STP 1253

Life Prediction Methodology for Titanium Matrix Composites

W. S. Johnson, J. M. Larsen, and B. N. Cox, editors

ASTM Publication Code Number (PCN)
04-012530-33

ASTM
100 Barr Harbor Drive
West Conshohocken, PA 19428-2959
Printed in the U.S.A.

Library of Congress Cataloging-in-Publication Data

Life prediction methodology for titanium matrix composites / W.S. Johnson, J.M. Larsen, and B.N. Cox. editors.
(STP ; 1253)
"ASTM publication code number (PCN) 04-012530-33."
Papers presented at the symposium held 22–24 March 1994 in Hilton Head, S.C.
Includes bibliographical references and indexes.
ISBN 0-8031-2039-7
1. Metallic composites—Testing—Congresses. 2. Titanium—Congresses. 3. Service life (Engineering)—Congresses. I. Johnson, W. S. (W. Steven) II. Larsen, James M. III. Cox, B. N. IV. Series: ASTM special technical publication; 1253.
TA481.L55 1996
620.1′89322—dc20 96-10845
CIP

Peer Review Policy

Each paper published in this volume was evaluated by three peer reviewers. The authors addressed all of the reviewers' comments to the satisfaction of both the technical editor(s) and the ASTM Committee on Publications.

The quality of the papers in this publication reflects not only the obvious efforts of the authors and the technical editor(s), but also the work of these peer reviewers. The ASTM Committee on Publications acknowledges with appreciation their dedication and contribution to time and effort on behalf of ASTM.

Printed in Chelsea, MI
April 1996

Foreword

The papers in the publication, *Life Prediction Methodology for Titanium Matrix Composites,* were presented at a symposium held 22–24 March 1994 in Hilton Head Island, South Carolina. The symposium was sponsored by ASTM Committee D30 on High Modulus Fibers and Their Composites, E8 on Fatigue and Fracture Mechanics, and NASA Langley Research Center. W. S. Johnson, Georgia Institute of Technology, J. M. Larsen, USAF Wright Laboratories, and B. N. Cox, Rockwell International Science Center, presided as symposium cochairmen and are coeditors of this publication.

Contents

Inelastic Material Behavior and Modeling

Fatigue

Overview

On 22–24 March 1994, a symposium entitled *Life Prediction Methodology for Titanium Matrix Composites* was held on Hilton Head Island, South Carolina. The symposium was sponsored by ASTM Committees D30 on High Modulus Fibers and Their Composites, E8 on Fatigue and Fracture Mechanics, and NASA Langley Research Center. Thirty-three papers were presented by researchers from government laboratories, universities, and industry. The volume represents a collection of those papers that were submitted and passed peer review by three knowledgeable referees.

The rationale for this symposium was to collect the current state of the art in life prediction methodology for Titanium Matrix Composites, TMCs. From the mid 1980's through the early 1990's several large technical programs pushed the development of TMCs and the development of the related life prediction methodology. Most visible among these programs were the National Aerospace Plane (NASP), the Integrated High Performance Turbine Engine Technology Program (IHPTET), and the Hi-Temp Program. These were large programs sponsored by various branches of the U.S. government. The NASP emphasized TMC for structural airframe applications on hypersonic vehicles, while the IHPTET (USAF) and Hi-Temp (NASA) programs developed TMCs for improved gas turbine engine performance. Of course researchers outside of the United States were also involved in TMC studies.

In order to develop accurate life prediction methodologies, a very through understanding of the mechanical damage mechanisms and environmental effects must be established. The envisioned usage conditions of these TMC materials are extreme. Temperature variations between 816°C and −130°C may be encountered on a hypersonic vehicle structure. Cyclic loads up to 60% of ultimate strength would be encountered routinely. Under these conditions the matrix material may not only undergo time dependent deformations but may undergo metallurgical changes as well. Many of the systems discussed in this volume are metastable beta-alloys of titanium. These materials may age with time at temperature in such a way that their mechanical properties, such as ultimate strength, yield strength, and modulus may increase over time. However, at elevated temperatures these materials become sensitive to oxidation, which causes embrittlement and enhanced susceptibility to cracking. Further complications arise because these composites that have fibers and matrices with significantly different coefficients of thermal expansions. Thus as temperature changes so does the internal stress state in the fibers and matrix. Because of this strong influence of thermally induced stresses, the approaches developed for thermal mechanical fatigue of metal that deal with the determination of mechanically induced stresses and strains are no longer applicable. Mathematical analysis must be developed to predict these complex internal stresses since they cannot be measured directly. Thus, the thermo-viscoplastic behavior of the constituents must be determined. Since both the fiber and matrix are very strong, the fiber/matrix interface is usually the weak link in the progression of mechanical failure. Understanding how the strength of the interface can change with time and how the interfacial strength is effected by the thermal residual stresses presents unique challenges on the microstructural scale. The interfacial strength play a significant role in the overall strength and fatigue resistance of TMCs.

This Special Technical Publication is organized into five sections that progress from the basic understanding to the life prediction methodology. The sections are as follows:

Interfacial Properties and Microstructure—Five papers are directed toward understanding the interplay between interfacial strength, residual thermal stresses and the effects of time at temperature.

Fiber Bridging Behavior—Four papers focus on the important role of fiber bridging in the progression of damage. Some papers present modeling approaches while others make detailed experimental observations of the bridging process.

Inelastic Material Behavior and Modeling—Seven papers deal with the complex aspects of experimentally determining inelastic material responses and how to model such behavior in a composite.

Fatigue—Seven papers discuss the various aspects of fatigue damage evolution.

Life Predictions—Six papers present various approaches to tying together all the basic understanding into a methodology that will allow one to predict total life of a TMC.

The collection of work presented here represents tremendous progress toward the characterization and modeling of Titanium Matrix Composites. Nearly all the work was conducted over a relatively short period of time, beginning in 1987. This body of understanding shows that TMCs can be attractive structural materials in stiffness and strength driven designs at elevated temperatures. Like any material, TMCs have their limitations in terms of usage temperature and life. In spite of their high costs, TMCs should find applications that demand their unique blend of mechanical properties and temperature tolerance.

We, the editors of this volume, would like to acknowledge the valuable support of our Steering Committee: Prof. Paul Bowen, University of Birmingham; Dr. Dave Buchanan, McDonnell-Douglas Aircraft; Dr. Rod Ellis, NASA Lewis Research Center; Prof. Tony Evans, Harvard University; and Dr. Ted Nicholas, USAF Wright Laboratory. Further we would like to thank the numerous peer reviewers that collectively spent hundreds of hours to ensure that each paper published in this volume is of a high quality. Lastly, our appreciation is extended to the ASTM staff for all their help in organizing the symposium, collecting the papers, assisting in the review process and publishing this volume.

Steve Johnson
Georgia Institute of Technology
Atlanta, GA 30320
Cochairman and Coeditor

Jim Larsen
USAF Wright Laboratory
Wright Patterson AFB, OH 45433
Cochairman and Coeditor

Brian Cox
Rockwell International Science Center
Thousand Oaks, CA 91364
Cochairman and Coeditor

Interface Properties and Microstructure

T. William Clyne,[1] Pascal Feillard,[1] and Alison F. Kalton[1]

Interfacial Mechanics and Macroscopic Failure in Titanium-Based Composites

REFERENCE: Clyne, T. W., Feillard, P., and Kalton, A. F., **"Interfacial Mechanics and Macroscopic Failure in Titanium-Based Composites,"** *Life Prediction Methodology for Titanium Matrix Composites, ASTM STP 1253,* W. S. Johnson, J. M. Larsen, and B. N. Cox, Eds., American Society for Testing and Materials, 1996, pp. 5–25.

ABSTRACT: Results are presented for Ti-6Al-4V containing silicon carbide (SiC) monofilaments having duplex (carbon plus TiB_2) coatings. Single fiber pushout testing has been used to study the debonding and frictional sliding characteristics under shear loading, with and without the residual radial compressive stress being reduced by applied in-plane tension. It is shown by finite element method (FEM) modeling that it is important to take account of thermal residual stresses when interpreting data from these tests. Prior heat treatments causing interfacial reaction tend to raise the resistance of the interface to debonding and sliding. This is correlated with data from tension testing of composites under axial and transverse loading, with continuous Poisson's ratio monitoring. The behavior under transverse loading is particularly sensitive to the mechanical response of the interface. The presence of interfacial reaction layers tends to inhibit the development of interfacial debonding and void formation under transverse loading, but causes embrittlement and leads to a reduced strain to failure.

The behavior under transverse load has also been studied with superimposed thermal cycling between 400 and 700°C. Strain histories have been monitored using scanning laser extensometry. Under these conditions, thermal stresses have a pronounced influence on the behavior. Many of the steady-state creep characteristics can be successfully modeled on the basis of matrix creep being controlled by volume-averaged stresses (predicted using the Eshelby method), with stress relaxation processes simulated via a variable stress-free temperature. However, the behavior tends to be influenced from a relatively early stage by interfacial damage development. Interfacial debonding and damage, promoted by a combination of opening mode stress from the applied load and shear stress from differential thermal contraction (particularly towards the specimen edges), soon starts to influence the strain history and has a strong influence on the rupture strain. In sharp contrast to the room temperature behavior, the heavily reacted specimens exhibited delayed onset of interfacial damage and a much longer lifetime, although the final strains to failure were similar. These results are considered in terms of stress fields and interfacial properties.

KEYWORDS: titanium, titanium matrix composites, life prediction, titanium alloys, fatigue (materials), modeling, thermal cycling, creep (materials)

It is now clear that the nature of the interface is of critical importance in controling the behavior of silicon carbide (SiC) monofilament-reinforced titanium, particularly under transverse loading. There have been several studies [*1–10*] of the inelastic deformation and failure characteristics of this class of composites under tensile loading. Several of these studies [*1,2,5–9*] have involved Poisson's ratio measurement during straining as a means of

[1]Reader, visiting scientist, and research student, respectively, Department of Materials Science and Metallurgy, University of Cambridge, Cambridge, CB2 3QZ, UK.

monitoring the inelastic processes taking place. The systems studied include Ti-15V-3Al-3Cr-3Sn (fully β) with Textron SCS-6 (carbon core, 3 μm carbon-rich surface layer) fibers [*2–6,8–10*], Ti-6Al-4V (α/β) with SCS-6 fibers [*1,3,4,10*] and Ti-6Al-4V with BP Sigma (tungsten core, 1 μm carbon + 1 μm TiB_2 layer) fibers [*7*]. In all of these cases, axial loading leads to matrix plasticity before final failure, which occurs soon after the fibers start to fracture. Under transverse loading, however, interfacial damage tends to dominate the behavior. This apparently occurs at relatively low macroscopic strains, particularly in locations where the applied load generates a strong Mode I component, but this does not immediately lead to macroscopic failure. Regimes have been identified in which interfacial debonding, local matrix plasticity, and global matrix plasticity occur. A general conclusion [*1,2,9*] is that the interface is inherently weak, but interfacial contact is maintained initially by the radial compressive stress resulting from differential thermal contraction.

In view of the significance of interfacial effects, particularly under transverse loading, several attempts [*1,7,10,11*] have been made to establish correlations between macroscopic properties and directly measured mechanical characteristics of the interface. (There is scope for altering the interfacial properties, not only by changing the fiber coating, but also by allowing various degrees of interfacial reaction to occur during fabrication and subsequent heat treatments.) However, very few such correlations have been successfully established. This is mainly because of limitations and interpretation difficulties associated with most interfacial test procedures. The one most easily applied to these materials is the single fiber push-out, or push-down, test [*12–18*]. Fiber pushout data generally indicate [*1,19–21*] that the interfacial shear strength (and frictional sliding stress) is raised substantially as interfacial reaction proceeds and the graphitic layer is consumed. The associated changes in the transverse tensile strength of the composite, on the other hand, appear to be small [*1,7*]. However, recent finite element method (FEM) studies [*22–24*] of the push-out test have confirmed that (*a*) the interface is loaded predominantly in shear (whereas the opening mode response probably controls the onset of damage in transverse tension testing), (*b*) thermal residual stresses have a very strong influence, and (*c*) loading conditions vary considerably along the length of the fiber. The recently-proposed [*24,25*] tensioned push-out test can in principle be used to explore the effect of Mode I loading, but it has not yet been systematically applied to these materials.

While the room temperature behavior of titanium-based metal-matrix composites (Ti MCCs) is important, interest really centers on the performance at high temperatures. A prime motivation for the incorporation of fibers lies in the dramatic improvement produced in creep resistance when loaded along the fiber direction. However, the creep behavior under transverse loading is of concern. While relatively little has been published [*26–29*] on creep of Ti MMCs, it is known that transverse creep rates can be relatively high, particularly if thermal cycling is imposed. Such cycling has little effect on the behavior when loaded along the fiber direction, but causes marked enhancement of creep under transverse loading [*29,30*]. This is consistent with a progressive transfer of load to the fibers under axial loading, but with matrix creep continuing in a quasi steady state for the transverse case.

It is anticipated that the variations in matrix stress state as the temperature is changed will affect the transverse creep behavior. Less attention has been paid to the fact that interfacial stress states will also be changing dramatically and that this is expected to affect the onset and development of damage processes and hence the creep rupture characteristics. In the present paper, an attempt is made to rationalize data from interfacial and macroscopic room temperature tests, together with results from isothermal and thermal cycling creep experiments.

Interfacial Testing

Material

Composite material was prepared at British Petroleum (BP) by vacuum hot-pressing stacks of aligned SiC monofilaments (coated with carbon/TiB_2-duplex layers, each about 1 μm thick) interleaved between 75-μm-thick Ti-6Al-4V foils, to produce unidirectional composite panels containing about 30% by volume of fibers. Post-fabrication heat treatments were carried out in sealed silica ampoules evacuated to about 10^{-5} torr. In view of the danger of residual oxygen penetrating along fiber/matrix interfaces [*31*], final cutting and machining operations (removing any material contaminated in this way) were carried out after the heat treatment. Composites were heat treated at 865°C for times ranging from 3 to 26 h. It has been confirmed [*7*] that the graphitic layer is progressively consumed at this temperature and has disappeared after the longest of the times employed. During this process, needles of titanium boride (TiB) grow from the coating into the matrix.

Single Fiber Push-out Testing

Wedge-shaped specimens were prepared from composite panels as described in previous publications [*7,21,32*]. Push-out testing was carried out on a number of fibers, the push-out load, and distance from the end of the wedge (and hence fiber aspect ratio = length/diameter) being noted for each fiber. Results from these tests are shown in Fig. 1 in the form of push-out stress as a function of aspect ratio for three heat treatments. The interfacial shear strength values shown for each specimen were obtained from the gradients of the push-out stress/aspect ratio plots, assuming the shear stress to be approximately constant in each case along the length of the fiber during the test. It can be seen that the shear strength rises as reaction proceeds, with a sharp rise apparently accompanying the final disappearance of the graphitic layer (between the 13- and 26-h heat treatments).

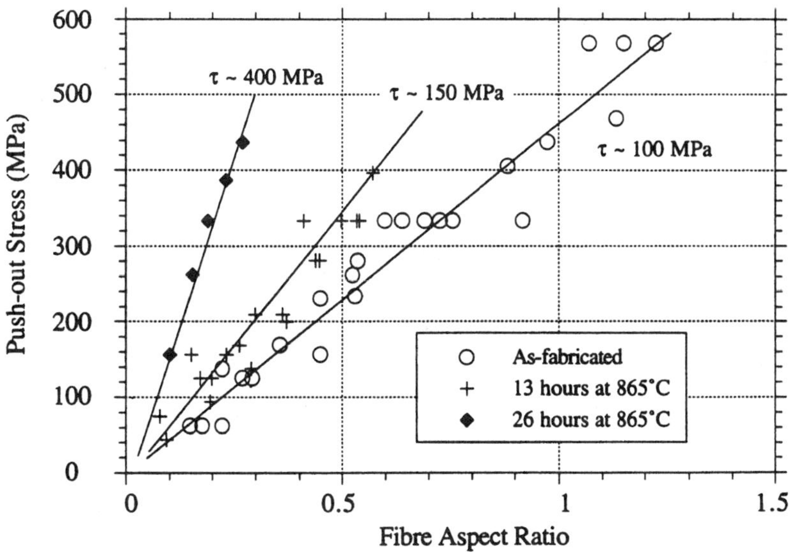

FIG. 1—*Single fiber push-out data* [7] *for specimens with different prior heat treatments.*

In fact, it has become clear from FEM calculations that the assumption of a uniform shear stress being generated along the length of the fiber during push-out testing is a poor one, even for short fibers. This is illustrated by the predictions shown in Fig. 2, giving distributions of interfacial shear stress under typical conditions, with and without thermal residual stress. (The value of 500 K is broadly consistent with the limited data [*33*] from X-ray measurement of thermal stresses.)

Clearly, the thermal stresses have a strong effect. Large variations in shear stress along the fiber length are generated, with the peak, where debonding should initiate, being located close to the bottom of the fiber. (This is predicted over a wide range of values for the applied axial compressive stress, σ_{zA}.) Such features should obviously be borne in mind, but they do not invalidate the deduction from experimental data that the interfacial shear strength rises as reaction proceeds. A further point worthy of note is that a large compressive radial stress also arises from differential thermal contraction (see following paragraphs). This confirms that all such testing normally yields information about how the interface responds to essentially pure Mode II conditions, with no opening mode loading.

Tensioned Push-Out Testing

Tensioned push-out testing has been carried out on composites in the as-fabricated condition, using the preparation procedures and operational details outlined previously [*25*]. A specimen of uniform thickness is required and the push-out load is measured with simultaneous applied in-plane tension (that is, normal stresses transverse to the fiber axis). It has been confirmed by FEM calculations [*24*] that the imposition of these tensile stresses (*a*) has little effect on the interfacial shear stresses and (*b*) strongly affects the interfacial radial

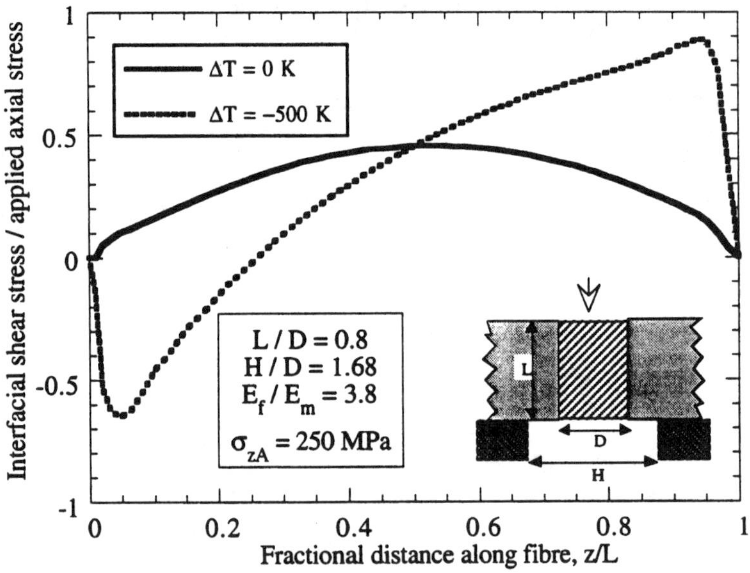

FIG. 2—*FEM predictions* [24] *showing the distribution of interfacial shear stress as a function of distance below the indenter, for typical conditions during push-out of SiC fibers in titanium, with and without thermal residual stress corresponding to an effective temperature decrease during fabrication of 500 K. The stress,* σ_{zA}, *is the applied normal compressive stress along the fiber* (z) *axis.*

stresses. The latter point is illustrated by Fig. 3 that shows for a typical case how the compressive radial stress from thermal contraction can be largely offset by the applied tension. It follows that this test can be used to examine the response of interfaces to loading states covering a range of mode mixities.

An example is shown in Fig. 4 of data obtained with the tensioned push-out test and interpreted using FEM modeling. This shows push-out load values obtained with a single specimen in the as-fabricated condition, converted to an average interfacial shear stress along the fiber length and plotted against the average interfacial radial stress. These data confirm that the application of an opening mode stress increases the ease of interfacial debonding/sliding, as expected. Extrapolation of the data to zero radial stress suggests that the interface will debond spontaneously at that point, with no shear stress or positive tension being necessary. This remains to be confirmed, although several workers [*1,2,9*] have pointed out that observations from macroscopic tests suggest that the interface in these composites (when a graphitic layer is present) seems to debond as soon as the residual compressive stress is offset by the applied load. Further work still required also includes the application of this test to reacted material in which the graphitic layer has been reduced in thickness, or eliminated.

Tensile Properties at Room Temperature

Specimen Preparation

Coupons for tension testing were machined from 1-mm-thick (6-ply) or 3-mm-thick (18-ply) unidirectional composite panels, using an electric-discharge machining technique to produce specimens with gage dimensions of 19 by 5 mm. The gage length was either parallel or transverse to the fiber axis. Gripping was with self-tightening friction grips and specimens were loaded to failure on a screw driven Schenk testing machine with a strain

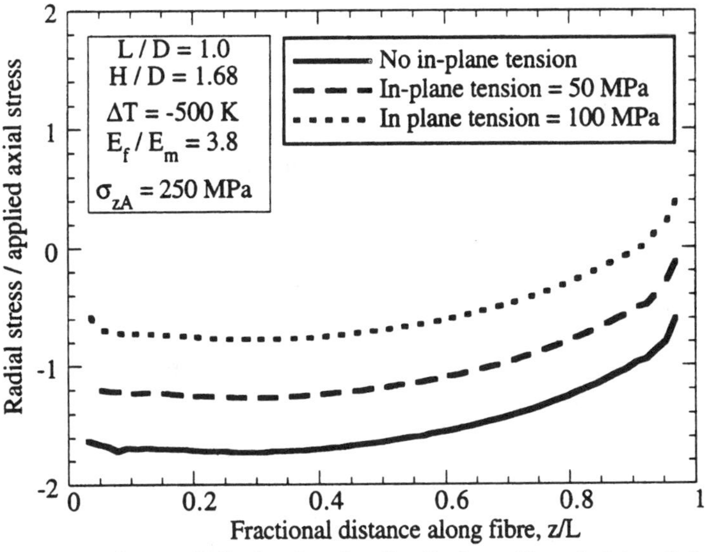

FIG. 3—*FEM predictions* [24] *showing the distribution of interfacial radial stress as a function of distance below the indenter, for typical conditions during push-out of SiC fibers in titanium, with and without applied in-plane tensile stresses.*

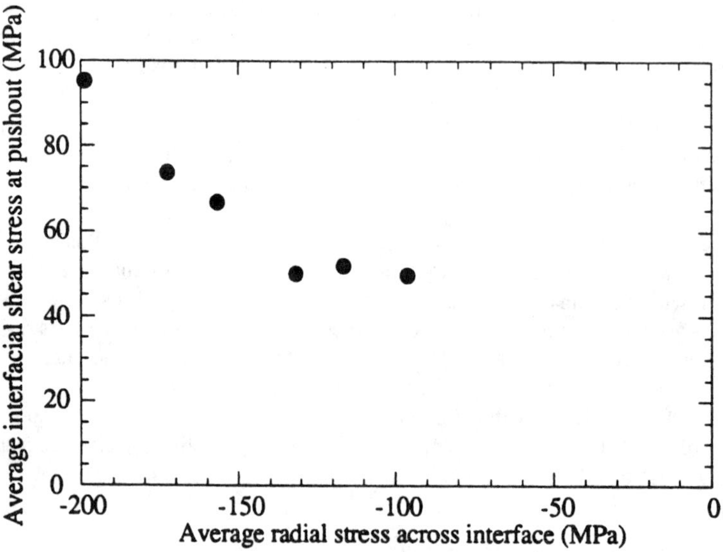

FIG. 4—*Experimental data* [25] *from tensioned push-out testing, plotted in the form of the average interfacial shear stress during frictional sliding as a function of the net average radial stress across the interface, taking account of thermal stresses* [24].

rate of 1.8 10^{-4} s^{-1}. Strain gages were applied to the specimen within the gage length. These were used to monitor strains in loading and longer transverse directions and, in some cases, in all three principal directions.

As-Fabricated Material

Data from axial loading of the composite are shown in Fig. 5. In the initial elastic regime, the experimental Young's modulus, E_1, and Poisson's ratio, ν_{12}, are close to the values predicted theoretically, using the Eshelby method [*27*] or simpler analytical expressions [*34*]. The inelastic behavior that is subsequently exhibited is consistent with matrix plasticity being the dominant feature. For example, the observed rise in Poisson's ratio is consistent with this.

Results from transverse loading experiments are presented in Fig. 6. Again the measured elastic constants in the early parts of the curves are broadly consistent with predicted [*27*] values. Beyond this region, however, the behavior is more complex. A sharp fall initially occurs in the Poisson's ratios, particularly in ν_{23}. This is attributed to interfacial debonding in regions where the fiber radial direction is parallel to the applied load that allows increments of extension in the loading direction without the corresponding lateral contractions (since matrix deformation is shielded by the fibers in this locality). This causes a fall in the Poisson's ratios. Subsequently, local and global matrix plasticity occur, probably while further debonding and voiding (gap formation) are occurring in interfacial regions. This is accompanied by a sharp rise in ν_{23}, while ν_{21} falls away to virtually zero. This reflects the fact that plastic flow is occurring under plane strain conditions, with all of the straining taking place within the 2–3 plane. That such behavior is expected can readily be demonstrated by examining volume-averaged matrix stresses in the elastic regime, again conveniently estimated using the Eshelby method. Predictions are presented in Fig. 7 that clearly

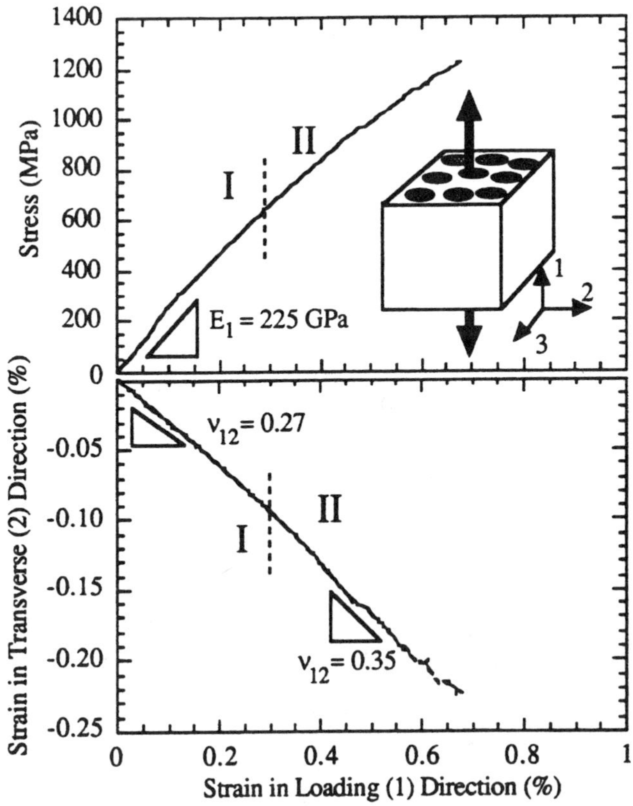

FIG. 5—*Experimental data from axial loading of the as-fabricated composite, up to final failure. The data indicate a transition from elastic behavior (Stage I) to a regime of matrix plasticity (Stage II). Predicted elastic values of the Young's modulus,* E_1*, and Poisson's ratio,* ν_{12}*, are as indicated (numerically and by means of the triangle gradients). The experimental value of* ν_{12} *in the plastic regime is also shown.*

show that the largest deviatoric stresses are in the 2–3 plane, provided the applied stress is greater than about 50 MPa.

The observation that interfacial debonding appears to start at an applied stress of around 170 MPa can be used to explore the local conditions under which this is stimulated, again using Eshelby calculations. The corresponding stress in the fibers along the loading direction is about 200 MPa, and this must also be the peak interfacial radial stress along the loading direction. The compressive radial stress at the interface from differential thermal contraction, on the other hand, is calculated to be about 150 MPa (for a 500 K temperature drop). These estimates are obviously approximate, but they do suggest that debonding occurs in opening mode soon after the net radial stress becomes tensile. More accurate analysis requires FEM modeling (and a reliable estimate of the appropriate temperature drop to employ). An idea of the extent of the region in which debonding is expected to occur can be obtained from the FEM stress field shown in Fig. 8. It can be seen that debonding is initially expected over only a small proportion of the fiber circumference, although the debonded region may grow as the applied load is increased. The details depend on the arrangement of fibers.

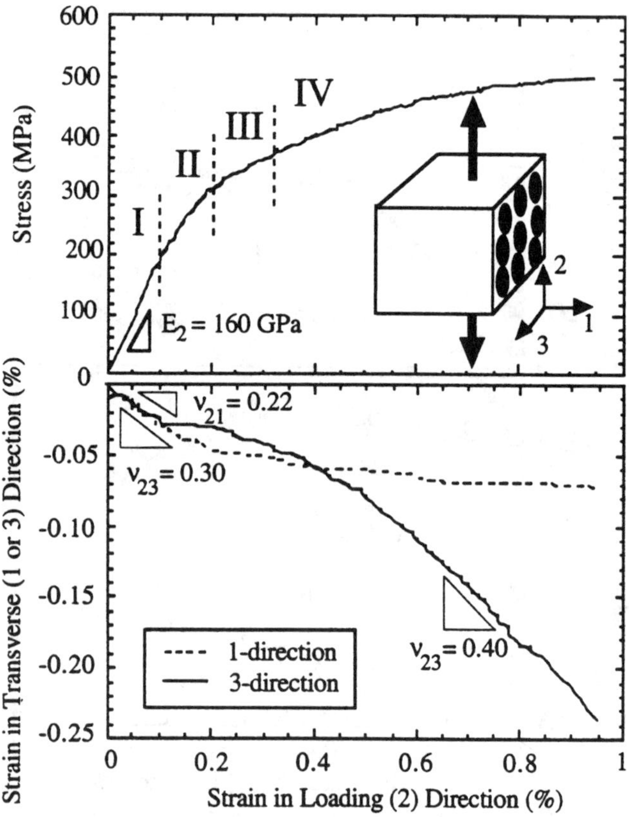

FIG. 6—*Experimental data from transverse loading of the as-fabricated composite, up to final failure. The results suggest the existence of four regimes, the last three marked by the onset of interfacial debonding, local plasticity, and global plasticity, respectively. Predicted elastic values of the Young's modulus,* E_2*, and the two Poisson's ratios,* ν_{21} *and* ν_{23}*, are as indicated. In the fully plastic regime, the observed value for* ν_{23} *is also shown. It can be seen that* ν_{21} *falls virtually to zero in this regime.*

Effect of Interfacial Reaction

Strain gage data from heavily reacted material are compared in Figs. 9 and 10 with those from composites in the as-fabricated state. Results are shown in Fig. 9 for axial loading. It can be seen that the Poisson's ratio changes during loading are very similar for the two cases. This is consistent with the nature of the interface having very little influence on the behavior. In neither case has interfacial debonding or sliding occurred to a significant degree. Failure occurs at a lower strain for the reacted material. This is probably because of cracking in the brittle reaction zone, perhaps assisted by general matrix embrittlement from diffusion of boron or carbon.

In contrast to the axially loaded case, it is clear from Fig. 10 that there is a significant difference between the behavior under transverse loading of as-fabricated and reacted materials. While only the variations in ν_{21} are available for comparison here, it is clear that the progressive decrease observed in the as-fabricated material from about 0.1% strain does not occur in the reacted material. Instead, the decrease in ν_{21} starts at a higher strain and is less

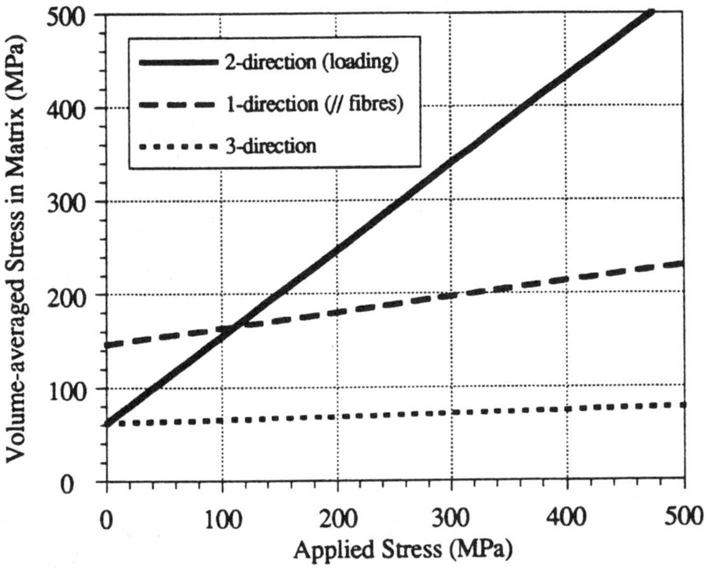

FIG. 7—*Eshelby method predictions for the dependence on applied stress in the 2-direction of the volume-averaged stresses within the matrix in the principal directions of a Ti-30SiC composite, assuming a thermal residual stress corresponding to a 500 K temperature decrease. Apart from the initial regime dominated by the thermal stresses, substantially larger deviatoric stresses are generated in the 2–3 plane than in either of the others.*

pronounced. The implication here is that the interfacial debonding that occurs in the as-fabricated material when the applied load approaches 200 MPa does not take place in the reacted material. The drop that does occur is probably associated with matrix plasticity (in the 2-3 plane). Again the strain to failure is reduced by the presence of the brittle reaction layer.

Thermal Cycling Creep Behavior

Measurement of Strain Histories during Creep

Specimens were subjected to thermal cycling under controlled atmosphere, while subject to a static, constant load transverse to the fibers. Various thermal histories were imposed (see following text). Material was used in as-fabricated and heat-treated states. The strain histories of the specimens were monitored using a scanning laser beam. Full details of the experimental set-up are given elsewhere [*35*].

Stress Fields and Stress Relaxation

It is well established that the creep behavior of fiber-reinforced MMCs is strongly affected by thermal stresses, particularly under thermal cycling conditions, in which case these stresses are continuously regenerated. An important feature of the behavior is that stress relaxation occurs readily, particularly at high temperature, which complicates the problem of predicting the changing stress field. This can be approached by identifying a temperature at which the composite is effectively free of thermal stress, T_{esf}. For example, calculations are shown in Fig. 11 of the average matrix stresses in the three principal

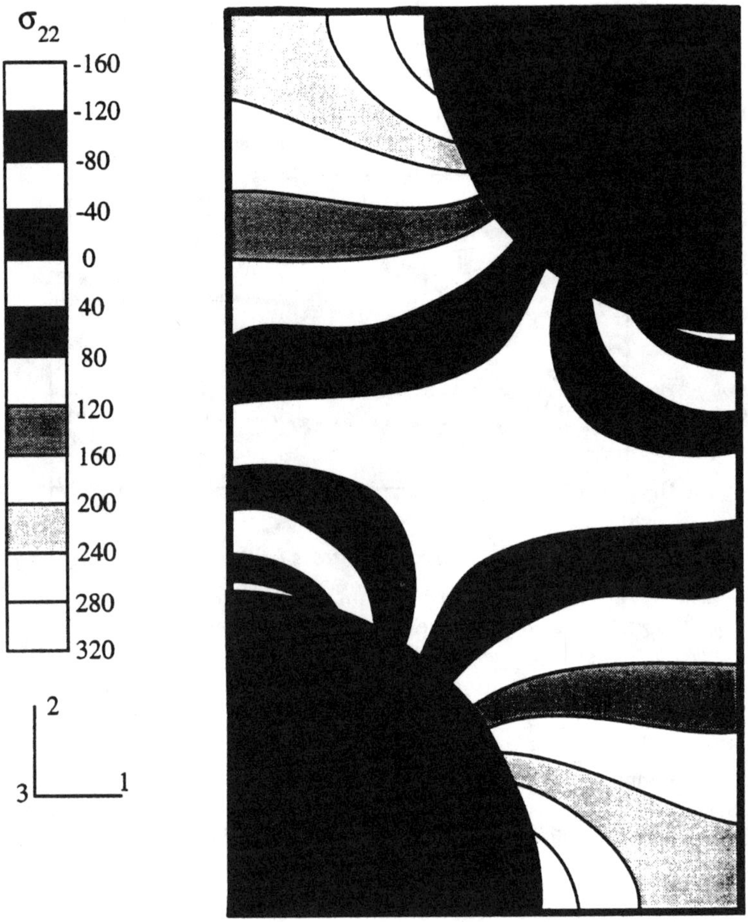

FIG. 8—*FEM contour plot for σ_{22} stresses (in MPa) resulting from a 500 K temperature decrease in a Ti-30SiC composite, with the fibers in an hexagonal array.*

directions as a function of temperature, assuming a T_{esf} value of 973 K (700°C), with a 20 MPa transverse load. The strong effects expected from temperature changes may be noted; for example, a drop in temperature of only 60 K below T_{esf} is sufficient for the stress to become more tensile parallel to the fibers (1-direction) than in the direction of applied load.

The composite creep rate at any point can be predicted from the stress state in the matrix, assuming it to obey similar creep laws to those applicable when no reinforcement is present, dictated by volume-averaged stresses. It is also possible to incorporate the effect of stress relaxation into this rationale by allowing the value of T_{esf} to change during the process. While the details of this stress relaxation are very difficult to measure or predict, the expected form of the variation can be identified. The general trend will always be for T_{esf} to approach the current temperature, T, at a rate that will increase with (*a*) higher values of T (faster diffusion kinetics) and (*b*) larger differences between T and T_{esf} (higher stress levels). An example of the effect of changes of this type is given in Fig. 12 that shows a specified thermal history together with a possible accompanying variation in T_{esf}. Information about the resultant matrix stress field is given in the form of the changing values of σ_2 and σ_1. For

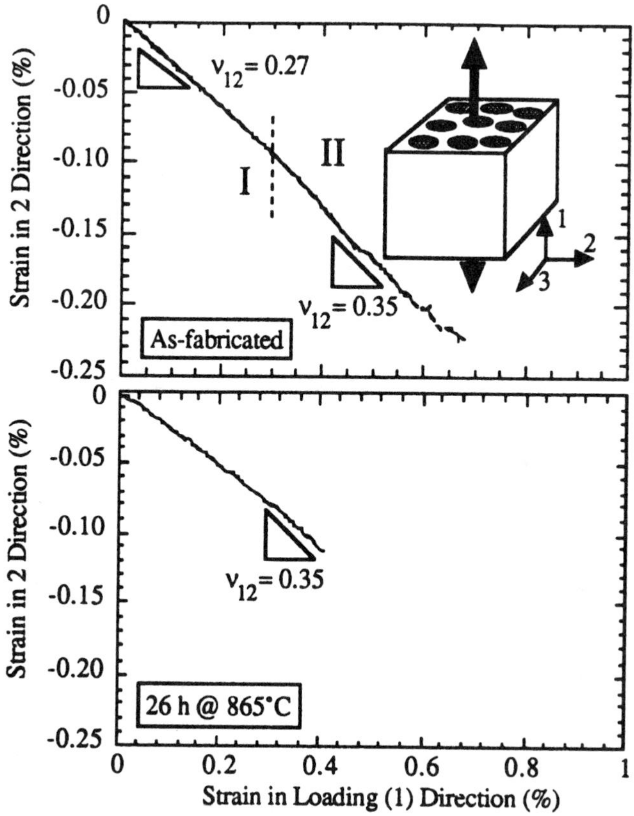

FIG. 9—*Experimental strain gage data* [7] *from axial loading at room temperature. Data are shown for composites in the as-fabricated state and after a substantial heat treatment of 26 h at 865°C.*

the case shown, the matrix stress in the loading (2) direction is greater than that in the fiber (1) direction only for a small part of the cycle, although this is the regime in which the temperature is high and in which much of the creep is expected to be concentrated. It can be seen, however, that there may be parts of the cycle where creep in the fiber direction (and therefore negative creep in the loading direction) could occur.

Steady-State Creep Behavior

Once the changing matrix stress field within the composite has been modeled in some way, it is possible to predict the creep behavior. The most tractable way to do this is on the basis of volume-averaged stresses. In practice, local variations may be of significance, in which case FEM modeling would be necessary; but for thermal cycling creep in particular, this represents a major computational effort, particularly since stress relaxation effects should be included. Modeling of creep by FEM techniques in which these are omitted [*36,37*] tends to predict a progressive load transfer to the fibers and a creep rate that correspondingly falls off to zero as the matrix stresses become low. This is realistic only for long fiber composites loaded parallel to the fibers. Unfortunately, while use of a volume-

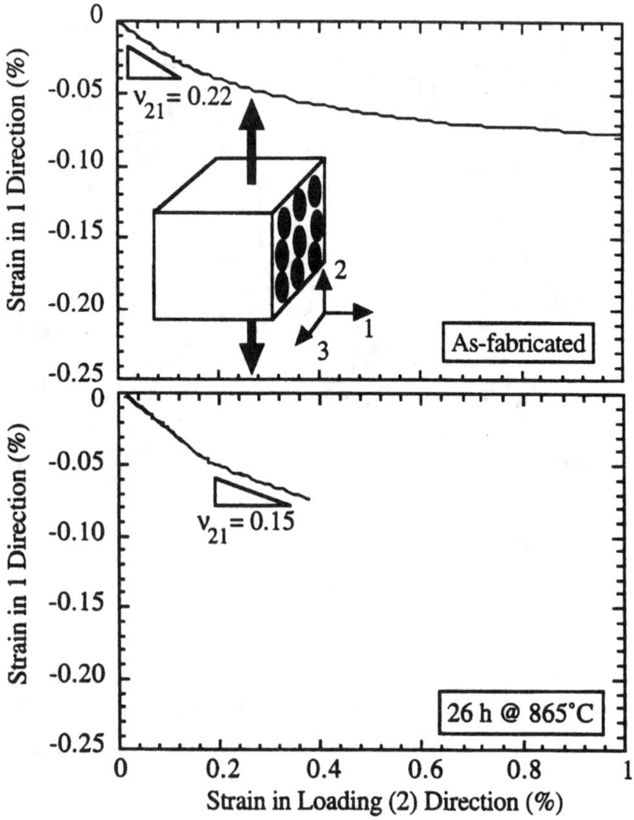

FIG. 10—*Experimental strain gage data* [7] *from transverse loading at room temperature. Data are shown for composites in the as-fabricated state and after a substantial heat treatment of 26 h at 865°C.*

averaged method, such as the Eshelby technique, reduces the amount of computation substantially, it also precludes any direct modeling of the stress relaxation processes, and the variation of T_{esf} can only be postulated empirically. Some results obtained in this way are presented in Fig. 13. Agreement between theory and experiment is not perfect, but some features are correctly predicted, including the approximate magnitude of the enhancement in creep rate induced by the thermal cycling. (This is obtained by comparing the average creep rate over the cycle with that during isothermal holding at the diffusional mean temperature of the temperature history, $\bar{T}$.)

A complication with such modeling for titanium composites concerns possible anisotropy in the matrix, from either crystallographic texture or grain shape effects [*38,39*]. This is illustrated by the data in Fig. 14, which gives a similar comparison to that in Fig. 13, but for the unreinforced matrix. No reverse creep, or enhancement of average creep rate, is expected in this case, since there should be no thermal stresses present. In practice, a certain amount of reverse creep is exhibited that may be attributed to inherent anisotropy. This is, however, almost certainly negligible when compared with the anisotropy introduced by the presence of ceramic fibers [*28*].

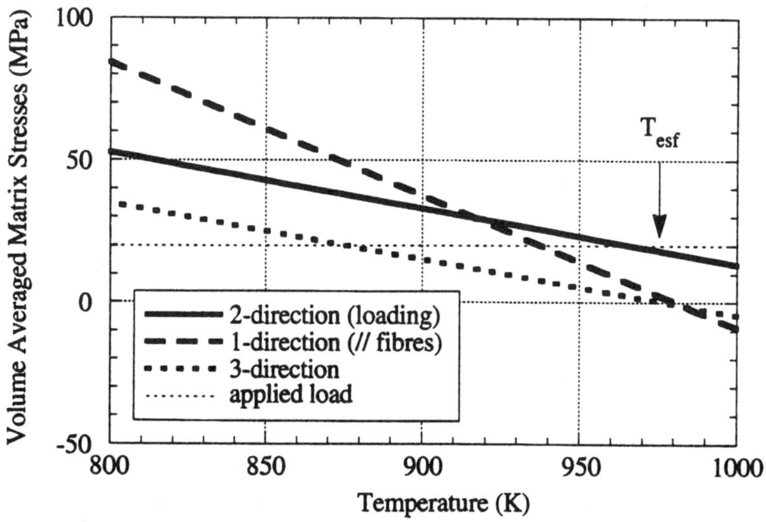

FIG. 11—*Calculations made using the Eshelby method of the dependence on temperature of volume-averaged matrix stresses in a Ti-30SiC composite, for an applied load of 20 MPa transverse to the fibers (2-direction) and a constant* T_{esf} *value of 973 K.*

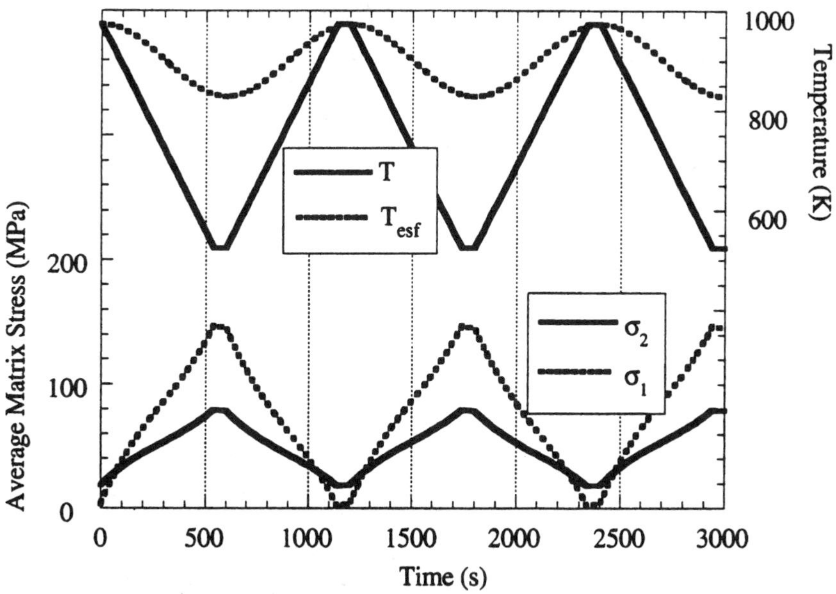

FIG. 12—*Calculations made using the Eshelby method showing the dependence on temperature of volume-averaged matrix stresses in a Ti-30SiC composite, for an applied load of 20 MPa in the 2-direction (transverse to the fibers) and a postulated variation of the effectively stress-free temperature,* T_{esf}, *as shown.*

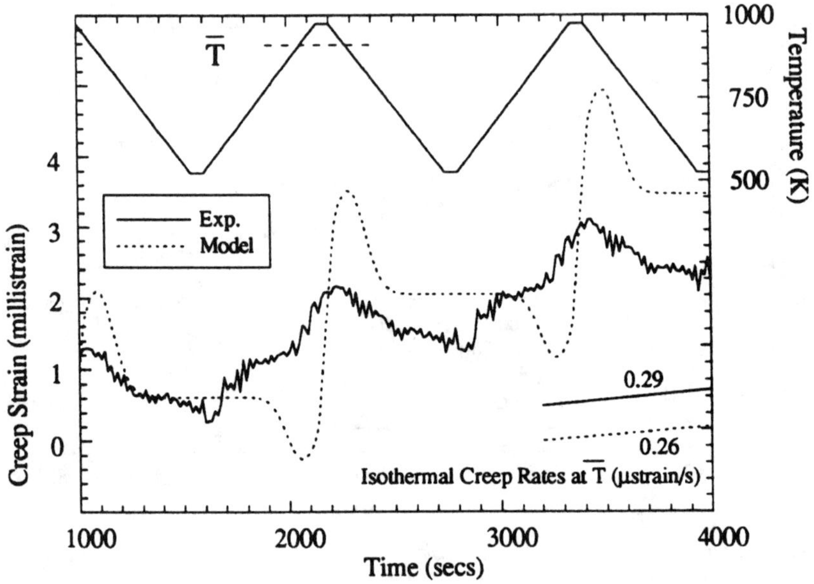

FIG. 13—*Comparison between measured and modeled creep strain data* [29] *for a composite under a transverse 20 MPa load. Data are shown for both a thermally cycled specimen and one held at the diffusional mean temperature.*

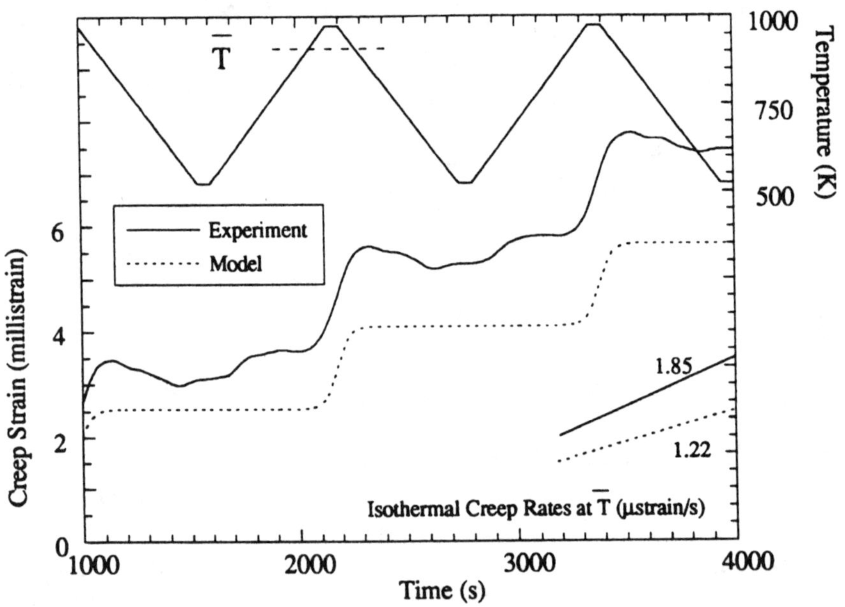

FIG. 14—*Comparison between measured and modeled creep strain data* [29] *for unreinforced Ti-6Al-4V, loaded parallel to the rolling direction. Data are shown for both a thermally cycled specimen and one held at the diffusional mean temperature.*

Interfacial Damage and the Onset of Failure

While some success has been obtained in modeling of the steady-state (Stage II) regime during thermal cycling creep under transverse load, it is in practice commonly observed with these materials that this regime is of short duration or is apparently never set up. While predicted and observed Stage II creep rates are usually lower than for the unreinforced matrix, a regime of accelerating damage and final failure often comes much earlier. The onset of damage tends to be very sensitive to the applied load. This is illustrated by the data shown in Fig. 15 that shows histories of total strain (including thermal expansion and contraction) for two different applied loads. For the 20 MPa load, a steady-state regime is identifiable, although it gives way after about 30 cycles to a regime of accelerating creep rate. For the load of 25 MPa, on the other hand, this acceleration occurs from the start and final rupture occurred after about 46 cycles, at a strain of about 12%. The damage that occurred during this accelerating regime was identified by microstructural examination as interfacial debonding and void formation.

This behavior is unsurprising in view of the deductions made from the room temperature testing. The compressive radial stress at the interface that is present at room temperature will presumably be virtually absent at temperatures around 600°C and above. It seems likely that even very modest applied loads may be sufficient to cause local debonding under these conditions. However, it may be noted that the development of damage seems to be very much aggravated by the imposition of thermal cycling. Further information on the significance of this was obtained by examining material with different interfacial structures (see following paragraphs).

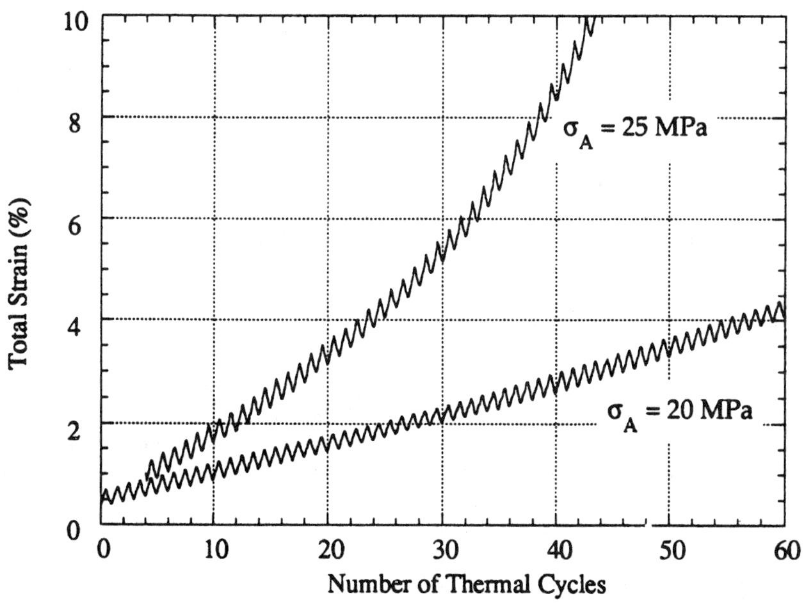

FIG. 15—*Complete strain histories for composite material, in the as-fabricated state, under two different applied transverse loads, as a function of the number of imposed thermal cycles. The temperature profile employed was of the same form as that shown in Figs. 13 and 14, but with higher heating and cooling rates of 100°C per min (so that each cycle was of about 10 min duration).*

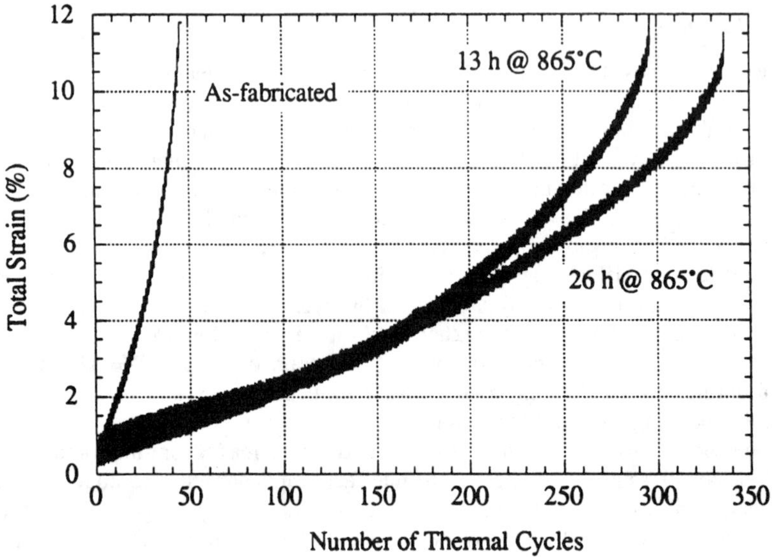

FIG. 16—*Complete strain histories for composite material, with and without prior heat treatments, under an applied transverse load of 25 MPa, as a function of the number of imposed thermal cycles. The thermal conditions during cycling were the same as in Fig. 15.*

The problem of rapid interfacial damage development in these materials under modest transverse loads, when combined with thermal cycling, would obviously be of concern to design engineers involved with these materials—although it should be recognized that the thermal excursions imposed in these experiments (to 700°C) are relatively severe. However, the damage is not associated with interfacial reaction between fiber and matrix during the tests—the rates of reaction are negligible below about 800°C. In fact, experiments with material that had been subjected to prior heat treatments in order to promote interfacial reaction showed considerably greater resistance to damage development and creep rupture. This is illustrated by the data in Fig. 16 that demonstrate that material in which the graphitic layer had been partially or completely consumed displays a much longer Stage II creep regime and considerably greater lifetime to rupture, although the final failure strain is similar for the three materials at about 12%.

It seems clear that the reacted materials are more resistant to interfacial damage. The higher interfacial bond strength, confirmed experimentally in shear and probably expected in opening mode as well, is evidently more significant under creep conditions than the associated embrittlement responsible for the reduced failure loads at room temperature. It is interesting to consider whether it is the greater resistance to shear or to opening that is of most significance here. It might seem likely that resistance to sliding would be of less importance, but study of fracture surfaces suggested that this might not be the case. In unreacted material, it was noticeable that the graphitic coating had debonded from the fibers near the edges of the specimen, while it remained adherent near the center, see Fig. 17. It seems likely that this resulted from damage progressively induced by reciprocating interfacial sliding under the influence of the thermal stresses as the temperature changed, although it should be recognized that ingress of oxygen along the interface from the specimen edge may also have played a part. It is important to note that both of these sources of damage would be eliminated with some loading geometries, such as the hoop wound tubes used by

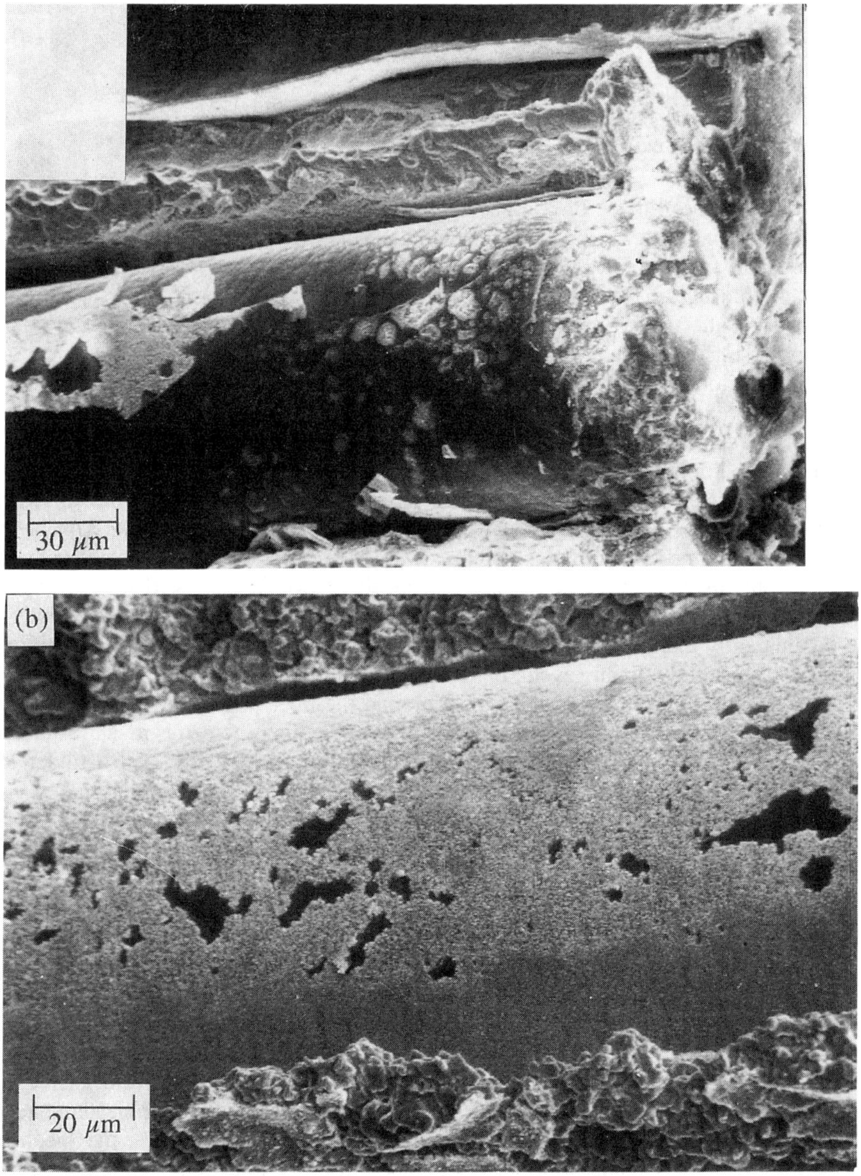

FIG. 17—*SEM micrographs of the fracture surface from an as-fabricated composite after thermally cycled transverse loading (25 MPa), taken from* (a) *near the specimen edge and* (b) *near the center.*

Dvorak et al. [*40,41*]. Tests with this specimen geometry might assist in identifying the critical interfacial properties under these loading conditions.

Summary and Conclusions

The following conclusions can be drawn from this work.

1. A study has been made of composites based on a Ti-6Al-4V matrix, reinforced with SiC monofilaments having C/TiB_2 coatings. This material has been tested in the as-fabricated state and after prior heat treatments that cause progressive consumption of the graphitic layer. Experiments have been carried out to measure interfacial properties, using the single fiber push-out test, with and without applied in-plane tension. These results have been correlated with data from tension testing at room temperature, with monitoring of Poisson's ratios, and creep testing with and without imposed thermal cycling.
2. As the graphitic layer is consumed, the resistance of the interface to debonding and frictional sliding under shear rises significantly.
3. Tests on as-fabricated material, with the compressive radial stress from differential thermal contraction partly offset by applied tension, indicated that the interfacial shear strength becomes small as this compression is removed. This is consistent with measurements made during transverse tensile loading, which suggest that interfacial debonding occurs as soon as the net radial stress becomes tensile. The reacted interface is apparently more resistant to such debonding, although further testing is required to quantify this fully. In any event, the strain to failure of the reacted material is lower, as a result of embrittlement of the region around the fiber.
4. Creep rates are very low when loaded parallel to the fibers. Creep is much faster under transverse loading, particularly when the specimen is thermally cycled under load. This behavior has been qualitatively explained in terms of the changing matrix stress state. Some success has been achieved in quantitative prediction of the steady-state strain history during cycling under transverse load, based on volume-averaged matrix stresses calculated by the Eshelby method and simulation of stress relaxation effects by means of a variable stress-free temperature. However, in many cases, the strain history is influenced by progressive interfacial debonding and damage that is difficult to incorporate into an analytical model.
5. During thermally cycled creep, the development of interfacial damage, and consequent final failure, are considerably delayed by prior interfacial reaction. The lifetime was raised by a factor of over seven in a presented example case. Creep rates were also much lower. This is attributed to greater resistance to shear or void formation or both. There is some evidence that the interface can initially become damaged by sliding in shear under thermal stresses, particularly near the edges of the specimens. Testing of hoop-wound specimens is planned in order to explore the significance of this and of the possibility of residual atmospheric oxygen having penetrated along the interface from the free surface during the tests.

Acknowledgments

Financial support has been provided by an NSF Fellowship, Corpus Christi College, Cambridge, an Overseas Research Studentship scholarship and Defence Research Agency Farnborough (for AFK) and by the French Ministry of Defence (for PF). The authors would also like to acknowledge useful contact and discussions with Dr. R. Shatwell and Mr. J.

Robertson, of BP Sunbury, and with Dr. M. C. Ward-Close, of DRA Farnborough. Dr. M. C. Watson of Cambridge Consultants and Dr. F. H. Gordon of Cambridge University were involved with some of the work described here.

References

[1] Wright, P. K., Nimmer, R., Smith, G., Sensmeier, M., and Brun, M., "The Influence of the Interface on Mechanical Behaviour of Ti-6Al-4V/SCS6 Composites," *Interfaces in Metal-Ceramic Composites,* R. Y. Lin, R. J. Arsenault, G. P. Martins, and S. G. Fishman, Eds., The Metals Society, Warrendale, PA, 1990, pp. 559–581.

[2] Jansson, S., Deve, H. E., and Evans, A. G., "The Anisotropic Mechanical Properties of a Ti Matrix Composite Reinforced with SiC Fibres," *Metallurgical Transactions,* Vol. 22A, 1991, pp. 2975–2984.

[3] Jeng, S. M., Yang, J.-M., and Yang, C. J., "Fracture Mechanisms of Fibre-reinforced Titanium Alloy Matrix Composites. Part II: Tensile Behaviour," *Materials Science and Engineering,* Vol. A138, 1991, pp. 169–180.

[4] Jeng, S. M., Yang, J.-M., and Yang, C. J., "Fracture Mechanisms of Fibre-reinforced Titanium Alloy Matrix Composites. Part III: Toughening Behaviour," *Materials Science and Engineering,* Vol. A138, 1991, pp. 181–190.

[5] Majumdar, B. S. and Newaz, G. M., "Inelastic Deformation of Metal Matrix Composites: Plasticity and Damage Mechanisms," *Philosophical Magazine,* Vol. 66, 1992, pp. 187–212.

[6] Majumdar, B. S., Newaz, G. M., and Ellis, J. R., "Evolution of Damage and Plasticity in Titanium-Based, Fibre-Reinforced Composites," *Metallurgical Transactions,* Vol. 24A, 1993, pp. 1597–1610.

[7] Watson, M. C. and Clyne, T. W., "Reaction-induced Changes in Interfacial and Macroscopic Mechanical Properties of SiC Monofilament Reinforced Titanium," *Composites,* Vol. 24, 1993, pp. 222–228.

[8] Gunawardena, S. R., Jansson, S., and Leckie, F. A., "Modeling of Anisotropic Behavior of Weakly Bonded Fibre Reinforced MMC's," *Acta Metallurgical Materials,* Vol. 41, 1993, pp. 3147–3156.

[9] Bonora, N., Costanzi, M., Newaz, G., and Marchetti, M., "Micro Damage Effects on the Overall Response of Long Fibre Metal Matrix Composites," *Composites,* Vol. 25, 1994, pp. 575–582.

[10] Fox, K. M., Strangwood, M., and Bowen, P., "The Influence of Thermal Treatments on Interfacial Properties and Fatigue Crack Growth Resistance in SCS-6/Ti/321S Composites," *Composites,* Vol. 25, 1994, pp. 684–691.

[11] Jeng, S. M., Yang, J.-M., and Yang, C. J., "Fracture Mechanisms of Fibre-reinforced Titanium Alloy Matrix Composites. Part I: Interfacial Behaviour," *Materials Science and Engineering,* Vol. A138, 1991, pp. 155–167.

[12] Marshall, D. B., "An Indentation Method for Measuring Matrix-fibre Frictional Stresses in Ceramic Composites," *Journal,* American Ceramics Society, Vol. 67, 1984, pp. C259–260.

[13] Marshall, D. B. and Oliver, W. C., "Measurement of Interfacial Mechanical Properties in Fibre-reinforced Ceramic Composites," *Journal,* American Ceramics Society, Vol. 70, 1987, pp. 542–548.

[14] Weihs, T. P. and Nix, W. D., "In situ Measurements of the Mechanical Properties of Fibres, Matrices and Interfaces in Metal Matrix and Ceramic Matrix Composites," *Metallic and Ceramic Composites,* S. I. Andersen, H. Lilholt, and O. B. Pedersen, Eds., Riso National Laboratory, Roskilde, Denmark, 1988, pp. 497–502.

[15] Shetty, D. K., "Shear Lag Analysis of Fibre Push-out (Indentation) Tests for Estimating Interfacial Friction Stress in Ceramic-Matrix Composites," *Journal,* American Ceramics Society, Vol. 71, 1988, pp. C107–109.

[16] Hsueh, C. H., Bright, J. D., and Shetty, D. K., "Interfacial Properties of SiC-Borosilicate Glass Composites Evaluated from Pushout and Pullout Tests," *Journal of Materials Science Letters,* Vol. 10, 1991, pp. 135–138.

[17] Singh, R. N. and Sutcu, M., "Determination of Fibre-Matrix Interfacial Properties in Ceramic Matrix Composites by a Fibre Push-out Technique," *Journal of Materials Science,* Vol. 26, 1991, pp. 2547–2556.

[18] Watson, M. C. and Clyne, T. W., "The Use of Single Fibre Pushout Testing to Explore Interfacial Mechanics in SiC Monofilament-reinforced Ti. Part I: A Photoelastic Study of the Test," *Acta Metal et Materials,* Vol. 40, 1992, pp. 135–140.

[19] Eldridge, J. I. and Brindley, P. K., "Investigation of Interfacial Shear Strength in a SiC Fibre/Ti-

24Al-11Nb Composite by a Fibre Push-out Technique,'' *Journal of Materials Science Letters,* Vol. 8, 1989, pp. 1451–1454.

[20] Yang, C. J., Jeng, S. M., and Yang, J. M., ''Interfacial Properties Measurement for SiC Fibre-reinforced Titanium Alloy Composites,'' *Scripta Metallurgica et Materials,* Vol. 24, 1990, pp. 469–474.

[21] Watson, M. C. and Clyne, T. W., ''The Use of Single Fibre Pushout Testing to Explore Interfacial Mechanics in SiC Monofilament-reinforced Ti. Part II: Application of the Test to Composite Material,'' *Acta Metal et Materials,* Vol. 40, 1992, pp. 141–148.

[22] Kallas, M. N., Koss, D. A., Hahn, H. T., and Hellman, J. R., ''Interfacial Stress State Present in a 'Thin Slice' Fibre Push-out Test,'' *Journal of Materials Science,* Vol. 27, 1992, pp. 3821–3826.

[23] Koss, D. A., Kallas, M. N., and Hellman, J. R., ''Mechanics of Interfacial Failure During Thin Slice Fibre Pushout Tests,'' *Proceedings,* Materials Research Society, Vol. 273, 1992, pp. 303–313.

[24] Kalton, A. F., Ward-Close, C. M., and Clyne, T. W., ''Development of the Tensioned Pushout Test for Study of Fibre-matrix Interfaces,'' *Composites,* Vol. 25, 1994, pp. 637–639.

[25] Watson, M. C. and Clyne, T. W., ''The Tensioned Push-Out Test for Measurement of Fibre/Matrix Interfacial Toughness under Mixed Mode Loading,'' *Materials Science and Engineering,* Vol. A160, 1993, pp. 1–5.

[26] Khobaib, M., ''Creep Behaviour of SCS-6/Ti-24Al-11Nb Metal Matrix Composites,'' *Titanium Aluminide Matrix Composites,* P. R. Smith, S. J. Balstone, and T. Nicholas, Eds., Wright-Patterson Air Force Base, Dayton, OH, 1991, pp. 450–461.

[27] Clyne, T. W. and Withers, P. J., *An Introduction to Metal Matrix Composites,* Cambridge University Press, Cambridge, UK, 1993.

[28] Gordon, F. H. and Clyne, T. W., ''The Effects of Texture and Microstructure on the Thermal Cycling Creep of Ti-6Al-4V with and without Fibre Reinforcement,'' *Proceedings,* 9th International Conference on Composite Materials (ICCM9), A. Miravete, Ed., Elsevier, Vol. 1, 1993, pp. 524–532.

[29] Gordon, F. H. and Clyne, T. W., ''Thermal Cycling Creep of Ti-6Al-4V/SiC Monofilament Composites Under Transverse Loading,'' *Residual Stresses in Composites: Modeling, Measurement and Effects on Thermomechanical Properties,* E. V. Barrera and I. Dutta, Eds., The Metals Society, Warrendale, PA, 1993, pp. 293–304.

[30] Colclough, A., Dempster, B., Favry, Y., and Valentin, D., ''Thermomechanical Behaviour of SiC-Al Composites,'' *Materials Science & Engineering,* Vol. A135, 1991, pp. 203–207.

[31] Hartley, M. V., ''The Effect of Isothermal Exposure on the Mechanical Properties of a Continuous Fibre Reinforced Titanium Matrix Composite,'' *Proceedings,* Ninth Risø International Symposium on Mechanical and Physical Behaviour of Metallic and Ceramic Composites, S. I. Andersen, H. Lilholt, and O. B. Pedersen, Eds., Risø National Laboratory, Roskilde, Denmark, 1988, pp. 383–390.

[32] Watson, M. C. and Clyne, T. W., ''Interfacial Mechanical Properties in Ti-SiC Monofilament Composites from Single Fibre Pushout Testing,'' *Interfacial Phenomena in Composite Materials, 91,* I. Verpoest and F. R. Jones, Eds., Butterworths, London, 1991, pp. 187–190.

[33] Brown, K. M., Hendricks, R. W., and Brewer, W. D., ''X-Ray Diffraction Measurements of Residual Stresses in SiC/Ti Composites,'' *Fundamental Relationships Between Microstructures and Mechanical Properties of Metal-Matrix Composites,* P. K. Liaw and M. N. Gungor, Eds., The Metals Society, Warrendale, PA, 1990, pp. 269–286.

[34] Clyne, T. W., ''A Compressibility-based Derivation of Simple Expressions for the Transverse Poisson's Ratio and Shear Modulus of an Aligned Long Fibre Composite,'' *Journal of Materials Science Letters,* Vol. 9, 1990, pp. 336–339.

[35] Furness, J. A. G. and Clyne, T. W., ''The Application of Scanning Laser Extensometry to Explore Thermal Cycling Creep of Metal Matrix Composites,'' *Materials Science & Engineering,* Vol. A141, 1991, pp. 199–207.

[36] Dragone, T. L. and Nix, W. D., ''A Numerical Study of High Temperature Creep Deformation in Metal-Matrix Composites,'' *Metal & Ceramic Matrix Composites: Processing, Modeling & Mechanical Behaviour,* R. B. Bhagat, A. H. Clauer, P. Kumar, and A. M. Ritter, Eds., The Minerals, Metals & Materials Society, Warrendale, PA, 1990, pp. 367–380.

[37] Dragone, T. L. and Nix, W. D., ''Steady State and Transient Creep Properties of an Aluminum Alloy Reinforced with Alumina Fibres,'' *Acta Metallurgica,* Vol. 40, 1992, pp. 2781–2791.

[38] Partridge, P. G., McDarmaid, D. S., and Bowen, A. W., ''A Deformation Model for Anisotropic Superplasticity in Two-phase Alloys,'' *Acta Metallurgica,* Vol. 33, 1985, pp. 571–577.

[39] Bowen, A. W., McDarmaid, D. S., and Partridge, P. G., ''Effect of High Temperature

Deformation on the Texture of a Two-phase Titanium Alloy,'' *Journal of Materials Science,* Vol. 26, 1991, pp. 3457–3462.

[*40*] Dvorak, G. J., Bahei-El-Din, Y. A., Macheret, Y., and Liu, C. H., ''An Experimental Study of Elastic-Plastic Behaviour of a Fibrous Boron-Aluminium Composite,'' *Journal of the Mechanics and Physics of Solids,* Vol. 36, 1988, pp. 655–687.

[*41*] Dvorak, G. J., ''Micromechanics of Inelastic Composite Materials: Theory and Experiment,'' *Journal of Engineering Materials and Technology,* Vol. 115, 1993, pp. 327–338.

Yutaka Kagawa,[1] *Chitoshi Masuda,*[2] *Chikara Fujiwara,*[3] *and Akira Fukushima*[3]

Effects of Reaction Layer on Interfacial Shear Properties and Strength of Fiber in Silicon-Carbide (SiC) Fiber-Reinforced Titanium Alloy Composite

REFERENCE: Kagawa, Y., Masuda, C., Fujiwara, C., and Fukushima, A., **"Effects of Reaction Layer on Interfacial Shear Properties and Strength of Fiber in Silicon-Carbide (SiC) Fiber-Reinforced Titanium Alloy Composite,"** *Life Prediction Methodology for Titanium Matrix Composites, ASTM STP 1253,* W. S. Johnson, J. M. Larsen, and B. N. Cox, Eds., American Society for Testing and Materials, 1996, pp. 26–42.

ABSTRACT: The effect of the interfacial thickness of the reaction layer on the interfacial shear properties and the tensile strength of double carbon-coated SCS-6 SiC fiber in Ti-15Mo-5Zr-3Al alloy matrix composite was examined. The major reaction layer thickness, that is, titanium-carbide (TiC) layer thickness, varied with heat-exposure temperature and time, respectively, and the resultant mean thickness of the reaction layer of the composite ranged from 0.4 to 1.7 μm. The critical interfacial toughness, G_i^c, and the mean shear sliding resistance, τ_s, were evaluated by the thin specimen pushout technique. Tensile strength of the silicon-carbide (SiC) fiber extracted from the titanium alloy matrix before and after the heat exposure was determined in relationship to the thickness of the reaction layer. The critical interface toughness, G_i^c, for the failure of the root of the reaction layer was $\approx$4 J/m^2, and the average shear sliding resistance of the interface, τ_s, was 102 to 118 MPa. The interfacial shear mechanical properties were adequate to prevent failure of the fiber due to the stress concentration caused by cracks that formed first in the reaction layer. The results showed that when the growth of reaction layer was within 1.7 μm, the mean strength of the extracted fiber was unaffected by the existence of the reaction layer because of weak bonding between it and the fiber. However, with the increase of the reaction layer thickness, the strength distribution of the extracted fiber tended to Weibull bimodal distribution.

KEYWORDS: silicon-carbide fiber-reinforced titanium alloys, composites, interfacial properties, fiber strength, reaction layer, debonding, frictional sliding, push-out tests, titanium, titanium matrix composites, life prediction, titanium alloys, modeling

Silicon-carbide (SiC) fiber-reinforced titanium alloy matrix composites are attractive for structural applications because of their high specific modulus and strength, and good stability at high temperatures [*1*]. One current area of research in SiC fiber-reinforced titanium (Ti)-based matrix composite systems is the optimization of the processing variables, including the secondary-forming process. The major problem during processing is

[1]Associate professor, Institute of Industrial Science, The University of Tokyo, Tokyo 106, Japan.
[2]Head, 4th Laboratory, Failure Physics Division, National Research Institute for Metals, Tokyo 153, Japan.
[3]Project engineer and research engineer, respectively, Materials Research Section, Engineering Research Department, Nagoya Aerospace Systems, Mitsubishi Heavy Industries, Ltd., Nagoya 455, Japan.

fiber degradation due to the unavoidable nature of active reaction behavior between the fiber and the matrix. To improve fiber degradation after incorporation into Ti-based matrices, a specially designed SiC fiber for Ti-matrix-based composites has been produced. The fiber has a carbon-rich double coating ≈3 μm thick with a fiber diameter of ≈140 μm (SCS-6)[4] and details of the fiber are reported in Refs *2* and *3*. Many studies of the interfacial reaction and its related phenomena in the Ti-based matrix composite reinforced by the SCS-6 fiber have been published [*4–12*]. Previous reports on the reaction behavior of the SCS-6 SiC fiber, Ti-matrix composite showed that the major reaction product is titanium-carbide (TiC) along the outer carbon coating layer and a trace of fine grain $Ti_xSi_y(C)$ (the majority reported is Ti_5Si_3) between the TiC and the matrix. The larger volume fraction of TiC compared to titanium silicate results from the presence of the carbon-rich coating layer on the fiber surface. The details of the microstructure of the reaction products between SiC fiber and various titanium alloy matrices were reported in Ref *12*. They clearly showed that the SCS-6 coating layer of the fiber acts as a diffusion barrier that reduces reaction rate at the interface during the consolidation and prevents damage of the fiber by the reaction.

Studies of the effect of the reaction layer on the fiber strength have been conducted for fiber-metal matrix systems. Metcalfe [*13*] showed that the strongly adhered reaction layer causes degradation of fiber strength due to the stress concentration effect. Metcalfe [*13*] also showed the relationship between the reaction zone thickness, x, and the fracture stress of the fiber. The work of Metcalfe was extended by Shorshorov et al. [*14*], who used a stress intensity factor approach to show the effect of the thickness of this layer on tensile strength of the fiber. DiCarlo [*15*] also derived a formula to predict the average fiber strength where the reaction zone controls the strength. Ochiai and Murakami [*16*] demonstrated a means of reducing the stress concentration effect of the cracked reaction layer on the strength of the fiber by a mechanism of crack deflection where the interfacial bonding is weak. Summarizing the previous papers, it seems important to control interfacial shear mechanical properties between the fiber and matrix to maintain the strength of the fibers in the matrix.

When evaluating the interfacial shear mechanical properties of the Ti-based matrix composites, various techniques were applied that include single-fiber pull-out tests [*17*], multiple fracture test [*18*], and push-out test [*19–22*]. The thin-sliced specimen fiber push-out technique is now used to measure interfacial shear property of Ti-based matrix composites, because it is applicable for various composite materials and require no special specimen. For example, Yang et al. [*19*] reported the effect of the matrices and thermal exposure on the shear debonding and sliding strength of the interface. Kagawa and Okura [*20*] and Kagawa [*21*] reported the procedures for obtaining interfacial shear debonding and sliding stress from a load-displacement curve of the single-fiber push-out test using the load-displacement relationship. Watson and Clyne [*22*] applied the push-out test to a SiC-Ti system and showed the effect of interfacial reaction conditions on the interfacial shear sliding stress.

Although, as summarized earlier, the reaction kinetics between the SCS-6 fiber and Ti-based matrices, and the interfacial shear mechanical properties of the composites have been relatively well studied, the effect of the formed reaction zone on the strength of the SCS-6 SiC fiber and interfacial shear mechanical properties for the SCS-6 SiC fiber-reinforced titanium matrix systems is not well known due to a lack of experimental evidence. In this paper, attention focuses on the effect of reaction zone thickness on the interfacial shear strength between fiber and matrix and the strength of the extracted fiber.

[4]Textron Corp., Lowell, MA.

Experimental Procedure

Composite Material and Specimen

The composite specimen was prepared by hot-isostatically pressing a stack of titanium alloy foils and monolayers of the SiC monofilaments. The SiC monofilament [*3,4*], (SCS-6 fiber) is a 144-μm diameter filament with a double-pass 3-μm silicon doped carbon coating added to increase fiber strength and enhance chemical compatibility with Ti-based matrices [*23*]. The matrix titanium alloy has a nominal composition of Ti-15Mo-5Zr-3Al. Detailed chemical analysis of the matrix indicated that it was composed of 16.00% molybdenum, 5.44% zirconium, 3.11% aluminum, 0.018% hydrogen, 0.098% oxygen, 0.0031% iron, 0.05% carbon, and the remainder titanium by weight. Properties of the fiber and matrix are listed in Table 1.

The composite panels were provided by Mitsubishi Heavy Industries Co., Ltd., Japan. The nominal volume fraction of the fiber was 32% by volume. The as-received composite consists of six fiber layers in through-the-thickness direction with a total thickness of 5 mm and the composite part was sandwiched between the titanium alloy foil. Figure 1 shows an optical photograph of a transverse cross section of the composite in an as-fabricated condition. The photograph reveals the relatively uniform fiber spacing and full consolidation of the matrix. Typical properties of the composite in the longitudinal direction were: Young's modulus ≈200 GPa and ultimate tensile strength ≈2000 MPa [*24*].

Specimens for heat-exposure were machined with a diamond cutting blade into a rectangular bar of 2 by 10 by 60 mm that contained fibers aligned uniaxially along the longest direction. Heat exposure for studying growth of the reaction layer at 1173 and 1253 K was performed in a vacuum of 0.13×10^{-3} Pa. Specimens were wrapped in a commercially pure titanium foil and held in the vacuum furnace at the temperature for 18, 54, and 21.6 ks, respectively, then vacuum cooled to room temperature. After the heat exposure, the reaction zone at the polished section perpendicular to the fiber axis, with or without etched surface, was observed by optical microscope. The etching was carried out by a mixture of H_2O-HF-HNO_3 dilute solution at 300 K. Chemical composition of the reaction layer was determined by wave-length-dispersive-type electron probe microanalyzer (EPMA).

Push-out Test

Fiber-matrix interfacial-shear mechanical properties, that is, interface toughness (critical shear debonding strain energy release rate), G_i^c, and shear sliding stress, τ_s, were evaluated by the thin specimen push-out method. In this study, the interface toughness was used because the shear debonding stress is believed to be essentially independent of a solution for the stress field [*25,26*], and the value can be used as a criterion of the fiber degradation due to formation of the reaction layer. Details of the in-house push-out test apparatus and its

TABLE 1—*Properties of fiber and matrix.*

Property	Fiber	Matrix
Young's modulus, GPa	406	9.9
Tensile yield stress, MPa	...	882
Tensile strength, MPa	4000	921
Elongation, %	1	5
Poisson's ratio	0.15	0.3

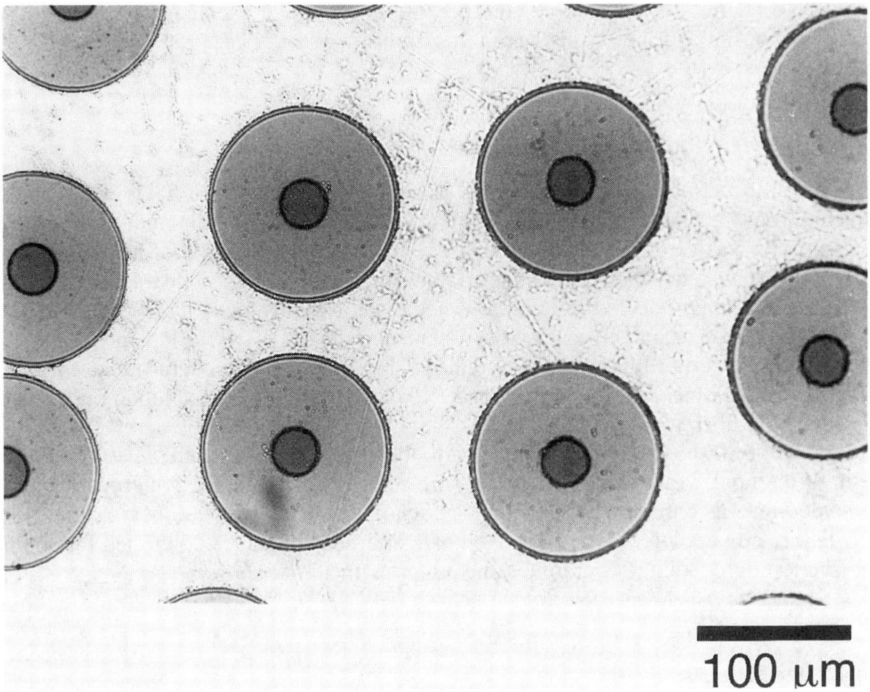

FIG. 1—*Optical micrograph of the polished section of a composite.*

application to SiC fiber-reinforced titanium alloy matrix composites have been reported [*20*].

The thin composite push-out specimens were prepared by the conventional metallurgical process, and the surfaces of the specimen perpendicular to the fiber axis were polished by diamond paste. Final polishing was done using 0.1 μm diamond paste. The final thickness of the specimen was $\approx$1 mm that corresponded to the aspect ratio of the fiber in the specimen being $\approx$7. The thickness of $\approx$1 mm was chosen because of the results of our preliminary experiment [*18*]. Before testing, all specimens were observed by optical microscope to identify interfacial damage incurred during the preparation process, and only undamaged samples (no debonded interface at the polished surface) were used for the test. The thin composite specimen was mounted on a 0.5-mm-diameter hole that was machined into a support block, allowing the fiber to be pushed out without resistance from the block. The sample support block was put on the load cell holder and the holder was secured to a three-axis translation table that allowed positioning of individual fibers in the composite beneath the indenter. Alignment of indenter and fiber was assessed by observation through an optical microscope ($\times$20) attached to the push-out test equipment.

During the push-out experiment, the load was applied at a constant displacement rate of 1.0×10^{-5} m/s. The push-out test was carried out in air at 293 K. Displacement of the indenter was continually monitored by magnetic type displacement meter attached to the arm of the indenter. A standard Pyramidal diamond indenter was used as the push-out indenter. The load-displacement behavior was stored in a digital memory scope with a sampling time of 10 ms. Following the push-out test, a true load-fiber surface displacement curve was obtained by substituting the penetration depth of the indenter obtained from another composite and the details were reported elsewhere [*20,21*].

Tensile Strength of the Extracted SiC Fiber

To verify the effect of reaction on the strength of the SiC fiber, fibers were carefully extracted from the composite after undergoing heat-exposure. The titanium alloy matrix of the composite was carefully dissolved in a boiling solution of 12 mol of hydrochloric acid (HCl) in 1 L of water. Thereafter, extracted fibers were fully rinsed in distilled water. Eighty fibers were randomly selected and tensile-tested in air at 297 K using an Instron-type testing machine with a constant crosshead displacement rate of 1.7×10^{-6} m/s. Before the experiment, as-received fiber was put into the HCl solution and the degradation of its tensile strength by this immersion was determined. No damage by the HCl solution on the tensile strength of the fiber was found.

Aluminum foil (0.1-mm thick) was added to the gripped section of the fiber, and the gage length of the fiber was set to 20 mm. The surface of extracted fibers, both before and after heat exposure, was observed by scanning electron microscope (SEM). Fibers extracted from an as-fabricated composite and as-received bare SiC fibers were tension tested for comparison. The testing conditions were the same as for extracted fibers.

Results and Discussion

Reaction Behavior

Figure 2 shows a typical example of optical micrographs of a polished and etched cross section of the composite with thermal exposure condition applied. The formation of a relatively thick reaction layer around the fiber is clearly recognizable after undergoing the heat exposure. Note that some parts of the reaction products were removed during the etching process, because the morphology of the reaction layer observed at the etched section differs in detail from the as-polished unetched specimen. The mean thickness of the reaction layer for the as-fabricated state was 0.4 μm, and the thickness varied with heat-exposure time and temperature, respectively. Growth in thickness of the reaction layer was noted following the heat exposure, and the thickness of the carbon coating layer decreased with the growth of the reaction layer. EPMA results of the titanium and carbon characteristic X-ray peaks are overlapped at the outer layer that is formed during consolidation or heat-exposure or both. The other elements (silicon, molybdenum, zirconium, and aluminum) were not detected in the reaction layer. This result suggests the existence of a relatively thick TiC reaction layer, that is, two layers adjacent to the SiC fiber are the carbon coatings of the SCS-6 fiber and the outer layer is the reaction product layer: the majority is assumed to be TiC. Hereafter, for convenience, the reaction layer is simply called the "TiC layer."

The randomly selected reaction zone thickness, that is, thickness of TiC layer, was measured from optical micrographs of the as-polished unetched specimen of 20 fibers. Figure 3 shows results of the reaction layer thickness, x, versus the square root of thermal-exposure time, t, for temperatures of 1173 and 1253 K. The solid line in the figure is obtained from the linear least squares of the reaction layer thickness versus the square root of time [*24*], that is

$$x = k\sqrt{t} \tag{1}$$

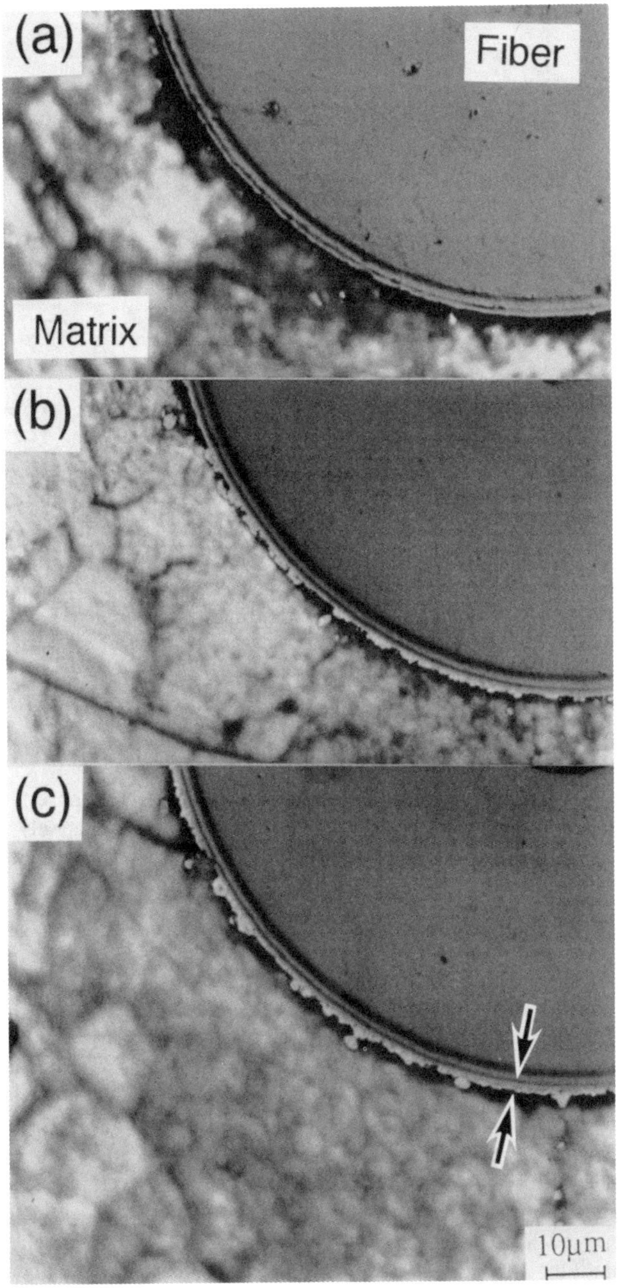

FIG. 2—*Polished and etched sections of a composite with heat-exposure condition:* (a) *as-fabricated,* (b) *1173 K, 18 ks, and* (c) *1173 K, 54 ks. Arrows in* c *indicate definition of the thickness of reaction layer.*

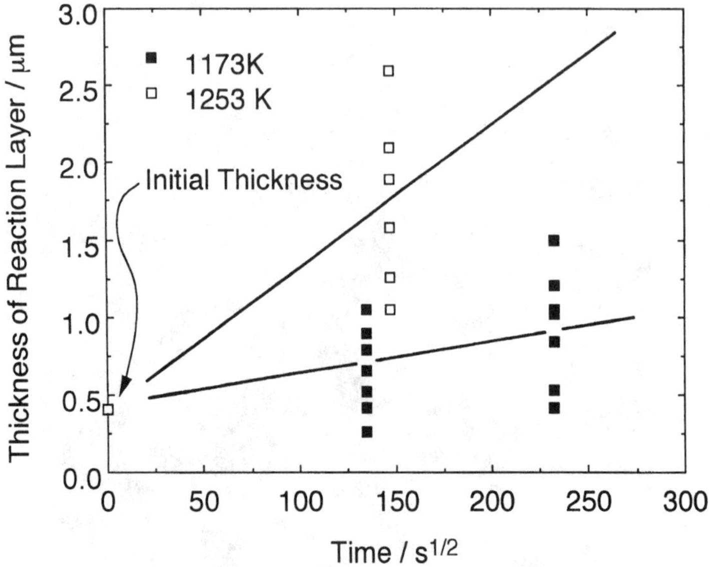

FIG. 3—*Plots of the mean reaction layer thickness increase versus heat-exposure time for temperatures at 1173 and 1253 K. Total thickness of the reaction zone is given by the sum of the increase of mean reaction layer thickness and initial reaction layer thickness (0.4 μm).*

where k is the constant.[5] The reaction rate constants, k in Eq 1, in this composite are 1.9×10^{-9} m/s$^{1/2}$ and 9.1×10^{-8} m/s$^{1/2}$ for the temperatures of 1173 and 1253 K, respectively. The measured reaction rate constant, k, values are consistent with those previously reported for the same kind of titanium alloy matrix composites [*25*]. The relatively slow reaction rate is probably related to the presence of the molybdenum-stabilized β phase [*25*].

Interfacial Shear Mechanical Properties

The push-out applied load-fiber surface displacement behavior was essentially independent of the thickness of the reaction layer and the curve agreed completely with previous reports [*20,21*]. The interfacial debonding between the coating layer and reaction layer occurs at the load level of approximately 20 N that corresponds to a shear stress of 133 MPa in the fiber. This means that the debonding at the interface began at an early stage. However, no visible relative slip at the interface between the fiber and matrix was recognized at this load, as previously reported [*20*]. Figure 4 shows a typical example of the interface debonding. Shear failure occurred at the root of the reaction layer, and this behavior was the same for all the composites tested. After the onset of interface debonding at the pushing surface, the applied load increased gradually, however, there was no fiber pushout at the back surface until the applied load had nearly reached maximum. The fiber pushout from the back surface began when the push-out load reached around maximum, and the load decreased with the increase in the push-out of the fiber from the back surface. The appearance

[5]Generally, the relationship between x and t is given by $x = kt^n$ where n is constant. In this study, $n = 0.5$ is applied because $n = 0.5$ is usually used for the reaction analysis of SiC(SCS-6) fiber-reinforced titanium alloy composites [*24*].

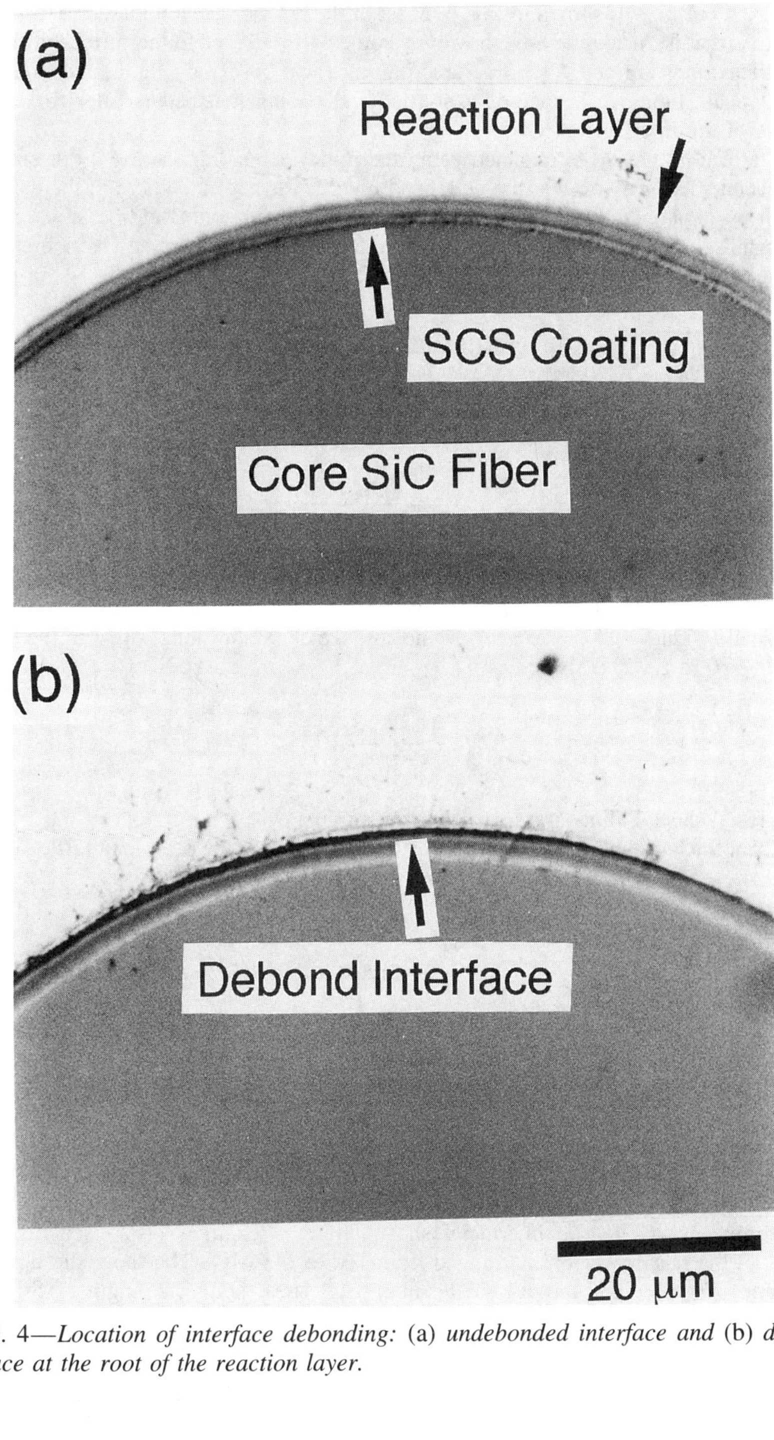

FIG. 4—*Location of interface debonding:* (a) *undebonded interface and* (b) *debonded interface at the root of the reaction layer.*

of the pushed out fiber is shown in Fig. 5. Most of the reaction layer adhered to the titanium alloy matrix, that is, inside the hole; however, some parts adhered to the surface of the push-out fiber. The rough surface of the fractured interface was observed (Fig. 6) and suggests the origin of shear sliding resistance against frictional sliding that occurs after the complete debonding of the interface.[6]

The interface toughness is obtained using the initial debonding load, P_f^*. For simplicity, the interfacial frictional stress at the debonded interface is neglected here in the calculation because it was assumed that the error attributed to this operation would be relatively small ($\pm 5\%$) against calculated G_i^c values [*18*]. The interface toughness in the absence of friction at the interface is given by [*26*]

$$G_i^c = \frac{1 - 2\nu_f \xi}{4\pi^2 R_f^3 E_f}(P_f^* + P_f^0)P_f^* \tag{2}$$

and

$$\xi = \frac{E_m \nu_f}{E_f(1 + \nu_m) + E_m(1 - \nu_f)} \tag{3}$$

where E_f and E_m are, respectively, Young's modulus of fiber and matrix, ν_f and ν_m are, respectively, Poisson's ratio of fiber and matrix, R_f is the radius of the fiber, and P_f^0 is the axial thermal residual load of the fiber. In the absence of axial residual stress in the fiber and setting $P_f^0 = 0$ in Eq 2, G_i^c becomes

$$G_i^c \approx \frac{(1 - 2\nu_f \xi)}{4\pi^2 R_f^3 E_f}(P_f^*)^2 \tag{4}$$

The interfacial shear sliding stress is obtained using the peak load, P_f^M because the push-out of fiber from the back surface of the composite occurred near the peak load [*20*]. Thus, τ_s is given by

$$\tau_s = \frac{P_f^M}{2\pi R_f L} \tag{5}$$

where P_f^M is the maximum load in the load-displacement curve and L is the specimen thickness ($\approx$1-mm thick). The thermally induced residual stress field in the composite specimen is not exactly obtained and the used interfacial shear sliding stress within this paper is defined as Eq 5.

The obtained G_i^c value and τ_s are plotted in Fig. 7 as a function of mean reaction layer thickness. The obtained mean G_i^c value was $\approx$4 J/m^2. The order of this value agrees with the interfacial shear strain energy release rate of the SCS-6 SiC fiber-reinforced β-Ti alloy matrix composites [*27*]. The interfacial shear sliding strength is also insensitive to the thickness of the reaction layer and ranged from 102 to 118 MPa. The shear sliding stress is comparable with previous reports on the interfacial shear sliding strength of SCS-6 SiC fiber-reinforced titanium alloy matrix composites [*10,20–22*].

[6]The effects of indenter shape and hole diameter on the measured result was not considered in this paper.

FIG. 5—*Appearance of the pushed-out fiber observed at the back surface.*

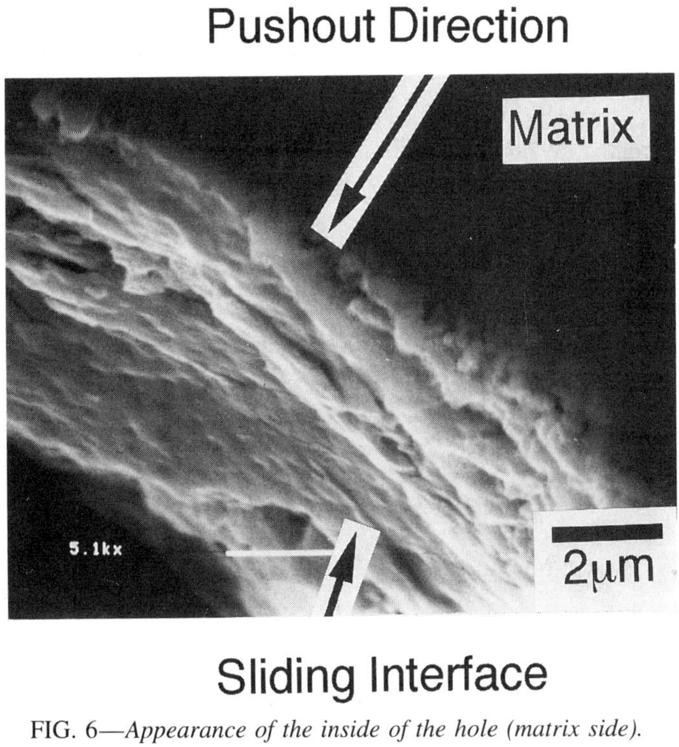

FIG. 6—*Appearance of the inside of the hole (matrix side).*

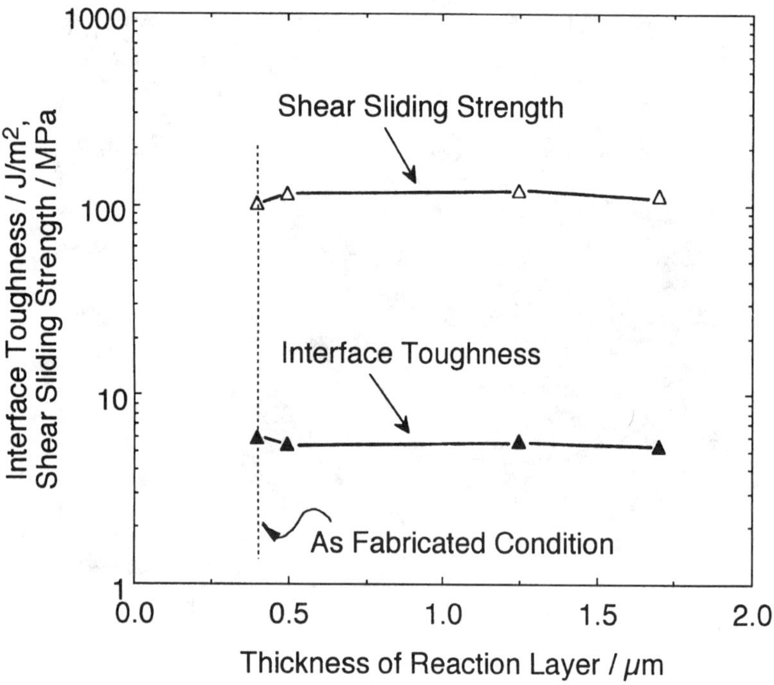

FIG. 7—*Plots of* G_i^c *and* τ_s *versus the mean thickness of the reaction layer.*

Effect of Reaction Layer on the Strength of Fiber

Figure 8 shows Weibull plots of the extracted fiber with the heat-exposure condition of the composite. The Weibull modulus was obtained using a two-parameter statistical distribution [*28*]. According to Weibull's statistical theory, the probability of failure, P_F, of the fiber subjected to nominal stress, σ_{fu}, is given as

$$P_F = 1 - \exp\left[-L\left(\frac{\sigma_{fu} - \sigma_u}{\sigma_0}\right)^{m_f}\right] \tag{6}$$

where m_f is the Weibull modulus of the fiber, L is the gage length of fiber, σ_u is the stress below which the fiber is assumed to have zero failure probability, and σ_0 is the normalizing parameter. The Weibull modulus of the fiber can be determined by rearranging Eq 6 into the form

$$\ln\ln\left[\frac{1}{(1 - P_F)}\right] = m_f\ln(\sigma_{fu} - \sigma_u) - m_f\ln\sigma_0 + \text{constant} \tag{7}$$

In Eq 7, the probability of failure, P_F, at the nth ranked sample from a total of N (=80 in this experiment) is obtained from mean rank method as

$$P_F = \frac{n}{N + 1} \tag{8}$$

The experimental data defining P_F as a function of σ_{fu} can then be a least squares fit to the linear relationship of Eq 7, thus defining m_f. The mean strength of the fiber, $\bar{\sigma}_{fu}$, is obtained

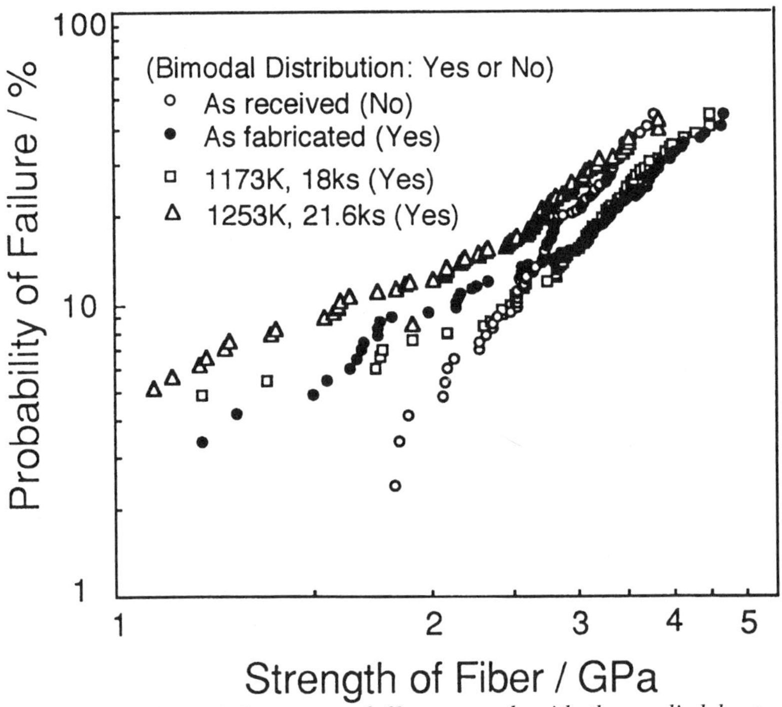

FIG. 8—*Weibull plots of the extracted fiber strength with the applied heat-exposure condition. The mode of distribution is also indicated in this figure.*

$$\bar{\sigma}_{fu} = \sigma_0^{-1/m_f}\Gamma\left(1 + \frac{1}{m_f}\right) \tag{9}$$

where

$$\Gamma(\zeta) = \int_0^{\infty} e^{-x}x^{\zeta-1}dx \tag{10}$$

The mean strength and Weibull modulus are shown in Fig. 9*a* and *b*, respectively, as a function of the mean thickness of the reaction layer. Regardless of the small difference in the mean strength with heat-exposure time, the mean strength of the fiber is unaffected by the presence of the reaction layer.[7] This result provides evidence that the fiber was not preferentially attacked by the reaction during composite consolidation. However, the distribution of fiber strength must be commented on. With the increasing thickness of the reaction layer, a tendency toward a bimodal distribution of fiber strength appeared.

In the tension testing of the extracted fiber, fracture of the reaction layer will occur prior to fracture of the fiber because the nominal fracture strain of the TiC layer is 0.4% [*29*] and that of SCS-6 fiber is 0.8%. After the onset of crack initiation at the reaction layer, three possible crack paths can be considered (Fig. 10).

If the reaction layer strongly adheres under the condition of no interfacial debonding (Fig.

[7]The other experimental result showed that the tensile strength of the SCS-6 fiber did not change after heat exposure [*21,29*].

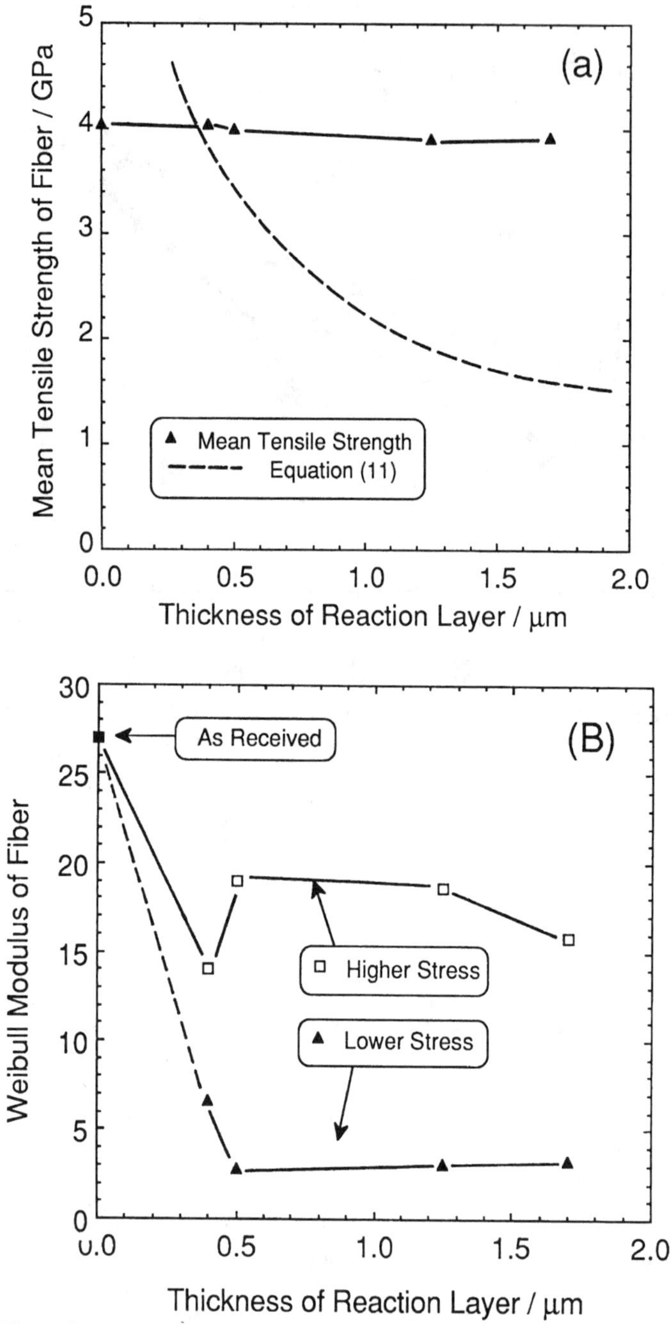

FIG. 9—*Plots of mean strength versus mean thickness of the reaction layer* (a) *and plots of the Weibull modulus versus mean thickness of reaction layer* (b). *(Two Weibull moduli (higher stress and lower stress) are ploted in the figure assuming bimodal distribution of extracted fiber strength).*

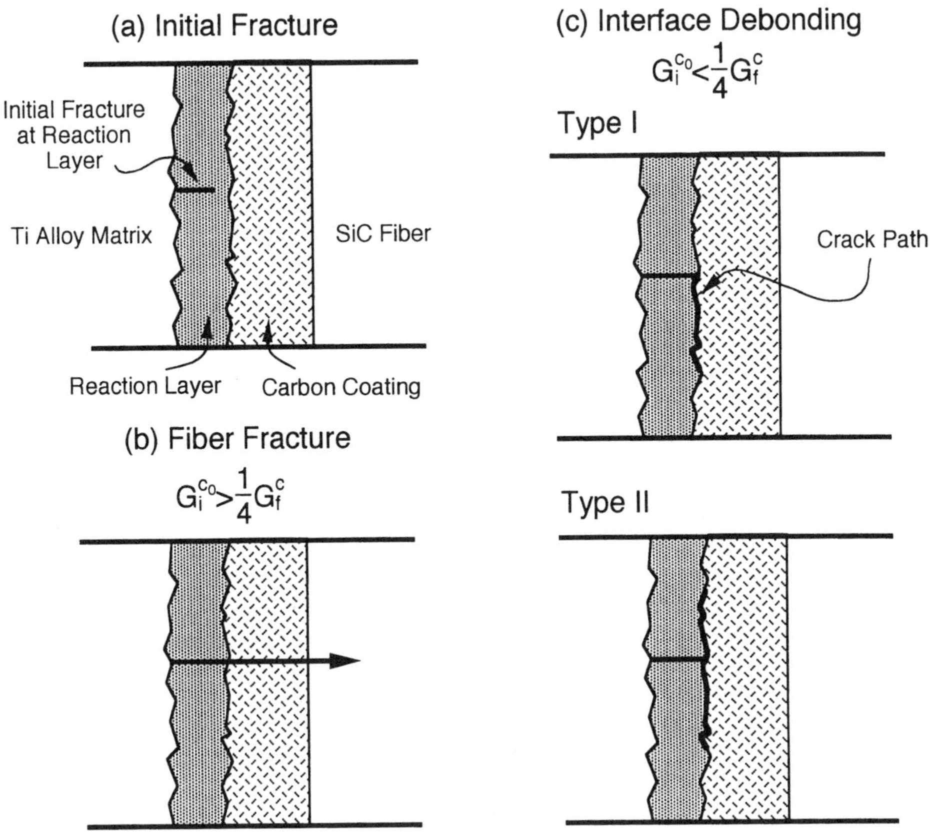

FIG. 10—*Modeling of the crack path in the extracted fiber:* (a) *initial cracking of the reaction layer,* (b) *interfacial undebonded model, and* (c) *interfacial debonding model (Type I and Type II).*

10*b*), assuming Young's modulus of the reaction layer (TiC $\approx$ 350 GPa) is nearly the same as that of the SiC fiber (400 GPa) and neglecting thermally induced stresses, the strength of the extracted fiber, σ_{fu}, is approximately given by [*30*]

$$\sigma_{fu} \approx \sqrt{\frac{G_f^c E_f}{\pi x}} \tag{11}$$

where G_f^c is the toughness of the fiber ($\approx$44 J/m^2) [*31*] and x is the thickness of the reaction layer. This equation means that the strength of the fiber is proportional to the inverse square root of the reaction layer thickness, $1/\sqrt{x}$. The calculation result of σ_{fu} as a function of x using Eq 11 is also shown in Fig. 9 by the broken line. The equation does not explain the experimental result. Thus, it is considered that the reaction layer is not a source of fracture for the extracted fiber.

In Fig. 10*a*, if the crack formed at the reaction layer is deflected along the interface (Fig. 10*c* Type I and Type II), the effect of stress concentration due to the fracture of the reaction layer should disappear and the tensile strength of the fiber should be independent of the thickness of reaction layer. Such a condition for interface debonding is provided by a minimum toughness of the interface, G_i^{co}. If the interface toughness, G_i^c, is smaller than G_i^{co}, the crack will be deflected along the weakest interface between fiber and matrix. The basic required condition for crack deflection at the interface is given by He and Hutchinson [*32*] as

$$G_i^{co} \leq \frac{1}{4} G_f^c \tag{12}$$

Assuming $G_f^c \approx 44$ J/m^2 [*31*], the calculated G_i^{co} is 11 J/m^2. In this experiment, the experimentally-obtained interface toughness was 4 J/m^2 and this value is smaller than the critical toughness of the interface for the onset of interface debonding prior to fracture of the fiber. Therefore, after the onset of fracture of the reaction layer, the interfacial shear failure occurs and this can significantly reduce the notch effect by blunting the crack along the debonded interface. In this experiment, G_i^c is independent of the thickness of the reaction layer, and therefore the formation of the reaction layer has no influence on the strength of the extracted fibers. This is because interfacial debonding (Fig. 10*c* Type I and Type II) is preferable to fiber fracture (Fig. 10*b*).

The increased scattering in strength, that is, the tendency toward a bimodal distribution, can not be explained by the interfacial shear mechanical properties. The result indicates that the effect of compositing is to increase the probability of failure at lower stress and not at higher stress. Bimodal distribution appears when the flow size increases or number of the flow changes or both. Figure 8 indicates that the initial flows get bigger with heat exposure or the bigger flow is introduced during the formation of the reaction layer or both. A typical SEM photograph of the surface of the extracted fiber is shown in Fig. 11. The extracted fiber surface in the figure shows a partially-separated part of the reaction layer between the SiC fiber and coating layer. This suggests some change in the SiC fiber surface after the formation of the reaction layer, however, more detailed studies are required to understand this bimodal distribution of the extracted fiber including the fiber damage during the extracting process.

Conclusion

The effect of the reaction layer thickness on the interfacial shear mechanical properties and tensile strength of the extracted SiC fiber has been studied. The interfacial toughness, shear sliding stress, and the strength of the extracted fiber were evaluated as a function of the reaction layer thickness.

The interfacial toughness and shear sliding stress are weakly dependent on the thickness of the reaction layer. The results of the SiC fiber tensile strength indicate that the mean fiber strength is the same up to a reaction layer thickness of 1.7 μm. Therefore, as soon as the reaction layer fractures, the interface between the reaction layer and the coating layer will tend to debond, and this debonding reduces the effect of the stress concentration of the fiber. The interface toughness is assumed to be enough to cause interfacial shear debonding along the fiber axis after the onset of cracking for a thick reaction layer.

The tendency toward a bimodal distribution of extracted fiber appears after the formation of a reaction layer. The result indicates that compositing increases the probability of failure at lower stresses and not at higher stresses. However, more detailed studies are required to understand this bimodal distribution after compositing.

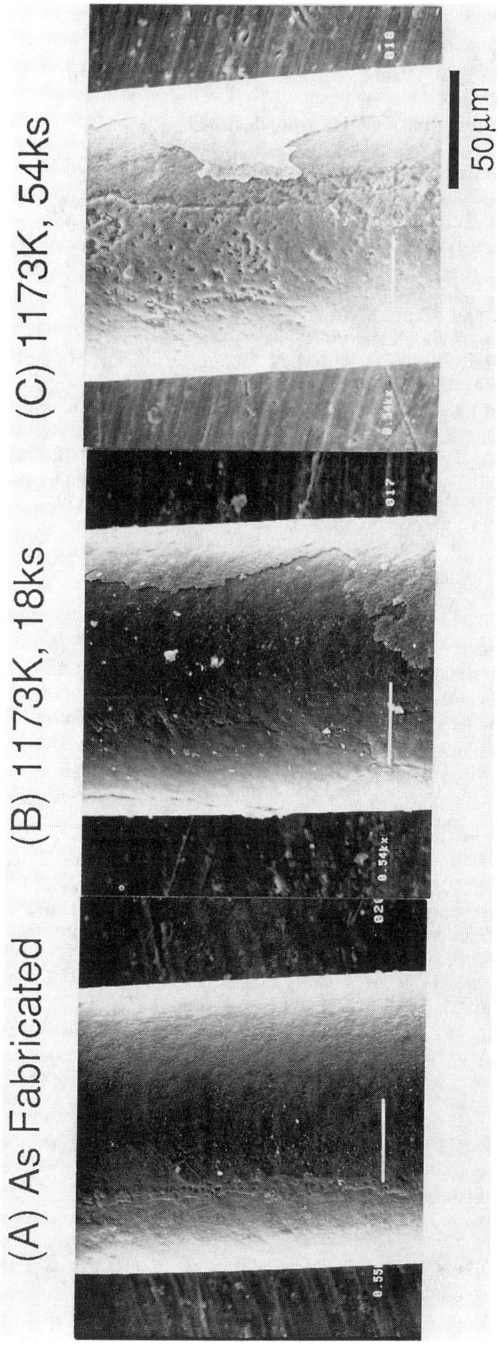

FIG. 11—*Typical SEM photographs of the surface of extracted fibers.*

References

[1] Smith, P. R. and Froes, F. H., *Journal of Metals,* Vol. 36, 1984, p. 19.
[2] Peticorps, Y. Le., Lahaye, M., Pailler, R., and Naslain, R., *Composite Science and Technology,* Vol. 32, 1988, p. 31.
[3] Ning, X. J., Pirouz, P., Lagerlof, K. P. D., and Dicarlo, J., *Journal of Materials Research,* Vol. 5, 1990, p. 2865.
[4] Rhodes, C. G. and Spurling, R. A., in *Recent Advances in Composites in the United States and Japan, ASTM STP 864,* J. R. Vinson and M. Taya, Eds., American Society for Testing and Materials, Philadelphia, 1985, p. 585.
[5] Rhodes, C. G., Ghosh, A. K., and Spurling, R. A., *Metallurgical Transactions A.,* Vol. 18A, 1987, p. 2151.
[6] Yang, J. M. and Jamg, S. M., *Journal of Metals,* Vol. 41, 1989, p. 56.
[7] Jones, C., Kiely, C. J., and Wang, S. S., *Journal of Materials Research,* Vol. 4, 1989, p. 327.
[8] Wei, W., in *Fundamental Relationships Between Microstructure & Mechanical Properties of Metal-Matrix Composites,* P. K. Liaw and M. N. Gungor, Eds., The Minerals, Metals, & Materials Society, Warrendale, PA, 1990, p. 353.
[9] Yang, J. M., Jeng, S. M., and Yang, C. J., *Materials Science & Engineering,* Vol. A138, 1991, p. 155.
[10] Wright, P. K., Nimmer, R., Smith, G., Senmeir, M., and Brun, M., in *Interfaces in Metal-Ceramics Composites,* R. Y. Lim, R. J. Arsenault, G. P. Martines, and S. G. Fishman, Eds., The Minerals, Metals, & Materials Society, Warrendale, PA, 1989, p. 559.
[11] Smith, P. R., Phodeo, C. G., and Revelos, W. C., in *Interfaces in Metal-Ceramics Composites,* R. Y. Lim, R. J. Arsenault, G. P. Martines, and S. G. Fishman, Eds., The Minerals, Metals, & Materials Society, Warrendale, PA, 1989, p. 35.
[12] Jansson, S., Deve, H. E., and Evans, A. G., *Metallurgical Transactions A,* Vol. 22A, 1991, p. 2975.
[13] Metcalfe, A. G., *Journal of Composite Materials,* Vol. 1, 1967, p. 356.
[14] Shorshorov, M. Kh, Ustinov, L. M., Zirlin, A. M., Olefirenko, V. I., and Vinogradov, L. V., *Journal of Materials Science,* Vol. 14, 1979, p. 1850.
[15] DiCarlo, J. A., in *Mechanical Behavior of Metal Matrix Composites,* J. E. Hack and M. F. Amateau, Eds., American Institute of Mining, Metallurgical, and Petroleum Engineers, Warrendale, PA, 1983, p. 1.
[16] Ochiai, S. and Murakami, Y., *Metallurgical Transactions,* Vol. 12A, 1981, p. 1155.
[17] Honda, K. and Kagawa, Y., *Journal of Japan Institute of Metals,* Vol. 58, 1994, p. 91.
[18] Kagawa, Y., Fujita, T., and Okura, A., *Acta Metallurgica & Materials,* Vol. 42, 1994, p. 3019.
[19] Yang, C. J., Jeng, S. M., and Yang, J. M., *Scripta Metallurgica,* Vol. 24, 1990, p. 469.
[20] Kagawa, Y. and Okura, A., *Journal of Japan Institute of Metals,* Vol. 56, 1992, p. 278.
[21] Kagawa, Y., ''Development and Evaluation of Advanced Inorganic Materials Mechanical Properties,'' Report of Scientific Research on Priority Areas, The Ministry of Education, Science and Culture, Tokyo, 1990, p. 151.
[22] Watson, M. C. and Clyne, T. W., *Acta Metallurgica & Materials,* Vol. 40, 1992, p. 141.
[23] Catalog (SCS-6 SiC Fiber), Textron Corp., Lowell, MA, 1992.
[24] Metcalfe, A. G., *Composite Materials, Vol. 1, Interfaces in Metal Matrix Composites,* A. G. Metcalfe, Ed., Academic Press, New York, 1974.
[25] Gundel, D. B. and Wawner, F. E., *Scripta Metallurgica,* Vol. 251, 1991, p. 437.
[26] Kernas, R. and Parthansarathy, T. A., *Journal,* American Ceramic Society, Vol. 74, 1991, p. 1585.
[27] Warren, P. D., Mackin, T. J., and Evans, A. G., *Acta Metallurgica & Materials,* Vol. 40, 1992, p. 1243.
[28] Weibull, W., *Journal of Applied Mechanics,* Vol. 18, 1951, p. 293.
[29] Soumelidis, P., Quenisset, J. M., Naslain, R., and Stoloff, N. S., *Journal of Materials Science,* Vol. 21, 1986, p. 895.
[30] Clyne, T. W. and Withers, P. J., *An Introduction to Metal Matrix Composites,* Cambridge University Press, Cambridge, UK, 1993, p. 173.
[31] Chan, K. S., *Metallurgical Transactions A,* Vol. 24A, 1993, p. 1531.
[32] He, M. Y. and Hutchinson, J. W., *International Journal of Solids and Structures,* Vol. 25, 1989, p. 1053.

David L. Davidson[1]

Fiber-Matrix Micromechanics and Microstructural Observations Under Tensile and Cyclic Loading

REFERENCE: Davidson, D. L., **"Fiber-Matrix Micromechanics and Microstructural Observations Under Tensile and Cyclic Loading,"** *Life Prediction Methodology for Titanium Matrix Composites, ASTM STP 1253,* W. S. Johnson, J. M. Larsen, and B. N. Cox, Eds., American Society for Testing and Materials, 1996, pp. 43–57.

ABSTRACT: A technique is described for measuring the displacements in fibers due to loading the composite. Fiber access was obtained by removing only very small pieces of the matrix, thereby minimizing the change in residual stresses. Fiber stresses were obtained near a hole, a slit, and a fatigue crack. It was found that fibers slipped within the matrix at a stress of 2 to 2.6 GPa near both the hole and slit; after cyclic loading the specimen with a hole, fiber stress at slip increased to 2.6 to 3.6 GPa. Fiber stresses bridging the wake of a fatigue crack were found to be approximately constant. The models used to compute a stress intensity factor at the tip of a bridged fatigue crack are reviewed, and it is shown that those assuming a constant fiber stress gave the best results.

KEYWORDS: fiber stress, stress concentration, fatigue crack growth, in situ measurements, stereoimaging, bridging zone fiber stress distribution, titanium, titanium matrix composites, life prediction, titanium alloys, fatigue (materials), modeling

A load applied in the fiber direction of continuous fiber metal matrix composite is shared by matrix and fiber. To determine analytically how the load is partitioned between fiber and matrix, it is necessary to assume elasticity and nonslippage between fiber and matrix. Fibers in these composites have a higher modulus than the matrix, and will, therefore, carry an increasingly higher load than the matrix which causes an increasingly larger shear stress at the fiber-matrix interface to occur. Thus, as the load is increased, either the fiber-matrix interface will break and interfacial slip will occur, or the fiber will break.

The methods developed for determining the point at which a fiber will separate from the matrix are: (1) the fiber pushout test [*1*] and (2) the fiber pullout test [*2*]. The value of these seemingly simple, direct methods of measurement is that the results should be easily interpreted, but experience has shown that the tests are more complex than they appear and that considerable analysis may be required to interpret the measurements [*3*]. One of the disadvantages of both of these testing methods is that the composite must be extensively altered to obtain specimens suitable for making the measurements.

The limitations in the fiber pushout and pullout tests have led to the development of an alternative solution for determining the response of fibers to loading. Ideally, any new method would circumvent the limitations imposed by the pushout and pullout tests, which is the need to cut thin slices from the material, thereby causing extensive material removal to expose the fibers for either pushing or the attachment of a tensile loading fixture. Removal

[1]Scientist, Southwest Research Institute, San Antonio, TX 78228.

of this much material for testing purposes alters the residual stress state of the material, which is important because it is the compressive radial residual stress that provides a force normal to the fiber-matrix interface that regulates the debonding and slippage of the fiber. This paper describes briefly a technique developed for the in situ observation of fibers that removes only a small amount of matrix, and, therefore, results in much less alteration of the residual stress state. Furthermore, the fibers of interest in a structure are those sampled, rather than fibers from a section remote to the notch or crack of interest.

Specimens prepared for the observation of fibers in situ were loaded through pins on the centerline of the specimen. The response of fibers to loading was determined by comparing photographs made at minimum load with those made at increasing load using the stereoimaging technique [*4*]. This allowed the visual system (eye and brain) to see the displacements in fiber and matrix caused by the application of load. Those displacements were measured using our system for automated photogrammetry, termed DISMAP [*5*]. The response of fibers to loading was observed in situ in specimens having a variety of stress concentrations, that is, holes, notches, slits, and fatigue cracks, and the results from those experiments are given following a description of the technique used to access the fibers.

Description of Technique

To reveal fibers for observation, the matrix overlaying the first row of fibers was removed, very carefully, exposing as small a portion of the fiber as possible, typically 0.060 by 0.15 mm, thereby perturbing the residual stresses in the composite to the minimum extent. The resulting hole in the matrix that exposes the fiber has been termed a "fiber port." Matrix removal was accomplished most successfully using a combination of ion etching and electropolishing, but some fiber ports were made using only electropolishing. The fiber port preparation sequence used is depicted in Fig. 1.

Ion etching was used to provide a trench with well-defined sides perpendicular to the fiber direction, Fig. 1 (top). The mask used to limit the region etched by ions was two parallel razor blades separated by 0.5 mm. Nitrogen ions at approximately 8 KeV were used to make a trench varying in depth from about 0.1 mm to almost 0, with the deepest part of the trench near the stress concentration. Each trench required several days of ion etching time. The resulting trench width was somewhat less than the 0.5-mm separation of the masks.

After ion etching, the specimen was masked off completely with stopoff lacquer, except for the trench emanating from the stress concentration. The sharp discontinuity of the ion-etched trench allowed the high-viscosity lacquer to be layed exactly to the edge of the trench, resulting in only a narrow region of matrix left exposed to the polishing solution.

Electropolishing was accomplished on all the titanium alloy matrix materials using nitric acid in ethanol, usually with a concentration of 25% by volume. The best polishing results were obtained with a potential of between 20 and 22 volt direct current (vdc) in a solution maintained at a temperature of -20 to -40°C. Matrix material was very carefully dissolved by using small time increments for electropolishing that varied from 30 s to 2.5 min. Between each increment, the specimen was examined under a microscope for evidence of fibers.

The specimen was first electropolished with the trench in the vertical position with the stress concentration at the top. This position caused the thickness of the viscous layer to increase with distance from the stress concentration, because gravity caused the heavier viscous layer to flow downward. Thus, the highest polishing rate occurred at the top of the trench, nearest the stress concentration where the viscous layer was the thinnest.

As soon as enough matrix was removed to reveal the first fiber, this portion of the specimen was masked with lacquer. Then, the specimen was turned 90° so that the trench

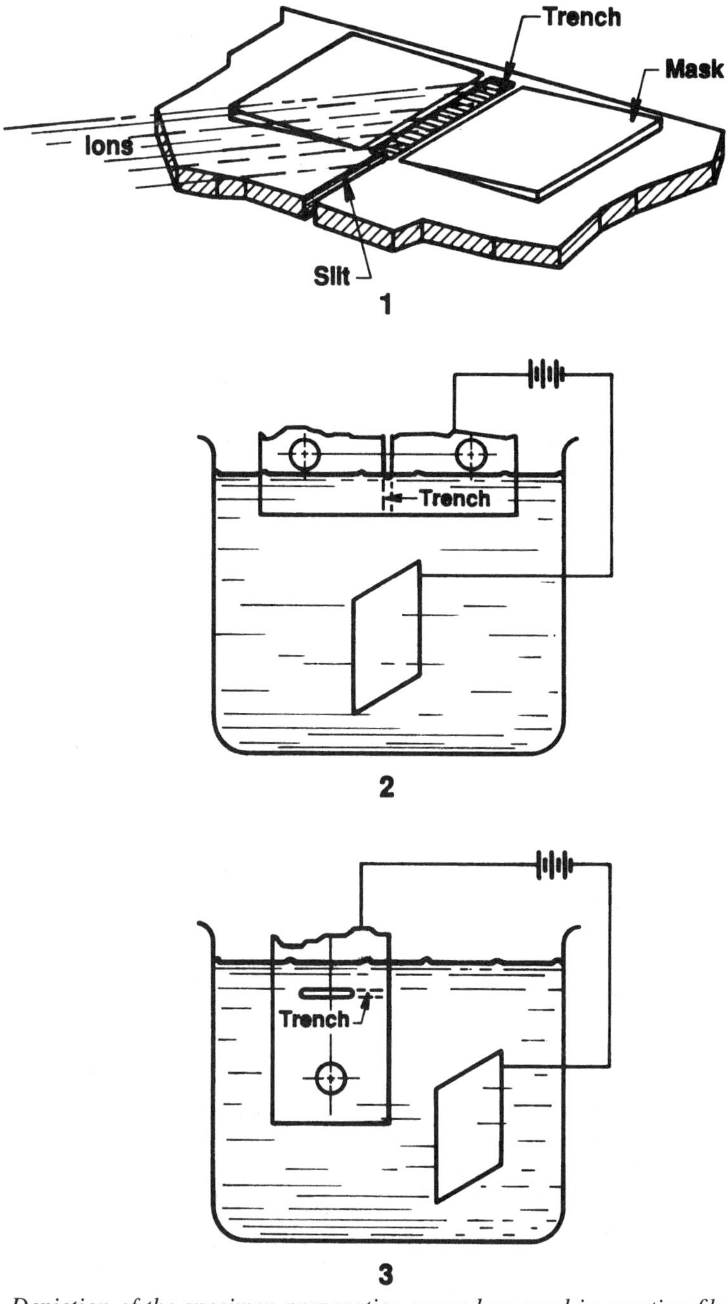

FIG. 1—*Depiction of the specimen preparation procedure used in creating fiber ports.*

was perpendicular to the force of gravity. This meant that the viscous layer was thicker in the lower part of the trench and drained uniformly along its length. Thus, polishing occurred fastest in the upper part of the trench. This manipulation of specimen position allowed the size of the fiber port to be controlled better than if one position alone had been used. Ports made using the method described ranged from 35 to 350 μm in length, with the typical port being elliptical with axes of 150 by 60 μm.

The surfaces of the fibers revealed through the fiber ports were relatively smooth and featureless, as viewed by secondary electrons in the scanning electron microscope (SEM), and would have caused difficulty during the observation and measurement of fiber response to loading. This problem was solved in two ways: (1) a light ion etch was used to texture the fiber surfaces, or (2) magnesium oxide particles were deposited from smoke on the fiber surfaces. Both of these treatments gave the fiber surface a fine texture when viewed using the secondary electron image in the SEM. Some of the fiber ports made using this process are shown in Fig. 2. Fibers in the case illustrated were decorated with MgO particles.

Loading of specimens in which fibers were exposed through fiber ports was accomplished using a special stage that fit within the specimen chamber of a SEM [*6*]. With this equipment, load was applied to both ends of the specimen using hydraulic actuators and the fibers were observed under high-resolution conditions. An illustration of the displacements measured in and near a fiber is shown in Fig. 3.

Results

Observations of fiber response were made under both unidirectional monotonic loading and after cyclic loading. All specimens used were straight sided, 25.4-mm wide by 54-mm long and 0.94-mm thick. The stress concentrations were located symmetrically about the centerlines of the specimen. Loading was through 6.3-mm holes located along the specimen long axis centerline about 30 mm apart. The specimens loaded monotonically had a Ti-15V-

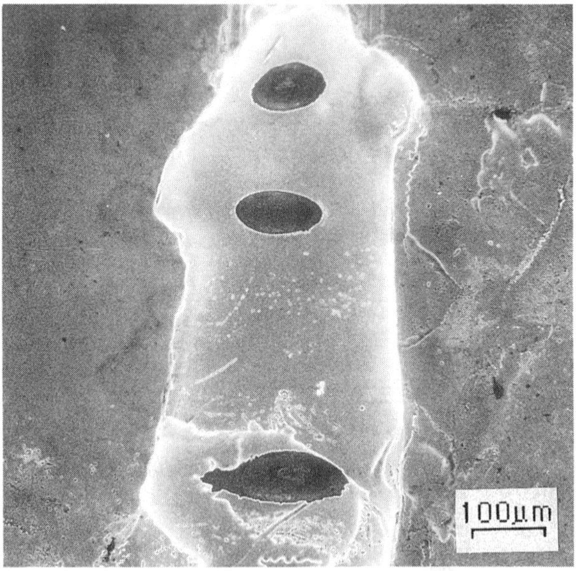

FIG. 2—*Illustration of fiber ports prepared using the procedure described in Fig. 1.*

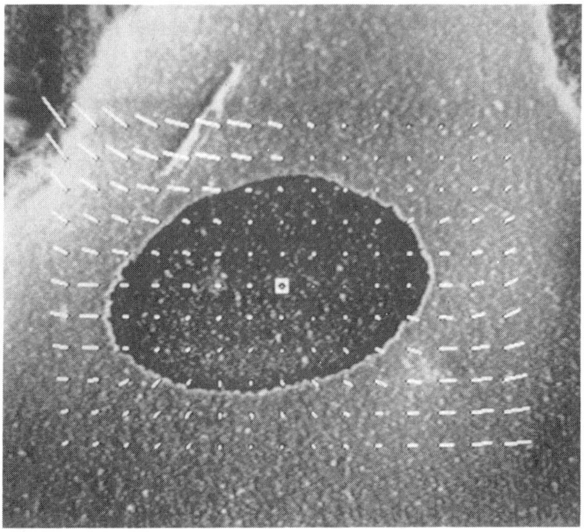

FIG. 3—*Displacements in a fiber and the surrounding matrix superimposed on a photograph of the fiber as seen through a fiber port. Displacements were measured each 11 μm, and the displacements are magnified by a factor of 15 for visualization.*

3Al-3Cr-3Sn (percent by weight) matrix reinforced with 35.5% by volume Textron Corp. SCS-6 fibers (≈150 μm diameter). The fatigue-cracked specimens were Ti-6Al-4V matrix reinforced with 42% by volume SCS-6 fibers. Both composites were manufactured by Textron Specialty Materials using the foil-fiber-foil method. Fiber architecture was different for the two composites used, although for both there were four fiber layers. The Ti-15-3-3-3 matrix specimens had two inner fiber layers transverse to the loading axis. Thus, only the two outer layers were parallel to the direction of loading. The Ti-64 matrix specimens had all four fiber layers oriented in the direction of loading.

Stress concentrations examined in the SCS-6/Ti-15-3 specimens were: (1) a "hole" 12.5 mm in diameter and (2) a "slit" 14.1 mm in length (1.61 mm longer on the side of the specimen having fiber ports) and 0.28 mm wide. Specimens of SCS-6/Ti-64 were *fatigue* cracked from centered slits approximately 8 mm long by about 0.5 mm wide. A complete description of the fatigue specimens can be found in Ref *7*.

The technique used to determine fiber stress was as follows: photographs of the fibers were made at minimum load through the fiber port holes, and then again with increasing load. Displacements were measured with DISMAP and converted into strains that were then multiplied by the modulus of the fiber (400 GPa) to obtain stress.

Hole Specimen Monotonically Loaded

During loading of hole specimen, the first fiber broke at a low load; it was probably damaged during specimen fabrication. The responses of the second fiber (0.39 mm from the hole edge) and third fiber (0.60 mm) were followed in detail. The fiber stresses shown in Figs. 4*a* and *b* are given as a function of the applied load. The figures show that the fiber (strain) stress increased with load to approximately (0.0056) 2.5 GPa, but either remained constant or decreased as load was further increased, then fiber (strain) stress again increased as the applied load increased.

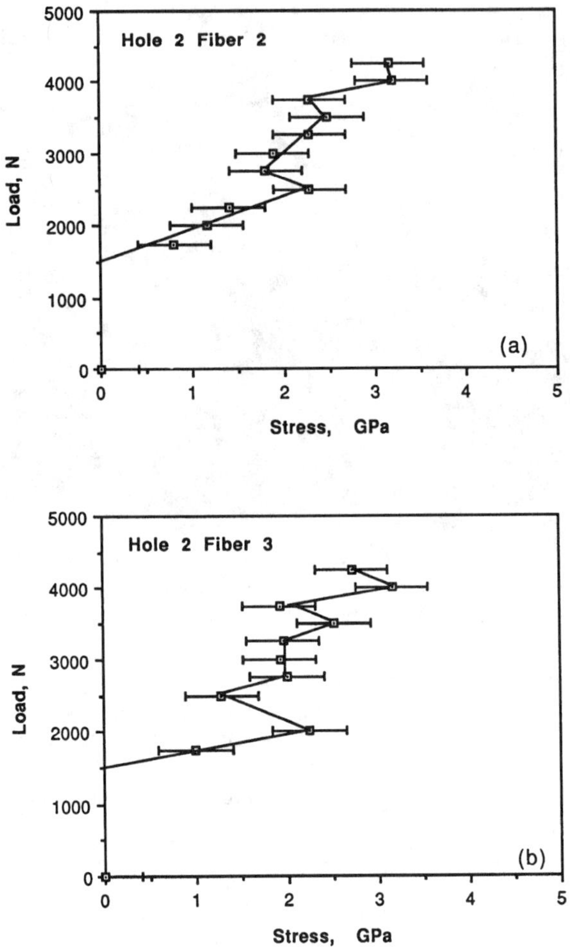

FIG. 4—(a) *Specimen with "hole": stress in* (a) *Fiber 2 and* (b) *Fiber 3, as a function of applied load.*

An abrupt change in slope of fiber stress versus load is interpreted as an indication that sufficient slip has occurred between the fiber and matrix to transfer part of the fiber stress to the matrix and other fibers. Slip may have initiated at a lower load, but it was insufficient to transfer enough load from the fiber to drastically alter its load-strain relationship. Thus, for loads sufficient to cause a fiber (strain) stress of about (0.0056) 2.5 GPa, considerable debonding between fiber and matrix is likely to have occurred.

Slit Specimen Monotonically Loaded

Stresses in the three fibers nearest the slit (0.17, 0.37, and 0.55 mm from the slit, respectively) were measured. The stress versus load history for these fibers, Figs. 5*a*, *b* and *c*, is similar to that found for fibers near the hole specimen, except that there is more of what might be interpreted as stick-slip behavior. Stress (strain) at the point of first slippage with

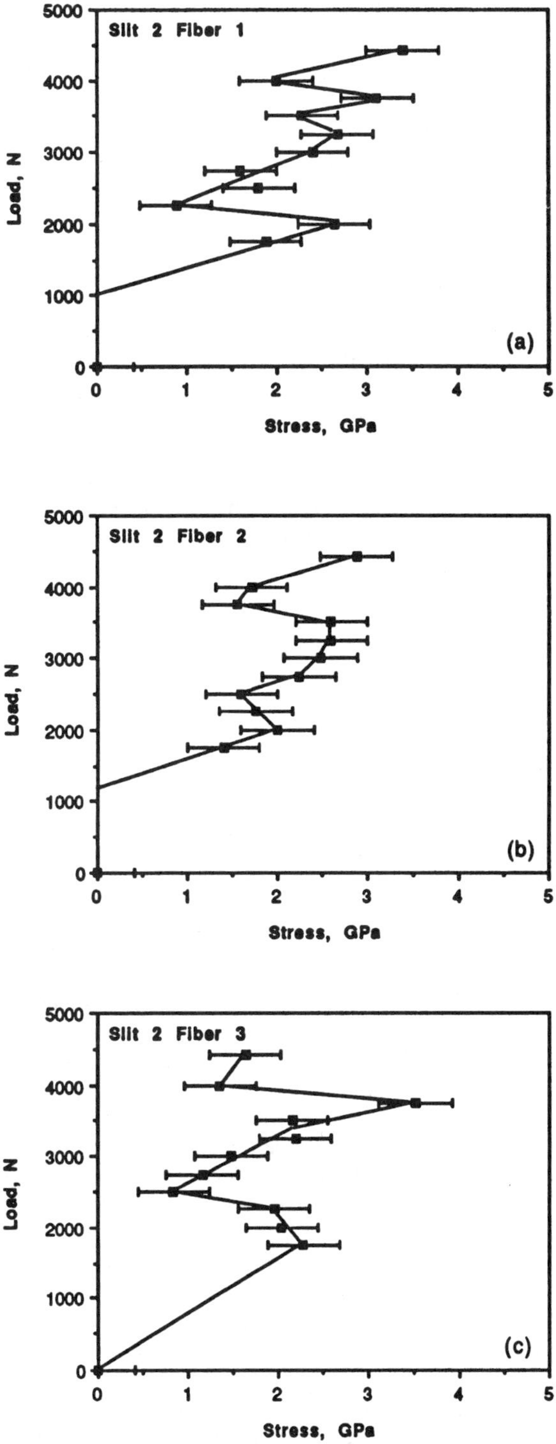

FIG. 5—*Specimen with "slit": stress in* (a) *Fiber 1,* (b) *Fiber 2, and* (c) *Fiber 3.*

increased loading varied between 2 GPa (0.0049) and 2.65 GPa (0.0066), which is about the same level as found for the hole specimen.

Irregularities in fiber slippage with increased loading could occur for at least two reasons: (1) surface irregularities on the interfaces between fiber coating and matrix or between the fiber and its coating could cause the fiber to slide easier at some points than at others, (2) sliding may be occurring at both interfaces (fiber/coating or coating/matrix), and (3) the load transfer between the various slipping fibers and their surrounding matrix is likely to be a complex interactive process.

Hole Specimen Cycled 30 Times

A second hole specimen was cycled 30 times between 150 and 4250 N. Cycling did not break any of the fibers that were being observed, although the loads should have been large enough to strain the fibers well beyond the point where slippage occurred between fibers and matrix. A number of load cycles were applied to bring to equilibrium some of the complex load redistribution between fibers and matrix and possibly smooth some of the interfacial roughness that might affect the load transfer. Strains in the first three fibers (0.13, 0.27, and 0.47 mm from the hole edge, respectively) were measured, as shown in Figs. 6*a*, *b*, and *c*. The same complex behavior found for fibers in the monotonically loaded hole is still evident for Fibers 2 and 3, but for Fiber 1 there may have been some change. For all the fibers, the stress at first unloading was 3.2 to 3.6 GPa, which is greater than the $\approx$2.2 GPa found for the monotonically loaded hole specimen. For both hole specimens, the number of stick-slip events observed was less than for the slit specimen.

Summary of Hole and Slit Specimen Results

The fiber stress levels at which slippage occurred between fiber and matrix are summarized in Table 1 and compared to the average stress that a fiber would have supported for the load applied when slippage began. The average stress was calculated on the basis of net section area, that is, ignoring the area of the stress concentration. The stress concentration factor tabulated in the last column was determined by dividing measured fiber stress by the average stress.

The stress concentration for the hole is very large compared to that expected at a hole in an isotropic material (about 4.3, as determined from Peterson [*8*]); however, the analysis by Bigelow [*9*] indicates that fiber stresses should be in the 1.1 to 1.8 GPa range for the applied stresses used in these experiments, which would give a first fiber stress concentration of 7 to 14, somewhat below the 14 to 17 found experimentally.

For the specimen with a slit, the stress concentration found (13 to 18) is about that expected based on elastic analysis [*8*] for an elliptical notch in a homogeneous solid with a height-to-width ratio of 8 ($K_{tg} \approx 20$). Also, the results of a three-dimensional finite element analysis by Johnson and Bigelow [*10*] for a slit of similar dimensions indicated that a stress concentration of 22 in the first fiber would be expected. Thus, the experimental results for the slit are about what was expected.

The increase in stress concentration caused by cycling the hole specimen evidently occurred due to the rearrangement of applied load between the fiber and matrix. Residual stresses in the matrix (tensile) and fibers (compressive) were relaxed due to cycling the hole specimen, and this resulted in the fibers carrying an increased load, thereby increasing the stress concentration. Fiber stresses may be smaller than measured for the uncycled hole and slit specimens because at minimum load the fibers were in compression. Conversely, if cycling redistributed the stresses as expected, the fiber stresses at minimum load would be

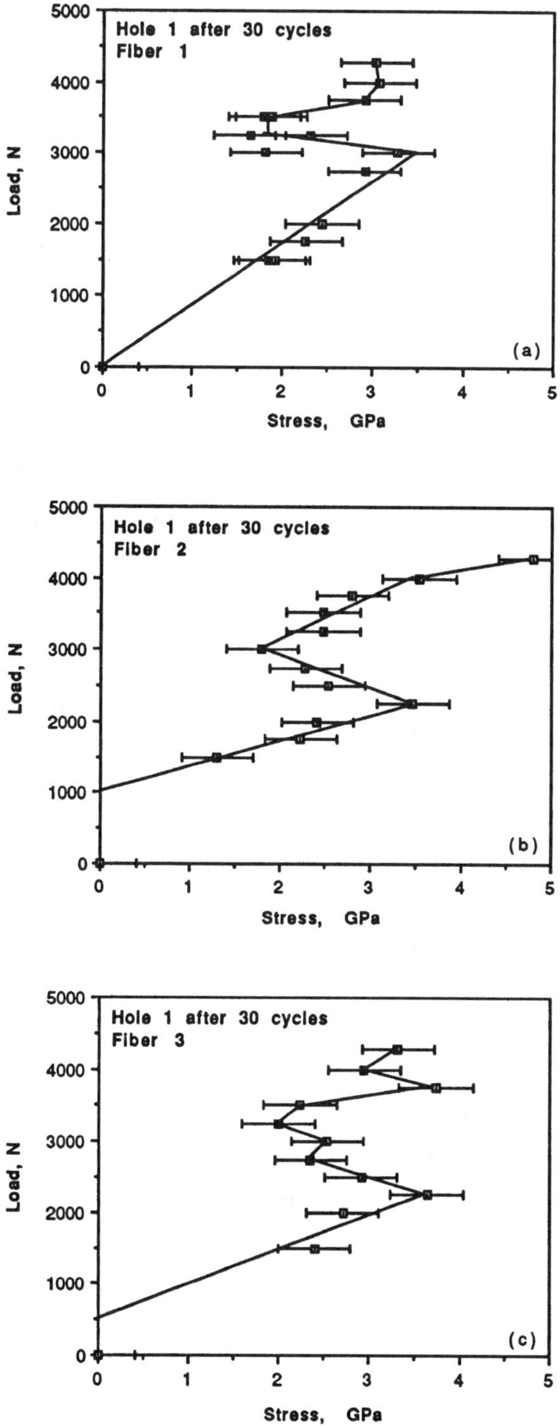

FIG. 6—*Specimen with "hole" cycled 30 times at a stress of 170 MPa:* (a) *Fiber 1,* (b) *Fiber 2, and* (c) *Fiber 3.*

TABLE 1—*Hole and slit results summary.*

Stress Concentration	Fiber	Load, N	Slip Stress, GPa	Average Stress, GPa	Stress Concentration Factor
Hole 2	2	2500	2.2	0.162	14
	3	2000	2.2	0.130	17
Hole 1	1	2000	2.6	0.130	20
(cycled 30 times	2	2200	3.5	0.145	25
at 170 MPa)	3	2200	3.6	0.142	25
Slit 2	1	2000	2.6	0.130	18
	2	2000	2.0	0.130	13
	3	1800	2.2	0.116	17

close to zero. Thus, the fiber stress concentrations for the cycled hole specimen could be more than given in the table. An analysis [*9*] of the first fiber stress concentration after 200 000 tension load cycles at a stress of 250 MPa indicated that the stress had increased by a factor of 2 (from 1.1 to 2.2 GPa) for an applied stress of 140 MPa. Experimentally, fiber stress increased by a factor of 1.4, which is lower than that predicted, but transfer of load from matrix to fiber was as predicted by the analysis.

Fatigue Specimens

Only two of the four specimens in which fatigue cracks were grown was prepared with fiber ports. Extensive fiber bridging was observed for these specimens at some levels of applied stress. Fiber strains were measured both ahead of and behind the crack tip. Fiber strains were measured only at maximum load as a function of crack length, applied stress level, and the extent of fiber bridging. Table V of Ref *7* lists all the measured fiber strains (converted to stress through multiplying by the modulus). These data are shown in Fig. 7 as a function of position relative to the crack tip. The scatter in these data are indicative of: (1) errors in the measurement of fiber strains (shown by the error bars), (2) the irregularities in stress caused by differences in fiber/matrix interface, (3) irregularities in fiber spacing within the matrix, and probably some other factors.

Fatigue cracks were grown at applied stresses between 118 and 175 MPa, and bridging lengths varied from 1 to 2.4 mm. Fibers stresses, taken separately for each combination of variables, or collectively as shown in the figure, do not show any effect of applied stress or location relative to the crack tip. Applied stress had such little effect on fiber stress because of differences in bridging length. The important result shown in the figure is the insensitivity of fiber stress to distance from the crack tip.

A simple method was devised to compute the expected fiber stresses, based on applied stress, crack length, and length of the bridging zone. The remotely applied load was assumed to partition in proportion to the cracked and uncracked cross-sectional areas, and fiber stress was assumed to be uniform in the bridging zone and support the entire load of the cracked area [*7*]. The fiber stresses thus computed (see Table V of Ref *7*), agreed with those measured reasonably well, as shown in Fig. 8. The reason fiber stresses are approximately constant in the bridging zone is that cycling has redistributed the stresses between fibers and matrix. This result is similar to that observed for the hole stress concentration that was load cycled, as described earlier.

Discussion

The levels of fiber stress and their distribution relative to the fatigue crack tip are not surprising when the stresses in fibers near the other stress concentration geometries are

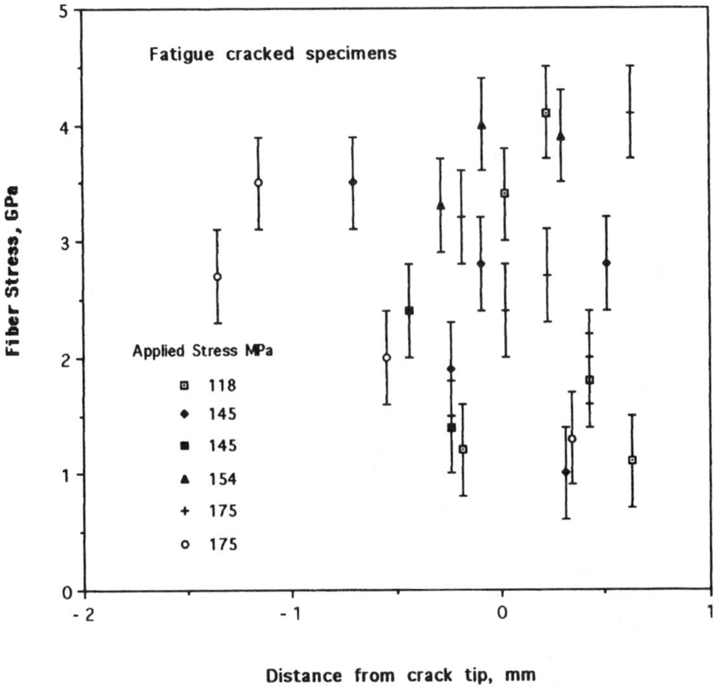

FIG. 7—*Fatigue cracked specimens: fiber stress as a function of distance from the crack tip. Symbols denote the gross section applied stress on the specimen. See Ref 7 for further details.*

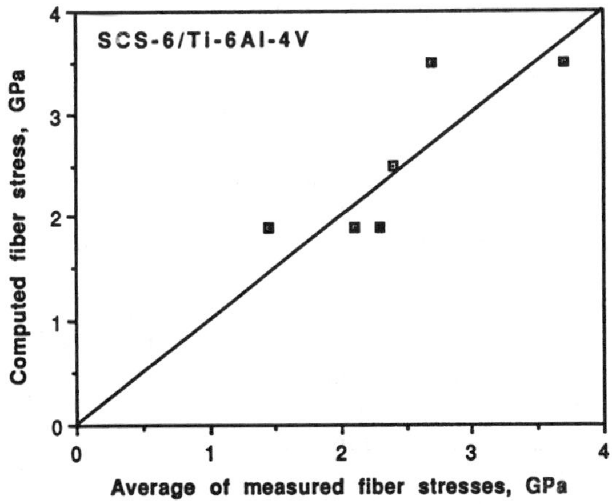

FIG. 8—*Comparison of the average of measured fiber stresses in the bridging zone of fatigue cracks to fiber stresses calculated assuming that stress in the bridging zone was uniform.*

considered. For both the hole and slit specimens, load application caused a redistribution of fiber stresses because of slippage between fiber and matrix. Whereas the amount of slippage was not determined for the hole and slit specimens, it was measured for the fatigue cracks [*7*]. Repeated cycling of the slipping interface damaged the fiber coating, so that when the surrounding matrix was removed chemically, spallation and cracking in the coating were evident. The damage zone was easily measured, and it was found that the extent of slip was a function of the crack opening displacement that increased with distance from the crack tip. This finding indicated that fiber stresses that would have been high because of their distance from the crack tip were lowered by an increase in slip distance between the fiber and matrix. Therefore, fibers bridging the crack tended towards constant stress in the equilibrium condition brought on by the application of many thousands of cycles. This result has been reinforced by a number of other investigators, as will be reviewed later.

Fiber coating damage that increased with distance from the crack tip has been reported also by Blatt et al. [*11*], using a nondestructive technique.

The results of the fiber stress measurements also indicate the mechanism of fiber fracture. The extensive damage to the interface caused by growing the fatigue crack lowered the load transfer characteristics of the interface in the damaged region to a level where significant load transfer within the slip zone is unlikely. Rather, load transfer to the fibers was occurring mainly beyond that point in the region of elastic interaction between fibers and matrix. Thus, the fibers were approximately free of load transfer within the slip zone, so that as distance to the crack tip increased with a longer bridging zone, the length of fiber subjected to a high stress increased. Since fiber strength is inversely proportional to the length being stressed, fiber fracture became more likely as distance from the crack tip increased. Thus, the length of the bridging zone would be expected to be limited under steady-state crack growth conditions.

As noted earlier, in addition to the direct experimental measurements presented by Davidson [*7*], there is also indirect evidence that fiber stress is constant, or nearly so, in the bridging region behind fatigue cracks. Ghosn et al. [*12*], successfully modeled the growth of fatigue cracks in single-edge-notched (SEN) specimens of SCS-6/Ti-15-3 using a "fiber pressure" approach that essentially assumed that the fibers within the bridging zone all carried the same stress, but there was a bending component added for the SEN specimen. Later work by these investigators [*13*] also included a center-cracked specimen loaded in tension for which there was no additional bending term. Fiber pressure was calculated as a function of applied stress, crack length, and bridging zone length. Conversely, when the shear lag model [*14*] was used, the value of interfacial shear strength (τ) required for agreement between model and experiment was ≈6.2 MPa, which is considerably below anything measured.

Bakuckas and Johnson [*15*] also attempted to use the shear lag theory to rationalize fatigue crack growth data using center-notched, unidirectional SCS-6/Ti-15-3 composites. They found that in order to obtain data collapse, the value of τ had to be varied by about a factor of 4, depending on the applied stress, crack length, and bridging zone length.

In another attempt to use the shear lag model to rationalize Davidson's fatigue crack data, Chan [*16*] assumed that the value of τ decreased as the distance from the crack tip increased due to interfacial damage. Consequently, the values of τ used in his model decreased by a factor of 5 as distance within the bridging region increased to ≈1 mm. Kantzos et al. [*17*] found that for SCS-6/Ti-15-3, measured values of τ decreased from ≈83 MPa to ≈55 MPa after three loadings using a repeated load pushout test that is evidence of interfacial damage and wear due to repeated sliding.

Fatigue crack growth rates through SCS-6/Ti-24Al-11Nb specimens with a central hole were measured by Jira and Larsen [*18*] and modeled by John et al. [*19*]. They concluded that the results were best modeled using a fiber stress that was constant in the bridging zone and that was dependent on applied stress and crack length. Measurement of crack opening displacements (COD) were made as a function of distance from the crack tip using a laser interferometric displacement gage. Those measurements best fit a COD profile computed by assuming that the fiber stress in the bridging zone was a constant ratio of the applied stress. Conversely, the value of τ that had to be assumed for the shear lag model to rationalize crack growth data was dependent on the level of applied stress, in general agreement with the results of Bakuckas and Johnson.

In contrast to the results given earlier is the work of Walls et al. [*20*] who grew fatigue cracks through unidirectional SCS-6/Ti-15-3 center-notched specimens and explained their results using the shear lag model, with $15 < \tau < 35$ MPa. However, they also noted that the stress required for interfacial sliding degraded with cycling.

The concept of a constant fiber stress within the bridging zone was extended from titanium matrix composites to ARALL, an aluminum alloy matrix composite reinforced with aramid fibers, by Davidson who used his model [*7*] to compute the crack-tip stress intensity factor (ΔK) for the composite from applied stress, crack length, and bridging distance. The model collapsed data from the composite to the fatigue crack growth rate line for the matrix material [*21*], and the results were found to be relatively insensitive to bridging length.

The collective results of all of the work referenced earlier leads to the conclusion that cyclic sliding between fiber and matrix degrades the strength of the interface. To get reasonable results using the shear lag model, this degradation must be included by allowing τ to vary with applied stress, crack length, and distance from the crack tip. However, predictive models for this degradation have not been developed. Also, the magnitudes of τ that must be used are lower than anticipated from pushout test results. Conversely, if a simple model is used to compute a constant value of fiber stress in the bridging region from applied stress, crack length, and bridging length, then the models give crack-tip stress intensity factors that collapse composite data onto the growth rate curves for the unreinforced alloy.

Finally, fiber stresses at the point of significant sliding ($\approx$2.2 GPa, Table 1) were used in the model of Dollar and Steif [*22*]. Their model for a fiber being pulled from a half space predicts that slip will begin when stress on the fiber reaches 1.22 σ_0 (σ_0 = the compressive residual stress normal to the fiber caused by differential thermal expansion). Using the measured value of 2.2 GPa would predict $\sigma_0 \approx$ 1800 MPa which does not compare well with the value of $\approx$300 MPa found in Cox et al. [*23*]. Conversely, using $\sigma_0 \approx$ 300 MPa, the fiber stress at slippage is predicted to be $\approx$0.16 of that measured. It is concluded that this model is inapplicable to the present results.

Summary and Conclusions

A technique has been developed that allows displacements to be measured in fibers within metal matrix composites that are being remotely loaded. Small amounts of matrix are removed, exposing a small part of the fiber surface. Displacements are measured using stereoimaging, from which strain is calculated, then stress, using fiber modulus.

1. The fiber stress concentration near a central hole stress concentration was found to be 14 to 17, which is somewhat higher than predicted by analysis (8 to 14). Fibers began to slip relative to the matrix at 2 to 2.6 GPa. Further loading caused fiber stress to decrease, then increase in a manner characteristic of stick-slip behavior.
2. After the specimen with a hole was cycled 30 times, fibers near the hole were found to sustain 2.6 to 3.6 GPa before slipping, an increase in stress concentration of 1.4 over

the uncycled hole, whereas, analysis of a similarly cycled hole specimen predicted an increase of two in stress concentration, but to a maximum stress of only 2.2 GPa.

3. The stress concentrations in fibers near the end of a slit were found to be 13 to 17, which compares well to values of 17 to 22 determined by analysis.
4. Fiber stresses measured in the bridging zone behind fatigue cracks did not exhibit any trend in magnitude with distance from the crack tip, although the scatter in the measurements was large. It was concluded that fiber stress within the bridging zone was constant. This result was consistent with the effect of load on fiber stress measured near the hole and slit stress concentrations.
5. Fatigue crack growth results were found by a number of investigators to be best modeled using a constant stress in the bridging zone.
6. The concepts developed for predicting fatigue crack growth rate through titanium alloy composites were shown also to predict accurately fatigue crack growth rates in the aluminum alloy/aramid composite, ARALL.
7. The shear lag model was found to be inappropriate for describing the growth of fatigue cracks through these composites, because the interfacial shear strength must be varied with applied stress and crack length to obtain a fit of the model to the experimental results.

Acknowledgments

Support for research on fatigue crack growth came from the Office of Naval Research, Contract No. N000-14-93-C-0040, Dr. S. G. Fishman, technical monitor, and support for determining fiber stresses near stress concentrations came from NASA Langley Research Center, Contract L-03581D, Dr. W. S. Johnson, technical monitor. John Campbell provided experimental assistance, and Jim Spencer provided the DISMAP measurements. This support and assistance is greatly appreciated.

References

[1] Marshall, D. B. and Oliver, W. C., "Measurement of Interfacial Mechanical Properties in Fiber Reinforced Ceramic Composites," *Journal,* American Ceramics Society, Vol. 70, 1987, pp. 542–548.

[2] Hsueh, C.-H., "Interfacial Debonding and Fiber Pull-Out Stresses of Fiber Reinforced Composites," *Materials Science and Engineering,* Vol. A123, 1990, pp. 1–11.

[3] Hsueh, C.-H., "Interfacial Friction Analysis for Fiber Reinforced Composites during Fiber Push Down (Indentation)," *Journal of Materials Science,* Vol. 25, 1990, pp. 818–828.

[4] Davidson, D. L. in *Scanning Electron Microscopy/1979/II,* Scanning Electron Microscopy, O'Hare, IL, 1979, pp. 79–86.

[5] Franke, E. A., Wenzel, D. J., and Davidson, D. L., "Measurement of Microdisplacements by Machine Vision Photogrammetry (DISMAP)," *Review of Scientific Instruments,* Vol. 62, 1991, pp. 1270–1279.

[6] Davidson, D. L. and Nagy, A., "A Low Frequency Cyclic Loading Stage for the SEM," *Journal of Physics F,* Vol. 11, 1978, pp. 207–210.

[7] Davidson, D. L., "The Micromechanics of Fatigue Crack Growth in Ti-6Al-4V Reinforced with SCS-6 Fibers," *Metallurgical Transactions A,* Vol. 23A, 1992, pp. 865–879.

[8] Peterson, R. E., *Stress Concentration Factors,* Wiley, New York, 1974, pp. 150 and 196.

[9] Bigalow, C. A., "A Micromechanics Based Strength Prediction Methodology for Notched Metal Matrix Composites," *Composites,* Vol. 24, 1993, pp. 113–121.

[10] Johnson, W. S. and Bigelow, C. A., "Elastic Plastic Stress Concentrations Around Crack Like Notches in Continuous Fiber Reinforced Metal Matrix Composites," *Test Methods and Design Allowables for Fibrous Composites, Second Volume, ASTM STP 1003,* C. C. Chamis, Ed., American Society for Testing and Materials, Philadelphia, 1989, pp. 261–275.

[11] Blatt, D., Karpur, P., Stubbs, D. A., and Matikas, T. E., "Observation of Interfacial Damage in

the Fiber Bridged Zone of a Titanium Matrix Composite,'' *Scripta Metallurgica and Materials,* Vol. 29, 1993, pp. 851–856.

[*12*] Ghosn, L. J., Kantzos, P., and Telesman, J., ''Modeling of Crack Bridging using a Unidirectional Metal Matrix Composite,'' *International Journal of Fracture,* Vol. 54, 1992, pp. 345–357.

[*13*] Ghosn, L. J., Telesman, J., and Kantzos, P., ''Specimen Geometry Effects on Fiber Bridging in Composites,'' *Fatigue '93,* Vol. II, J.-P. Bailon and I. J. Dickson, Eds., Engineering Materials Advisory Services, Ltd., Cradley Heath, UK, 1993, pp. 1231–1238.

[*14*] Marshall, D. B., Cox, B. N., and Evans, A. G., ''The Mechanics of Matrix Cracking in Brittle Matrix Composites,'' *Acta Metallurgica,* Vol. 33, 1985, pp. 2013–2021.

[*15*] Bakuckas, J. G. and Johnson, W. S., ''Application of Fiber Bridging Models to Fatigue Crack Growth in Unidirectional Titanium Matrix Composites,'' *Journal of Composite Technology and Research,* in press.

[*16*] Chan, K. S., ''Effects of Interface Degradation on Fiber Bridging of Composite Fatigue Cracks,'' *Acta Metallurgical Materials,* Vol. 41, 1993, pp. 761–768.

[*17*] Kantzos, P., Eldridge, J., Koss, D. A., and Ghosn, L. J., ''The Effect of Fatigue Loading on the Interfacial Shear Properties of SCS-6/Ti Based MMCs,'' *Proceedings,* Materials Research Society, Vol. 273, 1992, pp. 135–142.

[*18*] Jira, J. and Larsen, J. M., ''Crack Bridging Behavior in Unidirectional SCS-6/Ti-24Al-11Nb Composite,'' *Fatigue '93,* Vol. II, J.-P. Bailon and I. J. Dickson, Eds., Engineering Materials Advisory Services, Ltd., Cradley Heath, UK, 1993, pp. 1085–1090.

[*19*] John, R., Jira, J. R., Larsen, J. M., and Ashbaugh, N. E., ''Analysis of Bridged Fatigue Cracks in Unidirectional SCS-6/Ti-24Al-11Nb,'' *Fatigue '93,* Vol. II, J.-P. Bailon and I. J. Dickson, Eds., Engineering Materials Advisory Services, Ltd., Cradley Heath, UK, 1993, pp. 1091–1096.

[*20*] Walls, D. P., Bao, G., and Zok, F. W., ''Mode I Fatigue Cracking in a Fiber Reinforced Metal Matrix Composite,'' *Acta Metallurgica Materials,* Vol. 41, 1993, pp. 2061–2071.

[*21*] Davidson, D. L., ''Fatigue Crack Growth Through Composites with Continuous Fiber Reinforcement,'' *Proceedings,* ICCM-9, Vol. I, A. Miravete, Ed., University of Zaragoza and Woodhead Publishing, Ltd., Spain, 1993, pp. 571–577.

[*22*] Dollar, A. and Steif, P. S., ''Load Transfer in Composites with a Coulomb Friction Interface,'' *International Journal of Solids and Structures,* Vol. 24, 1988, pp. 789–803.

[*23*] Cox, B. N., Dadkhah, M. S., James, M. R., Marshall, D. B., Morris, W. L., and Shaw, M., ''On Determining Temperature Dependent Interfacial Shear Properties and Bulk Residual Stresses in Fibrous Composites,'' *Acta Metallurgica Materials,* Vol. 38, 1990, pp. 2425–2433.

David B. Marshall,[1] Winfred L. Morris,[1] Brian N. Cox,[1] and Dimitris A. Kouris[2]

The Role of Frictional Sliding in Transverse Failure of a Titanium Aluminide Composite

REFERENCE: Marshall, D. B., Morris, W. L., Cox, B. N., and Kouris, D., "**The Role of Frictional Sliding in Transverse Failure of a Titanium Aluminide Composite,**" *Life Prediction Methodology for Titanium Matrix Composites, ASTM STP 1253,* W. S. Johnson, J. M. Larsen, and B. N. Cox, Eds., American Society for Testing and Materials, 1996, pp. 58–65.

ABSTRACT: Circumferential frictional sliding between the matrix and fibers in a titanium aluminide matrix composite has been measured during transverse loading using a displacement mapping method based on differential image analysis. Sliding was detected over all of the interface before initiation of the cracks that eventually caused failure. The results are compared with recent analytical solutions for the interfacial stresses that account for near-neighbor fiber interactions. The location at which sliding initiated coincided with the predicted position of maximum shear stress, while the calculated magnitude of the shear stress when sliding began was equal to the previously measured resistance to frictional sliding in the longitudinal direction. The results indicate that circumferential sliding has a strong influence on the local stress concentrations that lead to failure.

KEYWORDS: interface, sliding, debonding, transverse failure, friction (materials), strain mapping, roughness, crack initiation, titanium, titanium matrix composites, life prediction, titanium alloys, fatigue (materials), modeling

Recent theoretical and experimental studies have suggested that circumferential sliding at debonded fiber-matrix interfaces is detrimental to the transverse strength of the composite [*1–3*]. In contrast, debonding and sliding is a prerequisite for enhanced longitudinal properties such as toughness in ceramic matrix composites [*4,5*] and resistance to fatigue crack growth in intermetallic matrix composites [*6–8*]. Therefore, it has been suggested [*1*] that optimal composite properties could be obtained with interfaces that debond readily and have low frictional sliding resistance longitudinally but large sliding resistance in the circumferential direction.

Circumferential sliding during transverse loading causes increased tensile hoop stress in the matrix adjacent to the fiber, at the location where cracking is induced normal to the applied load (Fig. 1). In composites containing fibers that are stiffer than the matrix, the stress at this location, $\sigma_\theta(\pi/2)$, is smaller than the applied stress if the interface does not slide: that is, the stress perturbation due to the fiber is compressive thus tending to inhibit the formation of strength-degrading transverse cracks. On the other hand, the hoop stress at $\theta = 0$ is tensile, thus tending to induce cracking parallel to the applied load. Such cracks have been observed in Ti_3Al/SiC composites [*1*]. However, because of their orientation, they do not lead to failure in unidirectional loading. If the interface is allowed to debond with

[1]Members of the Technical Staff, Rockwell International Science Center, Thousand Oaks, CA 91360.

[2]Professor, Department of Mechanical and Aerospace Engineering, Arizona State University, Tempe, AZ 85287–6106.

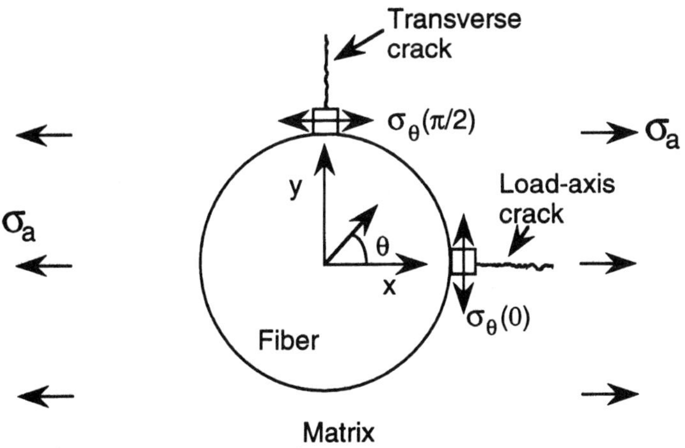

FIG. 1—*Stresses and cracking around a fiber subject to transverse loading.*

unrestricted sliding, these stresses change: the stress $\sigma_\theta(\pi/2)$ becomes larger than the applied stress thus favoring the formation of transverse cracks, which lead to failure, while the stress $\sigma_\theta(0)$ becomes compressive. The magnitudes of these changes in stress increase as the spacing of nearest neighbor fibers decreases (Fig. 2).

Analytical solutions have been obtained recently for the stress fields surrounding pairs of fibers [*3*], with the possibility of limited interfacial sliding being permitted by incorporating springs of various stiffness at the interface. The results indicate a continuous transition between the stress states mentioned earlier for bonded and freely sliding interfaces. The solutions could be used for guidance in designing interfaces that debond, yet have sufficient frictional sliding resistance to avoid transverse cracking. In this paper, we present experimental measurements of circumferential sliding (obtained by high-resolution strain mapping) and preliminary comparison with analysis.

Experiments

Transverse loads were applied to a titanium aluminide matrix composite as described in more detail elsewhere [*1*]. The composite, which was fabricated by Textron Specialty Materials, contained three rows of silicon-carbide fibers (SCS-6) in a two phase (α_2–β) matrix of Ti-25Al-10Nb-3V-1Mo (Super α_2). Beams with dimensions approximately 50 by 2 by 1 mm were cut from a sheet of the composite with the longest dimension normal to the fibers. The beams were polished on the sides containing the fiber ends and loaded in tension using a fixture attached to the stage of an optical microscope. High magnification micrographs were obtained from the polished sides of the beams during loading. The applied loads were measured with a load cell, and the corresponding strains were measured using strain gages attached to the sides of the specimens.

The in situ optical micrographs were analyzed using a high-resolution strain mapping technique [*9,10*]. This involved comparing images taken before and during loading and measuring relative displacements of corresponding image features. Measurements were obtained both stereoscopically and using a computerized image analysis system (HASMAP-High Accuracy Strain Mapping), both of which provide sensitivity of ~10 nm in differential displacement measurements from optical micrographs. The image analysis was facilitated by depositing magnesia (MgO) crystals on the surface of the specimen to provide additional sharp image detail.

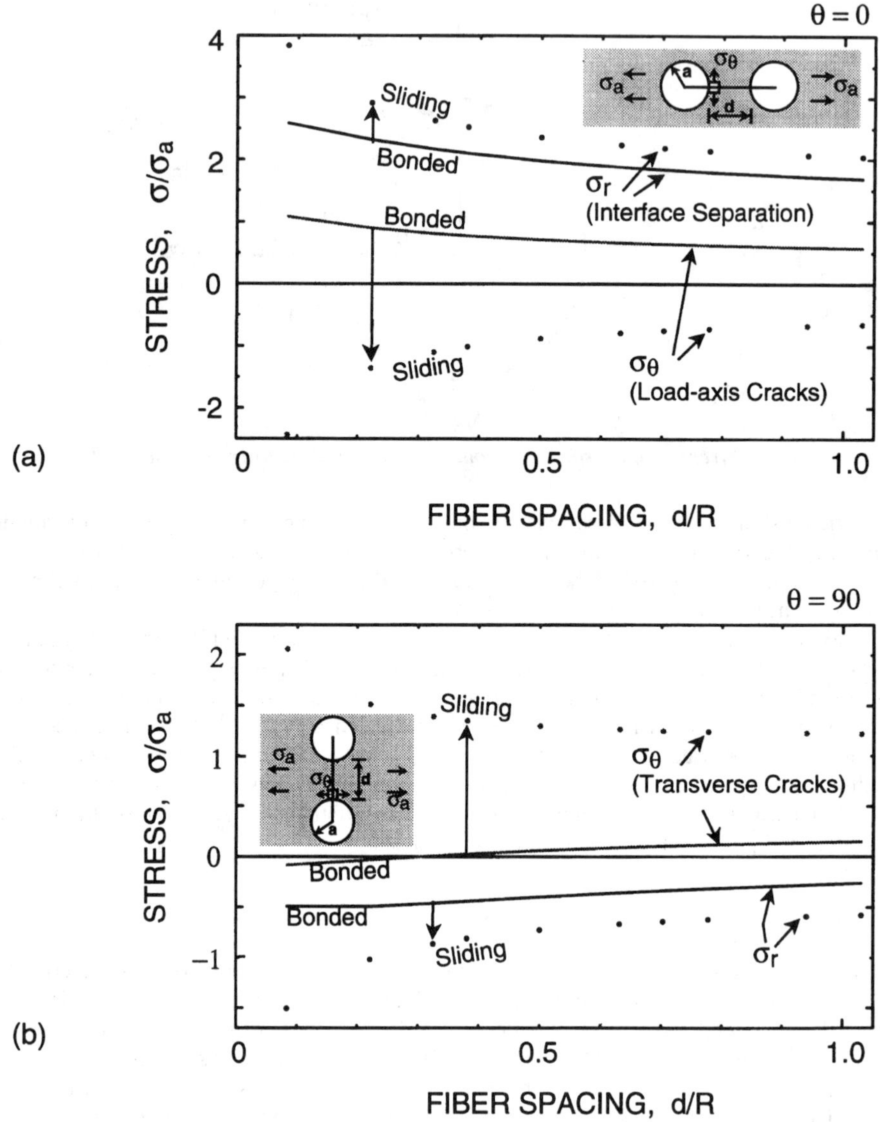

FIG. 2—*Hoop stresses in matrix adjacent to fiber, dependence on fiber spacing and interfacial sliding* [1].

Results and Analysis

Damage

The details of damage leading to failure are discussed in detail elsewhere [*1*]. The pertinent features, in order of appearance during loading, are: (1) separation of the fiber-matrix interface at the position, $\theta = 0$, (the separation initiated at applied stresses, σ_a, within the range 120 to 170 MPa, corresponding with the onset of nonlinearity in the stress-strain curve); (2) formation of "load-axis" radial cracks between some fibers (Fig. 1) at $\sigma_a \approx 150$ MPa; (3) formation of transverse radial cracks adjacent to some fibers at $\sigma_a \approx 160$ MPa; and (4) stable growth and linking of transverse cracks and debonded interfaces, leading to failure at $\sigma_a = 200$ MPa. Comparison of these results with stress field solutions for fully bonded and freely sliding interfaces suggested that significant circumferential sliding occurred between the onset of load-axis and transverse cracking at $\sigma_a \sim 150$ to 160 MPa [*1*].

In-Plane Surface Displacements

A map of in-plane surface distortions around a pair of closely spaced fibers at an applied load of 160 MPa is shown in Fig. 3. The arrows superimposed on the optical image are relative displacement vectors for corresponding image features located at the beginnings of the arrows within a reference micrograph at zero load and a second micrograph of the same area with the load applied. (The magnitudes of the displacement vectors are magnified by a factor of 40 compared with the dimensions on the micrograph.) The following deformations are readily discerned: tensile strains parallel to the applied load; Poisson's contraction from top to bottom of the micrograph; debonding and rotation of the right-hand fiber; and tensile strains in the matrix between the fibers in the direction normal to the applied stress.

At sufficiently high applied loads, the opening and sliding displacement discontinuities at the fiber-matrix interfaces can be readily measured using stereoscopic analysis. Results for the fiber shown in Fig. 4*a* at various stages of loading are shown in Fig. 4*b*.

Circumferential sliding displacements were first detected at a lower applied load (~100 MPa) than were radial opening displacements (120 MPa). Sliding began at the $\theta = 30°$ location and spread continuously in both directions with increasing load. This result is consistent with the calculated location of the maximum shear stress from the analysis of Ref

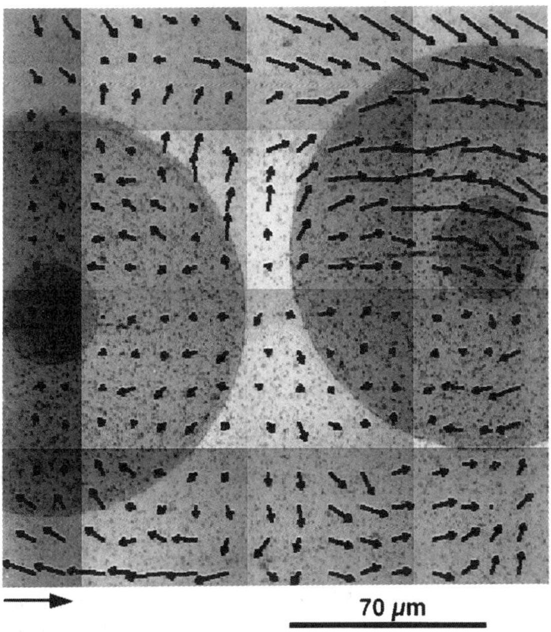

FIG. 3—*In-plane surface displacements (obtained by high-resolution differential image analysis) caused by applied transverse stress of 160 MPa. Reference arrow at bottom left represents displacement of 0.23 μm.*

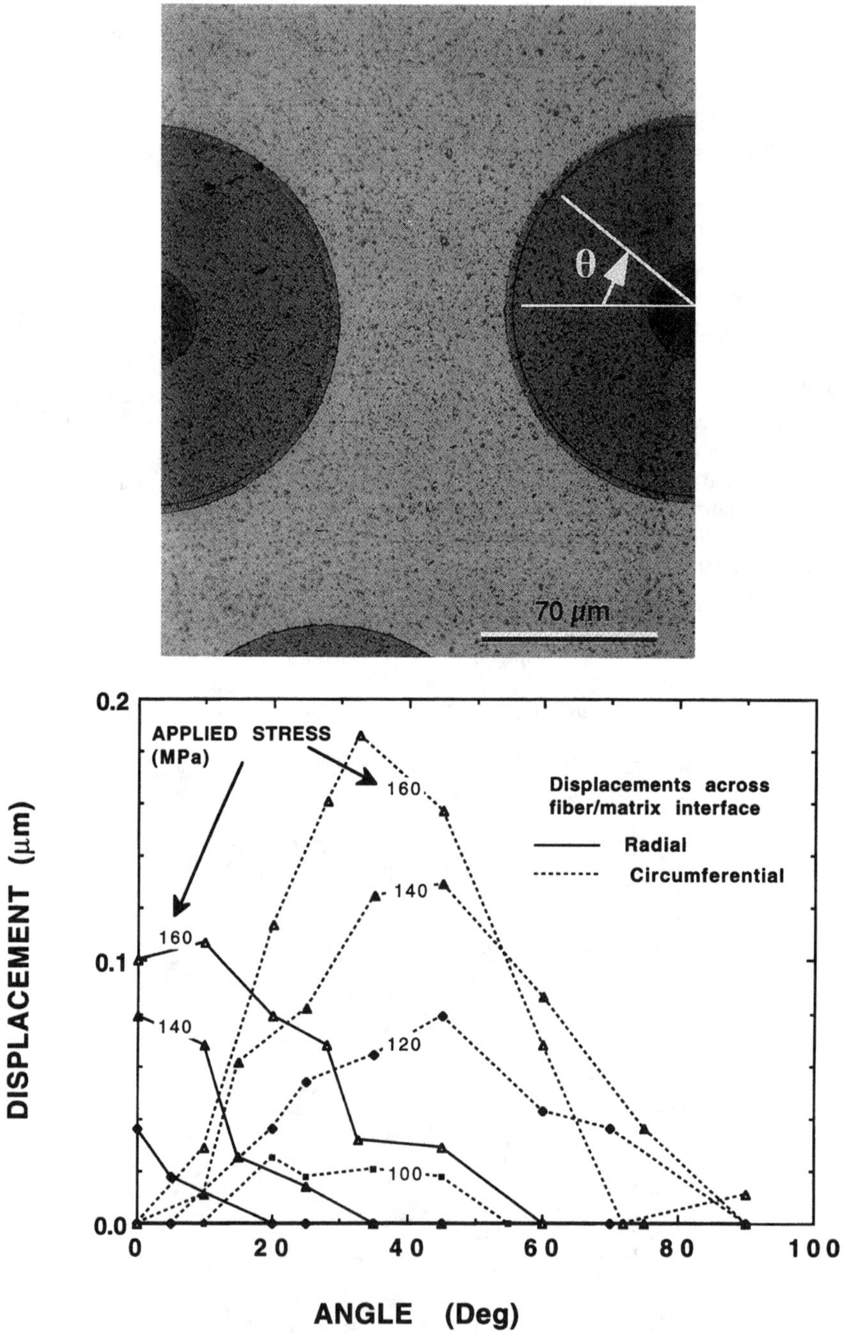

FIG. 4—(a) *Optical micrograph showing fiber used for measurements of* (b). (b) *Displacements across fiber-matrix interface in* (a), *measured by stereoscopic analysis of pairs micrographs obtained before and during loading. Data shown for various tensile stresses applied in the horizontal direction in* (a).

2 (Fig. 5). The radial opening displacements occurred first at $\theta = 0$ as expected and spread to higher angles with increasing load. By the stage at which transverse cracking was first observed in the composite (at a load of 160 MPa), circumferential sliding had extended around the entire interface and radial opening displacements were detected over the range, $0 < \theta < 60°$. However, the magnitudes of the opening displacements in Fig. 4*b* are smaller

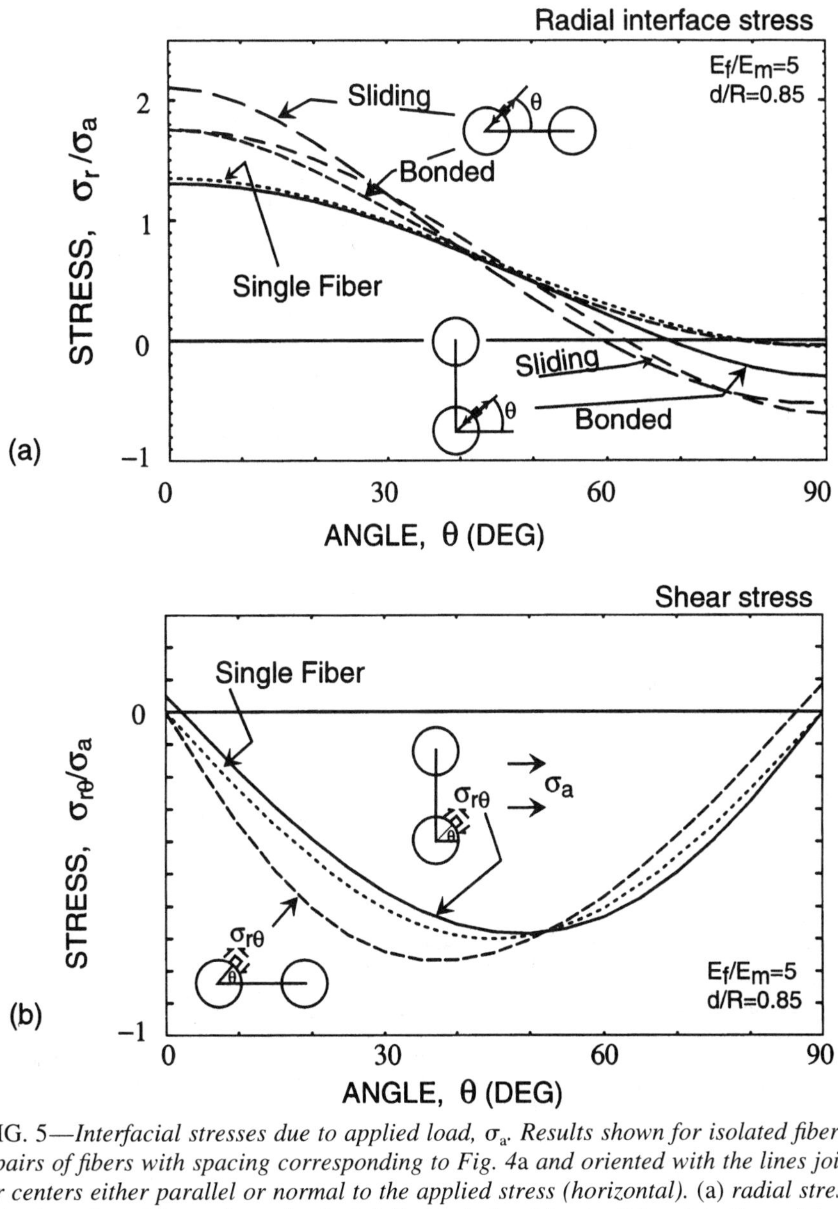

FIG. 5—*Interfacial stresses due to applied load, σ_a. Results shown for isolated fiber and for pairs of fibers with spacing corresponding to Fig. 4*a *and oriented with the lines joining their centers either parallel or normal to the applied stress (horizontal).* (a) *radial stresses: results for pairs of fibers shown for both fully bonded and freely sliding interfaces,* (b) *shear stresses* [1].

than the amplitude of the roughness that has been measured on the surfaces of SCS-6 fibers (~0.1 μm) [*11,12*]. Therefore, contact of asperities (and thus frictional resistance to sliding) is most likely maintained over much of this region.

The calculated interfacial shear stresses and radial stresses for pairs of fibers with the separation shown in Fig. 4*a* and with strongly bonded interfaces are shown in Fig. 5. Results for the radial stress at an interface that is free to slide but constrained to have zero normal displacement are also included in Fig. 5*a* for comparison. The radial stress at $\theta = 0$ from Fig. 5*a* is $\sim 2\sigma_a$, giving a value of ~240 MPa when interfacial separation began. This is close in magnitude to the residual compressive stress normal to the interface (which has been estimated from relaxation experiments to be ~260 MPa [*1*]), thus indicating that bonding at the interface is negligible and that the resistance to sliding is due entirely to friction. The largest shear stress occurs at $\theta \approx 35°$ and is equal to $\sim 0.7\sigma_a$. At the onset of sliding ($\sigma_a = 100$ MPa), this shear stress is approximately 70 MPa. This value is the same as the magnitude of the frictional stress for axial sliding, which has been evaluated independently from fiber-pulling experiments in this composite [*8*].

Conclusions

Differential image analysis revealed that, during transverse tensile loading, circumferential sliding occurred at the fiber-matrix interface well before failure of the composite. Such sliding, resisted by friction, is predicted to have an important influence on the local stresses surrounding the fiber; specifically, an increase in the tensile hoop stress that causes transverse radial cracks that eventually lead to failure. The circumferential sliding extended over the entire interface before transverse radial cracks were observed. Therefore, any analytical or numerical analysis of the local stresses for transverse cracking and failure must account for the circumferential sliding.

Acknowledgments

Funding for this work was supplied by the U.S. Air Force Office of Scientific Research under Contract AFOSR 90–0235.

References

[*1*] Marshall, D. B., Morris, W. L., Cox, B. N., Graves, J., Porter, J., Kouris, D., and Everett, R. K., "Transverse Strengths and Failure Mechanisms in Ti_3Al Matrix Composites," *Acta Metallurgica Materials,* Vol. 42, No. 8, 1994, pp. 2657–2673.

[*2*] Kouris, D., "Stress Concentration due to the Interaction Between Two Imperfectly Bonded Fibers in a Continuous Fiber Composite," *Journal of Applied Mechanics,* Vol. 60, No. 1, 1993, pp. 203–206.

[*3*] Kouris, D. and Tsuchida, E., "On the Elastic Interaction Between Two Fibers in a Continuous Fiber Composite under Thermal Loading," *Mechanics of Materials,* Vol. 12, 1991, pp. 131–146.

[*4*] Aveston, J., Cooper, G. A., and Kelly, A., "Single and Multiple Fracture," *Properties of Fiber Composites, Proceedings,* National Physical Laboratory, IPL Science and Technology Press Ltd., Surrey, UK, 1971, pp. 15–26.

[*5*] Marshall, D. B. and Evans, A. G., "Failure Mechanisms in Ceramic Fiber/Ceramic Matrix Composites," *Journal,* American Ceramics Society, Vol. 68, No. 5, 1985, pp. 225–231.

[*6*] Ghosn, L. J., Kantzos, P., and Telesman, J., "Modeling of Crack Bridging in a Unidirection Metal Matrix Composite," *International Journal of Fracture,* Vol. 54, 1992, pp. 345–357.

[*7*] Jira, J. R. and Larsen, J. M., "Fatigue of Unidirectional SCS-6/Ti-24Al-11Nb Composite Containing a Circular Hole (Parts I and II)," *Metallurgical Transactions,* Vol. 25A, No. 7, 1994, pp. 1413–1424.

[*8*] Walls, D. P., Bao, G., and Zok, F. W., "Mode I Fatigue Cracking in a Fiber Reinforced Metal Matrix Composite," *Acta Metallurgica Materials,* Vol. 41, No. 7, 1993, pp. 2061–2071.

[9] James, M. R., Morris, W. L., and Cox, B. N., ''A High Accuracy Automated Strain Field Mapper,'' *Experimental Mechanics,* Vol. 30, 1990, pp. 60–68.

[10] Marshall, D. B., Shaw, M. C., and Morris, W. L., ''Measurement of Debonding and Sliding Resistance in Fiber Reinforced Intermetallics,'' *Acta Metallurgica,* Vol. 40, No. 3, 1992, pp. 443–454.

[11] Jero, P. D., Kerans, R. J., and Parthasarthy, T. A., ''Effects of Interfacial Roughness on the Frictional Stress Measured using Push Out Tests,'' *Journal,* American Ceramics Society, Vol. 74, No. 11, 1991, pp. 2793–2801.

[12] Warren, T. L., Majumdar, A., and Krajcinovic, D., ''Atomic Force Microscope Imaging of the Surface Roughness of SCS6 and TiB_2 Coated SiC Fibers and Uncoated Sapphire Fibers,'' *Composites,* in press.

Partha Rangaswamy[1] *and N. Jayaraman*[2]

Issues Related to Prediction of Residual Stresses in Titanium Alloy Matrix Composites

REFERENCE: Rangaswamy, P. and Jayaraman, N., "**Issues Related to Prediction of Residual Stresses in Titanium Alloy Matrix Composites,**" *Life Prediction Methodology for Titanium Matrix Composites, ASTM STP 1253,* W. S. Johnson, J. M. Larsen, and B. N. Cox, Eds., American Society for Testing and Materials, 1996, pp. 66–82.

ABSTRACT: Recently, a detailed study of residual stresses on the as-processed SCS-6/Ti-24Al-11Nb $[0]_8$ composite, and SCS-6/Beta-21S composites in unidirectional $[0]_4$, cross-ply $[0/90]_s$, and quasi-isotropic $[0/\pm45/90]_s$ layups has been completed. In this study, residual stresses have been measured using X-ray diffraction ($\sin^2\psi$) technique. We have shown that the use of conventional unit cell models consisting of a quarter fiber surrounded by the matrix material to predict residual stresses for verification of experimental results is inadequate. Such models have predicted successfully the stresses at the fiber-matrix interface. However, experimental work to measure residual stresses have always been on surfaces far away from the interface region.

In this paper, the approach taken in extending the conventional unit cell model to the concept of multifiber models to predict average stresses are presented. In this process, several modeling issues have been identified. These issues are (1) use of conventional unit models for prediction of average surface residual stresses, (2) effect of orientation of the sub-surface plies on the residual stresses in the surface ply, (3) residual stresses in the interior plies, and (4) constituent material properties.

KEYWORDS: residual stress, finite element models, material removal effects, metal matrix composites, X-ray stress measurements, titanium matrix composites, coefficient of thermal expansion mismatch, titanium, life prediction, titanium alloys, fatigue (materials), modeling

It is well known that the origin of residual stresses in as-processed metal matrix composites (MMCs) are due to the coefficient of thermal expansion (CTE) mismatch between the reinforcement and the matrix material and the temperatures used in the processing of these composites. Other factors such as fiber orientation in a ply and stacking arrangement of these plies could also affect the nature and distribution of these stresses. Finally, one also has to take into account the basic constituent properties of the metal matrix and the reinforcement that could generally vary with temperature. The elastic-plastic behavior of the constituents, for example, may even exhibit time-temperature dependence. Also, one has to recognize the role of material properties and particularly, the fact that some of the constituent properties are sensitive to processing conditions and corresponding microstructural conditions. This is particularly true of those titanium alloy matrix materials that show potential to develop textures and therefore anisotropic properties. All of the preceding

[1]LANSCE/MST-5, Los Alamos National Laboratory, Los Alamos, NM 87545.

[2]Department of Materials Science and Engineering, University of Cincinnati, Cincinnati, OH 45221–0012.

factors could lead to a complex residual thermal stress-strain state both during processing and during service at elevated temperatures.

Residual stresses are measured generally on the surface of the MMC material using nondestructive techniques such as X-ray diffraction, neutron diffraction, etc. [*1–15*]. In many of these techniques, stress measurements are made by using a beam of finite size that covers several fiber diameters on the surface of a sheet specimen (Fig. 1). Generally, X-rays have very limited depth-of-penetration in metal specimens (less than about 10 μm). Therefore, the stresses measured are an average of all the stress gradients seen in the measuring direction in the specimen within the beam exposed volume. Residual stress measurements have been generally made in the direction of fiber (longitudinal stress, σ_L), transverse to the direction of fiber (transverse stress, σ_T), and at any other angle to the fiber (σ_ϕ) [*2–10*]. The stress profiles of the measured values are obtained by successive electropolishing of surface layers. Attempts have been made to verify these results using a conventional concentric cylinder model (CCM) and finite element solution for a unit cell model (UCM) (Fig. 2) [*12–26*]. Under these conditions, the normally used procedures of comparing the average stresses measured on the surface and the stress profiles obtained from electropolishing using X-ray diffraction technique with the CCM or UCM models are inadequate. The number and orientation of subsurface plies also have to be taken into consideration when predicting these stress profiles.

Recently, we have completed a detailed study of residual stresses on the as-processed SCS-6/Ti-24Al-11Nb $[0]_8$ composite, and SCS-6/Beta-21S composites in unidirectional $[0]_4$, cross-ply $[0/90]_s$, and quasi-isotropic $[0/\pm45/90]_s$ layups. The X-ray measurements were made in three different orientations on both surfaces of the composite: parallel (σ_L), transverse (σ_T), and 45° (σ_{45}) to the fiber direction as shown in Fig. 1. In this study, we used a beam size of about 3 by 7 mm for $\psi = 0°$ and 6 by 7 mm for $\psi = 60°$. The relatively large size of the beam encompasses an area covering at least 15 to 20 fibers that measures an average diameter of 140 μm each. Profiles of these average stress values versus depth (that is, after removing layers by electropolishing) from both surfaces of the as-received plate are shown in Fig. 3. To compare with the experimental results, finite element models (FEMs) were developed and designed to simulate real experimental conditions where the average

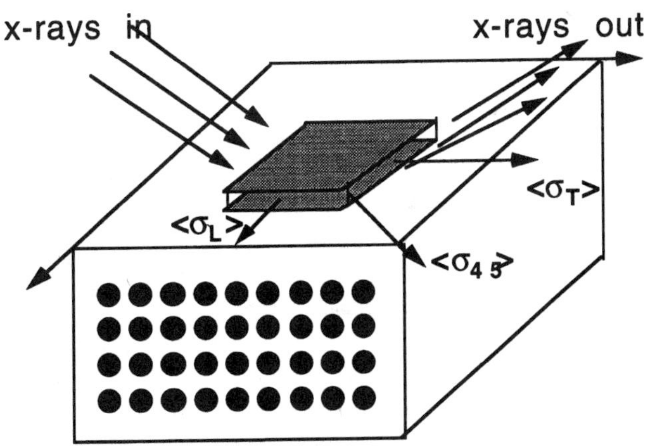

FIG. 1—*Schematic of a fiber-reinforced MMC showing the directions of stress measurement and the region on the surface of the composite where the X-ray beam is sampled.*

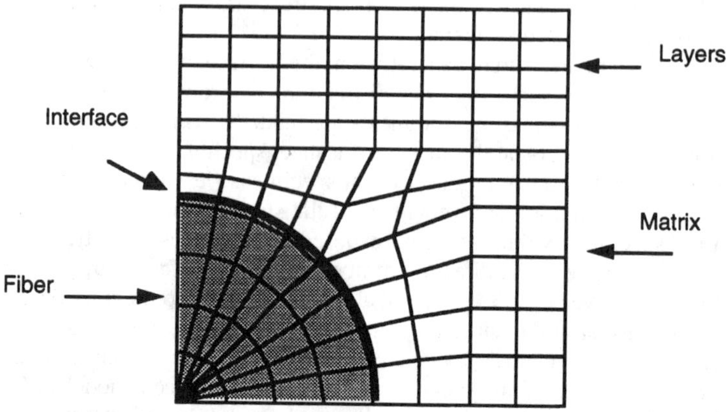

FIG. 2—*A schematic of a one-quarter fiber (¼ × 1) model showing the arrangement of matrix material in the form of layers. In conventional terminology, this is also known as "unit-cell finite element model."*

stresses can be calculated with little or no interference from the imposed surface boundary conditions in the CCM UCM models as also shown in Fig. 3. These results will be explained in detail in the later sections of this paper.

In this process, several modeling issues have been identified. These issues are (1) use of conventional unit models for prediction of average surface residual stresses, (2) effect of orientation of the subsurface plies on the residual stresses in the surface ply, (3) residual stresses in the interior plies, and (4) material properties.

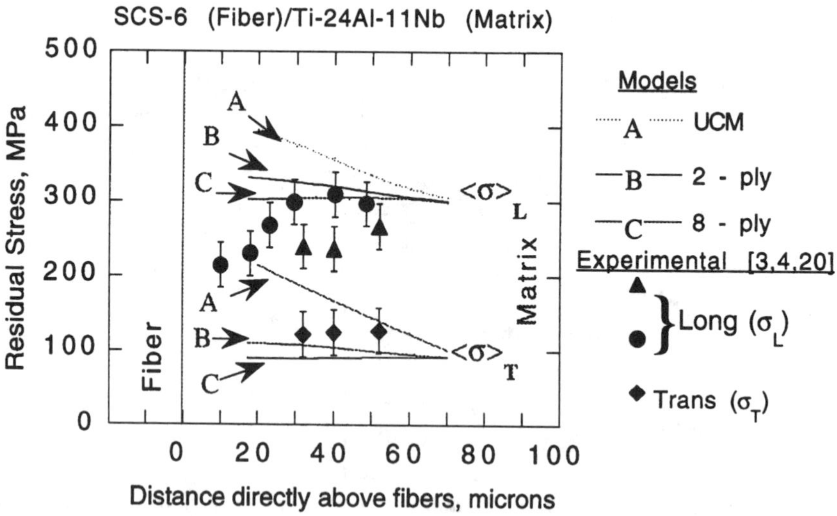

FIG. 3—*Comparison of longitudinal and transverse residual stress profiles between FEMs (UCM, two-ply and eight-ply) and experimental results.*

Prediction of Surface Residual Stresses in a Unidirectional MMC

The initial approach taken in developing the finite element models was based on first developing the UCM (Fig. 2), and adding to it additional fibers taking advantage of the symmetry boundary conditions offered by the finite element modeling technique. By choosing this approach, we could compare the stress prediction in the surface layers of the UCM and the multifiber models with the experimental results for verification. Without going into the details of the analysis that have been exhaustively explained in Refs *5* and *20*, it was found that beyond the one-fiber model (Figs. 4 and 5), development of the additional models such as two-fiber, three-fiber, etc., proved to give the same results of the one-fiber model because the additional models were repetitions of the one-fiber model and, therefore, similar results had to be expected. From these results, the one-fiber model was essentially representative of a two-ply model. Therefore the one-fiber model would henceforth be called a two-ply model. The two-ply model was extended to develop four-ply and eight-ply models of a

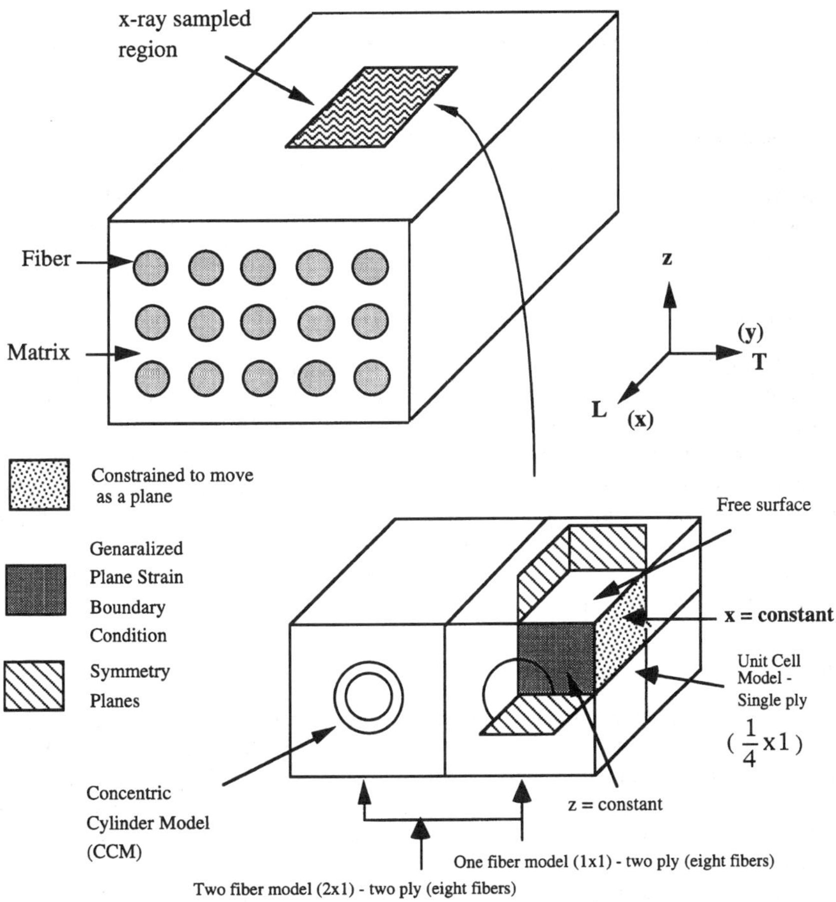

FIG. 4—*Schematic representation of concentric cylinder and UCMs of MMCs showing symmetry, generalized and constrained to move as a plane, boundary conditions. The extension of the UCM to a one- and two-fiber model is also shown.*

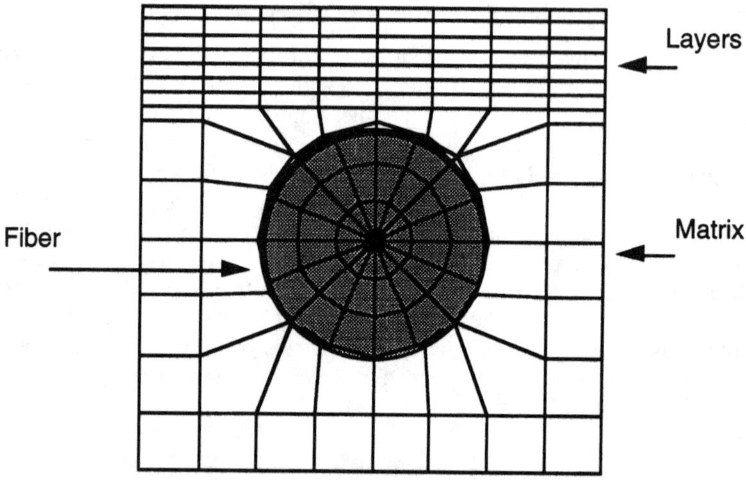

FIG. 5—*A schematic of a one fiber (1 × 1) model (two-ply) showing the arrangement of matrix material in the form of layers.*

unidirectional MMC (Figs. 4 through 7). The surface region of the models have been divided into layers to simulate material removal. Without going into the details of model development, codes, thermal loading procedures, boundary conditions, and layer removal techniques that have been explained in previous papers by the authors [*5,6,20*], we will briefly explain the finite element modeling procedure used in these analyses.

Boundary Conditions

All of the models were analyzed under symmetry boundary conditions on three mutually perpendicular faces and two out of the other three surfaces were subjected to generalized plane strain (constant strain) conditions. The third surface (the top surface) was constraint free. These boundary conditions are shown schematically in Fig. 4. Because of symmetry in the boundary conditions, a one-quarter fiber model (¼ × 1) actually predicts residual stress for a one-fiber model (single-ply). Similarly, a one-fiber model (1 × 1) predicts residual stress for four fibers (two plies, that is, two fibers in the top and bottom plies, respectively). The two-fiber model (2 × 1) predicts residual stress for eight fibers (two-ply, also, but four fibers in the top and bottom plies, respectively).

Thermal Loading Procedure

In this study, only development of thermal residual stresses have been considered. The stresses are assumed to be zero at a temperature of 950°C for this composite. Since the matrix material has low modulus, very low strength, and high ductility at temperatures close to the assumed processing temperature, the choice of the exact processing temperature is not very critical to the final solution. Residual stresses in the fiber alone due to its own manufacturing process, if any, were ignored. NIKE3D (nonlinear structural finite element code) [*27*] that is used to perform the finite element analysis uses a modified Newton-Raphson stiffness formula to solve the nonlinear equations. The thermal loading curve is divided into a number of increments and an iterative solution performed at each increment. The stiffness matrix is reformed at each increment. Small displacement theory is used. The

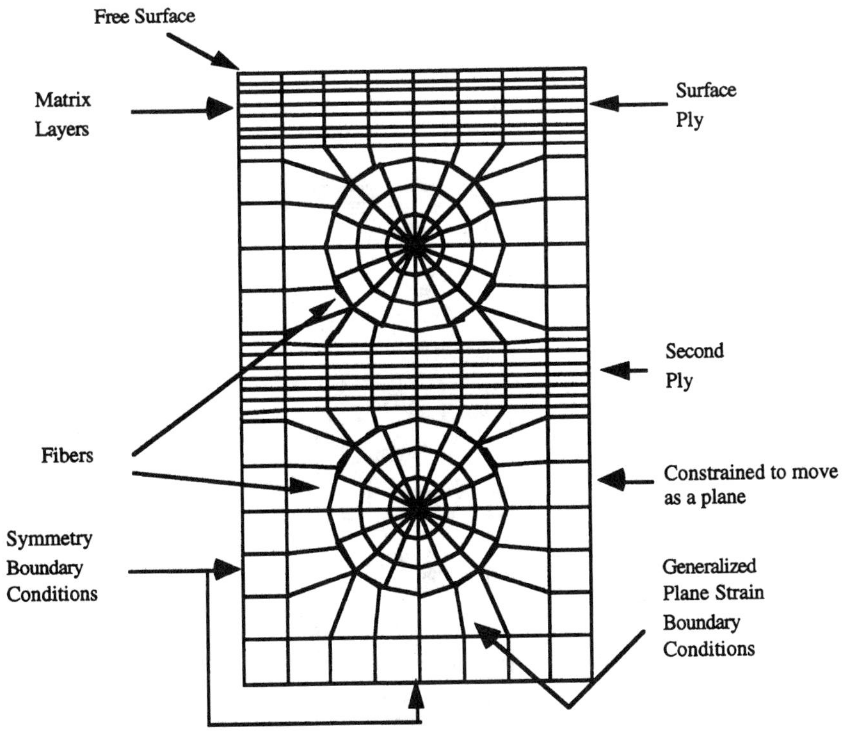

FIG. 6—*A view of the four-ply (1 × 4) FEM showing the matrix materials in the surface and the second internal ply arranged in the form of layers.*

plasticity algorithm makes use of a Von Mises yield surface model with associated plastic flow. Isotropic hardening was assumed in this analysis.

The initial analysis of residual stress was performed by dividing the cooling process into 100 steps for a total time of 1000 s. The FEM code has an option of incrementing through an automatic-time step-size control or fixed-time step-size control. The choosing of either option did not however have any significant effect on the results. About three to four iterations were generally required to obtain convergence at each time step.

Layer Removal Technique

In experimental studies [*2–4,20*], residual stress profiles in composite plates were measured by removing thin layers of matrix material. Material removal was done symmetrically on both sides of the composite plate in order to avoid inducing any bending effects. To simulate the material removal sequence, the surface regions of the composite models were modeled as layers. For example, Figs. 2, 5, 6, and 7 show the surface regions of the matrix material arranged in the form of layers.

The removal of material is simulated by using the birth/death option built into the finite element code or by identifying each layer of elements with different material identifier codes and deleting these materials in the restart procedure used after thermal loading of the composite to ambient temperature. The theory of material removal simulation by using the birth/death option in the finite element analysis code and examples of using this option to

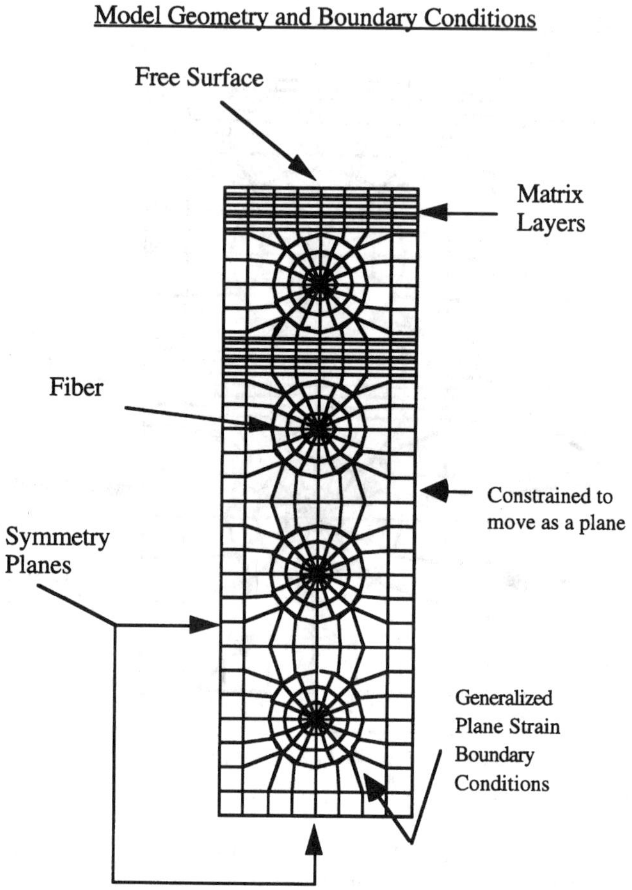

FIG. 7—*A view of the eight-ply (1 × 8) FEM showing the matrix materials in the surface ply and the second internal ply arranged in the form of layers.*

verify results of solutions obtained through analytical methods are explained in the appendix of Ref *5*.

Figure 8 shows the variation in the average stresses upon material removal in the longitudinal direction (fiber direction) and transverse direction for the UCM, two-ply, four-ply, and eight-ply models for the SCS-6/Ti-24Al-11Nb $[0]_8$ composite. Average stresses on the surface of the composite for all the components were computed by taking the average of individual elemental stress values on the outer surface. Such stress averaging was accomplished by using the Simpson's rule, a well-established numerical integration procedure. Before material removal, the average stresses predicted in the surface layer are the same for all the four models. This is indeed true (not shown) for all the layers in these models. But upon layer removal, the changes in stresses are highest in the UCM and the least in the eight-ply model. The longitudinal stresses in the layer closest to the fiber-matrix interface for the UCM, two-ply, four-ply, and eight-ply models are 396, 328, 316, and 302 MPa, respectively. Similarly, the transverse stresses in the layer closest to the fiber-matrix interface for the UCM, two-ply, four-ply, and eight-ply models are 214, 111, 107, and 91 MPa, respectively. These changes show the influence of the underlying fibers on the overall

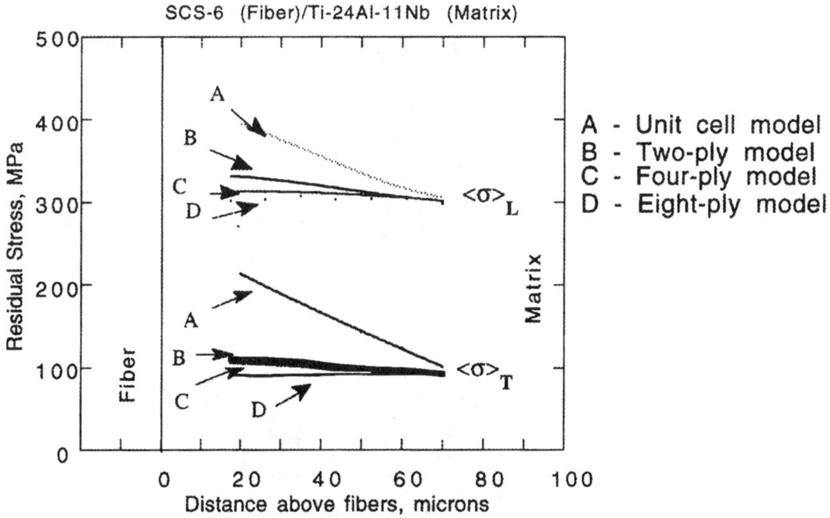

FIG. 8—*Comparison of longitudinal and transverse residual stress profiles between FEMs (UCM, two-ply, four-ply, and eight-ply).*

distribution of residual stresses in the composite. All these differences between the four models can be explained on the basis of volume fraction effects; that is, the differences are essentially due to different amounts of matrix material being removed in an otherwise apparently similar layer removal process used in all these models. In comparing the UCM and the two-ply model, due to the location of symmetric boundaries, twice the amount of material is removed and therefore these lead to a greater change in adjustment of residual stresses in the UCM. This reasoning can be extended to the four-ply and eight-ply models also. For the removal of an equivalent layer in the two-ply model, the increase in volume fraction will be greater as compared to the four-ply or the eight-ply model. In other words, the greater the change in volume fraction of the matrix, the greater will be the change in the residual stress. The results in Fig. 8 support this analysis. The predicted results from the preceding FEM models have been compared with experimental results and are shown in Fig. 3. In general as shown in the figure, the experimental results compare more favorably with the FEM results of the eight-ply (multiply) models than the conventional UCM. Similar observations have been made for the SCS-6/Beta21S $[0]_4$ composite. Conventionally, the UCM is used for most FEM analysis. The issue here clearly demonstrates the inadequacy of the conventional UCM to predict the changes in the residual stresses due to layer removal to obtain stress profiles.

Effect of Stacking Ply Orientation on the Surface Residual Stresses

It has been shown, thus far, that the underplies in a unidirectional composite have an effect on the surface residual stress distribution after the layer removal process. If indeed this is true, then the analysis of a cross-ply $[0/90]_s$ (Fig. 9) and an angle-ply $[0/45]_s$ model (Fig. 10) would also show an influence of the underlying plies on the stress/strain states on the surface of the composite.

The results for the longitudinal stresses and transverse stresses in the matrix material for a SCS-6/Beta-21S four-ply $[0]_4$ composite and the cross-ply $[0/90]_s$ models are shown in Figs.

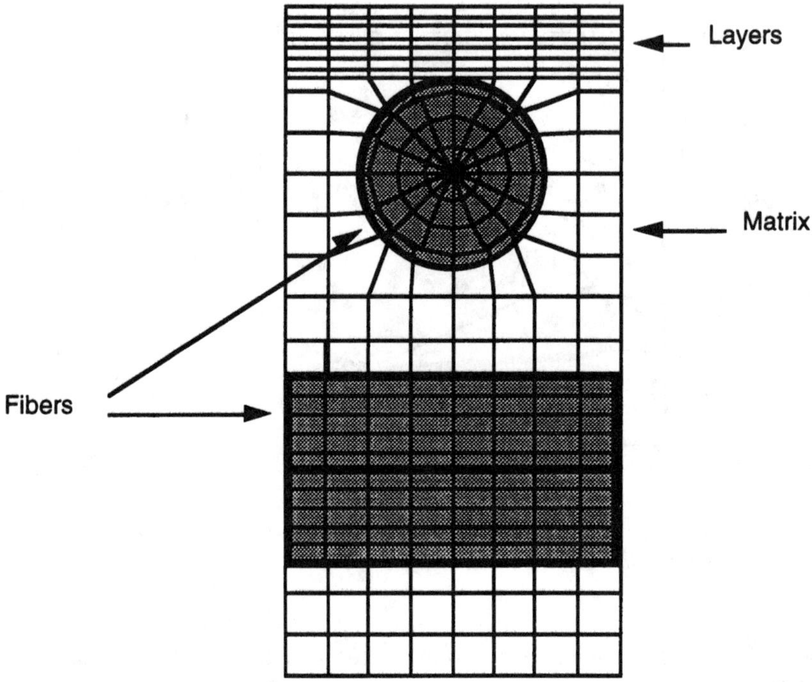

FIG. 9—*Schematic of a cross-ply [0/90] model showing the arrangement of matrix material in the form of layers.*

11*a* and *b*. The differences in the residual stress states between the unidirectional and the cross-ply composites are clearly contrasted in the residual stress plots for the longitudinal and transverse stresses. As seen in these figures, while the longitudinal and transverse stresses show similar trends for the cross-ply composite (Fig. 11*b*), these stresses are significantly different in the unidirectional composite (Fig. 11*a*). This difference between the two-ply layups indicates the strong influence of the underlying plies on the stress/strain states of the surface ply.

Results of the finite element models are compared with experimental results in Figs. 12*a* and *b* for the unidirectional and cross-ply composites, respectively. In these figures, the average longitudinal ($\langle \sigma_L \rangle$), and transverse stresses ($\langle \sigma_T \rangle$) from the finite element models after the layer-removal process and the residual stresses in the longitudinal (σ_L) and transverse (σ_T) directions obtained from experiments are plotted as a function of distance above the fibers. Figures 12*a* and *b* show very favorable comparisons in the trends in the variation of experimental and predicted residual stress profiles.

Residual Stresses in Internal Ply

Residual stress analysis in the internal (second) ply was conducted by removing symmetrically the surface ply (both matrix and the fiber layer) on both top and bottom surfaces of the model. This was done to simulate the corresponding experimental residual stress analysis conducted using the X-ray diffraction technique [*3,4,6,20*]. Finite element prediction of residual stresses in the internal (second) ply for the eight-ply model was analyzed, and the results are shown in Fig. 13. The longitudinal, transverse, and out-of-plane (σ_Z)

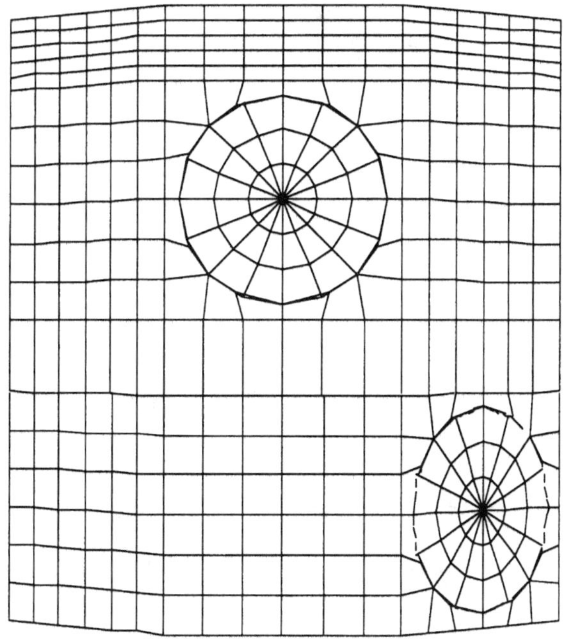

FIG. 10—*A view of the [0/45]$_s$ model showing the arrangement of the matrix material in the form of layers.*

stresses as a function of distance above the fiber plane are shown in Fig. 13. The average residual stresses in the surface-ply in the longitudinal and transverse directions are higher than the corresponding stresses in the second ply. The average out-of-plane stresses (σ_Z) in the surface-ply are consistently close to zero implying plane stress conditions. However, in the second-ply, the average σ_Z is gradually increasing from the outer surface layer towards the fiber-matrix interface layer. When the average elemental stress in the out-of-plane direction is not zero, it implies that the σ_Z gradient is very steep. The reasons for the steep stress gradient in the out-of-plane direction indicate the severity of the tri-axial stress state developed in the region between the two fiber plies. The FEM results seem to support the experimental observations.

In the measurement of residual stresses using the X-ray diffraction technique, normally one would expect for a biaxial plane stress condition on the surface that the $d_{\phi\psi}$ versus $\sin^2\psi$ plot would be linear (where $d_{\phi\psi}$ is the average interplanar spacing of the crystallographic planes, and ϕ and ψ are the specimen orientation parameters [*20*]). Any nonlinearity in this plot is normally attributed to severe stress gradients in the sampling volume and possible contribution from out-of-plane (σ_Z) stresses. Indeed, this was observed as shown in Fig. 13. In fact, to further prove this point using the FEM results, the $d_{\phi\psi}$ versus $\sin^2\psi$ was predicted for the second ply and a favorable comparison was seen between the predicted and experimentally observed plots, as seen in Figs. 14*a* and *b*.

Influence of Material Properties on Residual Stresses in MMCs

The material properties used in the FEMs are assumed isotropic, the constitutive behavior of the matrix material is thermal elastic/plastic, and the fiber is thermal elastic. For this type

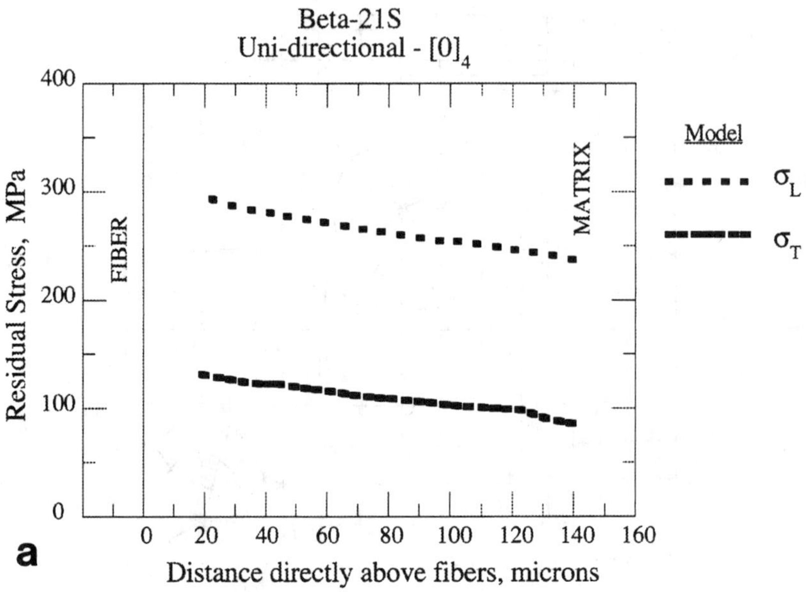

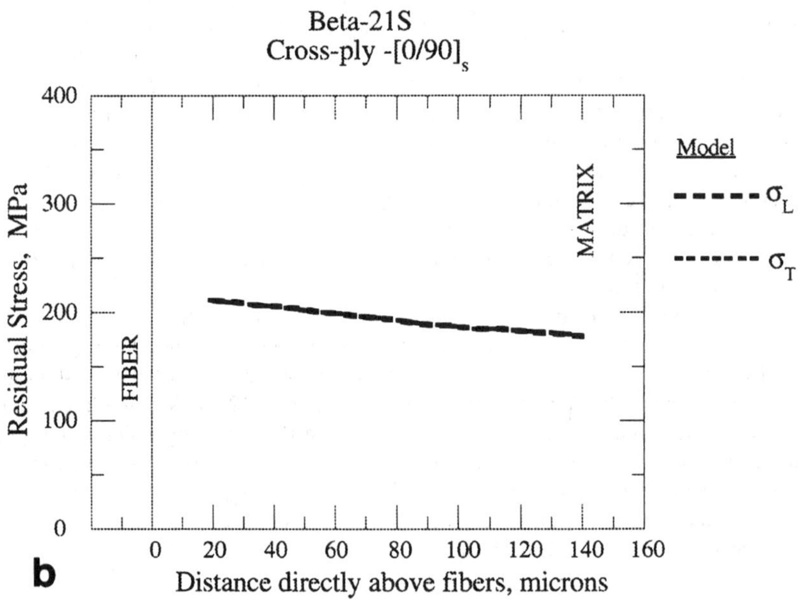

FIG. 11—*Comparison of matrix longitudinal and transverse residual stress profiles for the* (a) *$[0]_4$ (four-ply) and* (b) *$[0/90]_s$ (cross-ply), FEMs in the as-received condition.*

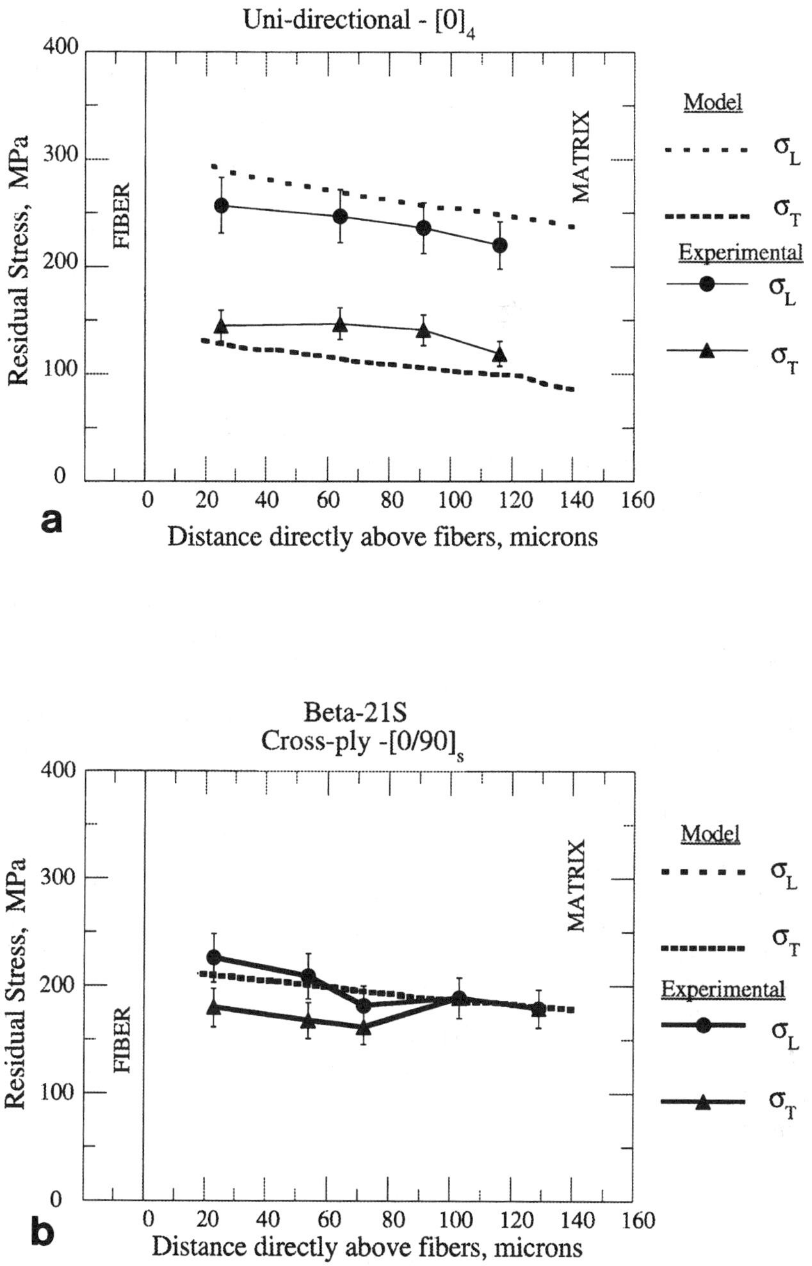

FIG. 12—*Comparison of matrix longitudinal and transverse residual stress profiles between* (a) *$[0]_4$ (four-ply) and* (b) *$[0/90]_s$ (cross-ply), FEMs and experimental results.*

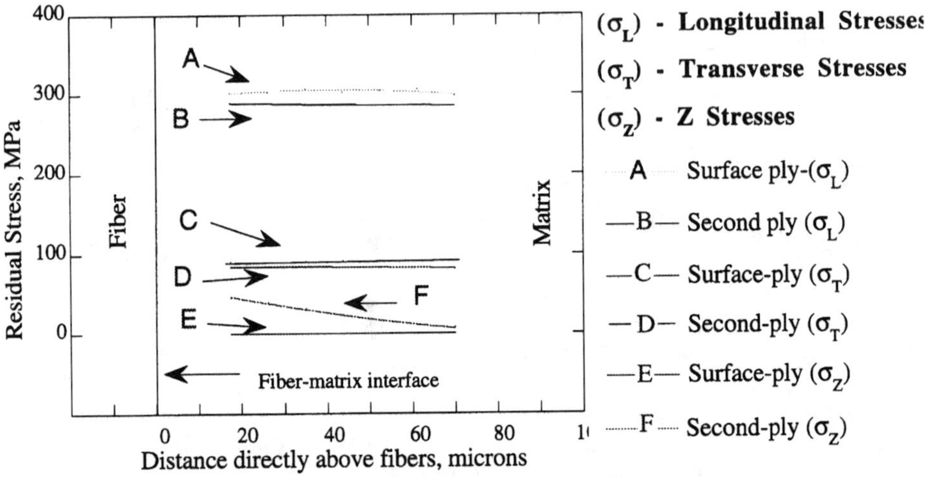

FIG. 13—*Comparison of matrix longitudinal* (σ_L), *transverse* (σ_T), *and out-of-plane* (σ_Z) *profiles between surface ply and second ply for the eight-ply* $[0]_8$ *unidirectional composite.*

of material model behavior, the constitutive material properties that are used are the Young's modulus (E_m), plastic modulus (E_p), Poisson's ratio (ν), yield strength (σ_y), and thermal coefficient of expansion (α) that are temperature dependent. Even though a more rigorous analysis would consider the thermal elastic/plastic/creep behavior of the matrix material, more sophisticated material models including creep effects were not considered mainly due to the fact that other researchers found the effect of creep to be insignificant on residual stresses in these materials [*12–19,22–26*].

For the SCS-6/Ti-24Al-11Nb and SCS-6/Beta-21S material systems used in this study, the influence of these constitutive material properties on the state of residual stresses could throw some light on the overall stress distribution. In comparing the residual stress profiles in the as-cooled and the material removal conditions for a two-ply model, it does appear for the same volume fraction and temperature of processing the mismatch in the thermal coefficient of expansion plays a more significant role in the stress distribution, than the constituent properties such as yield strength, elastic modulus, plastic modulus, and Poisson's ratio. Although, referring to Table 1, the material properties for the two matrices show major differences in yield strength, elastic modulus, plastic modulus, and thermal coefficient of expansion. Also, for a simple one-dimensional problem assuming thermal elastic behavior, the longitudinal stresses in the matrix and fiber along the direction of the fiber axis under iso-strain conditions are

$$\sigma_m = \frac{V_f E_m E_f \Delta\alpha \Delta T}{E_c} \tag{1}$$

$$\sigma_f = -\frac{V_m E_m E_f \Delta\alpha \Delta T}{E_c} \tag{2}$$

assuming $\Delta\alpha = \alpha_m - \alpha_f$ and $\Delta T = T_{\text{high}} - T_{\text{low}}$ where the key variables that directly influence the magnitude of the residual stresses are the CTE mismatch ($\Delta\alpha$), the fiber volume fraction (V_f), and the temperature change (ΔT) [*20*].

Substituting the appropriate material properties from Table 1 for the two matrix materials,

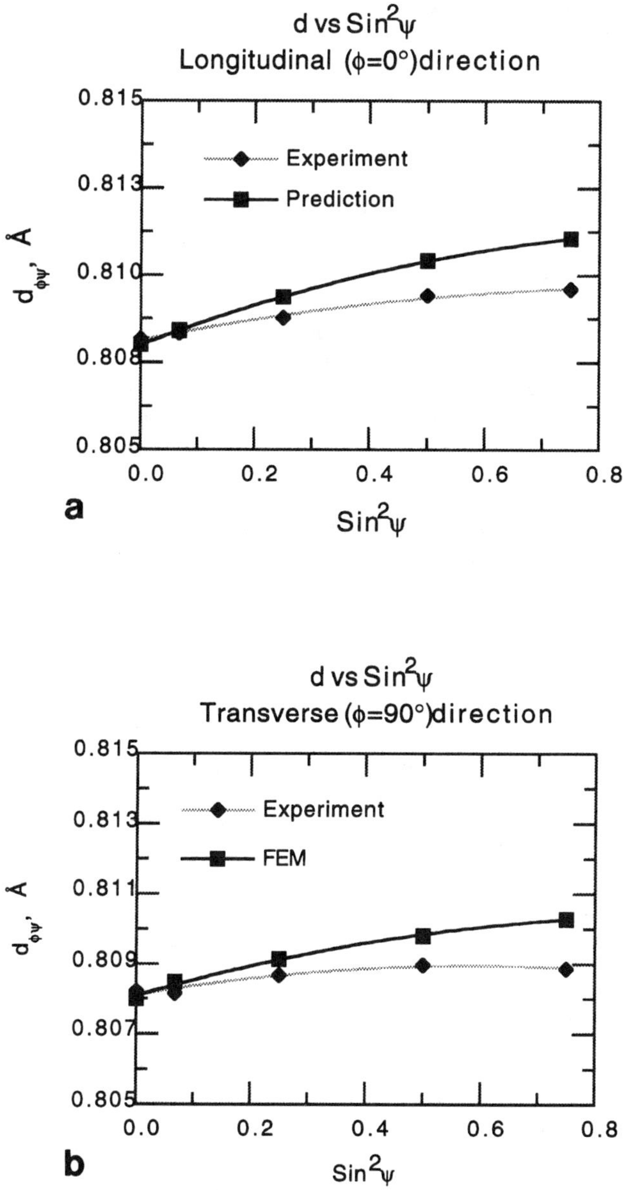

FIG. 14—*Comparison of* d *versus* $sin^2\psi$ *plots for the* (a) *longitudinal and* (b) *transverse directions between experimental results and predictions from FEMs in the second-ply region for SCS-6/Ti-24Al-11Nb MMC.*

it turns out that the matrix stresses for the Ti-24Al-11Nb and Beta-21S matrix systems are 369 and 350 MPa, respectively, using Eq 1. The finite element results from Table 2 show an average longitudinal stress of 292 and 283 MPa for the Ti-24Al-11Nb and Beta-21S matrix materials, respectively. The differences between the two material systems seem to be in fair agreement with the simple one-dimensional analytical model, considering that the constitutive model used in the finite element model is thermoelastic-plastic behavior of the matrix material and the material properties used are temperature dependent. Therefore, the slight increase in stresses in the Ti-24Al-11Nb matrix composite as compared to the Beta-21S matrix composite clearly underlines the impact of the mismatch in the thermal coefficients of the matrix phase and reinforcement phases for the same volume fraction and temperature change.

It is also interesting to note that the lower yield strength and Young's modulus of the Ti-24Al-11Nb matrix material could have contributed to the dropping of the transverse stresses because the Von Mises effective stress approaches the yield strength of the matrix material (Table 2), whereas the higher yield strength and Young's modulus of the Beta-21S matrix material has not caused any change in the longitudinal and transverse stresses because the effective stresses are comparatively small compared to the yield strength of the matrix material (Table 2). The transverse stresses in the layers close to the fiber are also strongly influenced by the hoop stresses that approach a maximum close to the fiber-matrix interface. Interestingly, this also explains why the stresses in the second ply region are lower than the surface ply regions. This is because of the increase in plastic deformation that is controlled by the Von Mises yield criteria in the FEM technique due to the presence of the tri-axial state of stresses in the second-ply region influenced by the presence of the two fibers, whereas in the surface ply, the tri-axial stress exists very close to one fiber only.

Summary

It is evident from this work that close attention should be paid to the experimental methods used in measurement and the models used for prediction of residual stresses in

TABLE 1—*Fiber and matrix material properties.*[a]

Temperature, °C	Poisson's Ratio	Elastic Modulus, GPa	Plastic Modulus, GPa	Yield Strength, GPa	Secant CTE, 10^{-6}/°C
		TITANIUM ALUMINIDE (Ti-24Al-11Nb) MATRIX PROPERTIES			
21	0.3	68.9	3.72	0.4990	12.15
149	0.3	67.6	4.21	0.4413	12.40
260	0.3	66.2	4.69	0.3902	12.69
427	0.3	64.8	5.38	0.3199	13.30
649	0.3	62.1	4.96	0.2599	14.85
760	0.3	51.0	2.48	0.2399	15.89
950	0.3	32.4	2.48	0.1103	18.72
		BETA-21S MATRIX PROPERTIES			
21	0.34	117	3.84	1.0500	9.940
316	0.34	101	5.40	0.7750	10.493
482	0.34	95.4	6.38	0.6900	10.824
566	0.34	78.1	16.96	0.4700	10.994
621	0.34	73.2	14.72	0.2890	11.108
650	0.34	70.6	0.0	0.2690	11.168
900	0.34	50.9	0.0	0.0940	11.685

NOTE—Ref *16*: T_{ref} = 900°C.
[a]SCS-6 fiber properties: ν = 0.33, E = 413.7 GPa, and α (secant CTE) = 4.86e − 06.

TABLE 2—*Two-ply model.*

Layer Number/ Distance from Fibers, μm	As-Cooled Condition $\langle\sigma\rangle_L$	$\langle\sigma\rangle_T$	$\langle\sigma\rangle_Z$	$\langle\sigma\rangle_E$ [a]	After Removing Each Layer $\langle\sigma\rangle_L$	$\langle\sigma\rangle_T$	$\langle\sigma\rangle_Z$	$\langle\sigma\rangle_E$ [a]
AVERAGE RESIDUAL STRESSES (MPA) IN Ti-24Al-11Nb MATRIX MATERIAL								
(1) 70.00	298	93	0	256	298	93	0	258
(2) 61.25	298	96	−9	263	300	97	0	262
(3) 52.50	298	97	−17	270	305	99	−1	270
(4) 43.75	298	94	−33	283	311	102	−1	275
(5) 35.00	298	91	−51	298	315	107	0	277
(6) 26.25	298	87	−72	316	319	110	0	281
(7) 17.50	298	89	−99	338	328	111	0	289
AVERAGE RESIDUAL STRESSES (MPA) IN BETA-21S MATRIX MATERIAL								
(1) 70.00	283	85	0	251	283	85	0	252
(2) 61.25	283	85	−3	254	289	89	−1	257
(3) 52.50	283	85	−10	259	296	94	1	261
(4) 43.75	283	85	−19	266	303	97	−1	269
(5) 35.00	283	85	−31	275	311	98	0	275
(6) 26.25	283	85	−44	285	319	100	0	283
(7) 17.50	283	85	−59	298	327	103	−1	290

[a] $\langle\sigma\rangle_E$ = Von Mises effective stress.

MMCs. Specifically, prediction models should be designed to simulate real experimental conditions. Many of the reported residual stress measurements in MMCs are done by the conventional X-ray diffraction technique. The stress profiles are usually obtained by successive electropolishing of surface layers. Under these conditions, it is shown here that the conventional UCMs are inadequate to predict the stress profiles resulting from the electropolishing. The number and orientation of subsurface plies have been shown to play a significant role in the residual stress profile of the surface ply.

Acknowledgments

The authors wish to thank the Ohio Super Computer Center (Steve Koehl and Larry Cooper) for a grant (PES 454) for running the Lawrence Livermore National Laboratory (LLNL) codes. The authors also wish to thank Robert Whirley of LLNL for helpful discussions on the codes and the personnel from WL/MLLN for their support, particularly Ted Nicholas, Jim Larsen, and Noel Ashbaugh. Many helpful discussions with Ken Wright of General Electric are also gratefully acknowledged.

References

[*1*] Tsai, S., Mahulikar, D., and Marcus, H. L., *Materials Science and Engineering,* Vol. 47, 1981, pp. 145–149.

[*2*] Rangaswamy, P., Revelos, W. C., and Jayaraman, N., "A Study of Residual Stresses in SCS-6/ Ti-24Al-11Nb Metal Matrix Composite," *Residual Stresses in Composites: Measurement, Modeling and Thermo-mechanical Modeling,* E. V. Barrera and I. Dutta, Eds., The Metallurgical Society of the American Institute of Mining, Metallurgical, and Petroleum Engineers, Warrendale, PA, 21–25 Feb. 1993.

[*3*] Jayaraman, N. and Rangaswamy, P., "Residual Stresses in Ti_3Al-SCS-6 Fiber Reinforced Metal Matrix Composites," *Proceedings,* Titanium Aluminide Composite Workshop, P. R. Smith, S. J. Balsone, and T. Nicholas, Eds., WL-TR-91-4020, Materials Behavior Branch, Wright Laboratory, Wright-Patterson AFB, OH, 1991.

[*4*] Rangaswamy, P., Revelos, W. C., and Jayaraman, N., "Residual Stresses in SCS-6/Ti-24Al-

11Nb Composite: Part 1—Experimental,'' *Journal of Composites and Technology Research,* Vol. 16, No. 1, Jan. 1994, pp. 47–53.

[*5*] Rangaswamy, P. and Jayaraman, N., ''Residual Stresses in SCS-6/Ti-24Al-11Nb Composite: Part II—Finite Element Modeling,'' *Journal of Composites and Technology Research,* Vol. 16, No. 1, Jan. 1994, pp. 54–67.

[*6*] Rangaswamy, P. and Jayaraman, N., ''Residual Stresses in SCS-6/Beta-21S Metal Matrix Composites,'' *Journal of Composites and Technology Research,* accepted for publication.

[*7*] Cox, B. N., James, M. R., Marshall, D. B., and Addison, R. C., Jr., *Metallurgical Transactions A,* Vol. 21A, 1990, p. 2701.

[*8*] James, M. R., ''Behavior of Residual Stresses during Fatigue of Metal Matrix Composites,'' *Residual Stresses-3,* H. Fujiwara, T. Abe, and K. Tanaka, Eds., Elsevier, New York, 1992, pp. 555–560.

[*9*] James, M. R., Bourke, M. A., Goldstone, J. A., and Lawson, A. C., ''Residual Stress Measurements in Continuous Fiber Titanium Matrix Composites,'' *Advances in X-Ray Analysis,* Vol. 36, 1993.

[*10*] James, M. R., ''Residual Stresses in Metal Matrix Composites,'' *Proceedings,* International Conference on Residual Stresses II, G. Beck, S. Denis, and A. Simon, Eds., Elsevier, New York, 1989, pp. 429–435.

[*11*] Brown, K. M., Hendricks, R. W., and Brewer, W. D., ''X-Ray Diffraction Measurements of Residual Stresses in SiC/Ti Composites,'' *Proceedings,* The Metals Society Fall Meeting, Indianapolis, IN, 1–5 Oct. 1989.

[*12*] Wright, P. K., Sensmeier, M. D., Kupperman, D., and Wadley, H., ''Thermal Stress Effects in Intermetallic Matrix Composites,'' NASA Contract No. NAS3-25790, Report No. 21, Lewis Research Center, Cleveland, OH, 1991.

[*13*] Saigal, A., Kupperman, D. S., and Majumdar, S., *Materials Science and Engineering,* Vol. A150, 1992, pp. 59–66.

[*14*] Majumdar, S., Singh, J. P., Kupperman, D., and Krawitz, A. D., ''Application of Neutron Diffraction to Measure Residual Strains in Various Engineering Composite Materials,'' *Journal of Engineering Materials and Technology,* Vol. 113, 1991, pp. 51–59.

[*15*] Kupperman, D. S., Majumdar, S., and Singh, J. P., ''Residual Strain in Advanced Composites,'' *Neutron News,* Vol. 2, 1991, pp. 15–18.

[*16*] Ashbaugh, N., Khobaib, M., John, R., et al., ''Mechanical Properties of Advanced Engine Materials,'' Final Report, Air Force Contract WL-TR-91-4149, Wright-Patterson AFB, OH, April 1992.

[*17*] Johnson, W. S., Lubowinski, S. J., and Highsmith, A. L., in *Thermal and Mechanical Behavior of Metal Matrix and Ceramic Matrix Composites, ASTM STP 1080,* J. M. Kennedy, H. H. Moeller, and W. S. Johnson, Eds., American Society for Testing and Materials, Philadelphia, 1990, pp. 193–218.

[*18*] Bigelow, C. A., Johnson, W. S., and Naik, A. R., ''A Comparison of Various Micromechanics Models for Metal Matrix Composites,'' *Mechanics of Composite Materials and Structures,* J. N. Reddy and J. L. Telpy, Eds., 1989.

[*19*] Bigelow, C., ''Thermal Residual Stresses in a Silicon-Carbide/Titanium [0/90] Laminate,'' *Journal of Composites and Technology Research,* submitted for publication.

[*20*] Rangaswamy, P., ''A Study of Residual Stresses in Titanium Metal Matrix Composites,'' Ph.D. thesis, University of Cincinnati, 1993.

[*21*] Uemura, M., Iyama, H., and Yamaguchi, *Journal of Thermal Stress,* Vol. 2, 1979, pp. 393–412.

[*22*] Vedula, M., Pangborn, R. N., and Queeney, R. A., *Composites,* Vol. 19, No. 1, 1988, pp. 55–60.

[*23*] Vedula, M., Pangborn, R. N., and Queeney, R. A., *Composites,* Vol. 19, No. 2, 1988, pp. 55–60.

[*24*] Wisnom, M. R., *Journal of Composite Materials,* No. 24, 1990, pp. 707–727.

[*25*] Nimmer, R. P., *Journal of Composites Technology & Research,* Vol. 12, No. 2, 1990, pp. 65–75.

[*26*] Kroupa, J. L., ''Elastic-Plastic FEM Analysis of MMC Subjected to Thermomechanical Fatigue,'' in *Proceedings,* Titanium Aluminide Composite Workshop, P. R. Smith, S. J. Balsone, and T. Nicholas, Eds., WL-TR-91-4020, Materials Behavior Branch, Wright Laboratory, Wright-Patterson AFB, OH, 1991.

[*27*] Maker, B. N., Ferencz, R. M., and Hallquist, J. O., ''NIKE3D: A Nonlinear, Implicit, Three Dimensional Finite Element Code for Solid and Structural Mechanics—User's Manual,'' University of California, Lawrence Livermore National Laboratory, Report UCRL-MA-105268, Jan. 1991.

Fiber Bridging Behavior

L. Neil McCartney[1]

Stress Transfer Mechanics: Models that Should be the Basis of Life Prediction Methodology

REFERENCE: McCartney, L. N., "**Stress Transfer Mechanics: Models that Should be the Basis of Life Prediction Methodology,**" *Life Prediction Methodology for Titanium Matrix Composites, ASTM STP 1253,* W. S. Johnson, J. M. Larsen, and B. N. Cox, Eds., American Society for Testing and Materials, 1996, pp. 85–113.

ABSTRACT: The initiation and growth of damage in composite materials are phenomena that precede the failure event where a material sample or component separates into two pieces. In fatigue, the damage grows slowly and leads to a gradual deterioration of mechanical properties. For the prediction of the fatigue behavior of unidirectional and laminated titanium composites, it is necessary to be able to take account of the effects on the thermoelastic constants of matrix cracking that is induced by fatigue stress cycling. The values of the thermoelastic constants for microcracked composites are determined by the way in which stress is transferred between fiber and matrix in unidirectional composites, and between neighboring plies in laminates, as a result of microcrack formation in the matrix.

A summary is given of the recent progress that has been made at the National Physical Laboratory (NPL) on the development of analytical stress transfer models for unidirectional and laminated composites. The models are each based upon just a single assumption concerning the stress field that leads to stress and displacement solutions for which the Reissner energy function, used in a variational calculation, has stationary values. The use of the Reissner function allows both applied traction and displacement conditions to be imposed on the fiber/matrix interface and external boundary. In contrast to other applications of variational techniques, the models provide both the stress and displacement distributions at every point in the composite. Thus, complete solutions can be derived that satisfy exactly the equilibrium equations, the interface conditions, the stress-strain relationships apart from one, and the boundary conditions involving tractions. The remaining stress-strain relationship and boundary conditions involving displacements are satisfied in an average sense. For unidirectional composites, both perfectly bonded and frictionally slipping interfaces (such that the interfacial shear stress is a constant) are considered; while for laminates, consideration is restricted to interfaces that remain perfectly bonded in the presence of transverse cracks. The analytical stress transfer models are of high quality to the extent that they represent the best models that can be derived subject to the single assumption on which they are based. The models do, therefore, offer life prediction methodologies the opportunity of confidently making use of analytical methods that obey the important principles of mechanics.

The estimation of effective stress intensity factors for bridged matrix cracks that occur in titanium composites subject to fatigue loading at modest load levels is an important element in the prediction of composite performance. Assuming that fibers remain intact, the paper makes use of the stress transfer models to predict the dependence of stress intensity factors (for long cracks) on the applied stress and crack separation, for bridged cracks in both unidirectional and laminated composites. The effects of crack interaction are determined, and the new model predictions are compared to those of shear-lag theory modified to take account of residual stresses arising from thermal expansion mismatch effects. It is also shown how predictions can be made for the dependence of the thermoelastic constants of unidirectional and laminated

[1]Head, Fracture and Performance Section, Centre for Materials Measurement and Technology, National Physical Laboratory, Teddington, Middlesex, UK, TW11 0LW.

titanium matrix composites on the nonuniform spacings of matrix cracks that can be encountered in practice.

KEYWORDS: unidirectional composites, cross-ply laminates, titanium matrix composites, stress transfer models, fiber bridging, stress intensity factors, fatigue crack growth, titanium, life prediction, titanium alloys, fatigue (materials), modeling

The estimation of the life of a composite, when subjected to fatigue or creep deformations, is perhaps the most difficult prediction that an engineer has ever to make. If material failure is defined to be the separation of a composite into two separate pieces, as in the case of metals, then it has to be recognized that composite failure is generally preceded by a multitude of localized damage effects that are strongly interactive, subject to statistical uncertainty, and that lead to an increasing deterioration in performance to the point of actual failure. If a practically useful life prediction design methodology is to be developed then it is vital that the initiation and growth of damage preceding failure is capable of being predicted with confidence. As predicting the complete failure of a composite is exceedingly difficult, a more pragmatic approach is to define composite failure as the loss of a specified amount of some property such as modulus or Poisson's ratio. The overall objective of this paper is, therefore, to outline some useful methods that can be recommended with confidence for use when predicting the growth and effects of fatigue induced damage on the mechanical performance of both fiber-reinforced unidirectional and cross-ply laminated titanium composites. The approach of the paper is to consider mainly models of stress transfer, and their use when predicting damage development, that have been shown to obey all the relevant laws of mechanics.

During fatigue loading (at modest stress levels) of unidirectional fiber-reinforced titanium composites, matrix cracking normal to the loading direction and the associated fiber/matrix debonding are important and highly interactive modes of damage that are encountered. This leads to the presence of cracks in the composite, normal to the loading direction, that are bridged by intact fibers. The presence of these cracks leads to a degradation of the properties of the composite, such as moduli, Poisson's ratios, and thermal expansion coefficients. It is thus important to understand the factors that determine the formation and growth of such cracks during fatigue loading. The stress intensity factors at the tips of bridged cracks are influenced strongly by the bridging fibers, and it is important to be able to predict the dependence of stress intensity factors on crack bridging forces. This leads to the need to be able to relate the effective stress, p, carried by a bridging fiber to the effective crack opening displacement, Δv. The quantitative knowledge of the $p - \Delta v$ relationship is the key to the understanding and prediction of the effect of bridging fibers on the stress intensity factors for bridged cracks, and, consequently, on crack growth rates during fatigue loading, and for calculating critical stress intensity factors for crack instability. The first objective of this paper is to describe a new theoretical stress transfer model that is capable of predicting the $p - \Delta v$ relationship in terms of the thermomechanical properties of the fiber and matrix of a composite. The model assumes that the fiber and matrix are deforming linearly so that resulting predictions can be used for fiber-reinforced titanium composites provided that extensive plasticity is not occurring. The second objective is to apply the stress transfer model to the prediction of stress intensity factors of bridged matrix cracks for the case when fibers remain intact during crack growth and to consider the point of instability of matrix cracks. The third objective is to indicate how the methodology may also be used to predict the effects of fatigue-induced transverse cracking on the thermoelastic constants of cross-ply laminates, allowing for the occurrence of nonuniform crack spacings.

A New Stress Transfer Model for Unidirectional Composites

Neither shear-lag methods (for example, *1* through *7*) nor variational methods based on the minimization of complementary energy (for example, *8* through *10*) provide solutions that enable the prediction of the distributions of all stress and displacement components throughout the composite system. Shear-lag theory neglects relevant stress and displacement components, one equilibrium equation and most stress/strain relationships, while variational methods based on complementary energy minimization do satisfy the equilibrium equations but do not necessarily satisfy the stress-strain relationships, nor do they predict the displacement distribution or allow one to examine the satisfaction of interface conditions involving displacement components. Pagano [*11–13*] has applied the Reissner variational principle [*14*] to unidirectional and laminated composites solving for the stress field and the weighted average displacements. The Reissner principle allows one to propose admissible solutions that involve both stress and displacement components. The Reissner variational procedure is based on the following energy function

$$J = \int_{v} \left[\frac{1}{2} \sigma_{ij} \left(\frac{\partial u_i}{\partial x_j} + \frac{\partial u_j}{\partial x_i} \right) - W(\sigma_{ij}) \right] dV - \int_{S_\tau} \hat{\tau}_j u_j \, dS \qquad (1)$$

where σ_{ij} is the stress tensor, u_i is the displacement vector, W is the strain energy function expressed in terms of stresses, and where $\hat{\tau}_j$ is the traction distribution applied to the region, S_τ, of the external surface, S, bounding the region, V, occupied by the composite. The remaining part of the external surface, S_u, is assumed to be subject to applied displacement conditions. The Reissner energy function, J, has a stationary value if the equations of elasticity are satisfied, namely

$$\frac{\partial \sigma_{ij}}{\partial x_i} = 0, \quad \sigma_{ij} = \sigma_{ji}, \quad \epsilon_{ij} = \frac{\partial W}{\partial \sigma_{ij}} \qquad (2)$$

where ϵ_{ij} is the strain tensor. The first two relationships of Eq 2 are the well-known equilibrium equations resulting from the linear and angular momentum balance equations of continuum mechanics, while the third relationship corresponds to the stress/strain equations.

McCartney [*15*] has recently modified an earlier stress transfer model [*16*] by making use of Nairn's representation for the stress field [*10*] (used in a variational calculation based on complementary energy), and then deriving the corresponding displacement field. Only one assumption is made in the "stress transfer" model where it is assumed that the axial stresses in the fibers and matrix do not depend upon the radial coordinate (it should, however, be noted that other assumptions are made when applying the model to the prediction of fatigue crack growth). The resulting solution has the following characteristics:

1. Equilibrium equations are satisfied exactly.
2. Interface conditions are satisfied exactly.
3. All stress-strain relationships are satisfied exactly apart from the axial relationship that is satisfied in an average sense (averaging across the cross-section of the fiber and of the matrix).
4. All external traction boundary conditions are satisfied exactly, and displacement conditions are satisfied in an average sense.
5. The problem reduces to solving an ordinary differential equation of fourth order.
6. The solution leads to stationary values of the Reissner energy function.
7. The model is the best possible that can be developed based upon the single fundamental assumption that axial stresses are independent of the radial coordinate.

In sharp contrast to the new stress transfer model just described, shear lag models are not able to match the new model with respect to any of the preceding criteria that can be used to characterize the quality of stress transfer models.

Application of Model to Stress Transfer for a Bridging Fiber

The model of stress transfer between fiber and matrix in a unidirectional composite just described has been applied successfully [*15*] to predicting stress transfer for broken fibers and for fibers bridging matrix cracks. The key quantities of interest are the distributions of stress and displacement in both fiber and matrix, taking account of thermal residual stresses, and of fiber/matrix debonding and frictional slip. Figure 1 shows a schematic diagram of a set of parallel matrix cracks normal to the principal loading direction in a unidirectional composite where the fiber/matrix interface has debonded and frictional slip is occurring. For a titanium matrix composite, such cracks are induced by fatigue loading; sometimes from some form of stress concentration such as a blunt notch or hole. The model to be discussed here represents a bridging silicon-carbide fiber as a linear elastic cylinder of radius, R, that is embedded in a concentric cylinder of titanium matrix that is also assumed to deform as a linear elastic solid. A representative volume element of the composite that will be the subject of stress analysis is also shown in Fig. 1. Cylindrical polar coordinates (r,z) are used such that the origin lies on the axis of the fiber at the midpoint between two adjacent matrix cracks that are separated by a distance, $2L$, as shown in Fig. 2. The outer radius of the matrix cylinder is $a = \sqrt{R/V_f}$, where V_f denotes the fiber volume fraction of the composite. Following manufacture, the fiber/matrix interfaces are assumed to be perfectly bonded. Hutchinson and Jensen [*17*], in their shear-lag analysis of stress transfer and crack bridging for both constant shear stress and Coulomb friction interface behavior, have considered two types of boundary condition applied to the outer surface of the matrix cylinder. First of all, on the surface $r = a$, the stress components, σ_{rr} and σ_{rz}, are assumed to be zero, defining a stress transfer problem (to be referred to as Problem 1) where there is no external lateral constraint. Secondly, the stress component, σ_{rz}, is assumed to be zero while the radial displacement, u_r, is assumed to have a value that corresponds to the lateral displacement that would arise in an uncracked composite subject to the same applied stress and temperature. Such boundary conditions will lead to nonzero values of the stress component, σ_{rr}, and to a nonzero effective transverse stress that will change with crack density. In this paper for radially constrained boundary conditions, the stress component, σ_{rz}, is assumed to be zero while the radial displacement, u_r, is assumed to have a value that leads to a zero value for the effective transverse stress, that is, the stress component, σ_{rr}, is averaged over the length, $2L$, of the composite lying between two neighboring matrix cracks. For this case (to be referred to as Problem 2), although its average over the length, $2L$, is zero, the distribution of the stress component, σ_{rr}, is nonzero indicating that there is localized lateral constraint that attempts to take account of the effect of surrounding composite material on the stress transfer between fiber and matrix.

Stress transfer calculations, using methods described in Ref *15*, have been performed for a typical titanium composite made of fibers and matrix having the following isotropic properties:

1. Fiber (SiC monofilament) where $E_f = 400$ GPa, $\nu_f = 0.25$, and $\alpha_f = 4.86 \times 10^{-6}/°C$
2. Matrix (Ti-6-4) where $E_m = 110$ GPa, $\nu_m = 0.34$, and $\alpha_m = 9.44 \times 10^{-6}/°C$

The values taken for the Young's moduli, E_f and E_m, and Poisson's ratios, ν_f and ν_m, correspond to those used by Bakuckas and Johnson [*18*] when assessing the relevance of

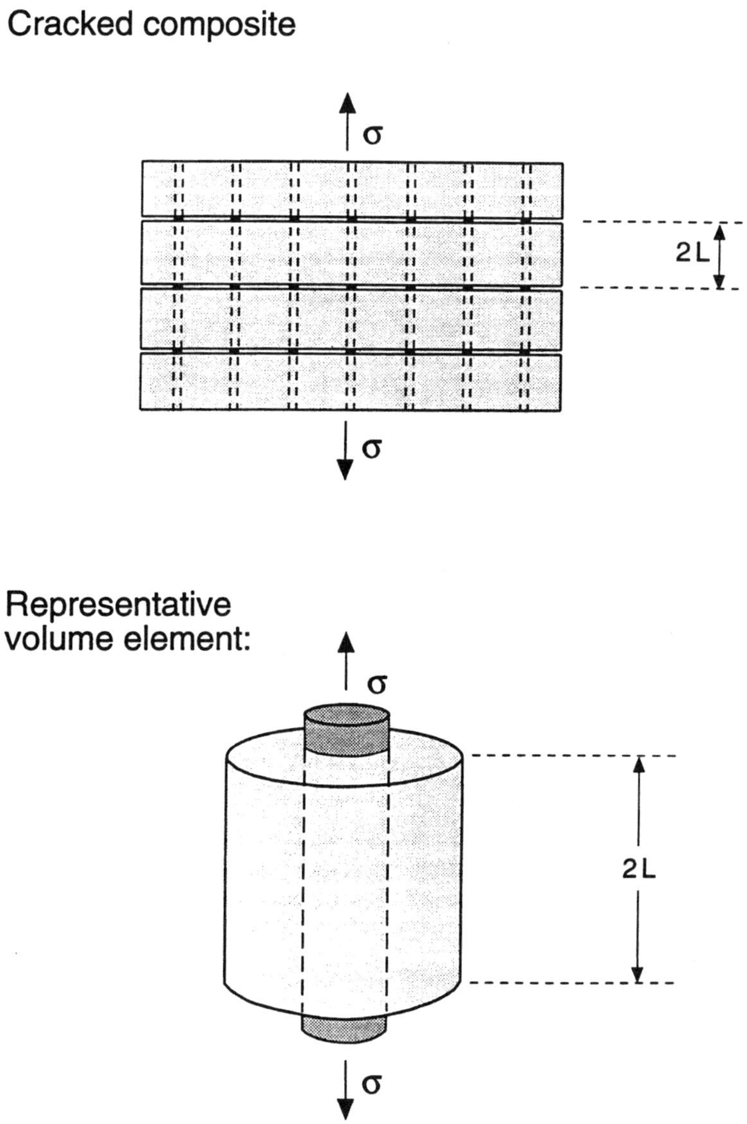

FIG. 1—*Schematic diagram of a unidirectional composite with an array of fatigue-induced matrix cracks. Also shown is the representative volume element used to solve the stress transfer problem.*

several fiber-bridging models to fatigue crack growth in titanium composites. The thermal expansion values, α_f and α_m, correspond to those used by Nimmer et al. [*19*] when carrying out a finite element analysis of a titanium composite. The fiber volume fraction of the composite is assumed to be 0.35 and the fiber radius is taken as 50 μm. The stress-free temperature of the composite is assumed to be 500°C so that ΔT, denoting the temperature difference between room temperature and the stress-free temperature, is -480°C.

The thermoelastic constants for an undamaged unidirectional composite have been calcu-

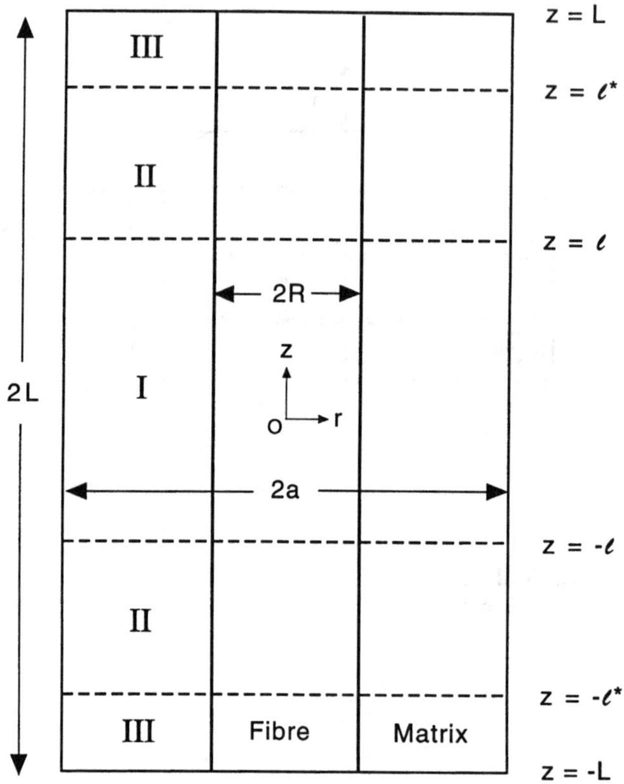

FIG. 2—*Diagram showing the region between two adjacent matrix cracks and identifying the three types of region exhibiting differing interface behavior.*

lated using a modified multiple concentric cylinder model of a unidirectional composite [*20*]. The values predicted, assuming that the composite is a transverse isotropic solid, are given by

$$E_A = 212\ \text{GPa} \qquad \nu_A = 0.302 \qquad \alpha_A = 6.57 \times 10^{-6}/°\text{C}$$
$$E_T = 167\ \text{GPa} \qquad \nu_T = 0.372 \qquad \alpha_T = 7.87 \times 10^{-6}/°\text{C}$$
$$\mu_A = 67.7\ \text{GPa}$$

These values will be used later when calculating the properties of cross-ply laminates. For titanium matrix composites, fatigue-induced matrix cracks with bridging fibers lead to fiber/matrix debonding and frictional slip that must be taken into account when predicting stress transfer. In this paper, it is assumed that stress transfer in the slip zone is characterized by an interfacial shear stress, τ, that is uniform at all points within the slip region. The value of τ may degrade during fatigue loading, or increase as a result of thermal exposure, but such effects are not considered here. One problem that arises, when assuming that the interfacial shear stress is a constant, is that the shear stress, σ_{rz}, is likely to be double valued at the points $r = R$, $z = \pm L$, since this shear stress must be zero at all points on the surfaces of the matrix crack and could be nonzero within the interface as $z \to \pm L$. Such behavior will require the stress field to have weak singularities at these points; features that are not easy to include in the stress transfer model under consideration, nor in finite element analysis. It is interesting to note that this problem does not arise if a Coulomb friction law is assumed for the slip region of the interface (see Ref *16*). To overcome this problem when using the constant interfacial shear stress model, a transition zone is introduced between the matrix crack and the constant shear stress zone. The geometry is illustrated in Fig. 2 where the region between two neighboring matrix cracks is divided into regions of three types. The use of the transition zone allows the model to be energetically consistent in the sense that no work is done on the stress-free surfaces of the problem. This would not be the case if the transition zone had been neglected, as a shear traction would arise from the model on the matrix crack surfaces, and would be doing work during deformation leading to spurious terms when carrying out energy balance calculations. This is particularly important when considering the fracture of brittle matrix composites.

For Region I, the fiber/matrix interface is either perfectly bonded, or has debonded but has not slipped, so that the stress components, σ_{rr} and σ_{rz}, and the displacement components, u_r and u_z, must be continuous across the interface, $r = R$, for all points, z, in the range, $-\ell \leq z \leq \ell$, that is

$$\sigma_{rr}(R-,z) = \sigma_{rr}(R+,z)$$

$$\sigma_{rz}(R-,z) = \sigma_{rz}(R+,z)$$

$$u_r(R-,z) = u_r(R+,z)$$

$$u_z(R-,z) = u_z(R+,z)$$

In Region I, it follows [*15*] that stress transfer is governed by the following fourth order ordinary differential equation

$$FR^4C''''(z) + GR^2C''(z) + HC(z) = 0 \tag{3}$$

where $C(z)$ is the function that determines the values of the stress and displacement components and is such that $C'(z)$ represents the interfacial shear stress distribution. The F, G, and H parameters are obtained from the micromechanical model and depend upon the properties of the fibers and matrix together with the fiber volume fraction.

For Regions II, $\ell \leq z \leq \ell^*$ and $-\ell^* \leq z \leq -\ell$, the stress components, σ_{rz} $(=\tau)$, σ_{rr}, and the displacement component, u_r, are continuous across the fiber/matrix interface at $r = R$, whereas the displacement component, u_z, can be discontinuous, that is

$$\sigma_{rr}(R-,z) = \sigma_{rr}(R+,z) \leq 0$$

$$\sigma_{rz}(R-,z) = \sigma_{rz}(R+,z)$$

$$u_r(R-,z) = u_r(R+,z)$$

$$\sigma_{rz}(R,z) = \tau$$

As the interfacial shear stress, τ, is constant in this region, the function, $C(z)$, has the form

$$C(z) = \tau z + k \tag{4}$$

where k is a constant.

For Regions III, the $\ell^* \leq z \leq L$ and $-L \leq z \leq -\ell^*$ transition zones, which allow the interfacial shear stress to tend to zero at $z = \pm L$ in a mechanically consistent manner, the stress components, σ_{rr}, σ_{rz} $(\neq\tau)$, and the displacement component, u_r, are continuous across the fiber/matrix interface at $r = R$, that is

$$\sigma_{rr}(R-,z) = \sigma_{rr}(R+,z)$$

$$\sigma_{rz}(R-,z) = \sigma_{rz}(R+,z)$$

$$u_r(R-,z) = u_r(R+,z)$$

The stress components in the matrix (for Region III only) are assumed to be held fixed at the values reached when the critical interfacial shear stress, τ, is first reached during loading. Such an assumption is also imposed for unloading during the execution of fatigue cycles. The stress in the fiber varies, dependent upon the load variations applied to the composite. The matrix part of Region III can be thought of as moving rigidly (having a frozen stress field) up and down the fiber as the load is cycled.

Complete stress transfer solutions have been found using the methods described in Ref *15* satisfying the interface conditions for Regions I, II, and III exactly. The precise functional relationships between the stress and displacement components and the function, $C(z)$, and its derivatives are given in Ref *15*.

The axial boundary conditions that must be satisfied are derived from the consideration of overall mechanical equilibrium and symmetry. Since the matrix is completely cracked on the $z = \pm L$ planes, it follows that $\sigma_{zz}(r, \pm L) = 0$ for $R \leq r \leq a$. Strictly speaking, σ_{zz} (when averaged over the fiber) should have the value, $\sigma/V_f = p$, where σ is the stress applied to the composite in the axial direction and where p is the effective crack bridging stress when regarded as being smoothed over the matrix crack surfaces. Because the stress transfer model assumes that the stress component, σ_{zz}, is independent of the radial coordinate, r, it follows that $\sigma_{zz}(r, \pm L) = \sigma/V_f$ for $0 \leq r \leq R$. Assuming that there is a regular array of equally spaced matrix cracks, symmetry considerations imply that $\sigma_{rz}(r, \pm L) = 0$ for $0 \leq r \leq a$. The preceding axial boundary conditions are satisfied exactly by the solution of the stress transfer model. For the predictions to be presented in this paper, the value used for the interfacial shear stress is $\tau = 100$ MPa that falls well within the range quoted by Bakuckas and Johnson [*18*] and corresponds to the value quoted by Clyne et al. [*21*] for a titanium composite in the as-received condition.

When carrying out stress transfer calculations, it is necessary to consider how the crack has been introduced. A damage-free composite is first loaded (compressively for a titanium composite) until the axial stress in the matrix is zero. The applied stress at such a point is denoted by σ_i. The crack can then be introduced without having to perturb the stress field. Mechanical loading is then applied to a level, σ, defining the relative applied stress, $\sigma - \sigma_i$. As soon as the loading of a cracked composite commences, interfacial shear stresses build up until the critical shear stress, τ, is reached at which point interfacial slipping takes place. Subsequent loading leads to the growth of the slip zone.

Although three separate regions of interface behavior have been introduced, nothing has been said about how their locations are determined. The transition zone Region III remains fixed in length during loading following the initiation of frictional slip. Prior to this stage of loading, Region I (indicating either perfect interface bonding or the absence of slip) occupies the whole of the region between the matrix cracks. When loaded, the maximum interfacial shear stress occurs at the point, $z = \ell^* < L$. Thus, frictional slip and the growth of Region II initiates at this point. From the mathematical representation of stress and displacement fields, it can be shown that the function, $C(z)$, characterizing stress transfer must at least be continuous and have continuous first and second derivatives in the region, $-L \leq z \leq L$. This ensures that all the stress components and the displacement component, u_r, are continuous in this region. The displacement component, u_z, depends upon the third derivative of $C(z)$ and thus is not necessarily continuous. However, the averages of u_z over cross sections of both the fiber and matrix are continuous, as they depend only on the derivative and integral of $C(z)$. The continuity of the function, $C(z)$, and its first two derivatives are sufficient to derive unique solutions to the equations governing stress transfer in Regions I, II, and III. In addition, they provide a condition for the value of the ℓ parameter that determines the boundary between Regions I and II. It should be noted that during loading the ℓ parameter decreases indicating the growth of the frictional slip zone, whereas the ℓ^* parameter does not vary during loading. It should also be noted that a consequence of the approach just described is that the interfacial shear stress will be continuous and bounded everywhere in the region, $-L \leq z \leq L$. The model does not therefore take account of the singularity that will exist at the point, $r = R$, $z = \ell$, and is therefore assuming that the interface is very weak such that either the interface may be regarded as always being debonded or that the stress singularity is weak enough for its effect to be negligible.

Once the boundary conditions described earlier have been used to derive a unique solution of the stress transfer problem, many predictions can be made concerning the stress and displacement distributions in the neighborhood of a matrix crack bridged by a fiber. Figures 3 through 7 show the results of predictions for the case when $\tau = 100$ MPa and when the crack density is 0.5 cracks/mm; a density for which the crack separation is large enough for there to be no interaction between the cracks, as seen from Figs. 3 through 7. For the boundary conditions applied, the solutions are symmetric about the plane, $z = 0$.

Figure 3 shows the distribution of the interfacial shear stress as a function of z/L for three applied stress levels when the crack density is 0.5 cracks/mm. The first corresponds to the point at which the critical interfacial shear stress, τ, of 100 MPa is first reached during loading. The critical shear stress is found at the $z = \ell^*$ location. Frictional slip in the interface commences when the applied stress is compressive having the value -115.9 MPa for Problem 1 (stress-free external surface) and the value -145 MPa for Problem 2 (radially constrained external surface); the fiber stress being compressive while the matrix stress is tensile. The second and third loading cases are for much larger applied stress values of 0.5 GPa and 1 GPa, respectively, where significant frictional slip has occurred. Figures 4 and 5 show the corresponding axial stress distributions in the fiber and matrix, respectively. Figures 6 and 7 show the effect of the external boundary condition on the interfacial shear and normal stresses, respectively. The dotted line shows the results when the external surface, $r = a$, of the system is stress-free (that is, Problem 1), while the solid line shows the corresponding results for an external surface subject to lateral constraint (Problem 2). It should be noted that for the stress-free case the normal stresses become tensile within Region III. As the interface in Region III has debonded, it cannot sustain a tensile normal stress with the result that the solution for a stress-free external surface is leading to inconsistency. For the constrained case, it remains compressive at all points on the interface and, therefore, is consistent with the presence of nonzero interfacial shear stresses in this region. As Fig. 6 indicates that the effect of applying constrained boundary conditions (that is, Problem 2) has a very slight effect on the interfacial shear stress distribution, the effects of such constraint will be neglected in the subsequent analysis. Thus, all subsequent results are derived for Problem 1 boundary conditions.

The continuous thick line in Fig. 8 shows the dependence of the length of the frictional slip zone on the relative applied stress, $\sigma - \sigma_i$, where $\sigma_i = -0.385$ GPa is the value of the applied stress for which an uncracked composite would have a zero axial stress in the matrix. For this curve, the crack density is 0.5 cracks/mm and the linear nature of the curve indicates that there is no interaction between the cracks. The continuous line does not extrapolate back to the origin as interfacial slip is initiated at a positive value (269.5 MPa for Problem 1) of the relative applied stress; in addition, the initial slip zone length is not zero. The thin, broken, and dotted lines show the corresponding curves for crack densities of 1, 2,

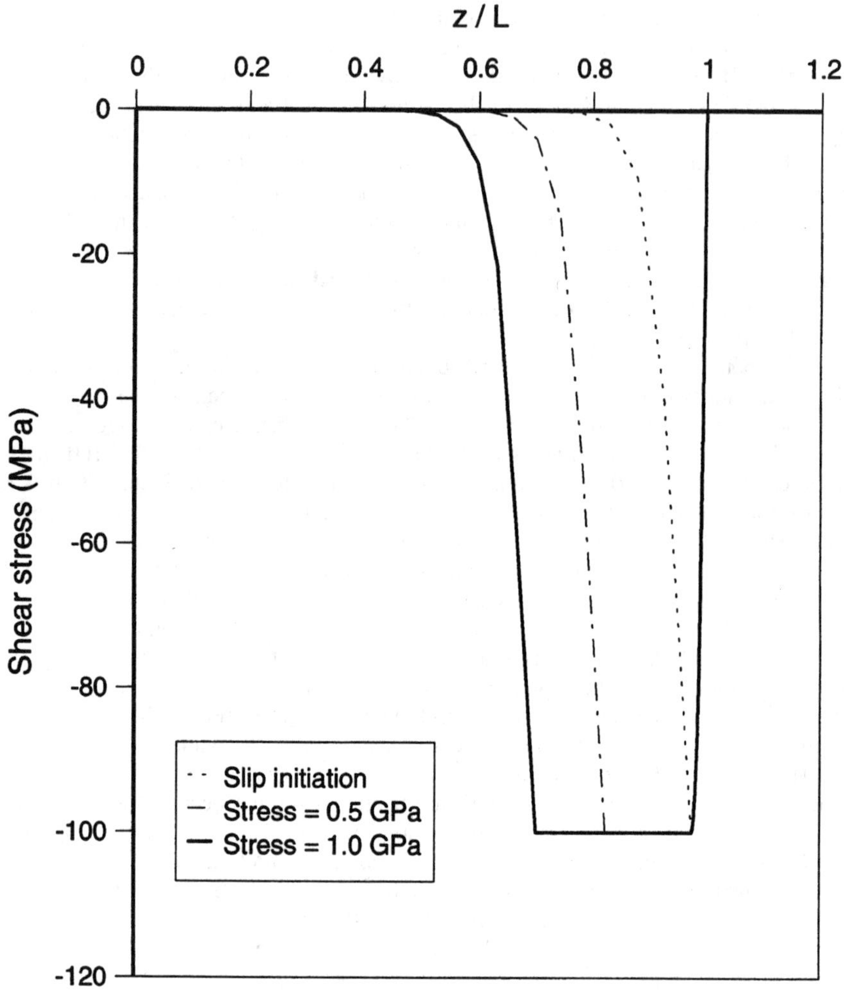

FIG. 3—*Comparison of the interfacial shear stress for three different loading conditions when the crack density is 0.5 cracks/mm.*

and 5 cracks/mm, respectively, and indicate that there are significant interaction effects only if the crack density or applied stress is large enough. The broken and dotted lines are tending towards limiting values of the slip-zone lengths that correspond to the limit imposed when applying the constant interfacial shear stress shear-lag model to two neighboring cracks. As the fiber radius is 50 μm, it follows that the limiting values of ℓ/R are 20, 10, 5, and 2 corresponding respectively to crack densities of 0.5, 1, 2, and 5 cracks/mm.

Figure 9 shows the dependence of the effective crack opening displacement on the relative applied stress, $\sigma - \sigma_i$, where $\sigma_i = -0.385$ GPa. This relationship is needed to understand and predict the influence of crack bridging forces on matrix crack-tip stress intensity factors. The thick continuous line shows the relationship for a crack density of 0.5 cracks/mm where crack interaction effects are so small that they can be neglected. The thin

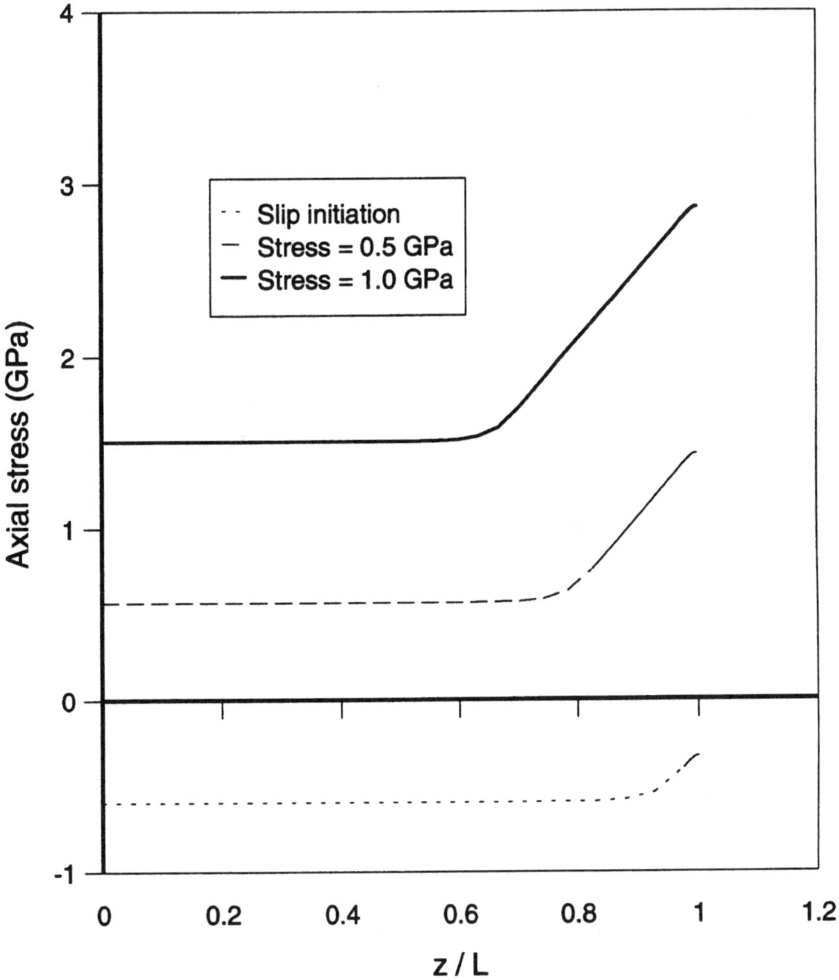

FIG. 4—*Comparison of axial fiber stress distribution for three different loading cases when the crack density is 0.5 cracks/mm.*

continuous, broken, and dotted lines show the corresponding results for crack densities of 1, 2, and 5 cracks/mm, respectively, indicating significant effects as the relative stress, $\sigma - \sigma_i$, increases.

To conclude this section, it is worth mentioning that finite element analysis (FEA), even for elastic solutions, would be presented with significant problems when solving Problems 1 and 2. The reason is that the point, $z = \ell$, defining the boundary between Regions I and II is not known in advance and must be found as part of the solution by iteration. For Problem II, an additional iteration would be required to determine the externally applied radial displacement consistent with a zero value for the radial stress averaged over $-L \leq z \leq L$. In contrast to FEA methods, the analytical approach described in this paper can readily deal with these problems using zero-finding routines.

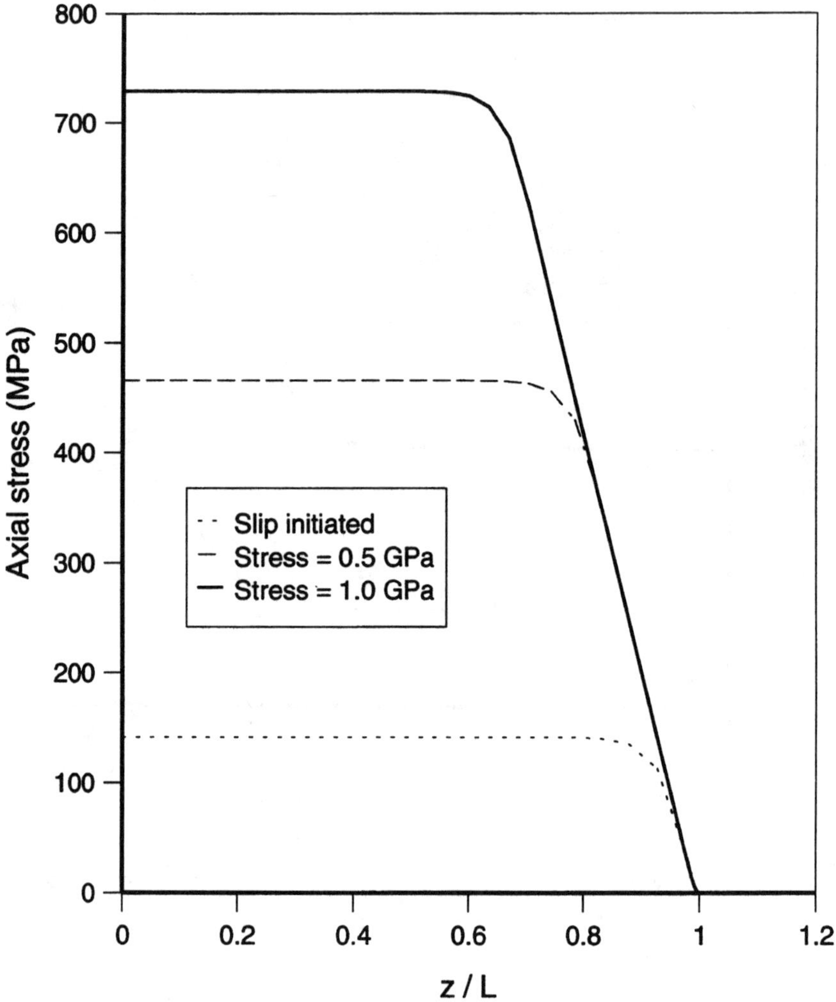

FIG. 5—*Comparison of axial matrix stress distribution for three different loading cases when the crack density is 0.5 cracks/mm.*

Stress Transfer for Cross-Ply Laminates

McCartney [*22,23*] has developed a generalized plane strain model of stress transfer for cross-ply laminates and has shown that the stress transfer problem reduces to solving a differential equation of the form (Eq 3) that is valid for unidirectional composites. Pagano[2] has shown that the approximations used, when deriving the solution for the generalized plane strain model of 0-90-0 laminates, are such that the Reissner energy function has a stationary value when carrying out a variational calculation. The conclusion is that the generalized plane strain model of stress transfer is the best possible that can be developed based upon the single fundamental assumption that axial stresses depend only on the axial

[2]Pagano, N.J., private communication, 1992.

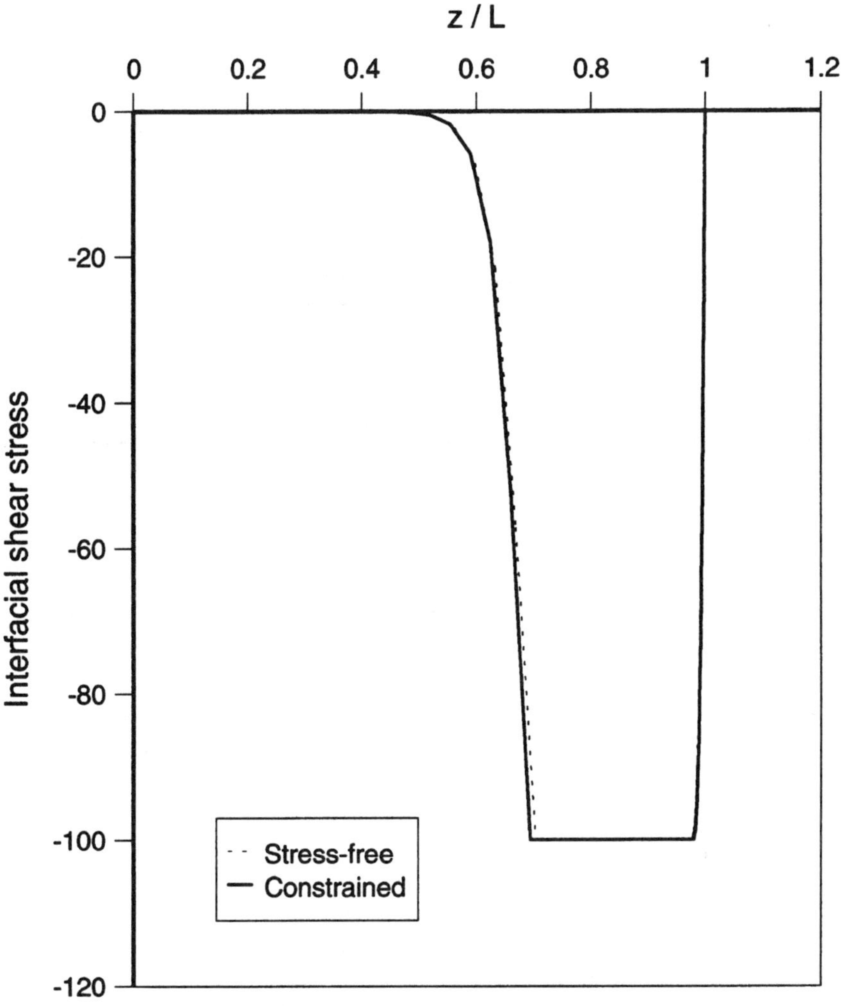

FIG. 6—*Comparison showing the effect of the external boundary condition on the interfacial shear stress distribution when* σ = *1 GPa and the crack density is 0.5 cracks/mm.*

coordinate. The model has the same characteristics as the new model for unidirectional composites described earlier, where averages are now taken over the cross-sections of the 0° and 90° plies, rather than over the fiber and matrix (see Refs *22* and *23* for details).

Fracture Mechanics for Bridged Cracks in Anisotropic Materials

When carrying out lifting calculations for unreinforced metals, four essential ingredients are needed. The first is a crack growth law that is usually assumed to have the form

$$\frac{dc}{dN} = A(\Delta K)^n \tag{5}$$

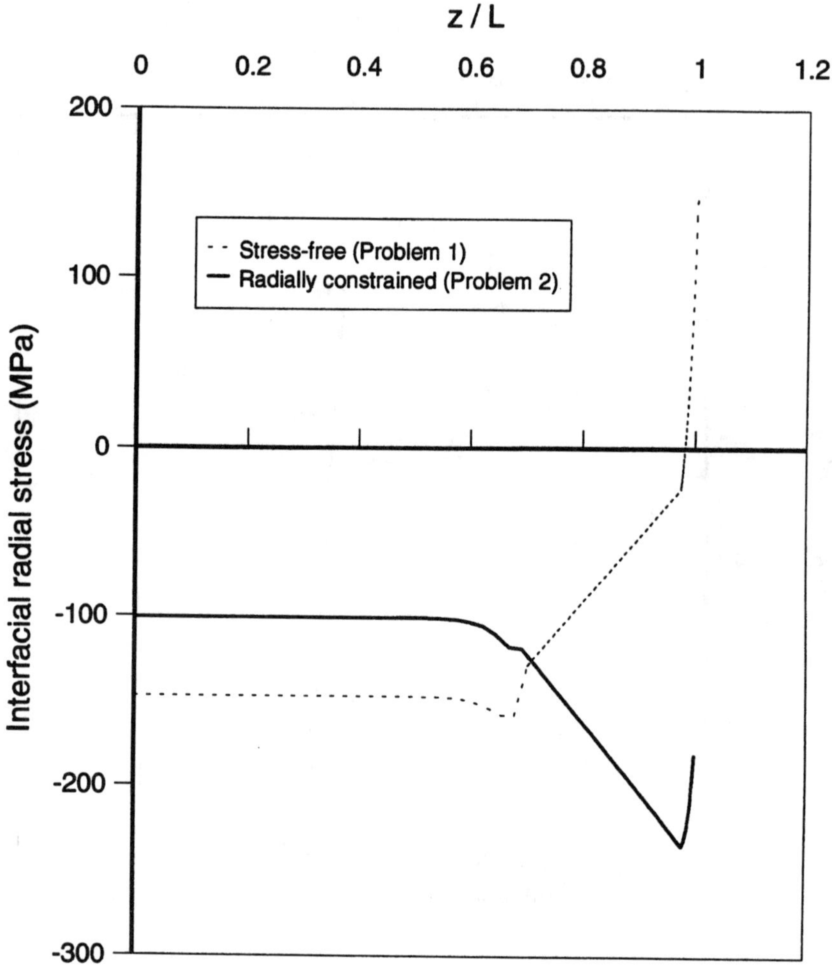

FIG. 7—*Comparison showing the effect of the external boundary condition on the interfacial normal stress distribution when* σ = *1 GPa and the crack density is 0.5 cracks/mm.*

where A and n are empirical constants, and where ΔK is the stress range intensity factor of the form

$$\Delta K = \Delta\sigma\sqrt{\pi c}Y \tag{6}$$

where $2c$ denotes the crack length and Y is a factor that takes geometrical effects into account; a topic not to be considered in this paper. The second ingredient, for loading cases where small-scale yielding conditions prevail, is a knowledge of the fracture toughness of the material, K_{Ic}, at which unstable fracture occurs. The third ingredient, not considered in this paper, is a knowledge of initial defect sizes that can be used, together with the instability condition and crack growth law, to predict fatigue lives. A fourth ingredient, again not considered in this paper, is the initiation of small fatigue cracks from microdefects. Analo-

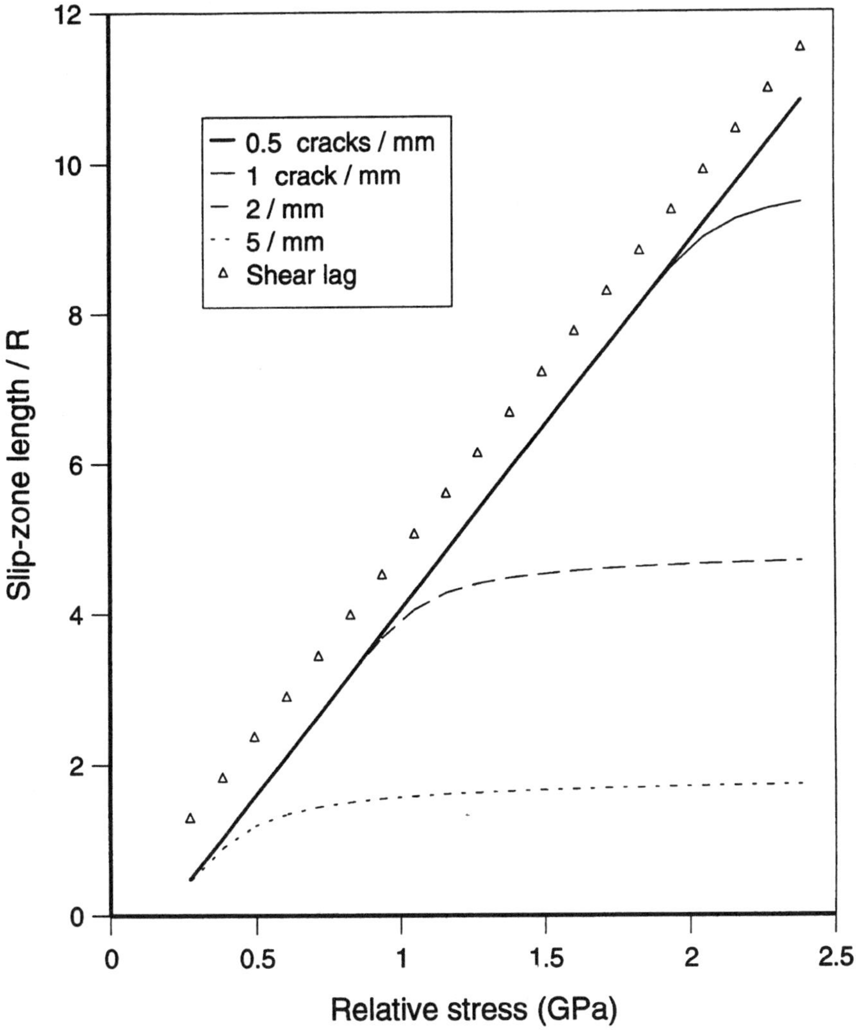

FIG. 8—*The dependence of the length of the frictional slip zone on the relative applied stress* ($\sigma - \sigma_i$) *for three crack densities.*

gous ingredients are required when predicting fatigue-induced matrix crack growth in fiber-reinforced metals.

For the case of matrix cracking in unidirectional and laminated composites under consideration in this paper, a crack growth law of the form

$$\frac{dc}{dN} = A(\Delta K_{\text{eff}}^{m})^{n} \tag{7}$$

is assumed where $\Delta K_{\text{eff}}^{m}$ is the effective stress range intensity factor for fatigue cracking in the matrix. This parameter, valid only for small-scale yielding conditions, will be influenced by the fibers bridging the matrix crack and by the fact that only the matrix is cracking during

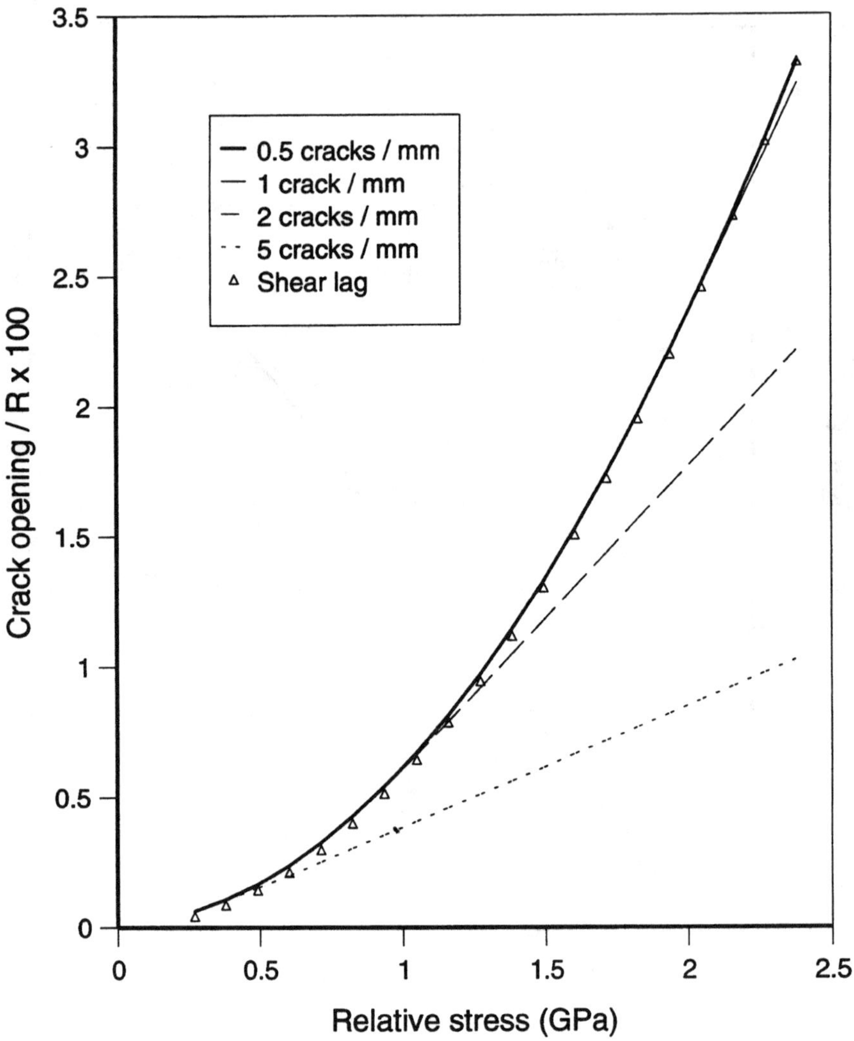

FIG. 9—*The dependence of the effective crack opening of a bridged crack on the relative applied stress* ($\sigma - \sigma_i$) *for three crack densities.*

fatigue loading. It is important that a rational approach is taken to defining the effective stress range intensity factor. As crack instability needs to be considered, it is useful first of all to consider the fracture of brittle matrix composites under monotonic loading conditions in order to define key fracture mechanics parameters.

It has been shown [*24*] that the fracture criterion for the unstable growth of bridged brittle matrix cracks has the following simple form

$$K^2 = \frac{2\gamma E_A}{\Omega} = K_{\mathrm{Ic}}^2 \tag{8}$$

that corresponds to the well-known Griffith fracture criterion when the material is isotropic so that $\Omega = 1$ (generalized plane stress) or $\Omega = 1 - \nu^2$ (plane strain), and when $E_A = E$, the Young's modulus for an isotropic solid. The parameter, K_{Ic}, is the effective Mode I fracture toughness of the composite. Equation 8 shows how K_{Ic} is related to the effective fracture energy, 2γ. To clarify the meaning of K_{Ic} and 2γ, it is useful to indicate their values for the case of unidirectional and laminated composites having brittle matrices for the case when only the matrix cracks leaving intact the fibers of unidirectional composites and the 0° plies of cross-ply laminates. As shown in Ref *24*, $\gamma = V_m\gamma_m$ when the matrix cracks in a unidirectional composite where $2\gamma_m$ is the fracture energy of the matrix and $V_m = 1 - V_f$ is the volume fraction of matrix. When the matrix cracks in a 0-90-0 cross-ply laminate, $\gamma = (a/h)\gamma_m$ where $2h$ is the laminate thickness and $2a$ is the thickness of the inner 90° ply. The effective fracture toughness of the composite is given [*24*] by

$$K_{Ic} = \sqrt{\frac{V_m E_A}{E_m}}\, K_{Ic}^m, \qquad K_{Ic}^m = \sqrt{2\gamma_m E_m} \tag{9}$$

where K_{Ic}^m is the fracture toughness of the matrix and E_m is Young's modulus for the matrix. For matrix cracks in titanium composites, the γ_m and K_{Ic}^m parameters should be interpreted as the effective fracture energy and fracture toughness of the titanium matrix that take account of plasticity effects. This identification is valid only if there is small-scale yielding at the tips of matrix cracks in the composite. Thus, Eq 9 can be used to determine the instability of matrix cracks growing during fatigue loading. At the point of instability for a unidirectionally reinforced composite, the matrix crack will instantaneously propagate across the remaining cross section of uncracked matrix in the plane of the crack. For a cross-ply laminate, the transverse matrix crack will instantaneously propagate across the remaining cross section of 90° ply in the plane of the crack. It is worth noting that the Eqs 8 and 9 are both valid when fibers fracture within the bridged zone but not if fiber fracture occurs ahead of the crack tip.

As fatigue crack growth can be thought of as resulting from energy dissipation (or energy absorption) in the plastic zones (see, for example, Ref *25*), it is reasonable to use energy concepts to define the effective stress range intensity factor, ΔK_{eff}^m, for the matrix. The Eq 9, which was derived for brittle matrix composites by energy balance arguments, is the basis for the following definition of the effective stress range intensity factor (also used by Bao and McMeeking [*26*])

$$\Delta K_{eff}^m = \sqrt{\frac{E_m}{V_m E_A}}\, \Delta K_c \tag{10}$$

where ΔK_c is the stress range intensity factor for the composite, regarded as a continuum, that will now be related to the stress intensity factor, K_c, for a bridged crack in a composite that is calculated using fracture mechanics principles.

When a matrix crack in a unidirectional composite is monotonically loaded, the resulting crack opening leads to frictional slip at the fiber/matrix interfaces. If now the stress is reduced by an amount, $\Delta\sigma$, as would occur during a fatigue loading cycle, then while $\Delta\sigma$ is relatively small, the composite responds elastically (that is, there is no relative slip at the fiber/matrix interfaces; a phenomenon not predictable when using the constant shear stress shear-lag model without transition zones). When $\Delta\sigma$ is larger, the maximum interfacial shear stress will reach the critical value, τ, that now acts in the opposite direction to that which was encountered during loading. To analyze the stress and displacement distributions at any stage during unloading, use is made of the linear superposition rule where the monotonic solution calculated for an applied stress of $\Delta\sigma$, for a temperature difference of $\Delta T = 0$, and for an interfacial shear stress of 2τ is subtracted from the monotonic solution where the applied stress is σ, $\Delta T \neq 0$, and the interfacial shear stress has the value, τ. For the case when $\Delta T = 0$, McMeeking and Evans [*27*] have given such a solution in the case of a shear-lag model based on a constant interfacial shear stress, τ. The same approach can be made for the new stress transfer model described in this paper, since the ℓ^* parameter is independent of the value of the applied stress. The following important result (derived for a shear lag model by McMeeking and Evans in the absence of thermal residual stresses) is established for the new model in the Appendix

$$\Delta K(\Delta\sigma, \Delta T) = 2K(\Delta\sigma/2, 0) \tag{11}$$

where $K(\sigma, \Delta T)$ is the stress intensity factor for a monotonically loaded crack when the applied stress has the value σ and $\Delta T \neq 0$. It is worth noting that although the stress intensity factor will depend linearly on the temperature difference, ΔT, the stress range intensity factor is independent of ΔT provided ΔT is held fixed during stress cycling. If the temperature were to be cycled while the stress was held fixed, then the stress intensity factor would not be given by Eq 11; a discussion of this topic is beyond the scope of this paper. The result (Eq 11), although derived for nonlinear $p - \Delta v$ relationships that result from the constant shear stress shear-lag model, is also valid for any linear $p - \Delta v$ relationship as would arise for unidirectional composites where interface slip has not occurred, and for laminates containing transverse cracks. Equations 7, 10, and 11 indicate that fatigue crack growth in unidirectional and cross-ply laminates is readily predicted if the stress intensity factor, $K(\sigma, \Delta T)$, is known. The method of determining this stress intensity factor will now be described.

Calculation of the Stress Intensity Factor

Now consider a large rectangular plate that is made of a unidirectionally fiber-reinforced composite material or of thin cross-ply material. The axes of material symmetry are parallel to the edges of the plate. On introducing a rectangular set of coordinate axes (x,y), the stress/strain/temperature relationships for the composite are of the form [*28,29*]

$$\epsilon_{xx} = a_{11}\sigma_{xx} + a_{12}\sigma_{yy} + \alpha_T\Delta T \tag{12}$$

$$\epsilon_{yy} = a_{12}\sigma_{xx} + a_{22}\sigma_{yy} + \alpha_A\Delta T \tag{13}$$

$$2\epsilon_{xy} = a_{66}\sigma_{xy} \tag{14}$$

The a_{11}, a_{12}, a_{22}, and a_{66} parameters are elastic constants for a transverse isotropic solid having the following values, where the axial direction is parallel to the y-axis:

1. For generalized plane stress conditions (thin sheets—unidirectional or laminates)

$$a_{11} = \frac{1}{E_T}, \qquad a_{22} = \frac{1}{E_A}$$
$$a_{12} = -\frac{\nu_A}{E_A}, \qquad a_{66} = \frac{1}{\mu_A} \tag{15}$$

2. For plane strain conditions (thick sheets—unidirectional only):

$$a_{11} = \frac{1 - \nu_T^2}{E_T}, \qquad a_{22} = \frac{1}{E_A}\left(1 - \nu_A^2\frac{E_T}{E_A}\right)$$
$$a_{12} = -\frac{\nu_A(1 + \nu_T)}{E_A}, \qquad a_{66} = \frac{1}{\mu_A} \tag{16}$$

where values of the constants, E_A, E_T, etc. for a typical unidirectional titanium composite have been given earlier.

It is convenient to introduce a dimensionless parameter, Ω, that fully characterizes the anisotropy of the material when considering crack opening distributions. The parameter, Ω, is defined [*28*] by

$$\Omega^2 = \frac{1}{4} E_A^2 a_{22}[2\sqrt{a_{11}a_{22}} + 2a_{12} + a_{66}] \tag{17}$$

For isotropic materials, it follows that $\Omega = 1$ for generalized plane stress and that $\Omega = 1 - \nu^2$ for plane strain where ν is Poisson's ratio for an isotropic solid. For a titanium composite having the fiber and matrix properties given earlier, it can be shown from Eq 17 that $\Omega = 1.0932$.

Frictionally slipping interfaces lead to nonlinear $p - \Delta v$ relationships that can be derived from micromechanical models and can be expressed [*28*] in the general form

$$p(x, c) = \sigma^* f(\Delta v/R) + \sigma_i, \Delta v = \Delta v(x, c) \tag{18}$$

where σ^* is a parameter having the dimensions of stress, R is the radius of bridging fibers, and σ_i is the value of the applied stress for which the axial matrix stress in an undamaged composite would be zero. The σ^* factor appearing in Eq 18 has been introduced so that f is a dimensionless function that is obtained from the micromechanical model. For the applications of this paper, it is convenient to select $\sigma^* = \tau$; the characteristic interfacial shear stress. The f function must be such that

$$f(0) = 0 \tag{19}$$

It is convenient to introduce the crack opening displacement $\Delta v_0 = v_0 R$ such that

$$f(v_0) = \frac{\sigma - \sigma_i}{\sigma^*} \tag{20}$$

It follows from analysis given in Refs *24* and *28* that, for an embedded bridged crack in a large panel, the crack opening distribution, $v(X) \equiv \Delta v(x,c)/R$, where $X = x/c$, may be determined from the nonlinear integral equation

$$v(X) = \mu \int_0^1 \{f(v_0) - f(v(t))\}\, k(t, X)\, dt, \qquad 0 \leq X \leq 1 \tag{21}$$

where

$$\mu = \frac{c}{c^*}, \qquad k(t, X) = \frac{1}{\pi} \ln \left| \frac{\sqrt{1 - t^2} + \sqrt{1 - X^2}}{\sqrt{1 - t^2} - \sqrt{1 - X^2}} \right| \tag{22}$$

and where

$$c^* = \frac{E_A R}{4\Omega\sigma^*} \tag{23}$$

The integral equation may be solved numerically by iteration (see Refs *24* and *28* for details) thus determining the crack opening distribution function $v(X)$. The corresponding value for the stress intensity factor for a bridged crack may be written

$$K = \sigma^* \sqrt{\pi c}\, \frac{2}{\pi} \int_0^1 \frac{f(v_0) - f(v(t))}{\sqrt{1 - t^2}}\, dt = (\sigma - \sigma_i)\sqrt{\pi c}\, Y \tag{24}$$

where

$$Y = \frac{2}{\pi} \int_0^1 \left(1 - \frac{f(v(t))}{f(v_0)} \right) \frac{dt}{\sqrt{1 - t^2}} \tag{25}$$

Solutions to the integral Eq 21 require extensive numerical computation especially when complex $p - \Delta v$ relationships are used. The results of such calculations will not be considered here as such results will find limited practical application. For fatigue cracking (under modest loads) in unidirectionally reinforced or cross-ply laminated plates, microcracks arise without accompanying fiber fracture, and for such cases a more useful simplified approach is possible. The approach is based on the assumption that fibers do not fail in the crack region and the matrix crack is long enough for the stress intensity to tend to a limiting value that is independent of crack length [*24,28*]. This assumption is reasonable for matrix crack formation in panels and plates but fiber fracture, and a more complex approach involving solutions of the integral Eq 21, will need to be considered for matrix crack formation in the presence of stress concentrations, for example, associated with a notch or hole.

Stress Intensity Factors in the Limiting Case of Long Cracks

It has been shown [*24,28*] that the equation determining the stress for the crack instability of long cracks in brittle matrix composites subject to monotonic loading is given by the simple relationship

$$\sigma \Delta v_0 - \int_0^{\Delta v_0} p(\Delta v) d(\Delta v) = \Omega \frac{K^2}{E_A} = 2\gamma \tag{26}$$

thus defining a relationship from which the stress intensity factor for a long bridged crack can be determined. It can be shown using Eqs 18, 20, 23, and 26 that the stress intensity factor may be calculated from the relationship

$$K^2 = \pi c^* (\sigma^*)^2 F(\sigma) \tag{27}$$

where

$$F(\sigma) = \frac{4}{\pi} \int_0^{v_0} [f(v_0) - f(v)]\, dv, \qquad \sigma = \sigma^* f(v_0) + \sigma_i \tag{28}$$

The values of the function, $F(\sigma)$, representing a dimensionless normalized square of the stress intensity factor, are thus readily calculated from micromechanical models. For example, in Fig. 10, the continuous line shows the dependence of the normalized square of the stress intensity factor, $K^2/\pi c^*(\sigma^*)^2$, as a function of the normalized relative stress, $(\sigma - \sigma_i)/\sigma^*$, for the case when the crack separation is 0.5 cracks/mm and there is no interaction between the cracks. The broken and dotted lines show the corresponding relationships for crack densities of 1, 2, and 5 cracks/mm. It should be noted that the results for crack densities of 0.5 and 1 cracks/mm appear coincident in Fig. 10. For relatively large crack densities, it is clear that crack interaction effects reduce the normalized stress intensity factor and such effects cannot be accounted for when using the shear-lag model.

Instability Condition for Long Fully Bridged Cracks

It follows from Eqs 27 and 28 that a fully bridged matrix crack (where fibers do not fracture in the bridged zone) growing under the influence of fatigue loading will become unstable when the following condition is satisfied

$$F(\sigma) = \frac{4}{\pi}\int_0^{v_0} \{f(v_0) - f(v)\}\, dv = \frac{K_{\rm Ic}^2}{\pi c^*(\sigma^*)^2} \quad (29)$$

where $K_{\rm Ic}$ is the fracture toughness of the composite defined by the first of the relationships (Eq 9). The result (Eq 29), valid for any crack bridging relationship, is regarded as the fracture criterion that determines (usually using numerical zero-finding techniques) the value of v_0, and hence of the applied stress, $\sigma = \sigma^* f(v_0) + \sigma_i$, for which a long pre-existing crack will grow unstably. It is clear from Eq 29 that the instability criterion for long cracks is independent of crack length. Thus, if fatigue cycling is in progress under a modest constant maximum load that prevents fiber fracture, then a crack will remain stable as it grows across

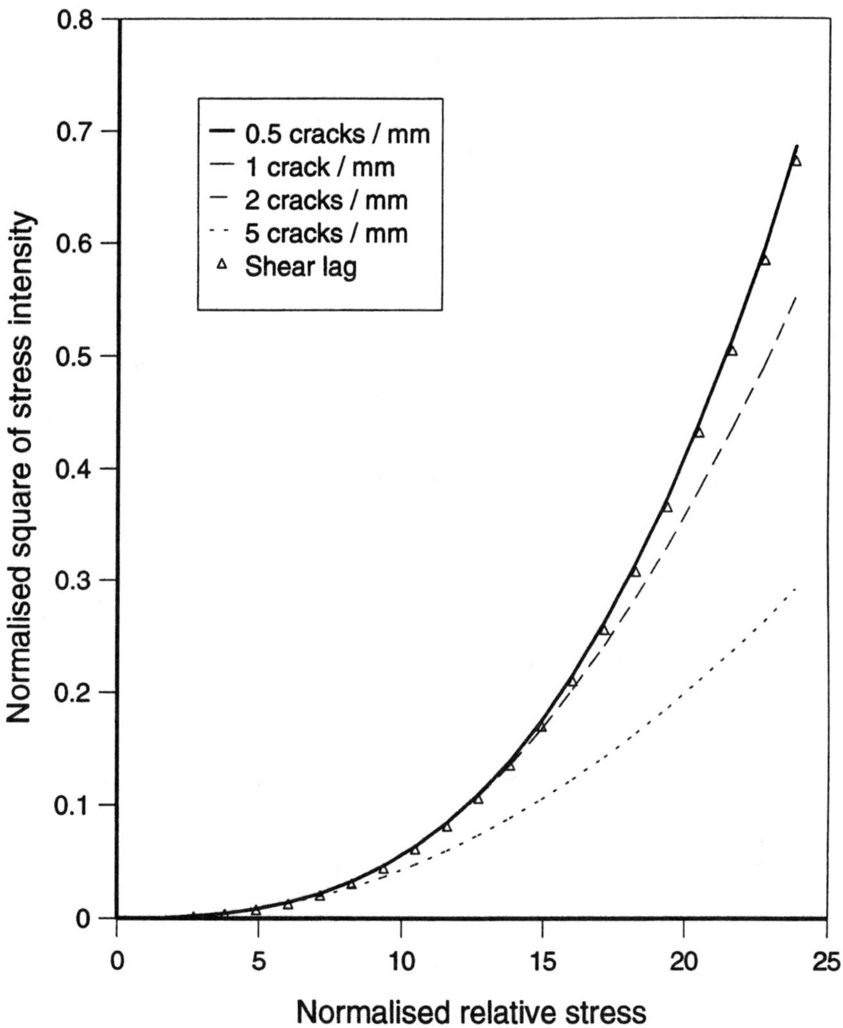

FIG. 10—*The dependence of the normalized square of the stress intensity factor* $K^2/\pi c^*(\sigma^*)^2$ *as a function of the normalized relative stress* $(\sigma - \sigma_i)/\sigma^*$ *for the case of long cracks and for three crack densities.*

the entire cross section of the composite. Instability may, however, be induced by stress transfer effects that will arise near the edges of the panel. The instability condition (Eq 29) for long cracks is very important if a cracked composite is likely to be subjected to intermittent overloads.

Comparison with Shear-Lag Theory

As is clear from the preceding discussion, shear-lag theory for a constant interfacial shear stress (see, for example, Ref *2*) is very approximate but is used extensively in the literature. It is worthwhile, therefore, to compare some of the predictions of the new model described in this paper with those of shear-lag theory (for unidirectional composites) modified to take into account the effects of residual stresses arising from thermal expansion mismatch effects (see, for example, Ref *29*). There are three parameters that are useful to compare. The first is the dependence on the applied stress of the extent of the debonded zone, in which the constant interfacial shear stress, τ, acts. From shear-lag theory, the length of this zone, when thermal residual stresses are present, is given [*29*] by

$$1 = \frac{R}{2\tau}\frac{V_m}{V_f}\frac{E_m}{E_A}(\sigma - \sigma_i) \tag{30}$$

where σ is the applied stress and σ_i is the applied stress (compressive for titanium composites) for which the matrix in an undamaged composite has a zero axial stress. Figure 8 compares the shear-lag relationship (shown as discrete points) with the new model predictions. It is seen that the shear-lag prediction will extrapolate to the origin and lies above the predictions of the new model. There is good agreement between shear-lag theory and the new model only if neighboring matrix cracks are far enough apart to avoid crack interaction effects.

The second parameter of interest is the relationship of the effective crack opening of a long matrix crack (with intact fibers) to the monotonic increasing applied stress, σ. For shear-lag theory, this is given by [*29*]

$$\Delta = \frac{R}{4\tau E_f}\left(\frac{V_m E_m}{V_f E_A}\right)^2 (\sigma - \sigma_i)^2, \quad f(v) = \sqrt{\frac{2E_f}{\tau}\frac{V_f E_A}{V_m E_m}}\sqrt{v}, \quad \text{when } \sigma^* = \tau \tag{31}$$

where 2Δ is the effective displacement discontinuity across the matrix crack for a continuum crack model that would be applied in a fracture mechanics calculation (see Ref *28* for extensive discussion of this aspect). The shear-lag prediction based on Eq 31 is shown in Fig. 9 as discrete points and is very close indeed to the new model predictions for the case when crack interaction effects are negligible. Significant deviations are seen when the crack density is two and five cracks/mm.

The third parameter of interest is the stress intensity factor for a long crack. It has been shown [*24*] that, when using the constant stress shear-lag model, the stress intensity factor, *K*, for a bridged crack in a unidirectional material is determined from the relationship

$$K^2 = \frac{4\pi(\sigma - \sigma_i)^3}{3\lambda^2\Omega} \tag{32}$$

where the parameter, λ, is given by

$$\lambda = 2\frac{V_f}{V_m}\sqrt{\frac{2\pi\tau}{R}\frac{E_f E_A}{E_m^2}} \tag{33}$$

It can be shown that for the shear-lag model the dimensionless function, $F(\sigma)$, appearing in Eq 27 for the normalized square of the stress intensity factor, is given by

$$F(\sigma) = \frac{2}{3\pi}\left(\frac{V_m E_m}{V_f E_A}\right)^2 \frac{(\sigma - \sigma_i)^3}{E_f \tau^2} \qquad \text{when } \sigma^* = \tau \tag{34}$$

In Fig. 10, the function, $F(\sigma)$, for shear-lag theory defined by Eq 34 is plotted as a function of the relative stress, $\sigma - \sigma_i$, in the form of discrete points that are very close to the predictions of the new model described in this paper provided that crack interaction effects are negligible. Again significant deviations are observed when the crack density is two and five cracks/mm.

The Effects of Accumulating Fatigue Damage

For structural panels subjected to fatigue loading at modest stress levels, a series of parallel cracks can be encountered leading to a degradation of the mechanical properties of the panel. The final objective of this paper is to indicate how micromechanical models have been used to predict the change of properties such as moduli, Poisson's ratios, and thermal expansion coefficients, as a function of crack density. Such calculations usually make the assumption that matrix cracks are uniformly spaced, but it has been shown [*30*] that transverse cracking in cross-ply laminates having nonuniform crack spacings can be treated.

Consider a titanium composite panel containing an array of $n + 1$ parallel matrix cracks labeled by the parameter, $i = 1 \ldots n + 1$, as illustrated in Fig. 11, such that $2l_i$ denotes the separation of the ith and $i + 1^{\text{th}}$ cracks and $2l$ denotes the distance between the first and $(n + 1)^{\text{th}}$ cracks. From the micromechanical models of unidirectional and laminated composites described in this paper, it is possible to calculate the effective thermoelastic constants of each segment of the panel lying between neighboring cracks. Thus, for that part of the panel lying between the ith and $i + 1^{\text{th}}$ cracks, the following thermoelastic constants can be calculated

$$E_A(l_i), E_T(l_i), \nu_A(l_i), \alpha_A(l_i), \alpha_T(l_i)$$

where E_A and E_T are the axial and transverse Young's moduli, respectively; ν_A, is the axial Poisson's ratio; and α_A and α_T are the axial and transverse thermal expansion coefficients, respectively.

The effective thermoelastic constants for a cracked panel of length, $2l$ containing $n + 1$ parallel cracks that may be spaced nonuniformly are denoted by

$$E_A(n), E_T(n), \nu_A(n), \alpha_A(n), \alpha_T(n)$$

where the argument, n, is used to indicate average properties when n randomly spaced cracks are present. By considering the whole assembly of n elements defined by the $n + 1$ cracks, it can be shown that the effective thermoelastic constants for the panel as a whole are given by the following addition formulae

$$E_T(n) = \sum_{i=1}^{n} \frac{l_i}{l} E_T(l_i) \tag{35}$$

$$\frac{1}{E_A(n)} = \sum_{i=1}^{n} \frac{\lambda_i}{E_A(l_i)} \tag{36}$$

$$\frac{\nu_A(n)}{E_A(n)} = \sum_{i=1}^{n} \lambda_i \frac{\nu_A(l_i)}{E_A(l_i)} \tag{37}$$

$$\alpha_A(n) = \sum_{i=1}^{n} \lambda_i \alpha_A(l_i) \tag{38}$$

$$\alpha_T(n) = \sum_{i=1}^{n} \lambda_i \alpha_T(l_i) \tag{39}$$

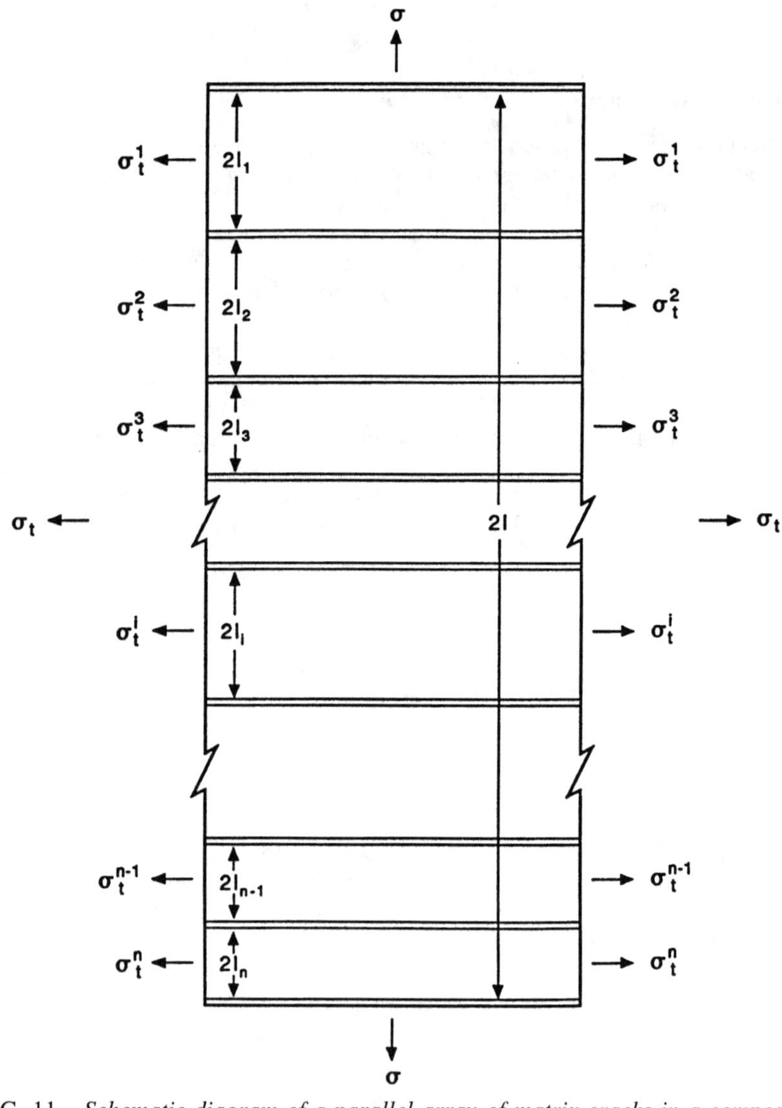

FIG. 11—*Schematic diagram of a parallel array of matrix cracks in a composite.*

where the parameters, λ_i, $i = 1 \ldots n$, are defined (consistent with Eq 35) by

$$\lambda_i = \frac{l_i}{l}\frac{E_T(l_i)}{E_T(n)}, \qquad \text{such that} \sum_{i=1}^{n} \lambda_i = 1 \tag{40}$$

It is emphasized that the Eqs 35 through 40 have been established only for 0-90-0 laminates. They are, however, expected to be valid also for unidirectional composites.

As an example, Fig. 12 shows the change in Young's modulus and Poisson's ratio as a function of crack density for the case of a cross-ply laminate where the 0° and 90° plies have the same thickness. The calculations were carried out for nonuniform crack spacings where additional cracks were inserted at nonuniformly distributed locations between existing cracks (see Ref *30* for details of technique used) in order to increase the crack density

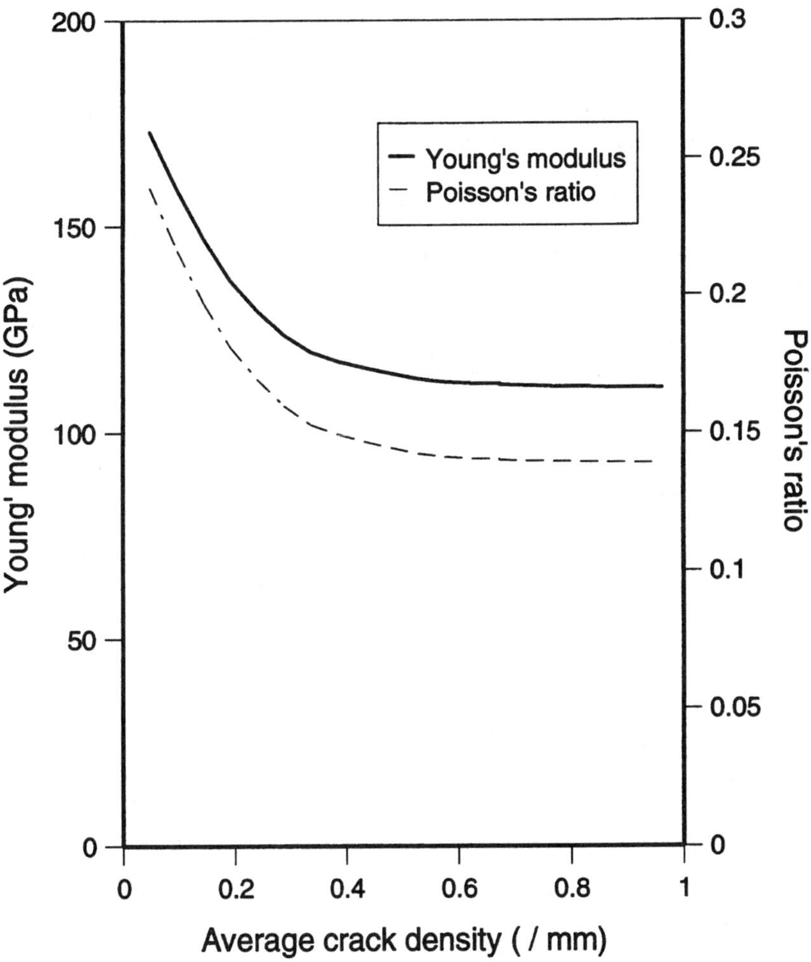

FIG. 12—*Dependence of Young's modulus and Poisson's ratio on average crack density for a cross-ply laminate having nonuniform distributions of transverse cracks in the 90° plies.*

incrementally. Similar effects will result when unidirectional composites contain arrays of parallel microcracks.

Discussion and Conclusions

This paper has described the features of two micromechanical models of stress transfer that occur when matrix cracks form in a fiber-reinforced titanium composite during fatigue loading. The first model of stress transfer for unidirectional composites has been developed for the case where interfacial frictional slip is assumed to be characterized by a constant interfacial shear stress, τ, while the second model of stress transfer applies to cross-ply laminates that are cracked in the transverse plies. Both these models are of similar quality in the sense that they are based on a single assumption, they satisfy all relevant laws of mechanics either exactly or in an average sense and lead to stress and displacement solutions for which the Reissner energy function has a stationary value. This means that the stress transfer models are the best that can be developed given the single assumption that was made. Consequently, the models are ideal for practical application, for example, to understand the behavior of fiber-reinforced composites, and to use the models with confidence as the basis of design methodology. Simple examples have been given in this paper of how the models can be used to understand the behavior of matrix cracks in fatigue for the case when loading levels are modest such that fiber fracture at the location of bridged cracks is avoided. Many more developments in this area are perfectly feasible. For example, the model for unidirectional composites can be applied to the case when frictional slip is governed by the Coulomb friction law instead of assuming a constant interfacial shear stress. Secondly, the fracture mechanics of small cracks can be investigated (where stress intensity factors are crack-length dependent) and the effects of fiber fracture taken into account.

The first important conclusion to be drawn from this paper concerns the nature of the boundary condition applied to the external boundary, $r = a$, of the representative volume element for that part of a fiber lying between two neighboring matrix cracks. In most analyses of this type, it is assumed that this external surface is stress-free and yet this leads to nonuniform radial displacements that do not reflect expected behavior in a cracked composite, and to tensile radial stresses that introduce inconsistency into the model. It is emphasized that such inconsistency does not arise when the model is radially constrained. The comparison given in Fig. 6, and other results not shown, indicates that the error introduced by assuming that external surface is stress-free is relatively small for fiber-reinforced titanium. This means that for titanium matrix composites at least one can neglect the effects of lateral restraint in stress-transfer calculations.

The second important conclusion concerns the effects of crack interaction when arrays of parallel cracks form in a unidirectional composite. The results of Figs. 8 through 10 indicate, for the constant interfacial shear stress model, that crack interaction effects are important for a titanium composite at relatively high applied stresses if the matrix crack density exceeds about one crack/mm.

The third conclusion concerns the performance of the relatively simple shear-lag model of stress transfer when modified to take account of residual stresses arising from thermal expansion mismatch. The results in Figs. 8 through 10 indicate that when interaction effects between neighboring matrix cracks are negligible, the modified shear-lag model predicts results that are more accurate than one should expect given the large number of approximations that are introduced when developing the model. This fact will increase significantly the confidence that can be placed in shear-lag models based upon the constant interfacial shear stress, τ, when applied to titanium matrix composites. It should be noted, however, that for other composite systems the shear-lag model may not perform as well as that demonstrated here for titanium composites.

The fourth and final conclusion concerns the effect of matrix cracking on the properties of panels of unidirectional and laminated fiber-reinforced titanium. Figure 12 shows that significant reductions in important properties such as modulus arise when a cross-ply laminate is damaged by the formation of arrays of transverse cracks, and similar results are found for matrix cracks in unidirectional composites.

To conclude, this paper has shown how high-quality micromechanical models of stress transfer can be used to predict the growth of matrix cracks during fatigue and to predict their effect on the thermoelastic constants of unidirectional and laminated composites. The methods described could relatively easily be used to predict the dependence of thermoelastic constants on the number of fatigue cycles executed. Such calculations, based on the models described in this paper, will be performed in the future. It is the author's opinion that, while the engineer would like to be able to predict strengths and fatigue strengths of composites, such predictions are likely to be very unreliable due to the complex nature of failure mechanisms in composites, and to the statistical factors that complicate the prediction process. A more pragmatic and reliable approach is to redefine the concept of failure to be a specified loss of some measurable property. As can be seen from the methods given in this paper, such an approach has simplicity and good potential for leading to reliable predictions that could be one basis for future design methodology.

Acknowledgment

The author would like to acknowledge that the work described in this paper was carried out as part of the Materials Measurement Programme, a program of underpinning research financed by the Department of Trade and Industry in the United Kingdom.

APPENDIX

Derivation of Expression for the Stress Range Intensity Factor

Consider the stress field that arises when the stress applied to a bridged crack is reduced, as would occur during fatigue cycling. Because the fiber/matrix interface is assumed to be debonded in the region of the matrix crack, and because frictional shear stresses are acting at the interface characterized by the uniform shear stress, τ, a stress reduction can lead to reversed frictional slip. The objective of this Appendix is to show how the stress field may be determined both before and after reversed slip occurs and to show how the value of the stress range intensity factor may be calculated.

Suppose that the maximum applied stress during a fatigue cycle is σ and that the stress field at the maximum stress, σ, is given by the function, $\sigma_{ij}(r, z, \sigma, \Delta T, \tau)$. The corresponding stress intensity factor for the bridged crack is denoted by $K(\sigma, \Delta T, \tau)$. It is convenient to decompose the stress field, σ_{ij}, as follows

$$\sigma_{ij}(r, z, \sigma, \Delta T, \tau) = \hat{\sigma}_{ij}(r, z, \sigma, \tau) + f_{ij}(r)\Delta T \tag{41}$$

where the first term is the stress field that results when $\Delta T = 0$, and where the second term represents the thermal stress distribution characterized by the function, f_{ij}, that is a function only of the radial coordinate, r. The thermal part of the stress field will remain unchanged during any stress change provided the temperature is fixed. Consider now a stress reduction from σ to the value, $\sigma - \Delta\sigma$. The stress field following any reduction in the applied stress may be determined by using the superposition principle. For the case when reversed frictional slip has not been initiated, the stress field following the stress reduction $\sigma \rightarrow \sigma - \Delta\sigma$ is given by

$$\sigma_{ij}(r, z, \sigma, \Delta T, \tau) = \hat{\sigma}_{ij}(r, z, \sigma, \tau) + f_{ij}(r)\Delta T - \hat{\sigma}_{ij}(r, z, \Delta\sigma, \infty) \tag{42}$$

where the term subtracted corresponds to the linear elastic solution for a bridged crack where the fiber/matrix interface is perfectly bonded (denoted by having an infinite value for the shear stress, τ). The stress intensity factor following the stress reduction, $\sigma \rightarrow \sigma - \Delta\sigma$, has the value $K(\sigma, \Delta T, \tau) - K(\Delta\sigma, 0, \infty)$ so that the stress range intensity factor for this case is given by $\Delta K = K(\Delta\sigma, 0, \infty)$. This expression for ΔK should be used only for those values of $\Delta\sigma$ for which the minimum value of the interfacial shear stress is greater than $-\tau$ so that the stress reduction, $\Delta\sigma$, is not large enough to initiate reversed frictional slip.

In order to determine the stress field for the case when reversed frictional slip occurs, it is first noted that in the reversed slip zone the frictional shear stress has the value, $-\tau$, so that it has been reduced by an amount 2τ relative to the value, τ, that arises when the maximum stress, σ, is acting. It then follows, using the superposition principle, that the stress field arising from the stress reduction, $\sigma \rightarrow \sigma - \Delta\sigma$, is given by

$$\sigma_{ij}(r, z, \sigma, \Delta T, \tau) = \hat{\sigma}_{ij}(r, z, \sigma, \tau) + f_{ij}(r)\Delta T - \hat{\sigma}_{ij}(r, z, \Delta\sigma, 2\tau) \tag{43}$$

It is noted that, because the stress field for a debonded fiber in the presence of a matrix crack is determined using the methods of linear elasticity theory, the following relationship must be satisfied by the function $\hat{\sigma}_{ij}$

$$\hat{\sigma}_{ij}(r, z, \sigma, \tau) \equiv \frac{1}{2}\hat{\sigma}_{ij}(r, z, 2\sigma, 2\tau) \tag{44}$$

The validity of this relationship has been verified by numerical methods, and it applies also to the corresponding displacement field. On using Eq 44, the result (Eq 43) may be expressed in the form

$$\sigma_{ij}(r, z, \sigma, \Delta T, \tau) = \hat{\sigma}_{ij}(r, z, \sigma, \tau) + f_{ij}(r)\Delta T - 2\hat{\sigma}_{ij}(r, z, \Delta\sigma/2, \tau) \tag{45}$$

with the result that the stress intensity factor following the stress reduction, $\sigma \rightarrow \sigma - \Delta\sigma$, is given by the value, $K(\sigma, \Delta T, \tau) - 2K(\Delta\sigma/2, 0, \tau)$. It then follows that the value of the stress range intensity factor, for the case when reversed frictional slip is occurring, is given by $\Delta K = 2K(\Delta\sigma/2, 0, \tau)$, corresponding to Eq 11 of the main text.

References

[1] Cox, H. L., "The Elasticity and Strength of Paper and Other Fibrous Materials," *British Journal of Applied Physics,* Vol. 3, 1952, pp. 72–79.

[2] Aveston, J., Cooper, G. A., and Kelly, A., "Conference on the Properties of Fibre Composites," National Physical Laboratory, IPC Science and Technology, Guildford, Surrey, UK, 1971, pp. 15–26.

[3] Aveston, J. and Kelly, A., "Theory of Multiple Fracture of Fibrous Composites," *Journal of Materials Science,* Vol. 8, 1973, pp. 352–362.

[4] Garrett, K. W. and Bailey, J. E., "Multiple Transverse Fracture in 90° Cross-Ply Laminates of a Glass Fibre-Reinforced Polyester," *Journal of Materials Science,* Vol. 12, 1977, pp. 157–168.

[5] Nayfeh, A. H., "Thermomechanically Induced Interfacial Stresses in Fibrous Composites," *Fibre Science and Technology,* Vol. 10, 1977, pp. 195–209.

[6] Parvizi, A., Garrett, K. W., and Bailey, J. E., "Constrained Cracking in Glass Fibre Reinforced Epoxy Crossply Laminates," *Journal of Materials Science,* Vol. 13, 1978, pp. 195–201.

[7] Nuismer, R. J. and Tan, S. C., ''Constitutive Relations of a Cracked Composite Laminate,'' *Journal of Composite Materials,* Vol. 22, 1988, pp. 306–321.
[8] Hashin, Z., ''Analysis of Cracked Laminates: A Variational Approach,'' *Mechanics of Materials,* Vol. 4, 1985, pp. 121–136.
[9] Hashin, Z., ''Analysis of Orthogonally Cracked Laminates under Tension,'' *Journal of Applied Mechanics,* Vol. 54, 1987, pp. 872–879.
[10] Nairn, J., ''A Variational Mechanics Analysis of the Stresses around Breaks in Embedded Fibres,'' *Mechanics and Materials,* Vol. 13, 1992, pp. 131–154.
[11] Pagano, N. J., ''Axisymmetric Micromechanical Stress Fields in Composites,'' *Proceedings,* IUTAM Symposium, J. N. Reddy and K. L. Reifsnider, Eds., Springer-Verlag, Blacksburg, VA, 1991, pp. 1–26.
[12] Pagano, N. J. and Brown, H. W., ''The Full-Cell Cracking Mode in Unidirectional Brittle-Matrix Composites,'' *Composites,* Vol. 24, 1993, pp. 69–83.
[13] Pagano, N. J., ''Stress Fields in Composite Laminates,'' *International Journal of Solids and Structures,* Vol. 14, 1978, pp. 385–400.
[14] Reissner, E., ''On a Variational Theorem in Elasticity,'' *Journal of Mathematics and Physics,* Vol. 29, 1950, pp. 90–95.
[15] McCartney, L. N., ''Stress Transfer Mechanics for Systems of Transverse Isotropic Concentric Cylinders, I—Basic Theory,'' National Physical Laboratory, Report CMMT(A)12, Feb. 1996.
[16] McCartney, L. N., ''New Theoretical Model of Stress Transfer Between Fibre and Matrix in a Uniaxially Fibre Reinforced Composite,'' *Proceedings,* Royal Society of London, Vol. A425, 1989, pp. 215–244.
[17] Hutchinson, J. W. and Jensen, H. M., ''Models of Fibre Debonding and Pull-Out in Brittle Composites with Friction,'' *Mechanics of Materials,* Vol. 9, 1990, pp. 139–163.
[18] Bakuckas, J. G., Jr. and Johnson, W. S., ''Application of Fibre Bridging Models to Fatigue Crack Growth in Unidirectional Titanium Matrix Composites,'' *Journal of Composite Technology and Research,* Vol. 15, 1992, pp. 242–255.
[19] Nimmer, P., Bankert, R. J., Russell, E. S., Smith, G., and Wright, P. K., ''Micromechanical Modeling of Fibre/Matrix Interface Effects in Transversely Loaded SiC/Ti-6-4 Metal Matrix Composites,'' *Journal of Composite Technology and Research,* Vol. 13, 1991, pp. 3–13.
[20] McCartney, L. N., ''Analytical Method of Calculating the Thermoelastic Constants of a Multi-Phase Unidirectional Composite,'' National Physical Laboratory, Report DMM(A)57, March 1992.
[21] Clyne, T. W., Feillard, P., and Kalton, A. F., ''Interfacial Mechanics and Macroscopic Failure in Titanium-Based Composites,'' in this volume.
[22] McCartney, L. N., ''Theory of Stress Transfer in a 0-90-0 Cross-Ply Laminate Containing a Parallel Array of Transverse Cracks,'' *Journal of the Mechanics and Physics of Solids,* Vol. 40, 1992, pp. 27–68.
[23] McCartney, L. N., ''The Prediction of Cracking in Biaxially Loaded Cross-Ply Laminates having Brittle Matrices,'' *Composites,* Vol. 24, 1993, pp. 84–92.
[24] McCartney, L. N., ''Mechanics of Matrix Cracking in Brittle-Matrix Fibre-Reinforced Composites,'' *Proceedings,* Royal Society of London, Vol. A409, 1987, pp. 329–350.
[25] Lindley, T. C. and McCartney, L. N., ''Mechanics and Mechanisms of Fatigue Crack Growth,'' *Developments in Fracture Mechanics,* 2nd ed., G. G. Chell, Ed., Applied Science Publishers, London, 1981, Chapter 5.
[26] Bao, G. and McMeeking, R. M., ''Fatigue Crack Growth in Fibre-Reinforced Metal-Matrix Composites,'' Vol. 42, 1994, pp. 2415–2425.
[27] McMeeking, R. M. and Evans, A. G., ''Matrix Fatigue Cracking in Fibre Composites,'' *Mechanics of Materials,* Vol. 9, 1990, pp. 217–227.
[28] McCartney, L. N., ''Mechanics for the Growth of Bridged Cracks in Composite Materials: I—Basic Principles,'' *Journal of Composite Technology and Research,* Vol. 14, 1992, pp. 133–146.
[29] McCartney, L. N., ''Mechanics for the Growth of Bridged Cracks in Composite Materials: II—Applications,'' *Journal of Composite Technology and Research,* Vol. 14, 1992, pp. 147–154.
[30] McCartney, L. N., ''The Prediction of Non-Uniform Cracking in Biaxially Loaded Cross-Ply Laminates,'' National Physical Laboratory, Report DMM(A)142, July 1994.

James M. Larsen,[1] *Jay R. Jira,*[1] *Reji John,*[2] *and Noel E. Ashbaugh*[2]

Crack-Bridging Effects in Notch Fatigue of SCS-6/TIMETAL 21S Composite Laminates

REFERENCE: Larsen, J. M., Jira, J. R., John, R., and Ashbaugh, N. E., "**Crack-Bridging Effects in Notch Fatigue of SCS-6/TIMETAL 21S Composite Laminates,**" *Life Prediction Methodology for Titanium Matrix Composites, ASTM STP 1253,* W. S. Johnson, J. M. Larsen, and B. N. Cox, Eds., American Society for Testing and Materials, 1996, pp. 114–136.

ABSTRACT: Fatigue tests of middle-hole tension specimens of SCS-6/TIMETAL 21S composite (silicon-carbide fibers reinforcing a matrix of Ti-15Mo-2.6Nb-3Al-0.2Si alloy) were performed on three laminate architectures: unidirectional, cross ply, and quasi-isotropic. Specimens were tested over a range of stress levels, and fatigue damage was documented in situ by macrophotography and direct–current electric potential drop measurements. Typically, failure evolved by the formation of a few dominant cracks at the notch that propagated into the composite matrix and, in many instances, were substantially affected by unbroken fibers bridging the cracks. Fractographic and failure mode characterization revealed key differences in the effectiveness of crack bridging in the three laminates. A shear-lag crack-bridging model was shown to correlate crack growth data in the laminates based on an empirical value of fiber/matrix interfacial shear stress. Crack-bridging stress distributions were predicted using the shear lag model and verified by comparing the predicted crack opening displacement profiles with measurements made using a laser interferometric displacement gage system. Implications of the results are discussed with respect to the potential use of these materials in practical structural applications.

KEYWORDS: crack bridging, crack propagation, fatigue (materials), fracture mechanics, mechanical properties, metal matrix composites, titanium matrix composites, titanium, titanium alloys, life prediction, modeling

Continuous-fiber, titanium alloy matrix composites (TMCs) appear to offer promise for aerospace structures that require high specific strength and stiffness at elevated temperatures [*1,2*]. The most attractive applications for these materials in turbine engines are principally as unidirectional rings and rods. Typical aircraft needs call for panel and beam structures and the more balanced properties offered by composite laminates containing off-axis plies. In many aircraft configurations, assembly requires the use of fasteners in loaded holes that may be expected to serve as sites for the initiation of fatigue cracks. The problem of notched fatigue in unidirectional TMCs has been studied by a number of investigators [*3–10*], but relatively little research has dealt with notch fatigue of composite laminates of practical architectures [*5,7,11*]. Moreover, experimental and analytical studies applicable to crack growth in TMCs have also primarily addressed unidirectional composites [*12–31*]. Although effects of fiber bridging on fatigue cracking of titanium matrix composites have been the subject of several investigations [*15–31*], a consensus regarding the applicability of

[1]Group leader and material research engineer, respectively, Materials Directorate, Wright Laboratory, WL/MLLN, Wright-Patterson AFB, OH 45433–7817.
[2]Associate and senior research engineer, respectively, University of Dayton Research Institute, Dayton, OH 45469–0128.

various crack-bridging models is yet to emerge [*21,24,27*]. As first noted by Bakuckas and Johnson [*21*], many of the shear-lag models of crack bridging appear to give inconsistent predictions of crack growth rates and the related crack opening displacements. The present work on TMC laminates examines the ability of crack-bridging models to correlate crack growth behavior in practical composite architectures and explores the underlying mechanisms involved in crack bridging in these materials.

Experimental Procedure

Materials

The SCS-6/TIMETAL 21S[3] composite panels examined in this investigation were fabricated by Textron Specialty Materials Division (SMD) by hot isostatic pressing alternating layers of woven fiber mat and matrix foil. The fiber mat contained the Textron SCS-6 fiber, which is produced by chemical vapor deposition on a carbon monofilament core. The fiber has a nominal diameter of 142 μm and is β-phase silicon carbide (SiC) with a radially columnar grain structure. A double-pass, carbon-rich outer coating is applied to the fiber to control reactions with titanium alloy matrices and to limit fiber damage during normal handling.

Three composite architectures were produced: four-ply unidirectional ($[0]_4$), cross-ply ($[0/90]_S$), and eight-ply quasi-isotropic ($[0/\pm45/90]_S$). Following consolidation, the composite was aged in vacuum for 8 h at 621°C to stabilize the matrix. As shown in Fig. 1, the microstructure of the aged matrix is composed of transformed beta grains containing fine acicular alpha-phase precipitates. The relatively uniform fiber distribution shown in the figure is representative of all three laminates. The properties of the constituent materials and the various laminates at room temperature (23°C) are listed in Table 1.

Fatigue Testing

Middle-hole tension specimens, MH(T), of nominal dimensions, 150 by 19 mm, were fabricated by abrasive water-jet machining of the composite panels. In all cases, the length of the specimen was parallel to the plies containing [0] fibers (the surface plies in the laminates). Circular holes, 4.75 mm in diameter, were machined in the middle of the specimens by ultrasonic abrasive drilling, producing a theoretical stress concentration factor (K_t) equal to 3.3 [*32*] based on the remote section stress and assuming the material to be homogeneous and orthotropic.

The fatigue experiments were performed in air at room temperature at a frequency of 20 Hz and a stress ratio ($R = \sigma_{min}/\sigma_{max}$) of 0.1 using a servohydraulic fatigue machine with hydraulic friction grips. Instrumentation for all tests included direct-current electric potential measurement and a macrophotography system. During the fatigue tests, photographs were taken periodically to obtain a visual record of the damage accumulation in the notch region. The projected crack lengths from the photographs were used to calibrate the electric potential measurements of crack lengths that were used for subsequent crack growth analysis.

In addition to tests of the composite, fatigue crack growth tests were performed on the fiberless Timetal 21S that had been consolidated from matrix alloy foils into 0.9-mm-thick panels using the composite material processing and aging conditions. The tests were conducted in accordance with ASTM Test Method for Measurements of Fatigue Crack Growth Rates (E 647–93) using both conventional compact tension specimens, C(T), and single-

[3]Registered trademark of Timet Corporation, Henderson, NV.

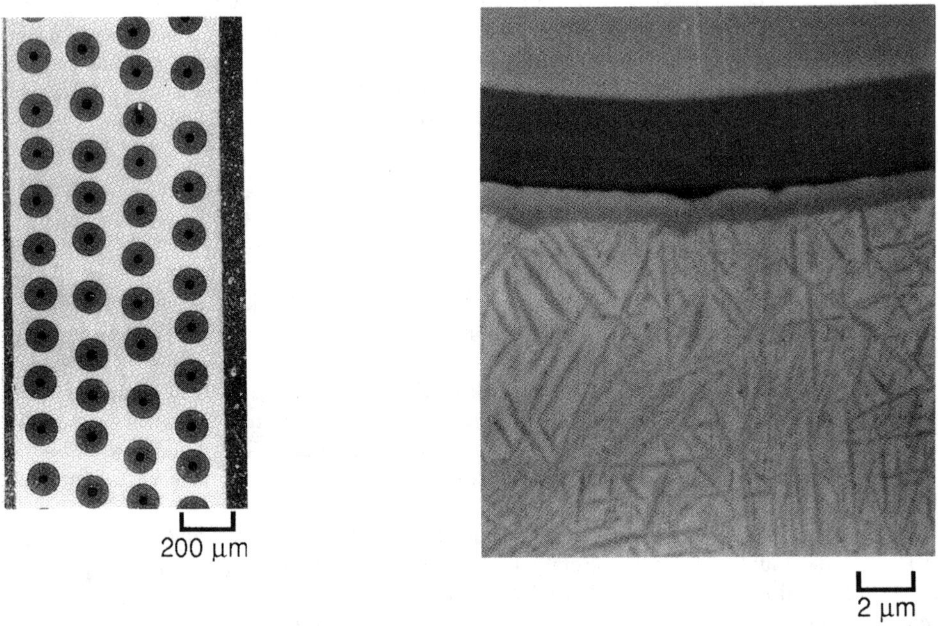

FIG. 1—*Microstructure SCS-6/TIMETAL 21S composite:* (left) *Cross section of four-ply, unidirectional composite and* (right) *fiber-matrix interface region.*

TABLE 1—*Room temperature properties of the constituent materials and the composite laminates.*

	SCS-6 Fiber	Fiberless TIMETAL 21S	SCS-6/ TIMETAL 21S $[0]_4$	SCS-6/ TIMETAL 21S $[0/90]_S$	SCS-6/ TIMETAL 21S $[0/\pm45/90]_S$
Fiber volume fraction	...	...	0.334	0.345	0.348
[0] Fiber volume fraction	...	...	0.334	0.172	0.087
Young's modulus in fatigue,[a] GPa	393.0	114.1	208	157	130
Poisson's ratio	0.25	0.34	...	...	...
Orthotropic modulus, GPa	...	...	178.2	...	...

[a]Young's modulus for the laminates was measured after the off-axis fiber matrix interfaces had separated under fatigue loading.

edge tension specimens, SE(T), with clamped ends [*33*], and load levels were selected to avoid specimen buckling. Crack length and crack closure behavior were determined for both specimen geometries from crack–mouth–opening compliance measurements using an extensometer.

Stress Intensity Factor for Cracking in MH(T) Specimens

As shown schematically in Fig. 2, the typical cracking geometry observed in the MH(T) specimens with unidirectional fibers involved the initiation and propagation of two cracks on either side of the hole. The cracks initiated at four symmetric locations approximately 65° from the axis of loading. These cracks tended to propagate initially in the direction of the hole radius, but soon assumed an orientation perpendicular to the axis of loading. The solution for the stress intensity factor, K, for these dual, kinked cracks was determined by finite element analysis under a separate investigation [*34*]. In the analysis of a specimen of width, $2W$, and hole radius, r, the crack length, $2a$, was defined as the length projected on a plane normal to the load axis. It was shown that the stress intensity factor for each of the dual cracks in the range of $a/W \geq r/W + 0.025$ is accurately described by multiplying the Newman and Raju K solution [*35*] for planar, symmetric cracks from a hole by a factor of 0.80. This modified equation was used for all specimens with four cracks from the hole. The standard solution [*35*] was used for cases in which only two cracks grew from the hole. This crack geometry, in which cracks initiated at the 90° location, was typical of the $[0/90]_S$ and $[0/\pm 45/90]_S$ laminates.

Crack-Bridging Analysis

The customary approach to analysis of Mode I matrix fatigue cracks propagating normal to the fiber direction in a unidirectional composite defines a remotely applied stress intensity factor range, ΔK_a. The influence of crack bridging by unbroken fibers in the crack wake is described by an effective bridging stress intensity factor range, ΔK_b, which reduces ΔK at the matrix crack tip according to the relationship

$$\Delta K_{\text{tip}} = \Delta K_a - \Delta K_b \tag{1}$$

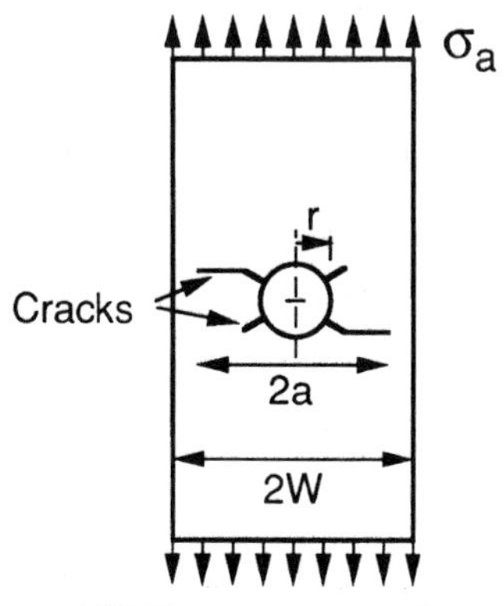

FIG. 2—*Middle-hole tension, MH(T), specimen indicating typical cracking mode for unidirectional SCS-6/TIMETAL 21S composites.*

In addition to bridging, the effective stress intensity factor range in the matrix material, ΔK_m, is further reduced by a shielding factor, F_s, such that

$$\Delta K_m = F_s \Delta K_{\text{tip}} \tag{2}$$

where various forms of F_s have been proposed [*12–14,36*] having values in the approximate range of $0.5 \leq F_s \leq 1.0$ for the current materials. For reasons discussed later, the form of Marshall, Cox, and Evans [*12*] was selected, giving

$$F_s = E_m/E_c \tag{3}$$

where E_m = elastic modulus of the composite matrix and E_c = elastic modulus of the composite in the direction of loading.

The physical process of crack bridging is believed to restrict both the maximum and minimum values of K in fatigue. K_{max} is limited by the stress of bridging fibers, and K_{min} is limited by a crack closure effect due to the fibers as they attempt to return to their original, unloaded position [*14,15*]. Thus, in addition to reducing ΔK_{tip}, bridging alters the effective stress ratio, $R_{m,\text{eff}}$, at the matrix crack tip. $R_{m,\text{eff}}$ may be expected to be greater than the remotely applied stress ratio, $R_a \equiv \sigma_{a,\text{min}}/\sigma_{a,\text{max}}$. In titanium matrix alloy composites, the process of fabrication by hot isostatic pressing results in significant residual stresses in the matrix at room temperature, with a significant component of tensile stress parallel to the fiber. This axial stress, which may relax significantly due to fatigue cycling, also serves to alter $R_{m,\text{eff}}$. Since $R_{m,\text{eff}}$ was uncertain due to the potential for matrix stress relaxation and degradation of fiber/matrix interfaces, a matrix crack growth relationship that was independent of R was used. This was accomplished using a closure-corrected effective ΔK approach discussed later.

The Shear Lag Model of Crack Bridging in Unidirectional Composites

The analysis of fiber bridging of Mode I cracks propagating normal to the fiber direction in unidirectional composites having weak fiber/matrix interfaces has been widely addressed in the literature. A particularly extensive set of solutions has been reported by Cox and Lo [*17,18*]. Using the nomenclature of these authors for a bridged crack propagating along the x axis, the general problem is defined as follows.

The form of the relationship between bridging stress, $p(u)$, and the crack opening displacement, $u(x)$, in fatigue is given by

$$|p(u) - p_r(x)| = \beta(|u(x) - u_r(x)|)^{1/2} \tag{4}$$

where $p_r(x)$ and $u_r(x)$ are the values for the last fatigue reversal. Based on the shear lag analysis of McCartney [*13*], the bridging stiffness parameter, β, is given by

$$\beta = \frac{2\eta f E_c}{(1 - f)E_m} \left[\frac{\tau E_f}{\rho} \right]^{1/2} \tag{5}$$

where η is equal to $\sqrt{2}$ for fatigue [*14*], f is the fiber volume fraction, E_f is the fiber modulus, τ is the shear stress of the fiber-matrix interface, and ρ is the fiber radius. For a Mode I crack perpendicular to the fiber direction, the crack opening profile is related to the applied load by the equation

$$u(X) = \frac{4a}{\pi E'} \int_X^1 \left\{ \int_0^t \frac{h_1(s/t, at/W)}{\sqrt{t^2 - s^2}} \times (\sigma_a(as) - p[u(as)])ds \right\} \frac{h_1(X/t, at/W)}{\sqrt{t^2 - X^2}} tdt \tag{6}$$

where $X \equiv x/a$, h_1 is a function of specimen geometry, and E' is the effective elastic modulus for a transversely isotropic material. From the solution to this equation, which must be solved numerically, the effect of $u(x)$ on the stress intensity factor at the crack tip, K_{tip}, follows directly from

$$K_{\text{tip}} = K_a - K_b = 2\sigma_a \sqrt{\frac{a}{\pi}} \int_0^1 \frac{h_1(s, a/W)}{\sqrt{1 - s^2}} \times \left[\frac{\sigma_a(sa) - p[u(sa)]}{\sigma_a(sa)} \right] ds \tag{7}$$

where σ_a is the remote stress. A comprehensive set of numerical solutions to these equations has been reported by Cox and Lo [*17*], who have developed canonical functions of best fit to represent the solutions [*18*]. These functions provide an expedient method for evaluation of realistic crack-bridging problems without resolving the integral equations (as noted in the Appendix, these equations were modified slightly to correct typographical errors in the original reference). The functions are presented in terms of $\Delta K_{\text{tip}}/\Delta K_a$ that, when combined with the ΔK solution for the MH(T) specimens, provided a general solution for the basic problem of crack bridging in the current experiments. In the discussion to follow, this solution is used in a parametric analysis to examine the influence of crack bridging.

Results

Crack Growth in Fiberless TIMETAL 21S

The results of crack growth tests performed on the fiberless matrix alloy are shown in Fig. 3 that plots the data in terms of both the applied and effective stress intensity factor ranges, ΔK_m and $\Delta K_{m,\text{eff}}$, respectively. The subscript, m, is used to designate data in the matrix alloy, and the effective ΔK is defined in the conventional manner as $\Delta K_{m,\text{eff}} = K_{m,\max} - K_{m,\text{cl}}$, where $K_{m,\text{cl}}$ is the stress intensity factor corresponding to crack-wake closure determined from measurements of crack-mouth-opening compliance. As shown in the figure, use of

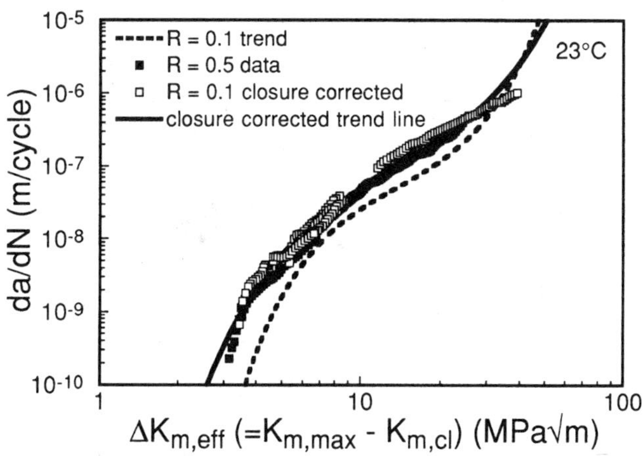

FIG. 3—*Crack growth rate data,* da/dN-ΔK *and* ΔK_{eff}, *for unreinforced (monolithic) TIMETAL 21S. Also shown is a least squares fit to the data.*

$\Delta K_{m,\text{eff}}$ as a correlating parameter consolidated the crack growth rate data for stress ratios of 0.1 and 0.5. An analytical expression representing these data over the range of growth rates was obtained by performing a least square regression of the form [37]

$$\log\left(\frac{da}{dN}\right) = C_1(\sinh(C_2(\log(\Delta K_{m,\text{eff}}) + C_3))) + C_4 \tag{8}$$

where C_i are the constants of regression, and the units of da/dN and $\Delta K_{m,\text{eff}}$ are m/cycle and MPa$\sqrt{\text{m}}$, respectively. The correlation coefficient for this regression was 0.984, and the values of the coefficients were: $C_1 = 0.7158$, $C_2 = 3.0207$, $C_3 = -1.1139$, and $C_4 = -7.069$.

Crack Growth in SCS-6/TIMETAL 21S Laminates

Constant-load-amplitude fatigue tests of MH(T) specimens were conduced over a range of stress levels from crack-arrest conditions to near the respective laminate's notched tensile strength. For all of these tests, crack growth data in the form of half-crack length (a) in the composite matrix versus fatigue cycles (N) were reduced to the form of da/dN versus ΔK_a. As shown in Fig. 4, the data for the three laminates exhibited crack growth behavior that differed significantly from crack growth in the unreinforced matrix alloy and the predictions for unbridged cracks (dashed curve) in the composite based on Eqs 2 and 3. In each laminate, the trends in da/dN for tests conducted at the lowest stresses displayed crack growth rate decreasing, or holding approximately constant, with increasing ΔK_a. At the highest stresses, the da/dN-ΔK_a data tended to be parallel to the data from crack growth in the matrix alloy; the data from the $[0/90]_S$ and $[0/\pm 45/90]_S$ tests tended to align near the curve representing crack growth in the fiberless matrix alloy, while the $[0]_4$ data were offset from the matrix crack growth curve. Due to differences in crack growth behavior for the three laminates, analyses of these results are most conveniently treated separately. The

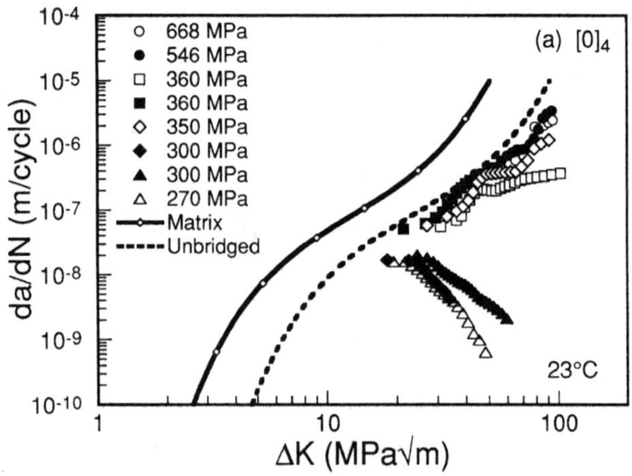

FIG. 4—*Crack growth rate data,* da/dN-ΔK, *at various stress levels for three laminates compared to the trend line for unreinforced (monolithic) TIMETAL 21S and the predicted trend for unbridged laminates:* (a) $[0]_4$, (b) $[0/90]_S$, *and* (c) $[0/\pm 45/90]_S$.

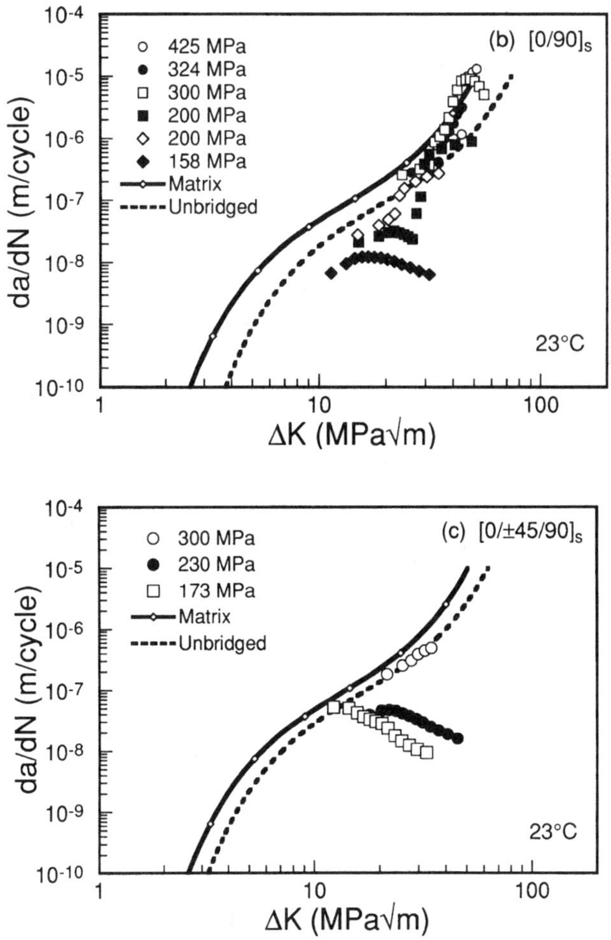

FIG. 4—*(Continued)*

unidirectional material will be addressed first, followed by a discussion of the applicability of this analysis to $[0/90]_S$ and $[0/\pm45/90]_S$ laminates.

Crack Bridging in $[0]_4$ Composite

As shown in Fig. 4*a*, there is a distinct transition in crack growth behavior in the $[0]_4$ tests as the maximum remote stress is increased from 300 to 350 MPa. The form of the data suggests that crack bridging was significant at the lower stress, while the extent of bridging was limited at the higher stresses. This presumption was confirmed by chemical removal of the matrix material from the outer ply of selected specimens from tests interrupted prior to failure. Matrix dissolution was accomplished using a solution of 10% bromine in methanol saturated with tartaric acid, which gently removed the matrix alloy while leaving the fiber/matrix reaction zone (the light sheath surrounding the fibers) relatively unattacked. Figure 5 shows a specimen tested with σ_{max} = 300 MPa in which all of the fibers bridging the fatigue cracks were intact; a higher magnification view of the region of the upper crack is shown in

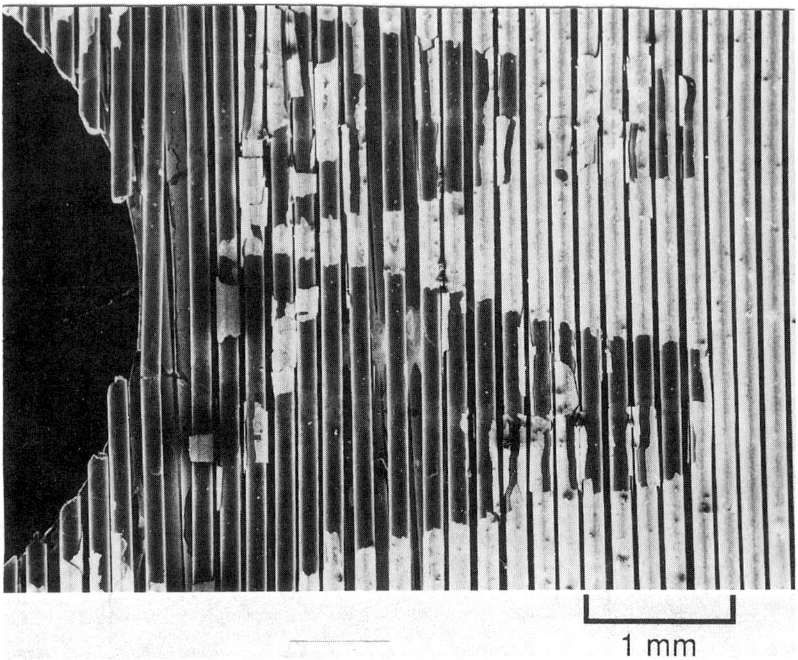

FIG. 5—*SEM photograph of fatigue-cracked region in* $[0]_4$ *composite after chemical removal of the outer layer of matrix. This specimen was removed from testing after 1 483 000 fatigue cycles with* $\sigma_{max} = 270$ *MPa.*

Fig. 6. As shown in Fig. 7 for a test with $\sigma_{\mathrm{max}} = 360$ MPa, all of the fibers behind the crack tip were broken. Thus, these two cases represent the limiting conditions of fully bridged and unbridged cracks, respectively. In all cases examined, the bridged cracks exhibited, at most, only an occasional broken fiber in the crack wake, while specimens tested under conditions that produced unbridged cracks were characterized by fracture of essentially all fibers behind the crack tip. Typically, no broken fibers were observed ahead of the crack tip. This visual evidence of fiber bridging and fiber fracture was in complete agreement with measurements of crack opening displacement profiles to be presented later.

Analytical Correlation of Crack Growth Rates in the Composite

Assuming that the *da/dN*-ΔK data of the monolithic TIMETAL 21S material (Fig. 3) represent a material property curve, crack growth rates in the composite should be related to crack growth in the matrix alloy by Eq 8 using the substitution

$$\Delta K_{m,\mathrm{eff}} = F_s\,\Delta K_{\mathrm{tip}} = F_s(\Delta K_a - \Delta K_b) \tag{9}$$

where ΔK_b represents the reduction in ΔK_{tip} due to the combination of bridging pressure and crack closure. For maximum stress levels sufficient to fail bridging fibers, the contribution from crack-bridging pressure was negligible, although effects of crack closure due to contact of broken fibers in the crack wake on unloading should exist at some level. Assuming that $\Delta K_b = 0$ for an unbridged crack in a composite, the difference in crack growth rates in the composite versus the fiberless matrix would be due solely to the correction factor, F_s. Of the various forms of F_s, the one proposed by Marshall, Cox, and Evans (Eq 3) produced the best correlation with the unbridged crack growth rate data and was, therefore, selected for use in

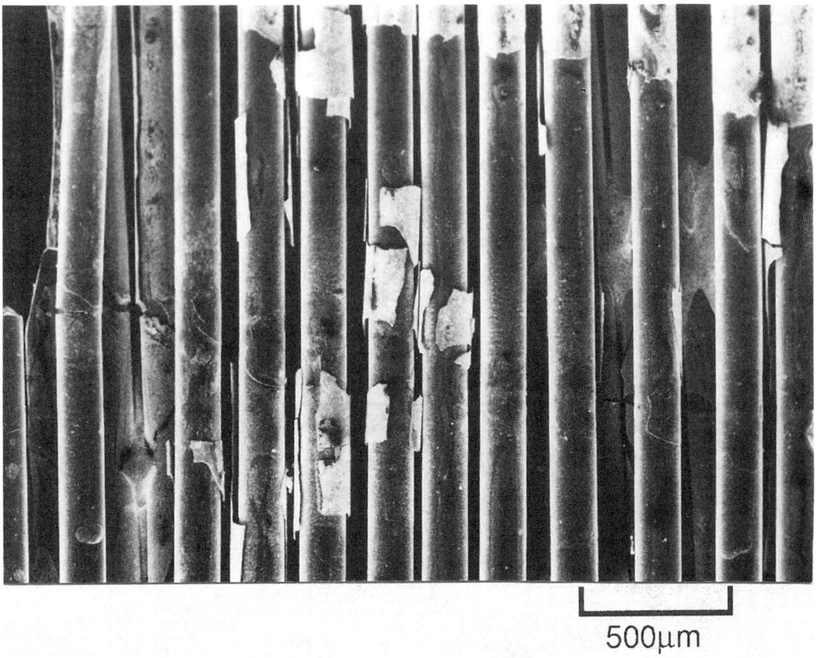

FIG. 6—*Higher magnification view of the lower crack showing unbroken fibers along the matrix crack near the notch.*

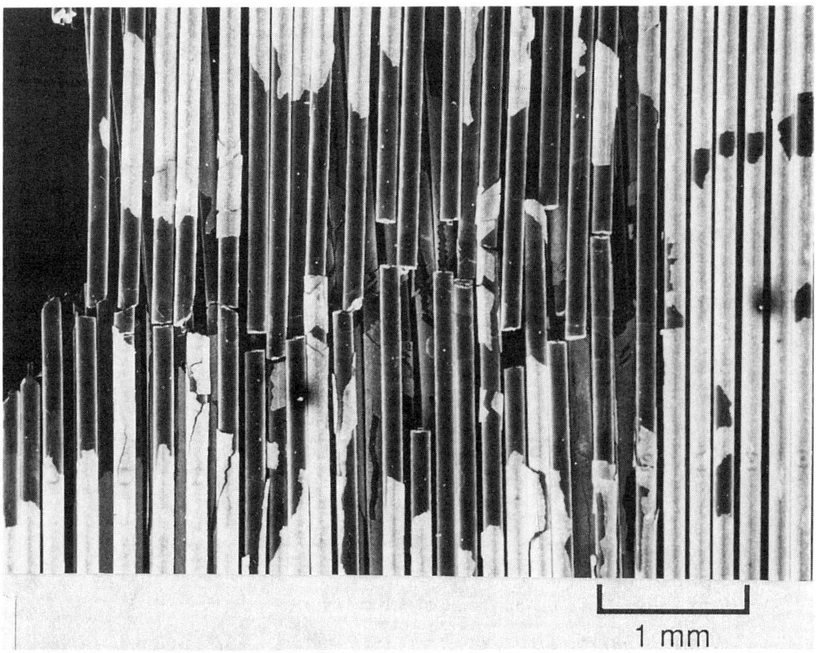

FIG. 7—*SEM photograph of fatigue-cracked region in $[0]_4$ laminate after chemical removal of the outer layer of matrix revealing broken fibers along the entire crack length. This specimen was removed from testing after 45 000 fatigue cycles with σ_{max} = 360 MPa.*

the analysis of the bridged crack growth rate data. Use of the other forms of F_s over-predicted crack growth rates of the unbridged specimens by as much as an order of magnitude.

In the equations for crack bridging, the only term that was not measured directly was the fiber/matrix interfacial shear stress, τ. Using the bridging analysis of Cox and Lo [*18*] in conjunction with Eqs 8 and 9, effective values of τ were calculated for the fully bridged cracks to obtain agreement between the experimentally measured and predicted crack growth rates in the $[0]_4$ composite. In this sense, τ was treated as a fitting parameter, and any error in the analysis or measured material parameters would be incorporated in τ. Figures 8*a* and *b* plot the calculated values of τ versus the normalized bridged crack length and the normalized crack propagation life, respectively ($N_i \equiv$ cycles to initiate a crack of length 300

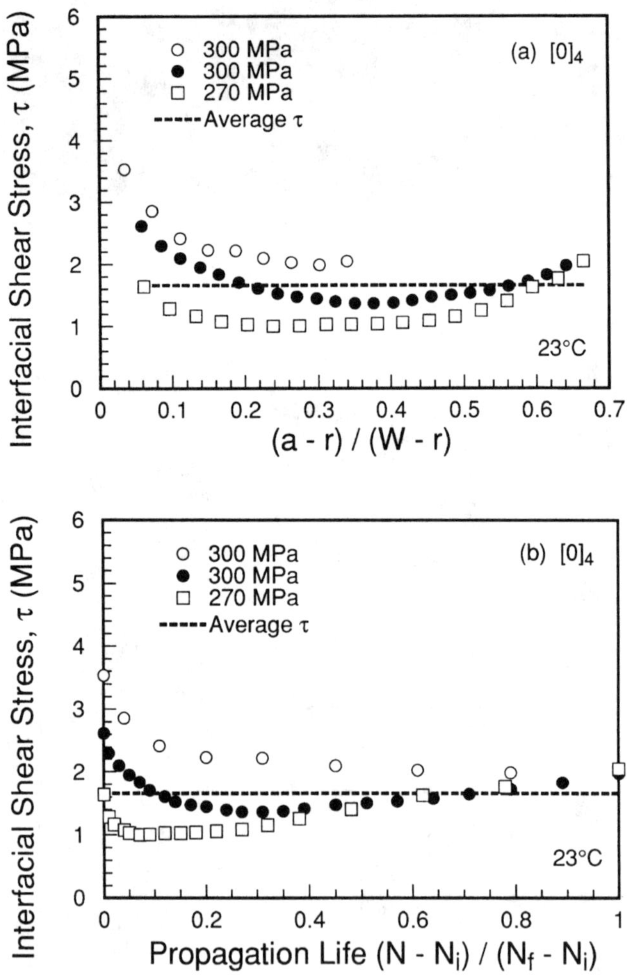

FIG. 8—*Calculated interfacial shear stress (τ) for applied stress levels exhibiting fully bridged crack growth in SCS-6/TIMETAL 21S $[0]_4$ composite:* (a) *verses normalized bridged crack length and* (b) *verses normalized propagation life.*

μm from the hole edge). As shown in Fig. 8*b*, the calculated values of τ quickly degraded from the initial value and became nominally constant, or slightly increasing thereafter. As the cracks extended, most of the fibers in the crack wake had experienced thousands of fatigue cycles, and it is likely that the approximately constant values of τ represent a stabilized effective interfacial shear stress that had degraded from its pristine value, as reported for other SCS-6 fiber composites by Walls et al. [*26*]. Note that for these three nominally-equivalent tests, the calculated values of τ varies by a factor of two or more in some instances, indicating the level of test-to-test variability inherent in the data from this material.

As a test of the ability of the shear lag analysis of crack bridging to correlate crack growth rate data from the bridged fatigue cracks, the average value of τ was used in the bridging analysis to "re-predict" the crack growth rate data. As shown by the solid curves in Fig. 9, these correlations were in good agreement with the experimental data, although the calculated crack growth rates based on the constant value of τ indicated the development of a steady-state crack growth rate toward the end of the test, while the actual crack growth rates continued to decrease. For stress levels sufficient to fail the bridging fibers, ΔK_b is zero, and the predicted crack growth rates overlap the dashed curve. The data fall slightly below the curve, suggesting that processes not accounted for in the analysis are at work. As noted earlier, crack closure effects due to interference by broken fibers in the crack wake were not discernible and were therefore not included; this omission may be partially responsible for the difference between predicted and actual growth rates in the unbridged specimen. In addition, it is possible that limited crack bridging occurred in some instances, although no direct evidence of crack bridging was found.

Crack Bridging in [0/90]$_S$ Composite

As shown in Fig. 4*b*, many of the features of crack growth behavior in $[0]_4$ composite are present in the $[0/90]_S$ data. There appears to exist a critical stress for crack bridging (≈200 MPa). For maximum stress levels less than this nominal value, crack growth rates are approximately independent of ΔK_a, while data from tests conducted with maximum stresses

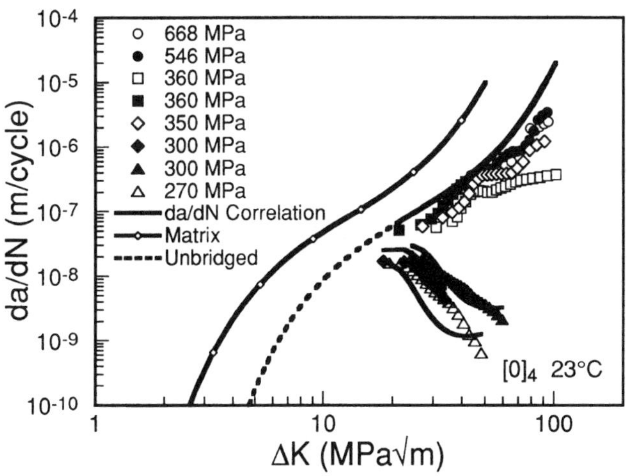

FIG. 9—*Comparison of crack growth rate data for the $[0]_4$ laminate from Fig. 4*a *with the predicted trends using the average τ shown in Fig. 8.*

greater than approximately 200 MPa exhibit unbridged crack growth. The unbridged data fall in the region of the dashed-line prediction made using Eqs 2 and 3 with E_c defined as the stabilized elastic modulus measured in fatigue. In other words, E_c represents the behavior of the composite after the fiber/matrix interfaces in the [90] plies have failed, increasing the specimen compliance. The inferences regarding fiber bridging and fracture are supported by the micrographs presented in Fig. 10, which show metallographic sections through the [0] ply of $[0/90]_S$ specimens tested at maximum stress levels of 158 and 324 MPa. At 158 MPa, all fibers in the crack wake are intact, while at 324 MPa, all fibers in this region are broken. The observed failure of bridging fibers is also consistent with calculations of fiber stresses to be discussed later.

As was done for the $[0]_4$ material, the fiber-bridging analysis was used to obtain estimates of the effective fiber-matrix interfacial stress, τ, for the $[0/90]_S$ tests of fully bridged cracks. In this instance, only the 158 MPa data were used, as this was the only specimen to exhibit fully bridged behavior throughout the test. The results are presented in Fig. 11 and indicate an initial decline in τ with increasing crack length that is more pronounced than in the unidirectional composite. The figure also shows an average τ for $[0/90]_S$ composite (solid line) as compared to the dashed-line representation of the average τ determined from the $[0]_4$ tests. The extent to which specimen-to-specimen variability contributed to the difference in average τ between these two materials is not known. Based on the variability in τ for the $[0]_4$ material shown in Fig. 8, however, inherent material variability may be a significant contributing factor.

Figure 12 presents a comparison between the experimental and calculated crack growth rates using the average value of τ for the $[0/90]_S$ and $[0]_4$ tests, the dashed and solid curves, respectively. The correlations between the experimental and calculated crack growth rates are within a factor of two using the average τ from the $[0/90]_S$ data, while use of the average τ from the $[0]_4$ tests predicts somewhat higher crack growth rates. As shown in the figure, the crack growth rates for the unbridged test conditions are in reasonable agreement with the unbridged prediction shown by the dashed line. In contrast to the $[0]_4$ behavior, where all of the unbridged data fell below the predicted crack growth rates, the unbridged data for $[0/90]_S$ tend to be centered about the predicted behavior.

Crack Bridging in $[0/\pm45/90]_S$ Composite

As shown in Fig. 4*c*, crack growth behavior in the quasi-isotropic, $[0/\pm45/90]_S$, composite exhibited trends similar to the unidirectional and cross-ply results. There was a critical bridging stress, 230 MPa $> \sigma_{crit} >$ 300 MPa, below which fiber-bridging effects were significant, and above which the effects of fiber bridging appeared to be absent. Although no definitive analysis exists for crack bridging by [$\pm$45] fibers, the experimental evidence suggests that the contribution of these fibers is significant, since the critical stress for crack bridging in $[0/\pm45/90]_S$ is greater than the critical stress for the $[0/90]_S$ composite, which had double the volume fraction of [0] fibers. Clearly, a traditional shear lag analysis for [0] bridging fibers is not generally applicable to the complexities present in the $[0/\pm45/90]_S$ architecture. Therefore, no additional analysis of these data will be presented.

Crack Opening Displacements During Crack Growth in $[0]_4$ Composite

As shown in Fig. 4, the crack growth behavior of the $[0]_4$ composite changed from fully-bridged to unbridged when σ_a increased from 300 to 360 MPa. The unbridged behavior implies that, at σ_a = 360 MPa, the maximum fiber stresses exceeded the tensile failure strength, *S*, from the beginning of the test. Previous studies [*19,20,24,27*] have shown that

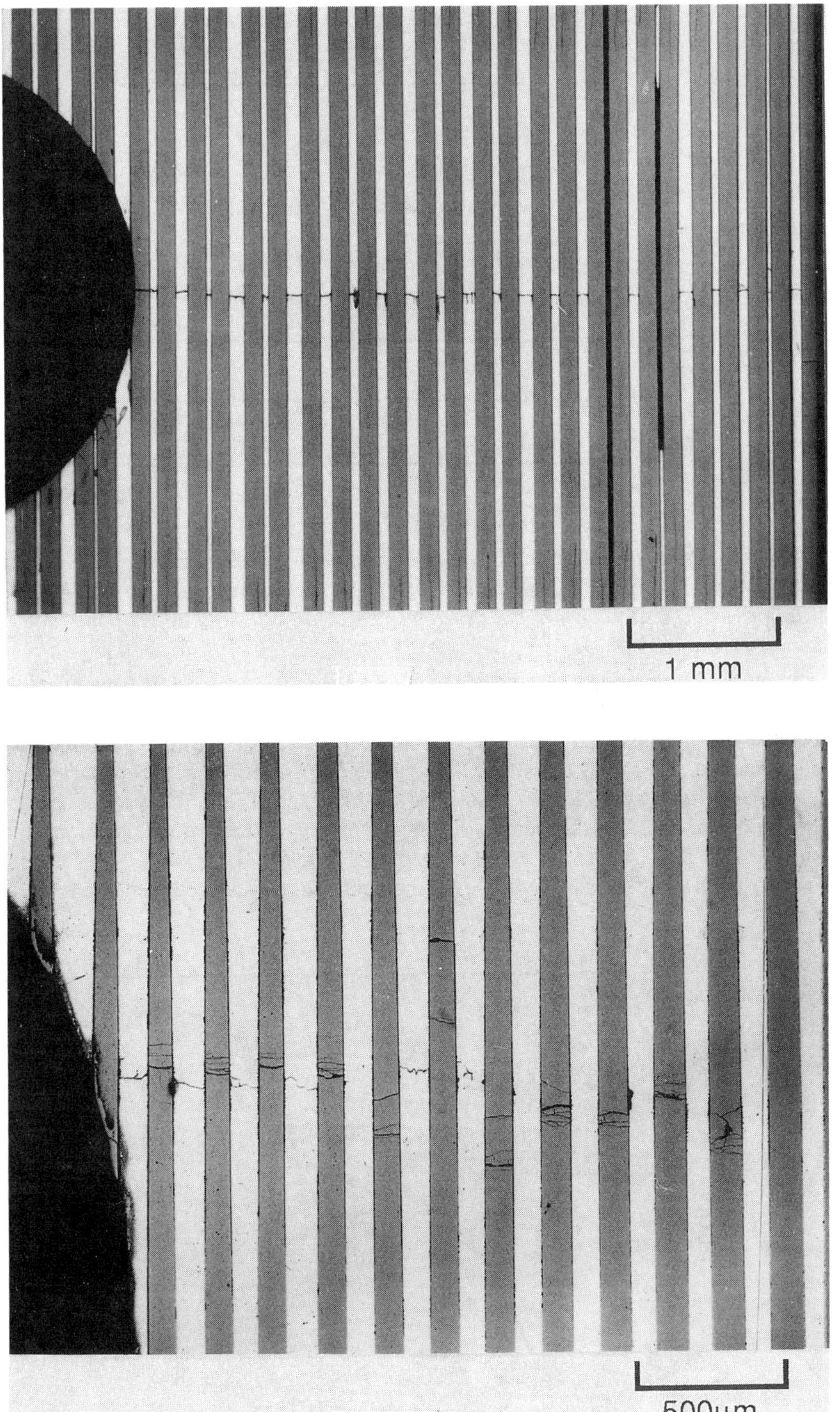

FIG. 10—*Metallographic section through the outermost layer of fibers in* $[0/90]_S$ *laminates tested at different stress levels:* (top) *specimen that was removed from testing after 488 400 fatigue cycles with* σ_{max} = *158 MPa. Fibers are unbroken along the length of the matrix crack; and* (bottom) *specimen that was removed from testing after 8058 fatigue cycles with* σ_{max} = *324 MPa. Fibers are broken along the length of the matrix crack.*

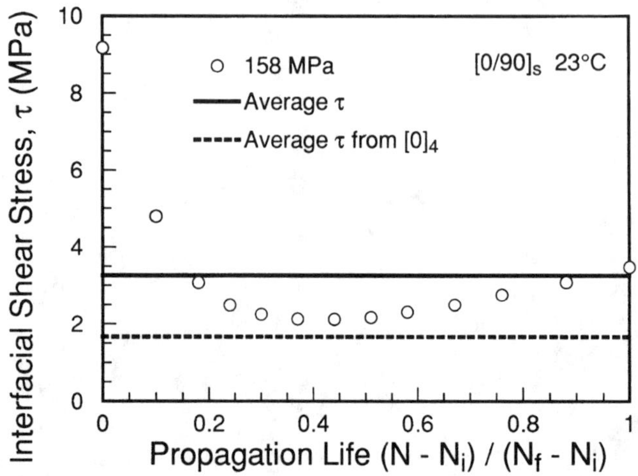

FIG. 11—*Calculated interfacial shear stress (τ) verses normalized propagation life for an applied stress level that exhibited fully bridged crack growth in SCS-6/TIMETAL 21S $[0/90]_S$ composite. Also shown is the average τ calculated from the $[0]_4$ laminate.*

measurements of crack-opening-displacement profiles provide an independent method of verifying the extent of bridging and the magnitude of the bridging-fiber stresses in the crack wake. Hence, during this study, the crack-opening-displacement profiles were measured for tests conducted at σ_a = 300 and 360 MPa using a laser interferometric displacement gage (IDG). The laser IDG system has a resolution of the order of 0.01 mm [*38*]. The measured data are plotted as half-crack-opening-displacement range (Δu) versus location along the crack in Figs. 13 and 14 for σ_a = 360 and 300 MPa, respectively.

Using Eq 6, the half-crack-opening-displacement range was predicted for the test con-

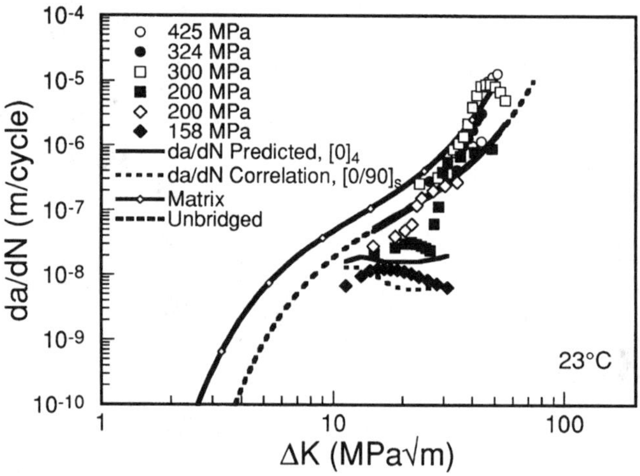

FIG. 12—*Crack growth rate data for the $[0/90]_S$ laminate from Fig. 4*b *with the trends predicted using the average τ from the $[0]_4$ laminate (Fig. 8). Also shown is the correlation using the average τ calculated from $[0/90]_S$ laminate.*

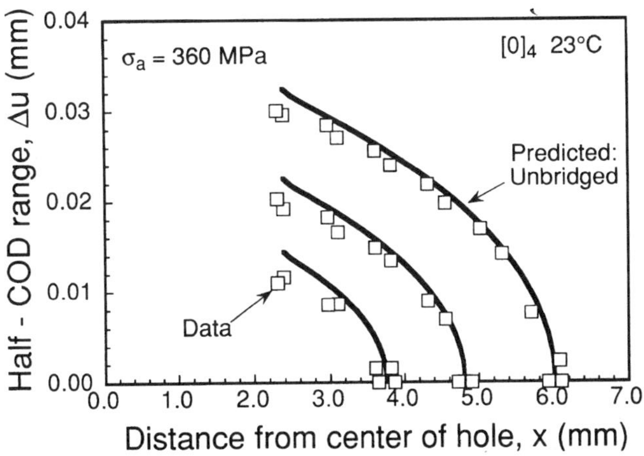

FIG. 13—*Comparison of measured and predicted half-crack-opening-displacement profile for a [0]$_4$ lay-up during crack growth in the absence of fiber bridging.*

ducted at σ_a = 360 MPa and compared with the measured data in Fig. 13 (Δu was calculated by replacing $\sigma_a(x)$ and $p(x)$ in Eq 6 with $\Delta\sigma_a(x)$ and $\Delta p(x)$, respectively). The unbridged profile was predicted assuming $\Delta p(x) = 0$. As shown in Fig. 13, the predictions correlate well with the measured data for different crack lengths, confirming that fiber bridging was not operative at σ_a = 360 MPa. Using Eqs 4 through 7, the fully bridged displacement profile was predicted and is shown in Fig. 14 for σ_a = 300 MPa. The average τ (1.66 MPa) from Fig. 8 was used to predict $\Delta u(x)$. The prediction correlates well with the data from measurements of the four cracks emanating from the hole, validating the magnitude of the bridging stresses predicted for the [0]$_4$ layup using the shear lag model. For the bridged crack, the maximum Δu at the edge of the hole was only about 5 μm, while for the

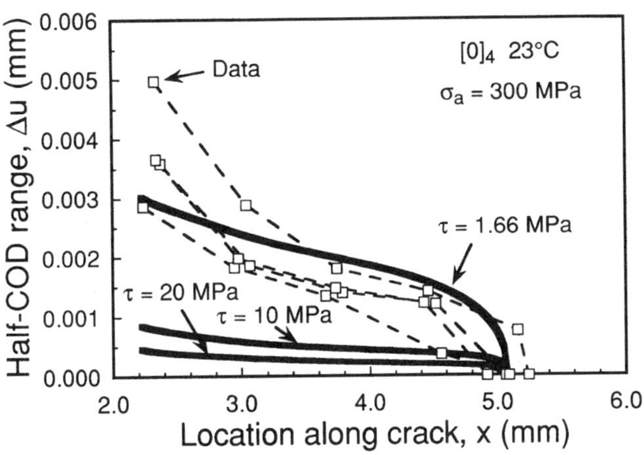

FIG. 14—*Comparison of measured and predicted half-crack-opening-displacement profiles for a [0]$_4$ lay-up. Data from the four cracks emanating from a hole are in good agreement with the shear-lag model (SLM) predictions for fully bridged cracks.*

unbridged crack, the maximum Δu was about 30 μm. The difference in Δu for these two cases highlights the significant shielding effect of bridging fibers.

Bridging Fiber Stresses During Crack Growth in $[0]_4$ Composite

An important parameter involved in the analysis of the crack growth in the composite is the bridging stress distribution, $p(x)$. Knowing $p(x)$, the bridging fiber stress, σ_f, can be determined using the relationship

$$\sigma_f = p(x)/f \tag{10}$$

Comparing σ_f with the tensile strength of the fibers, S, the onset of failure of the fibers (leading to composite fracture) can be predicted. The analysis conducted in previous sections was based on the applied and bridging stress ranges, that is, $\Delta\sigma_a$ and Δp, respectively. $\Delta p(x)$ can be related to $p(x)$ assuming that the effective stress ratio in the fibers is the same as that for $\Delta\sigma_a$. Hence

$$p(x) = \Delta p(x)/(1 - R) \tag{11}$$

As discussed earlier, $\Delta p(x)$ and $\Delta u(x)$ were determined for the $[0]_4$ specimen tested at σ_a = 300 MPa. The predicted σ_f for τ = 1.66 MPa is shown in Fig. 15, which corresponds to the $\Delta u(x)$ shown in Fig. 14. The shear lag model predicts that fiber bridging stress range along the crack wake increases from zero at the crack tip to about 1500 MPa at the edge of the hole. As shown in Fig. 15, the maximum σ_f occurs at the hole edge, that is, $x = r$. Hence, the stress in the fiber at the hole edge should be tracked to determine if the fibers will eventually fail under the applied loading conditions.

Using Eqs 10 and 11, the bridging fiber stresses, σ_f, at the edge of the hole were calculated as a function of the crack length for the test conducted at σ_a = 300 MPa. Figure 16 shows the predicted line for R = 0.1 corresponding to the applied stress ratio. The predicted fiber stress at the hole increases from 1067 to 1410 MPa for the specimen tested at 300 MPa. The predicted σ_f is less than 50% of the reported value for the quasi-static tensile

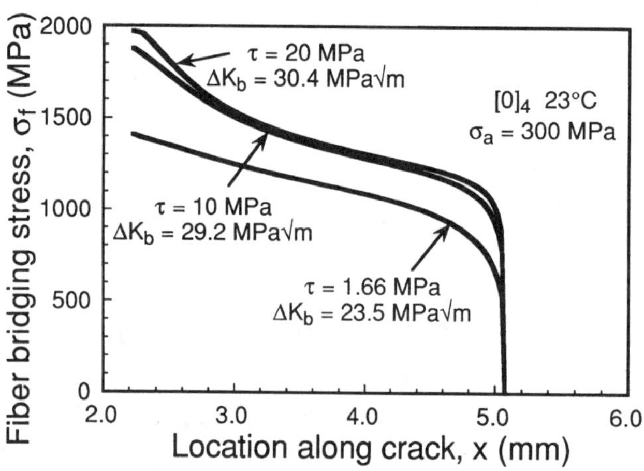

FIG. 15—*Shear-lag model (SLM) predictions of bridging stress along the crack wake for* a = *5 mm.*

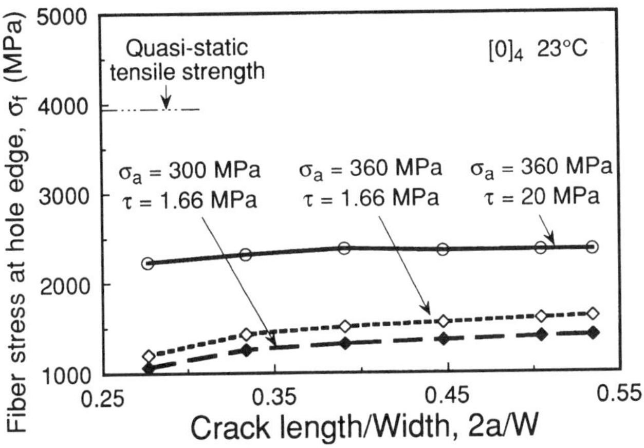

FIG. 16—*Predicted variation of the fiber stress at the hole edge with increasing crack length in the* $[0]_4$ *composite.*

strength of 3943 MPa [*39*] for this composite. As shown in Fig. 4, the specimen tested at 360 MPa exhibited no crack bridging, implying that a 20% increase in applied stress resulted in fiber failure from the beginning of test. It should be noted that similar experiments conducted by the present authors on nominally equivalent composite material revealed a fiber-bridging stress equal to the fiber strength [*40*], indicating that the strength of the bridging fibers in the current material was anomalously low.

Discussion

As revealed by the experimental results, phenomena affecting crack bridging in titanium composite laminates are often complex, and details of the physical processes may be partially obscured by variability inherent in the data. However, the traditional shear lag analysis of crack bridging exhibited considerable ability to correlate crack growth in the composite, based on knowledge of crack growth behavior in the matrix alloy and an empirical choice of the fiber/matrix interfacial shear stress, τ. Moreover, the value of τ obtained from the crack growth rate data was in good agreement with independent measurements of crack-opening-displacement-range profiles.

As shown in Fig. 4, the growth rates of the unbridged cracks in the unidirectional composite were approximately an order of magnitude lower than in the fiberless matrix alloy. One possible explanation for this difference is that a significant level of crack closure occurred in the composite due to interference of broken fibers in the crack wake. Detailed examination of load versus crack-opening-displacement data from these tests, however, suggested that crack closure occurred near the minimum applied load, indicating that $\Delta K_{\rm tip} \approx \Delta K_a$ for these tests. In addition, the predicted unbridged crack-opening-range profiles shown in Fig. 13 would agree with the experimental data only if $\Delta K_{\rm tip} \approx \Delta K_a$. The suggested negligible levels of crack closure would be possible if little fiber pullout occurred prior to fiber fracture or if repeated contact of the fracture surfaces forced the broken fibers to retract, or both. In any event, the apparent absence of appreciable levels of crack closure in the unbridged cracks eliminates this mechanism as a significant explanation for the reduced composite crack growth rates.

Another factor that could influence the crack growth rates in the composite is the presence of residual stress in the matrix alloy, which would tend to modify the effective stress ratio in the matrix material. Although the existence of significant tensile residual stresses in the matrix of titanium composites is well documented [*41–43*], these stresses should be significantly reduced by cyclic plasticity during fatigue, particularly at the higher stress levels. Although residual stresses were not treated explicitly, as discussed earlier, effects of residual stress were addressed through use of $\Delta K_{m,\text{eff}}$ as a correlating parameter that essentially eliminated stress ratio effects.

It should be noted that it is common practice to use as a baseline material property the da/dN-ΔK_a data for crack growth in the matrix alloy, while the analysis of bridged fatigue cracks incorporates a crack closure effect by bridging fibers through the parameter, η, in Eq 5. In the present work, crack growth rates in the matrix alloy are assumed to be controlled by $\Delta K_{m,\text{eff}}$ in both the composite and fiberless material.

As shown earlier, the unbridged crack growth behavior in each of the laminates correlated relatively well with predictions made using the shielding factor (F_s in Eq 3) proposed by Marshall, Cox, and Evans (MCE) [*12*]. Much of the recent literature on crack bridging has tended to discount this factor, choosing either a value of 1.0 (for example, McMeeking and Evans [*14*]) or the energy-based approach of McCartney [*13*] that, for the current material, predicts a value near unity. Based on the agreement between the MCE prediction and the experimental data for unbridged crack growth, the factor, F_s, given by Eq 3 was chosen for analysis of fully bridged crack growth. This shielding factor alone accounts for much of the difference between crack growth rates in the composite and the fiberless matrix material.

Use of the shear lag model to describe bridged crack growth in laminates was justified by the ability to correlate crack growth rate data over a range of stress levels, as well as the accuracy of the predicted crack opening profiles. The solution of Cox and Lo [*17,18*] correlated the data reasonably well from both $[0]_4$ and $[0/90]_S$ composites. This approach appeared to have some empirical applicability for $[0/\pm45/90]_S$, although the presence of the ±45 plies in this material obviously makes it a special case. While there is evidence that fiber/matrix interfacial shear stress degrades during repeated sliding of a bridging fiber in the wake of a fatigue crack, the current results suggest that τ degrades to a stable value in a relatively small number of fatigue cycles and remains approximately constant thereafter. As shown in Fig. 8, τ appears to decrease initially and then increase prior to failure. Although these variations in τ may correspond to specific physical processes, the trend may also be an artifact of the assumed quadratic relationship between bridging pressure and crack opening assumed in Eq 4.

An unusual finding of the analysis was the low value of interfacial shear stress, τ, that was needed to correlate the data. This average value of 1.66 MPa is well below the measurements made by a variety of micromechanical tests of single-fiber titanium composites and is in contrast with some of the values suggested for cracks in other SCS-6/titanium matrix composites [for example, *14,17,26*]. The reasons for the low values of τ are traceable to a number of factors. First, as stated earlier, a number of authors have chosen values of F_s in Eq 2 near unity, while the present findings indicate that this factor should be close to 0.5. As a result, the calculated values of the bridging pressure (and τ) are substantially reduced. Secondly, because the crack growth rate data in the composite specimens often approached threshold (ΔK_{th}) growth rates, it was important that the corresponding crack growth behavior of the fiberless matrix alloy be well defined in this region. The bridging analysis is highly sensitive to uncertainties in matrix crack growth behavior near ΔK_{th}, where the crack growth rate curve becomes very steep and the Paris law is inaccurate. To address this issue, extensive data were obtained at near-threshold growth rates, and these data were represented by Eq 8. Note that use of the Paris law will result in higher values of τ. Finally, the value of τ determined in the present work represents the stabilized, degraded behavior that would exist after hundreds of thousands of cycles. It is not surprising that this value is substantially below measurements made on pristine fibers, or bridging fibers subjected to only a small number of fatigue cycles.

The 1.66 MPa value of τ is in excellent agreement with the independent measurements of crack-opening-displacement range in the current work (Fig. 14) and other research on this material [*40*]. The effect of using higher values of τ is shown in Fig. 14. When τ increases from 1.66 to 20 MPa, the predicted $\Delta u(x)$ profile is substantially lower than the measured values. Accordingly, the predicted σ_f also increases with increase in τ, as shown in Fig. 15, but the increase in σ_f is only about 40% for a twelve-fold increase in value of τ. ΔK_b calculated for the different values of τ are also shown in Fig. 15. The higher value of ΔK_b at $\tau = 20$ MPa is consistent with the assumption of $F_s \approx 1.0$, as discussed earlier. Based on Figs. 4 and 14, we conclude that the lower values of τ determined during this study correlate well with both crack growth and $\Delta u(x)$ data, unlike the higher values of τ reported in Refs *14, 17,* and *26*.

Figure 16 shows that the σ_f predicted using $\tau = 20$ MPa and $\sigma_a = 360$ MPa is still significantly lower than the tensile strength of the fibers. Hence, the use of higher values of τ cannot explain the observed unbridged crack growth behavior at $\sigma_a > 300$ MPa.

When considering the potential use of TMCs in structural applications, the present findings highlight a number of issues and opportunities. A key concern is the extreme sensitivity of fiber bridging to applied stress and the increase in crack growth rates that occurs when the strength of the bridging fibers is exceeded. As shown in Fig. 4, an increase in σ_a of about 20% changed the crack growth behavior from fully bridged to completely unbridged. Such differences in the character of crack growth can have serious design implications, because an overload-induced loss of bridging would dramatically reduce the damage tolerance of the material. Although the value of the critical stress to permit bridging, σ_{crit}, is dependent on the unbridged length and would increase significantly as the notch dimension (that is, initial unbridged length) decreases, the hole size used here is representative of typical applications for the panel-form of this material. Equivalent center-hole specimens of $[0]_8$ SCS-6/Ti-24Al-11Nb tested previously exhibited large-scale fiber bridging at all stresses up to the material's notched tensile strength of approximately 825 MPa [*24,27*], and the bridging fiber stress achieved the expected strength of approximately 4000 MPa. Hence, σ_{crit} is highly material dependent. Although it is unclear exactly what characteristics of these two composites produced the different bridging behaviors, it should be possible to tailor future TMCs to optimize bridging behavior and the associated material's fatigue performance.

Conclusions

1. Using the closure-corrected $\Delta K_{m,\mathrm{eff}}$ as a crack driving force parameter, consolidated stress ratio effects in the fiberless matrix alloy and was an important aspect of the analysis of bridged crack growth behavior.
2. The shear-lag model for crack bridging was in reasonable agreement with the experimental findings for large-scale bridging when an average value of the interfacial shear stress, τ, was used as a fitting parameter.
3. The effective value of interfacial shear stress operating in the bridging region of a fatigue crack in SCS-6/TIMETAL 21S was determined from the crack growth rate data to be 1.66 MPa. This value was independently confirmed by precise measurements of crack-opening-displacement profiles.
4. Calculated crack opening profiles were in excellent agreement with measured profiles for both bridged and unbridged cracks.

5. The critical stress for crack bridging in the $[0]_4$ composite was less than 50% of the material's notched tensile strength, and the corresponding fiber stress in the bridging region was less than half of the reported fiber strength. This behavior is, however, highly material dependent, since large-scale bridging has been found to persist up to the material's notch strength for equivalent specimens of unidirectional SCS-6/TIMETAL 21S and SCS-6/Ti-24Al-11Nb composite.

Acknowledgments

This research was performed in the Materials Directorate, Wright Laboratory, Wright-Patterson Air Force Base, under Research Project 2302P101 of the Air Force Office of Scientific Research (AFOSR). The authors from the University of Dayton Research Institute were supported by an on-site contract (Nos. F33615-91-C-5606 and F33615-94-C-5200) with the Materials Directorate. The authors gratefully acknowledge the assistance of Mr. R. A. Kleismit in conducting the experiments and Ms. A. Hutson, Mr. G. Clemons, and Ms. J. B. Moran in data reduction and analysis.

APPENDIX

In applying the solution of Cox and Lo [*18*], three typographical errors were discovered and confirmed by the original authors. The analysis presented here used the canonical solutions from the cited reference with the following corrections:

1. Equation 2 in Ref *18* should read as shown in Eq 5 in the present paper (this form was used in Cox and Lo [*17*].
2. Equation 5 in Ref *18* should read

$$X \equiv \frac{\Delta\sigma_a}{w} \cdot \frac{\pi E'}{4\beta^2} \text{(cyclic loading)} \tag{12}$$

where the cyclic component of this expression is contained in β.

3. The sign of a_{66} in Eq A1 of Ref *18* should be (+).

References

[*1*] Larsen, J. M., Williams, K. A., Balsone, S. J., and Stucke, M. A., "Titanium Aluminides for Aerospace Applications," *High Temperature Aluminides and Intermetallics,* S. H. Whang, C. T. Liu, D. P. Pope, and J. O. Stiegler, Eds., The Minerals, Metals, and Materials Society, Warrendale, PA, 1990, pp. 521–556.

[*2*] Larsen, J. M., Russ, S. M., and Jones J. W., "Possibilities and Pitfalls in Aerospace Applications of Titanium Matrix Composites," *Characterization of Fiber Reinforced Titanium Metal Matrix Composites,* Report AGARD-R-796, NATO Advisory Group for Aerospace Research and Development (AGARD), Specialized Printing Services Ltd., Loughton, Essex, UK, 1994, pp. 1.1–1.21.

[*3*] Harmon, D. and Saff, C., "Damage Initiation and Growth in Fiber Reinforced Metal Matrix Composites," *Metal Matrix Composites: Testing, Analysis, and Failure Modes, ASTM STP 1032,* W. S. Johnson, Ed., American Society for Testing and Materials, Philadelphia, 1989, pp. 237–250.

[*4*] Bain, K. R. and Gambone, M. L., "Fatigue Crack Growth of SCS-6/Ti-64 Metal Matrix Composite," *Fundamental Relationships Between Microstructure and Mechanical Properties of*

Metal Matrix Composites, P. K. Liaw and M. N. Gungor, Eds., The Minerals, Metals, and Materials Society, Warrendale, PA, 1990, pp. 459–469.

[*5*] Naik, R. A. and Johnson, W. S., "Observations of Fatigue Crack Initiation and Damage Growth in Notched Titanium Matrix Composites," *Composite Materials: Fatigue and Fracture, Third Volume, ASTM STP 1110,* T. K. O'Brien, Ed., American Society for Testing and Materials, Philadelphia, 1991, pp. 753–771.

[*6*] Newaz, G. M. and Majumdar, B. S., "Crack Initiation Around Holes in a Unidirectional MMC Under Fatigue Loading," *Engineering Fracture Mechanics,* Vol. 42, No. 4, 1992, pp. 699–711.

[*7*] Hillberry, B. M. and Johnson, W. S., "Prediction of Matrix Fatigue Crack Initiation in Notched SCS-6/Ti-15-3 Metal Matrix Composites," *Journal of Composites Technology and Research,* Vol. 14, No. 4, 1992, pp. 221–224.

[*8*] Walls, D. and Zok, F., "The Effect of Notches on the Fatigue Behavior of a Ti-SiC Composite," *Proceedings,* Advanced Metal Matrix Composites for Elevated Temperatures Conference, American Society for Metals, Metals Park, OH, 1992, pp. 101–107.

[*9*] Jira, J. R. and Larsen, J. M., "Crack Bridging Behavior in Unidirectional SCS-6/Ti-24Al-11Nb Composite," *FATIGUE 93, Vol. 2,* J.-P. Bailon and I. J. Dickson, Eds., Engineering Materials Advisory Services, Ltd., UK, 1993, pp. 1085–1090.

[*10*] Jira, J. R. and Larsen, J. M., "Fatigue of Unidirectional SCS-6/Ti-24Al-11Nb Composite Containing a Circular Hole," *Metallurgical and Materials Transactions A,* Vol. 25A, 1994, pp. 1413–1424.

[*11*] Herrmann, D. J., Ward, G. T., Lawson, E. J., and Hillberry, B. M., "Prediction of Matrix Fatigue Crack Initiation from Notches in Titanium Matrix Composites," in this volume.

[*12*] Marshall, D. B., Cox, B. N., and Evans, A. G., "The Mechanics of Matrix Cracking in Brittle-Matrix Fiber Composites," *Acta Metallurgica,* Vol. 33, No. 11, 1985, pp. 2013–2021.

[*13*] McCartney, L. N., "Mechanics of Matrix Cracking in Brittle-Matrix Fiber-Reinforced Composites," *Proceedings,* Royal Society of London A, Vol. A409, 1987, pp. 329–350.

[*14*] McMeeking, R. M. and Evans, A. G., "Matrix Fatigue Cracking in Fiber Composites," *Mechanics of Materials,* Vol. 9, No. 3, 1990, pp. 217–227.

[*15*] Cox, B. N. and Marshall, D. B., "The Determination of Crack Bridging Forces," *International Journal of Fracture,* Vol. 49, 1991, pp. 159–176.

[*16*] Sensmeier, M. D. and Wright, P. K., "The Effect of Fiber Bridging on Fatigue Crack Growth in Titanium Matrix Composites," *Fundamental Relationships Between Microstructure and Mechanical Properties of Metal Matrix Composites,* The Metallurgical Society, Inc., Warrendale, PA, 1990, pp. 441–450.

[*17*] Cox, B. N. and Lo, C. S., "Load Ratio, Notch, and Scale Effects for Bridged Cracks in Fibrous Composites," *Acta Metallurgica et Materialia,* Vol. 40, No. 1, 1992, pp. 69–80.

[*18*] Cox, B. N. and Lo, C. S., "Simple Approximations for Bridged Cracks in Fibrous Composites," *Acta Metallurgica et Materialia,* Vol. 40, No. 7, 1992, pp. 1487–1496.

[*19*] Davidson, D. L., "The Micromechanics of Fatigue Crack Growth at 25°C in Ti-6Al-4V Reinforced with SCS-6 Fibers," *Metallurgical Transactions A,* Vol. 23A, 1992, pp. 865–879.

[*20*] Ghosn, L. J., Kantzos, P., and Telesman, J., "Modeling of Crack Bridging in a Unidirectional Metal Matrix Composite," *International Journal of Fracture,* Vol. 54, 1992, pp. 345–357.

[*21*] Bakuckas, J. G., Jr. and Johnson, W. S., "Application of Fiber Bridging Models to Fatigue Crack Growth in Unidirectional Titanium Matrix Composites," NASA Technical Memorandum 107588, NASA Langley Research Center, Hampton, VA, July 1992.

[*22*] Bowen, P., Cardona, D. C., and Ibbotson, A. R., "Micromodelling of Crack Growth in Fiber Reinforced Composites," *Materials Science and Engineering A,* Vol. 153, 1992, pp. 628–634.

[*23*] Bao, G. and McMeeking, R. M., "Fatigue Crack Growth in Fiber-Reinforced Metal-Matrix Composites," *Acta Metallurgica et Materialia,* Vol. 42, No. 7, 1994, pp. 2415–2425.

[*24*] John, R., Jira, J. R., Larsen, J. M., and Ashbaugh, N. E., "Analysis of Bridged Fatigue Cracks in Unidirectional SCS-6/Ti-24Al-11Nb Composite," *FATIGUE 93, Vol. 2,* J.-P. Bailon and I. J. Dickson, Eds., Engineering Materials Advisory Services, Ltd., UK, 1993, pp. 1091–1096.

[*25*] Ghosn, L. J., Telesman, J., and Kantzos, P., "Specimen Geometry Effects on Fiber Bridging in Composites," *FATIGUE 93, Vol. 2,* J.-P. Bailon and I. J. Dickson, Eds., Engineering Materials Advisory Services, Ltd., UK, 1993, pp. 1231–1238.

[*26*] Walls, D. P., Bao, G., and Zok, F. W., "Mode I Fatigue Cracking in a Fiber Reinforced Metal Matrix Composite," *Acta Metallurgica et Materialia,* Vol. 41, No. 7, 1993, pp. 2061–2071.

[*27*] John, R., Kaldon, S. G., and Ashbaugh, N. E., "Applicability of Fiber Bridging Models to Describe Crack Growth in Unidirectional Titanium Matrix Composites," *Titanium Metal Matrix Composites II,* WL-TR-93-4105, P. R. Smith and W. C. Revelos, Eds., Wright-Patterson AFB, OH, 1993, pp. 233–250.

[28] John, R., Jira, J. R., Larsen, J. M., and Ashbaugh, N. E., "Fatigue Crack Growth in Unidirectional SCS-6/Ti-24Al-11Nb Composite Containing a Circular Hole," *Metallurgical Transactions A*, to be submitted for publication, 1996.

[29] Covey, S. J., "A Study of Fiber Volume Fraction Effects in Notched Unidirectional SCS-6/Ti-15V-3Cr-3Al-3Sn Composite," NASP Contractor Report 191165, NASA Lewis Research Center, Cleveland, OH, 1993.

[30] Cardona, D. C., Barney, C., and Bowen, P., "Modeling and Prediction of Crack Arrest in Fiber Reinforced Composites," in this volume.

[31] Bowen, P., "Characterization of Crack Growth Resistance under Cyclic Loading in the Presence of an Unbridged Defect in Fiber-Reinforced Titanium Metal Matrix Composites," in this volume.

[32] Greszczuk, L. B., "Stress Concentrations and Failure Criteria for Orthotropic and Anisotropic Plates with Circular Openings," *Composite Materials: Testing and Design (Second Conference), ASTM STP 497,* American Society for Testing and Materials, Philadelphia, 1972, pp. 363–381.

[33] Blatt, D., John, R., and Coker, D., "Stress Intensity Factor and Compliance Solutions for a Single Edge Notched Specimen with Clamped Ends," *Engineering Fracture Mechanics,* Vol. 47, No. 4, 1994, pp. 521–532.

[34] John, R. and Johnson, D. A., "Weight Function for Two Cracks Emanating on Each Side of a Hole," to be submitted for publication, 1996.

[35] Newman, J. C., Jr., "A Nonlinear Fracture Mechanics Approach to the Growth of Small Cracks," *Behavior of Short Cracks in Airframe Components, AGARD Conference Proceedings No. 328,* NATO Advisory Group for Aerospace Research and Development, Specialized Printing Services, Ltd., Loughton, Essex, UK, 1982, pp. 6.1–6.26.

[36] Chan, K. S. and Davidson, D. L., "Fatigue Crack Growth in Fiber Reinforced Metal-Matrix Composites," *Fatigue of Advanced Materials,* R. O. Ritchie, R. H. Daushardt, and B. N. Cox, Eds., MCE Publications, Birmington, UK, 1991.

[37] Miller, M. S. and Gallagher, J. P., "An Analysis of Several Fatigue Crack Growth Rate (FCGR) Descriptions," *Fatigue Crack Growth Measurement and Data Analysis, ASTM STP 738,* Hudak, Jr. and Bucci, Eds., American Society for Testing and Materials, Philadelphia, 1981, pp. 205–251.

[38] Hartman, G. and Nicholas, T., "An Enhanced Laser Interferometer for Precise Displacement Measurements," *Experimental Techniques,* 1987, Vol. 2, pp. 24–26.

[39] Gambone, M. L. and Wawner, F. E., "The Effect of Elevated Temperature Exposure of Composites on the Strength Distribution of the Reinforcing Fibers," *Intermetallic Composites III,* Materials Research Society, Pittsburgh, 1994, pp. 111–118.

[40] Jira, J. R., John, R., Larsen, J. M., and Kleismit, R. A., "Analysis of Bridged Fatigue Cracks in SCS-6/TIMETAL® 21SComposite," *Materials Science and Engineering,* to be submitted for publication, 1996.

[41] Cox, B. N., James, M. R., Marshall, D. B., and Addison, R. C., Jr., "Determination of Residual Stresses in Thin Sheet Titanium Aluminide Composites," *Metallurgical Transactions A,* Vol. 21A, 1990, pp. 2701–2707.

[42] Rangaswamy, P., Revelos, W. C., and Jayaraman, N., "Residual Stresses in SCS-6/Ti-24Al-11Nb Composite: Part I—Experimental," *Journal of Composites Technology and Research,* Vol. 16, No. 1, 1994, pp. 47–53.

[43] Rangaswamy, P. and Jayaraman, N., "Residual Stresses in SCS-6/Ti-24Al-11Nb Composite: Part II—Finite Element Modeling," *Journal of Composites Technology and Research,* Vol. 16, No. 1, 1994, pp. 54–67.

Dian Zheng[1] *and Hamouda Ghonem*[1]

High Temperature/High Frequency Fatigue Crack Growth in Titanium Metal Matrix Composites

REFERENCE: Zheng, D. and Ghonem, H., "**High Temperature/High Frequency Fatigue Crack Growth in Titanium Metal Matrix Composites,**" *Life Prediction Methodology for Titanium Matrix Composites, ASTM STP 1253,* W. S. Johnson, J. M. Larsen, and B. N. Cox, Eds., American Society for Testing and Materials, 1996, pp. 137–163.

ABSTRACT: An experimental study was conducted to examine the effects of high temperature/high frequency loading on the crack growth behavior in titanium composites. Fatigue crack growth tests were performed on the unidirectional SM1240/TIMETAL-21S composite in air environment at three different temperatures: 24, 500, and 650°C with a loading frequency of 10 Hz and a stress ratio of 0.1. Tests were also carried out in vacuum and on aged specimens. In all the tests, the crack length was measured continuously while the crack opening displacement (COD) was measured at discrete positions along the crack length. Results show that, at all temperatures, the crack progresses in two sequential stages; a stage in which the crack growth rate decreases as the crack length increases followed by a stage characterized by an accelerated crack growth rate. The first stage corresponds to a crack-bridging condition and is the focus of the analysis in this paper. In additional to COD measurements, the microstructural features associated with the fracture process during this stage was examined by scanning electron microscopy with emphasis being placed on the matrix crack morphology, debonding location, temperature, and environmental influences on the fiber coatings. An analytical procedure, employing concepts of the theory of fracture mechanics and micromechanics analysis of the stress state along bridging fibers, was applied in order to calculate the fiber-bridging stress, the frictional shear stress, and the effective stress intensity factor. Results of this work suggest that the fiber-bridging stress and the frictional shear stress are variables along the crack length and are test temperature dependents. Furthermore, it was concluded that both the crack-tip opening displacement and the effective stress intensity factor can be used as measurements of the crack-tip driving force in the crack-bridging stage. Based on the experimental and analytical results collected in this study, a bridging damage mechanism under the loading condition specified here is suggested. This mechanism argues that the effect of temperature on the high frequency crack growth process is established through the temperature-related modification of the frictional shear stress and the resulting adjustment of the internal stress state in the debonded regions of the bridging fibers.

KEYWORDS: fatigue crack growth rate, crack bridging, crack opening displacement, bridging fiber stress, frictional shear stress, thermal aging, titanium, titanium matrix composites, life prediction, titanium alloys, fatigue (materials), modeling

An important factor in the development of continuous fiber-reinforced metal matrix composites (FRMMC) is the understanding of the fatigue crack growth mechanisms and governing factors. The focus on the crack growth stage stems from the fact that the different

[1]Research associate and professor, respectively, Mechanics of Materials Laboratory, Department of Mechanical Engineering, University of Rhode Island, Kingston, RI 02881.

methods used to fabricate this class of composites inevitably lead to the formation of microscopic defects. The presence of these defects in applications involving load reversals and high temperature environment lead to a significant reduction in the crack initiation life of the corresponding composite. Furthermore, the damage tolerance procedure employed in aerospace applications, for which FRMMC are candidate materials, is based on fracture mechanics concepts aimed at describing the growth behavior of an existing crack in the material under consideration. In the case of titanium matrix composites, fatigue crack growth occurs by mechanisms involving fiber-matrix debonding, matrix cracking, and fiber breakage. For load levels that are not sufficient to produce immediate fiber breakage at the crack tip, the fatigue crack growth is dominated by matrix fracture and fiber bridging. This mode of damage is facilitated by the fact that the silicon-carbide (SiC) type fibers used to reinforce titanium matrices are coated with one or more carbon-rich layers designed to accommodate fiber/matrix debonding and sliding that occur during the matrix cracking. In a bridging crack condition, the crack tip is shielded by the amount of the load carried by the bridging fibers. As a result, the crack-tip driving force is decreased leading to a crack growth rate lower than that of the matrix material when subjected to the same externally applied load. Efforts have been made to identify the different factors influencing the crack bridging process with the aim of determining the crack-tip driving force. These efforts have largely followed models that employ concepts of linear fracture mechanics coupled with micromechanical interpretation of balance of forces acting on the debonded region of bridging fibers [*1–8*]. While the majority of these models may not be directly applicable to fatigue crack growth in composites with ductile matrices, they do, however, identify the crack opening displacement range, $\Delta\delta$, the fiber bridging stress range, Δp, and the fiber/matrix frictional shear stress, τ_s, as being the critical governing parameters of the fatigue crack bridging process. Efforts have been made to study each of these parameters and their form of correlations with mechanical and material properties of the composite. In this, the crack opening displacement, $\Delta\delta$, particularly that of the near-crack-tip region, is viewed, on the basis of the experience gained from studying the fracture processes of high strength monolithic materials, to be a measure of the crack-tip driving force. In addition, $\Delta\delta$ has the advantage, in spite of its expected small magnitude in bridging cracks, of being a measurable parameter. This provides the possibility of establishing an experimentally based methodology to analyze the crack-bridging behavior. A number of studies have been carried out in this regard. For example, Davidson [*5*], using a stereo-imaging technique, measured the $\Delta\delta$ caused by cyclic loading of discontinuously grown cracks in SCS-6/Ti-6Al-4V and SCS-6/Ti-14Al-21Nb. The crack-tip driving force and the slip length were then correlated with $\Delta\delta$ measurements in order to compute crack growth rates using fracture mechanics concepts. Telesman et al. [*9*] used a loading stage mounted inside a scanning electron microscope (SEM) chamber to measure the $\Delta\delta$ profiles of cracks propagating in SCS-6/Ti-24Al-11Nb. These measurements were then used to develop an experimentally based approach to calculate the fatigue crack growth driving force for bridged cracks. Hartman and Nicholas [*10*] developed an automated laser interferometric displacement gage system that was then used by Jira and Kleismet [*11*] to measure the $\Delta\delta$ at different positions along a bridged crack length in SCS-6/Ti-24Al-11Nb. John et al. [*12*] utilized these measurements to examine the validity of different crack bridging models applied to unidirectional titanium matrix composites. Aside from these studies, knowledge related to the evolution of $\Delta\delta$ as a function of the crack length is nonexistent. Equally absent in the literature is the influence of high temperature on the $\Delta\delta$ in titanium matrix composites.

The second important crack-bridging parameter is the fiber bridging stress, Δp, that has a direct influence on the crack-tip shielding and the consequent effect on the reduction of the crack-tip driving force. The calculation of Δp, which is not amenable to experimental measurements, has relied on micromechanical concepts dealing with the balance of forces acting on the debonded region of bridging fibers [5,7,12,13]. Once Δp is determined, the effective stress intensity factor, ΔK_{eff}, of the corresponding bridged crack could be calculated by considering the crack to be propagating in a matrix material with an imposed closure pressure distributed along the crack length. Davidson [5] estimated Δp through the computed fiber stress by measuring the fiber strain. The work of Telesman et al. [6] and John et al. [12] calculated Δp based on either an elementary strength of materials approach or an approach assuming that σ_b has a power-law form. These types of Δp calculations are not associated with parameters such as the frictional shear stress, τ_s, or the crack opening displacement, $\Delta\delta$. Furthermore, the distribution of the fiber-bridging stress obtained from various techniques are quite different. In the work of Davidson, for example, an assumption was made that Δp is approximately constant. The simple shear-lag analysis, however, indicates that Δp is proportional to the square of the crack opening displacement, $\Delta\delta$. This implies that Δp vanishes at the crack tip and the bridging fibers at the crack tip carry no load [8,13]. Chiang et al. [8] applied a modified shear-lag model and showed that Δp attains a finite value at the crack tip. Although a significant amount of research has been made in relation to the fiber-bridging stress distribution in metal matrix composites, no single calculation of Δp distribution for high temperature conditions exists. Moreover, the available information for Δp distribution at room temperature does not discuss the effect of the crack length, for example, on the nature of the distribution.

Among the three bridging parameters mentioned earlier, the frictional shear stress, τ_s, has been the parameter most studied [4,13–15]. This is due to the recognition that the mechanics of the fiber/matrix interface region is critical for the understanding of the composite behavior. Different methods of measurement have been applied to determine the magnitude of τ_s. The most common of these is the use of fiber push-out/pull-out tests to measure τ_s in terms of the stress required to cause fiber/matrix slippage in as-received specimens of the composite under investigation [16–20]. These methods however view τ_s as being a single material constant. This view has been utilized in a variety of bridging models to predict the crack-tip driving force, the debonding length, and the crack opening displacement. The existence of a gap between the results of these models and corresponding experimental data has focused the attention on examining the characteristics of τ_s and the validity of extending its single value as measured from pre-fatigued specimens to conditions under which fiber/matrix interfacial damage accumulates [5,6,13]. Recent work in this area has shown that repeated sliding on a fiber/matrix interface may lead to deterioration of that interface either by fracture of fiber coatings [21,22], wear of asperities [17], or plastic relaxation of the friction stress on the interface [17]. Indeed, available fiber push-out/pull-out tests have showed reduced interface frictional stress in metal, intermetallic, and ceramic composites after a few loading/unloading cycles. However, push-out tests performed on fatigued composite specimens have shown either further decrease or limited increase in the value of τ_s when compared with that obtained from as-received specimens. Furthermore, the variation in debonding mechanisms along the interface as discussed by Bakuckas and Johnson [13] indicates that the value of τ_s is apparently a function of the fiber location along the debond length and crack length. These results led Chan [14], Bukuckas and Johnson [13], and Ghonem et al. [23] to suggest that τ_s is not a material constant but should rather be viewed as a variable that depends on loading conditions (applied stress level, temperature, and frequency), debonding mechanisms, viscoplastic deformation of surrounding matrix, and environmental attack of the fiber/matrix interface region. Eldridge [18] measured the frictional shear stress in titanium matrix composites at different high temperatures in a vacuum-controlled environment. This work showed that the increase in temperature causes a decrease in the value of τ_s. In contrast, pull-out tests by Marshall et al. [24] and push-out tests by Clyne [25] both applied to SM1240/TIMETAL-21S, at room temperature, have showed that high temperature aging prior to testing leads to an increase in the as-received τ_s. These contradictory results imply that τ_s is a condition-sensitive parameter. Examining the state-of-the-art database for τ_s indicates that the distributions of τ_s along the bridging zone under both room (ambient) temperature and elevated temperatures are not available.

The objective of the present study is to identify the evolution of $\Delta\delta$, Δp, and τ_s in terms of both crack length and test temperature. This has been achieved using an experimental approach to measure $\Delta\sigma$ that was then used through an analytical approach to determine both Δp and τ_s. This paper discusses the main characteristics of these three parameters and describes their interdependent relationships as well as their individual and combined effects on the crack-bridging mechanism at selected elevated temperatures.

Material and Experimental Procedure

The material used in this study is the SM1240/TIMETAL-21S composite. The matrix material is the metastable β titanium alloy, TIMETAL-21S. The composition of this alloy (in percent by weight) is 0.1 Fe, 16.0 Mo, 3.06 Al, 2.9 Nb, 0.2 S, 0.22 C, 0.12 O, 0.005 N, with the balance being titanium. The SM1240 (SiC) fiber (100 μm in diameter) consists of a 10-μm-diameter tungsten-cored SiC monofilament with dual vapor-deposited coatings of carbon and boron-rich titanium diboride at the surface. The thickness of each of these layers is about 1 μm. The composite material is made of eight plies of TIMETAL-21S foils, each 112 μm in thickness, with continuous SM1240 fibers having approximately a 35% by volume fraction and oriented along the 0° direction. Table 1 lists the basic properties of both the fiber and the matrix material. The composite was consolidated using vacuum hot pressing employing pre-established critical processing parameters. All consolidated plates, measuring 900 by 100 by 1 mm, received a stabilization heat treatment (621°C/8 h) in a dynamic vacuum furnace. Figure 1 is a chemically etched cross section of the as-fabricated composite showing a hexagonal fiber arrangement and traces of the foil-to-foil interfaces. The average center-to-center fiber spacing is approximately 150 μm.

The plates were machined into rectangular specimens measuring 75 by 10 mm with a center hole of 1.575 mm in diameter. Notches were cut at opposite horizontal edges of the center. In addition to testing at room temperature, two other temperatures have been investigated, 500 and 650°C. The temperature of 500°C is the expected application temperature while 650°C represents a near upper limit use of this type of Sigma fiber [26]. In addition to tests in air environment, one vacuum test (10^{-7} torr) was performed at 650°C/10 Hz. High temperature tests were carried out using a two-zone resistance furnace and allowing a ± 4°C maximum temperature variation along the 40-mm specimen gage length.

TABLE 1—*Mechanical properties of the fiber and the matrix.*

Property	24°C	500°C	650°C
E_m, GPa	98.2	72.4	47.6
ν_m	0.35	0.35	0.35
$\sigma_{Y,m}$, MPa	1043	552	245
$\sigma_{U,m}$, MPa	1089	641	289
$\epsilon_{F,m}$, %	>3.9	>10.5	50 to 55
E_f, GPa	390	390	390
ν_f	0.3	0.3	0.3
$\sigma_{U,f}$, MPa	3340	3340	3340
$\epsilon_{F,f}$, %	1	1	1

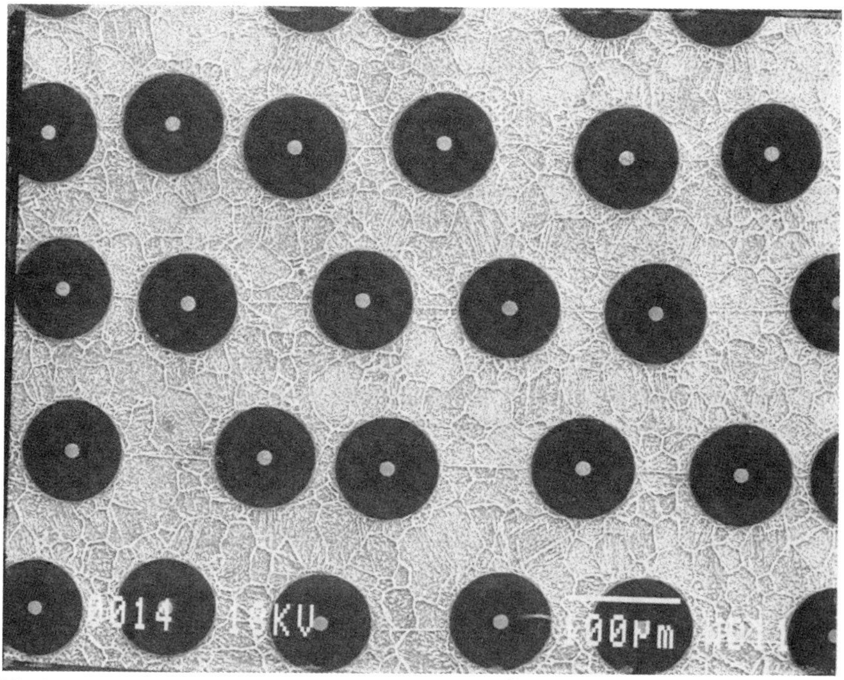

FIG. 1—*A polished and etched cross section of the as-received SM1240/TIMETAL-21S composite specimen.*

All specimens were gripped by a flat-ended hydraulic grip and tested using a servohydraulic material testing machine equipped with computer control and data acquisition capabilities. Tests were carried out under constant load range with a stress ratio of 0.1. In all cases, the applied stress range was 273 ± 3 MPa.

As the objective of this work was to identify the role of temperature on the crack growth mechanism without introducing significant time-dependent deformation effects, the loading frequency was selected to be 10 Hz for all test conditions performed here. This selection is based on results obtained from previous work on TIMETAL-21S fiberless laminates [*27*] that showed that frequencies higher than 10 Hz produce insignificant oxidation damage and creep deformation at temperatures below 650°C. Frequencies lower than 10 Hz, on the other hand, were shown to induce effects related to viscoplastic deformation and oxidation at temperatures up to 650°C. In addition, a 500°C test was carried out in air using a loading frequency of 20 Hz. To assess the role of the fiber/matrix interface aging during the high temperature crack growth testing, several specimens were aged for durations ranging from 40 to 600 h at 650 and 720°C. These specimens were then subjected to crack growth testing at room temperature.

In all the tests, the crack length was measured continuously using the d-c potential drop technique supported by optical measurements using traveling microscopes. Furthermore, the crack opening displacement, $\Delta\delta$, was measured by calculating the difference between the maximum and minimum values of the displacement-load loops of the measuring position near the crack tip. This loop that corresponds to a full loading cycle was obtained using a linear array laser interferometric displacement gage system with a resolution of 10 nm. As this technique depends on the fringe pattern obtained from a laser reflection of two micro-hardness indents placed above and below the crack surfaces at the location of interest, the COD measurements were collected at discrete crack lengths with the condition that a measurement be performed only when the crack tip advances a distance of at least 600 μm from the previous measuring position. This was to isolate the fringe pattern of the current measuring position from reflection interference caused by indents of a neighboring measuring position. No COD measurements were carried out for vacuum tests.

Several tests were stopped before the specimen final separation occurred in order to identify related fiber/matrix damage features. In each of these cases, one of the specimen's surfaces parallel to the fiber direction was ground to a depth sufficient to uncover the near-surface fiber array. This provided an opportunity to observe possible fiber bridging and the extent of fiber/matrix debonding in relation to the crack path.

Post-test examinations included fractographic analysis of the fracture surface in order to identify features of the fracture mechanisms and changes taking place in the fiber/matrix interface region. The fiber/matrix interfacial growth at several temperatures and for different exposure durations was also studied using scanning electron microscopy.

Results and Analysis

The purpose of the analysis carried out in this section is to determine the evolution characteristics of the parameters governing the crack-bridging mechanism, in particular, the crack growth rate, the fiber-bridging stress, the frictional shear stress, and the crack-tip driving force. The success in identifying the characteristics of these parameters for the test temperatures used here will lay the ground for a proposed high temperature, crack growth damage mechanism of crack bridging to be described in the next section.

The Crack Growth Rate

The curves representing the crack growth rate versus the applied stress intensity factor for all test conditions are shown in Fig. 2. The common feature of these curves is the presence of an initial crack-bridging stage signified by the continuous decrease in the crack growth rate as the applied stress intensity factor, ΔK_a (or the crack length), increases. In this case, unlike the monolithic materials, ΔK_a cannot therefore be treated as a crack-tip driving force. This crack-bridging stage is followed by a stage that consists of repeated events of crack growth acceleration and retardation leading ultimately to crack instability and final failure of the test specimen. In no test condition performed here was full crack arrest observed.

Focusing on the initial crack-bridging stage, which occupies not less than 30% of the fatigue life of any of the test specimens, it is observed that for a particular crack length the crack growth rate decreases as the temperature increases thus signifying the dependency of the crack-tip shielding force on temperature. At 500°C, it was observed that the crack growth rates at 10 Hz (air) and 20 Hz (air) are identical. The role of environment is noticeable in the difference between the 650°C crack growth rates in air and in vacuum. The slope of the crack growth rate curve corresponding to the first crack-bridging stage, which could be considered a measure of the build up rate of the crack-tip shielding force, is lowest in the 24°C test and highest in the 650°C air test. This observation reinforces the fact that the crack-tip shielding force increases as temperature increases. The end of the crack-bridging stage is marked by the lowest crack growth rate attained in each of the tests. This rate ranges from 5.6×10^{-6} mm/cycle in the room temperature test to 2.4×10^{-7} mm/cycle in the 500°C. The crack length at the end of this stage, which marks a transition from decelerated to accelerated crack growth rate, decreases as the temperature increases. This length is, however, similar in the 500° and 650°C (vacuum) tests.

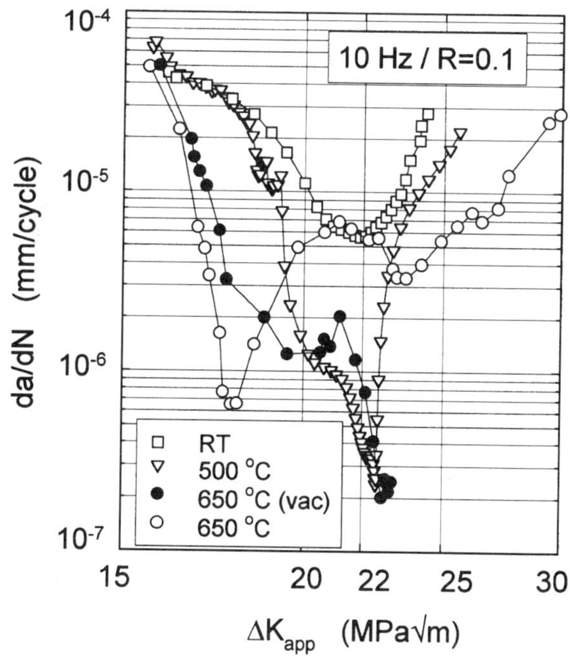

FIG. 2—*Fatigue crack growth rate,* da/dN, *in the composite versus the applied stress intensity factor,* ΔK_a, *for various temperatures at 10 Hz.*

As will be detailed later, the analysis in this paper will focus only on the initial crack-bridging stage of each of the tests carried out here. This stage is hereafter termed the crack-bridging stage.

Fractography

The different fracture processes encountered in each of these crack growth stages are evident in the appearance of the fracture surface features. For example, Fig. 3, which belongs to the 500°C test, shows two different fracture zones: the first for the crack-bridging stage and the second for the repeated crack growth acceleration/retardation stage. The first zone has excessive fiber core damage that may have resulted from the application of a large number of fatigue cycles and prolonged thermal exposure. The pull-out fibers in this zone are relatively short. The second zone shows no fiber core damage and displays longer fiber pull-out length. An example of the unstable fracture zone is shown in Fig. 4; it has a high degree of matrix deformation surrounding the pull-out fibers.

The matrix cracking mode for all tests was observed to be transgranular. A typical example of the fracture surface features associated with this mode is shown in Fig. 5. They consist of cleavage facets; the size of which seems to depend on their location on the fracture surface and the corresponding crack growth rate; the higher the growth rate, the larger the cleavage facet size. Furthermore, foil-to-foil interfaces along the fracture surface are difficult to identify and did not interrupt the continuity of the cleavage planes. Isolated colonies of fatigue striations were found on fracture surfaces of some of these tests but did not represent a dominate fracture mode in any of the test cases.

In examining the interfacial region of the bridging fibers, it was interesting to observe

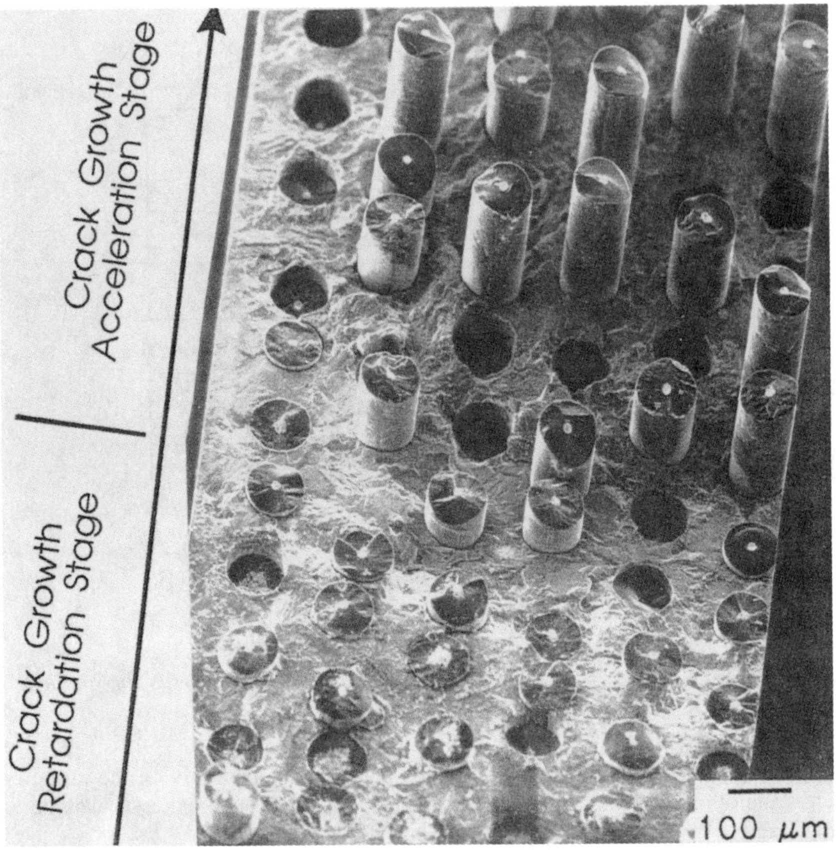

FIG. 3—*Typical view of fracture surface obtained from 500°C/10 Hz test showing distinctive regions that related to the crack bridging and the crack acceleration stages.*

that, regardless of the test temperature, debonding always occurred within the carbon-coating layer or between the carbon and the TiB_2 layers, see also Refs *28* and *29*. Figure 6 shows an example of a debonding crack in the carbon layer immediately under the fracture surface of a 24°C test. The debonding that occurred at these locations is possibly due to the compliant nature of the carbon layer.

The surface morphology of the bridging fibers in the crack-bridging stage are distinct for different temperatures. At 24°C, a patch-like residual carbon layer can be observed (Fig. 7), while at 500° and 650°C, this type of layer is absent (Fig. 8). It is possible that the presence of this layer at room temperature is a source of increased fiber/matrix friction during the frictional sliding of the bridging fibers. This may in fact be the cause of the groove-shaped traces observed on the debonded surface of pull-out fibers at 24°C, Fig. 9. At high temperature, this extra carbon layer would be gradually oxidized to form CO/CO_2 and would evaporate, thus making the fiber/matrix interfaces smoother than that at 24°C. This speculation is supported by the fact that the carbon patch layer does exist on the surfaces of the pull-out fibers within the final failure stage of the 500° and 650°C tests (Fig. 10). The insufficient exposure time during the rapid fracture stage results in no oxidation of the preceding residual carbon layer along the fibers in this region. It is worth mentioning that the surface morphology of the bridging fibers at 650°C in vacuum is similar to the one at 24°C (Fig. 11).

FIG. 4—*View of final fracture region at low magnification (650°C/10 Hz). Arrow indicates the crack growth direction and the lower part of the fracture surface is the final fracture region.*

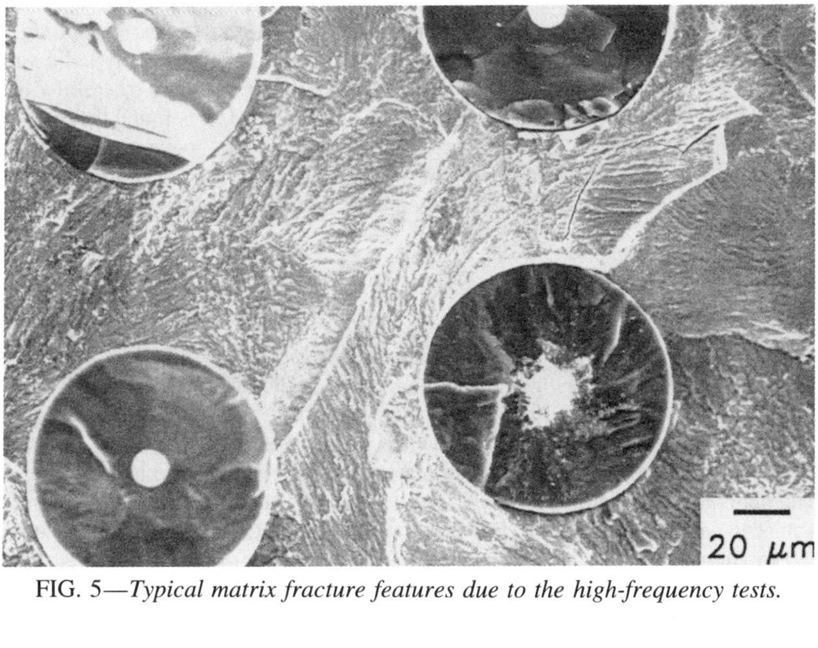

FIG. 5—*Typical matrix fracture features due to the high-frequency tests.*

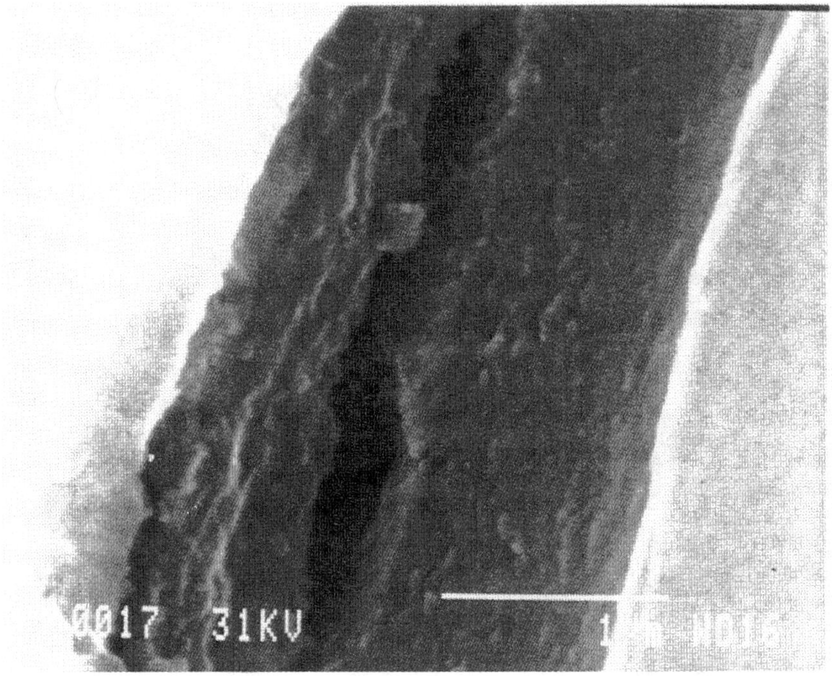

FIG. 6—*An SEM micrograph (1 mm below the fracture surface) showing the location of the interfacial debonding occurring within the carbon coating layer of the fiber (650°C/10 Hz).*

The debonding length of fibers along the crack length was examined by removing the outermost matrix layer in several fractured specimens; Fig. 12 shows the debonding length for 650°C. It can be seen that fiber breakage occurred in the bridging fibers indicating that the fiber pull-out length could be proportional to the debonding length. By carefully measuring the fiber pull-out length from the fracture surface using SEM, it is found that the average fiber pull-out length in the crack-bridging stage increases as the temperature increases (Table 2). In fact, when comparing the trend of these lengths with that of the crack growth rate, one could conclude that the higher the crack growth rate, the lower the average fiber pull-out length.

The Crack Opening Displacement, $\Delta\delta(x)$

The range of the crack opening displacements, $\Delta\delta$, were obtained at positions, x, measured from the center of the bridged crack length during the 24, 500, and 650°C (air) tests. These discrete data points for different (half) crack lengths, in addition to those obtained by setting $\Delta\delta$ at the crack tip at zero, are shown in Figs. 13*a* through *c*. The smooth curves connecting the points in these curves were generated by using a spline curve-fitting procedure. The general trend of these curves is that, for the same crack length, $\Delta\delta(x)$ increases as the distance from the crack tip increases; a trend similar to that observed in unbridged cracks. The influence of bridging fibers on the crack opening displacement is, however, apparent in the fact that the measured values of $\Delta\delta(x)$ are much smaller than those of monolithic materials [*4*] and in the weak dependency of $\Delta\delta(x)$ on the increase of the crack

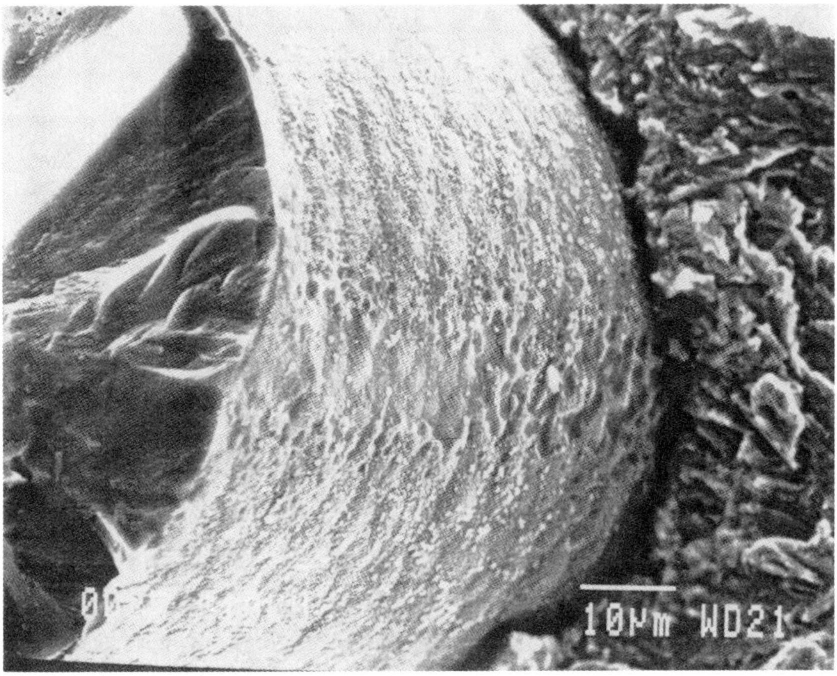

FIG. 7—*Surface morphology of the fiber in the bridging region at 24°C.*

length. In the case of the 24°C test, for example, the advance of the half crack length from 1.39 to 1.88 mm leads to a 0.3-μm increase in the $\Delta\delta$ measured at the notch mouth. The increase in the test temperature produces, however, higher values of $\Delta\delta$. This is evident in the 24°C test results when compared with those of 650°C.

Fiber-Bridging Stress, $\Delta p(x)$

In this study, the bridging crack length is considered to be a short crack falling within the transient region mentioned by Chiang et al. [*8*]. Here, the stress and strain fields are complex on both sides of the crack tip, and the discrete bridging traction of the bridging fibers supposedly control the crack-tip driving force. This section focuses on determining the fiber-bridging stress and its relationship with temperature and crack length.

With reference to the previous section, the measured crack opening displacement, $\Delta\delta(x)$, is equal to the difference between the crack opening displacement, $\Delta\delta_a(x)$, that is proportional to the externally applied stress, $\Delta\sigma_a$, and the crack opening displacement, $\Delta\delta_b(x)$, that is related to the fiber bridging stress, $\Delta p(x)$. $\Delta\delta(x)$ can thus be written as

$$\Delta\delta(x) = \Delta\delta_a(x) + \Delta\delta_b(x) \quad (1)$$

where on the basis of the linear elastic fracture mechanics concepts, both $\Delta\delta_a(x)$ and $\Delta\delta_b(x)$ can be written as [*30*]

$$\Delta\delta_a(a) = \frac{4\Phi}{E_c}\Delta\sigma_a\sqrt{a^2 - x^2} \quad (2)$$

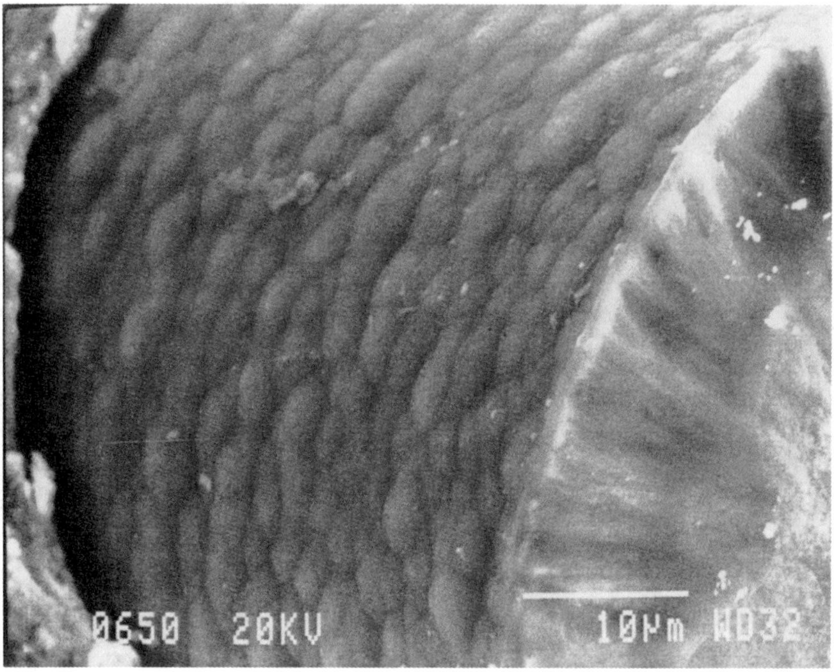

FIG. 8—*Surface morphology of the fiber in the bridging region at 650°C.*

and

$$\Delta\delta_b(x) = \frac{4\Phi}{E_c\pi}\int_{a_0}^{a} \Delta p(t) \ln \left| \frac{\sqrt{a^2 - x^2} + \sqrt{a^2 - t^2}}{\sqrt{a^2 - x^2} - \sqrt{a^2 - t^2}} \right| dt \tag{3}$$

Φ is a correction factor accounting for the finite width [*31*] and is given as

$$\Phi = \ln \left(\frac{1 + \sin(\pi a/W)}{\cos(\pi a/W)} \right) \bigg/ \frac{\pi a}{W} \tag{4}$$

By setting

$$\int_{a_0}^{a} \Delta p(t) \ln \left| \frac{\sqrt{a^2 - x^2} + \sqrt{a^2 - t^2}}{\sqrt{a^2 - x^2} - \sqrt{a^2 - t^2}} \right| dt = F(x) \tag{5}$$

Equation 3 could thus be written as

$$F(x) = \pi\Delta\sigma_a\sqrt{a^2 - x^2} - \frac{\pi E_c \Delta\delta(x)}{4\Phi} \tag{6}$$

Equation 6 is a typical Fredholm equation of the first kind that can be solved numerically by the least-square method using the singular value decomposition technique [*32,33*]. This solution, yields values of $\Delta p(x)$ as a function of the crack length, a. Results for the three

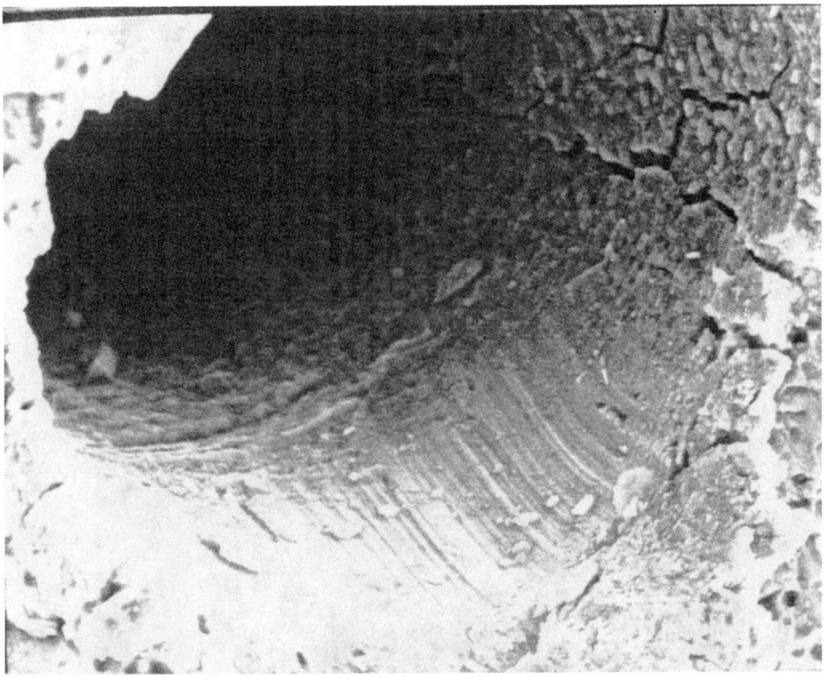

FIG. 9—*Evidence of sliding grooves on the debonded surface at 24°C.*

temperature levels of 24, 500, and 650°C are shown in Figs. 14*a* through *c*, respectively. A feature of these distributions is the presence in all test cases of a minimum finite value of Δp at the crack tip. Also, the average values of $\Delta p(x)$ increase proportionally to both the increase in the temperature and the crack length. Considering the fiber-bridging stresses to be the principal contributor to the shielding of the crack-tip driving force, this result explains the decrease in the crack growth rate associated with the increase in both temperature and crack length. Furthermore, for the same test temperature, the difference between the maximum and minimum fiber-bridging stress at a particular crack length decreases as the crack length increases. This tendency of $\Delta p(x)$ to approach a uniform level is also influenced by the temperature as can be seen in Fig. 14.

Once the fiber-bridging stress distribution, $\Delta p(x)$, is determined, the discrete fiber traction, ΔS_f, acting on the individual bridging fibers can be calculated as

$$\Delta S_f(x) = \Delta p(x)/V_f \tag{7}$$

where V_f is the fiber volume fraction. The maximum value of ΔS_f is reached at the end of the crack-bridging stage and acts on the first fiber located at the crack mouth. These values for the three temperatures are 776 MPa (24°C), 873 MPa (500°C), and 877 MPa (650°C). Figure 15 shows the evolution of ΔS_f calculated at the crack mouth as a function of the crack length and at different temperatures. This figure shows that the dependency of ΔS_f on temperature is not significant while its dependency on the crack length decreases as the temperature increases. Figure 15 moreover, indicates that while ΔS_f increases with increasing crack length and temperature, its value seems to converge to an upper limit of about 900 MPa. Since these crack-mouth bridging fibers are most likely to be the first to fracture at the

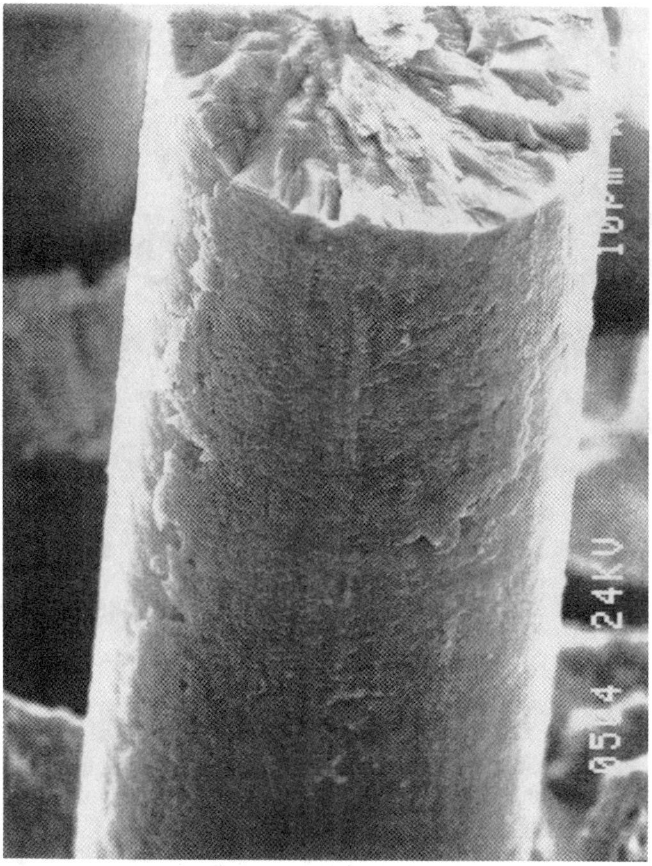

FIG. 10—*Surface morphology of the fiber in the final fracture region at 650°C showing intact parts of the carbon coating layer.*

end of the crack bridging stage, the 900 MPa would represent the fracture strength of the Sigma fibers. This value is much lower than the monotonic tensile strength of this type of fiber that is estimated as the 3000 MPa level [*34*]. This significant discrepancy could be primarily related to the fact that the fatigue fracture strength of SiC fibers is lower than its monotonic fracture strength. The mechanical damage induced in the fiber surface due to load reversibility and fiber/matrix frictional wear effects could also be seen as another factor accelerating the fibers failure at a lower traction stress.

Frictional Shear Stress, $\tau_s(x)$

When the tip of the matrix crack bypasses a fiber during crack bridging, neglecting the details of the fiber/matrix interfacial debonding process, the loading part of the fatigue cycle is assumed to produce a maximum debonding length, L_{max}, along the bridging fiber. The subsequent unloading of the cycle will introduce, within L_{max}, a reversed frictional sliding, L_R, along which the interface degradation would expect to alter the average frictional sliding stress, τ_s. The change in τ_s, could, in turn, affect the reversed sliding length, L_R. However, once τ_s is stabilized within a few loading cycles [*6,12*], L_{max} and L_R could also reach stable

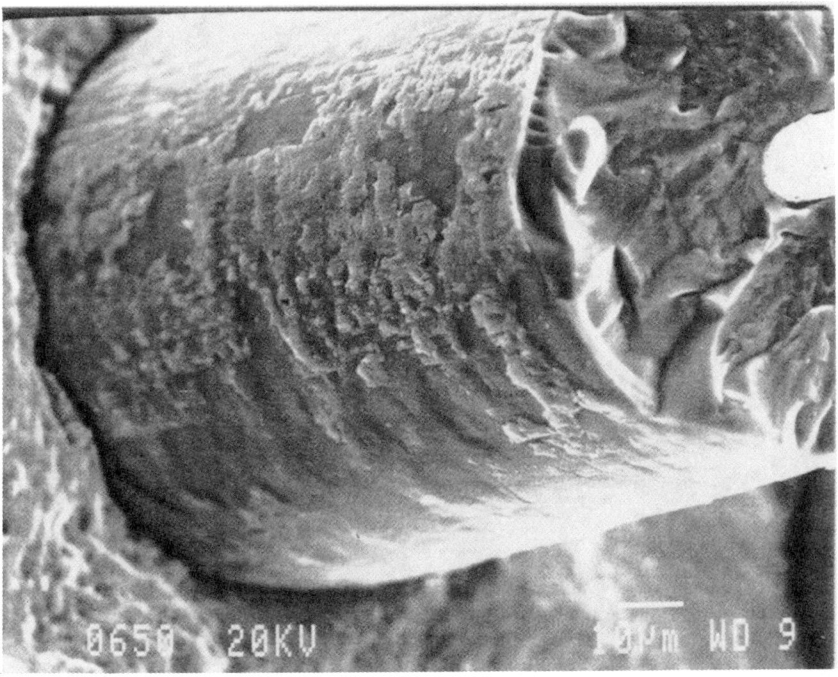

FIG. 11—*Surface morphology of the fiber in the bridging region at 650°C (vacuum).*

lengths. This steady-state condition is maintained until the crack tip advances and involves a new fiber. Focusing on this condition for an arbitrary location, x, along the crack length, a, a geometric model, based on the work of Budiansky, Hutchinson, and Evans [*35*], has been developed by the authors [*36*] to identify the stress states in three regions along the loading axis of the fiber/matrix interface. These regions are the bonded zone ($z > L_{max}$), the initial debonded zone ($L_R < z \leq L_{max}$), and the frictional sliding zone ($0 \leq z \leq L_R$), see Fig. 16. This model takes into account the influence of the axial fiber and matrix residual stresses, σ_f^R and σ_m^{R0}, respectively, as well as effects of the mean load. The formulation of this model leads to the determination of $\tau_s(x)$ in terms of the crack opening displacement range, $\Delta\delta(x)$. This is expressed as

$$\tau_s(x) = \frac{(B_1 + B_2)}{\Delta\delta(x) - B_3} \tag{8}$$

where B_1, B_2, and B_3 are explicit functions of material properties, fiber volume fraction, fiber-bridging stress, $\Delta p(x)$; applied stress, $\Delta\sigma_a$; residual stresses of fiber, σ_f^{R0} and matrix, σ_m^{R0}; as well as the initial frictional shear stress, τ_0. The input requirements to execute the preceding expression, in particular, the values of σ_f^{R0} and σ_m^{R0}, were obtained from data generated in the work of Ghonem et al. [*37*] on the same composite material; these values are listed in Table 3. Furthermore, τ_0 was measured using a room temperature fiber push-out procedure carried out on thin composite specimens that were previously aged (100 h) in air at elevated temperatures for different durations. Results of this work, detailed in Ref *38*, show that τ_0 has the values 67, 34, and 22 MPa for the 24, 500, and 650°C tests, respectively (these results are being examined by the authors in order to confirm their validity when both aging and push-out procedures are performed in high temperature/

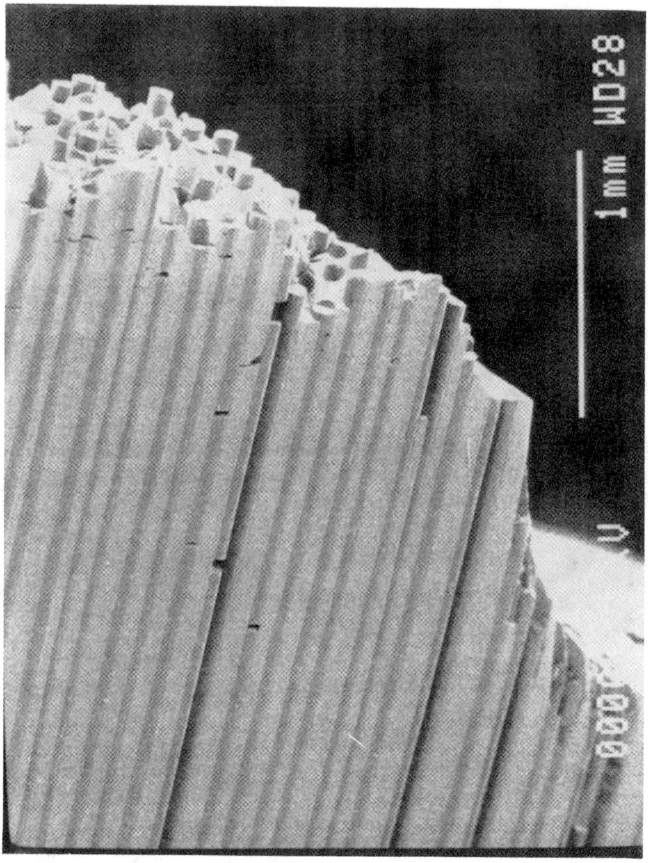

FIG. 12—*Ground and polished surface showing fiber pull-out lengths in the case of a 650°C/10 Hz test.*

TABLE 2—*Average fiber pull-out length.*

Temperature, °C	Average Fiber Pull-Out Length, μm
24 (air)	61
500 (air)	86
650 (air)	107
650 (vacuum)	96

vacuum conditions). Now, with the use of $\Delta\delta(x)$ profiles, the frictional shear stress, $\tau_s(x)$, for various crack lengths at the different test temperatures can be determined from Eq 8. Results for three different crack lengths at the three temperatures under consideration are shown in Fig. 17. It can be seen that the frictional shear stress, τ_s, does not have a single value but is a variable along the crack length. The relative variation of $\tau_s(x)$ in terms of the temperature, T, and the half crack length, a, can be written as follows

$$\tau_s(x) = \tau_0(T) \exp \{-\beta a^{-2} T^{-0.05}[1 - (x/a)^{\alpha}]\} \tag{9}$$

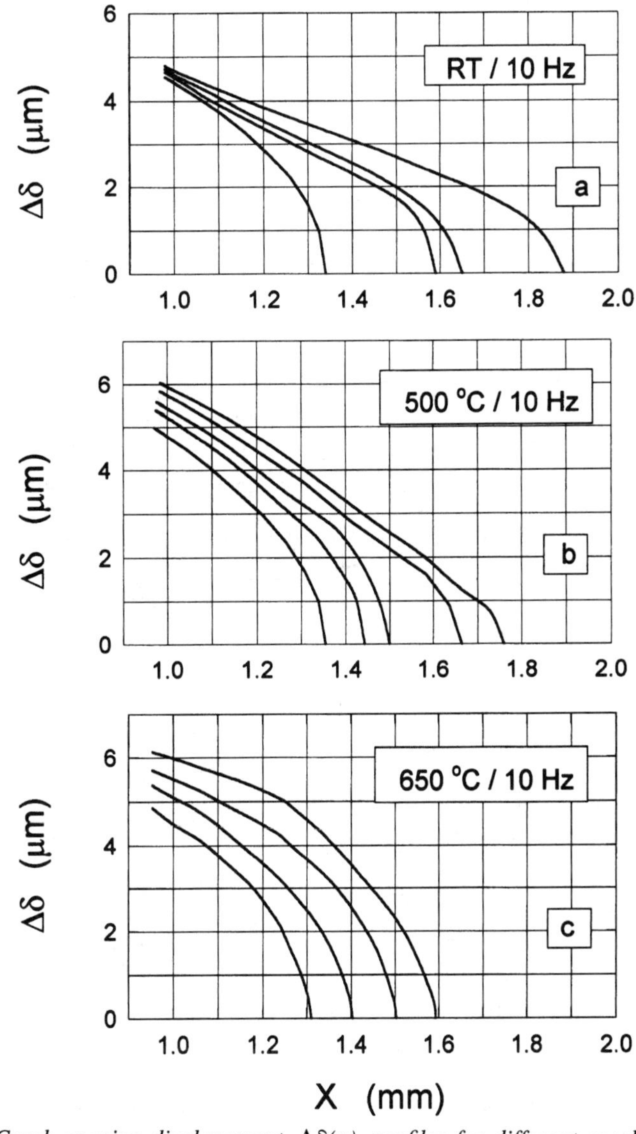

FIG. 13—*Crack opening displacement,* Δδ(x)*, profiles for different crack lengths at* (a) *24°C,* (b) *500°C, and* (c) *650°C.*

where β is a material-related parameter that could encompass factors such as matrix material properties, fiber diameter, and fiber volume fraction. In the current study, β is approximately equal to 1.09. The exponent, α, is the prime controlling factor of Eq 9 and depends on both the crack length and the temperature via the following expression

$$\alpha = (C_1 + C_2T)a^{-4.145} \tag{10}$$

where C_1 and C_2 are constants equal to 54.33 and 0.0253, respectively.

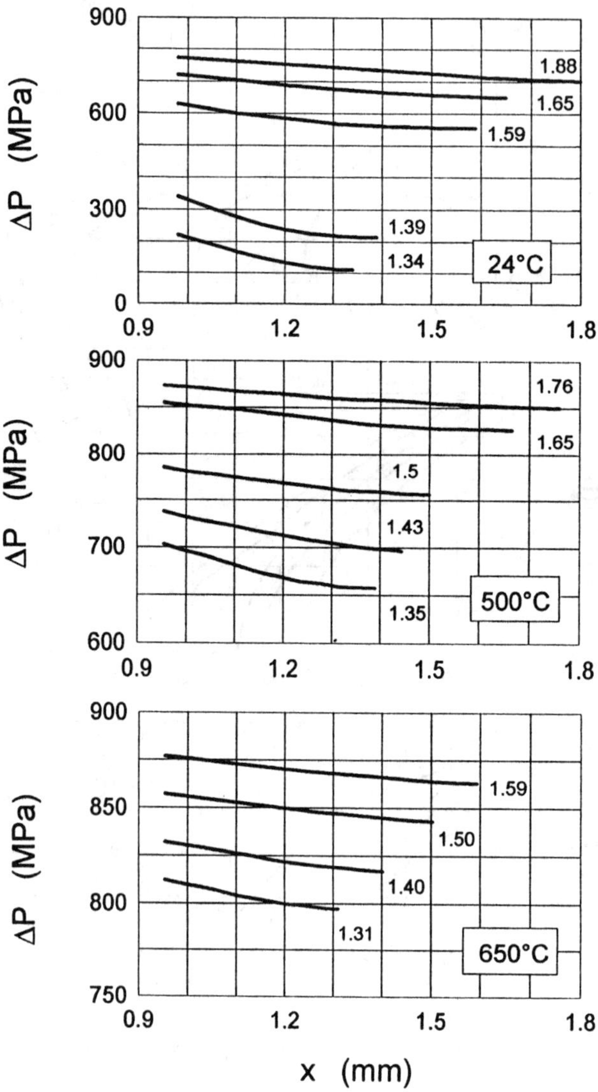

FIG. 14—*Fiber-bridging stress distributions for different crack lengths at* (a) *24°C,* (b) *500°C, and* (c) *650°C. X is the distance measured from the crack tip to the center of the hole.*

Crack-Tip Driving Force

It has been shown previously that ΔK_a cannot be considered an effective crack-tip driving force due to the presence of a crack-tip shielding force exerted by the fiber traction. A correction of ΔK_a that yields an effective stress intensity factor, ΔK_{eff}, can be calculated by subtracting from ΔK_a the shielding component produced by fiber-bridging stress. This is expressed as

$$\Delta K_{\text{eff}} = Y\Delta\sigma_a\sqrt{\pi a} - 2Y\sqrt{a/\pi}\int_{a_0}^{a}\frac{\Delta p(x)}{\sqrt{a^2 - x^2}}dx \tag{11}$$

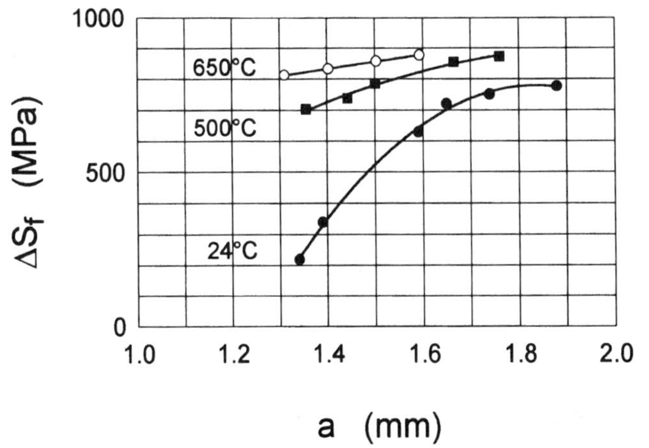

FIG. 15—*Fiber-bridging traction at the crack mouth as a function of crack length (measured from the center of the hole) and temperature.*

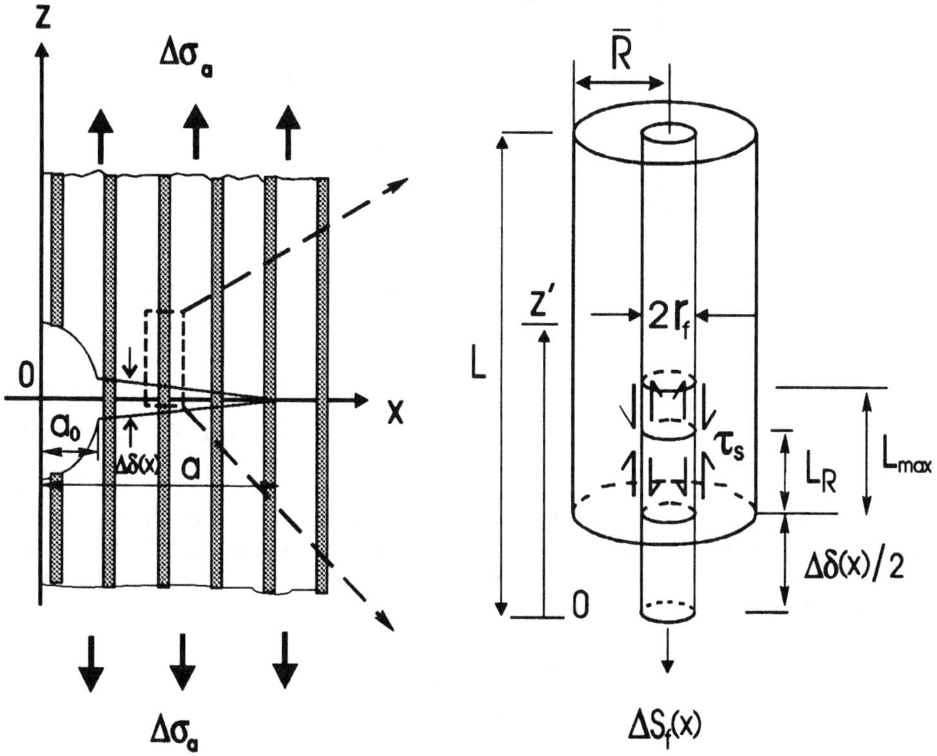

FIG. 16—*A geometric configuration of the proposed micromechanical model.*

TABLE 3—*Thermal residual stress of the matrix.*

Temperature, °C	24	500	650
σ_m^{R0}, MPa	200	47	21

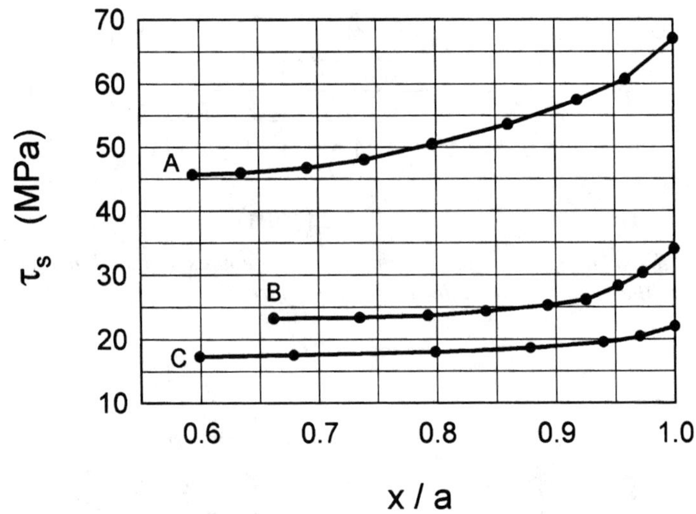

FIG. 17—*Frictional shear stress for different crack lengths and temperatures:* (a) *1.65 mm/24°C,* (b) *1.665 mm/500°C, and* (c) *1.593 mm/650°C.*

where a_0 is the initial half crack length and Y is a geometric factor for finite width correction

$$Y = (\sec(\pi a/2W)^{1/2}$$

The use of Eq 11 to calculate ΔK_{eff} depends only on the knowledge of $\Delta p(x)$ that, in turn, was calculated on the basis of the experimentally determined $\Delta\delta(x)$. This equation in conjunction with results shown in Fig. 14, was applied to various bridged crack lengths for the three test temperatures. Results in the form of ΔK_{eff} versus the crack growth rate, da/dN, are plotted in Fig. 18. It shows that da/dN is proportional to ΔK_{eff}, and for the same ΔK_{eff}, da/dN increases as the temperature increases. The trends of these results are similar to those observed in monolithic materials. The curves shown in Fig. 18 could be expressed in a Paris-type equation written as follows

$$\frac{da}{dN} = C(T)\,\Delta K_{\text{eff}}^{n(T)} \tag{12}$$

where both the coefficient, C, and the exponent, n, are temperature dependents.

The influence of crack length and test temperature on the effective stress intensity factor are further examined in Fig. 19. It shows that ΔK_{eff} is inversely proportional to both a and T. For comparison between the behavior of ΔK_{eff} and that of the crack-tip opening displacement, see Ref *38*.

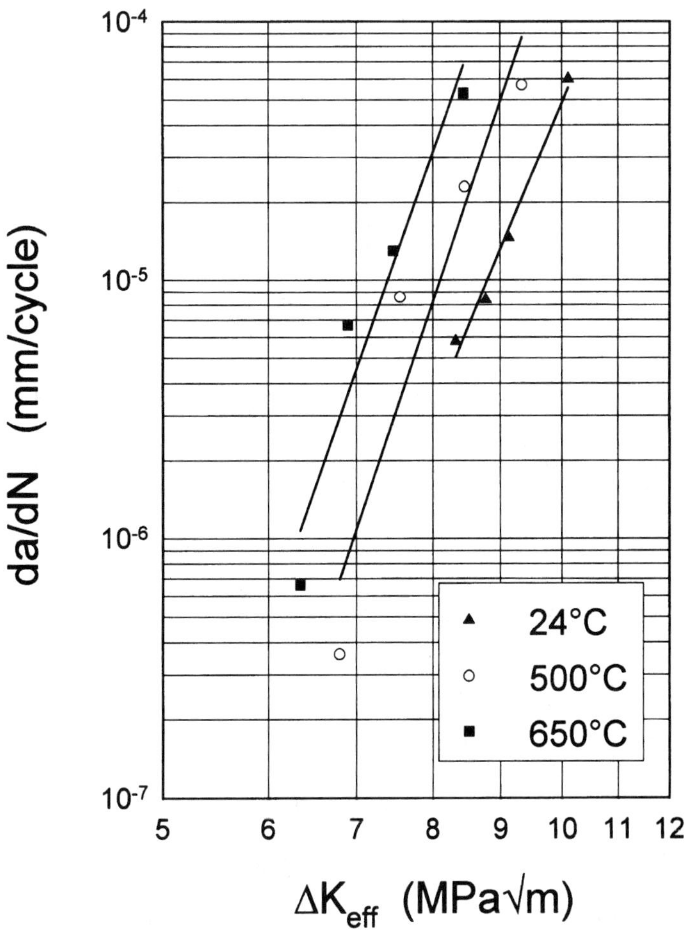

FIG. 18—ΔK_{eff} *versus crack growth rate for different temperatures.*

Discussion

The work in this paper has focused on the influence of high temperature on the bridging fatigue crack growth in the SM-1240/TIMETAL-21S composite when subjected to high frequency loading. The basic conclusion is that, in the range from 24 to 65°C; an increase in the test temperature leads to a decrease in the fatigue crack growth rate. This result was supported by experimental and analytical work identifying the evolutionary behavior of the fatigue crack opening displacement and the distributions of both the fiber-bridging stress and the frictional shear stress. Results of this work show that the temperature-related reduction in the crack growth rate is primarily due to events taking place in the debonding zone of bridging fibers leading to an increase in the crack-tip shielding. In an attempt to formulate, on the basis of these results, a damage mechanism capable of interpreting the observed response of the bridging crack to the temperature, one should recall the following important observations:

1. Debonding of bridging fibers, regardless of the test temperature, occurs within the carbon coating layer or between the carbon and the TiB_2 layers. The occurrence of this

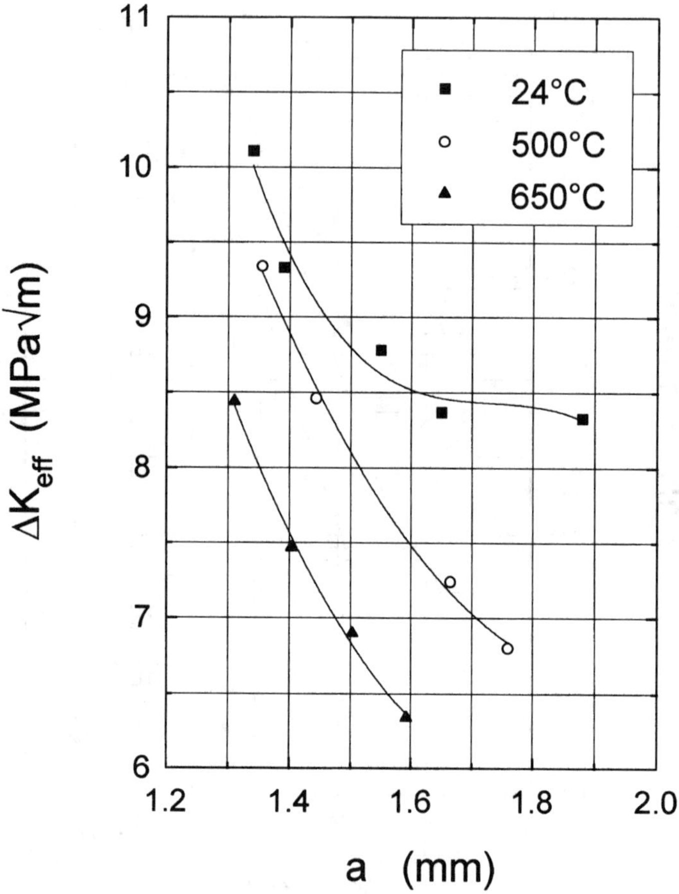

FIG. 19—ΔK_{eff} *versus crack length for different temperatures.*

at room temperature is possibly due to the compliant nature of the carbon layer. At high temperatures, however, this may be also due to the degradation of this layer by means of carbon diffusion outward through the TiB_2 layer or the formation of new phases such as TiC at TiB_2/C interface. The work of Osborne and Ghonem [*29*] on the interfacial kinetics of this composite has demonstrated, however, that the thickness change of the TiB_2 layer as well as the outward diffusion of the carbon are insignificant up to 650°C.

2. The average fiber pull-out length increases as the temperature increases. This result in addition to the fact that the crack opening displacement, particularly at the crack mouth, increases with an increase in temperature, means that the debonded length along a bridging fiber, that is, the Mode II crack, is temperature dependent.
3. The role of environmental effects on the crack growth behavior at 500°C is negligible. This is due to the fact that crack growth rates of the 10- and the 20-Hz tests are identical. At 650°C, the role of environment seems to play a role due to the observation that there exists a difference between the crack growth rates of the corresponding air and vacuum tests. This influence is primarily through the modification of the surface morphology of the debonded area in the bridging fibers to alter the frictional shear stress.

On the basis of these observations, a crack-bridging damage mechanism could be established. Here, an assumption is made that the initial debonding under the same external loadings is approximately identical in all temperatures. Two important factors will, however, influence the ensuing events. They are: (a) the Mode II fracture toughness, K_{II}, of the fiber carbon coating layer, and (b) the magnitude of the radial stress acting on the boundary of the would-be-debonded and debonded regions. It is documented that the carbon becomes more brittle as the temperature increases up to 1000°C [*39,40*]. This implies that for the same applied ΔK_{II} acting at the tip of the initially debonded region (Mode II crack) on the bridging fiber, a relatively longer debonded length within the more brittle carbon layer would be generated at higher temperatures. The relative movement between the two surfaces of this Mode II crack is governed primarily by the frictional shear stress, τ_s, acting along these surfaces. τ_s is, in fact, a function of both the surface roughness characteristics and the clamping force provided by the radial stress, σ_r, acting perpendicular to the frictional surfaces. One could then express τ_s as where τ_0 is the initial frictional shear stress of the frictional surfaces, and μ is the coefficient of friction. Tamin and Ghonem [*41*] have shown, on the basis of a time-dependent finite element study performed on the same composite, that the radial stress, σ_r, decreases as the temperature increases. It is, therefore, possible to assume that τ_s, being a variable, decreases with the increase in temperature. A relatively longer debonding length with a reduced frictional shear stress between its surfaces would then permit, for the same applied ΔK_{II}, a larger surface movement that is translated as a larger crack opening displacement of the main bridging crack. This larger movement between the surfaces of the debonded region would, in turn, result in a higher shear stress and consequently higher ΔK_{II}, at the corresponding Mode II crack tip. At the steady stage where the debonding length is stable, a combination of larger relative displacement of the frictional surfaces and higher (Mode II) crack-tip shear stress necessitates, for achieving a balance of axial forces, a higher fiber traction. For higher temperature test conditions, this traction, when integrated along the bridged crack length, produces a crack-tip shielding stress higher than that in lower temperature test conditions and thus a lower crack growth rate. This mechanism is therefore based on the notion that the temperature-related reduction in the crack growth rate is traced to the associated stress state variation with respect to temperature. In this, temperature-related effects in the form of phase changes or growth of the fiber/matrix interface are assumed to play a minor role. In order to examine the validity of this concept, a specimen was aged for 40 h at 650°C. This duration is equal of that required to complete the crack growth test at this temperature. After aging, the specimen was tested at 24°C. The resulting crack growth rate versus ΔK_a is shown in Fig. 20 and compared to the results of the nonaged 24°C test. The similar crack growth rates suggest that the fiber/matrix interfacial-related modification has little influence on the bridging crack growth behavior.

Conclusions

The experimental and analytical studies in this work have been performed in order to examine high temperature fatigue crack damage mechanisms in titanium metal matrix composites. The SM1240/TIMETAL-21S composite was used as a test material. The results of this work show that the fatigue crack growth rate of the composite at all temperatures consists of a deceleration crack growth stage followed by the acceleration crack growth stage. The present work has focused on the understanding of the controlling parameters and the damage events occurring in the initial crack bridging stage. The major conclusions of these efforts are as follows:

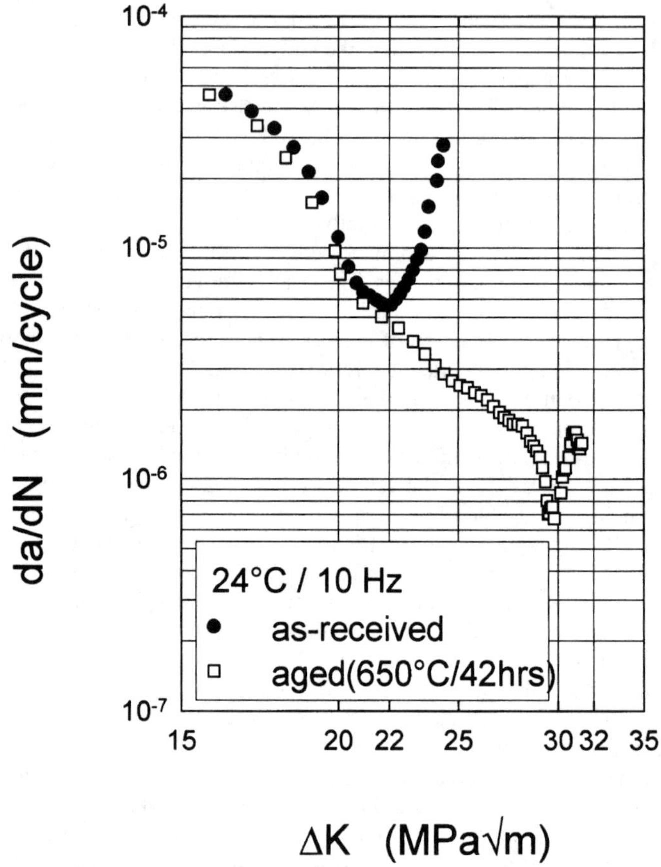

FIG. 20—*The crack growth rates for as-received and aged specimens (24°C/10 Hz).*

1. For the 10-Hz fatigue loading condition, based on the presentation of the crack growth rate versus applied stress intensity factor, the fatigue crack growth rate decreases with the increase of the temperature.
2. The crack opening displacement, $\Delta\delta(x)$, at a position farthest from the crack tip increases as the crack length or the temperature increases, while the crack opening displacement near the crack tip, $\Delta\delta_T$, is found to decrease as both the crack length and the temperature increase.
3. The fiber-bridging stress, $\Delta p(x)$, is a variable along the crack length with a decreasing distribution from the crack mouth to the crack tip. At the crack tip, $\Delta p(x)$ is a finite value. Its magnitude increases as the crack length or the temperature increases, the longer the crack length or the higher the temperature, the more uniformly distributed is the fiber-bridging stress, $\Delta p(x)$.
4. The frictional shear stress, $\tau_s(x)$, is not a single value; instead it varies along the crack length with the lowest value at the crack mouth.
5. The present study indicates that both the crack-tip opening displacement, $\Delta\delta_T$, and the effective stress intensity factor, ΔK_{eff}, are measurements of the crack-tip driving force in the condition of crack bridging.

6. The temperature influence is expressed through the alternation of the internal stress state which, in turn, affects the frictional shear stress. The effect of environment is shown by further modifying the surface morphology of the sliding region, which also directly changes the frictional shear stress.

Acknowledgments

The composite material used in this study was manufactured by United Technologies, Pratt & Whitney, East Hartford, Connecticut, as part of a joint research program between Pratt & Whitney, Wright-Patterson AFB Materials Laboratory, and the University of Rhode Island. The authors acknowledge the contributions of Dr. M. Thompson of United Technologies Research Center and Dr. T. Nicholas of Wright-Patterson AFB towards this program. This research was supported by U.S. Air Force Office of Scientific Research under Grant F49620. Dr. W. Jones is the program manager.

References

[*1*] Marshall, D. B., Cox, B. N., and Evans, A. G., ''The Mechanics of Matrix Cracking in Brittle-Matrix Fiber Composites,'' *Acta Metallurgica,* Vol. 33, 1985, pp. 2013–2021.

[*2*] McCartney, L. N., ''Mechanics of Matrix Cracking in Brittle-Matrix Fiber-reinforced Composites,'' *Proceedings,* Royal Society of London, Vol. 409A, 1987, pp. 329–350.

[*3*] McMeeking, R. M. and Evans, A. G., ''Matrix Fatigue Cracking in Fiber Composites,'' *Mechanics of Materials,* Vol. 9, 1990, pp. 217–227.

[*4*] Ghosn, L., Kantzos, P., and Telesman, J., ''Modeling of Crack Bridging in a Unidirectional Metal Matrix Composite,'' *International Journal of Fracture,* Vol. 54, 1992, pp. 345–357.

[*5*] Davidson, D. L., ''The Micromechanics of Fatigue Crack Growth at 25°C in Ti-6Al-4V Reinforced with SCS-6 Fibers,'' *Metallurgical Transactions,* Vol. 23A, 1992, pp. 865–879.

[*6*] Telesman, J., Ghosn, L. J., and Kantzos, P., ''Methodology for Prediction of Fiber Bridging Effects in Composites,'' *Journal of Composites Technology & Research,* Vol. 15, 1993, pp. 234–241.

[*7*] Walls, D. P., Bao, G., and Zok, W., ''Mode I Fatigue Cracking in a Fiber Reinforced Metal Matrix Composite,'' *Acta Metallurgica Materials,* Vol. 41, 1993, pp. 2061–2071.

[*8*] Chiang, Y.-C., Wang, A. S., and Chou, T.-W., ''On Matrix Cracking in Fiber Reinforced Ceramics,'' *Journal of the Mechanics and Physics of Solids,* Vol. 41, 1993, pp. 1137–1154.

[*9*] Telesman, J., Kantzos, P., and Brewer, D., ''In Situ Fatigue Loading Stage Inside Scanning Electron Microscope,'' *Lewis Structures Technology–1988,* NASA CP 3003, NASA Lewis Research Center, Cleveland, OH, Vol. 3, May 1988, pp. 161–172.

[*10*] Hartman, G. A. and Nicholas, T., ''An Enhanced Laser Interferometer for Precise Displacement Measurements. Experimental Techniques,'' *Society of Experimental Mechanics,* Vol. 11, No. 2, 1987, pp. 24–26.

[*11*] Jira, J. R. and Kleismet, R., ''Crack Opening Displacement Measurements in Unidirectional Metal Matrix Composites,'' to be submitted for publication.

[*12*] John, R., Kaldon, S. G., and Ashbaugh, N. E., ''Applicability of Fiber Bridging Models to Describe Crack Growth in Unidirectional Titanium Matrix Composites,'' *Titanium Metal Matrix Composites II, Proceedings,* Titanium Metal Matrix Composites II Workshop, June, 1993, WL-TR-93-4105, Materials Directorate, Wright-Patterson AFB, OH, pp. 270–290.

[*13*] Bakuckas, J. G., Jr. and Johnson, W. S., ''Application of Fiber Bridging Models to Fatigue Crack Growth in Unidirectional Titanium Matrix Composites,'' *Journal of Composites Technology & Research,* Vol. 15, 1993, pp. 242–255.

[*14*] Chan, K. S., ''Effects of Interface Degradation on Fiber Bridging of Composite Fatigue Cracks,'' *Acta Metallurgica Materials,* Vol. 41, 1993, pp. 761–768.

[*15*] Thouless, M. D., Sbaizro, O., Sigl, L. S., and Evans, A. G., ''Effect of Interface Mechanical Properties on Pullout in a SiC-Fiber-Reinforced Lithium Aluminum Silicate Glass-Ceramic,'' *Journal,* American Ceramics Society, Vol. 72, 1989, pp. 525–532.

[*16*] Marshall, D. B. and Oliver, W. C., ''Measurement of Interfacial Mechanical Properties in Fiber-Reinforced Ceramic Composites,'' *Journal,* American Ceramic Society, Vol. 70, 1987, pp. 542–548.

[17] Mackin, T. J., Warren, P. D., and Evans, A. G., "Effects of Fiber Roughness on Interface Sliding in Composites," *Acta Metallurgica Materials,* Vol. 40, 1992, pp. 1251–1257.

[18] Eldridge, J., "Fiber Push-Out Testing of Intermetallic Matrix Composites at Elevated Temperatures," *Proceedings,* Materials Research Society Symposium, Vol. 273, D. Miracle, J. Graves, and D. Anton, Eds., Pittsburgh, 1992, pp. 325–330.

[19] Marshall, D. B., Shaw, M. C., Morris, W. L., and Graves, J., "Interfacial Properties and Residual Stresses in Titanium and Titanium Aluminide Matrix Composites," *Proceedings,* Workshop for Titanium Matrix Composites, P. R. Smith and W. C. Revelos, Eds., Wright-Patterson AFB, OH, 1991, pp. 329–347.

[20] Watson, M. C. and Clyne, T. W., "The Use of Single Fiber Pushout Testing to Explore Interfacial Mechanics in SiC Monofilament-Reinforced Ti—II. Application of the Tests to Composite Material," *Acta Metallurgica Materials,* Vol. 40, 1992, pp. 141–148.

[21] Kantzos, P., Ghosn, L., and Telesman, J., "The Effect of Degradation of the Interface and Fiber Properties on Crack Bridging," *HITEMP Review 1992,* Vol. II, NASA Lewis Research Center, Cleveland, OH, 1992, pp. 32-1–32-14.

[22] Warren, P. D., Mackin, T. J., and Evans, A. G., "Design, Analysis and Application of An Improved Push-Through Test for the Measurement of Interface Properties in Composites," *Acta Metallurgica Materials,* Vol. 40, 1992, pp. 1243–1249.

[23] Ghonem, H., Zheng, D., Tamin, M. N., and Zhang, T., "Investigation of High Temperature Crack Bridging Mechanisms in SM1240/TIMETAL-21S," Report MML-94-2, Mechanics of Materials Laboratory, Department of Mechanical Engineering, University of Rhode Island, April 1994.

[24] Cox, B. N. and Marshall, D. B., "The Determination of Crack Bridging Forces," *International Journal of Fatigue,* Vol. 49, 1991, pp. 159–176.

[25] Clyne, T. W. and Withers, P. J., *An Introduction to Metal Matrix Composites,* Cambridge University Press, London, 1993.

[26] Schulte, K. and Minoshima, K., "Mechanisms of Fracture and Failure in Metal Matrix Composites," *Metal Matrix Composites—Processing, Microstructure and Properties,* N. Hansen et al., Eds., RISO National Laboratory, Roskilde, Denmark, 1991, pp. 129–147.

[27] Ghonem, H., Wen, Y., Zheng, D., Thompson, M., and Linsey, G., "Effects of Temperature and Frequency on Fatigue Crack Growth in Ti-β21S Monolithic Laminate," *Materials Science and Engineering,* Vol. 161, 1993, pp. 45–53.

[28] Warwick, C. M. and Smith, J. E., "Interfacial Reactions in Titanium Alloys Reinforced with C/TiB_2-Coated SiC Monofilament," *Metal Matrix Composites—Processing, Microstructure and Properties,* N. Hansen et al., Eds., RISO National Laboratory, Roskilde, Denmark, 1991, pp. 735–740.

[29] Osborne, D. and Ghonem, H., "On the Interfacial Reaction Kinetics of TIMETAL-21S Reinforced with SM1240 and SCS-6 Continuous Fibers," Report MML-95-1, Mechanics of Materials Laboratory, Department of Mechanical Engineering, University of Rhode Island, March 1995.

[30] McCartney, L. N., "General Solution of a Certain Mixed Boundary Value Crack Problem," *International Journal of Engineering Science,* Vol. 21, 1983, pp. 131–142.

[31] Sneddon, I. N. and Lowengrub, M., *Crack Problems in the Classical Theory of Elasticity,* Wiley, New York, 1969.

[32] Marti, J. T., "An Algorithm for Computing Minimum Norm Solutions of Fredholm Integral Equations of the First Kind," *SIAM Journal on Numerical Analysis,* Vol. 15, 1978, pp. 1071–1076.

[33] Baker, C. T. H., *The Numerical Treatment of Integral Equations,* Oxford University Press, Oxford, UK, 1977.

[34] Le petitcorps, Y., Lahaye, M., Pailler, R., and Naslain, R., "Modern Boron and SiC CVD Filaments: A Comparative Study," *Composites Science and Technology,* Vol. 32, 1988, pp. 31–55.

[35] Budiansky, B., Hutchinson, J. W., and Evans, A. G., "Matrix Fracture in Fiber-Reinforced Ceramics," *Journal of the Mechanics and Physics of Solids,* Vol. 34, 1986, pp. 167–189.

[36] Zheng, D. and Ghonem, H., "A Model for High Temperature Fatigue Crack Bridging in Metal Matrix Composites," Report MML-95-2, Mechanics of Materials Laboratory, Department of Mechanical Engineering, University of Rhode Island, July 1995.

[37] Ghonem, H., Wen, Y., and Zheng, D., "An Interactive Simulation Technique to Determine the Internal Stress States in Fiber Reinforced Metal Matrix Composites," *Materials Science and Engineering,* Vol. A177, 1994, pp. 125–134.

[38] Zhang, T. and Ghonem, H., "Time-Dependent Fatigue Crack Growth in Titanium Metal Matrix

Composites,'' *International Journal of Fatigue and Fracture of Engineering Materials,* Vol. 18, No. 11, 1995, pp. 1249–1262.

[*39*] Buesking, K. W., Young, J. L., and Reynolds, G. H., ''Status of Advanced Filament Development for Titanium Alloy and Titanium Intermetallic Matrix Composites,'' *Proceedings,* Workshop for Titanium Matrix Composites, P. R. Smith and W. C. Revelos, Eds., Wright-Patterson AFB, OH, 1991, pp. 25–64.

[*40*] Blackman, L. C. F., *Modern Aspects of Graphite Technology,* Academic Press Inc., London, 1970.

[*41*] Tamin, M. N., Zheng, D., and Ghonem, H., ''The Determination of the Stress State Variation during High Temperature Crack Bridging Process in MMCs,'' *Damage of Metal Matrix Composites,* 3rd Symposium on Advances in Fatigue, Lifetime Predictive Techniques, Montreal, Canada, May 1994.

Diana C. Cardona,[1] *Craig Barney,*[2] *and Paul Bowen*[2]

Modeling and Prediction of Crack Arrest in Fiber-Reinforced Composites

REFERENCE: Cardona, D. C., Barney, C., and Bowen, P., **"Modeling and Prediction of Crack Arrest in Fiber-Reinforced Composites,"** *Life Prediction Methodology for Titanium Matrix Composites, ASTM STP 1253,* W. S. Johnson, J. M. Larsen, and B. N. Cox, Eds., American Society for Testing and Materials, 1996, pp. 164–181.

ABSTRACT: This paper addresses the modeling of crack growth resistance of titanium-based metal matrix composites under cyclic loading by considering the conditions necessary to promote the crack arrest of dominant cracks that are often observed in these materials. The presence of dominant cracks facilitates the application of fracture mechanics parameters such as the stress intensity factor, K, and the stress intensity factor range, ΔK, to describe crack growth resistance under monotonic and cyclic loading, respectively. Using such fracture mechanics parameters, weight function methods have been used to model, assess, and predict the onset of crack arrest for cracks growing in both uniform and nonuniform stress fields. The model requires the experimental determination of crack arrest conditions for a particular testpiece geometry, for a given loading condition, and where the specific number of intact bridging fibers is also determined. It can then be used to predict the influence of crack shape and crack size for a variety of testpiece geometries and loading conditions. It is clear that corner cracks and semi-elliptical cracks are potentially more damaging than equivalent through-thickness cracks. Particular attention has also been taken to model the behavior of cracks growing in selectively reinforced composites, and the importance of the thickness of any unreinforced cladding is considered.

KEYWORDS: titanium matrix composites, titanium, fatigue crack propagation, life prediction, crack arrest determination, modeling, fatigue (materials), cracks, titanium alloys, maximum safe stress, monolithic cladding, fiber-reinforced composites

The aim of lifing is to identify the safe life of a component under given design stresses and after which it must be removed to maintain the integrity of a structure or an engine. Fiber-reinforced composites with outstanding strength and stiffness are becoming available increasingly. These are tailored for envisaged applications by judicious selection of matrix materials and the use of very high-strength monofilament fibers. Materials such as titanium metal matrix composites (Ti MMCs) are expected to require different damage tolerance concepts and different empirical engineering approaches compared with traditional monolithic alloys. From present understanding, it appears that the lifing of Ti MMC components is a complex issue and a number of issues need to be solved before the material can be used and certified "safe" for component applications.

Recent work has shown that under cyclic loading if no fiber fracture occurs in Ti MMCs, then cracks arrest eventually (defined where the crack increment per cycle is less than 10^{-8} mm/cycle) [*1*]. The aim of this paper is to address the conditions required for crack arrest

[1]Composites technologist, Rolls-Royce plc, Derby DE24 8BJ, UK.

[2]Research student and professor of Mechanical Metallurgy, respectively, School of Metallurgy and Materials/IRC in Materials for High Performance Applications, University of Birmingham, Edgbaston, Birmingham B15 2TT, UK.

and to present a methodology to predict the behavior and onset of crack arrest for cracks of different sizes and shapes, when growing in both selectively and fully reinforced plates, under cyclic loading.

Lifing of Conventional Materials

Fracture mechanics can quantify the effects of cracks under carefully controlled conditions, and it is often used to predict instability during component service life. The stress intensity factor (K) is used as one primary characterization parameter, and it describes the magnitude of the elastic crack tip stress field.

A fracture mechanics based approach has been used extensively to analyze fatigue crack growth in conventional materials. Its main advantage is that data on small testpieces can then be applied to structural components rather well, regardless of geometry or loading conditions, provided that the cracks are subjected to the same driving force, ΔK, where ΔK is the stress intensity factor range. Fatigue damage is a complex process that may depend both on intrinsic variables such as material properties and microstructure, and on extrinsic factors such as applied stress, temperature, and the fabrication process. Mode I crack growth in monolithic alloys can be characterized relatively easily using fracture mechanics parameters, at least for a range of variables where growth can be represented by the Paris-Erdogan relationship. This relates the crack increment per cycle, da/dN, and the stress intensity factor range, ΔK, by

$$\frac{da}{dN} = C(\Delta K)^m \tag{1}$$

where C and m are empirical material constants. The applied stress intensity range (ΔK_{nom}) can be related to applied stress range ($\Delta\sigma_{\text{nom}}$) by

$$\Delta K_{\text{nom}} = \Delta\sigma_{\text{nom}} \sqrt{\pi a} \tag{2}$$

for a central crack of length $2a$, strictly derived under biaxial tension in an infinite sheet (but also accurate for uniaxial tension). Analogous expressions for finite testpiece size, different crack geometry, and loading conditions are available [*2,3*].

Thus, for monolithic materials, the crack growth rate per cycle can be linked to the applied stress range, and, after integration of equations such as Eq 1, nondestructive examination intervals can be established to prevent growth of the crack to a critical (unstable) size.

Understanding and quantifying the severity of cracks are important parts in the development of life prediction methodologies. Development of stress intensity factors for through-thickness, embedded, surface, and corner cracks in structural components using analytical, numerical (finite element analysis), and analytical methods has continued for the past few decades, but most analyses have concentrated on growth in uniform stress fields. The standard analytical solutions are inadequate for treating complexities at geometrical discontinuities, cracks in self-equilibrating residual/thermal stress fields, and cracks in emerging composite materials where various toughening mechanisms (such as fiber bridging) may reduce the effective stress intensity factor range. (Such comments are equally true when considering effects of crack closure on fatigue life, but the influence of crack closure is generally most significant at near threshold crack growth rates. In this work, threshold is defined as a growth rate of $1c^{-8}$ mm/cycle. This near threshold region is not represented by the Paris-Erdogan relationship and to date it is rarely taken into account.) Usually, such complex problems require more than the use of standard analysis methods. Although numerical methods can be used, they require multiple crack length steps, and this leads to high costs in both manpower and computational resources.

Weight function methods are a versatile way to calculate stress intensity factors for a variety of (complex) loading conditions. They allow the evaluation of stress intensity factors by the simple integration of the weight function, $m(a,x)$, and the stress distribution, $\sigma(x)$

$$K = \int_0^a \sigma(x)\, m(x,a)\, dx \tag{3}$$

where a is the crack length and x is the position in the direction of crack growth (usually taken as through the testpiece width) [*4*].

The strength of this method is the separation in Eq 3 of the crack line stress (or stress in the uncracked body), $\sigma(x)$, and the cracked geometry that is contained completely in the weight function, $m(a,x)$.

Provided the loading mode is the same (for example, Mode I loading) superposition of the stress intensity factors due to different stress profiles can be used, by an algebraic summation of stress intensity factors. This is very useful, as complicated loading conditions can be readily characterized by the addition of several simple solutions and hence this method can achieve computational efficiency without compromising the accuracy of the solutions. Based on this approach, equations are now available for rapid calculation of stress intensity factors in the presence of nonuniform stress distributions, and for different crack shapes both for monolithic materials and for composite materials (provided $\sigma(x)$ profiles can be obtained). In many cases, the validity of the approach has been confirmed by independent finite element analysis [*5–8*].

Crack Growth in Fiber-Reinforced Titanium MMCs in the Presence of an Unbridged Defect

In Ti MMC materials, fatigue loading often results in the propagation of a dominant matrix crack. Fiber failure does not take place ahead of the advancing but stable matrix fatigue crack, but, instead, local debonding of the interface of those fibers ahead of the crack tip is observed. A typical fatigue crack growing in a Ti MMC material is shown in Fig. 1. Provided the interfacial bonding between the matrix and the fibers is not too strong (and

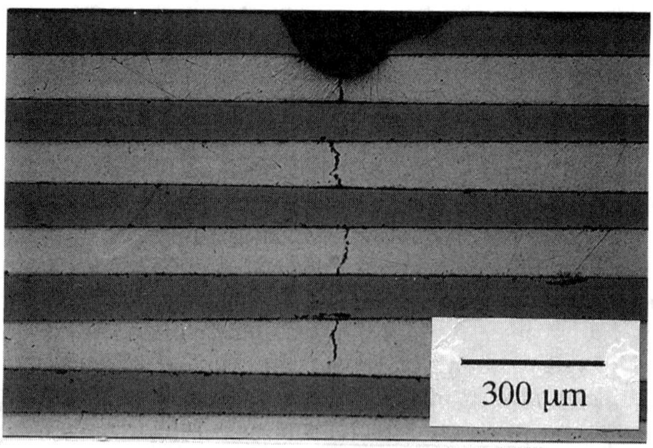

FIG. 1—*Metallographic section through a Mode I crack after crack arrest.*

allows fiber debonding ahead of the advancing matrix crack), stable crack growth can take place, with intact fibers bridging behind the crack tip and in its wake (as shown in Fig. 1). This crack bridging zone forms an important source of toughening by crack tip shielding. This operates by creating a zone where the crack tip is protected from the nominal stress intensity factor (defined by the product of the applied stress and the total crack length). Fibers "tie together" the crack faces behind the crack tip and these bridges exert compressive stresses on the crack faces (to restrict crack opening displacements) until they fracture and pull out. Therefore, in the presence of crack bridging, crack extension may be self-arresting. This will depend critically upon the number of intact fibers bridging the crack surfaces in the wake of the crack.

Dominant cracks that result from fatigue loading in Ti MMC makes the use of fracture mechanics parameters for crack characterization much more self-evident. Studies have considered the quantitative use of stress intensity based parameters for the assessment of crack bridging in unidirectionally reinforced composite materials [*9–17*] The fracture behavior of Ti MMCs has been analyzed usually by replacing discrete bridging fiber forces by continuum crack closure tractions. These models appear to be suitable to model crack bridging effects, and the application of any particular model depends only upon the type and accuracy of information available prior to modeling. Provided that some reference experimental crack growth resistance data on a particular composite are available, then the discrete fiber (plate) bridging model [*12,18,19*] has some advantages. One application of the model can be seen in Fig. 2, where a comparison between experimental data and predictions are shown to be in close agreement for a crack growth rate regime where the crack growth resistance varies markedly with crack extension (increase in nominal stress intensity factor range). The experimental data used to obtain the "effective stress partitioning" between the

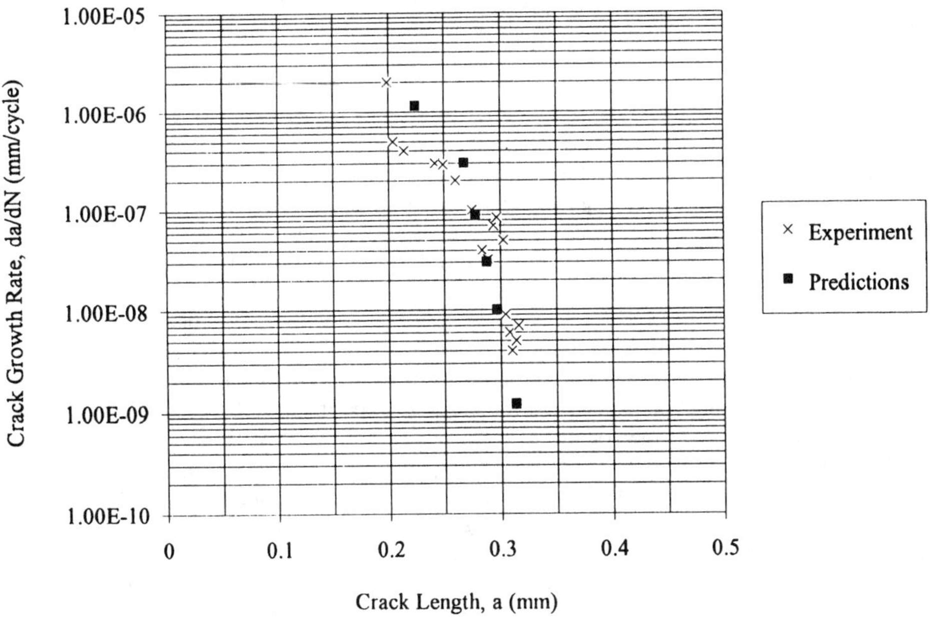

FIG. 2—*Crack growth resistance curves:* da/dN *versus crack length for* $\Delta P = 67N$, *initial* $\Delta K_{app} = 7.5\sqrt{MPa\ m}$, R $= 0.5$, *and* $a_0/W = 0.05$. *Comparison of experimental and theoretical predictions.*

fibers and the matrix that are a necessary input to the model were obtained from the crack growth behavior of the material at a higher crack growth rate and with reference also to crack growth rates in monolithic material (processed in the same way as the composite). Weight function methods are used to model the effect of discrete fibers (plates) and to derive suitable "effective stress partitioning," $\sigma(x)$, distributions. Once this effective stress partitioning has been determined, the effective crack growth resistance curve for the material can then be evaluated for a wide variety of loading conditions and crack shapes, crack sizes, and testpiece geometries. An explanation for the occurrence of crack arrest, of course, can also be quantified in terms of when the predicted effective stress intensity range at the crack tip is reduced below the long crack threshold, ΔK_{th}, for the matrix material. The model can be used for any material provided an experimental crack growth resistance curve is available for that material. This allows the model to generalize the stress partitioning between fibers and matrix and to include local effects such as residual stresses and interfacial strength, but experimental crack growth resistance curves for both the composite and monolithic matrix material must be available at the temperature of interest.

It is clear that the relationship between the nominal applied stress intensity range and the crack growth resistance of the composite is dependent upon many factors including the testpiece geometry and fiber distribution. Factors such as the interfacial strength, the number of fibers bridging the crack in the crack wake and initial unbridged crack size and shape may also be critical [*12,18–21*], and this indicates the potential complexities inherent to models that attempt to predict crack growth resistance.

Crack Arrest in Fiber-Reinforced Composites in the Presence of an Unbridged Defect

In the presence of fiber bridging, crack extension in Ti MMCs may become self-arresting [*1,20,21*]. This depends critically upon the number of intact fibers bridging the crack surfaces. A methodology for evaluating crack arrest experimentally has been developed, and it is reported in detail elsewhere [*22,23*]. To establish crack arrest conditions, a number of similar testpieces of a single composite system are required. All must be of similar dimensions, and they must all contain an unbridged defect of closely similar shape and size. The easiest crack shape to utilize is a through-thickness defect and, for most tests, an a_0/W = 0.25 value (where a_0 is the through thickness depth of the unbridged defect, and W is the testpiece width) has been used [*21*]. For a single mean stress, an initial stress intensity factor range, $\Delta K_{initial}$, is chosen. If the crack grows to failure, then the initial stress intensity factor range should be reduced on the next testpiece, by approximately 10%, and the test is repeated until crack arrest (defined by $da/dN \leq 10^{-8}$ mm/cycle) is established. If the initial stress intensity factor range promotes crack arrest, then for the next testpiece, the $\Delta K_{initial}$ is increased by approximately 10% and the sequence is repeated. In this manner, a crack arrest failure transition can be established. In principle from experimental observations, the modeling must address carefully the conditions under which the first row of fibers to be breached by the matrix fatigue crack will not fail [*20,23*]. Previous work has suggested to a good engineering approximation that the value of the maximum nominal stress intensity factor, K_{max}, just after the first row of fibers can be used to rank the propensity of the composite to fiber failure and hence eventually to crack arrest. It should be noted, that the crack increment per cycle, da/dN, chosen to give crack arrest of $da/dN \leq 10^{-8}$ mm/cycle is broadly similar to that used to define ΔK_{th} in monolithic materials. Clearly, in composite materials, its physical basis is to reduce the effective stress intensity factor range, ΔK_{eff}, in the matrix regions (the continuous phase) below ΔK_{th} for the monolithic matrix material alone. This approach is considered in detail, and its experimental basis is reviewed. It also provides a convenient method with which to compare the crack growth resistance of a wide range of composites and microstructural conditions. Such comparisons are not easy to produce in single crack growth resistance tests.

Experimental

The material used for the majority of the tests considered here was in the form of a consolidated six-ply plate of 1-mm thickness of a titanium Ti-6Al-4V alloy matrix unidirectionally reinforced with continuous Sigma (SM1240) silicon carbide fibers. The fibers were coated with dual carbon and titanium boride layers and were an approximate diameter of 100 μm and the fiber volume fraction was approximately 31%. Specimens of 4.5-mm width were machined from the plate by electro discharge machining (EDM).

To date, tests on this composite have been carried out on through-thickness, single-edge-notched specimens (such as shown in Fig. 1) and corner-cracked testpieces. All tests have been carried out using three-point-bending conditions with an overall span to width ratio (S/W) of 13.3. Through-thickness notches of 1100 μm depth, were cut using a 180-μm-thickness diamond blade to give an initial unbridged crack length to width ratio, a_0/W, of $\approx$0.25. For corner-cracked specimens, the notch surface length, c, was 450 μm, and the notch depth was 450 μm. This gave a surface crack length (c) to depth (a_0) ratio of both $c/a_0 \approx 1$ and $a_0/W \approx 0.1$. Crack growth was monitored using a d-c potential difference technique and crack growth rates were calculated using the seven-point secant method [*21,23,24*].

Tests were carried out using ESH servohydraulic testing machines fitted with 1 and 10 kN load cells, at frequencies of 0.5 and 10 Hz, respectively. Tests were carried out at room temperature. A commercially available PAC LOCAN acoustic emission analyzer was used to characterize fiber fracture in situ during some of the tests. The effect of load ratio, R (where $R = P_{min}/P_{max}$ and P_{min} and P_{max} are minimum and maximum applied load over the fatigue cycle, respectively), on crack growth rates, da/dN, has been considered. An R ratio of 0.1 (at a frequency of 0.5 Hz), and R ratios of 0.5 and 0.7 (at a frequency of 10 Hz) were studied. A constant load range, ΔP (which can also be considered to be a constant surface stress range, $\Delta\sigma$), was used in each test. This allows the nominal applied stress intensity range, ΔK_{app}, to increase with increase in fatigue crack length. Crack paths and the incidence of surface fiber failure were studied optically on polished metallographic sections [*23*].

Results

Experimental Characterization of Crack Arrest Conditions

The experimental observations for a range of load ratios on the crack arrest conditions of Ti-6Al-4V/SM1240 six-ply composites are shown in Figs. 3, 4, and 5 and are summarized in Table 1 [*23*]. (Note that the crack arrest condition for $R = 0.1$ tests has been to be a growth rate $da/dN \leq 10^{-7}$ mm/cycle, because these early tests were performed at a cyclic frequency of 0.5 Hz). Fiber fracture in situ was monitored only for tests performed at $R = 0.7$, and these events are shown in Fig. 5. In general, from the form of the crack growth resistance curves, Figs. 3, 4, and 5, and from the acoustic emission events, Fig. 5, it can be deduced that crack arrest occurs only if relatively few fibers or no fibers fail during crack growth. For R ratios of 0.1 and 0.5 (Table 1), the initial level of K_{max} (applied to the unbridged defect) appears to be of use in predicting the onset of crack arrest. Such K_{max} values are conservative for tests performed under higher $R = 0.7$ conditions, but they would still appear to have some merit. To include effects of "bridging scale lengths" and based on a sound physical description of the process, the value of "K_{max} passed the first row of bridging fibers" has been suggested. Such nominal K_{max} values must now be treated with

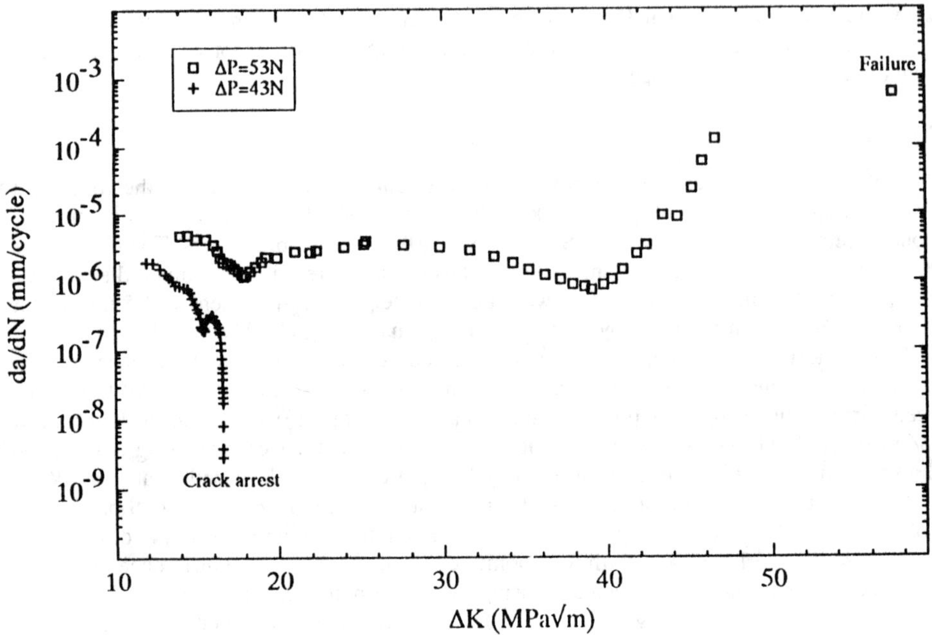

FIG. 3—ΔK *versus* da/dN *for* R = *0.5*, *υ* = *10 Hz, and for initial unbridged through-thickness defects.*

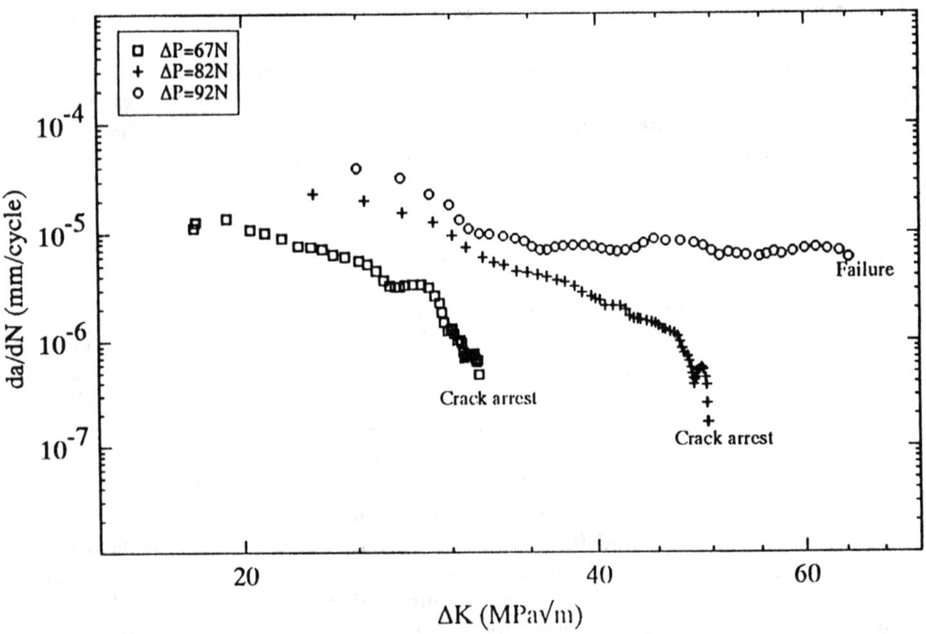

FIG. 4—ΔK *versus* da/dN *for* R = *0.1*, *υ* = *0.5 Hz, and for initial unbridged through-thickness defects.*

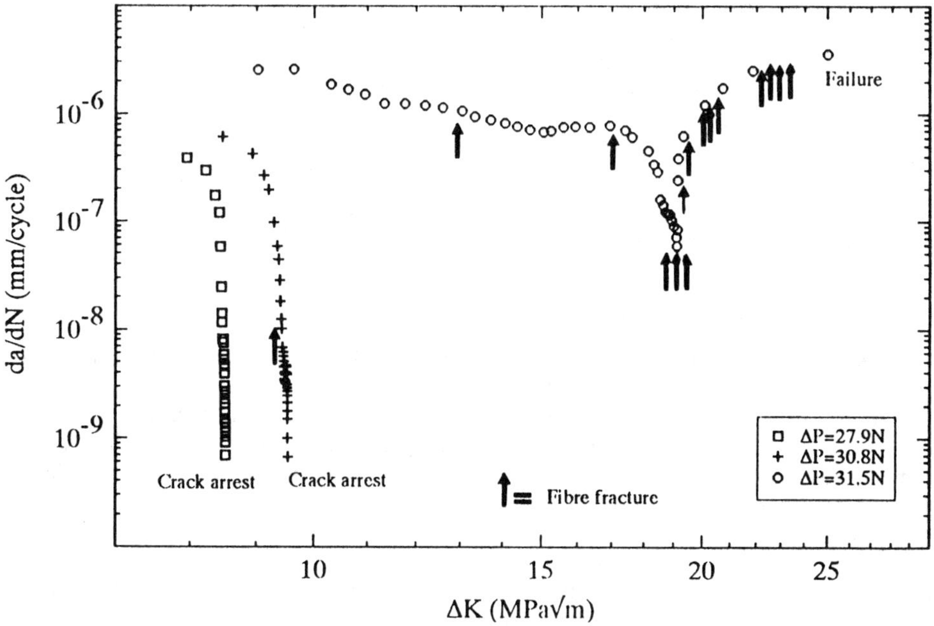

FIG. 5—ΔK *versus* da/dN *for* R = *0.7,* υ = *10 Hz, and for initial unbridged through-thickness defects.*

TABLE 1—*Summary of the effect of mean stress on the outcome of the test and final crack bridged length* [23].

Stress Ratio	Initial ΔK, MPa$\sqrt{m}$	Initial K_{max}, MPa$\sqrt{m}$	K_{max} after First Row of Fibers, MPa$\sqrt{m}$	Test Result	Final Bridged Crack Length, mm
0.1	17.8	19.8	21.7	crack arrest	0.94
0.1	22.3	24.8	27.5	crack arrest	1.23
0.1	25.0	27.8	30.2	catastrophic failure	NA[a]
0.5	12.1	24.2	25.8	crack arrest	0.57
0.5	13.8	27.6	29.8	catastrophic failure	NA
0.7	8.0	26.5	29.0	crack arrest	0.12
0.7	8.5	28.3	31.0	crack arrest	0.21
0.7	9.0	30.1	33.0	catastrophic failure	NA

[a]NA = not available.

caution, but they do emphasize the fact that fibers must be breached before they can fail. A complete description of the condition for fiber failure would require the measurement of K_{max} values or crack opening displacements or both under load, and ongoing studies are considering such approaches. It must also be noted that the crack arrest condition will derive directly from matrix ΔK_{eff} values being reduced below ΔK_{th} values in general, and therefore a nominal K_{max} value should strictly be considered to describe the onset of fiber fracture alone. This K_{max} value will also be specific to the number of bridging fibers interacting with the crack front, and, since in the experiments performed earlier uniform through-thickness crack growth is observed, it is reasonable to assume that six fibers (over the testpiece thickness of 1 mm) bridge the crack front "simultaneously." For nonuniform crack growth and for part-through-thickness defects (such as corner cracks) of reduced crack front size, fewer fibers may be breached for any given crack growth increment. Under such conditions, the requirements for crack arrest are likely to be more severe, and this is shown in Fig. 6, where to date crack arrest has been defined only for an initial K_{max} value of 13.8 MPa$\sqrt{m}$ (an initial K_{max} value of 20.6 MPa$\sqrt{m}$ produces catastrophic failure). Similar observations have also been made for semi-elliptical cracks [*25*].

Prediction of Crack Arrest Conditions

Based on the value of K_{max} beyond the first row of fibers, which gives crack arrest at a low R ratio (R = 0.1) for the Ti-6Al-4V/SM1240 composites, predictions of crack arrest conditions are now given in Fig. 7 for a range of crack shapes, crack sizes, and loading conditions (for a crack depth increment of 140 μm in all cases). For simplicity, only semicircular (part-through thickness) and quarter-circular (corner) cracks were considered. These predictions are presented here in terms of the nominal or applied maximum stress at the surface of the component or testpiece σ_{max} per bridging fiber divided by the crack front length (in millimeters). Based on the most probable number of fibers bridging over the estimated crack front (Table 2) actual maximum nominal stress levels to give crack arrest are given for this range of crack shapes in Fig. 8.

In essence, the model employed to date predicts crack arrest conditions only as a function of the maximum nominal stress intensity factor, K_{max}, passed the first row of bridging fibers,

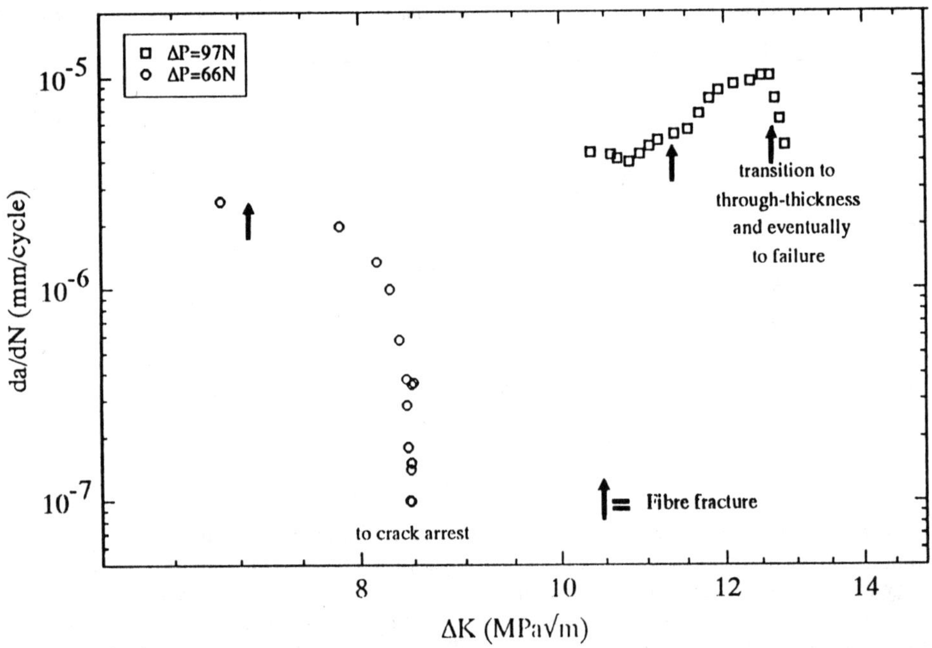

FIG. 6—ΔK *versus* da/dN *for* R = *0.5, v = 10 Hz, and for initial unbridged corner (part-through-thickness) defects.*

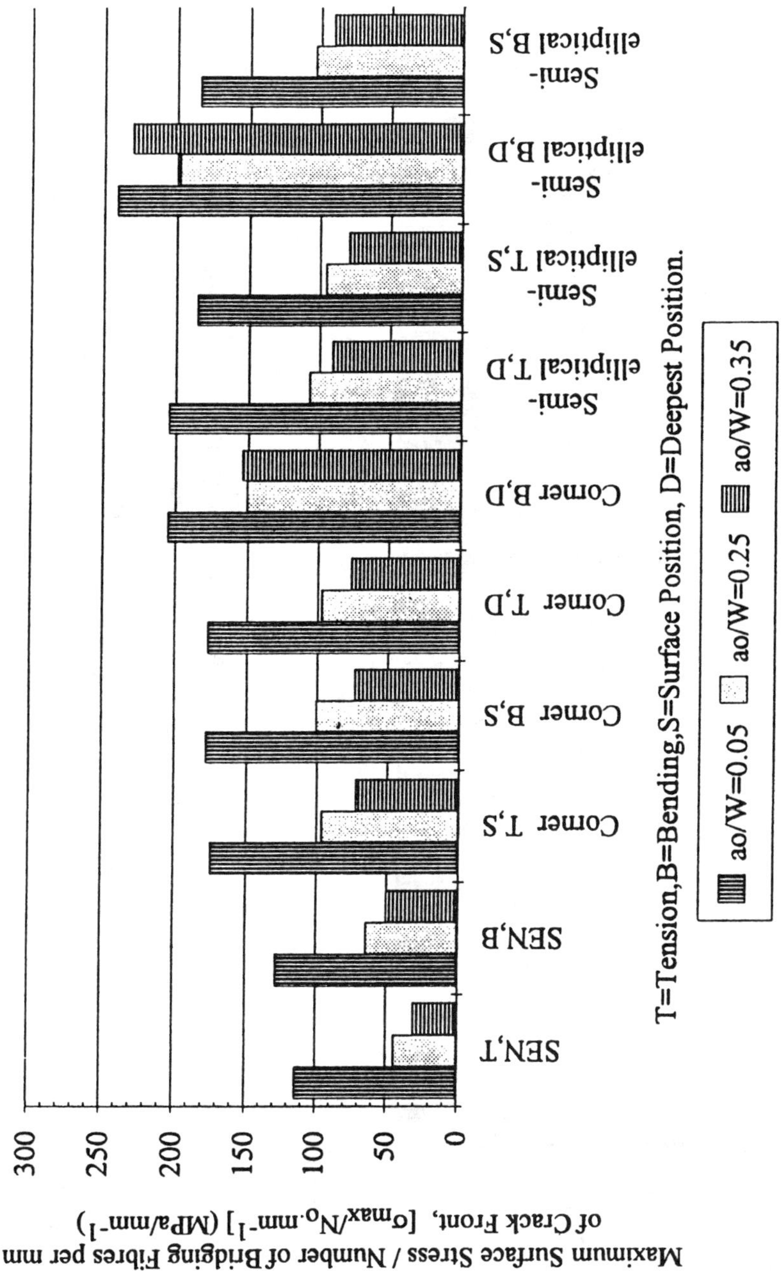

FIG. 7—*Normalized maximum surface stress predicted to allow crack arrest for a range of crack shapes and crack sizes in bending and in tension.*

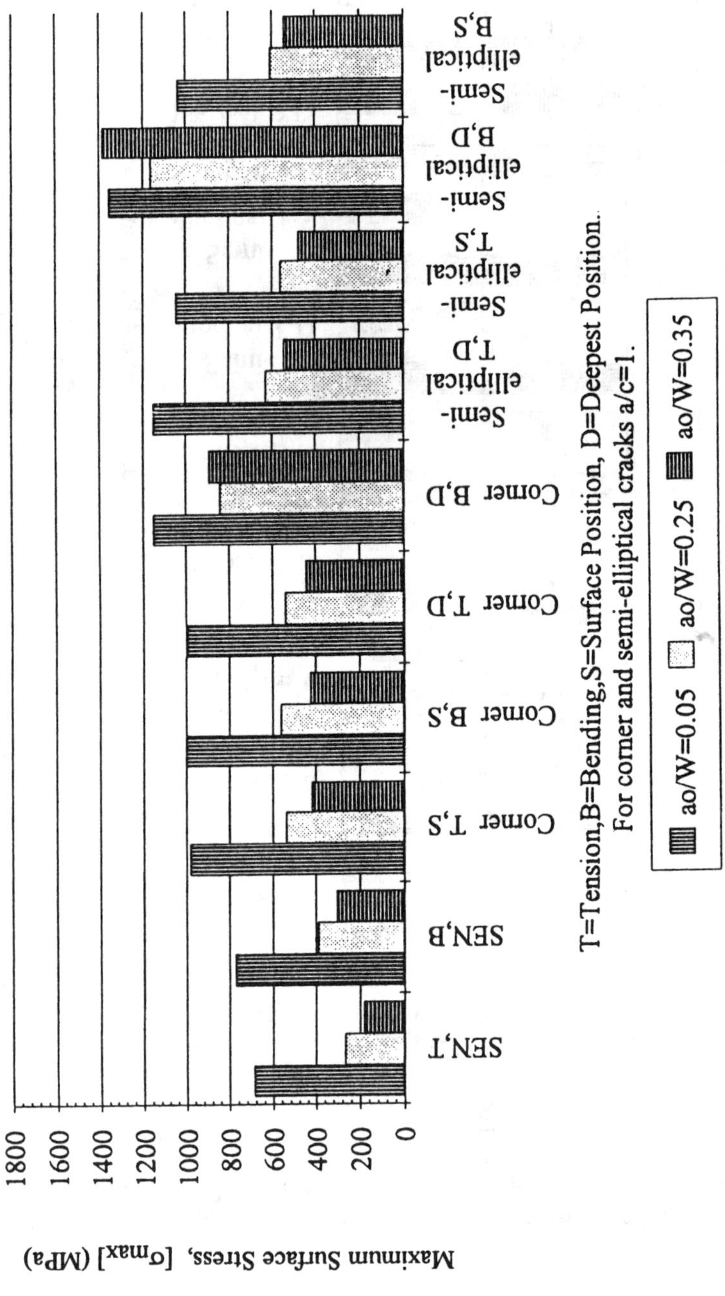

FIG. 8—*Maximum surface stress predicted to allow crack arrest for a range of crack shapes and crack sizes in bending and in tension. See Table 2 for number of intact fibers bridging over the crack front.*

TABLE 2—*Number of fibers interacting with the crack front. Here* $a_0/W = 0.05$.

Crack Shape, $a_0 = 200\ \mu m$ and $(a/c = 1)$	No Cladding, Initial Crack Size			Cladding Size = Initial Crack Size	
	a_0	$5a_0$	$7a_0$	a_0	$5a_0$
Through-thickness SEN (crack front length = 1 mm)	6	6	6	6	6
Corner crack (crack front length = $\pi(a_0 + 140\ \mu m)/2$	3	10	14	1	2
Semi-elliptical crack (crack front length = $\pi(a_0 + 140\ \mu m)$	6	21	29	2	4

and the absolute number of bridging fibers "over the crack front length." This methodology must be handled with care as the value of K_{max} passed the first row of bridging fibers, which leads to crack arrest varies for different composites. Examples are given in Table 3 [*22*]. (It should be noted that the crack arrest conditions for composites other than Ti-6Al-4V/SM1240 should be treated with caution because they have not been validated as thoroughly).

The model can be extended to predict the behavior of cracks growing in nonuniform stress distributions. This is required for testpieces that are selectively reinforced, that is, those that possess regions of composite material and regions of monolithic material. Again, it is essential to know the distribution of stress and the maximum value of stress intensity factor up to which crack arrest would be expected for the composite system. The value of maximum stress intensity factor at the crack tip just after the first row of bridging fibers can be evaluated using weight functions to account for the nonuniformity of the stress profiles. Such factors are inherent to the predictions made in Fig. 9 for a range of composites containing given depths of monolithic cladding. In Fig. 9, the influence of the composite system (see Table 3) and cladding thickness can be clearly seen.

TABLE 3—*Comparison of experimental observations of crack growth behavior related to* K_{max} *after the first row of breached fibers* [22].

Material	K_{max} after First Row of Fibers, MPa$\sqrt{m}$	Test Result
Ti-6Al-4V/SCS6	22.6	crack arrest
	28.8	catastrophic failure
β21S/SCS6	39.6	crack arrest
	40.6	catastrophic failure
Ti-15-3/SCS6	35.0	crack arrest
	42.2	catastrophic failure
Ti-6Al-4V/SCS6 (0/90 laminate)	15.0	crack arrest
	19.8	catastrophic failure

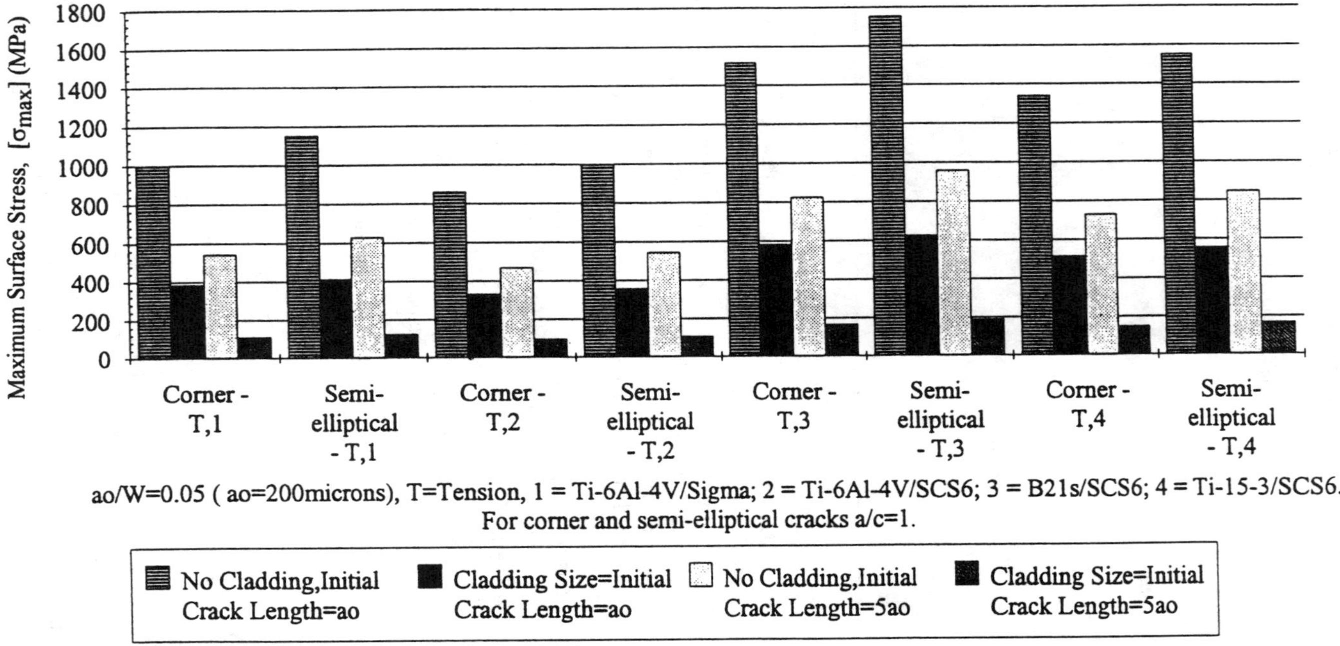

FIG. 9—*Maximum surface stress predicted to allow crack arrest for part-through-thickness cracks in tension. The influence of crack shape, crack size, composite system, and cladding thickness is demonstrated. See Table 2 for number of intact fibers bridging over the crack front.*

Discussion

Fatigue crack propagation in Ti MMCs has been observed to occur by the growth of dominant cracks that often follow a Mode I direction. Therefore, the use of linear elastic fracture mechanics parameters to characterize crack propagation seems to be appropriate. In addition, under some conditions, cyclic loading will result in crack arrest. These conditions can be obtained from experimental observations. If these can be quantified in terms of nominal stress intensity parameters, then crack arrest can be predicted for any crack shape and loading configuration. A simple approach that can predict the onset of crack arrest has been presented here. It is based on the local maximum stress intensity passed the first row of bridging fibers. The number of fibers interacting with the crack tip front (through-thickness) is also an essential factor and must be taken into account.

Experimental Characterization of Crack Arrest

Great care is required to ensure that the experimental conditions for crack arrest are established accurately. Moreover, some degree of uncertainty exists when determining the transition from catastrophic failure to crack arrest, and a nominal K_{max} value should be considered only as a useful engineering approximation with which to characterize crack arrest. Also, the crack does not always arrest after the first row of fibers has been breached (except for the case of $R = 0.7$, see Table 1), and, therefore, the precise micromechanism of crack arrest may involve several rows of fibers in general. Nevertheless, the parameter does appear to be of some use in providing an explanation for the behavior observed experimentally.

Crack Growth from an Initial Unbridged Through-Thickness Defect

The concept of "K_{max} criterion passed the first row" of fibers in predicting fiber failure and hence eventual crack arrest is generally supported by the results shown in Table 1. However, predictions of crack arrest at lower stress ratios of $R = 0.1$ and $R = 0.5$ are conservative with respect to predictions of crack arrest at the highest stress ratio of $R = 0.7$. Here, it is appropriate to note that although the threshold stress intensity range for matrix alloy will be expected to decrease modestly as the mean stress ratio increases, the degree of clampback (traction from intact fibers) required to reduce the effective stress intensity range to below this level will be reduced markedly. For example, at $R = 0.7$, initial ΔK_{app} values of ≤ 8.5 MPa$\sqrt{m}$ will give crack arrest, whereas at $R = 0.1$, initial ΔK_{app} values of ≤ 22 MPa$\sqrt{m}$ will promote crack arrest. The bridged crack length at arrest also reflects such factors (Table 1), because it increases as the stress ratio decreases. Thus, it may be suggested that larger repeated unloading cycles may lead to easier fiber fracture:

1. as a result of cumulative damage, such as interfacial wear that can plausibly reduce fiber strength, or
2. at lower stress ratios the crack is observed to grow further prior to crack arrest than at higher stress ratios. These longer cracks result in a reduction of testpiece stiffness that itself produces larger fiber displacements (under load control tests) and plausibly increases fiber fracture,
3. or both.

Crack Growth from Constrained Corner Defects

These tests have highlighted further the critical role of the number of intact bridging fibers acting over the crack front in promoting crack arrest for a given crack length increment. It is crucial to appreciate that a smaller number of bridging fibers should be involved for a given crack increment of a constrained crack. More important in this context is that because the initial unbridged defect (corner crack) size is usually small, then a given fixed crack increment (dictated by the requirement to grow passed a fiber) will increase the nominal stress intensity factor markedly compared to the effect of the same fixed crack increment on a larger initial defect size (such as a deep unbridged through-thickness crack) provided that the initial ΔK_{app} is similar [*23*]. Moreover, the crack arrest criterion suggested by Table 1 is specific to a given number of bridging fibers acting over the crack front because the growth of a through-thickness defect samples six fibers "simultaneously" (over a crack front of 1 mm through-thickness) with depth increase, and each fiber may "screen" each other fiber from the "full" applied ΔK_{max} level. This is in contrast to the growth of a part-through-thickness crack that conceptually could breach and stress each bridging fiber singly. This situation would be expected to promote fiber failure and hence less severe initial ΔK_{app} levels will be tolerated if crack arrest is to be achieved. Figure 6 supports this proposed mechanism and these initial observations have stimulated much current work to extend crack arrest predictions to a range of crack shapes and sizes. In the present example (Fig. 6), the behavior of the corner-cracked testpiece is consistent with a reduced number of fibers bridging the (reduced) crack front of between three and four (scaled to the crack arrest conditions pertaining to the through-thickness crack where six fibers over the crack front of 1 mm are assumed to bridge the crack simultaneously). This number of bridging fibers would appear to be reasonable in view of the initial unbridged defect size of $a_0 = c = 450$ μm (and since the number of fibers per millimeter is six), but note that the analysis method for the most probable number of fibers bridging the crack must be based on a comparison of equivalent length of crack front between the part-through-thickness crack and that of the reference (through-thickness) crack. In general, the actual volume fraction of reinforcement will be an important parameter. All predictions in this paper are based on a single volume fraction of fibers of 31% and, for the fiber size considered, this represents approximately six fibers per millimeter. (The most probable distance required to grow past a fiber has been taken as 140 μm throughout the analyses primarily for illustrative purposes. A lower limiting value to this distance would be 100 μm, and an upper limiting value would be 165 μm).

Prediction and Modeling of Crack Arrest

Figure 7 shows the predictions of the maximum allowable value of nominal or applied stress at the surface of the component or testpiece, σ_{max}, per number of bridging fiber over the crack front (in mm) that will result in crack arrest. The virtue of presenting predictions in terms of the maximum allowable nominal stress in this way is that the figure can be used to begin to define a "crack arrest map" for different situations. Specific examples are given in Fig. 8 and where the influence of crack shape and crack size on maximum allowable stresses that will give crack arrest can be seen clearly. In general, Fig. 8 is encouraging because at small crack sizes, high surface stresses can be tolerated. Note that the figure is based on the most probable number of fibers bridging the crack front, but it assumes that fibers are not isolated singly during nonuniform crack growth. Further experimental work on the growth of part-through-thickness cracks is required to validate this assumption.

Selectively Reinforced Composites

The effect of a cladding of monolithic material covering the composite layer is of interest for potential applications. It may be assumed that a cladding layer of monolithic material will usually be added to the composite section, either to facilitate joining or to allow machining tolerances, or for protection of the composite during handling or a combination thereof. Therefore, fatigue crack propagation in this (nonuniform) stress environment has to be considered carefully. Provided that the stress partitioning between composite and monolithic cladding layers can be evaluated then predictions of crack arrest can still be made (see Fig. 9). The problem that accompanies the presence of a cladding layer of monolithic material is that in a component, surface crack initiation in the monolithic material is likely to occur. This crack will tend to propagate towards the composite regime, where "only" as it interacts with bridging fibers will crack arrest be a possibility. Therefore, the cladding size will determine the size of the crack passed the first row of bridging fibers, and the crack shape will determine the maximum number of fibers that may bridge the crack. The predictions in Fig. 9 in terms of the effect of cladding thickness require some clarification and are for part-through-thickness cracks only. Simple stress partitioning (within the composite layers and the monolithic cladding) based on an equi-strain assumption (for the testpiece loaded in tension) has been included in the analysis. Also by definition, the cladding layers of increased thickness have a reduced effective volume fraction of reinforcement. Thus, allowable stresses will be reduced both because of the stress partitioning (equi-strain assumption) and because the crack will be of increased size when it reaches the first row of fibers. The only factor that may mitigate against the adverse effect of cladding thickness predicted in Fig. 9 is that the growth of a part-through-thickness crack through a thick cladding layer may itself contribute significantly to the required testpiece (component) lifetime. Further modeling and experimental work is ongoing to evaluate the damage tolerance of such testpieces.

For simplicity, the analysis shown in Fig. 9 does not include the residual stresses that would result from cooling from consolidation temperatures due to the different thermal expansion coefficients for the composite layers and the monolithic cladding, but such nonuniform stress distributions are easy to incorporate using the weight function approach.

In practice, it can be seen from Fig. 9 that corner cracks are slightly more damaging than semi-elliptical cracks as at any one time, passed the first "row" of bridging fibers, fewer fibers will interact with the crack tip (see Table 2). Sequential fiber failure may occur in either case unless very low values of stress were applied on the surface of the testpiece. It is also clear that a smaller cladding thickness in general will result in higher allowable stresses to meet the requirements for crack arrest. The potential for composite optimization can be clearly inferred from Fig. 9 (and Table 3). The increased stresses allowable for the metastable β matrix alloys for any given testpiece gometry are a direct result of the higher K_{max} conditions that still promote crack arrest (Table 3).

In summary, one aim of this work is to develop a universal fail-safe map for Ti MMCs with matrix cracks bridged by fibers, and to use such concepts within a lifing methodology. Attempts at this task should result in important empirical tools for transferring data obtained from laboratory testpieces, to sub-scale components, and ultimately to full-size components. Further work must also establish precise failure criteria for actual components. Indeed, the growth of Mode I cracks may not result in catastrophic failure of an actual component, but this must at least be considered in the overall lifing philosophy. Ongoing studies are further addressing such issues.

Conclusions

A simple approach for the prediction of crack arrest for Ti MMCs reinforced unidirectionally with continuous fibers has been presented. It is based on the maximum nominal stress intensity factor that occurs passed the first row of bridging fibers and the absolute number of bridging fibers acting over the crack front. The model can take into account the effect of different crack shapes, cracks sizes, testpiece geometry, and loading conditions. The model can predict the potency of part-through-thickness cracks compared with equivalent through-thickness cracks, and the model predictions are consistent with experimental observations.

The methodology has also been used to predict the onset of the crack arrest of cracks growing in selectively reinforced material. Such an approach may be useful in establishing limits for the thickness of any cladding layers, and it highlights the potential potency of part-through-thickness cracks in such materials.

Acknowledgments

Support for Mr. Barney by a SERC CASE award with BP Research Limited, Sunbury, UK, is gratefully acknowledged. The authors would like to thank Rolls-Royce plc and the controller of Her Majesty's Stationary Office for permission to publish this paper.

References

[*1*] Bowen, P., "Characterization and Micromodeling of Fatigue Crack Growth in Unidirectional Continuous Fiber Reinforced Titanium Alloy Composites," *The Processing Properties and Applications of Metallic and Ceramic Materials,* M. H. Loretto and C. J. Beevers, Eds., MCE Publications, Birmingham, UK, Sept. 1992, pp. 697–818.

[*2*] Wu, X. R. and Carlsson, A. J., *Weight Functions and Stress Intensity Factor Solutions,* first ed., Pergamon Press, Oxford, UK, 1991.

[*3*] Murakami, Y., *Stress Intensity Factors Handbook,* Vols. 1 and 2, Pergamon Press, Oxford, UK, 1987.

[*4*] Bueckner, H. F., "A Novel Principle for the Computation of Stress Intensity Factors," *Zeitschrift Angewandte Mathematik und Mechanik,* Vol. 50, 1970, p. 129.

[*5*] Sih, G. C. and Lee, Y. D., "Review of Triaxial Crack Border Stress and Energy Behavior," *Theoretical and Applied Fracture Mechanics,* Vol. 12, No. 1, 1989, p. 1.

[*6*] Glinka, G. and Shen, G., "Universal Features of Weight Functions for Cracks in Mode I," *Engineering Fracture Mechanics,* Vol. 40, No. 6, 1991, p. 1135.

[*7*] Zhao, W., Wu, X. R., and Yan, M. G., "Weight Function Method for Three Dimensional Crack Problems—I. Basic Formulation and Application to an Embedded Elliptical Crack in Finite Plates," *Engineering Fracture Mechanics,* Vol. 34, No. 3, 1989, p. 593.

[*8*] Pickard, A. C., *The Application of 3-Dimensional Finite Element Methods to Fracture Mechanics and Fatigue Life Prediction,* Engineering Materials Advisory Services, Ltd., Warley, UK, 1986, p. 97.

[*9*] Cardona, D. C. and Bowen, P., *International Journal of Fracture,* to be submitted.

[*10*] Walls, D., Bao, G., and Zok, F., "Effects of Fiber Failure on Fatigue Cracking in a Ti/SiC Composite," *Scripta Metallurgica and Materials,* Vol. 25, 1991, pp. 911–916.

[*11*] Bakuckas, J. G., Jr. and Johnson, W. S., NASA TM 107588, *Journal of Composites and Technology Research,* to be submitted.

[*12*] Bowen, P., Cardona, D. C., and Ibbotson, A. R., "Micromodeling of Crack Growth in Fiber Reinforced Composites," *Materials Science and Engineering,* Vol. A153, 1992, pp. 628–634.

[*13*] McMeeking, R. M. and Evans, A. G., "Matrix Fatigue Cracking in Fiber Composites," *Mechanics of Materials,* Vol. 9, 1990, p. 217.

[*14*] Ghosn, J. L., Kantzos, P., and Telesman, J., "Modeling of Crack Bridging in a Unidirectional Metal Matrix Composite," *International Journal of Fracture,* Vol. 54, 1992, p. 345.

[*15*] Hoagland, R. G., Rosenfield, A. R., and Hahn, G. T., *Metallurgical Transactions,* Vol. 3, 1972, p. 123.

[*16*] Cox, B. N., Marshall, D. B., and Thouless, M. D., "Influence of Statistical Fiber Strength

Distribution on Matrix Cracking in Fiber Composites,'' *Acta Metallurgica,* Vol. 37, No. 7, 1989, p. 1933.
[*17*] Cox, B. N. and Marshall, D. B., ''Crack Bridging in the Fatigue of Fibrous Composites,'' *Fracture Engineering and Materials Structures,* Vol. 14, No. 8, 1991, p. 847.
[*18*] Cardona, D. C., Barney, C., and Bowen, P., ''Micromodeling of Effective Stress Intensities for Bridge Cracks in Fiber Reinforced Titanium Metal Matrix Composites,'' *Journal of Composites,* Vol. 24, 1993, p. 122.
[*19*] Cardona, D. C. and Bowen, P., ''Micromodeling of Fatigue Crack Growth in Fiber Reinforced Titanium Matrix Composites,'' *Theoretical Concepts and Numerical Analysis of Fatigue,* Engineering Materials Advisory Services, Ltd., Birmingham, May 1992, p. 505.
[*20*] Barney, C., Cardona, D. C., and Bowen, P., ''Predictions of Crack Arrest under Cyclic Loading in Continuous Fiber Reinforced Metal Matrix Composites,'' *Proceedings,* ICCM/9, Vol. 1, 1993, pp. 541–548.
[*21*] Bowen, P., ''Crack Growth under Cyclic Loading in Fiber Reinforced Titanium Metal Matrix Composites,'' *Characterization of Fiber Reinforced Titanium Composites,* AGARD-R–796, Advisory Group for Aerospace Research and Development, France, 1993, pp. 18.1–18.16.
[*22*] Bowen, P., ''Characterization, Modeling and Predictions of Crack Growth Resistance in Fiber Reinforced Titanium Matrix Composites,'' *Proceedings,* ICCM/9, Vol. 1, 1993, pp. 585–593.
[*23*] Barney, C., Cardona, D. C., and Bowen, P., ''Crack Arrest under Cyclic Loading in Continuous Fiber Reinforced Ti-6Al-4V Alloy Metal Matrix Composites,'' *Proceedings,* Third Japan International SAMPE Symposium, Vol. 2, Japan, Dec. 1993, pp. 1830–1835.
[*24*] Cotterill, P. J., ''Fatigue Crack Growth at Elevated Temperature in a 9% Cr-1% Mo Steel,'' Ph.D. thesis, University of Cambridge, UK, 1989.
[*25*] Ibbotson, A. R., ''Fatigue Crack Growth in Continuous Fiber Reinforced Titanium Alloy Matrix Composites,'' Ph.D. thesis, The University of Birmingham, UK, 1994.

Inelastic Material Behavior and Modeling

Mohammad Khobaib,[1] Reji John,[1] and Noel E. Ashbaugh[1]

Sustained Load Behavior of SCS-6/ TIMETAL 21S Composite

REFERENCE: Khobaib, M., John, R., and Ashbaugh, N. E., "**Sustained Load Behavior of SCS-6/TIMETAL 21S Composite,**" *Life Prediction Methodology for Titanium Matrix Composites, ASTM STP 1253,* W. S. Johnson, J. M. Larsen, and B. N. Cox, Eds., American Society for Testing and Materials, 1996, pp. 185–207.

ABSTRACT: TIMETAL 21S matrix composite reinforced with silicon-carbide (SCS-6) fibers is a potential candidate material for the high temperature structural components of advanced aircraft. However, a considerable amount of mechanical property characterization is required before this material can be used for aerospace structures. This paper discusses the characterization of the sustained load behavior of SCS-6/TIMETAL 21S composite under simulated service conditions. TIMETAL 21S, reinforced with approximately 35% SCS-6 fibers (by volume) was used for this study. The tension creep specimens were fabricated from $[0]_4$, $[90]_4$, $[0/90]_S$, and $[0/\pm45/90]_S$ laminates. Sustained load creep tests were conducted under cold grip conditions at 650, 760, and 815°C at various stress levels. The creep response for $[90]_4$, $[0/90]_S$, and $[0/\pm45/90]_S$ layups exhibited generic creep curves with three distinct stages: (1) an early region of primary creep, (2) a linear region similar to a steady stage, and (3) a relatively short region similar to a tertiary stage. In contrast, the creep response of the $[0]_4$ layup showed hardly any tertiary region and the failure was abrupt just after the secondary stage. The total strain to failure for all orientations, except the $[90]_4$, was near the ultimate strain for fiber failure. Accelerated rate of damage accumulation was observed with increasing stress and temperature that resulted in higher creep strain under these conditions. A simple creep model based on the relaxation of matrix stress in the primary and secondary creep stages was developed to predict the creep response of the unidirectional SCS-6/TIMETAL 21S composite at 650°C. The model predictions correlated well with the data for stresses less than 700 MPa. The optical and scanning electron microscopic (SEM) analyses of the failure surfaces showed multiple crack initiation sites near fiber/matrix interface. Two distinct zones of failures were observed in all cases. The flat zone, with little or no fiber pull-out, represented the stable crack growth region. The rugged zone, with significant fiber pull-out, was caused by overload failure. In all cases, the damage (crack) initiated from the edge. For $[90]_4$ and $[0/90]_S$, the crack initiation sites were observed to be close to the 90° fiber/matrix interface, while in the case of $[0/\pm45/90]_S$, the crack initiated predominantly from the 45° fiber matrix interface. The fibers that were damaged during the fabrication process provided the crack initiation site for the $[0]_4$ orientation. These edge cracks enhanced the environmental access to the fiber/matrix interface that resulted in progressive damage of the interface and the fiber. Data plotted on a Larson-Miller Parameter (LMP) chart enabled a direct comparison of the creep resistance of all the layups tested. The creep resistance of the $[0]_4$ layup was significantly greater than that of the other layups. The creep rupture life of specimens with 0° fibers was governed by the failure of unidirectional fibers. An empirical model was developed to predict the creep rupture life of all layups containing 0° fibers based on the LMP approach assuming that only the 0° fibers carry the load for the majority of the life.

KEYWORDS: metal matrix composite, silicon-carbide fibers, sustained load, creep (materials), laminates, fiber orientation, damage accumulation, fiber/matrix interface, rupture lives,

[1]Research engineer, associate research engineer, and senior research engineer, respectively, The University of Dayton Research Institute, Dayton, OH 45469–0128.

titanium, titanium matrix composites, life prediction, titanium alloys, fatigue (materials), modeling

The emerging technology in advanced aerospace structures such as the hypersonic vehicle structural component design require light-weight materials possessing high stiffness and high strength with improved environmental resistance at elevated temperatures. TIMETAL 21S[2] matrix composite reinforced with silicon-carbide (SCS-6) fibers is a potential candidate material for such applications. The SCS-6/TIMETAL 21S composite provides improved strength, lower density, and high temperature capability needed to permit the design of advanced aerospace material components. However, a considerable amount of mechanical property characterization is required before this material can be used for aerospace structures. Since most of the structures will be subjected to sustained stress at elevated temperatures during service, the design community will require an understanding of the creep behavior of SCS-6/TIMETAL 21S composite.

While SCS-6 fibers possess excellent creep resistance, TIMETAL 21S is susceptible to creep, especially at the usable load and temperature of these composites. In addition, the off-axis plies of the fibers offer a potential debit to the creep resistance of composite laminates containing off-axis fibers. Hence, detailed information on the creep behavior of various layups of this composite is required. This paper discusses the characterization of the sustained load behavior of various layups of the SCS-6/TIMETAL 21S composite.

Background

Several experimental investigations [*1–6*] have been conducted to understand the creep deformation of titanium metal matrix composites. A number of micromechanical models are also available to predict the creep response of metal matrix composites reinforced with continuous fibers [*7–10*]. Experimental investigation and analysis of results from the creep behavior of a unidirectional composite reinforced with continuous fibers suggest that the longitudinal creep can be described by an iso-strain model [*11,12*]. Transverse stresses that develop on a micromechanical level during creep deformation were neglected. The iso-strain concept was used by McDanels et al. [*13*] to predict the steady-state creep of unidirectional copper reinforced with tungsten fiber, where both the matrix and fiber creep at a relatively constant rate. Soliman [*14*] adapted the virtual displacement analysis of creep proposed by deSilva [*15*] to predict the secondary and primary creep behavior of a graphite-epoxy composite.

Usually, the creep rate of the reinforcing fiber is several orders of magnitude less than that of the matrix, under identical conditions. To achieve a steady state, the matrix must relax to a very low stress value in order to exhibit a creep rate equal to that of the fiber. Thus, in beryllium-reinforced epoxy and boron-reinforced aluminum composites, the matrix contribution to creep resistance has been frequently neglected by assuming that fibers carry the entire load [*14*]. A uniaxial model was proposed by McLean [*10*] to predict the creep behavior of the composite based on elastic noncreeping fibers in a creeping matrix. Schwenker et al. [*16,17*] used this creep relaxation model to correlate the unidirectional creep data of the SCS-6/Ti-6Al-4V composite. These analyses assumed that the matrix exhibited secondary-state behavior in the composite. Wright [*18*] proposed a two-bar model to predict the creep behavior of the unidirectional composite laminate of SCS-6/Ti-6Al-4V composite based on the primary creep behavior of the matrix. Ohno et al. [*19*] used both matrix stress relaxation and fiber bundle strength in their model to describe the conditions of

[2]Registered trademark of Timet Corporation, Henderson, NV.

creep rupture in unidirectional SCS-6/Ti-15-3. McMeeking et al. [*20*] used a probabilistic model for fiber fracture and shear lag analysis to describe the creep behavior of a unidirectional composite. Foye [*21*] pioneered the use of a finite element method to analyze the composite deformation. The elastic and plastic deformation in the cross-plied composite laminate was determined [*22,23*] by coupling the laminate theory with finite element analysis. The VISCOPLY code, which accounts for the orthotropic properties of the composite, was used to predict the time-dependent response of the metal matrix composite using the laminated plate theory [*24*]. Teply and Dvorak [*25*] proposed a generalized deformation model to predict the creep response of composites. However, the model did not account for inelastic deformation and imperfect fiber/matrix interface effects. Recently, Boyle and Ashbaugh [*26*] used the finite element analysis to predict the creep response of $[0]_8$ SCS-6/Ti-24Al-11Nb and $[0]_4$ SCS-6/TIMETAL 21S using the primary and secondary creep behavior of the neat matrix.

This paper describes the experimental results of an investigation of the effect of stress level and temperature on the creep behavior of various layups of SCS-6/TIMETAL 21S. An analytical approach to predict the accumulation of creep strain in the unidirectional SCS-6/TIMETAL 21S composite is presented. The application of the Larson-Miller parameter to characterize the creep rupture lives of various layups at different stress levels and temperatures is also discussed.

Material and Experimental Procedure

TIMETAL 21S reinforced with approximately 35% by volume of continuous SCS-6 fiber was used in this study. All tests were conducted using straight-sided specimens, about 110 mm long and 12.5 mm wide, in a horizontal servohydraulic test machine with closed-loop computer control. Specimens were cut from various panels of $[0]_4$, $[90]_4$, $[0/90]_S$, and $[0/\pm45/90]_S$ layups. The specimens were tested using a rigid grip system in which only the middle 50 mm of the specimen was heated. Radiant-heat quartz lamps were used to achieve the desired temperature. The grips were maintained at room temperature. Sustained load tests were conducted at 650, 760, and 815°C at various stress levels. A number of tests were continued to failure to obtain the rupture lives, while others were stopped after about 100 h. After testing, each specimen was analyzed for damage mechanisms such as surface/edge crack, oxidation, and fiber pull-out using optical and scanning electron (SEM) microscopes. Additional creep tests were also conducted on neat (fiberless) TIMETAL 21S to aid in the understanding of the rate of matrix relaxation and increase of fiber stress during creep deformation of the composite. The creep strain reported in the following sections was calculated as the difference between the current total strain and the total strain when the applied stress reached the maximum value. The typical ramp time from zero to maximum load varied between 75 to 130 s.

Creep Response of the $[0]_4$ Laminate

The creep behavior of the $[0]_4$ laminate of SCS-6/TIMETAL 21S for different stress levels at 650°C is shown in Fig. 1. The figure shows the response during the early stages of creep deformation. As shown in Fig. 1, the creep response exhibited classical primary and secondary creep behaviors. An increase in primary creep strain with increasing stress is evident. Figure 2 shows creep behavior of a $[0]_4$ composite tested to failure. The applied stress level was about 70% of the tensile strength at 650°C. The tertiary stage was negligible, and the failure occurred abruptly after an extended period of secondary stage, as shown in Fig. 2. The maximum measured creep strain prior to failure was only about 0.002

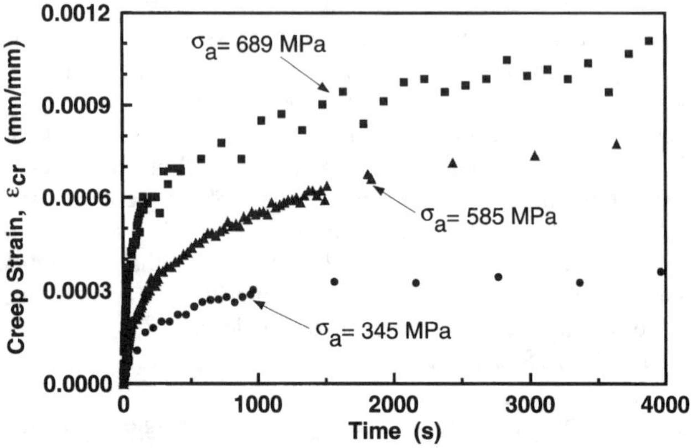

FIG. 1—*Effect of stress during early stages of creep behavior of the unidirectional SCS-6/TIMETAL 21S at 650°C.*

mm/mm. Even the specimen tested at 689 MPa eventually failed at ≈673 000 s. This stress was only about 58% of the tensile strength at 650°C, and the fibers were not expected to creep or fail under this combination of load and temperature. However, the specimens tested at stress levels ≥690 MPa failed after some time-dependent deformation. The rupture occurred with little or no tertiary stage at a total strain less than 0.01 mm/mm, which is the failure strain of the SCS-6 fibers.

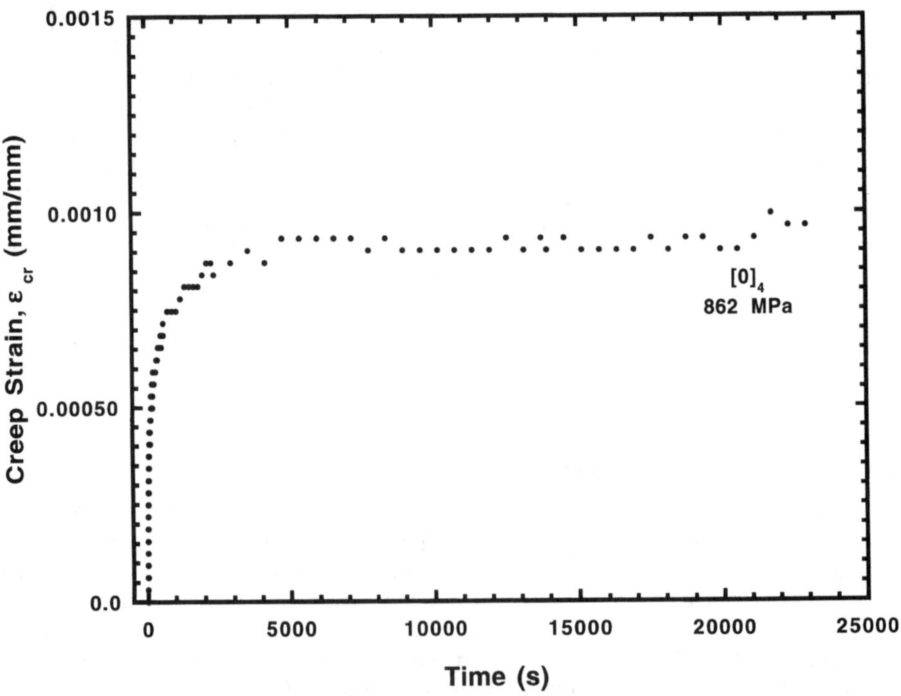

FIG. 2—*Creep behavior of the $[0]_4$ SCS-6/TIMETAL 21S tested to failure at 650°C.*

Figure 3 is a SEM photograph from a specimen tested at 689 MPa showing the evidence of multiple fiber breakage. A number of cracks were found to initiate from surface and edge flaws. The frequency of cracks was higher in the vicinity of the fracture. On continued exposure under sustained load, some of these cracks were observed to interlink and form a dominant crack, which ultimately led to specimen failure. In a previous investigation [*4*], nearly 10% of the first layer fibers were found to be cracked at the onset of the secondary stage of creep. The optical analysis of the creep ruptured surfaces always showed two distinct fracture zones. The oxidized zone with relatively flat surface represented the time-dependent slow growth region, while the remaining fracture surface was dominated by fiber pull-out indicating fast fracture. This zone appeared to be oxidized. This is in contrast to the larger jagged zone of the overload region that exhibited evidence of extensive fiber pull-out. Similar mechanisms were also observed in the specimens tested at 760 and 815°C.

Table 1 shows the initial and maximum possible constituent stresses and strains prior to initiation of fiber failure for the specimens tested at 650°C. The volume fraction of the specimens was calculated based on the distribution of fibers and the thickness of the specimen. The constituent stresses were calculated using the rule of mixtures. The maximum possible stresses and strains were computed assuming that the matrix had relaxed to zero stress and that the applied load was carried completely by the fibers. The Young's modulus of the matrix and fiber was 78 and 370 GPa, respectively. The maximum initial matrix stress is 303 MPa, which is less than the yield strength of 350 MPa at 650°C. For stress levels up to 689 MPa, the initial and the maximum fiber stresses are well below the failure strength of 3800 MPa. The maximum strains indicate that the creep deformation prior to fiber failure is

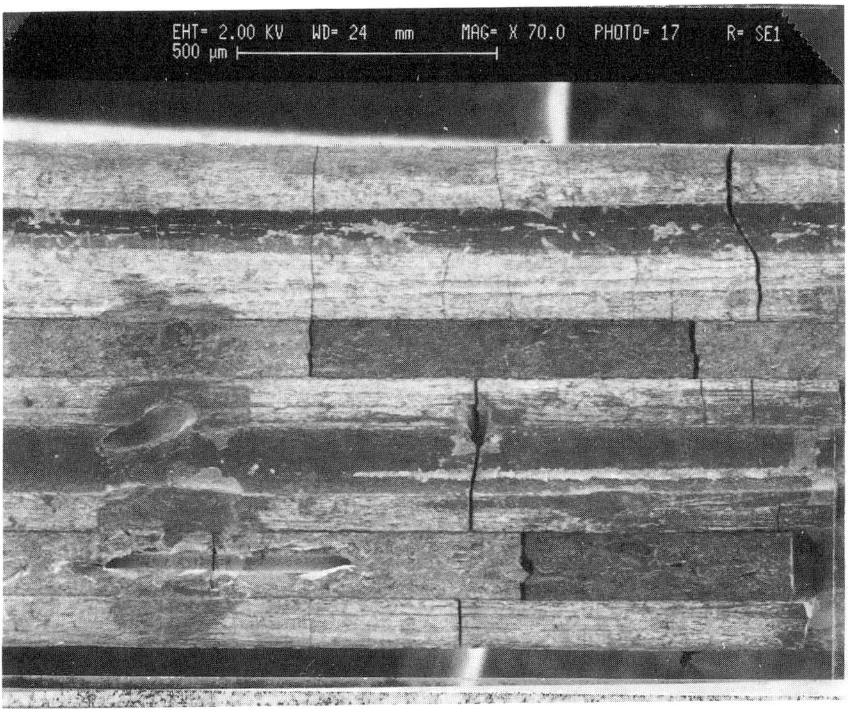

FIG. 3—*Evidence of multiple fiber breakage during creep of the* $[0]_4$ *SCS-6/TIMETAL 21S.*

TABLE 1—*Initial and maximum possible stress/strain states of $[0]_4$ SCS-6/TIMETAL 21S tested at 650°C.*

Composite Stress, MPa	Volume Fraction of Fibers, %	Modulus of Composite, GPa	Initial Matrix Stress, MPa	Initial Fiber Stress, MPa	Maximum Fiber Stress, MPa	Initial Composite Strain, mm/mm	Maximum Composite Total Strain, mm/mm	Maximum Composite Creep Strain, mm/mm
345	36	182.68	147	699	962	0.001 89	0.002 60	0.000 71
585	38	190.16	240	1138	1523	0.003 08	0.004 12	0.001 04
689	34	177.17	303	1439	2029	0.003 89	0.005 48	0.001 59

TABLE 2—*Composite stress corresponding to initial matrix yield.*

Temperature, C	Modulus of Matrix, GPa	Modulus of Fiber, GPa	Modulus of Composite, GPa	Yield Stress of Matrix, MPa	Composite Stress at Initial Matrix Yield, MPa
482	90	376	190.1	810	1710.9
650	78	370	180.2	350	808.6
760	70	365	173.3	120	297.0
815	64	362	168.3	110	289.3

extremely small; about 0.002 mm/mm. Table 2 shows the moduli of the fiber, the matrix, and the composite; the matrix yield stress, and the composite stress corresponding to initial matrix yield at different temperatures. The specimens tested at stress levels >810 MPa at 650°C and all the stress levels at 760 and 815°C during this study resulted in yielding of the matrix when the composite was loaded to the maximum stress.

Modeling of Unidirectional Creep Behavior at 650°C

This section discusses the mathematical representation of the creep behavior of neat TIMETAL 21S and the prediction of strain accumulation in the unidirectional SCS-6/TIMETAL 21S composite at 650°C. As discussed earlier, the creep response of SCS-6/TIMETAL 21S was dominated by the creep of the matrix, TIMETAL 21S, since the SCS-6 fibers exhibit negligible creep at 650°C. Under sustained loading of the composite, the matrix starts creeping due to the initial stress. The fibers retard the extension of the matrix, leading to an increase of the fiber stress with a simultaneous decrease of the matrix stress. Hence, the creep rate of the matrix decreases with increasing time resulting in a continuously decreasing creep rate of the composite, as shown in Figs. 1 and 2. The primary- followed by secondary-type creep behavior is exhibited by the composite prior to significant fiber failures. The modeling of the creep behavior of the composite requires an accurate representation of the creep behavior of the neat (fiberless) matrix.

Representation of Creep Behavior of Neat (Fiberless) TIMETAL 21S—Figure 4 shows the creep response of neat TIMETAL 21S at 650°C for stress levels ranging from 51.7 to 172 MPa. The maximum creep strain at rupture was higher than 0.15 mm/mm. At high stress levels, very little secondary stage was evident. We propose Eq 1 to represent the primary and near-secondary creep response of the neat matrix behavior at 650°C.

$$\epsilon_{cr} = \frac{\epsilon_{cr,lim}}{2}\left[\left(\frac{t}{t_{lim}}\right)^{p} + \left(\frac{t}{t_{lim}}\right)^{q}\right] \quad (1)$$

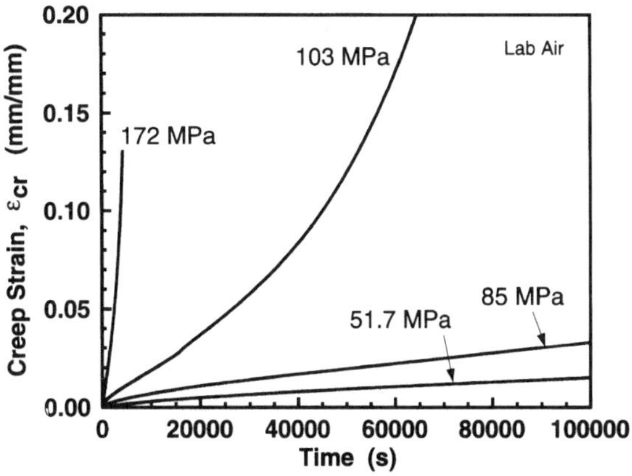

FIG. 4—*Creep behavior of the neat (fiberless) TIMETAL 21S at 650°C.*

in which, ϵ_{cr} = creep strain in the matrix, t = time, $\epsilon_{cr,lim}$ = limiting matrix creep strain, t_{lim} = time at $\epsilon_{cr,lim}$, and p and q are constants. Typically, the neat matrix creep rupture strain was approximately 0.15 mm/mm as shown in Fig. 4. In contrast, the total strain in the unidirectional composite is limited by the fiber failure strain of about 0.01 mm/mm. Hence, the matrix creep, up to a limiting creep strain, $\epsilon_{cr,lim}$ = 0.015 mm/mm, was represented using Eq 1. Figure 5 shows a plot of ϵ_{cr} versus the ratio, t/t_{lim}, for all applied stress levels. Interestingly, the data for all the stress levels coalesce into a single curve. A least-squares fit of Eq 1 to the data yielded p = 0.6112 and q = 0.9052. The first and second terms in Eq 1 describe the primary and near-secondary creep behavior of the neat matrix. If q = 1, the second term in Eq 1 will describe the conventional secondary (steady-state) behavior.

Equation 1 can be used to describe the matrix creep behavior at any stress level if the dependence of t_{lim} with respect to the applied stress, σ_m, is known. Figure 6 shows a plot of the measured t_{lim} versus σ_m. The solid line in Fig. 6 is the power law given by the following equation

$$t_{lim} = 6.7526 \times 10^{11}(\sigma_m)^{-3.9656} \tag{2}$$

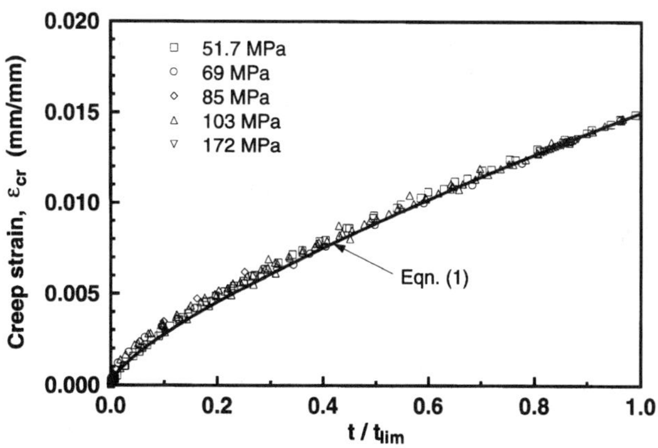

FIG. 5—*Creep strain versus normalized time for the neat TIMETAL 21S at 650°C.*

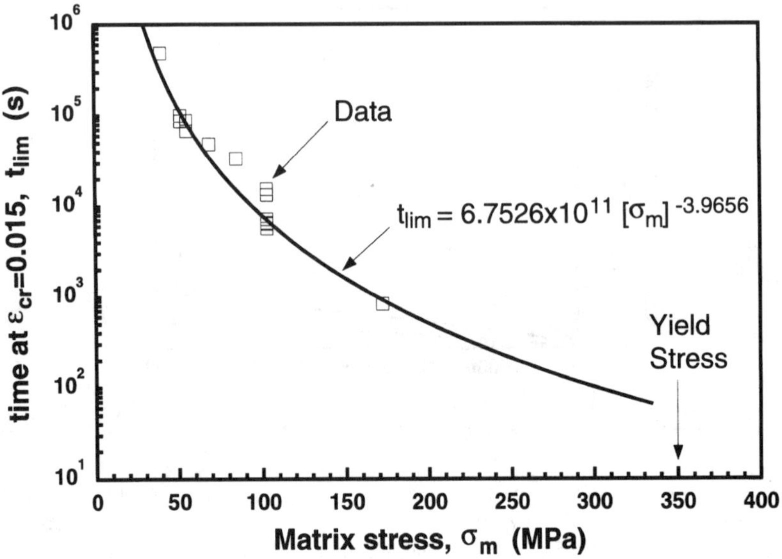

FIG. 6—*Stress dependency of* t_{lim} *(time at* $\epsilon_{cr} = 0.015$*) for the neat TIMETAL 21S at 650°C.*

in which $t_{\rm lim}$ is in seconds and σ_m is in MPa. Using Eqs 1 and 2, the creep behavior of the neat matrix can be satisfactorily described up to $\epsilon_{cr} = 0.015$ mm/mm for any stress level at 650°C. The application of the preceding matrix creep equations to predict the creep behavior of the unidirectional composite is discussed next.

Prediction of Creep Behavior of Unidirectional SCS-6/TIMETAL 21S—Equations 1 and 2 are used in conjunction with the two-bar model proposed by Wright [*18*] to predict the creep behavior of $[0]_4$ SCS-6/TIMETAL 21S. During sustained loading of the composite subjected to a stress, σ_a, the constituent stresses are continuously redistributed according to the following equation given by iso-strain based rule of mixtures

$$\sigma_a = \sigma_m(1 - V_f) + \sigma_f V_f \quad (3)$$

where σ_m = matrix stress, σ_f = fiber stress, and V_f = volume fraction of the fibers. The fibers were assumed to deform elastically and the matrix was represented using Eqs 1 and 2.

An incremental solution scheme [*18,27*] was used to determine the accumulation of creep strain with increasing time. The composite strain at the end of the time interval, Δt, is given by

$$\epsilon'_{cr} = \epsilon_{cr} + \Delta\epsilon_{cr}/2 \quad (4)$$

where ϵ_{cr} = composite creep strain at the middle of the time interval and $\Delta\epsilon_{cr}$ = assumed increment of matrix creep strain after Δt. At $t = 0$, ϵ_{cr}, and $\Delta\epsilon_{cr}$ were set equal to 0.0. Using the iso-strain condition, the composite strain, matrix stress, and fiber stress were derived [*18*] as

$$\epsilon_c = [\sigma_a + \{(1 - V_f)\epsilon'_{cr}E_m\}]/E_f \quad (5)$$

$$\sigma_m = [\epsilon_c - \epsilon'_{cr}]E_m \quad (6)$$

$$\sigma_f = \epsilon_c E_f \quad (7)$$

where E_m and E_f are the matrix and fiber moduli, respectively. In the next step, the creep strain rate was calculated as $\dot{\epsilon}_{cr}$ at time $= t + \Delta t/2$ using Eqs 1 and 2. Note that the current matrix stress, σ_m given by Eq 6 was substituted in Eq 2. The new increment of creep strain was evaluated as

$$\Delta\bar{\epsilon}_{cr} = \dot{\epsilon}_{cr}\Delta t \tag{8}$$

If $\Delta\bar{\epsilon}_{cr} \neq \Delta\epsilon_{cr}$, $\Delta\epsilon_{cr}$ was updated to $\Delta\bar{\epsilon}_{cr}$ and the analysis from Eqs 4 to 8 was repeated until $\Delta\bar{\epsilon}_{cr} = \Delta\epsilon_{cr}$, which becomes the current creep strain increment. After convergence, ϵ_{cr} and t were increased by $\Delta\epsilon_{cr}$ and Δt, respectively, and the preceding procedure was repeated for the next time increment. The flowchart for the preceding procedure is given in Ref *18*.

Using Eqs 1 to 8, the creep response of the composite was predicted for different applied stresses and compared with the experimental data in Fig. 7. The predictions correlate well with measured data during the early stages of creep deformation. The primary response and the stress dependency of the creep behavior are predicted satisfactorily by the model up to 4000 s.

As discussed earlier, during sustained loading of the composite, the fiber and matrix stresses are continually redistributing throughout the creep deformation. The predicted variation of the matrix and fiber stresses with increasing time is shown in Figs. 8 and 9, respectively. σ_m is plotted versus log(t) in Fig. 8. For all σ_a, σ_m decreased rapidly to about 50 MPa in 1000 s. Hence, the rapid decrease in the composite strain rate during the early stages of creep deformation can be attributed to the decrease in matrix stress. The matrix is entirely relaxed to less than 10% of the initial stress in about 4000 s. The corresponding rapid increase of fiber stress to stable magnitude is shown in Fig. 9. As expected, the stable values of fiber stresses in Fig. 9 correspond to the maximum fiber stresses reported in Table 1. For $t > 4000$ s, the load is carried almost completely by the fibers. Hence, the failure of the composite is controlled by the fiber failure. Figure 9 and Table 1 show that the maximum fiber stress is about 2000 MPa for $\sigma_a = 689$ MPa.

Figure 10 compares the predicted and measured response of $[0]_4$ SCS-6/TIMETAL 21S for $\sigma_a = 690$ MPa. The model did not predict creep rupture, but the composite failed at ≈673 000 s. After about $= 10\,000$ s, the data showed a gradual increase in creep strain until close to failure. As shown in Fig. 8, the matrix relaxed completely in about 10 000 s.

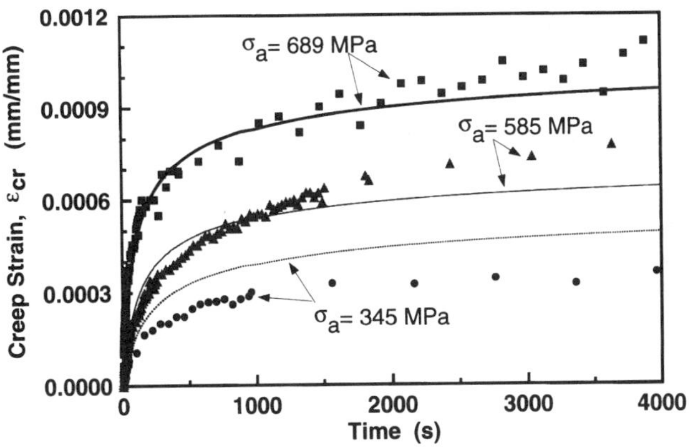

FIG. 7—*Predicted and measured response during early stages of creep of the $[0]_4$ SCS-6/TIMETAL 21S at 650°C.*

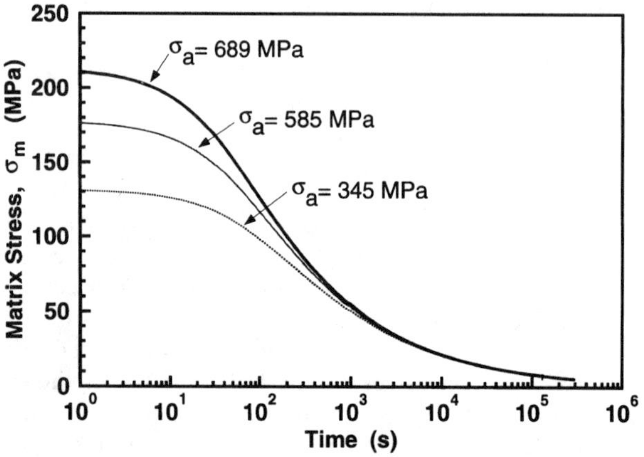

FIG. 8—*Predicted relaxation of matrix stress during early stages of creep of the $[0]_4$ SCS-6/TIMETAL 21S at 650°C.*

Hence, the additional strain that eventually led to failure of the composite can be attributed to the accumulation of strain due to time-dependent fiber breakages leading to a gradual increase of the stress in the remaining in-tact fibers. The abrupt failure of the composite at total strain levels less than the failure strain of the fiber (=0.01 mm/mm) implies that the strength of the fibers in the composite could be degrading with increased exposure to temperature under stress.

Discussion—The maximum possible creep strain in the composite prior to fiber failure subjected to different stress levels was reported in Table 1. The maximum possible creep strains for the maximum σ_a was slightly less than 0.002 mm/mm. As shown in Figs. 4 and 5, these values of creep strain are less than 15% of $\epsilon_{cr,lim}$. Hence, an accurate prediction of the primary creep of composite would require an accurate representation of the matrix creep behavior in the very early stages of the creep deformation of the matrix. Hence, the common practice of only using the secondary creep behavior of the matrix to predict the early creep deformation of the composite would lead to erroneous results.

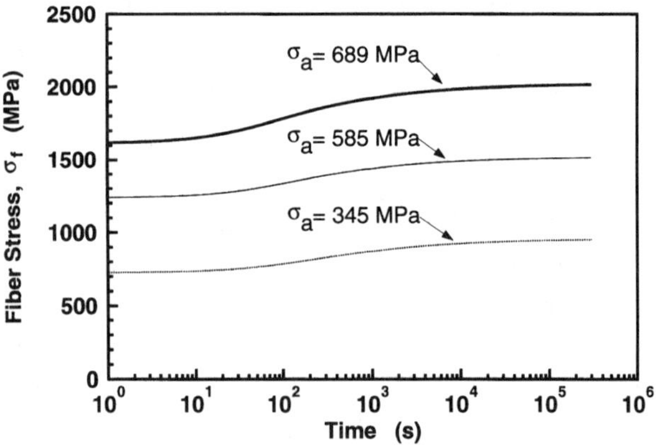

FIG. 9—*Predicted increase of fiber stress during early stages of creep of the $[0]_4$ SCS-6/TIMETAL 21S at 650°C.*

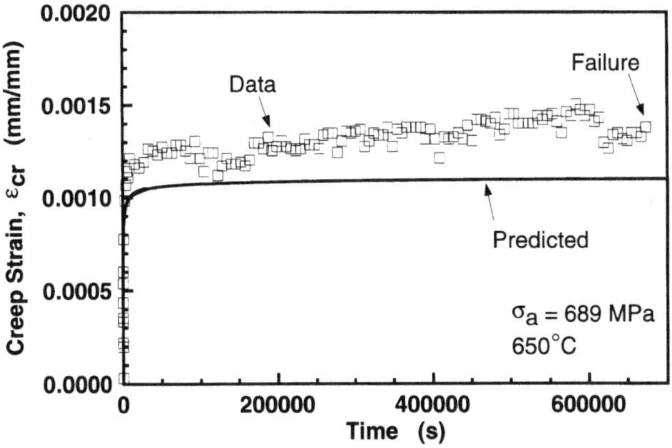

FIG. 10—*Predicted and measured creep response of the $[0]_4$ SCS-6/TIMETAL 21S tested at applied stress of 689 MPa at 650°C.*

Based on the initial matrix stresses reported in Table 1 and the predicted matrix stress relaxation history in Fig. 8, we can conclude that the matrix was subjected to high stress levels during the primary creep deformation of the composite. Therefore, additional matrix creep data for high values of σ_m are required to determine the constants in Eqs 1 and 2.

The composite stress and matrix stresses corresponding to the yield stress of the matrix at different temperatures were reported in Table 2. At 650°C, σ_a = 689 MPa did not result in matrix yielding consistent with the preceding analysis. The effect of residual stresses on the creep behavior of the composite was ignored. Three-dimensional finite element analysis [*28*] has shown that for SCS-6/TIMETAL 21S the residual stresses become negligible at 650°C.

As discussed earlier, the matrix is subjected to near-yield stresses initially. Hence, the composite can be expected to creep during the ramping of the applied load from zero to maximum load at temperature.

Creep Response of the $[0/90]_S$ Laminate

The creep behavior of $[0/90]_S$ SCS-6/TIMETAL 21S at 650°C is shown in Fig. 11. The $[0/90]_S$ laminates exhibit classical creep behavior and consist of all the three stages: primary, secondary, and tertiary. The overall creep behavior of the $[0/90]_S$ laminate is similar to that of the $[0]_4$ laminate. The primary zone is longer than that of the $[0]_4$ laminate due to extensive matrix deformation in the presence of 90° fiber. The increase in primary strain and the continued extension with time in the secondary stage is partly due to the continued debonding of 90° fiber and matrix cracking. Figure 12 shows evidence of a crack initiating from the fiber/matrix interface near the debonded 90° fiber. Such damage initiates at several locations during the course of the test and results in substantial deformation, as shown in Fig. 12. With continued exposure under load, some of these cracks link together and create a dominant crack that leads to final failure. The process is similar to that of transverse (that is, $[90]_4$) creep, but in the $[0/90]_S$ composite, the crack initiating from matrix/fiber interface near the debonded 90° fiber is bridged by 0° fiber. Such bridging extends the period of matrix relaxation along the 0° fiber loading.

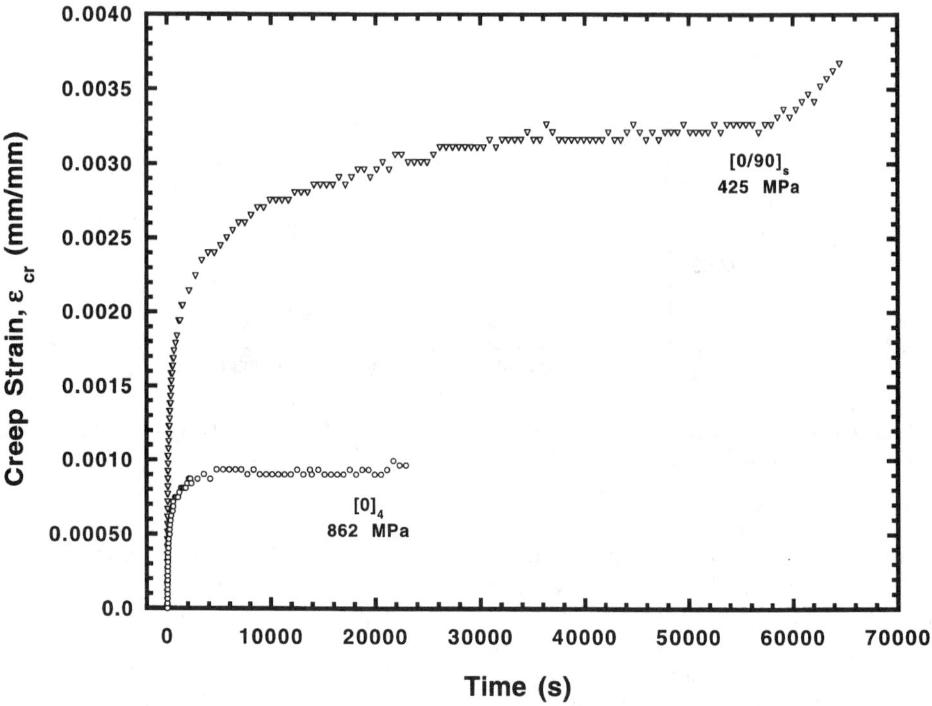

FIG. 11—*Typical creep behavior of the* $[0/90]_S$ *laminate of SCS-6/TIMETAL 21S at 650°C.*

Creep Response of the $[0/\pm45/90]_S$ Laminate

Several tests were conducted at three stress levels for the $[0/\pm45/90]_S$ laminate of SCS-6/TIMETAL 21S. The typical creep behavior of $[0/\pm45/90]_S$ laminate is shown in Fig. 13. Similar to the other laminates, larger creep strain with higher creep stress is quite evident. The tertiary stage is also significant.

The $\pm45°$ fibers in the creep of the $[0/\pm45/90]_S$ composite contribute to more extensive deformation in matrix as compared with $[0/90]_S$ specimen resulting in a larger primary zone. The load is primarily picked up by the ±45 fibers all through the primary and secondary stage. Depending upon the temperature and load, ±45 fibers can significantly align towards the tensile (loading) axis resulting in an increase of the load-carrying capability of the composite. Figure 14 shows an optical micrograph of a $[\pm45]_S$ specimen creep tested at 650°C at a stress of 34.5 MPa. During the course of the test, the fibers rotated towards the tensile axis to nearly $\pm24°$. Such realignment or rotation of $\pm45°$ fibers towards the tensile axis was also reported by Toth [*29*], for the creep of $[\pm45]_S$ laminates of aluminum reinforced with boron fibers. In the boron/aluminum composite, the shear deformation of the matrix caused the $\pm45°$ fibers to rotate from their original orientation to an angle of $\pm28°$, and the creep rates improved by several orders of magnitude [*29*]. The resistance to the rotation of fiber is provided by the matrix. The resistance, or the frictional force, enables these fibers to take up higher load, and causes continued matrix relaxation with time. The continued debonding and rotation of $\pm45°$ fibers toward the tensile axis occurs all over the specimen resulting in extensive matrix deformation and multiple crack initiation in the

FIG. 12—*Matrix crack initiating from a 90° fiber during creep of the [0/90]$_S$ laminate.*

matrix. Figure 15 clearly shows the evidence of extensive deformation and cracking of matrix around ±45° fiber. As stated earlier, the load is initially transferred to ±45° fibers that continuously try to orient towards the tensile axis, and such rotation is opposed by the frictional force of matrix. In this process, the fiber/matrix interface initiates cracking in the matrix as shown in Fig. 15. This provides environmental access to 0° fiber through accelerated diffusion on continued exposure. Some 0° fibers are also degraded and eventually break with time on continued exposure under load. The linking of several cracks leads to the formation of a dominant crack and eventual failure. For [0/90]$_S$ and [0/±45/90]$_S$ laminates, the growth of a dominant crack has a longer life as compared to the creep behavior of [0]$_4$ orientation. This is due to partial crack bridging by ±45° and 90° fibers. Such bridging leads to the development of a noticeable tertiary stage.

Creep Response of the [90]$_4$ Laminate

The creep behavior of the [90]$_4$ composite is clearly distinct from the other three layups. All the three stages of creep are exhibited at all stress levels as shown in Fig. 16. The matrix behavior of neat TIMETAL 21S at 69 MPa is also shown in this figure. These plots appear closer to matrix behavior than the behavior shown by the other composite layups discussed earlier. The classical creep behavior of [90]$_4$ is easy to understand because the transverse fibers act as defects and do not contribute to any creep resistance. In fact, the debonded transverse fibers act like holes in the matrix and result in a significant debit in creep strength [*30*]. Figure 17 shows the debonding of the 90° fiber that occurs in the early stage of primary creep. The rupture life of the [90]$_4$ layup was less than one half that of the neat matrix [*30*].

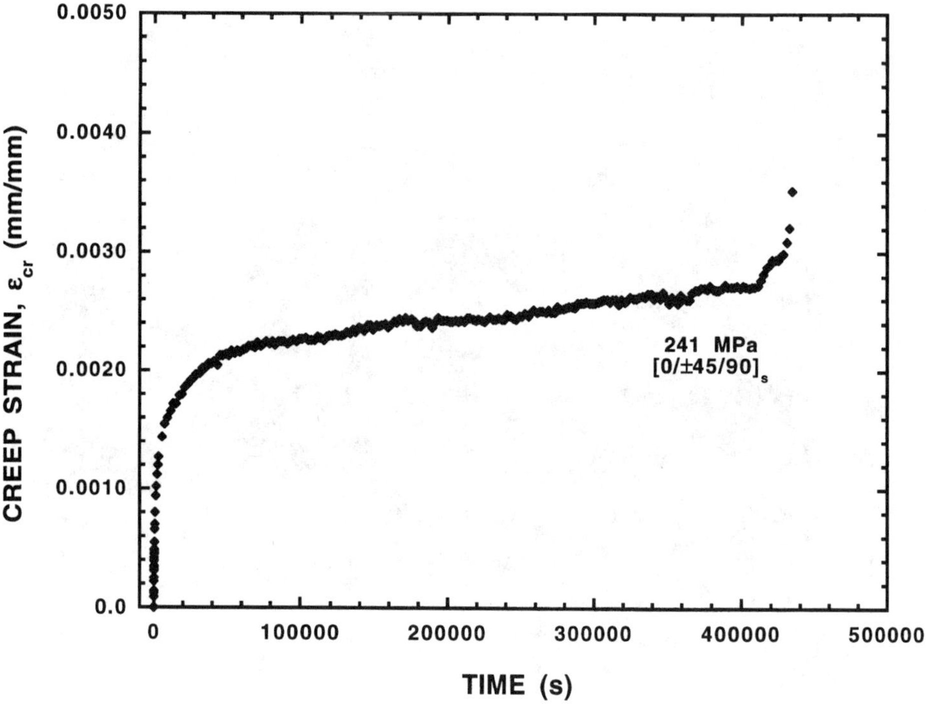

FIG. 13—*Typical creep behavior of the* $[0/\pm45/90]_S$ *laminate of SCS-6/TIMETAL 21S at 650°C.*

Hence, the case of unidirectional layups would require extremely low allowable transverse stresses during service.

Effect of Temperature on Creep Behavior

The effect of temperature on the creep behavior of the $[0]_4$ laminate of SCS-6/TIMETAL 21S is shown in Fig. 18. Three tests were conducted at 650, 760, and 815°C at a stress of 587 MPa. Like most materials, the creep strain is observed to increase with increasing temperature. The increase in creep strain is related to increased matrix relaxation as well as faster reaction kinetics, such as debonding, environmental degradation, and related factors at higher temperatures. On continued exposure, failure occurs due to the synergistic effect of load and temperature. Similar effects of increasing temperature were observed for $[0/90]_S$, $[0/\pm45/90]_S$, and $[90]_4$ laminates.

Life Prediction of Composite Laminates

During this investigation, a large number of tests at different stresses and temperatures were continued to failure to obtain the creep rupture lives. The data for all the layups are reported in Table 3. Over the years, rupture life has been used extensively as an engineering design parameter. Several creep parameters [*31–34*] have been effectively used to determine the creep-rupture life. The Larson-Miller parameter (LMP) [*31*] has been used extensively to characterize the rupture behavior of various materials.

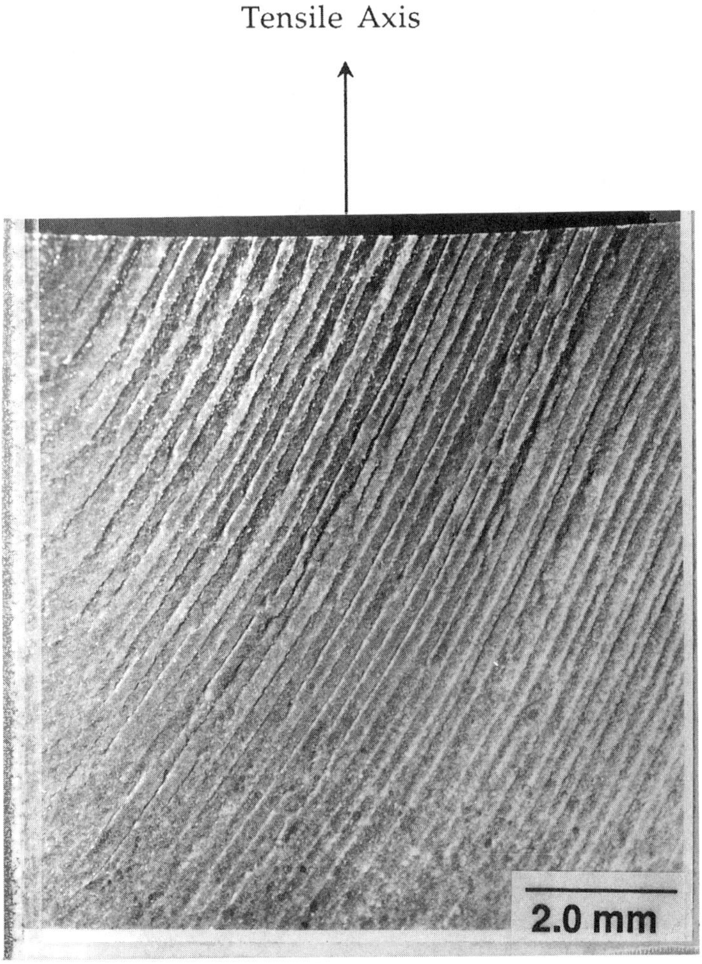

FIG. 14—*Rotation of the* $\pm 45°$ *fibers towards the loading axis.*

In this study, the application of the LMP to characterize the creep rupture life of different layups of SCS-6/TIMETAL 21S for a variety of stress levels and temperatures was investigated. The generalized equation for creep in most materials is expressed as

$$\dot{\epsilon} = \dot{\epsilon}_0(T,\sigma) \exp\left(\frac{-G(T,\sigma)}{KT}\right) \tag{9}$$

where $\dot{\epsilon}$ is strain rate, $\dot{\epsilon}_0$ is a pre-exponential factor, G is Gibb's free energy of activation, T is absolute temperature, σ is applied stress, and K is Boltzmann's constant. In general, both $\dot{\epsilon}_0$ and G depend on temperature and stress [*35–36*]. Larson and Miller [*31*] expressed the preceding rate equation as

$$\dot{\epsilon} = \dot{\epsilon}_0 \exp\left(\frac{-Q(S)}{KT}\right) \tag{10}$$

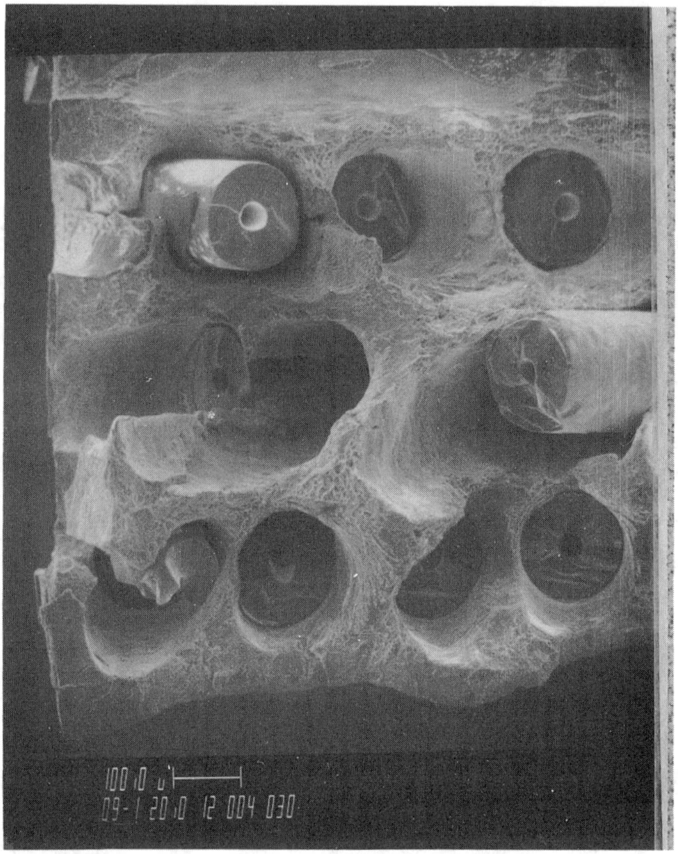

FIG. 15—*Extensive deformation around the ±45° fibers during creep of the [0/±45/90]$_S$ laminate.*

where $Q(S)$ is the activation energy for the process under consideration. Both Eqs 9 and 10 can be written in a simple form, as shown by Eq 11

$$\dot{\epsilon} = \dot{\epsilon}_0 \exp\left(\frac{-H}{KT}\right) \tag{11}$$

where H is the activation enthalpy and $\dot{\epsilon}_0$ contains an entropy term. Equation 11 can be rearranged for any specific stress as

$$\frac{H}{K} = T\left(\ln\frac{\dot{\epsilon}_0}{\epsilon} + \ln t\right) \tag{12}$$

where t denotes time during creep. Larson and Miller [*31*] showed empirically that Eq 12 can be written as

$$\log \sigma = A + Bm \tag{13}$$

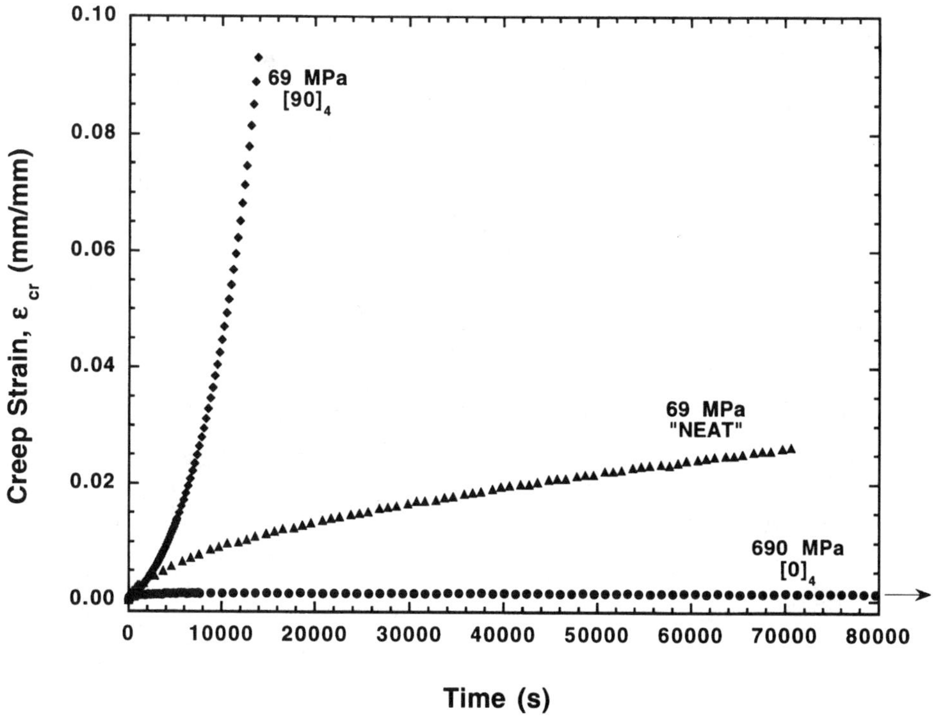

FIG. 16—*Comparison of creep behavior of the* $[90]_4$ *laminate of SCS-6/TIMETAL 21S with that of the "neat" matrix and the* $[0]_4$ *laminate at 650°C.*

where

$$m = T(C + \log t) \tag{14}$$

where T is temperature in Kelvin, t is time in hours, C is a material constant, and A and B are the intercept and slope, respectively, in the log σ versus m plot. In Eq 13, m is the well known Larson-Miller parameter. The value of C has been determined to be between 18 and 22 for a wide variety of metals and alloys. The variation in the value of C occurs due to the wide range of temperatures, the strain range to failure, and the inherent material creep property. For most cases, the value of C fluctuates around 20, and, for short strain range, $C = 20$ produces good correlation with the data. As discussed earlier, the strain at failure for all the $[0]_4$, $[0/90]_S$, and $[0/\pm 45/90]_S$ specimens were close to 0.01 mm/mm. Hence, in this study, C was assumed equal to 20.

A direct comparison of the creep resistance of various laminates from creep strain versus time plots was difficult because conducting creep tests at one fixed stress and temperature for all laminates was not practicable due to the large difference in their creep behavior. However, when all the creep rupture data were represented on a Larson-Miller plot, distinct trends become evident. Figure 19 shows a direct comparison of the creep behavior of various layups in the temperature range from 650 to 815°C. The significant difference in the creep resistance for various laminates can be seen in the figure. The $[0]_4$ layup exhibited the highest creep resistance while the $[90]_4$ layup showed the least creep resistance. The neat

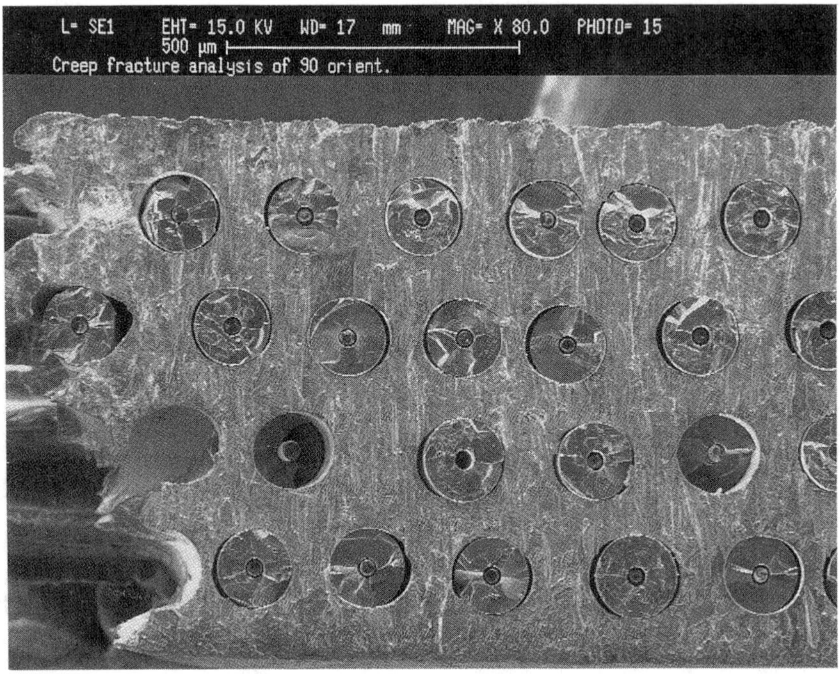

FIG. 17—*Debonded at 90° fiber during creep of the* $[90]_4$ *laminate.*

matrix data are also shown in the figure and highlights the debit in creep strength of the $[90]_4$ layup due to the introduction of 90° fiber. Such debit in creep strength implies that any realistic use of a unidirectional layup requires the use of a high creep-resistant matrix and fiber/matrix interface.

Analysis of the creep data obtained from all the layups suggested that the fiber stress controls creep rupture life. As discussed earlier in Fig. 8, that the matrix relaxes completely in less than 5% of creep life and the load is carried by 0° fibers only during most of the creep life. Assuming that the matrix and 90° lamina have lost all the load-carrying capability, the stress in the 0° fibers for various laminates was determined from the following equation that is based on load equilibrium

$$\sum_{i=1}^{n} \frac{B}{n} W V_f g(\theta_i) \sigma_f = B W \sigma_a \tag{15}$$

where B is the specimen thickness, W is the specimen width, n is the number of plies (lamina) in the laminate, V_f is the volume fraction of the fiber, σ_f is the 0° fiber stress, σ_a is the composite stress, θ_i is the angle of Lamina i, and $g(\theta_i)$ is the lamina orientation factor ranging from 1.0 for the 0° layup to 0.0 for the 90° layup. The values of σ_f for $[0]_4$, $[0/90]_S$, [0/90/0], and $[0/\pm45/90]_S$ laminates were calculated using Eq 15 and plotted versus the LMP in Fig. 20. The calculated values of σ_f and the LMP are also reported in Table 3. Remarkably, the creep rupture data for all the laminates with 0° fibers coalesce to a master line with negligible scatter. Only a limited amount of data is required from creep tests with the $[0]_4$ layup to generate the master σ_f versus LMP plot. As shown in Fig. 20, the proposed

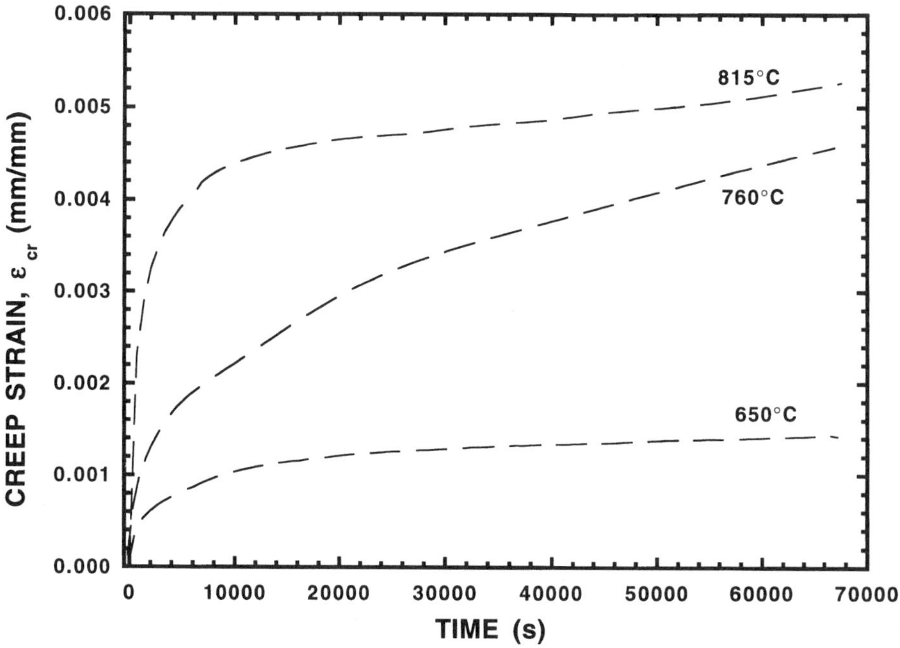

FIG. 18—*Effect of temperature on creep response of the unidirectional SCS-6/TIMETAL 21S.*

modified LMP approach offers a useful tool for the designer to accurately predict the creep strength/rupture life of any laminate of the SCS-6/TIMETAL 21S composite containing 0° fibers subjected to various combinations of stresses and temperatures.

Summary and Conclusions

The creep behaviors of the $[0]_4$, $[0/90]_S$, $[0/\pm45/90]_S$, and $[90]_4$ laminates of SCS-6/TIMETAL 21S at different temperatures was characterized. Little or no tertiary creep was observed for the $[0]_4$ layup and failure was abrupt, occurring after the secondary stage. The $[90]_4$, $[0/\pm45/90]_S$, and $[0/90]_S$ layups produced generic creep curves exhibiting the primary, secondary, and tertiary stages of creep. Specimens with $[0]_4$ orientation showed the highest creep resistance over other layups. The creep life of all the layups containing the 0° fiber was found to be governed by the failure of the unidirectional fibers. The creep strain to failure for $[0]_4$, $[0/90]_S$, and $[0/\pm45/90]_S$ was lower than the failure strain of fibers (1%), while the $[90]_4$ composite laminate elongated more than 10%. The $[90]_4$ layup showed a significant reduction in creep strength compared to the neat matrix. Creep strain increased with increasing stress and increasing temperature for all the laminates. In all the cases, multiple matrix cracks initiated from the surface and edges, and some fibers on the edges showed evidence of crack growth from pre-existing cracks introduced during machining. Failure of all layups occurred due to the growth of a dominant crack.

The environment played a major role in the damage evolution, and for the case of $[0]_4$ orientation, the degradation of the outer layer of fibers was life limiting. For $[0/90]_S$ layups, the debonding of the 90° fiber and crack initiation near the fiber/matrix interface played a major role in the creep behavior. The crack initiation site in the $[0/\pm45/90]_S$ laminate was

TABLE 3—*Creep rupture life, composite and fiber stresses, and LMP for all layups.*

Layup	Temperature, °C	Composite Stress, MPa	Rupture Life, h	Calculated Fiber Stress, MPa	Larson-Miller Parameter (LMP)
$[0]_4$	650	862.0	6.9	2463	19.23
	650	825.0	173.2	2357	20.52
	760	672.0	22.2	1920	22.05
	760	621.0	161.0	1774	22.93
	815	517.5	111.0	1479	23.98
[0/90/0]	650	586.5	277.0	2513	20.72
	815	379.5	63.9	1626	23.72
$[0/90]_S$	650	425.0	18.0	2429	19.61
	650	400.0	93.7	2286	20.02
	650	379.5	190.0	2169	20.78
	815	345.0	9.0	1971	22.79
	815	310.0	116.7	1771	24.00
$[0/\pm45/90]_S$	650	241.5	121.6	2276	21.78
	650	220.0	138.0	2073	22.56
	760	200.0	194.4	1885	23.02
$[\pm45]_S$	650	103.0	5.8	...	19.16
	650	69.0	122.2	...	20.38
	760	55.2	277.7	...	23.18
	815	34.5	92.2	...	24.23
$[90]_4$	650	86.2	1.0	...	18.49
	650	77.6	1.4	...	18.60
	650	69.0	3.9	...	19.00
	650	48.3	9.7	...	19.37
	650	44.8	14.4	...	19.53
	650	41.4	44.3	...	19.98
	650	34.5	117.3	...	20.37

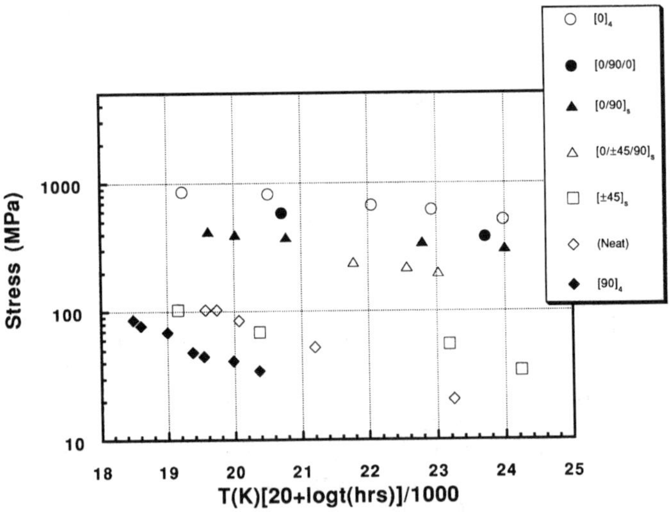

FIG. 19—*Plot of applied stress versus Larson-Miller parameter (LMP) for all layups of SCS-6/TIMETAL 21S tested at different stresses and temperatures.*

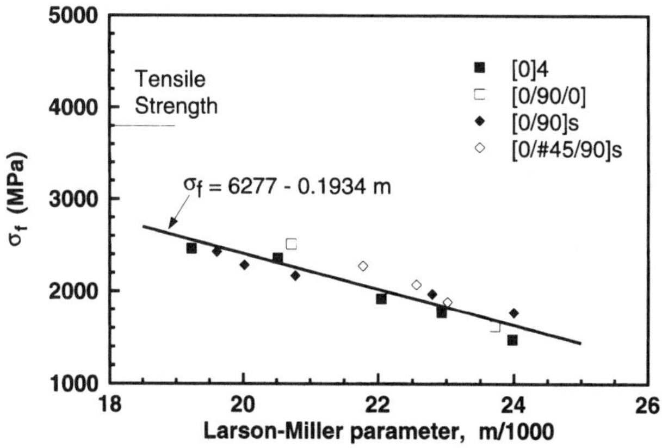

FIG. 20—*Plot of equivalent fiber stress versus Larson-Miller parameter (LMP) for all layups of the SCS-6/TIMETAL 21S tested at different stresses and temperatures.*

observed near the matrix/±45° fiber interface. Extensive matrix deformation and evidence of cracking was found around the ±45° fibers.

A simple two-bar model was successfully used to predict the creep behavior of the unidirectional SCS-6/TIMETAL 21S composite laminate at 650°C. The model predicted that the matrix relaxes to less than 10% of the initial stress in about 4000 s. At anticipated service stress levels and temperatures, the initial stress in the matrix is at the yield surface. Comparison of the analysis and data indicates that the in situ fiber strength could be degrading due to exposure to temperature.

The applied stress versus the Larson-Miller parameter (LMP) plot offered a direct comparison of the creep lives of various layups. The LMP approach was used assuming that the composite behaved like an effective monolithic material. The modified LMP approach based on fiber stress can be used as an engineering tool to predict the creep rupture lives of various layups containing 0° fibers under different conditions of load and temperature. The use of the modified LMP approach is entirely empirical and needs further research.

Acknowledgments

This research was conducted at the Materials Directorate, Wright Laboratory (WL/MLLN), Wright-Patterson Air Force Base, Ohio. The authors from the University of Dayton Research Institute were supported under Contract F33615-91-C-5606, which was funded under Task 2302P101 of the Air Force Office of Scientific Research. The authors acknowledge the assistance of D. Garner and D. Knapke in conducting the experiments and preparing some of the graphs.

References

[*1*] Gambone, M. L., *Fatigue and Fracture of Titanium Aluminides, Vol. II,* WRDC-TR-89-4145, Wright-Patterson AFB, OH, Feb. 1990.

[*2*] Evans, D. J., "Study of Creep Damage Mechanisms and Steady State Creep Regime in SCS-6/Ti-6Al-4V Composites," M. S. thesis, University of Dayton, OH, 1990.

[*3*] Khobaib, M., "Creep Behavior of SCS-6/Ti-24Al-11Nb Composite," *Titanium Aluminide Composites,* WL-TR-91-4020, P. R. Smith, S. J. Balsone, and T. Nicholas, Eds., Wright Laboratory, Wright-Patterson AFB, OH, Feb. 1991, pp. 450–466.

[4] Khobaib, M., "Damage Evolution in Creep of SCS-6/Ti-24Al-11Nb Metal-Matrix Composites," *Journal of Reinforced Plastics and Composites,* Vol. 12, March 1993, pp. 296–310.
[5] Schuyler, D. R., *Feasibility of Aluminide Metal Matrix Composites (MMC) for 1400°F Applications,* NARDC-91066-60, Naval Air Development Center, Warminster, PA, 1990.
[6] Schwenker, S. W., Evans, D. J., and Eylon, D., "Longitudinal Creep Behavior and Damage in SCS-6/Ti-6Al-4V Metal Matrix Composites," *Proceedings,* Seventh World Conference on Titanium, San Diego, CA, 1992, pp. 2593–2600.
[7] Kelly, A. and Street, K. N., "Creep of Discontinuous Fiber Composites II. Theory for the Steady-State," *Proceedings,* Royal Society of London, Vol. A328, 1972, pp. 283–293.
[8] Mileiko, S. T., "Steady State Creep of a Composite Material with Short Fibers," *Journal of Material Science,* Vol. 5, 1970, pp. 254–261.
[9] Taya, M., "Creep Behavior of Metal Matrix Composites," *Metal Matrix Composites: Mechanisms and Properties,* R. K. Everett and R. J. Arsenault, Eds., Academic Press, Boston, MA, 1991, pp. 189–215.
[10] McLean, M., "Mechanisms and Models of High Temperature Deformation of Composites," *Materials and Engineering Design: The Next Decade,* B. F. Dyson and D. R. Hayhurst, Eds., Institute of Metals, London, 1989, pp. 287–294.
[11] Branca, T. R., "Creep of a Uniaxial Metal Matrix Composite Subjected to Axial and Normal Lateral Loads," T&AM Report 341, University of Illinois at Urbana-Champaign, June 1971.
[12] Gato, S. and McLean, M., "Role of Interfaces in Creep of Fiber Reinforced Metal-Matrix Composites-1," *Acta Metallurgica Materials,* 1991, pp. 153–164.
[13] McDanels, D. L., Singorelli, R. A., and Weaton, J. W., "Analysis of Stress Rupture and Creep Properties of Tungsten Fiber-Reinforced Copper Composites," NASA-TN-D-4173, Technical Note, NASA Lewis Research Center, Cleveland, OH, 1967.
[14] Soliman, F. Y., "Creep and Rupture of Graphite-Epoxy Composites," *Composite Materials: Testing and Design, ASTM STP 460,* American Society for Testing and Materials, Philadelphia, 1969, pp. 254–270.
[15] de Silva, A. R. T., "A Theoretical Analysis of Creep in Fiber Reinforced Composites," *Journal of the Mechanics and Physics of Solids,* Vol. 16, 1968, pp. 169–186.
[16] Schwenker, S. W., Roman, I., and Eylon, D., "Creep Behavior of SCS-6/Ti-6Al-4V Unidirectional Composites," *Advanced Composites,* Feb. 1993, pp. 1169–1176.
[17] Schwenker, S. W., Evans, D. J., and Eylon, D., "Longitudinal Creep Behavior and Damage in SCS-6/Ti-6Al-4V Metal Matrix Composites," *Proceedings,* Seventh World Conference on Titanium, F. H. Froes and I. L. Caplan, Eds., The Metallurgical Society of the American Institute of Mining, Metallurgical, and Petroleum Engineers, 1993, pp. 2593–2600.
[18] Wright, K., "Creep Behavior and Modeling of SCS-6/Titanium MMC," *Titanium Matrix Components,* WL-TR-92-4035, P. R. Smith and W. C. Revelos, Eds., Wright-Patterson AFB, OH, April 1992, pp. 251–275.
[19] Ohno, N., Toyoda, K., Okamoto, N., Miyake, T., and Nishide, S., "Creep Behavior of Unidirectional SCS-6/Ti-15-3 Metal Matrix Composite at 450°C," *Journal of Engineering Materials and Technology,* 1994, pp. 116–125.
[20] Du, Z. Z. and McMeeking, R. M., "Creep Models for Metal Matrix Composites With Long Brittle Fibers," *Journal of the Mechanics and Physics of Solids,* Vol. 43, No. 5, 1995, pp. 701–726.
[21] Foye, R. L., "Theoretical Post-Yielding Behavior of Composite Laminates: Part 1—Inelastic Micromechanics," *Journal of Composite Materials,* Vol. 7, 1973, pp. 178–193.
[22] Foye, R. L., "Creep Analysis of Laminates," *Composite Reliability, ASTM STP 580,* American Society for Testing and Materials, Philadelphia, PA, 1975, pp. 381–395.
[23] Adams, D. F., "Inelastic Analysis of a Unidirectional Composite Subjected to Transverse Loading," *Journal of Composite Materials,* Vol. 4, 1970, pp. 310–328.
[24] Bahei-El-Din, Y. A., "Plasticity Analysis of Fibrous Composite Laminates Under Thermomechanical Loads," *Thermal and Mechanical Behavior of Ceramic and Metal Matrix Composites, ASTM STP 1080,* J. M. Kennedy, H. H. Moeller, and W. S. Johnson, Eds., American Society for Testing and Materials, Philadelphia, 1990, pp. 20–39.
[25] Teply, J. L. and Dvorak, G. J., "Bounds on Overall Instantaneous Properties of Elastic-Plastic Composites," *Journal of the Mechanics and Physics of Solids,* Vol. 36, No. 1, 1988, pp. 29–58.
[26] Boyle, M. and Ashbaugh, N. E., unpublished work, University of Dayton Research Institute, Dayton, OH, 1993.
[27] Mendelson, A., *Plasticity: Theory and Applications,* Robert E. Krieger Publishing Company, Malabar, FL, 1986.

[*28*] Kroupa, J. L. and Neu, R. W., "The Nonisothermal Viscoplastic Behavior of a Titanium Matrix Composite," *Composites Engineering,* accepted for publication.
[*29*] Toth, I. J., "Creep and Fatigue Behavior of Unidirectional and Cross-plied Composites," *Composite Materials: Testing and Design, ASTM STP 460,* American Society for Testing and Materials, Philadelphia, PA, 1969.
[*30*] John, R., Khobaib, M., and Smith, P. R., "Prediction of Creep-Rupture Life of Unidirectional Titanium Matrix Composites Subjected to Transverse Loading," *Metallurgical and Materials Transactions,* submitted for publication.
[*31*] Larson, F. R. and Miller, J., "A Time-Temperature Relationship for Rupture and Creep Stresses," *Transactions,* American Society of Mechanical Engineers, Vol. 74, July 1952, pp. 765–775.
[*32*] Orr, R. L., Sherby, O. D., and Dorn, J. E., "Correlations of Rupture Data for Metals at Elevated Temperatures," *Transactions,* American Society for Metals, Vol. 46, 1954, pp. 113–119.
[*33*] Monkman, F. C. and Grant, N. J., "An Empirical Relationship Between Rupture Life and Minimum Creep Rate in Creep Rupture Tests," *Proceedings,* American Society for Testing and Materials, Philadelphia, Vol. 56, 1956, pp. 593–620.
[*34*] Manson, S. S., Succop, G., and Brown, W. F., Jr., "The Application of Time-Temperature Parameters to Accelerated Creep-Rupture Testing," *Transactions,* American Society for Metals, Vol. 51, 1959, pp. 911–934.
[*35*] Basinski, Z. R., "Activation Energy for Creep of Aluminum at Sub-atmospheric Temperatures," *Acta Metallurgica,* Vol. 5, 1957, pp. 684–686.
[*36*] Pharr, G. M. and Nix, W. D., "Application of the Methods of Dislocation Dynamics to Describe Plastic Flow in Both B.C.C. and F.C.C. Metals," *Acta Metallurgica,* Vol. 27, 1979, pp. 433–440.

Golam M. Newaz,[1] *Bhaskar S. Majumdar,*[2] *and Frederick W. Brust*[3]

Inelastic Deformation Mechanisms in SCS-6/Ti 15-3 Metal Matrix Composite (MMC) Lamina Under Compression

REFERENCE: Newaz, G. M., Majumdar, B. S., and Brust, F. W., "**Inelastic Deformation Mechanisms in SCS-6/Ti 15-3 Metal Matrix Composite (MMC) Lamina Under Compression,**" *Life Prediction Methodology for Titanium Matrix Composites, ASTM STP 1253,* W. S. Johnson, J. M. Larsen, and B. N. Cox, Eds., American Society for Testing and Materials, 1996, pp. 208–230.

ABSTRACT: An investigation was undertaken to study the inelastic deformation mechanisms in $[0]_8$ and $[90]_8$ Ti 15-3/SCS-6 lamina subjected to pure compression. Monotonic tests were conducted at room temperature (RT), 538°C, and 650°C. Results indicate that mechanical response and deformation characteristics were different in monotonic tension and compression loading whereas some of those differences could be attributed to residual stress effects. There were other differences because of changes in damage and failure modes. The inelastic deformation in the $[0]_8$ lamina under compression was controlled primarily by matrix plasticity, although some evidence of fiber-matrix debonding was observed. Failure of the specimen in compression was due to fiber buckling in a macroscopic shear zone (the failure plane). The inelastic deformation mechanisms under compression in $[90]_8$ lamina were controlled by radial fiber fracture, matrix plasticity, and fiber-matrix debonding. The radial fiber fracture was a new damage mode observed by MMCs. Constitutive response was predicted for both the $[0]_8$ and $[90]_8$ laminae, using AGLPLY, METCAN, and Battelle's Unit Cell FEA model. Results from the analyses were encouraging.

KEYWORDS: inelastic deformation mechanisms, metal matrix composites, compression, constitutive response, titanium, titanium matrix composites, life prediction, titanium alloys, fatigue (materials), modeling

Titanium-based metal matrix composites (MMCs) offer excellent potential for high temperature applications that require a combination of high temperature strength, stiffness, and toughness. Current advances in processing of this class of materials have made them technologically viable for their use in various engine and aerothermal applications. Considerable research to date has been conducted in characterizing silicon-carbide (SiC) fiber-reinforced titanium composites [*1–9*]. The references reflect only a small number of representative publications from a much larger pool.

An important feature of MMCs is that there is significant difference in the coefficient of thermal expansion between the fiber and matrix; the matrix has a larger coefficient of thermal expansion. Consequently, large compressive radial stresses build up at the interface between the matrix and the fiber as the composite is cooled from the processing temperature. Of course, fiber volume fraction affects the magnitude of the residual stresses. On the other hand, the axial and circumferential stresses in the matrix are both tensile in nature. This

[1]Professor, Wayne State University, Detroit, MI 48202.
[2]Senior scientist, Universal Energy Systems, Inc., Dayton, OH 45433.
[3]Senior scientist, Battelle Memorial Institute, Columbus, OH 43201.

combination of stresses sets up yielding or near yielding conditions in the matrix immediately adjacent to the fiber. A consequence of such high stresses is that microyielding may occur well before macroscopic yield is observed when the MMC is loaded mechanically. In addition, the strength of the chemical bond between the fiber and the matrix has been shown by many researchers [*5,7,9*] to be weak in the composite. Also, the presence of the reaction zone adds another dimension to the complex deformation characteristics in these composites under applied loading.

The loading mode that has received most attention by researchers in the field is tensile loading. Although this is certainly a dominant loading mode for the application of fiber-reinforced MMCs, compressive loading for these structural materials is a definite possibility for various components. Compression on one surface is a reality in bending. Furthermore, in the case of reversed fatigue loading, the composite will experience compressive loading in part of the cycle.

It is also anticipated that structural design engineers will require design allowables for tensile, compressive, and shear properties. Compressive loading has not received much attention to date for high temperature composites. Past studies on boron/aluminum composites concentrated on their longitudinal compression behavior [*10*]; however, their transverse compression behavior was not studied. Understanding of the evolution of damage and failure modes under compressive loading of the MMC at high temperature is seriously lacking. Considerable research is needed to clarify and resolve the micromechanics issues for compressive response. It is also necessary to complement experimental efforts with analytical ones to predict the constitutive response of the MMC under compressive loading since such efforts are not under consideration in many of the major high temperature programs.

Often, damage accumulates in local regions in a structure due to high stresses and strains. If the damage reaches a critical level, it can lead to failure of the structure. Constitutive models are essential to evaluate local stress and strain fields developed due to global structural loading. However, rational constitutive models can be only developed if the mechanisms for material deformation and damage development can be clarified and used to develop the models. Analytical constitutive models often require an assumption of ideal microstructures and may simplify the problem to a level where realism is compromised, particularly for heterogeneous materials like composites. A computational framework on the other hand provides much versatility and can be effectively utilized provided deformation mechanisms are properly accounted for. For MMC structures subjected to compressive loading (a situation that exists even in the case of simple bending), constitutive modeling under compressive loading is a clearly defined need. Clarification of damage modes and deformation mechanisms are equally important compared with constitutive modeling and are integrally related. The material deformation characteristics and modeling for monotonic loading constitutes the basic requirements for structural analysis.

Inelastic deformation is critical in that its initiation and evolution contributes to the failure of a structural part. Proper constitutive models can only be developed when actual inelastic deformation mechanisms can be ascertained and incorporated into modeling.

In recognition of the need to clarify the inelastic deformation mechanisms under compressive loading, a brief investigation was undertaken to monitor the initiation and progression of damage in the SCS-6/Ti 15-3 MMC for both $[0]_8$ and $[90]_8$ laminae subjected to monotonic compressive loads, particularly at elevated temperature. This small effort was only part of a much larger effort for damage evaluation under cyclic loading in titanium-based MMCs that is covered in a separate report. This information will be extremely valuable for constitutive and damage modeling for this class of materials. Results were also compared with tension response and corresponding deformation mechanisms.

Experimental Aspects

The material tested was an eight-ply unidirectional SCS-6/Ti 15-3 composite, approximately 1.99 mm thick, with a fiber volume fraction of approximately 0.35. The SCS-6 (SiC) fiber diameter is approximately 140 μm, and it contains alternating outer layers of carbon and silicon that protect the fiber from damage during handling and consolidation. The Ti 15-3 alloy is a metastable body-centered cubic (bcc) beta titanium-alloy, the bcc phase being stabilized by vanadium. The material is metastable because hexagonal close-packed (hcp) alpha phase, which is the stable room temperature phase of pure titanium, precipitates when the material is held at temperatures as low as 400°C for long periods of time.

The $[0]_8$ and $[90]_8$ rectangular test specimens were machined from the unidirectional panel using an electric-discharge machining (EDM) technique, with specimen dimensions of 2.0 by 10.2 by 104.1 mm. The gage length was 25.4 mm. The specimens were mechanically polished after EDM machining to minimize any damage associated with the machining. All specimens were tested in the as-fabricated condition, that is, no heat-treatment was performed prior to the testing. Tests were conducted at room temperature, 538°C (1000°F), and 650°C (1200°F).

Specimens were gripped using friction grips, and loaded on a servohydraulic testing machine at a strain rate of approximately 0.004/s. Compression testing was done using a Battelle developed compression fixture as shown in Fig. 1 that uses buckling guides. The longitudinal strains were measured using an extensometer.

A few tests were conducted without buckling guides, particularly for $[0]_8$ specimens. Longitudinal macrobuckling of the specimen is shown in Fig. 2. A failure mode typically not observed in MMCs, that is delamination, was also observed. Premature buckling may not allow evaluation of proper mechanical response and properties at higher stress or strains. Therefore, lateral restraint for unidirectional higher strength MMCs are essential to develop a good understanding of key deformation mechanisms under pure compressive loading.

Following mechanical testing, most of the compression specimens were not loaded to failure. A number of specimens were polished at the edges. A fairly complex polishing procedure has been developed at Battelle for evaluating MMCs to preserve the characteristics of deformation. Repeated polishing was performed on a few tested and untested samples to confirm that any cracking, debonding, or slip bands was due to deformation alone, and not associated with the polishing procedure or manufacturing. Specimens were etched using Kroll's reagent, which was found to be effective in revealing slip bands both at surface and subsurface locations. Metallographic specimens were examined optically and key deformation characteristics were photographed.

The overall test matrix is given in Table 1. The study did not allow for any statistical property evaluation.

Mechanical Behavior Evaluation

Longitudinal Lamina

The typical stress-strain response including unloading for a $[0]_8$ specimen is shown in Fig. 3. The compression test result without buckling guides is also shown in the same figure. The elastic modulus in tension and compression were similar, ranging between 175 to 179 GPa. Also, the corresponding tension data is presented in the same plot. It may be noted that the compression capability of the $[0]_8$ lamina is significantly higher than the tension case both in terms of (1) the onset of nonlinearity and (2) the higher ultimate strength (when premature buckling is avoided). The compression specimen was not loaded to failure. However, exper-

FIG. 1—*Compression test setup.*

imental efforts suggest that the compression strength will be quite high. In the absence of restraint for buckling, it is clear that premature buckling takes place resulting in no useful information in the nonlinear deformation regime. The failure mode for compression without buckling guides will be discussed in the next section. The unloading line is parallel to the elastic loading line for the $[0]_8$ composite and indicate that the deformation response is controlled primarily by plastic deformation. The mechanical responses at 538 and 650°C are included in Fig. 4 along with room temperature response. Table 2 provides data on the effect of temperature on the proportional limit and modulus. The decrease in modulus and proportional limit are due to the decrease in flow properties of the matrix at elevated temperature.

The onset of nonlinear deformation at room temperature (RT) in compression (~2200 MPa) was significantly higher than in tension (~980 MPa). This difference is due to a residual tensile stress in the matrix; calculations show [*11*] that for this system the axial

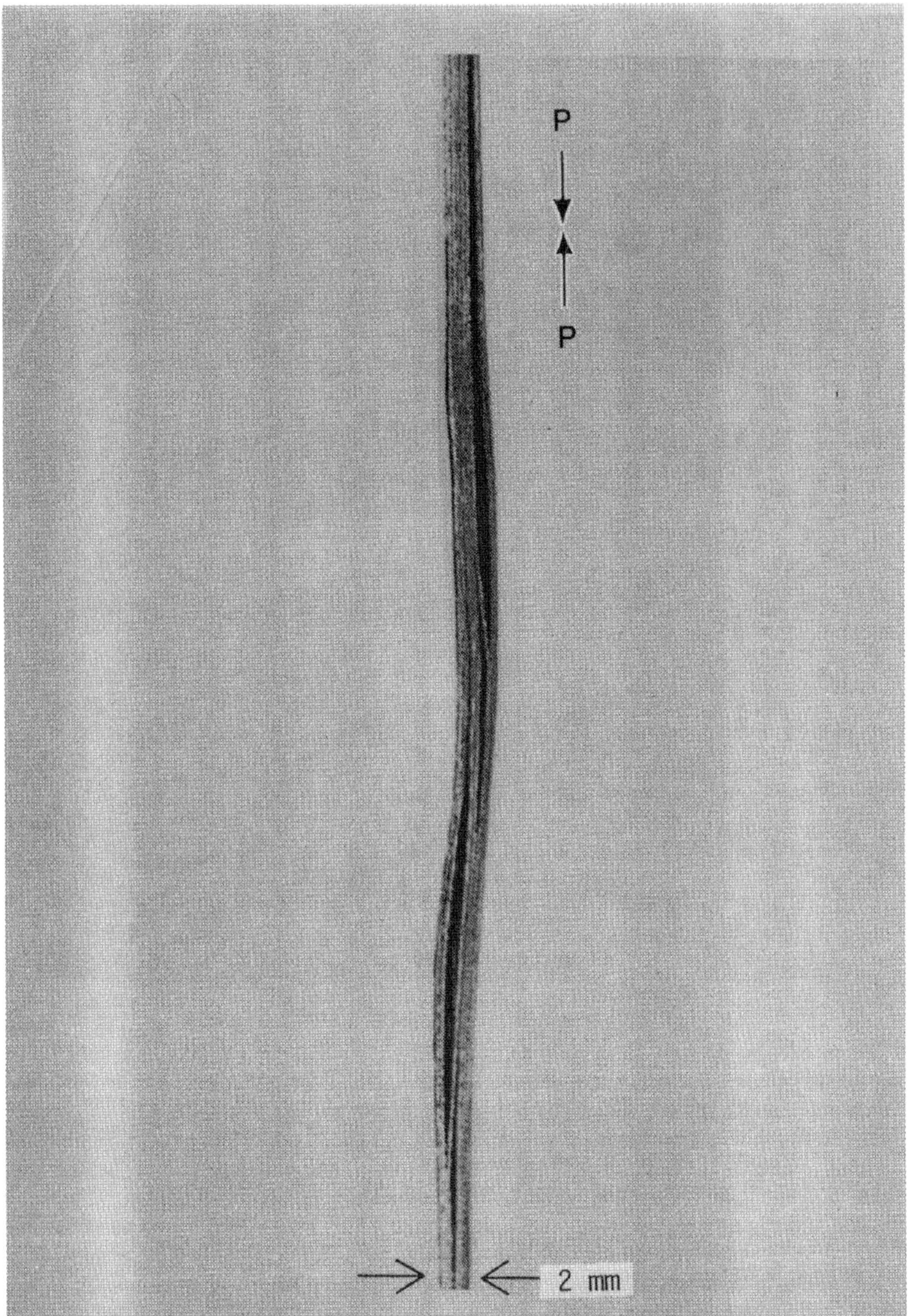

FIG. 2—*Macrobuckling of unidirectional specimen without restraints, subjected to compression.*

TABLE 1—*Compression test matrix (SCS-6/Ti 15-3).*

Layup	Number of Specimens	Temperature	Comments
$[0]_8$	1	RT[a]	stress-strain, edge replication
	2	RT	intermediate unloading
	1	RT	without guides
	1	538°C	stress-strain, unloading
	1	650°C	stress-strain, unloading
$[90]_8$	1	RT	stress-strain
	2	RT	intermediate unloading
	1	RT	edge replication
	1	538°C	stress-strain, unloading
	1	650°C	stress-strain, unloading

[a]RT = room temperature.

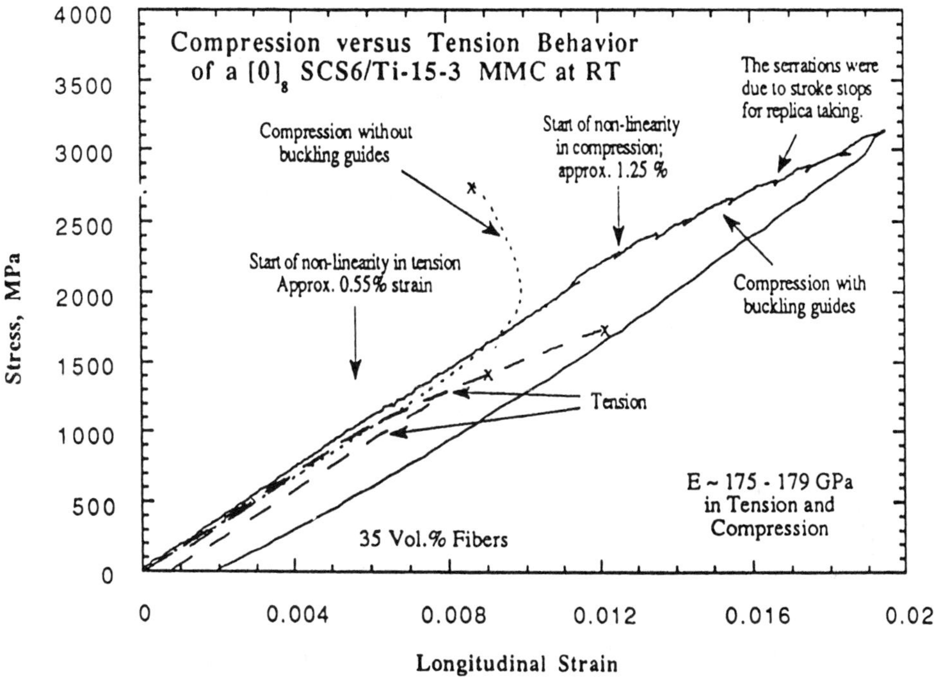

FIG. 3—*Typical stress-strain response of $[0]_8$ lamina in compression. Also includes tension results for the same lamina.*

residual stresses at RT are approximately 412 MPa in the matrix at the fiber-matrix interface, and −1000 MPa in the fiber. Assuming an elastic one-dimensional response, the residual axial stress can be calculated using the formula

$$S_r = [(|S_c| - |S_t|)/2) \cdot E_m/E_c$$

where S_c and S_t are the measured proportional limits in compression and tension, and E_m and E_c are matrix and composite moduli, respectively. Using this formula, the tensile axial residual stress in the matrix is calculated to be 335 MPa.

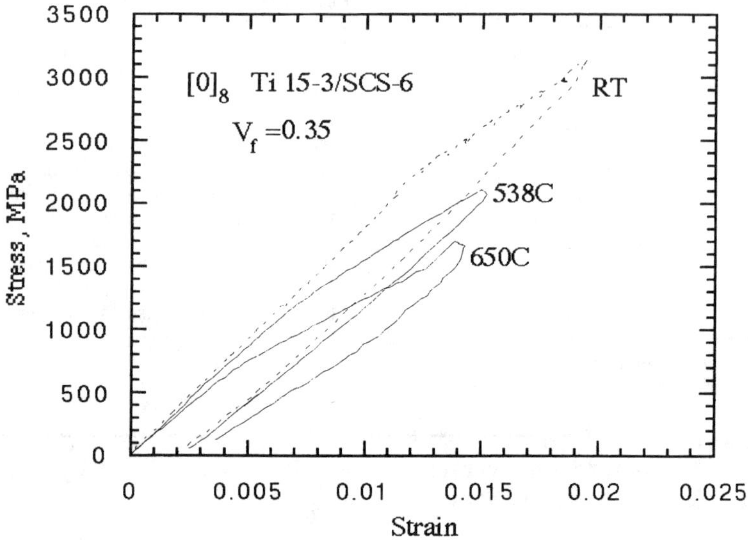

FIG. 4—*Mechanical response at elevated temperature for $[0]_8$ lamina in compression.*

TABLE 2—*$[0]_8$ SCS-6/Ti 15-3 properties in compression.*

Temperature	Young's Modulus, GPa	Proportional Limit	
		Stress, MPa	Strain
RT[a]	179	2200	0.012
538°C	170	1200	0.007
650°C	150	600	0.004

[a]RT = room temperature.

Transverse Lamina

Figure 5 shows the typical stress-strain response of the $[90]_8$ lamina for the three different temperatures, namely, room temperature, 538°C (1000°F), and 650°C (1200°F). In each case, the specimens were unloaded in the nonlinear regime. There is a progressive loss in elastic modulus and proportional limit as a function of temperature as given in Table 3. The distinct knee observed in tension for the onset of nonlinearity in the $[90]_8$ lamina (Fig. 5) is clearly absent in the case of compression. The onset of nonlinearity in the tension stress-strain curve was attributed to fiber-matrix debonding—a dominant damage mechanism for the $[90]_8$ MMC lamina. Note the large work hardening period for the MMC, extending from approximately 0.5% strain to approximately 1% strain. This period was much longer than for the matrix material.

The RT and 538°C compression stress-strain results are compared in Fig. 6 with those of the tension case to accentuate the difference in the response of the same SCS-6/Ti 15-3 composites under different loading conditions. This plot also contains the pure matrix stress-strain response in tension at RT and at 538°C. There are two important observations that can be made from this comparison: (1) the initial elastic or tangent moduli under tension and compression are similar, and (2) the proportional limits in compression are higher than for their corresponding tension counterparts. The higher proportional limit can be explained by the presence of a beneficial tensile residual stress that must be overcome in compression

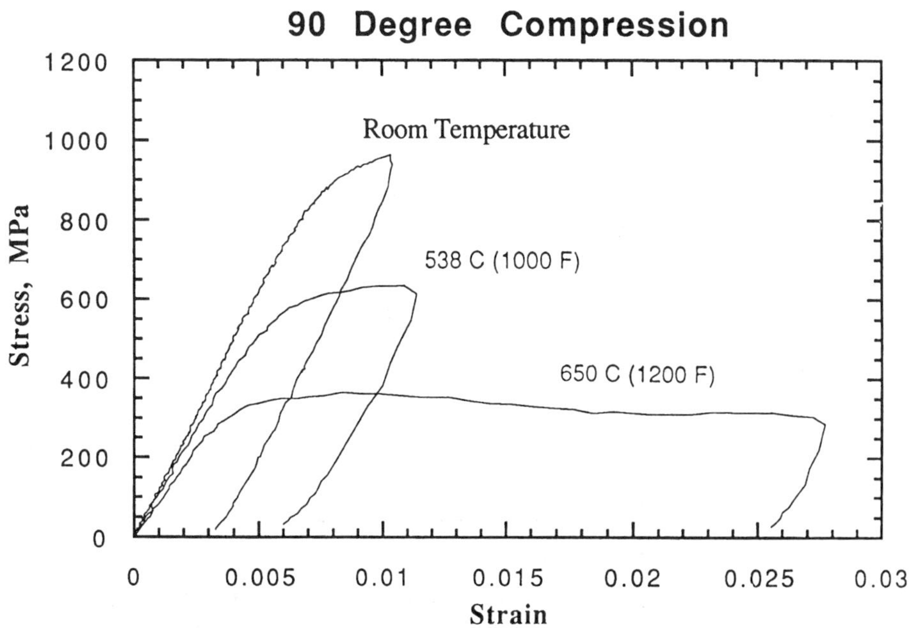

FIG. 5—*Mechanical response of $[90]_8$ lamina in compression.*

TABLE 3—*$[90]_8$ SCS-6/Ti 15-3 properties in compression.*

		Proportional Limit	
Temperature	Young's Modulus, GPa	Stress, MPa	Strain
RT[a]	125	697	0.006
538°C	108	428	0.004
650°C	92	276	0.003

[a]RT = room temperature.

before a compressive proportional limit is reached. A matrix average radial residual compressive stress of ~190 MPa at RT, estimated from the proportional limits in tension and compression, compares well with a prediction of ~200 MPa using finite element analysis (FEA).

The unloading characteristics for the $[90]_8$ specimens under compression are quite different than those observed in tension. The initial unloading line is parallel to the initial loading line in compression signifying plastic deformation at work. The offset strains in each case for all temperatures is significantly large (Fig. 5). Further unloading results in a somewhat more compliant behavior as zero load is approached.

Microstructural Evaluation

Longitudinal Lamina

Optical photomicrographs of replicas taken from polished specimens show the presence of extensive matrix plasticity in the lamina (Fig. 7). Fiber-matrix debonding is also evident at higher strain levels within the ply. These observations were only made at RT. The failure

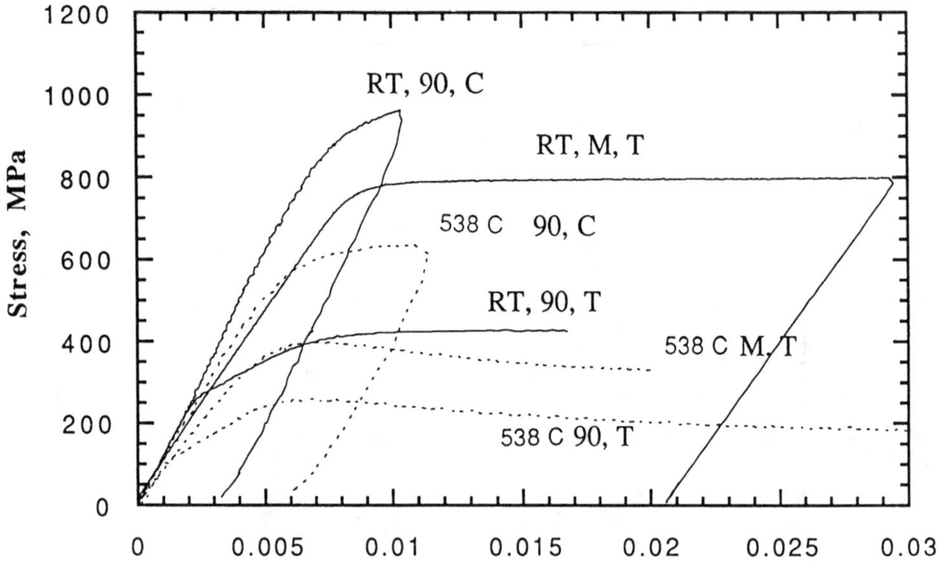

FIG. 6—*Comparison of compression and tension response of* $[90]_8$ *lamina RT and 538°C. Matrix behavior in tension is also shown (T = tension, C = compression, and M = matrix).*

due to fiber buckling in an unrestrained test is shown in Fig. 8. The fibers break in a shear plane within the gage section. This failure is indicative of a buckling instability due to compressive load.

Transverse Lamina

Optical photomicrographs of specimens polished after testing revealed the nature of deformation characteristics under compressive loading quite clearly. The initial deformation mechanism under monotonic compressive loading is found to be matrix plasticity along the loading axis between fibers as shown in Fig. 9 for a specimen tested at room temperature. This appears to be a dominant mechanism controlled by high strains between the fibers within a ply along the loading line during compression and less plasticity between the adjacent plies. This partitioning of plasticity was likely responsible for the high work hardening that was observed for the compression specimen (Fig. 6). It may be recalled that in the case of tension, there was only a knee associated with debonding, and the smooth behavior under compression as shown in Fig. 6 was absent. The slip lines are considerably more prominent compared to the similar slip lines observed in the tension case at similar locations [7,9]. In the case of tension, the initial microplasticity is followed by fiber-matrix debonding at the fiber poles at the onset of departure from the linear elastic response (knee). In the case of compression, the nonlinearity is a smooth one (Fig. 6) without any abrupt change from linearity. Debonding, as observed in the tension case at the two poles of a fiber, is absent in the compression case, as the matrix is in compression at the poles of the fiber (top and bottom points on the fibers in Fig. 9).

Careful observation also reveals a very interesting damage condition in the composite under compression. There are radial cracks within the fibers oriented primarily along the compression loading axis that form after the initial slip lines have developed in the matrix as mentioned earlier. Reaction zone cracks (RZC) are also observed. These damage modes are

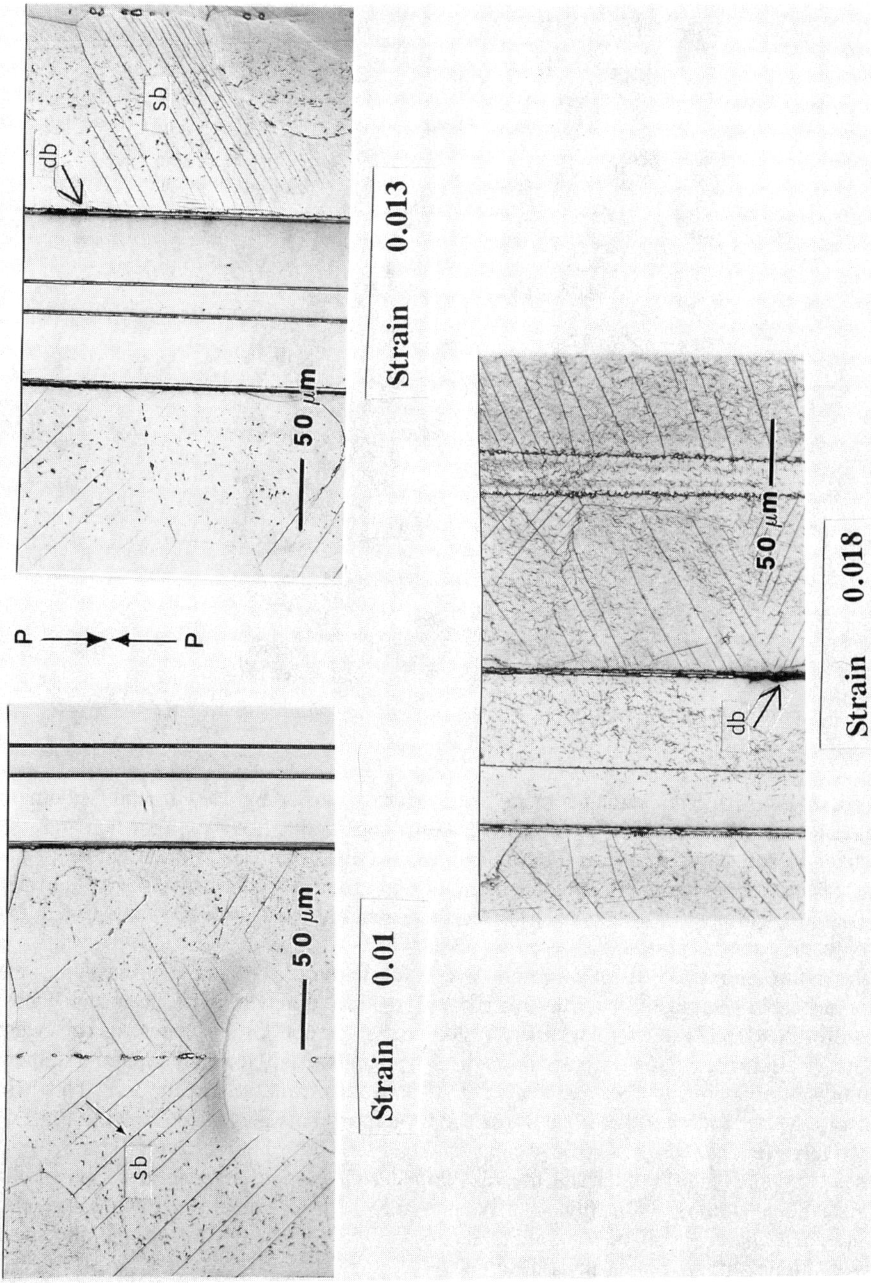

FIG. 7—*Plastic slip bands (sb) and debonding (db) in* $[0]_8$ *lamina at RT in compression.*

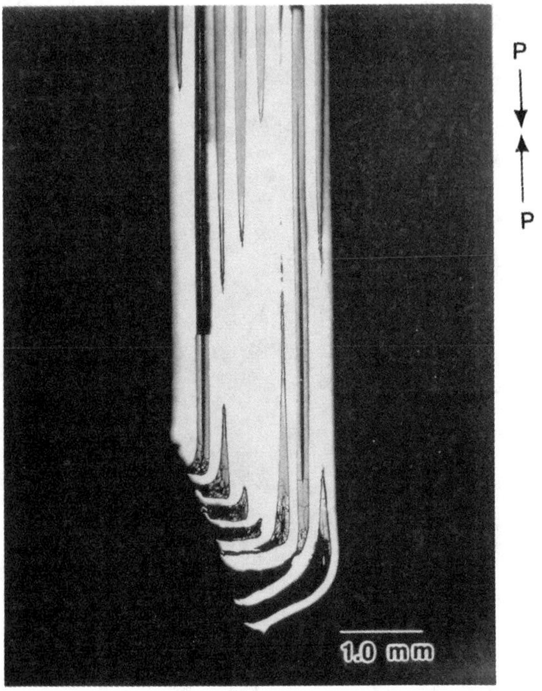

FIG. 8—*Fiber buckling in [0]$_8$ lamina in compression at RT.*

depicted in Fig. 10. This damage mode was reported earlier by Newaz and Majumdar [*12–14*]. Radial fiber cracking is indicative of the low radial fiber strength that may be related to the soft carbon core in the SCS-6 fiber and the columnar grain structure of the chemical vapor deposition (CVD) deposited SiC with grain boundary facets oriented radially. These cracks are not observed in as-received materials and were not observed in the tension loading mode.

In the nonlinear regime, further compressive deformation of the composite results in the development of bulk plasticity in the matrix with extensive networks of shear slip bands. This is illustrated in Fig. 11 for specimens tested at 538 and 650°C. These micrographs were taken under polarized light to capture the active slip bands. Also, it may be noted that there is significant debonding of fiber and matrix at 90° locations from the loading axis. This is in contrast to the debonding at the poles for the tension case. This is clearly shown in Fig. 12. Qualitatively, the debonding in compression can be attributed to high local tensile transverse strains at the fiber-matrix interface that develop during compressive deformation of the specimen. Shear strains at the fiber-matrix interface are also expected. However, precise computational analysis will be required to determine the contribution of normal and shear strains to the debonding phenomena observed.

The transverse compression response is compared with the transverse tension response in Table 4 in terms of the leading inelastic deformation mechanisms. Although, the radial cracking of fiber in compression is a concern, it may be noted that the most debilitating damage, that is, one that occurs at a fairly low strain level, is fiber-matrix debonding under tension. Radial fiber cracking is observed in compression at a strain level of 0.007 whereas debonding under tension initiates around a strain level of 0.003.

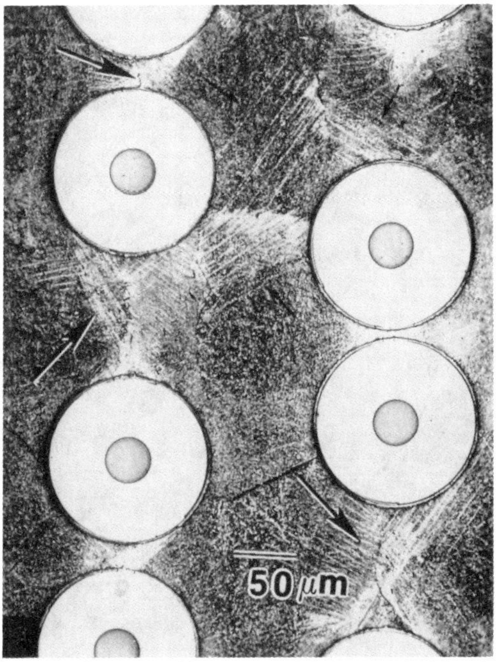

FIG. 9—*Matrix plasticity between fibers in Ti 15-3/SCS-6 $[90]_8$ lamina under compression at RT. Loading is up and down.*

The sequence and location of damage development under tension and compression can explain the different stress-strain responses of the composite. Furthermore, this specific behavior with regard to the "knee" can be rationalized based on the effect of debonding orientation on the mechanical response of the composites in tension and compression as illustrated in Fig. 13. Reaction zone cracks and debonding that are perpendicular to load axis as in the case in tension are anticipated to have a larger effect on compliance.

Analytical Considerations

Within the scope of the compression investigation, an effort was made to determine the suitability of a few constitutive models to predict the stress-strain response under monotonic compression. Our effort was only confined to the transverse constitutive response.

A unique set of micromechanics equations have been developed by NASA researchers for high temperature metal matrix composites [*15*]. These equations, coupled with a set of thermoviscoplastic nonlinear multifactor relationships developed by these researchers [*16*], are consolidated into a computational code titled METCAN. It treats the material nonlinearity at the constituent level, where the material's nonlinear time-temperature-stress dependence is modeled using power-law type multifactor interaction relationships. Although the model can account for interface damage in a limited way, by changing the mechanical properties of the interphase region (the region consisting of the reaction-zone and the interfaces), it cannot predict a priori what the interphase property should be. METCAN was used to predict the stress-strain behavior for both $[0]_8$ and $[90]_8$ SCS-6/Ti 15-3 laminae under both room and elevated temperatures. These results are presented in Fig. 14. The

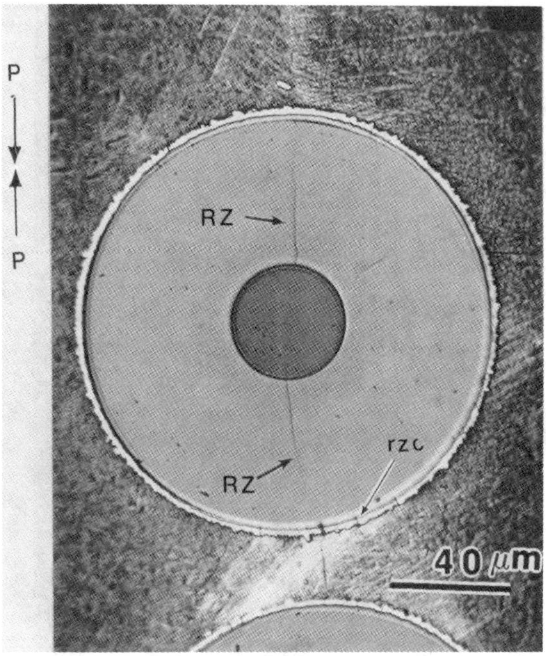

FIG. 10—*Radial cracks in SCS-6 fiber in a [90]$_8$ Ti 15-3/SCS-6 lamina under compression at RT. Loading is up and down.*

correlations are quite good and are encouraging. METCAN is not suitable for unloading and, therefore, no attempt was made to predict the unloading response.

Also, the AGLPLY code was used (continuum plasticity-based model by Dvorak and Baheiel-Din [*17*], which is based on a vanishing fiber diameter model, and can be used effectively to predict lamina and laminate responses from the elastic-plastic responses of the matrix and fiber materials. The primary drawback of the model is that it cannot account for damage. Comparison of the AGLPLY predictions with experimental data for the [0]$_8$ MMC at RT is given in Fig. 15. In the predictions, the [90]$_8$ MMC are compared with experimental data in Fig. 16. The figure also includes experimental data obtained by Lerch [*18*] at NASA Lewis using a 32-ply MMC without any buckling guides. While the experimental data from the two investigations are in agreement, the AGLPLY predictions are not good in the inelastic regime of deformation. Rather, AGLPLY appeared to mimic the matrix behavior. The inadequacy of AGLPLY prediction may be related to the assumption of vanishing fiber diameter that averages out the microstructure and cannot account for the way plasticity is partitioned between different regions of the MMC.

Battelle's finite element computational unit cell model developed by Brust et al. [*11*] was also used to predict the stress-strain response of [0]$_8$ and [90]$_8$ laminae for the RT case. The following analytical aspects were involved in the model: (1) two-dimensional generalized plane-strain conditions; (2) periodic boundary conditions; (3) classical thermal plasticity for the matrix, whose property was assumed to be independent of position; (4) thermo-elastic properties of the fiber; and (5) the fiber-matrix interface was modeled as a discrete interface. For the case of simulating constraints due to the buckling guides, transverse stiffening was modeled using springs with appropriate constants. The results are shown in Figs. 17 and 18.

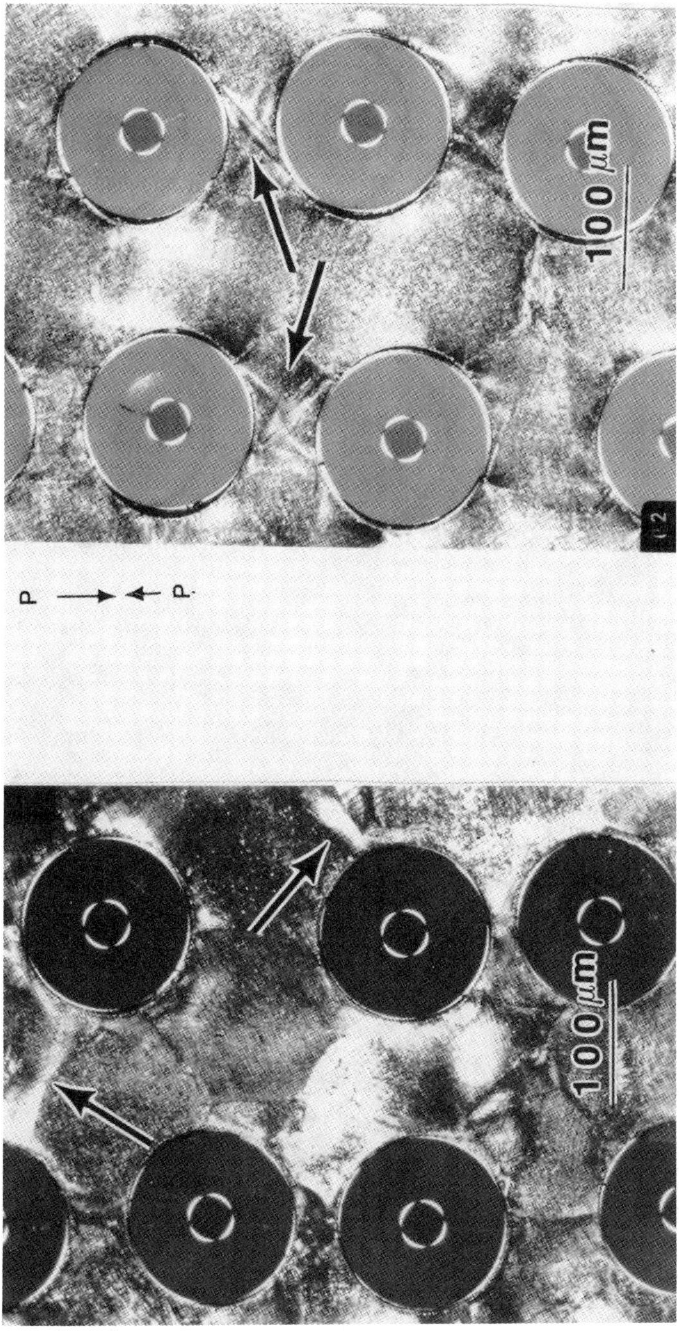

a) 538°C

b) 650°C

FIG. 11—*Matrix plasticity in the form of concentrated shear slip bands in* $[90]_8$ *Ti 15-3/SCS-6 lamina under compression.*

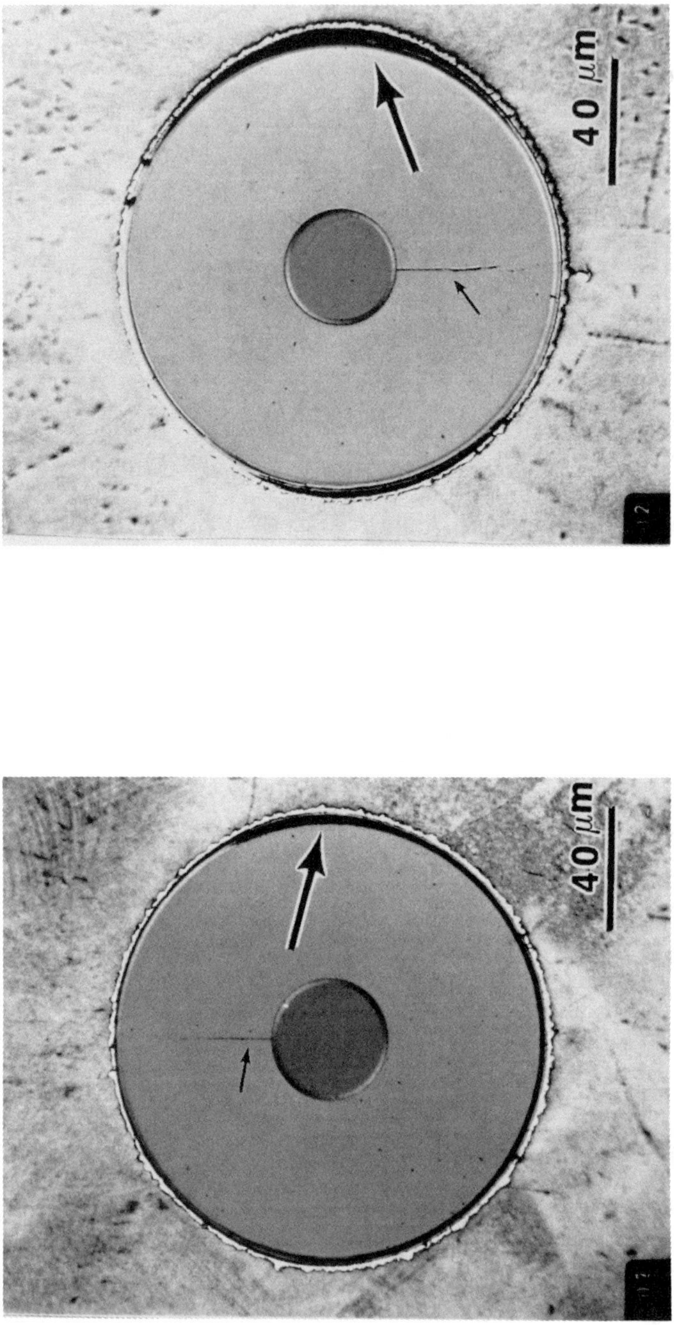

a) 538°C

b) 650°C

FIG. 12—*Debonding (large arrow) and radial fiber cracking (small arrow) in* $[90]_8$ *Ti 15-3/SCS-6 lamina. Loading is up and down.*

TABLE 4—*Deformation mechanisms in tension and compression, $[90]_8$ SCS-6/Ti 15-3.*

Inelastic Deformation Mechanisms	Tension		Compression	
	Strain	Stress, MPa	Strain	Stress, MPa
Plastic slip band	0.002	225	0.002[a]	225[a]
Debonding	0.003 to 0.005	280 to 340	0.006 to 0.008	700 to 900
Reaction zone cracks	0.005	350	0.006	700
Radial fiber cracking	...	...	0.006 to 0.008	700 to 900
Extensive Matrix plasticity	>0.006	400	>0.01	900

[a]Data not retrievable at earlier strain/stress levels because of loss of replica.

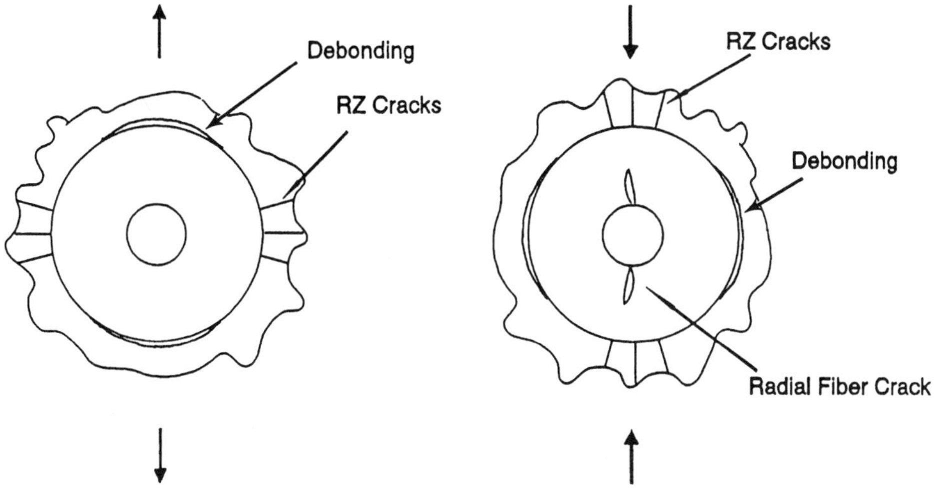

FIG. 13—*Orientation of damage (cracking) in tension and compression.*

The correlation of experimental and predicted results are quite good. Notably, the unloading response was captured for the more complex $[90]_8$ stress-strain response. The compliance change late in unloading at low loads requires further clarification.

Further modeling was undertaken to explain the compliance change on unloading. Figure 19 compares the predicted and experimental response to one cycle of load for a 90° system loaded in compression. (Note from Fig. 19 that the model was loaded to a higher initial strain value before unloading for emphasis.) Note that during unloading (near-zero applied stress), the curve hooks back. This phenomenon is observed in all tests. At maximum load, the fiber and matrix have failed at Point A. Interestingly, upon unloading, the ''curve back'' illustrated in Fig. 19 occurs because the interface at Point B fails. Figure 20 shows the progression of damage for a monotonically loaded case (no unload). Initially, damage develops 90° from the load direction (near Point A) first as fiber debonds from matrix (Fig. 20). Then, as loading progresses, the damage location shifts to near the top (Point B). This is captured using simple classical damage mechanics theories [*19*]. Indeed, this predicted damage evolution and ultimate failure are observed experimentally with shear bands and cracks developing near this predicted location. We emphasize that this computational micromechanics model and the damage model are very simple and robust to use, and the computational effort is minimal.

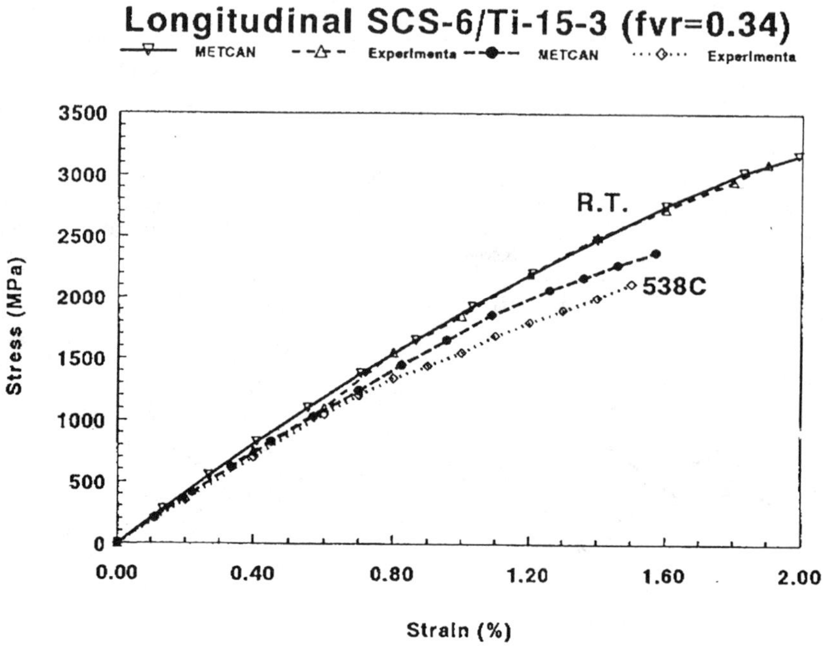

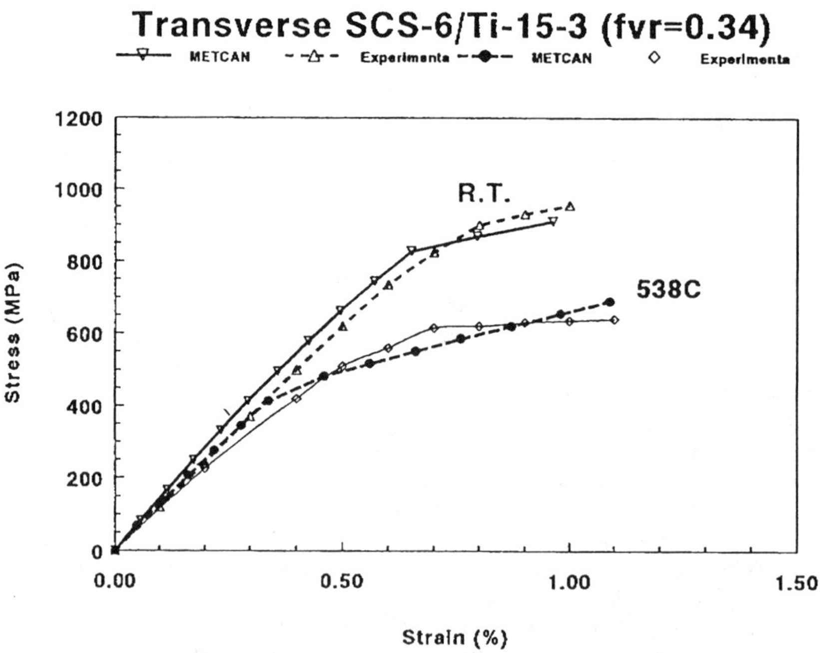

FIG. 14—*METCAN analysis results for* (top) *[0]$_8$ and* (bottom) *[90]$_8$ stress-strain response in compression.*

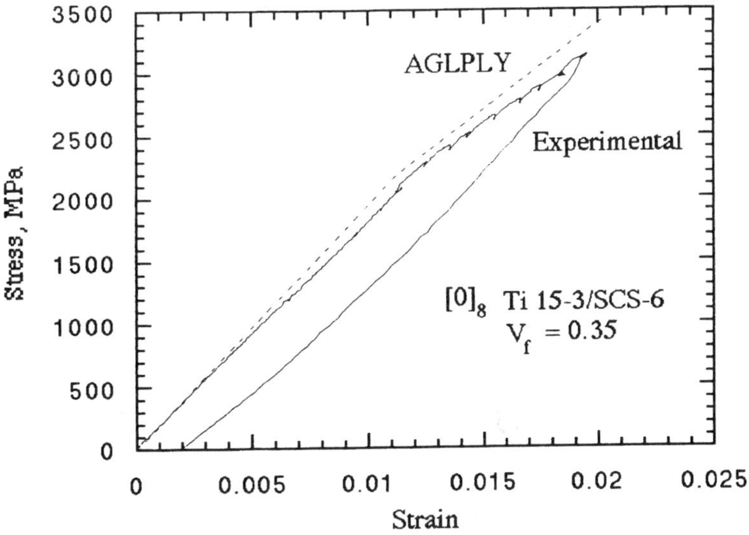

FIG. 15—*AGLPLY analysis results for* $[0]_8$ *stress-strain response in compression.*

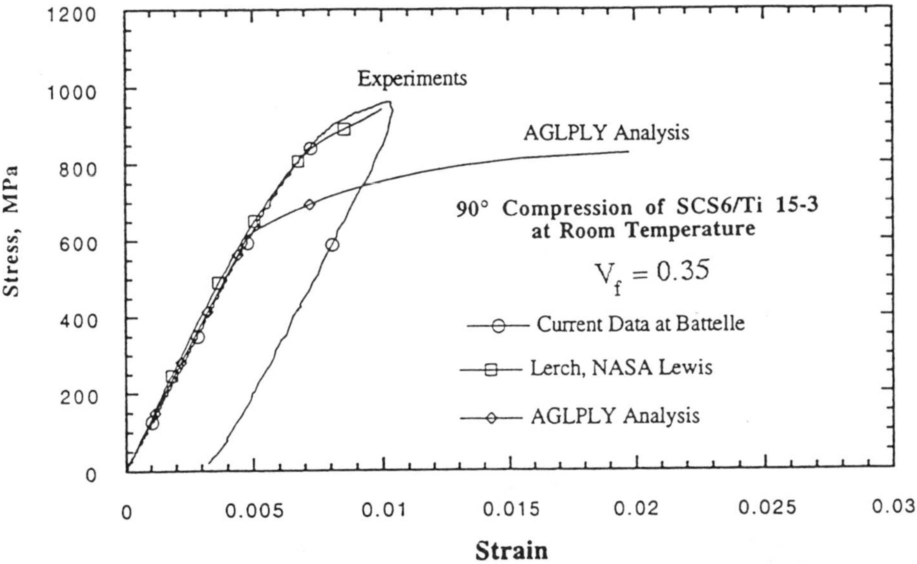

FIG. 16—*AGLPLY analysis results for* $[90]_8$ *stress-strain response in compression.*

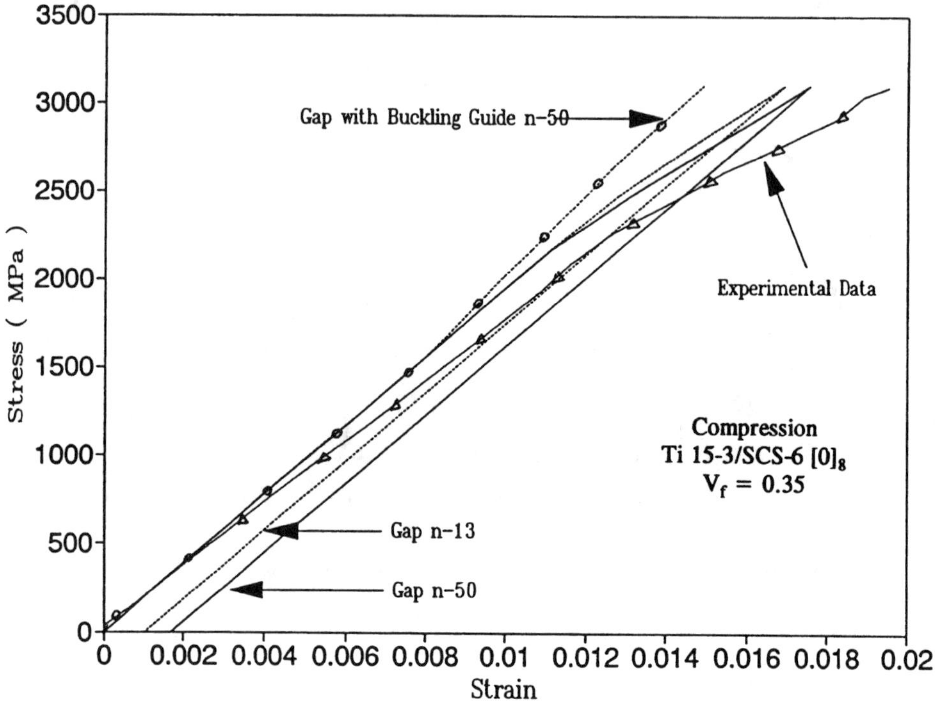

FIG. 17—*Battelle's unit cell computational micromechanics results for $[0]_8$ lamina at RT. Values of* n *represent Ramberg-Osgood exponent for matrix experimental data.*

Experimental observation after complete unloading shows separation not only around Location A, but also around Location B in these titanium-based composites (Fig. 21).

Conclusions

Based on the current investigation, the following conclusions can be reached.

1. Inelastic deformation in the $[0]_8$ lamina under compression is controlled primarily by matrix plasticity. Some fiber-matrix debonding is also observed.
2. Premature buckling can result if restraints (buckling guides) are not used for $[0]_8$ lamina in compression.
3. Macroscopic instability resulting in fiber failure by buckling within a shear zone in the gage section of the specimens was observed in longitudinal compression for the $[0]_8$ lamina.
4. The mechanical response and deformation characteristics in SCS-6/Ti 15-3 $[90]_8$ lamina are different in monotonic tension and compression. The characteristic early knee associated with fiber-matrix debonding in the stress-strain response of the $[90]_8$ lamina under tension is absent in the case of the monotonic compressive response.
5. There is a marked decrease in proportional limit and increased ductility as the temperature is increased for both the $[0]_8$ and the $[90]_8$ SCS-6/Ti 15-3 laminae in compression. These can be rationalized in terms of matrix flow properties.

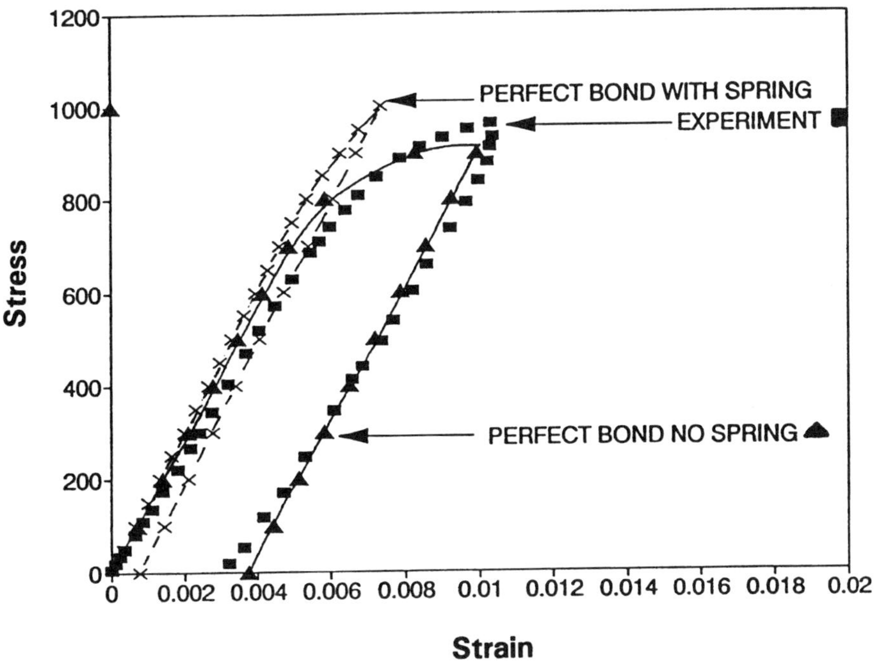

FIG. 18—*Battelle's unit cell computational micromechanics results for [90]$_8$ lamina at RT. Lateral constraint was modeled using springs.*

6. Radial fiber fracture at the carbon core in SCS-6 fibers is a distinct damage mechanism observed under compressive loading in the [90]$_8$ lamina. The transverse compression capability of the SCS-6 fibers are anticipated to be low relative to the longitudinal direction due to the columnar structure of the SCS-6 fiber. Extensive matrix plasticity and debonding are also dominant mechanisms under compression. These mechanisms are observed at elevated temperature as well.
7. AGLPLY analysis predicts the [0]$_8$ stress-strain response quite well, but does not predict the [90]$_8$ response as well. Also, computational results using NASA's METCAN and Battelle's Unit Cell Model adequately predict the stress-strain response of both the [0]$_8$ and [90]$_8$ laminae under compression.
8. The compliance change late on transverse unloading was computationally modeled to be due to debonding of fiber and matrix at the top and bottom of the fiber. This was confirmed experimentally as well.

Acknowledgments

Funding for this work was provided by NASA HITEMP NRA (Dr. Bradley Lerch, program monitor) and AFOSR (Dr. Walter Jones, program monitor). Current efforts at Battelle are continuing under the AFOSR contract. Dr. J. R. Ellis from NASA-LeRC and Dr. Theodore Nicholas of Wright Laboratories is acknowledged for valuable discussions. Mr. Norman Frey and Mr. Glenn Foster are recognized for their excellent efforts in testing and microscopy. Mr. S. Naboulsi and Dr. S. Mital are acknowledged for computational results including METCAN analysis.

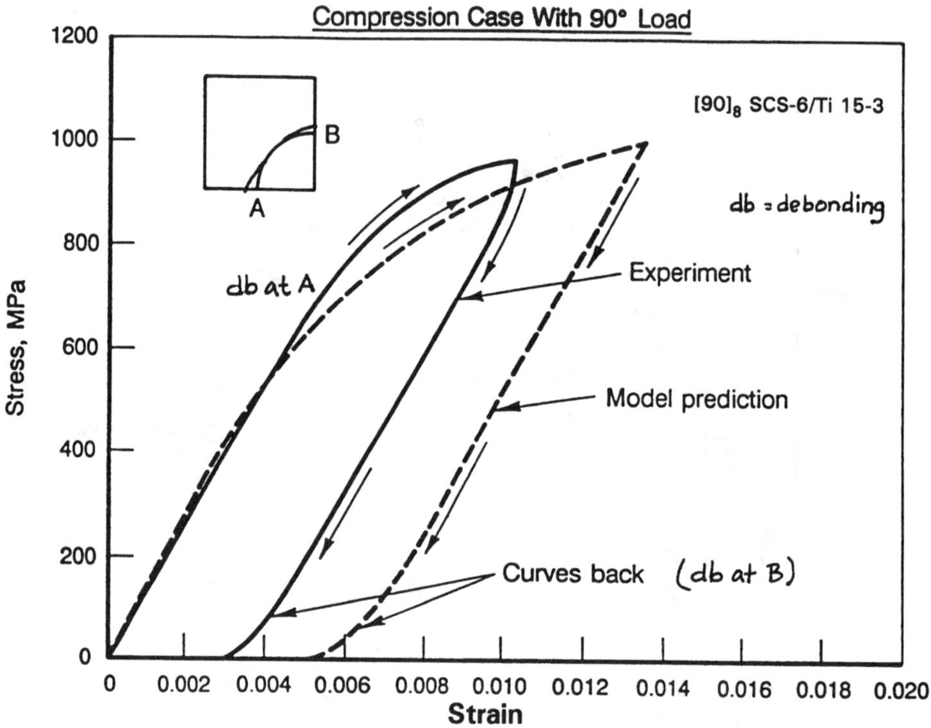

FIG. 19—*Unit cell model prediction of compression behavior in transverse MMC lamina that captures actual unloading feature.*

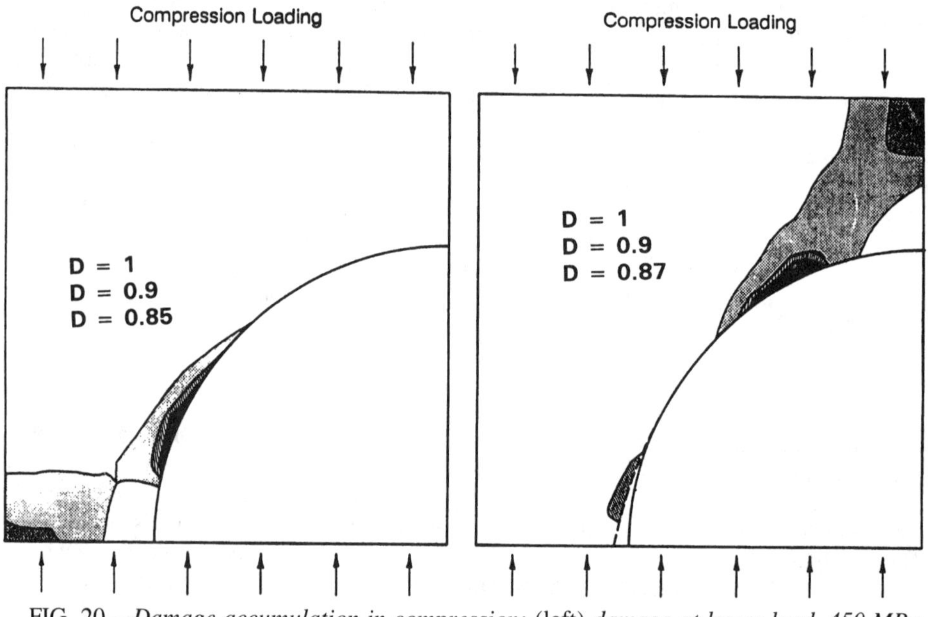

FIG. 20—*Damage accumulation in compression:* (left) *damage at lower load, 450 MPa, and* (right) *damage at higher load, 900 MPa.*

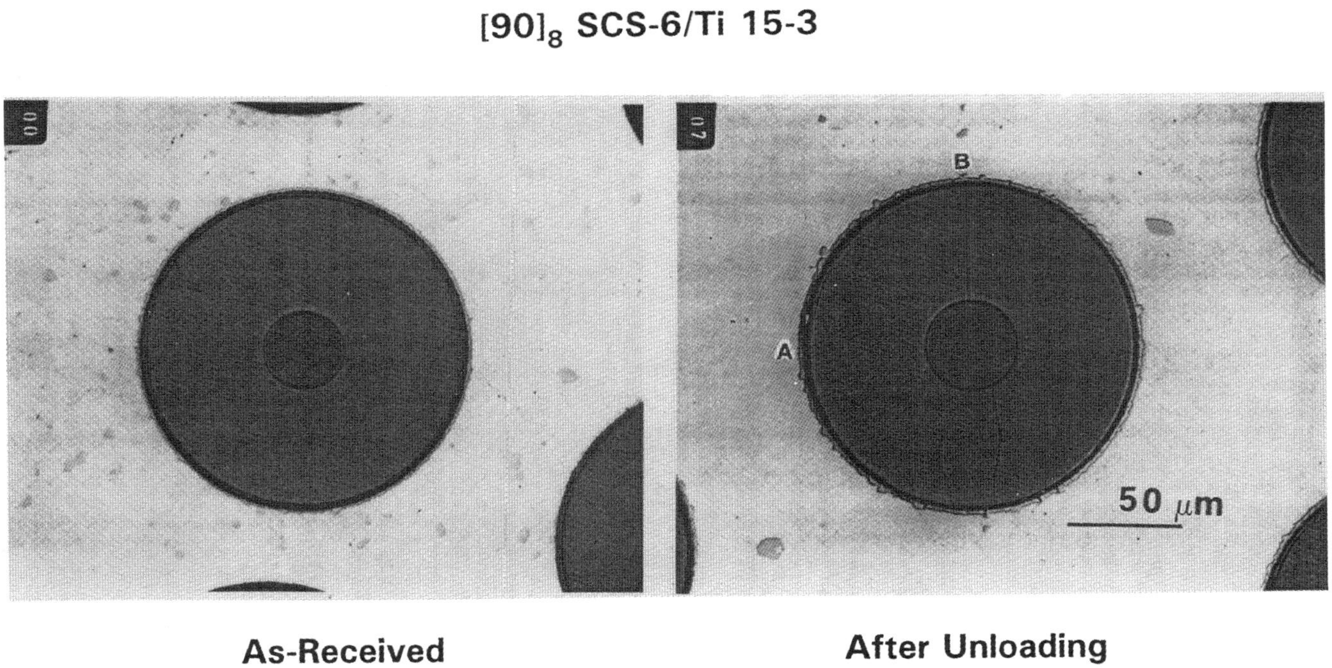

FIG. 21—*Complete debonding after unloading shift of debonding location from Location A (created during loading) to Location B (created during unloading);* (left) *as-received and* (right) *after unloading.*

References

[1] Sun, C. T., ''Modeling Continuous Fiber Metal Matrix Composite as an Orthotropic Elastic-Plastic Material,'' *Metal Matrix Composites: Testing, Analysis, and Failure Modes, ASTM STP 1032,* W. S. Johnson, Ed., American Society for Testing and Materials, Philadelphia, 1989, pp. 148–160.

[2] Lerch, B. A. and Saltsman, J. F., ''Tensile Deformation Damage in SiC Reinforced Ti-15V-3Cr-3Al-3Sn,'' NASA Technical Memorandum 103620, NASA, Cleveland, OH, April 1991.

[3] Majumdar, B. S. and Newaz, G. M., ''Thermomechanical Fatigue of a Quasi-Isotropic Metal Matrix Composite,'' *Composite Materials: Fatigue and Fracture, 3rd Volume, ASTM STP 1110,* T. K. O'Brien, Ed., American Society for Testing and Materials, Philadelphia, 1991, pp. 732–752.

[4] Newaz, G. M., Majumdar, B. S., and Brust, F. W., ''Thermal Cycling Response of Quasi-Isotropic Metal-Matrix Composites,'' *ASME Journal of Engineering Materials and Technology,* April 1992, pp. 156–161.

[5] Johnson, W. S., Lubowinski, S. J., and Highsmith, A. L., ''Mechanical Characterization of Unnotched SCS-6/Ti-15-3 Metal Matrix Composites at Room Temperature,'' *Thermal and Mechanical Behavior of Metal Matrix and Ceramic Matrix Composites, ASTM STP 1080,* Kennedy, Moeller, and Johnson, Eds., American Society for Testing and Materials, Philadelphia, 1990, pp. 193–218.

[6] Nimmer, R. P., Bankert, R. J., Russell, E. S., Smith, G. A., and Wright, K. P., ''Micromechanical Modeling of Fiber/Matrix Interface Effects in Transversely Loaded SiC/Ti-6-4 Metal Matrix Composites,'' *Journal of Composites Technology and Research,* Vol. 13, No. 1, 1991, pp. 3–13.

[7] Majumdar, B. S. and Newaz, G. M., ''Inelastic Deformation in Metal Matrix Composites: Plasticity and Damage Mechanisms,'' *Philosophical Magazine,* London, Vol. 66, No. 2, 1992, pp. 187–212.

[8] Gunawardena, S. R., Jansson, S., and Leckie, F. A., ''Transverse Ductilities of Metal Matrix Composites,'' *Proceedings,* AD-22, ASME Winter Annual Meeting, Atlanta, American Society of Mechanical Engineers, Dec. 1991, pp. 23–30.

[9] Newaz, G. M. and Majumdar, B. S., ''Deformation and Failure Mechanisms in MMC,'' *Proceedings,* AD-22, ASME Winter Annual Meeting, Atlanta, American Society of Mechanical Engineers, 1991, pp. 55–66.

[10] Lamothe, R. M. and Nunes, J., ''Evaluation of Fixturing for Compression Testing of Metal Matrix and Polymer Matrix Composites,'' *Compression Testing of Homogeneous Materials and Composites, ASTM STP 808,* R. Chait and R. Paprino, Eds., American Society for Testing and Materials, Philadelphia, 1983, pp. 241–253.

[11] Brust, F. W., Majumdar, B. S., and Newaz, G. M., ''Constitutive and Damage Response of Ti 15-3/SCS-6 MMC,'' *Proceedings,* ASTM Conference on Fatigue and Fracture, Pittsburgh, PA, 4–5 May 1992.

[12] Newaz, G. M. and Majumdar, B. S., ''Failure Modes in Transverse MMC Lamina Under Compression,'' *Journal of Materials Science & Letters* (UK), Dec. 1992, accepted for publication.

[13] Newaz, G. M. and Majumdar, B. S., ''Inelastic Deformation Mechanisms in a Transverse MMC Lamina Under Compression,'' *Proceedings,* Fracture and Damage, AD-Vol. 27, ASME Winter Annual Meeting, Anaheim, American Society of Mechanical Engineers, 1992, pp. 77–84.

[14] Majumdar, B. S. and Newaz, G. M., ''Inelastic Deformation Mechanisms in MMC: Compression and Fatigue,'' HITEMP Review, NASA Conference Publication 10104, NASA, Cleveland, OH, 1992, p. 49–1.

[15] Hopkins, D. A. and Chamis, C. C., ''A Unique Set of Micromechanics Equations for High-Temperature Metal Matrix Composites,'' *Testing Technology of Metal Matrix Composites, ASTM STP 964,* Di Giovanni and Adsit, Eds., American Society for Testing and Materials, Philadelphia, 1988, pp. 159–176.

[16] Chamis, C. L. and Hopkins, D. A., ''Thermoviscoplastic Nonlinear Constitutive Relationships for Structural Analysis of High-Temperature Metal Matrix Composites,'' *Testing Technology of Metal Matrix Composites, ASTM STP 964,* Di Giovanni and Adsit, Eds., American Society for Testing and Materials, Philadelphia, 1988, pp. 177–196.

[17] Dvorak, G. J. and Bahei-El-Din, Y. A., *Journal of Applied Mechanics,* Vol. 49, 1982, pp. 327–335.

[18] Lerch, B. A., private communication, NASA-LeRC, Cleveland, OH, Dec. 1992.

[19] Lemaitre, J. and Chaboche, J. L., *Mechanics of Solid Materials,* Cambridge University Press, Cambridge, UK, 1990.

Steven M. Arnold,[1] Atef F. Saleeb,[2] and Michael G. Castelli[3]

A Fully Associative, Nonlinear Kinematic, Unified Viscoplastic Model for Titanium-Based Matrices

REFERENCE: Arnold, S. M., Saleeb, A. F., and Castelli, M. G., "**A Fully Associative, Nonlinear Kinematic, Unified Viscoplastic Model for Titanium-Based Matrices,**" *Life Prediction Methodology for Titanium Matrix Composites, ASTM STP 1253,* W. S. Johnson, J. M. Larsen, and B. N. Cox, Eds., American Society for Testing and Materials, 1996, pp. 231–256.

ABSTRACT: Specific forms for both the Gibb's and complementary dissipation potentials are chosen such that a complete (that is, fully associative) potential-based multiaxial, unified viscoplastic model is obtained. This model possess one tensorial internal state variable that is associated with dislocation substructure, with an evolutionary law that has nonlinear kinematic hardening, and both thermal- and strain-induced recovery mechanisms. A unique aspect of the present model is the inclusion of nonlinear hardening through the use of a compliance operator, derived from the Gibb's potential, in the evolution law for the back stress. This nonlinear tensorial operator is significant in that it allows both the flow and evolutionary laws to be fully associative (and therefore easily integrated) and greatly influences the multiaxial response under nonproportional loading paths. In addition to this nonlinear compliance operator, a new, consistent, potential-preserving, internal strain unloading criterion has been introduced to prevent abnormalities in the predicted stress-strain curves that are present with nonlinear hardening formulations during unloading and reversed loading of the external variables. Specification of an experimental program for the complete determination of the material functions and parameters for characterizing a metallic matrix, for example, TIMETAL 21S, is given. The experiments utilized are tensile, creep, and step-creep tests. Finally, a comparison of this model and a commonly used Bodner-Partom model is made on the basis of predictive accuracy and numerical efficiency.

KEYWORDS: viscoplasticity, nonlinear hardening, isothermal model, deformation, multiaxial response, correlations, prediction, titanium, titanium matrix composites, life prediction, titanium alloys, fatigue (materials), modeling

Nomenclature

Invariants

Ω	Complementary dissipation potential
Φ	Gibb's complementary potential
F	Bingham-Prager threshold function
G, Ψ	Normalized second invariant functions
J_2	Second invariant of effective deviatoric stress tensor

[1]Research engineer, National Aeronautics and Space Administration, Lewis Research Center, Cleveland, OH 44135.

[2]Professor, Department of Civil Engineering, University of Akron, Akron, OH 44325.

[3]Research engineer, NYMA, Inc., NASA Lewis Research Center Group, BrookPark, OH 44142.

I_2	Second invariant of internal deviatoric stress tensor
H_1	Invariant material functions

Stresses

σ_{ij}	Cauchy stress tensor
S_{ij}	Deviatoric stress tensor
Σ_{ij}	Effective deviatoric stress tensor
α_γ	Internal state variables (stress-like)
α_{ij}	Internal (or back) stress tensor
a_{ij}	Deviatoric internal stress tensor
κ, κ_0	Drag stress and initial drag stress
Y	Yield stress

Strains

e_{ij}, ϵ^I_{ij}, ϵ^R_{ij}, ϵ^T_{ij}	Total, inelastic, reversable, and thermal strain tensors, respectively
$\mathcal{A}_\xi$	Conjugate internal state variables (displacement-like)
$\mathcal{A}_{ij}$	Internal strain tensor

Material Parameters

C_{ijkl}, E_{ijkl}	Elastic compliance and stiffness tensor, respectively
B_0, B_1, R_α	Hardening and thermal recovery material parameters
μ, κ	Material parameters
β	Denotes the extent of strain-induced recovery
n, p, q	Material exponents
$\underline{g}(G)$, $f(F)$	Material functions
$\bar{E}(\sigma_{ij})$, $\bar{H}(\alpha_\gamma)$, $\underline{Z}(T)$	Material functions
η, $\hat{E}$, $\hat{G}$	Coefficient of thermal expansion, Young's modulus, and shear modulus, respectively

Miscellaneous

T_{ref}	Reference temperature
$sgn(\ )$	Sign function: takes on 1 or -1 depending upon the sign of argument
Q_{ijkl}, L_{ijkl}	Internal compliance and internal stiffness operators, respectively
δ_{ij}	Kronecker delta function
$\langle\ \rangle$	Macauley bracket
$Hv[\ ]$	Heaviside unit function
$\dot{(\cdot)}$	Time derivative (or rate) notation
$\dot{\mathcal{S}}$	Instantaneous state trajectory
$\dot{\mathcal{G}}$, $\dot{\mathcal{D}}$	Components of the state trajectory, $\dot{\mathcal{S}}$

A number of titanium matrix composite (TMC) systems are currently being researched and evaluated for high temperature air frame and propulsion system applications. As a result, numerous computational methodologies for predicting both deformation and life are under development. An integral part of these methodologies is an accurate and computationally efficient constitutive model for the metallic matrix constituent. Furthermore, because of the proposed elevated operation temperatures for which these systems are designed, the required constitutive models must account for both time-dependent and time-independent deformations. To accomplish this, we will employ a recently developed complete potential-based framework [1] utilizing internal state variables that was put forth for the derivation of reversible and irreversible constitutive equations. This framework, and consequently the resulting constitutive model, is termed complete because the existence of the total (integrated) form of the Gibbs complementary free energy and complementary dissipation potentials are assumed a priori. The specific forms selected here for both the Gibbs and complementary dissipation potentials result in a fully associative, multiaxial, isothermal, unified viscoplastic model with nonlinear kinematic hardening. Thus, this model constitutes one of many models in the GVIPS (Generalized VIsoplasticity with Potential Structure) class of inelastic constitutive equations that can be constructed using the generalized framework of Arnold and Saleeb [1].

The particular unified GVIPS model of interest in this study possesses one tensorial internal state variable (that is, the back or internal stress) that is associated with dislocation substructure and an evolutionary law that has nonlinear kinematic hardening and both thermal and strain-induced recovery mechanisms. A unique aspect of the present model is the inclusion of nonlinear hardening through the use of a compliance operator (derived from the Gibb's potential) in the evolution law for the back stress. This nonlinear tensorial operator is significant in that it allows both the flow and evolutionary laws to be fully associative [2] (and therefore easily integrated) and greatly influences the multiaxial response under nonproportional loading paths [1,3,4]. In addition to this nonlinear compliance operator, a new consistent, potential preserving, internal unloading criterion is introduced in this paper to prevent abnormalities in the predicted stress-strain curves, which are present with nonlinear hardening formulations, during unloading and reversed loading. This new criterion is motivated by considering the trajectory of the state point through the state space during unloading and hinges upon the introduction of internal state discontinuity inequalities for the conjugate back (or internal) strain rate.

A primary objective of the present study is to specify material functions and characterize the associated material parameters for the current kinematic, isothermal GVIPS model given TIMETAL 21S,[4] an advanced titanium-based matrix commonly used in TMCs. Although both long- and short-term behavior is important, capturing the short-term (or transient) behavior and rate sensitivity of the material is of primary importance given that the applications of interest are primarily those involving material processing and propulsion systems. Furthermore, to illustrate the effectiveness of the present model, a comparison of the GVIPS model and a particular Bodner-Partom (BP) model discussed in Ref 5 (of the commonly used BP models [6–8] in the literature) is made on the basis of predictive accuracy and numerical efficiency.

The paper begins by briefly summarizing the complete potential structure, followed by a multiaxial and uniaxial statement of the proposed GVIPS model, including a discussion on the recent addition of a potential preserving internal unloading criterion. A discussion regarding the characterization of the proposed model is then followed by numerous results illustrating the predictive accuracy and numerical efficiency of the model.

Complete Potential Structure

Here, the basic thermodynamic framework put forth by Arnold and Saleeb [1] is summarized. Expressions for the Gibb's thermodynamic and the complementary dissipation poten-

[4]TIMETAL 21S is a registered trademark of TIMET, Titanium Metals Corporation, Toronto, OH 43964.

tial functions are assumed in terms of a number of state and internal variables characterizing the changing internal structure of the material. For conciseness, the discussion is limited to a case involving (1) small deformations (in which the initial state is assumed to be stress free throughout), (2) an initially isotropic material, (3) isothermal conditions, and (4) the specialized potential framework discussed in Ref *1*. A Cartesian coordinate reference frame and index notation are utilized (repeated Roman subscripts imply summation).

Given the Gibb's potential in the following form

$$\Phi = \Phi(\sigma_{ij},\alpha_\gamma,T,\epsilon^I_{ij}) \tag{1}$$

and assuming a priori that the inelastic strain is an independent parameter (and not an internal state variable), for example

$$\Phi = \bar{E}(\sigma_{ij}) - \sigma_{ij}\epsilon^I_{ij} + \bar{H}(\alpha_\gamma) - Z(T) - \frac{\sigma_{kk}}{3}\eta(T - T_0) \tag{2}$$

an expression for the total strain rate can be obtained by differentiating, that is

$$\dot{e}_{ij} = \frac{d}{dt}\left(\frac{-\partial\Phi}{\partial\sigma_{ij}}\right) = C_{ijrs}\dot{\sigma}_{rs} + \dot{\epsilon}^I_{ij} + \delta_{ij}\frac{\eta}{3}\dot{T} \tag{3}$$

as well as the rate of change of the conjugate internal variables ($\mathcal{A}_\xi$),

$$\dot{\mathcal{A}}_\xi = \frac{d}{dt}\left(\frac{-\partial\Phi}{\partial\alpha_\xi}\right) = Q_{\xi\gamma}\dot{\alpha}_\gamma \tag{4}$$

where

$$C_{ijrs} = \frac{-\partial^2\Phi}{\partial\sigma_{ij}\partial\sigma_{rs}} = \frac{-\partial^2\bar{E}(\sigma_{ij})}{\partial\sigma_{ij}\partial\sigma_{rs}} \tag{5}$$

$$Q_{\xi\gamma} = \frac{-\partial^2\Phi}{\partial\alpha_\xi\partial\alpha_\gamma} = \frac{-\partial^2\bar{H}(\alpha_\gamma)}{\partial\alpha_\xi\partial\alpha_\gamma} \tag{6}$$

are the external and internal compliance operators, respectively. Note the three terms in Eq 3 may then be identified as the reversible, irreversible (inelastic), and thermal expansion components of the total strain rate, respectively. Thus

$$\dot{e}_{ij} = \dot{\epsilon}^R_{ij} + \dot{\epsilon}^I_{ij} + \dot{\epsilon}^T_{ij} \tag{7}$$

where

$$\dot{\epsilon}^R_{ij} = C_{ijrs}\dot{\sigma}_{rs} \tag{8}$$

and

$$\dot{\epsilon}^T_{ij} = \delta_{ij}\frac{\eta}{3}\dot{T} \tag{9}$$

but $\dot{\epsilon}^I_{ij}$ (the inelastic strain rate) is defined separately using the concept of a dissipation potential $\Omega(\sigma_{ij},\alpha_\gamma,T)$.

Given

$$\Omega = \Omega(\sigma_{ij},\alpha_\gamma,T) \tag{10}$$

and using the Clausius-Duhem inequality; the flow law becomes

$$\dot{\epsilon}^I_{ij} = \frac{\partial \Omega}{\partial \sigma_{ij}} \tag{11}$$

and the evolutionary laws for the thermodynamic conjugate internal state variables

$$\dot{\mathcal{A}}_\gamma = -\frac{\partial \Omega}{\partial \alpha_\gamma} \tag{12}$$

Utilizing Eq 4, the internal constitutive rate equations for the internal state variables is obtained

$$\dot{\alpha}_\gamma = L_{\gamma\xi}\dot{\mathcal{A}}_\xi \tag{13}$$

where

$$L_{\gamma\xi} = [Q_{\gamma\xi}]^{-1} = \left[\frac{-\partial^2\Phi}{\partial\alpha_\xi\partial\alpha_\gamma}\right]^{-1} \tag{14}$$

Thus, Eqs 11 and 12 represent the flow and evolutionary laws, for an assumed $\Omega = \Omega(\sigma_{ij},\alpha_\gamma,T)$, and Eq 13 is the internal constitutive rate equation, given a Gibb's potential Φ, wherein both potentials are directly linked through the internal state variables, α_γ.

Viscoplastic Constitutive Model

A complete multiaxial statement of a GVIPS model can be derived by using the preceding framework, given a specific form for both the Gibb's potential, Φ, and the complementary dissipation potential, Ω. Form invariance (objectivity) of these potentials, and material symmetry considerations, requires that they depend only on certain invariants of their respective tensorial arguments (that is, an integrity basis [*9*]). In the spirit of von Mises and because of the deviatoric nature of inelastic deformation, only the quadratic invariant will be considered at this time in specifying the dissipation potential. Similarly, only the linear elastic strain energy contribution will be considered in specifying the Gibb's potential, with the internal state groupings being functions of the respective quadratic invariants. Finally, although Eq 10 indicates that an unlimited number of internal state variables can be specified, here our attention will be restricted to a GVIPS model with a single "independent," evolving, internal state variable. This internal state variable, a_{ij}, is taken to be a second-order symmetric traceless tensor that represents the internal (or back) stress associated with dislocation substructure. Two additional scalar state variables, one representing the drag strength (κ) and taken to be nonevolving and the other a yield stress (Y) that is uniquely assumed to implicitly evolve with internal stress are considered. Both variables are associated with dislocation density and are included in anticipation of the nonisothermal

GVIPS model [*10*]. However, in the forthcoming non-isothermal formulation instead of κ being taken as an evolving internal state variable with its own associative evolution law as in Ref *3*, κ will be taken as an explicit function of temperature only.

Consequently, the Gibb's potential may be written as

$$\Phi = -\frac{1}{2} C_{rskl}\sigma_{rs}\sigma_{kl} - \sigma_{ij}\epsilon^{I}_{ij} + H_1(G) - Z(T) - \frac{\sigma_{kk}}{3}\eta(T - T_0) \tag{15}$$

and the dissipation potential as

$$\Omega = \mu \int f(F)dF + R_\alpha B_0 \int g(G)dG \tag{16}$$

where

$$F = \left\langle \frac{\sqrt{J_2}}{\kappa} - Y \right\rangle \tag{17}$$

$$Y = \langle 1 - \beta\sqrt{G} \rangle \tag{18}$$

$$G = \frac{I_2}{\kappa^2} \tag{19}$$

$$H_1(G) = -B_0(G + B_1 G^p) \tag{20}$$

$$I_2 = \frac{3}{2} a_{ij}a_{ij} \tag{21}$$

$$J_2 = \frac{3}{2} \Sigma_{ij}\Sigma_{ij} \tag{22}$$

$$\Sigma_{ij} = S_{ij} - a_{ij} \tag{23}$$

$$S_{ij} = \sigma_{ij} - \frac{1}{3}\sigma_{kk}\delta_{ij} \tag{24}$$

$$a_{ij} = \alpha_{ij} - \frac{1}{3}\alpha_{kk}\delta_{ij} \tag{25}$$

Note that in the preceding expression for the dissipation potential, the stress dependence, both external and internal, enters through the scalar functions, F and G, in the form of effective (Σ_{ij}) and internal (a_{ij}) deviatoric stresses, respectively. Furthermore, the function, F, acts like a threshold surface, because when $F < 0$, no inelastic strain can occur. Clearly, this threshold value is dictated by the magnitude of both the drag strength (κ) and yield stress (Y). A unique aspect of this model is that the internal variable representing the yield stress is specifically taken as a special scaled function of the back stress and drag strength. Consequently, this allows for (1) the model to possess features of a model with three internal variables yet without any additional computational cost, and (2) the presence of an induced strain recovery term (as opposed to the common ''ad-hoc'' introduction of such terms) in the evolution of the back strain (that is, the associated conjugate variable, $\mathcal{A}_\gamma$). It is important to realize that the product, κY, constitutes the radius of the threshold surface (see Fig. 1*a*), thereby dictating from physical arguments, that both Y and κ be always positive valued. Furthermore, given the specified form of Y in Eq 18, it is clear that the material parameter ratio (β/κ) will dictate the limit value (that is, when $Y = 0$) of the internal stress (α_{ij}), or the cut-off limit for dynamic recovery.

By selecting the preceding scalar functions, a general yet "complete" potential-based model, with "associated" flow and evolutionary laws can be constructed. The second invariants, J_2 and I_2, are also scaled for tension. These invariants could just as easily have been scaled for shear by replacing the coefficient 3/2 with 1/2, and modifying the definition of the magnitude of the inelastic strain rate that follows. Also, it should be stated that the linear forms of F and Y in Eqs 17 and 18 were chosen in order to allow algebraic manipulation and analytical solution of the resulting expressions (for example, inversion of the flow law), so as to ease the characterization stage of the model, as discussed subsequently in the next two sections. Note, however, that the quadratic forms of F and Y do provide some additional interesting features in the model as discussed previously by Arnold et al. [*3*].

Finally, it is interesting to note the close similarity between the present assumed form of the threshold surface, F, of Eq 17 to that used by Lemaitre and Chaboche [*11*] in the context of inviscid plasticity. The significant difference between the two forms is that Lemaitre and Chaboche [*11*] utilized a nonassociated format; the form similar to Eq 17 was taken to affect

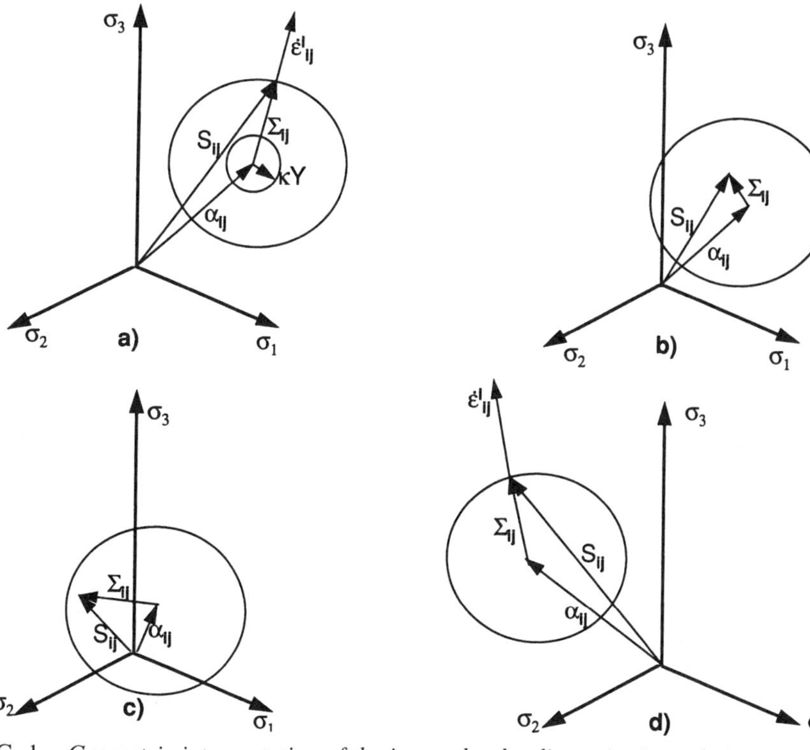

FIG. 1—*Geometric interpretation of the internal unloading criterion relative to positions of the threshold surfaces.*

only the directions of inelastic and internal state, whereas a classical J_2 form was employed to control yielding (that is, the consistency condition). In the present fully associative case, the same threshold surface, F, is used to affect both yielding and the direction of inelastic and internal state.

Taking the appropriate derivatives of both Φ and Ω as indicated in Eqs 2 through 14, one obtains the multiaxial isothermal specification particular to the present constitutive model. Here the decomposition of the total strain rate is that of Eq 7, where the reversible strain rate is given by Eqs 8 and 9 and the irreversible (or inelastic) component is defined by the following flow law

$$\dot{\epsilon}_{ij}^{I} = \frac{3}{2}\|\dot{\epsilon}_{ij}^{I}\|\frac{\Sigma_{ij}}{\sqrt{J_2}} \qquad \text{if} \quad F \geq 0 \tag{26}$$

or

$$\dot{\epsilon}_{ij}^{I} = 0 \qquad \text{if} \quad F < 0 \tag{27}$$

where

$$\|\dot{\epsilon}_{ij}^{I}\| = \sqrt{\frac{2}{3}\dot{\epsilon}_{ij}^{I}\dot{\epsilon}_{ij}^{I}} = \frac{\mu f(F)}{\kappa} \tag{28}$$

The internal constitutive rate equation is always given by

$$\dot{a}_{ij} = L_{ijrs}\dot{\mathcal{A}}_{rs} \tag{29}$$

while the evolutionary law for the back strain rate is given by

$$\dot{\mathcal{A}}_{kl} = \dot{\epsilon}_{kl}^{I} - \frac{3}{2}\frac{\beta}{\kappa}\|\dot{\epsilon}_{ij}^{I}\|\frac{a_{kl}}{\sqrt{G}}Hv[Y] - \frac{3R_{\alpha}B_0\, g(G)}{\kappa^2}a_{kl} \qquad \text{if} \quad a_{ij}\Sigma_{ij} \geq 0 \tag{30}$$

during internal loading and

$$\dot{\mathcal{A}}_{rs} = Q_{rslm}E_{lmnp}\left(\dot{\epsilon}_{np}^{I} - \frac{3}{2}\frac{\beta}{\kappa}\|\dot{\epsilon}_{ij}^{I}\|\frac{a_{np}}{\sqrt{G}}Hv[Y] - \frac{3R_{\alpha}B_0\, g(G)}{\kappa^2}a_{np}\right) \qquad \text{if} \quad a_{ij}\Sigma_{ij} < 0 \tag{31}$$

during internal unloading, as indicated by the inequality, $a_{ij}\Sigma_{ij} < 0$. The internal stiffness operator is defined as

$$L_{ijrs} = [Q_{ijrs}]^{-1} = \frac{\kappa^2}{3B_0(1 + B_1 pG^{p-1})}\left(I_{ijrs} - \frac{3B_1(p-1)G^{p-2}}{\kappa^2(1 + B_1 pG^{p-1}(6p-5))}a_{rs}a_{ij}\right) \tag{32}$$

with $I_{ijrs} = \delta_{ir}\delta_{js}$.

Equation 31 constitutes a new consistent, potential preserving, internal unloading criterion that will prevent the classical abnormalities [*11*] in the cyclic response associated with nonlinear hardening formulations, upon unloading and reversed loading of the "external" variables. This new criterion is motivated by considering the trajectory of the state point within the state space during unloading and hinges upon the introduction of an internal state discontinuity inequality for the conjugate internal (or back) strain rate. Mathematically, this is accomplished by the introduction of a dissipation potential that is distinctly different when the material is undergoing internal loading versus unloading. The geometric interpretation of this criterion (discontinuity inequality) in the π–plane is illustrated in Fig. 1, where the situation of the deformed material, $a_{ij} \neq 0$, is depicted. Note that the projection of the applied stress vector, σ_{ij}, on the π–plane is, of course, S_{ij}, the deviatoric stress and that the projection of the cylindrical surfaces of constant inelastic strain rate (F = constant) appear as concentric circles in this plane, although only one is illustrated in Fig. 1. The current effective stress vector, $\Sigma_{ij} = S_{ij} - a_{ij}$ emanates from the translated center of the nested circles, and the corresponding inelastic strain rate vector, $\dot{\epsilon}^I_{ij}$, is collinear with it. Now, if the applied stress is reduced to a value less than the internal stress, the effective stress will change sign and unloading internal to the material will take place, thus resulting in a readjustment (state recovery, that is, shift toward the origin of the π–plane) in internal stress, a_{ij}, as shown in Fig. 1*b*. If the stress reversal were continued, the material would also continue to rapidly readjust until once again internal reloading commenced, as depicted in Fig. 1*c*.

Justification for this criterion stems from the work of Orowan [*12*] and others, and has indicated that upon stress reversals dislocations are remobilized and consequently rapid rearrangement of these dislocations is possible within the wake of the previous load path. To describe this rapid motion of dislocation remobilization (readjustment of internal stress) during material unloading, we introduced, in the spirit of Robinson [*13*] and Onat [*14*], distinct regions within the state space in which the "rate" of the conjugate internal state variable (internal strain) changes discontinuously. Currently, until more exploratory tests can be performed to fully describe the affected regions of the state space, the internal (back) strain rate during internal unloading is taken to be proportional to the back strain rate during loading, through the product of the external stiffness and internal compliance operators, see Eq 30. Thus, implying that the internal stress and strain during internal unloading are related by the external stiffness tensor; the accuracy of such an assumption can be verified by conducting a number of transient stress dip tests [*15*]. Furthermore, an examination of the evolution of the conjugate of the internal stress (that is, the internal strain, $\mathcal{A}_{ij}$), clearly shows that Eq 30 possesses, as typically assumed in the literature [*16,17*] competitive mechanisms consisting of a hardening term (which accounts for strengthening mechanisms) and two state recovery terms (which account for softening mechanisms). The first state recovery term evolves with inelasticity; it is strain induced and is commonly called dynamic recovery; whereas, the second term, interchangeably called static or thermal recovery, evolves with time and is thermally induced. Again, it is important to note that as the internal stress evolves and finally reaches a limit value (dictated by the material parameters, β and κ) the yield stress will vanish, as will the strain-induced recovery term. The added flexibility provided by the dynamic recovery term is particularly advantageous under conditions where thermal recovery is relatively small (that is, low homologous temperatures).

Finally, the preceding expressions are further specialized by assuming the nonlinearity of inelasticity and the thermal recovery process to be represented by power law functions, that is

$$f(F) = F^n \tag{33}$$

$$g(G) = G^q \tag{34}$$

Uniaxial Simplification and State Space Representation

If we consider only a uniaxial state of stress, the deviatoric applied and internal stress tensors become

$$S_{ij} = \frac{1}{3}\begin{bmatrix} 2\sigma & 0 & 0 \\ 0 & -\sigma & 0 \\ 0 & 0 & -\sigma \end{bmatrix}$$

and

$$a_{ij} = \frac{1}{3}\begin{bmatrix} 2\alpha & 0 & 0 \\ 0 & -\alpha & 0 \\ 0 & 0 & -\alpha \end{bmatrix}$$

such that

$$F = \left\langle \frac{\sigma - \alpha}{\kappa} - Y \right\rangle \quad (35)$$

$$Y = \left\langle 1 - \beta \frac{\alpha}{\kappa} \right\rangle \quad (36)$$

$$G = \frac{\alpha^2}{\kappa^2}$$

and the flow and evolutionary laws, respectively, become

$$\dot{\epsilon}^I = \frac{\mu}{\kappa} F^n \frac{(\sigma - \alpha)}{|\sigma - \alpha|} \quad (37)$$

and

$$\dot{\alpha} = \frac{3}{2}\mathcal{H}\left(1 - \frac{2}{3}\mathcal{C}\alpha^2\right)\left\{\dot{\epsilon}^I - \beta\|\dot{\epsilon}^I\|Hv[Y]\alpha - \frac{2R_\alpha B_0}{\kappa}\left(\frac{\alpha}{\kappa}\right)^{2q+1}\right\} \quad \text{if} \quad (\sigma - \alpha)\alpha \geq 0 \quad (38)$$

$$\dot{\alpha} = -3\hat{G}\left\{(1 + \beta\alpha Hv[Y])\|\dot{\epsilon}^I\| + \frac{2R_\alpha B_0}{\kappa}\left(\frac{\alpha}{\kappa}\right)^{2q+1}\right\} \quad \text{if} \quad (\sigma - \alpha)\alpha < 0 \quad (39)$$

Utilizing the concept of the state space representation in Refs *14, 15,* and *18*, we can examine and illustrate geometrically the various features of the present unified viscoplastic model and discuss its characterization. The pertinent state space is constructed by combining and rewriting Eqs 7, 37, and 38, in standard vector form as follows, and plotting the corresponding $\mathcal{G}$ fields or relaxation trajectories. That is

$$\dot{\mathcal{S}} = \dot{\mathcal{G}}(\sigma,\alpha) + \dot{\mathcal{D}}(\dot{e})$$

or, for example, during loading

$$\begin{Bmatrix} \dot{\sigma} \\ \dot{\alpha} \end{Bmatrix} = \left\{ \frac{3}{2}\mathcal{H}\left(1 - \frac{2}{3}\mathcal{C}\alpha^2\right)\left[\begin{matrix} -\hat{E}\,\dot{\epsilon}^I \\ \dot{\epsilon}^I - \beta\|\dot{\epsilon}^I\|Hv[Y]\alpha - \dfrac{2R_\alpha B_0}{\kappa}\left(\dfrac{\alpha}{\kappa}\right)^{2q+1} \end{matrix}\right]\right\} + \begin{Bmatrix} \hat{E}\dot{e} \\ 0 \end{Bmatrix} \tag{40}$$

where

$$\|\dot{\epsilon}^I\| = \frac{\mu}{\kappa}F^n \tag{41}$$

$$\mathcal{H} = \frac{\kappa^2}{3B_0\left(1 + B_1 p\left(\dfrac{\alpha}{\kappa}\right)^{2p-2}\right)} \tag{42}$$

$$\mathcal{C} = \frac{3B_1(p-1)\left(\dfrac{\alpha}{\kappa}\right)^{2p-4}}{\kappa^2\left(1 + B_1 p\left(\dfrac{\alpha}{\kappa}\right)^{2p-2}(6p-5)\right)} \tag{43}$$

and $\dot{e}$ is the total uniaxial strain rate. These relaxation trajectories are obtained from Eqs 37 and 40 by imposing a constant total strain rate (that is, $\dot{e} = 0$), thereby implying that $\dot{\mathcal{S}} = \dot{\mathcal{G}}(\sigma,\alpha)$.

Figure 2 illustrates the state space representation associated with Eqs 37, 38, and 39, and the material parameters given in Table 1. Immediately, a number of qualitative features of the first and third quadrants of the state space can be deduced from Eqs 37 and 40. For example, the locus of steady-state or "fixed" points in creep, denoted by Line *ab* in Fig. 2 is obtained by imposing the condition, $\dot{\mathcal{S}} = 0$, on Eq 40. Mathematically, such a condition is obtained when

$$\sigma = (1 - \beta Hv[Y])\alpha + \kappa\left[1 + \left[\frac{2R_\alpha B_0}{\mu(sgn(\dot{\epsilon}^I) - \beta Hv[Y]\alpha)}\left(\frac{\alpha}{\kappa}\right)^{2q+1}\right]^{1/n}\right] \tag{44}$$

and is denoted as the "locus of steady-state creep." The associated steady-state creep rate is

$$\dot{\epsilon}^I = \frac{\dfrac{2R_\alpha B_0}{\kappa}\left(\dfrac{\alpha}{\kappa}\right)^{2q+1}}{(sgn(\dot{\epsilon}^I) - \beta Hv[Y]\alpha)} \tag{45}$$

Second, the "locus of maxima in stress" (that is, $\dot{\sigma} = 0$) during relaxation ($\dot{e} = 0$) occurs when

$$\dot{\epsilon}^I = -\frac{\dot{\sigma}}{\hat{E}} = 0 \tag{46}$$

This is satisfied when

$$\sigma = \kappa + \alpha(1 - \beta Hv[Y]) \tag{47}$$

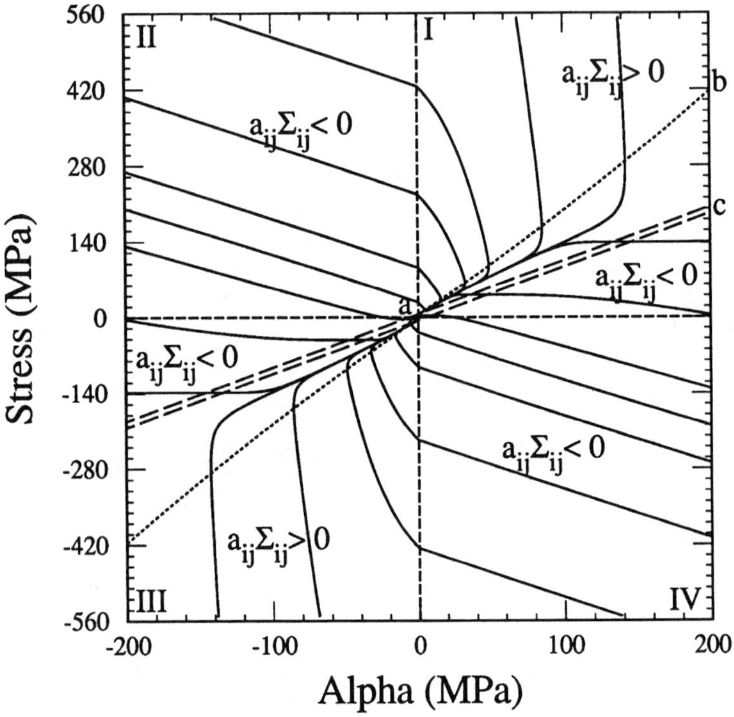

FIG. 2—*State space associated with the material parameters in Table 1.*

TABLE 1—*Isothermal, GVIPS, material parameters representing TIMETAL 21S at 650°C.*

Constants	Units	TIMETAL 21S
	VISCOPLASTIC	
κ	MPa	5.86
μ	MPa/s	5.52×10^{-9}
n	...	3.3
B_0	MPa	5.86×10^{-4}
B_1	...	0.05
p	...	1.8
q	...	1.35
R_α	1/s	0.1×10^{-5}
β	...	0.01
	ELASTIC	
$\hat{E}$	MPa	80 671
ν	...	0.365

and is denoted by Line *ac* in Fig. 2. The points on Line *ac* do not correspond to fixed points in the state space, as do those associated with Line *ab*, since the internal stress continues to evolve, that is

$$\dot{\alpha} = -\frac{3}{2}\mathcal{H}\left(1 - \frac{2}{3}\mathcal{C}\alpha^2\right)\left\{\frac{2R_\alpha B_0}{\kappa}\left(\frac{\alpha}{\kappa}\right)^{2q+1}\right\} \tag{48}$$

Note that the inelastic strain is zero along and within the zone denoted by Line *ac*, as in this region, $F < 0$. Finally, Fig. 2 illustrates clearly that the second and fourth quadrants as well as the lower portions of quadrants one and three contain distinctly different relaxation trajectories resulting from the imposition of the internal unloading criterion.

Characterization

The most important, and often times most difficult, aspect of modeling the behavior of a given material at elevated temperature is obtaining the required material functions, for example, $f(F)$ and $g(G)$, and material parameters. The difficulty associated with this process typically stems from not only the variety in mathematical forms for the material functions (for example, power law, exponential, hyperbolic sine, etc.), but also the fact that given the material functions there is not a unique set of material parameters for any given load path. Therefore, numerous iterations and difficult compromises are required before a final set of material parameters, for the assumed material functions, can be obtained.

With the present isothermal GVIPS model, nine material parameters need to be determined, as given in Table 1: three are associated with the flow law (that is, κ, μ, and n); three with the nonlinear hardening operator, (that is, B_0, B_1, and p); two with the internal thermal recovery term (that is, R_α and q); and one with the dynamic recovery term (that is, β). Both elastic and inelastic material parameters given in Table 1 correspond to the reference temperature (T_{ref}) of 650°C. This temperature was chosen as the reference temperature because at 650°C the material exhibits cyclically neutral behavior and thus supports the use of a purely kinematic model. Here, a brief outline of the characterization process is given, with a more complete discussion regarding both exploratory and characterization testing given by Arnold [*15*] and Castelli et al. [*19*]. All tests addressed are uniaxial isothermal experiments, thus implying that the multiaxial material constants are typically generalized from their uniaxial counterparts. This need for generalization is precisely why a consistent multiaxial theory, such as that developed from a potential formulation, is imperative. The available tests for use in characterizing the current model are: three relaxation tests, three creep tests, one multiple-step creep test, and three tensile curves at different total strain rates.

Initially, the yield stress is unity and, therefore, the radius of the threshold surface is determined by the drag strength and represents that value of stress below which only elastic behavior is "observed"; that is, no measurable inelastic behavior occurs. The actual numerical value employed will obviously be non-unique as it will be dependent upon the sensitivity of the experimental equipment or the definition of inelasticity employed, or both. A preferred approach to arrive at a value for κ is to conduct a sequence of creep or relaxation probing tests to determine the maximum value of stress for which no time-dependent behavior ($\dot{\epsilon}^I = 0$) is observed. An alternative approach and the one taken here (given the availability of three relaxation test results) is to use the minimum stress value attained at the end of a 24-h period as our maximum allowable drag strength (that is, $\kappa \leq 7$ MPa) since all relaxation trajectories in the state space tend toward the origin (that is, $\sigma = \kappa$, $\alpha = 0$, see Fig. 2).

Given κ and taking $\beta \approx 0$ (because at T_{ref} the thermal recovery mechanism dominates), the material parameters, μ and n, may be obtained from either variable strain rate tests or stress transient dip tests, as discussed in Arnold [*15*]. Here, given a multiple step creep test that is a type of variable strain rate test, data pairs can be obtained at the beginning and end of each step in stress, for example, $(\sigma_1, \dot{\epsilon}_1^I)$, $(\sigma_2, \dot{\epsilon}_2^I)$, respectively. Assuming that the internal stress, α, is equal to α^* both before and after the step in stress (that is, Points 1 and 2), the following relationship may be obtained from eliminating α^* from the corresponding flow laws at Points 1 and 2 (see Eq 36); that is

$$\left[\frac{\sigma_1 - \sigma_2}{\kappa}\right] = \left(\frac{\kappa}{\mu}\dot{\epsilon}_1^I\right)^{1/n} - \left(\frac{\kappa}{\mu}\dot{\epsilon}_2^I\right)^{1/n} \tag{49}$$

where the only unknowns are μ and n. Employing a sufficient number of data pairs and a nonlinear regression scheme, optimal values for μ and n can be found.

Next, assuming that hardening dominates the early portion of a creep test (primary creep stage), the thermal recovery term in the evolutionary law may be neglected. Consequently, the internal stress rate of Eq 38 can be expressed as

$$\dot{\alpha} = \frac{3}{2}\mathcal{H}\left(1 - \frac{2}{3}\mathcal{C}\alpha^2\right)\dot{\epsilon}^I \tag{50}$$

given that $\beta \approx 0$. Furthermore, an expression for $\dot{\alpha}$ in terms of $\dot{\epsilon}^I$, $\ddot{\epsilon}^I$, $\dot{\sigma}$, and α can be obtained from differentiation of the flow law in Eq 37, that is

$$\dot{\alpha} = \frac{\dot{\sigma} - \dfrac{\kappa^2}{\mu n}\left(\ddot{\epsilon}^I + \dfrac{2\dot{\epsilon}^I|\dot{\sigma}|}{(\sigma - \alpha)}\right)\left(\dfrac{\kappa}{\mu}\dot{\epsilon}^I\right)^{(1-n)/n}}{\left[1 + \dfrac{\kappa^2}{\mu n}\dfrac{2\dot{\epsilon}^I}{(\sigma - \alpha)}\left(\dfrac{\kappa}{\mu}\dot{\epsilon}^I\right)^{(1-n)/n}\right]} \tag{51}$$

Thus, equating Eqs 50 and 51, assuming that κ, μ, and n are known, considering data pairs $(\sigma,\epsilon^I,\dot{\epsilon}^I,\ddot{\epsilon}^I,\alpha)$ taken from primary creep data, and using again a nonlinear regression scheme, the material parameters, B_0, B_1, and p, can be determined. Given multiple creep curves, a number of sets of B_0, B_1, and p may be obtained and then a weighted averaged can be computed to give a single optimum set.

Finally, the material parameters associated with thermal recovery (R_α and q) may be determined using the data pairs $(\sigma_s,\dot{\epsilon}_s^I)$ associated with the constant strain rate portion of the creep curve, and the corresponding internal stress (that is calculated from the flow law) at this assumed steady state (α_s),

$$\alpha_s = \frac{\sigma_s - \kappa\left[1 + \left(\dfrac{\kappa}{\mu}\dot{\epsilon}_s^I\right)^{1/n}\right]}{1 - \beta H v[Y]} \tag{52}$$

Upon obtaining these data pairs, one may take the natural logarithm of both sides of Eq 45 and determine R_α and q from the ordinate intercept and slope, respectively, of the resulting ln-ln plot.

The procedure just outlined can easily be applied to obtained an initial set of required material parameters. This set must, however, be further optimized to obtain a final set of material parameters that will allow accurate correlation of the model over a "wide range of loading." In the present case, the authors are primarily interested in the short-term response ($\leq$7200 s) and small strain ($\approx$1%) regime of the titanium alloy TIMETAL 21S. The resulting optimized material parameter set for TIMETAL 21S is given in Table 1.

The corresponding correlation for, (1) the tensile response of TIMETAL 21S over three orders of magnitude in total strain rate (that is, $\dot{e} = 8.33 \times 10^{-4}$, 8.33×10^{-5}, $8.33 \times$

10^{-6}), (2) short-term creep at three stress levels (that is, σ = 72, 110, and 128 MPa), and (3) relaxation at 238 MPa are shown respectively in Figs. 3, 4, and 5, wherein the symbols denote experimental data and the lines are the model correlations. In the opinion of the authors, the model does a very good job, given a single set of constants, for the wide variety of loading conditions examined. It is important to remind the reader that if one were to focus on only one type of loading condition, superior correlations could be achieved for that class of loading; however, predictions of other classes of loading would severely suffer. For example during the preceding characterization process, excellent tensile correlation was obtained but at the expense of poor creep and relaxation behavior for a given set of material parameters. Conversely, when the model was calibrated for creep response, poor tensile and relaxation behavior were similarly predicted.

GVIPS Predictions

The present GVIPS model, characterized to represent the behavior of TIMETAL 21S at 650°C, is now exercised and its predictive capability assessed. The assessment begins by considering relaxation tests performed at stress levels above and below those used in the characterization of the model. The corresponding short-term (that is, ≤800 s) and mid-term (that is, ≤7200 s) relaxation responses are shown in Figs. 5 and 6, respectively. Again, it is apparent that overall agreement with short-term behavior is good while that of mid-term is only reasonable. Note how the initial stress rates are accurate at the higher stress levels and yet diminish more rapidly than do the actual experimental values.

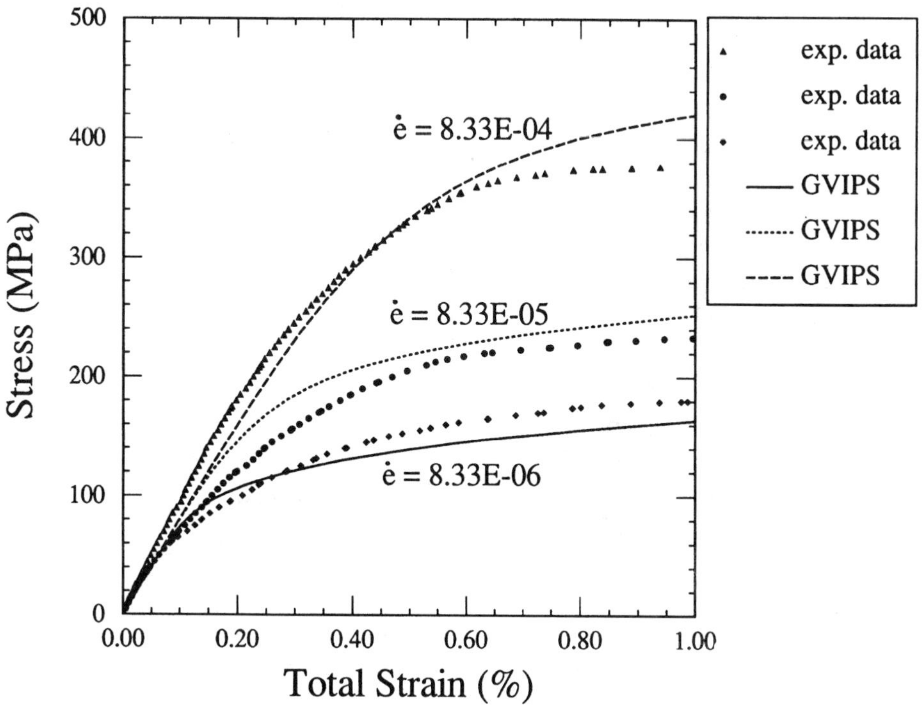

FIG. 3—*GVIPS correlation with experimental tensile data* [20] *considering the effect of strain rate.*

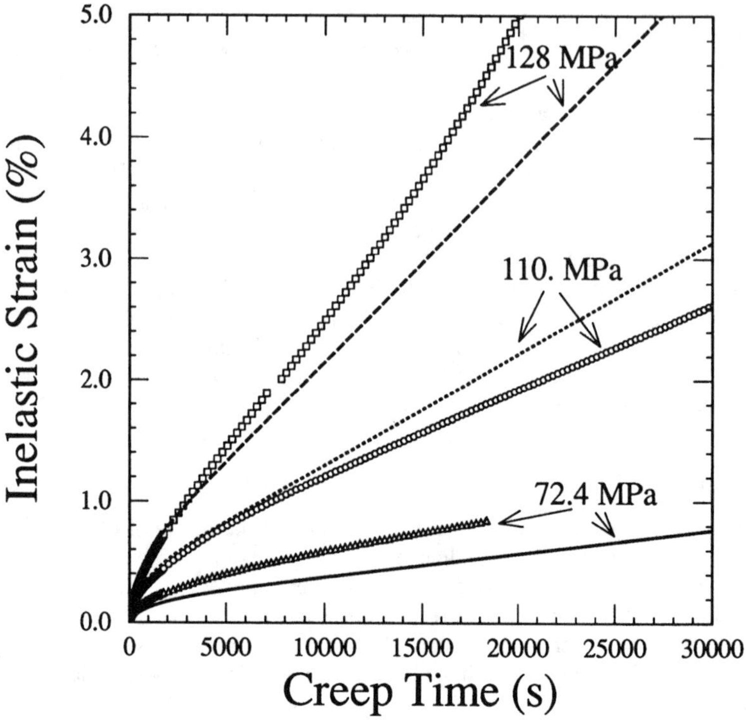

FIG. 4—*GVIPS correlation with experimental data of TIMETAL 21S under creep.*

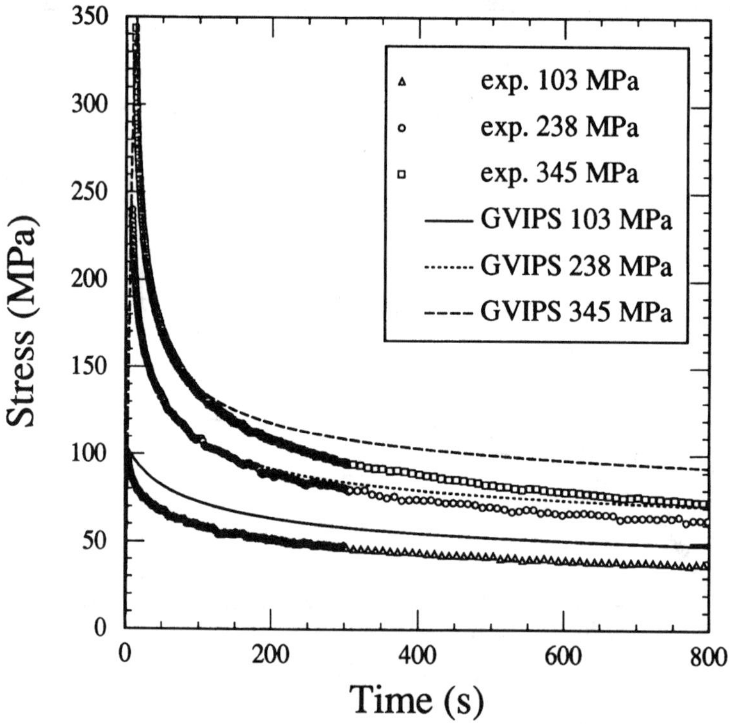

FIG. 5—*Short-term relaxation response of TIMETAL 21S: GVIPS simulation versus experimental results.*

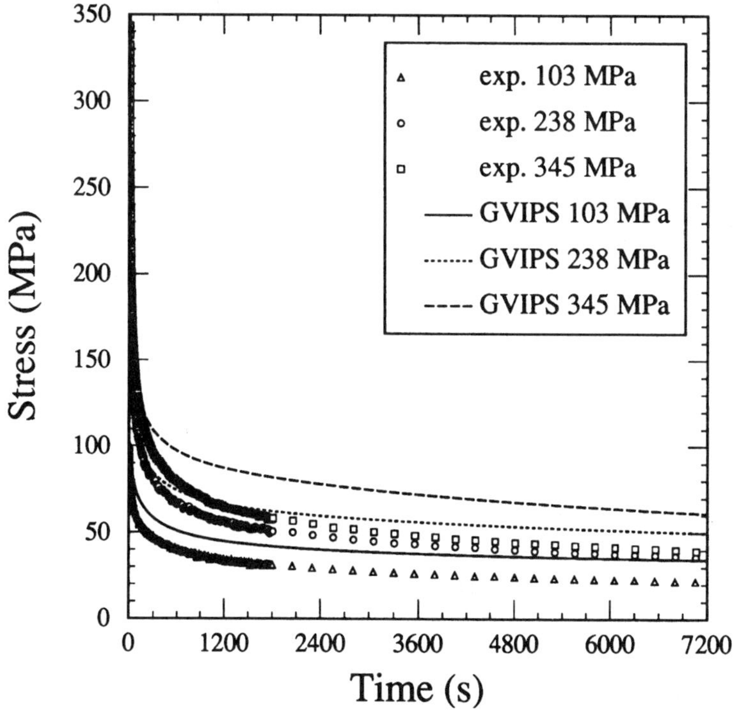

FIG. 6—*Mid-term relaxation response of TIMETAL 21S: GVIPS simulation versus experimental results.*

Figure 7 illustrates the cyclic behavior under strain control given a total strain rate of 8.33×10^{-5}, wherein the symbols represent the experimental results and the solid line the GVIPS prediction. Again, the agreement is adequate with approximately a 20% over-prediction of the maximum stress at ± 0.01 strain. Although not shown in Fig. 7, note that two cycles are required with the GVIPS model before the hysteresis loop will close.

A multiple step creep test is considered next, where the resulting inelastic strain versus time (at constant stress) is shown in Fig. 8. Here the GVIPS (solid line) simulation is seen to significantly under-predict, the accumulated inelastic strain as compared with the experimental data (denoted by the symbols). This is not surprising, however, given the fact that the model under-predicts the creep behavior at 72 MPa (see Fig. 4) and that this test was performed under constant load and not constant stress as was the simulation. As a consequence of the constant load-controlled creep test, the experimental results have both additional geometric effects and possibly creep damage (thereby leading to tertiary creep) included in them. Neither of these factors have been included in the numerical simulation. The influence of these factors is particularly evident during the third step, where the inelastic strain rate during the experiment is significantly greater than that simulated (see $\sigma = 128$ MPa in Fig. 4).

Lastly, a classic plasticity-creep interaction experiment was performed and simulated to illustrate the need for a unified viscoplastic formulation. Here, the material is subjected to an overload prior to performing a creep test at a lower stress value, as shown in Fig. 9. Typically, as is found here, the initial creep response following the overload is significantly reduced when compared to that produced from a pure creep test at this lower stress amplitude, Fig. 10. Once again, comparing the numerical simulation (of the GVIPS model)

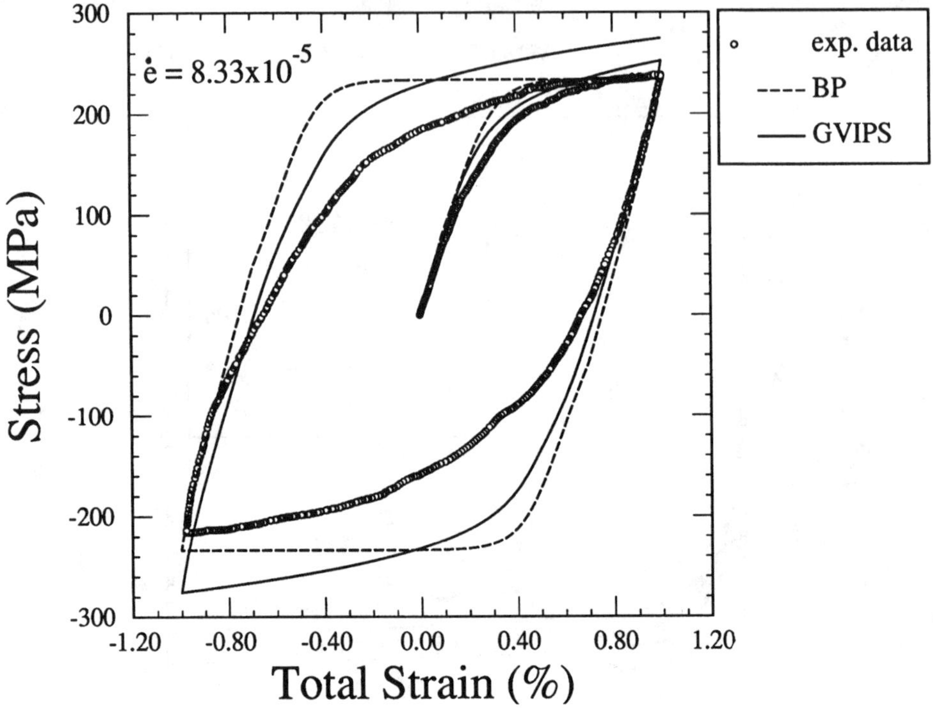

FIG. 7—*GVIPS and BP simulations versus cyclic behavior of TIMETAL 21S* [20].

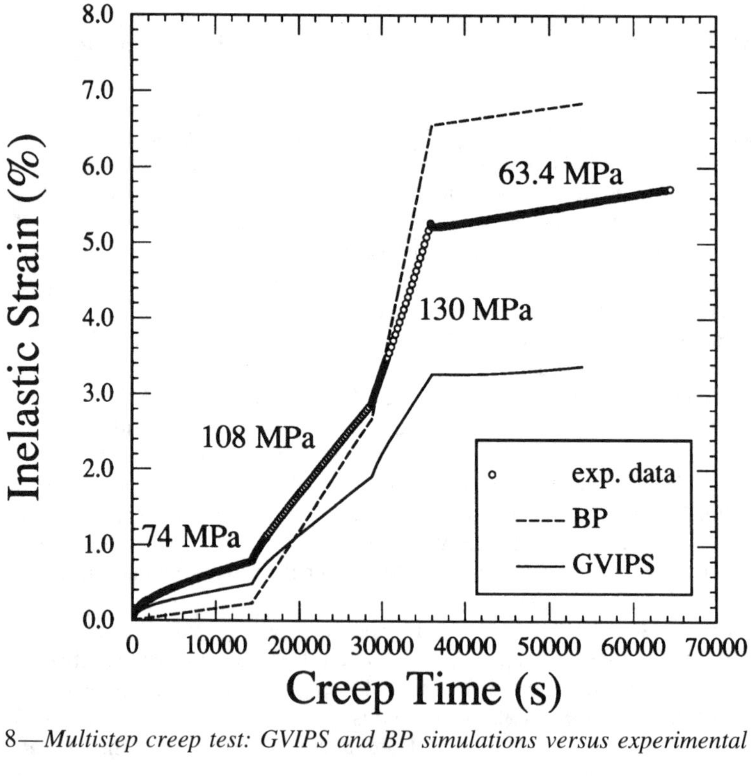

FIG. 8—*Multistep creep test: GVIPS and BP simulations versus experimental data.*

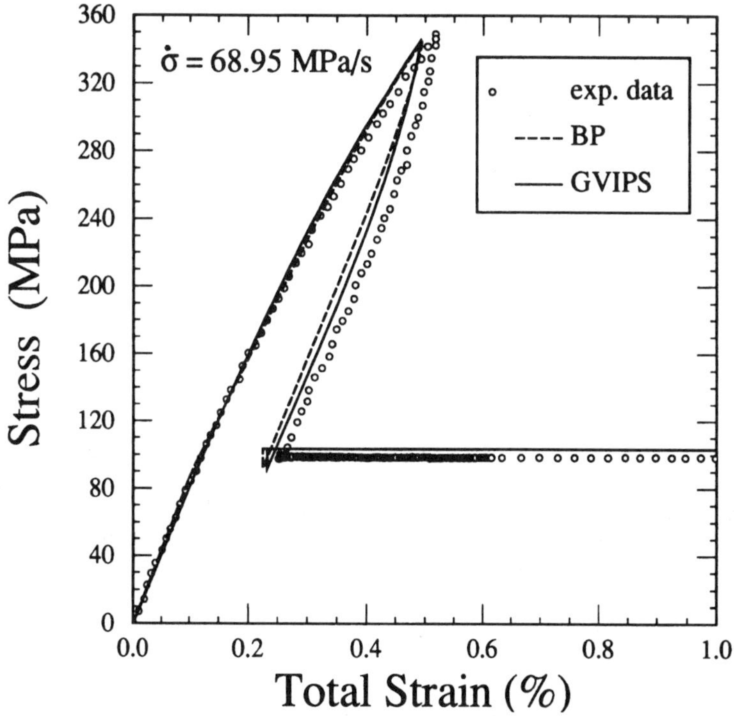

FIG. 9—*Plasticity–creep interaction test: stress-strain response simulation versus experimental data.*

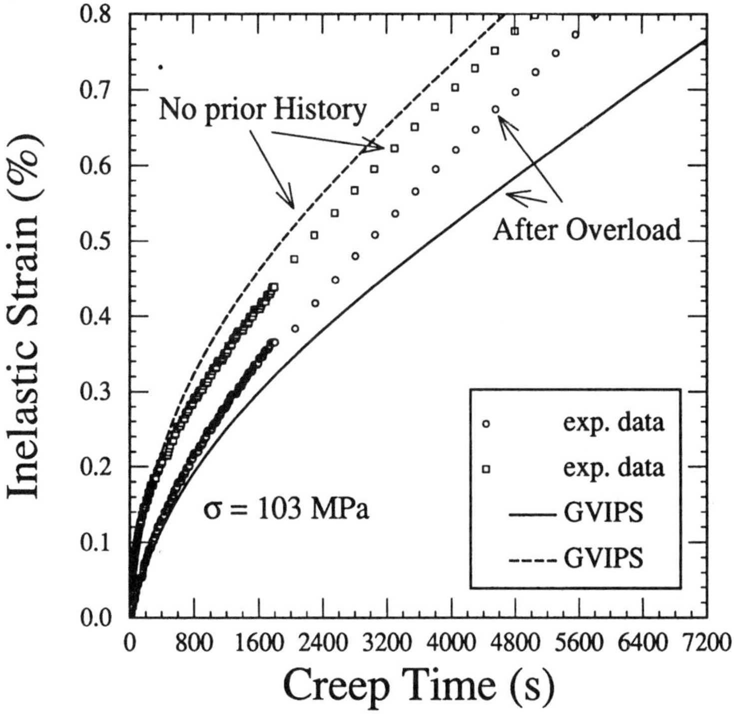

FIG. 10—*Plasticity–creep interaction test: illustrating the influence of prior history on inelastic response.*

to that of the experimental results one observes reasonable overall agreement. Although, the amount of total accumulated creep strain at the end of the test, that is, 72 000 s, is significantly under estimated, the initial reduction in primary creep rate, due to the overload, is extremely accurate during the first 800 s and good over the first hour of creep loading, thus fulfilling our primary objective.

GVIPS versus BP Model

Reviewing Figs. 3 through 10 it is evident that the present GVIPS model in conjunction with the material parameters of Table 1 does a very good overall job of simulating the short-term and small strain behavior of TIMETAL 21S at 650°C. In an attempt to assess the relative quality of these correlations and predictions as well as the numerical efficiency of the present model, an alternative and commonly employed unified viscoplastic model was implemented into our in-house Inelastic Deformation Analysis Code (IDAC). The full multiaxial, nonisothermal formulation of the Bodner-Partom (BP) model, given in Refs *5* and *21* was chosen as the alternative model because it too has been characterized for the titanium matrix, TIMETAL 21S. The specific tests employed to correlate this model are unknown to the authors, consequently, all loading conditions investigated are assumed to be predictions.

Figures 11 through 15 illustrate the tensile, creep, cyclic, and relaxation behavior predictions of the BP model at 650°C. The corresponding GVIPS simulations are also included in these figures for ease of comparison. Examining Fig. 11, it is apparent that at the higher

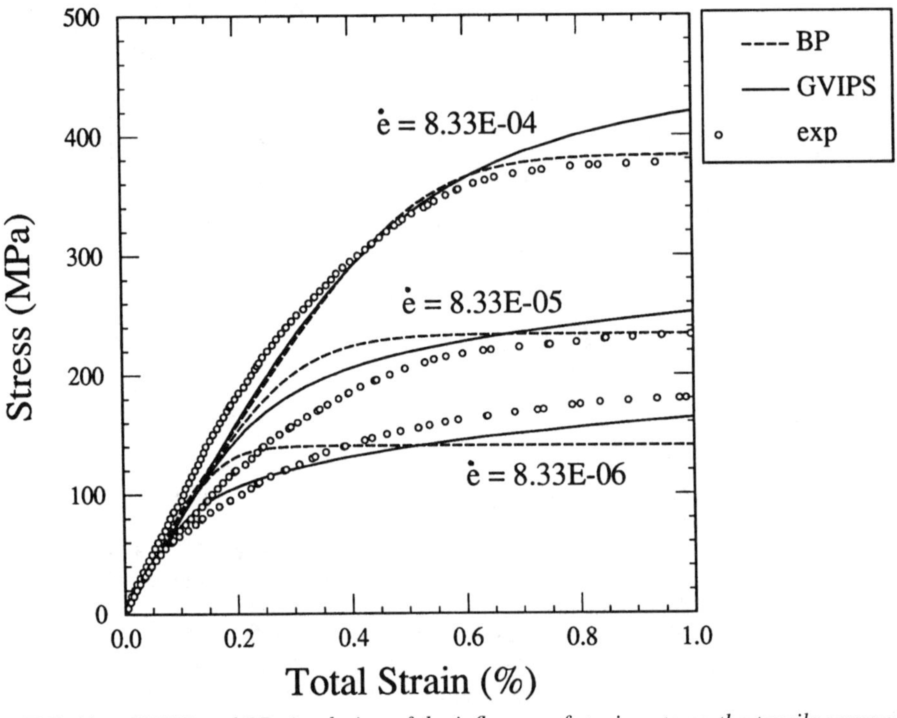

FIG. 11—*GVIPS and BP simulation of the influence of strain rate on the tensile response.*

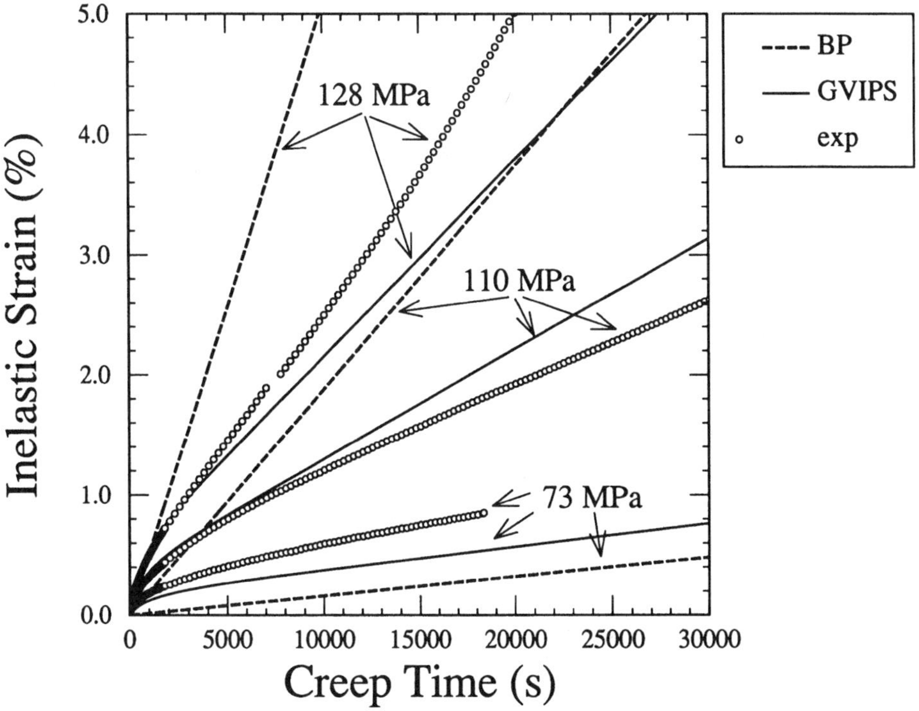

FIG. 12—*GVIPS and BP simulation of the creep response.*

strain rates (8.33×10^{-5} and 8.33×10^{-4}) the BP model does a better job of predicting the saturated stress values at 0.01 strain than does the GVIPS model. However, at the lower strain rates, the BP model does a poorer job in representing the knee (or transient portion) of the tensile curves. This poorer simulation by the BP model of the time-dependent behavior is further illustrated in Figs. 12, 14, and 15, wherein the creep and relaxation behavior is compared with the GVIPS simulations. Clearly, both the primary and secondary creep rates differ greatly and disagree with the experimental observations. Similarly, the short- and mid-term BP relaxation response is inaccurately predicted in Figs. 14 and 15 as compared with the more accurate GVIPS simulations. This inferior performance in creep and relaxation is not unexpected as the BP model is known for its ability to predict the rate-dependent tensile and cyclic response of a material and not the time-dependent creep and relaxation behavior. The significance of this failure to adequately capture the time-dependent behavior is further emphasized in both the multistep creep test shown in Fig. 8 and the plasticity-creep interaction test shown in Figs. 9 and 16 wherein the BP model poorly represents the primary creep (transient or short-term response) following a change in the applied loading.

The ability of the BP model to simulate the rate-dependent cyclic behavior is shown in Figs. 7 and 13, where both the BP and GVIPS responses are in reasonably good agreement, except for the predicted maximum stress amplitudes achieved, where they differ between 10 and 30% of one another. Also, note that in Fig. 13 a numerical problem was observed in the BP model during unloading at the fastest strain rate of 8.33×10^{-4}. No special numerical techniques where implemented to attempt to overcome this difficulty as suggested in the work of Kroupa [5], because a strict comparison in the numerical integration of these two models was desired.

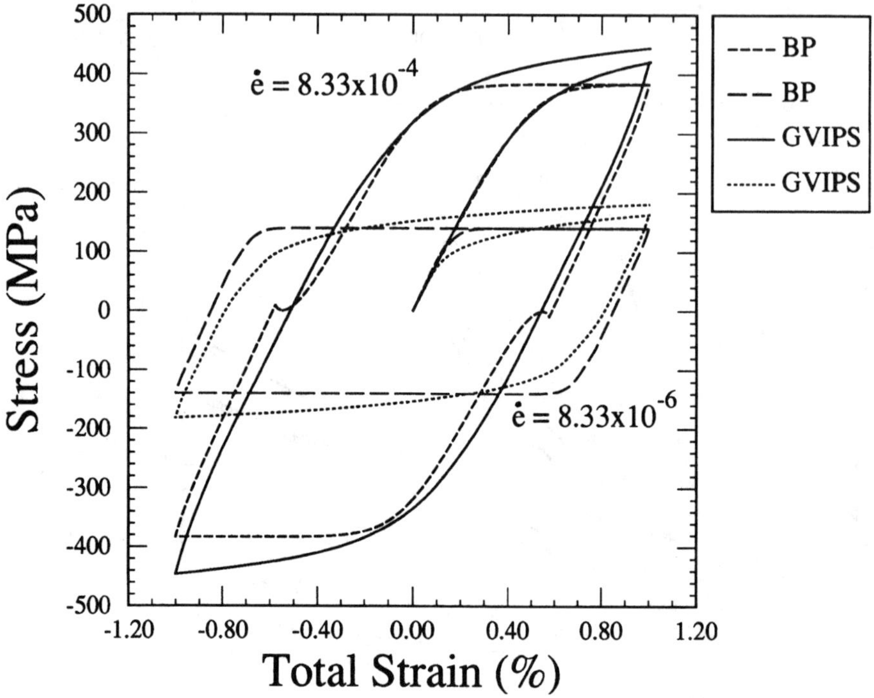

FIG. 13—*GVIPS and BP simulation of cyclic behavior with two orders of magnitude difference in total strain rate.*

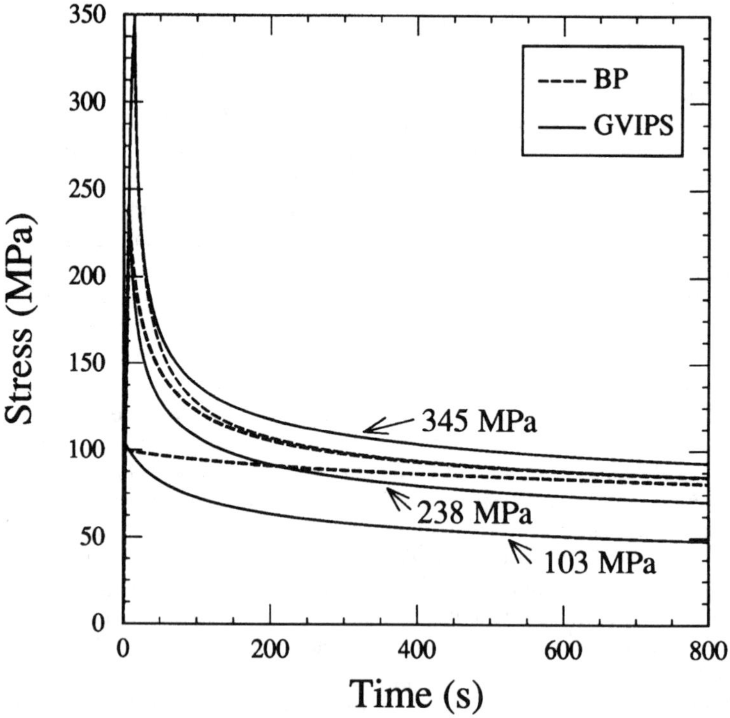

FIG. 14—*GVIPS and BP simulation of short-term relaxation response.*

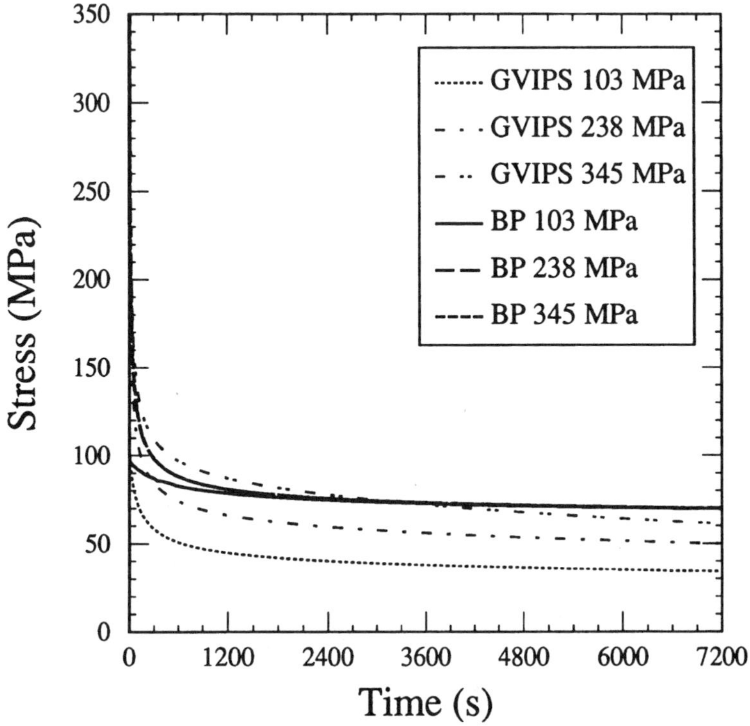

FIG. 15—*GVIPS and BP simulation of mid-term relaxation response.*

Finally, a comparison of the numerical efficiency of both models is made, relative to the total number of integration steps required to complete the previous simulations. This comparison is completely legitimate as both multiaxial models were implemented into the same computer code, integrated using the same self-adapting fouth order Runge-Kutta and Adams Bashforth predictor-corrector integration algorithm, and executed on the same Sun IPX workstation. Table 2 contains the number of total integration steps required to perform each loading history discussed. Clearly, the GVIPS model is extremely easy to integrate (due to the use of power law material functions as opposed to exponential ones and the potential structure itself) when compared with the BP model and therefore would provide significant computational savings in a large-scale structural finite element analysis. Note that in the case of tensile and cyclic behavior both models give similar accuracy and yet the GVIPS formulation provides at least a factor of 2 improvement in numerical efficiency over the BP model. Furthermore, in the case of creep and relaxation where the simulation accuracy of GVIPS is clearly superior to the BP model, the numerical efficiency is also superior, particularly in the case of creep where numerical stiffness can become a problem [*22*]. Thus, the GVIPS (generalized viscoplasticity with potential structure) formulation provides not only accurate predictive capability, but also superior numerical performance over other nonpotential formulations, the BP model being an example.

Summary

A fully associative, multiaxial, isothermal, nonlinear kinematic hardening viscoplastic model has been presented. It contains, in general, three internal state variables (two scalars

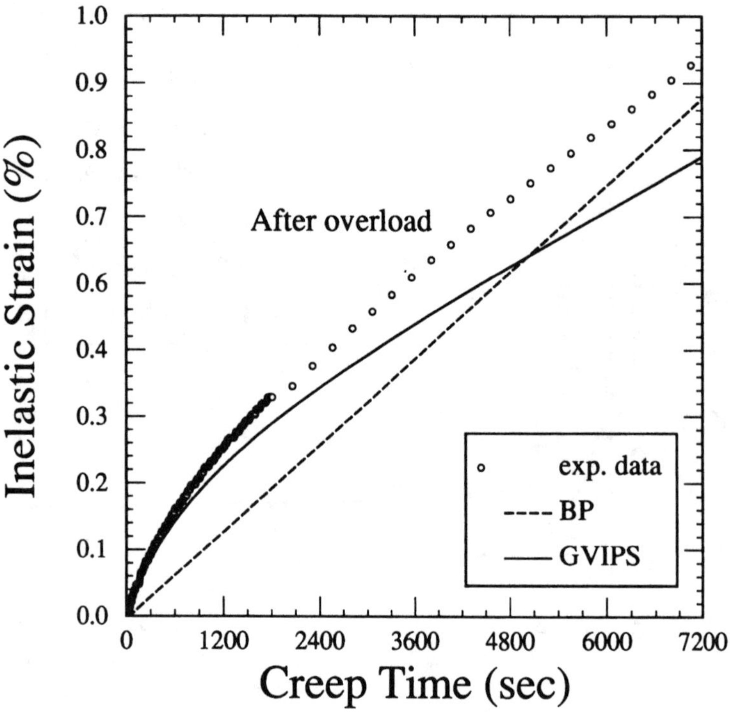

FIG. 16—*GVIPS and BP simulation of inelastic (creep) response following overload shown in Fig. 9.*

TABLE 2—*Numerical comparison of the GVIPS and BP models.*

Type of Test	Integration Steps		Speed-Up
	GVIPS	BP	
Tensile, 1/s			
$\dot{e} = 8.33 \times 10^{-4}$	170	338	1.99
$\dot{e} = 8.33 \times 10^{-5}$	196	619	3.16
$\dot{e} = 8.33 \times 10^{-6}$	231	992	4.29
Cyclic, 1/s			
$\dot{e} = 8.33 \times 10^{-4}$	761	15 007	19.72
$\dot{e} = 8.33 \times 10^{-5}$	828	2 727	3.29
$\dot{e} = 8.33 \times 10^{-6}$	938	4 937	5.26
Creep, MPa			
$\sigma = 72.4$	335	758	2.26
$\sigma = 110$	359	12 203	34
$\sigma = 128$	364	32 621	89.6
Relaxation, MPa			
$\sigma = 345$	372	624	1.68
$\sigma = 238$	365	519	1.42
$\sigma = 103$	358	382	1.07
Four-step creep test	1189	6 631	5.57
Plasticity/creep	476	8 839	18.5

and one tensor) and both thermal and strain-induced recovery mechanisms. The two, nonevolving, scalar internal state variables are associated with dislocation density and are defined as the drag and yield stress. The evolving tensorial variable known as the internal (or back) stress is a second-order, traceless, symmetric tensor and is associated with dislocation substructure. A unique aspect of the present model is the inclusion of nonlinear hardening through the use of a compliance operator, derived from the Gibb's potential, in the evolution law for the back stress. This nonlinear tensorial operator is significant in that it allows both the flow and evolutionary laws to be fully associative (and therefore easily integrated) and greatly influences the multiaxial response under nonproportional loading paths. In addition to this nonlinear compliance operator, a new consistent, potential preserving, internal strain unloading criterion has been introduced to prevent abnormalities in the predicted stress-strain curves, which are present with nonlinear hardening formulations, during unloading and reversed loading of the external variables. The resulting isothermal unified viscoplastic model was then characterized for the titanium-based alloy, TIMETAL 21S, at 650°C. Results illustrated the good overall correlation and predictive capabilities of the model for a wide range of loading conditions, that is, tensile, cyclic, creep, and relaxation tests. The proposed model was then compared with a commonly accepted and employed version of the Bodner-Partom viscoplastic model and was found to be preferable both in its predictive capabilities and numerical performance for the conditions investigated.

Acknowledgments

Thanks be to God. We would also like to acknowledge the efforts of Dr. T. E. Wilt in the development of IDAC and implementation of both the GVIPS and Bodner-Partom models and to thank him for his assistance in performing a number of the simulations presented in this paper.

References

[1] Arnold, S. M. and Saleeb, A. F., "On the Thermodynamic Framework of Generalized Coupled Thermoelastic-Viscoplastic-Damage Modeling," *International Journal of Plasticity,* Vol. 10, No. 3, 1994, pp. 263–278.

[2] Saleeb, A. F. and Wilt, T. E., "Analysis of the Anisotropic Viscoplastic-Damage Response of Composite Laminates-Continuum Basis and Computational Algorithms," *International Journal of Numerical Methods and Engineering,* Vol. 36, 1993, pp. 1629–1660.

[3] Arnold, S. M., Saleeb, A. F., and Wilt, T. E., "A Modeling Investigation of Thermal and Strain Induced Recovery and Nonlinear Hardening in Potential Based Viscoplasticity," NASA TM-106122, 1993; *Journal of Engineering Materials and Technology,* Vol. 117, No. 2, 1995, pp. 157–167.

[4] Saleeb, A. F., Seif, Y., and Arnold, S. M., "Fully-Associative Viscoplasticity with Anisotropic and Nonlinear Kinematic Hardening," *International Journal of Plasticity,* submitted for publication.

[5] Kroupa, J. L., "Implementation of a Nonisothermal Unified Inelastic-Strain Theory Into ADINA 6.0 for a Titanium Alloy—User Guide," WL-TR-93-4005, Materials Directorate, Wright Laboratory, Cleveland, OH, 1993.

[6] Bodner, S. R. and Partom, Y., "Constitutive Equations for Elastic-Viscoplastic Strain Hardening Materials," *Journal of Applied Mechanics,* Vol. 42, 1975, p. 385.

[7] Chan, K. S., Bodner, S. R., and Lindholm, U. S., "Phenomenological Modeling of Hardening and Thermal Recovery in Metals," *Journal of Engineering Materials and Technology,* Vol. 110, 1988, pp. 1–8.

[8] Stouffer, D. C., Ramaswamy, V. G., Laflen, J. H., Van Stone, R. H., and Williams, R., "A Constitutive Model for the Inelastic Multiaxial Response of Rene'80 at 871C and 982C," *Journal of Engineering Materials and Technology,* Vol. 112, 1990, pp. 241–246.

[9] Spencer, A. J. M., *Continuum Physics,* Vol. 1, A. C. Eringen, Ed., Academic Press, London, 1971, p. 240.

[10] Arnold, S. M., Saleeb, A. F., and Castelli, M. G., "A Fully Associative, Nonisothermal, Non-Linear Kinematic, Unified Viscoplastic Model for Titanium Based Matrices," *Thermo-mechanical Fatigue Behavior of Materials: Second Volume, ASTM STP 1263,* M. J. Verrilli and M. G. Castelli, Eds., Philadelphia, 1995.
[11] Lemaitre, J. and Chaboche, J. L., *Mechanics of Solid Materials,* Cambridge University Press, New York, 1990.
[12] Orowan, E., "Causes and Effects of Internal Stresses, Internal Stresses and Fatigue of Metals," *Proceedings,* Symposium on Internal Stresses and Fatigue in Metals, Detroit, MI, G. M. Rassweiler and W. L. Grvise, Eds., Elsevier Publishing, New York, 1959.
[13] Robinson, D. N., "A Unified Creep–Plasticity Model for Structural Metals at High Temperature," ORNL TM-5969, Oak Ridge National Laboratory, Oak Ridge, TN, 1978.
[14] Onat, E. T. and Fardshisheh, F., "Representation of Creep of Metals," ORNL TM–4783, Oak Ridge National Laboratory, Oak Ridge, TN, 1972.
[15] Arnold, S. M., "Effects of State Recovery on Creep Buckling Induced by Thermomechanical Loading," Ph.D. thesis, University of Akron, Akron, OH, 1987.
[16] *Unified Constitutive Equations for Plastic Deformation and Creep of Engineering Alloys,* A. K. Miller, Ed., Elsevier Applied Science, New York, 1987.
[17] Freed, A. D., Chaboche, J. L., and Walker, K. P., "A Viscoplastic Theory with Thermodynamic Considerations," *Acta Mech,* Vol. 90, 1991, pp. 155–174.
[18] Onat, E. T. and Fardshisheh, F., "Representation of Creep, Rate Sensitivity and Plasticity," SIAM, *Journal of Applied Mathematics,* Vol. 25, No. 3, 1973, pp. 522–538.
[19] Castelli, M. G., Arnold, S. M., and Saleeb, A. F., "Specialized Deformation Tests for the Characterization of a Viscoplastic Model: Application to a Titanium Alloy," NASA TM-106268, 1994.
[20] Ashbaugh, N. E. and Khobaib, M., University of Dayton Research Institute, Dayton, OH, unpublished data.
[21] Neu, R. W., "Nonisothermal Material Parameters for the Bodner-Partom Model," MD-Vol. 43, *Material Parameter Estimation for Modern Constitutive Equations,* L. A. Bertram, S. B. Brown, and A. D. Freed, Eds., ASME Book No. H00848-1993, 1993, pp. 211–226.
[22] Arnold, S. M., "Quantification of Numerical Stiffness for a Unified Viscoplastic Constitutive Model," *Journal of Engineering, Materials and Technology,* Vol. 112, 1990, pp. 271–276.

Cliff J. Lissenden,[1] *Carl T. Herakovich,*[2] *and Marek-Jerzy Pindera*[3]

Inelastic Deformation of Titanium Matrix Composites Under Multiaxial Loading

REFERENCE: Lissenden, C. J., Herakovich, C. T., and Pindera, M.-J., **"Inelastic Deformation of Titanium Matrix Composites Under Multiaxial Loading,"** *Life Prediction Methodology for Titanium Matrix Composites, ASTM STP 1253,* W. S. Johnson, J. M. Larsen, and B. N. Cox, Eds., American Society for Testing and Materials, 1996, pp. 257–277.

ABSTRACT: A theoretical model capable of predicting the thermomechanical response of continuously reinforced, titanium matrix composite (TMC) laminates, subjected to multiaxial loading has been developed. The model is based on micromechanics and employs classical lamination theory to determine inelastic response. The constitutive relationships for each lamina are determined from a micromechanics analysis that is performed numerically using the finite element method. Matrix viscoplasticity, thermal stresses, and damage to the fiber/matrix interfacial zone are explicitly included in the model.

The representative cell of the micromechanical model is considered to be in a state of generalized plane strain, enabling a quasi two-dimensional analysis to be performed. Constant strain triangular elements are formulated with elasto-viscoplastic constitutive equations. Interfacial debonding is incorporated into the model through interface elements that are based on a constitutive model that includes normal and tangential debonding.

Theoretical predictions are compared with the results of an experimental program conducted on SCS-6/Ti-15-3 unidirectional, $[0_4]$, and angle-ply, $[\pm 45]_S$, tubular specimens. Multiaxial loading included increments of axial tension, compression, torque, and internal pressure. Loadings were chosen in an effort to distinguish inelastic deformation due to damage from matrix plasticity and separate time-dependent effects from time-independent effects. Results show that fiber/matrix debonding can be a major factor in the effective stress-strain response and that significant room temperature creep can occur at relatively low applied stress levels.

KEYWORDS: micromechanics, viscoplasticity, interfacial debonding, room-temperature creep, silicon carbide/titanium, lamination theory, finite element analysis, titanium, titanium matrix composites, life prediction, titanium alloys, fatigue, materials, modeling

Metal matrix composites (MMCs) are candidate materials for the next generation of jet engines, hypersonic flight vehicles and missiles. These applications will present very severe thermomechanical loading environments and that is the reason these advanced materials are necessary. Before a new composite material system can be used for a structural component, analyses must be performed on two basic levels. First, the basic thermomechanical response of the material must be characterized at the micro-level. Then the geometry of the component and anticipated loading must be considered to design the component at the macro-level.

[1]Assisted professor, Department of Engineering Science and Mechanics, Penn State University, Unversity Park, PA; formerly visiting assistant professor, Department of Engineering Mechanics, University of Kentucky, Lexington, KY 40506–0046.

[2]Henry L. Kinnier professor and director of Applied Mechanics, Department of Civil Engineering and Mechanics, University of Virginia, Charlottesville, VA 22903–2442.

[3]Associate professor, Department of Civil Engineering and Applied Mechanics, University of Virginia, Charlottesville, VA 22903–2442.

Most current work involves the characterization of MMCs at the micro-level. This is quite a formidable task because advanced MMCs being developed for elevated temperature applications are susceptible to damage development in addition to matrix plasticity at relatively low stress levels in comparison with anticipated service stress levels.

The theoretical model described herein is aimed specifically at titanium (Ti) matrix composites reinforced with continuous silicon-carbide (SiC) fibers; however, the model is valid for any continuous fiber composite with a viscoplastic matrix and possible fiber/matrix interfacial debonding. Numerous forms of damage have been detected experimentally in MMCs. The most significant of these, at least for SiC/Ti in the presence of transverse tensile loading, is fiber/matrix interfacial debonding. The interface in SiC/Ti composites is weak and debonds at relatively low stress levels due to coatings applied to the SiC fiber that are required to control the reaction zone and preserve their strength. Other damage mechanisms will not be considered at this point in order to focus attention on interfacial debonding.

Experimental Evidence of Damage

Fiber/matrix interfacial debonding was first detected experimentally by Johnson et al. [*1*] using the edge replica technique. Tension tests were conducted on a number of different SiC/Ti laminates. A distinct knee was observed in the stress-strain response of those laminates containing off-axis plies. Edge replicas taken under load revealed interfacial debonding had occurred and was responsible for the observed knee. The knee in the initial loading cycle occurred at a higher stress level than in subsequent cycles suggesting that the initial knee was associated with the breaking of a chemical bond, while subsequent knees were associated with overcoming residual stresses created during processing. Nimmer et al. [*2*] reported similar findings for a Ti-6-4 matrix composite. Other researchers [*3–5*] observed interfacial debonding along with other damage mechanisms through careful microstructural evaluations before loading, at interrupted load levels, and after loading.

Modeling Interfacial Damage

Composites with imperfect interfacial bonding were modeled in Refs *6* through *11*. The evolution of damage in MMCs was first studied by Adams [*12*] using a finite element analysis. Interfacial debonding and matrix crack growth were studied by effectively removing an element from the analysis once the element stresses exceeded a specified failure strength.

Aboudi [*13*], Wisnom [*14*], and Achenbach and Zhu [*15,16*] all used different interface failure criteria as well as different micromechanical models to predict the effects of interfacial debonding on MMCs including matrix plasticity. Nimmer [*17*] and Brust et al. [*18*] used finite element analysis and assumed that the interfacial strength was associated solely with the compressive radial residual stresses created during fabrication. Once the residual stresses at the interface were overcome, the interface failed. Various forms of the Needleman interfacial debonding model [*19*] have been used in finite element analyses [*20–23*].

The current paper presents an overview of a theoretical model [*21,24*] capable of predicting the inelastic response of symmetrically laminated MMCs in the presence of interfacial debonding. Model predictions are compared with experimental results obtained from the multiaxial tests conducted on $[0_4]$ and $[\pm 45]_S$ SiC/Ti tubular specimens.

Theoretical Model

Micromechanics

Consider a composite comprised of continuous fiber reinforcement aligned in the 1-direction and arranged in a doubly periodic array. The representative volume element (RVE) is chosen to be one rectangular unit cell that contains a single fiber embedded in matrix material. In the presence of an imperfect fiber/matrix interface, this unit cell is herein called a debonding unit cell model (DUCM).

The effective response of the RVE is obtained through homogenization of the local stress and strain fields. Local stress and strain fields are volume averaged over the entire RVE in accordance with

$$\bar{\sigma}_{ij} = \frac{1}{V}\int_V \sigma_{ij}(x_k)dV = c_f\bar{\sigma}^f_{ij} + c_m\bar{\sigma}^m_{ij} \tag{1}$$

$$\bar{\epsilon}_{ij} = \frac{1}{V}\int_V \epsilon_{ij}(x_k)dV - \alpha_{ij} = c_f\bar{\epsilon}^f_{ij} + c_m\bar{\epsilon}^m_{ij} - \alpha_{ij} \tag{2}$$

where a bar indicates effective or homogenized variables, c denotes a volume fraction, sub- and superscripts (f and m) represent fiber and matrix quantities or variables, respectively, and α_{ij} is defined as the following integral over the fiber-matrix interfaces.

$$\alpha_{ij} = \frac{1}{2V}\int_S ([u_i]n_j + [u_j]n_i)dS \tag{3}$$

following Boyd et al. [*25*]. The quantity

$$[u_i] = u_i^f|s - u_i^m|_S \tag{4}$$

is the jump discontinuity in the displacement field at the damaged interface. The effective stress-strain constitutive relationship is described by

$$\bar{\sigma}_{ij} = C^*_{ijkl}[\bar{\epsilon}_{kl} - \bar{\epsilon}^{pl}_{kl} - \bar{\epsilon}^t_{kl}] \tag{5}$$

where C^*_{ijkl} denotes the current effective stiffness tensor, $\bar{\epsilon}_{kl}$ is the total effective strain, $\bar{\epsilon}^{pl}_{kl}$ is the effective plastic strain, and $\bar{\epsilon}^t_{kl}$ is the effective thermal strain.

Viscoplastic Material Model

The elastic-viscoplastic constitutive model of Bodner and Partom [*26*] is used to represent the metallic matrix. A review of the model is given by Bodner in Ref *27*. The total strain rate is

$$\dot{\epsilon}_{ij} = \dot{\epsilon}^{el}_{ij} + \left\{\frac{D_o^2}{J_2}\exp\left[-\left(\frac{Z^2}{3J_2}\right)^n\left(\frac{n+1}{n}\right)\right]\right\}^{1/2} s_{ij} \tag{6}$$

where J_2 is the second invariant of the deviatoric stresses, s_{ij}, and

$$Z = Z_1 + (Z_o - Z_1)\exp\left[-m(W_p/Z_o)\right] \tag{7}$$

where W_p is the plastic work. There are five material parameters in the model: Z_o is associated with the yield stress of the material in simple tension; Z_1 is proportional to the ultimate stress; m determines the work-hardening of the material; the rate sensitivity is controlled by the parameter n; and the constant, D_o, defines the limiting strain rate for the material. Material parameters have been fit from tension test data on fiberless, as-fabricated, matrix material.

Interfacial Constitutive Model

The interface constitutive model used is that of Tvergaard [*28*] with slight modifications. It is based on the cohesive zone model developed by Needleman [*19*] that describes the process of void nucleation from initial debonding through complete decohesion. The basic premise of the model is that with increasing interfacial separation the traction across the interface increases to a maximum and then decreases until it vanishes, so that complete decohesion occurs as illustrated in Fig. 1. The response of the interface is specified in terms of a critical strength, σ_o, and the work of separation per unit area. Dimensional considerations introduce a characteristic length, δ, into the model.

Consider an interface supporting a traction field, **T**, which, in general, has normal and shear components. Two adjacent material points, initially on opposite sides of the interface, are considered, and the interfacial traction depends only on the discontinuity in the displacement field at the interface [**u**]. At each point along the interface

$$u_n = \mathbf{n} \cdot [\mathbf{u}], \quad u_t = \mathbf{t} \cdot [\mathbf{u}], \quad u_b = \mathbf{b} \cdot [\mathbf{u}] \tag{8}$$

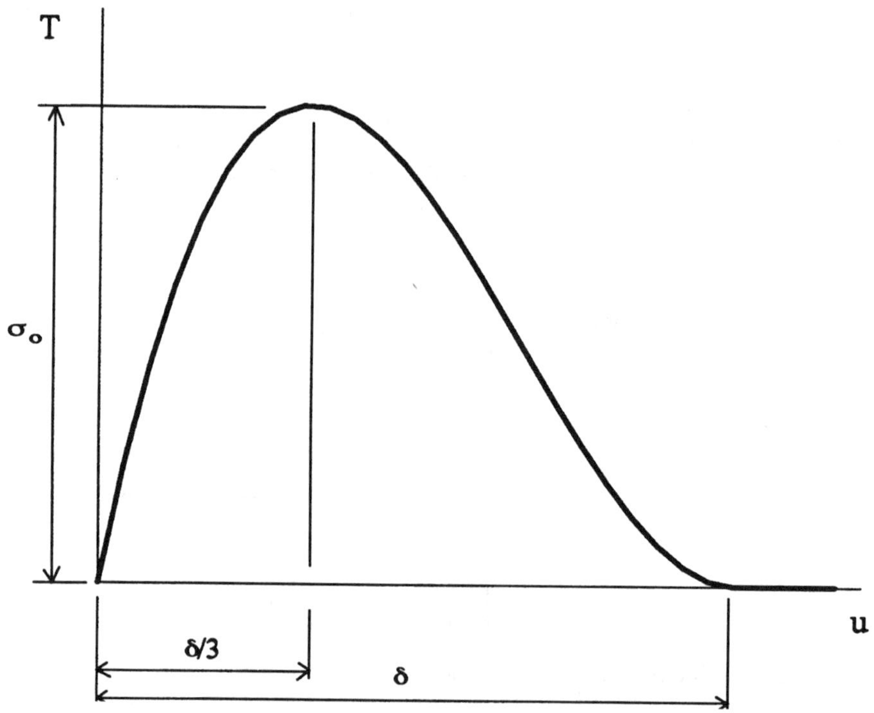

FIG. 1—*One-dimensional interfacial traction-displacement relationship.*

and

$$T_n = \mathbf{n} \cdot \mathbf{T}, \quad T_t = \mathbf{t} \cdot \mathbf{T}, \quad T_b = \mathbf{b} \cdot \mathbf{T} \tag{9}$$

where **n**, **t**, and **b** are unit vectors that form a right-hand coordinate system (Fig. 2). Positive u_n corresponds to increasing normal separation; u_t is tangent to the interface and normal to the fiber direction; while u_b is tangent to the interface and parallel to the fiber direction. Once complete decohesion has occurred at a point along the interface, tangential tractions are accounted for by Coulombic friction.

The condition of the interfacial bond is described by the resultant nondimensionalized displacement, λ

$$\lambda = \begin{cases} [\eta^2 + \chi^2]^{1/2} & \text{for } u_n \leq 0 \\ [\xi^2 + \eta^2 + \chi^2]^{1/2} & \text{for } u_n > 0 \end{cases} \tag{10}$$

where

$$\xi = u_n/\delta_n, \quad \eta = u_t/\delta_t, \quad \text{and} \quad \chi = u_b/\delta_b \tag{11}$$

are the interfacial displacements normalized with respect to the appropriate characteristic lengths. Normal compressive tractions are not considered to be detrimental to the interface. In the most general case, there are three characteristic lengths: δ_n, δ_t, and δ_b. A perfect

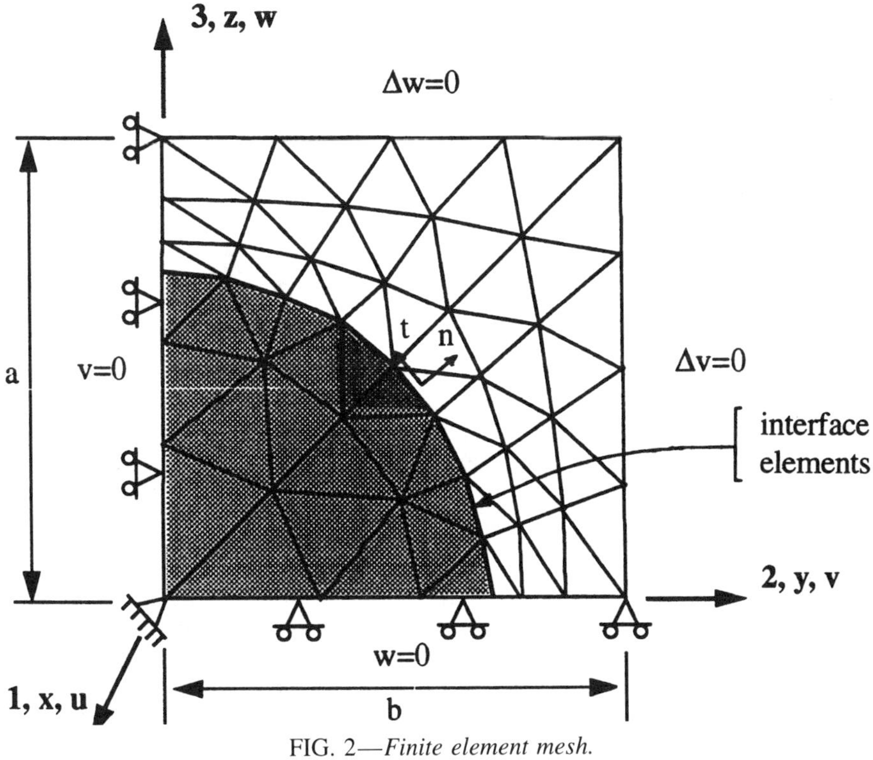

FIG. 2—*Finite element mesh.*

interfacial bond is represented by $\lambda = 0$. Any value of λ between zero and one represents imperfect bonding and $\lambda \geq 1$ corresponds to the no-bond condition. Clearly, complete bond failure at a point occurs when $\lambda = 1$ (Fig. 1).

A scalar interaction function, $F(\lambda)$, will be used to characterize the traction-displacement relationship

$$F(\lambda) = \frac{27}{4}\sigma_o(\lambda^2 - 2\lambda + 1) \tag{12}$$

For monotonically increasing loads, the current value of λ is always the maximum value, λ_{max}. However, for unloading and reloading conditions, the current value of λ may be less than the maximum value, λ_{max}, that may have occurred at some previous time in the load history. In order to prevent the interface from repairing itself, it is necessary to set the current value of λ in Eq 12 equal to the current maximum.

For $\lambda_{max} < 1$, the interfacial tractions are defined by

$$T_n = \begin{cases} 10^6\xi & \text{for } u_n \leq 0 \\ F(\lambda_{max})\xi & \text{for } u_n > 0 \end{cases} \tag{13}$$

$$T_t = \alpha F(\lambda_{max})\eta \tag{14}$$

$$T_b = \alpha F(\lambda_{max})\chi \tag{15}$$

The normal traction, T_n, has been made much stiffer than in Ref *28* for normal compression to minimize interpenetration of the fiber and matrix phases.

For $\lambda_{max} \geq 1$, which signifies loading after the interface has failed and is only capable of transmitting compressive and frictional tractions, the interfacial tractions are defined by

$$T_n = \begin{cases} 10^6\xi & \text{for } u_n \leq 0 \\ 0 & \text{for } u_n > 0 \end{cases} \tag{16}$$

$$T_t = \begin{cases} -\operatorname{sgn}(du_t)\mu T_n & \text{for } |T_t| \geq \mu|T_n| \text{ and } u_n \leq 0 \\ -4\mu T_n\eta & \text{for } |T_t| < \mu|T_n| \text{ and } u_n \leq 0 \\ 0 & \text{for } u_n > 0 \end{cases} \tag{17}$$

$$T_b = \begin{cases} -\operatorname{sgn}(du_b)\mu T_n & \text{for } |T_b| \geq \mu|T_n| \text{ and } u_n \leq 0 \\ -4\mu T_n\chi & \text{for } |T_b| < \mu|T_n| \text{ and } u_n \leq 0 \\ 0 & \text{for } u_n > 0 \end{cases} \tag{18}$$

where μ is the coefficient of kinetic friction at the interface, sgn(·) is the algebraic sign of the term in parentheses, and du_t, du_b are the current tangential interfacial displacement increments.

Lamination Theory

Inelastic behavior of laminated composites can be predicted using the nonlinear lamination theory [*29*]. The assumptions required for the nonlinear lamination theory (NLLT) are the same as those for the classical lamination theory (CLT). The development of CLT can be found in many composite mechanics textbooks, for example, Ref *30*. Thus, only the final form of the equations will be presented here.

Define resultant mechanical forces and moments acting on a laminate to be

$$\{N,M\} = \int_{-H}^{+H} \{\sigma\}_x(1,z)\, dz \tag{19}$$

where z is the coordinate direction through the thickness of the laminate having a thickness of $2H$. Equivalent thermal force and moment resultants are defined as

$$\{N^t,M^t\} = \int_{-H}^{+H} [\bar{Q}]\{\epsilon^t\}_x(1,z)\, dz \tag{20}$$

where $[\bar{Q}]$ is the transformed reduced stiffness. Inelastic pseudo-force and moment resultants represent the effects of damage and plasticity at any instant and are defined by

$$\{N^{in},M^{in}\} = \int_{-H}^{+H} [\bar{Q}]\{\epsilon^{in}\}_x(1,z)\, dz \tag{21}$$

The laminate extensional, coupling, and bending stiffnesses, respectively, are defined as

$$[A,B,D] = \int_{-H}^{+H} [\bar{Q}](1,z,z^2)\, dz \tag{22}$$

With these definitions, the overall laminate constitutive relationships can be written in the familiar compact form

$$\begin{Bmatrix} \bar{N} \\ \bar{M} \end{Bmatrix} = \begin{bmatrix} A & B \\ B & D \end{bmatrix} \begin{Bmatrix} \epsilon^{\circ} \\ \kappa \end{Bmatrix} \tag{23}$$

where

$$\begin{aligned} \bar{N} &= N + N^t + N^{in} \\ \bar{M} &= M + M^t + M^{in} \end{aligned} \tag{24}$$

Implementation

The viscoplastic and interfacial debonding models are implemented numerically using finite element analysis. Consider a laminate with continuous fiber reinforcement. Individual plies are assumed to be in a state of generalized plane strain [*31*]. The stress and strain fields are then independent of the axial 1-coordinate, but displacement can occur in all three coordinate directions. Each displacement is dependent on the 2- and 3-coordinates, and the displacement in the 1-direction is also linearly dependent on the 1-coordinate. Including the 2- and 3-dependence of the 1-displacement allows axial shear deformation to occur. Thus, response can be modeled in a generic plane.

Due to symmetry, only a quarter of the RVE need be analyzed. This region is discretized using 68 constant strain solid elements and eight interface elements as shown in Fig. 2. The current model is restricted to thermal and in-plane mechanical loading of flat symmetric laminates. The analysis of a laminate is initiated by using the micromechanical model to determine the thermomechanical properties of the laminae. These properties, along with the loading specifications, are used as input to the nonlinear lamination theory to determine both laminate and laminae stresses and strains. The strains in each lamina are input back into the micromechanical model to determine the inelastic strains. Because the strains in the material principal coordinate system are, in general, different for each lamina in the laminate and because the interfacial constitutive relationships are history-dependent, each lamina in a half-laminate has its own unit cell associated with it.

Experimental Program

Materials

Three $[0_4]$ and three $[\pm 45]_S$ SiC/Ti thin-walled tubular specimens having a nominal fiber volume fraction of 0.40 and a nominal ply thickness of 0.2032 mm (0.008 in.) were tested on an axial-torsional load stand. The tubes were fabricated (by Textron Specialty Materials) using the hot isostatic pressing (HIP) technique. The laminated tubes consisted of the titanium alloy, Ti-15V-3Cr-3Sn-3Al (commonly designated Ti-15-3), reinforced with SCS-6 silicon-carbide fibers. Molybdenum wires, 0.0508 mm (0.002 in.) in diameter were woven between the SiC fibers to help keep the fibers aligned properly.

The $[0_4]$ tubes (radius to thickness ratio, $r/t = 23.4$) were nominally 177.8 mm (7 in.) long and had an outer diameter of 38.1 mm (1.5 in.). Steel plugs, 50.8 mm (2 in.) long, were bonded to the inner surface of the ends of the tubes to facilitate a uniform loading of the tube during testing. Thus, the gage length was 76.2 mm (3 in.). The $[\pm 45]_S$ tubes ($r/t = 62.5$) were nominally 304.8 mm (12 in.) long and had an outer diameter of 101.6 mm (4 in.). Tapered steel collars 101.6 mm (4 in.) long were bonded both inside and outside the ends of each tube, leaving a gage length of 101.6 mm (4 in.). Strains were measured by strain gage rosettes.

Loadings

Three types of multiaxial loadings, designated I, II, and III, were applied to the tubes. All tests were conducted at room temperature. A detailed test matrix is provided in Ref *24*. Type I tests (Fig. 3*a*) comprised a sequence of axial tension tests to increasingly higher load levels, where at the peak of each individual cycle, a biaxial stress state was introduced by applying an increment of torque. The torque was then reduced to zero while the tensile load was held constant, and the cycle was completed by removal of the tensile load. Type I loading was modified by the addition of two hold periods (Points J and K in Fig. 3*a*) during which current load levels were held constant. This loading is designated Type IA and was applied to $[\pm 45]_S$ tubes only.

Type II tests (Fig. 3*b*) consisted of a sequence of torsional tests to increasingly higher load levels, where at the peak of each individual cycle, a biaxial stress state was introduced by applying an increment of axial tension. The tension was then reduced to zero while the torsional load was held constant, and the cycle was completed by removal of the torsional load.

Type III tests (Fig. 3*c*) required the presence of only the circumferential (transverse, for a [0] tube) stress component initially. This required that the longitudinal stress component due to internal pressure be eliminated by the application of axial compression. Due to the test setup, internal pressure and axial compression had to be applied incrementally for each cycle (A-B). At the peak internal pressure of each cycle, an increment of axial tension was applied (B-D) and removed (D-E), then an increment of torque was applied (E-F) and removed (F-G). Finally, the internal pressure was unloaded (G-H).

Experimental Results and Comparison with Theory

The finite element model (DUCM) was used to predict the material response for the loading cases applied to the tubular specimens. Thermal residual stresses were considered in

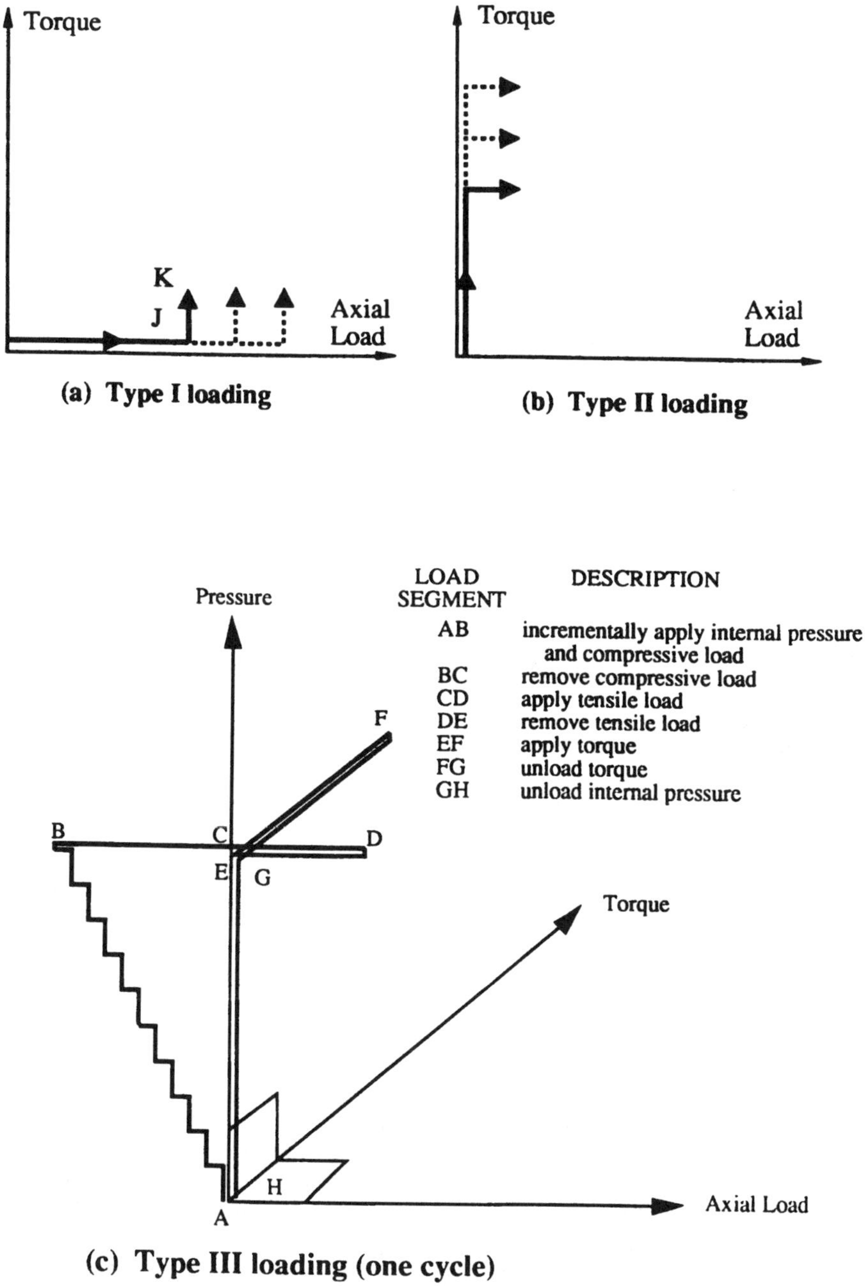

FIG. 3—*Multiaxial loadings.*

TABLE 1—*Constituent properties.*

	Thermoelastic Properties			Viscoplastic Parameters				
Material	E, GPa (Msi)	ν	α, 10^{-6}/°C (10^{-6}/°F)	D_0, 1/s	Z_0, MPa (ksi)	Z_1, MPa (ksi)	m	n
SiC	400 (58)	0.25	4.99 (2.77)	0	...	...	...	...
Ti	91 (13.2)	0.36	9.25 (5.14)	10 000	965 (140)	1172 (170)	1700	7

the DUCM predictions by applying an effective uniform change in temperature of −816°C (−1500°F) prior to mechanical loading. The constituent properties and viscoplastic material parameters used in the study are given in Table 1.

Unidirectional [0] SiC/Ti

The interface properties required for DUCM were chosen from a trial-and-error curve-fitting procedure that considered both axial shear and transverse tensile loading. The following properties were selected for unidirectional SCS-6/Ti-15-3:

$\sigma_0 = 103.4$ MPa (15 ksi),
$\alpha = 3.3$, $\delta = 254$ μm (10×10^{-6} in.),
and $\mu = 0$.

Predicted initial elastic moduli are compared with experimentally determined values in Fig. 4. The predicted and experimental axial (Fig. 4*a*) and shear (Fig. 4*c*) moduli agree very well, while the predicted transverse modulus (Fig. 4*b*) is approximately 7% higher than the average of Rosettes 1, 3, and 4 (R1, R3, and R4, respectively).

Type I Loading—Dominant Axial Tensile Load—The response of a $[0_4]$ tube to Type I loading was linear elastic to failure. DUCM predicted that the axial tensile response of a unidirectional composite is independent of the condition of the interface.

Type II Loading—Dominant Axial Shear Load—The inelastic shear deformation of a $[0_4]$ tube subjected to Type II loading under rotation control was nonuniform around the tube [*32*]. Two rosettes R2 and R3 located at midlength of the tube registered quite different shear strains as shown in Fig. 5. The large uneven horizontal spacing between experimental data points in the response of R3 during Cycle 13 (Fig. 5*b*) suggests that the inelastic deformation after the knee occurs in an unstable manner. This rapid, nonuniform inelastic shear deformation was accompanied by a sudden increase in acoustic activity.

The DUCM prediction is compared with experimental results from Rosettes R2 and R3 in Fig. 5*a* and *b*, respectively. Predictions assuming a perfect interfacial bond (PB) and an interface with no bond (NB) are also shown. Good agreement between the prediction and experimental results is observed up to a shear stress of 276 MPa (40 ksi) at which point the predicted response displays an unstable loss of stress, whereas the experimental response exhibits only a large instantaneous reduction in stiffness (that is, a knee). The model predicts sudden debonding over more than half of the fiber-matrix interface. It also predicts that matrix plasticity has not yet occurred. The predicted maximum von Mises stress ($\sqrt{1.5 s_{ij} s_{ij}}$, where s_{ij} is the deviatoric stress) in the matrix is reduced from 696 MPa (101 ksi) to 614 MPa (89 ksi) by the sudden debonding. However, the maximum von Mises stress prior to debonding is near the yield stress of the matrix, 758 MPa (110 ksi).

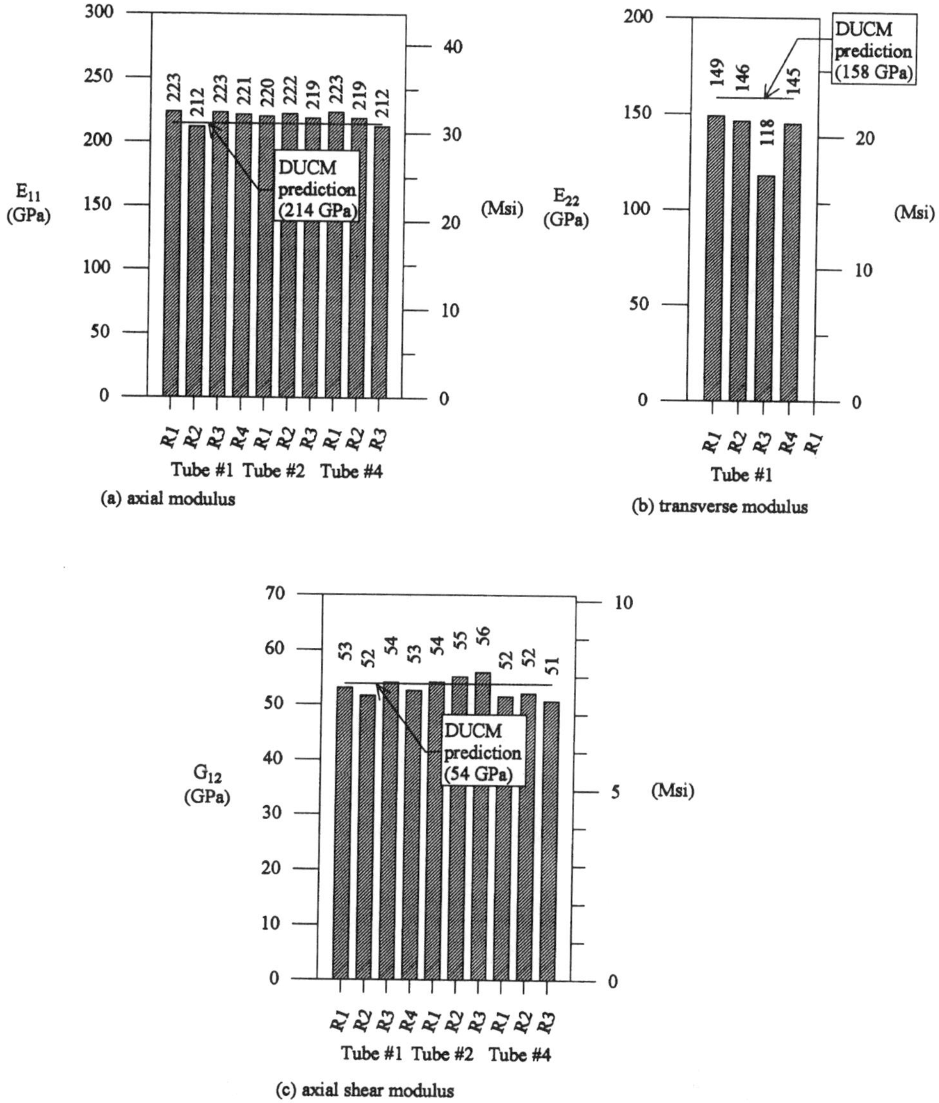

FIG. 4—*[0_4] initial elastic moduli.*

The predicted and experimental responses diverge after the predicted sudden debonding because one of the key model assumptions is no longer valid in the actual test; that being the assumption that debonding is macroscopically uniform. Thus, considering the interface properties to be equivalent to the average of all interfaces, as was done here, does not result in predictions that agree well with the experimental data. It appears that the RVE must include more interfaces so that nonuniform debonding can be modeled, or that a statistical approach be used.

Type III Loading—Dominant Transverse Tensile Load—The inelastic transverse defor-

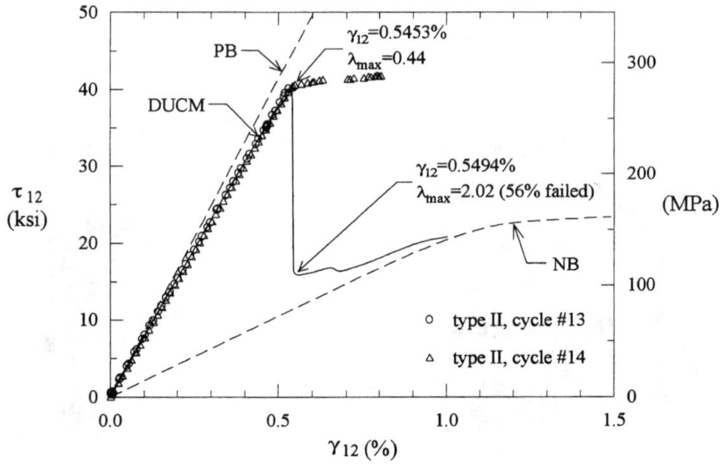

(a) rosette #2

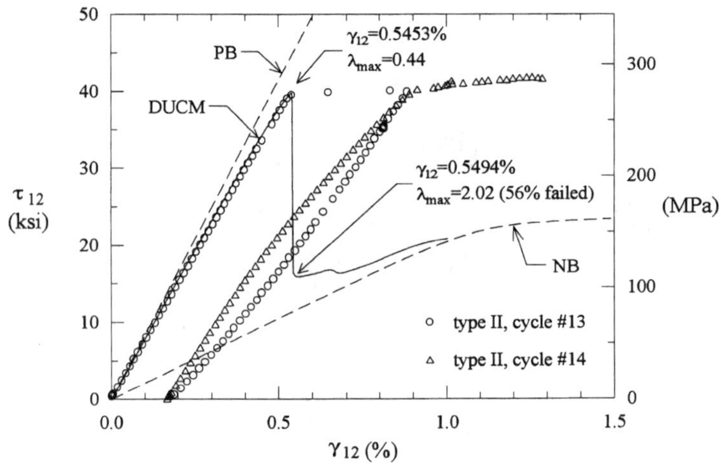

(b) rosette #3

FIG. 5—*[0_4] axial shear response—Type II loading* (a) *Rosette 2 and* (b) *Rosette 3.*

mation of a [0_4] tube under Type III loading was also nonuniform around the tube, as shown by the results from two different rosettes (Fig. 6). The DUCM prediction, along with predictions for the extreme conditions of PB and NB are compared with results from both rosettes in Fig. 6. For clarity, the NB prediction has been offset to start at the same point as results from the second experimental cycle (Cycle 15) shown in the figure. The predicted response for NB illustrates the effects of the thermal residual stresses. The distinct knee predicted to occur at approximately 124 MPa (18 ksi) is associated with overcoming the compressive residual stresses at the interface. DUCM predicts the knee associated with overcoming residual stresses to occur at a higher stress, 158 MPa (23 ksi), because there is also a bond strength associated with the interface. A distinct knee was not observed

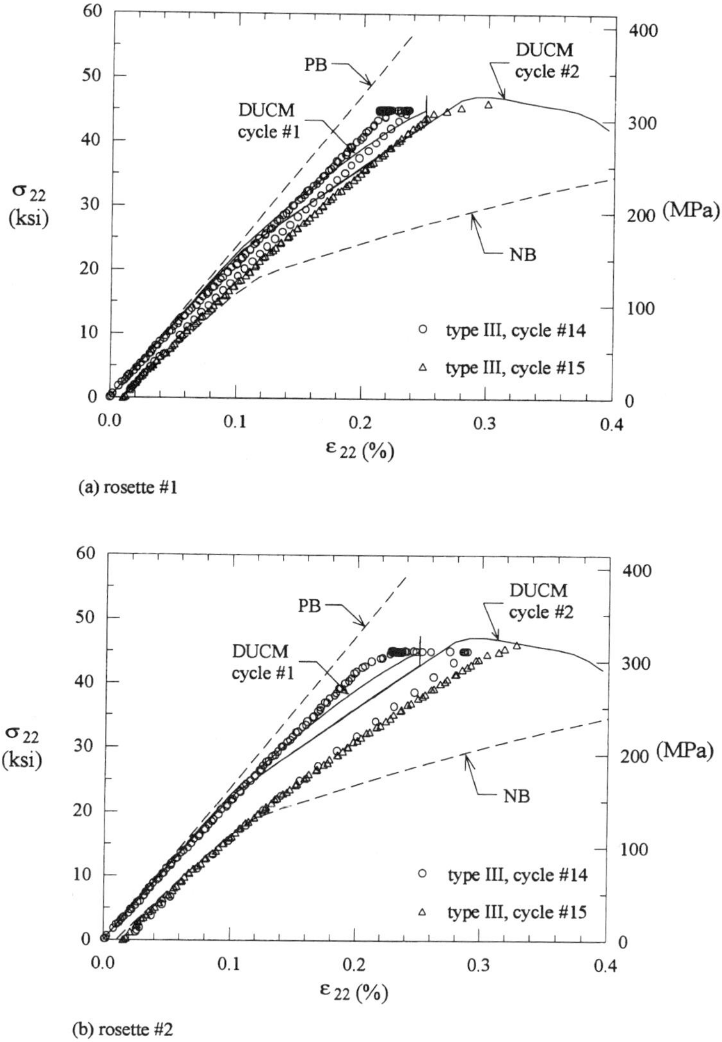

FIG. 6—*[0₄] transverse response—Type III loading* (a) *Rosette 1 and* (b) *Rosette 2.*

experimentally in the Type III tests until a higher stress level, 276 MPa (40 ksi), probably because of the additional restraint to circumferential deformation provided by the fixtures bonded to the tube.

The vertical spike present in the DUCM prediction at the transverse strain of 0.25% (Figs. 6*a* and *b*) was created by an increase and then a decrease in the predicted overall transverse stress during axial and shear loading and unloading, while the transverse strain was held constant. This appears to be caused by changes in the predicted local stress fields.

Both rosettes indicate that permanent transverse strains were present after the first cycle (Cycle 14) shown in Figs. 6*a* and *b*. DUCM predicted local plastic strain in the matrix adjacent to the interface, but no significant overall plastic strain for an applied strain of 0.25%. The predicted unloading response does not follow the predicted initial loading response (Cycle 1) due to the interfacial damage that occurred during loading. Subsequently, the predicted reloading response (Cycle 2) retraced the unloading response.

The predicted and experimental axial strain-transverse strain response for Type III loading is shown in Fig. 7. For clarity, only data from two transverse loading cycles are presented. The agreement between theory and experiment is very good except that DUCM did not predict permanent strains after loading to the transverse strain of 0.25% and then unloading, while a small permanent transverse strain was observed experimentally at the end of Cycle 14.

The minor Poisson's ratio, ν_{21}, determined by the slope of the axial strain-transverse strain diagram (Fig. 7), is observed to be degraded by interfacial damage. This is evident for both the experimental results and the first cycle predicted by DUCM. The second predicted cycle exhibits a negative Poisson's ratio at high transverse strains that is not expected to be observed physically. The predicted negative Poisson's ratio, like the stress drop predicted by DUCM at large transverse strains (Fig. 6), is due to the decrease in interfacial traction for an increase in interfacial displacement (Fig. 1). In composites having a distribution of interface strengths, once a fiber becomes debonded the surrounding material restricts the opening of the interface and the stress drop created by the interface model does not occur [*22,33*].

Angle-Ply [±45]$_S$ SiC/Ti

The experimentally determined axial and shear moduli shown in Fig. 8 varied by 28 and 12%, respectively, at different locations around the tube due to material nonuniformity in the form of fiber-rich and fiber-depleted regions [*34*]. Also, the average experimental axial and shear moduli were, respectively, 27 and 19% below their predicted values for perfect bonding, that is, E_{xx} = 158 GPa (23.0 Msi) and G_{xy} = 74 GPa (10.8 Msi). Figure 9 shows the global and material principal stresses predicted by a linear elastic laminated tube analysis [*35*] for a [±45]$_S$ tube subjected to thermal, axial, and torsional loads applied independently. The figure shows that cooling the composite tube during processing creates tensile transverse stresses, σ_{22}, in all plies. Because transverse tension may cause debonding, the thermal stresses associated with processing can initiate debonding. The presence of interfacial debonding prior to mechanical loading has been verified through a microstructural evaluation [*36*].

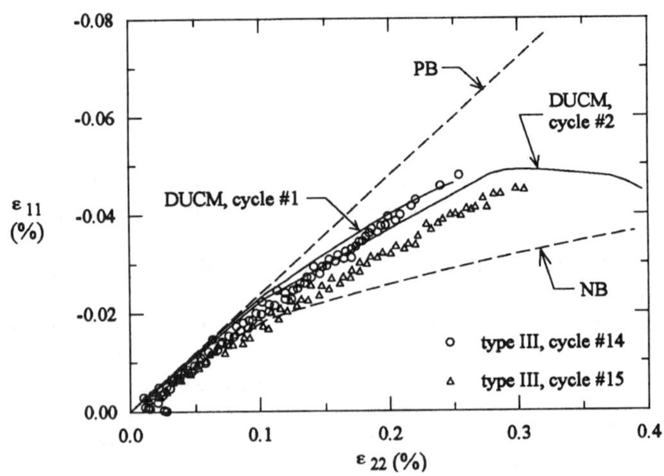

FIG. 7—*[0$_4$] axial strain versus transverse strain—Type III loading.*

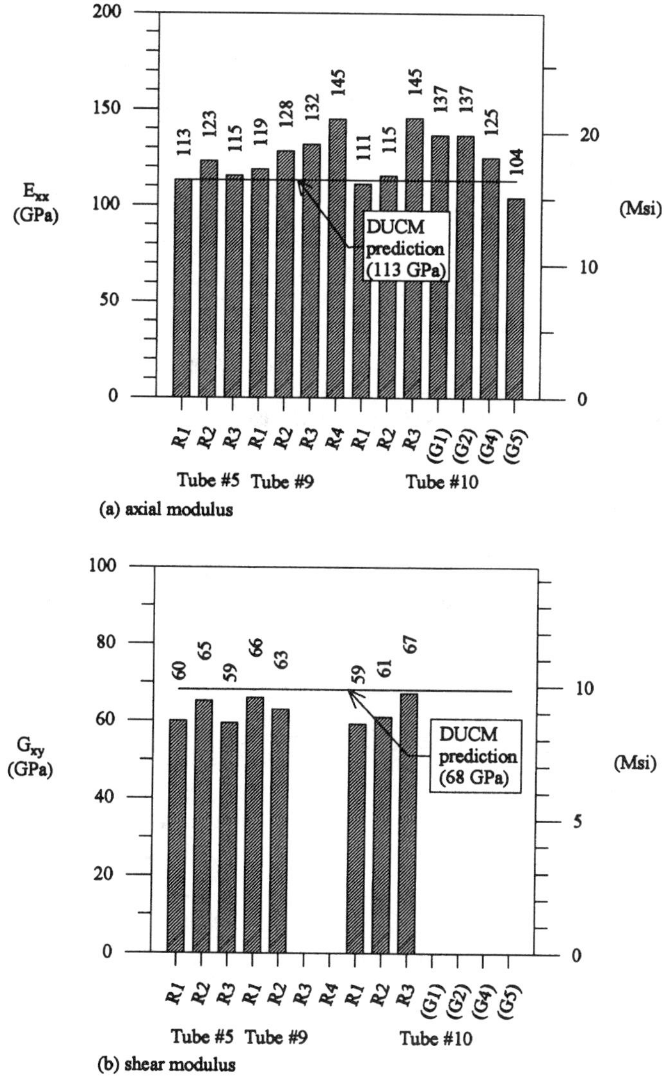

FIG. 8—*$[\pm 45]_S$ initial elastic moduli.*

For $[\pm 45]_S$ tubes, the interface properties required for DUCM were taken to be representative of the statistical average of all interfaces. Through a curve-fitting procedure, the following values were chosen:

$\sigma_0 = 103$ MPa (15 ksi),
$\alpha = 2.0$,
$\delta_n = 254$ μm (10×10^{-6} in.),
$\delta_t = \delta_b = 10\delta_n$,
and $\mu = 0$.

Type I Loading—Dominant Tensile Load—Low-level acoustic emissions (AEs) indicat-

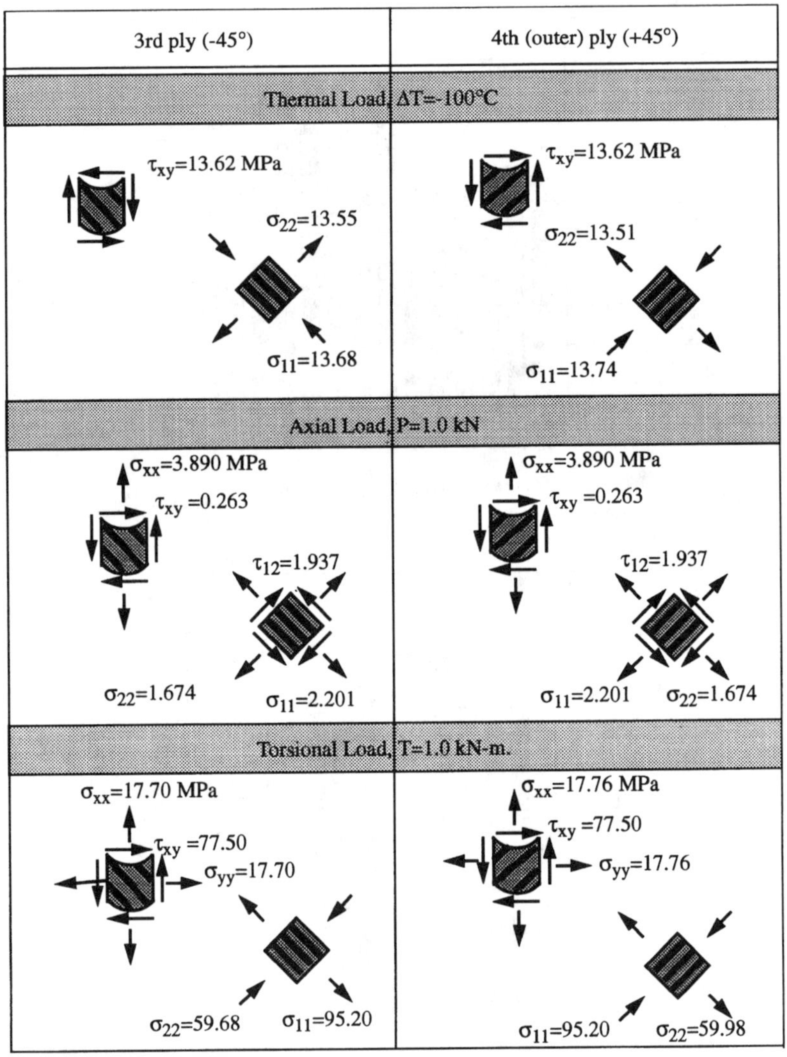

FIG. 9—*[±45]$_S$ tube global and material principal stresses.*

ing some form of damage initiation were recorded at an axial stress of 83 MPa (12 ksi). The intensity of the AE increased monotonically with monotonically increasing axial stress. In subsequent cycles, AE was not heard until the maximum stress level of the previous cycle was surpassed; this is known as the Kaiser effect [*37*]. The acoustic activity is attributed to interfacial debonding. (Recovery of fibers from failed specimens indicated essentially no fiber breakage.) The experimental proportional limit of 138 MPa (20 ksi) was only 18% of the predicted axial stress required to initiate yielding in a perfectly bonded composite. Damage was also evidenced by the degraded unloading stiffness. In fact, after 15 load cycles, the axial modulus was degraded approximately 50%. Further, permanent deformations were evident for stresses well below those predicted for a perfectly bonded composite. Additional details related to these experimental findings may be found in Ref *38*.

Experimental results from a $[\pm 45]_S$ tube subjected to Type I loading are compared with the DUCM prediction as well as the PB and NB extreme conditions in Fig. 10. The experimental data lie within the extremes predicted for PB and NB. The stiffness was degraded with each load cycle and approached the NB prediction as the maximum applied stress level in each cycle increased indicating that initially there was an imperfect mechanical bond present. The comparison between experimental data and DUCM prediction is inexact because load control was used in the experimental investigation, while DUCM relies upon displacement loading. At the higher applied load levels, creep deformation may be

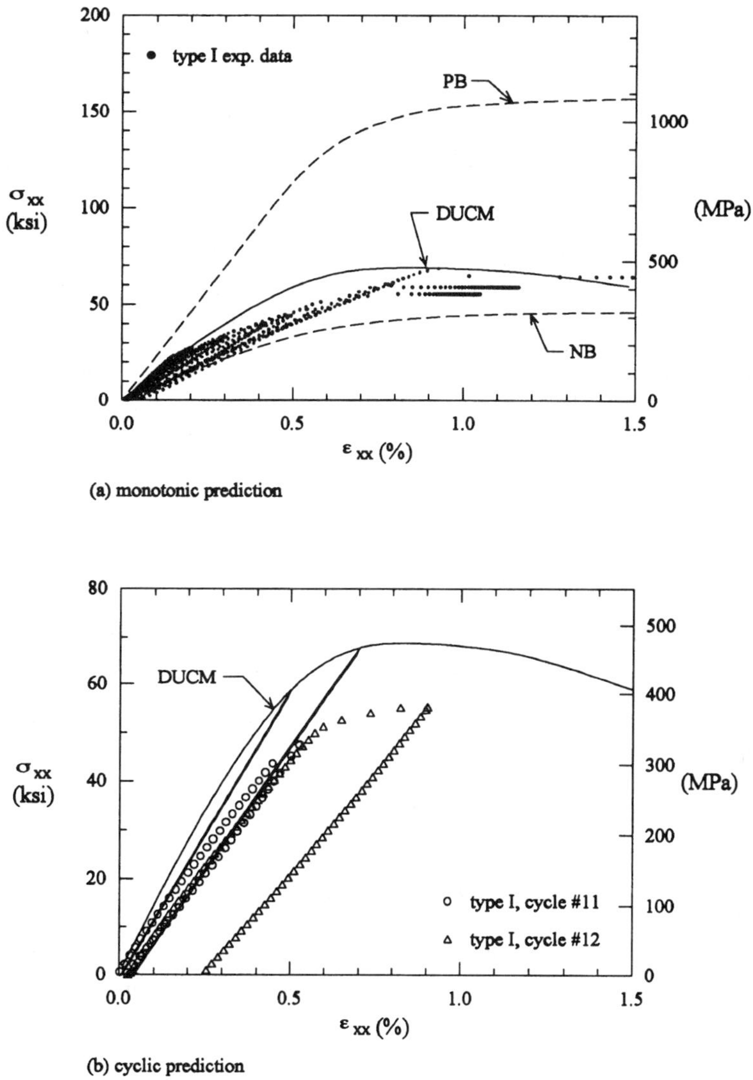

FIG. 10—*$[\pm 45]_S$ axial response—Type I loading* (a) *monotonic prediction and* (b) *cyclic prediction.*

present, as suggested by the data points that lie on a horizontal line and are spaced closer and closer together for increasing axial strain in Fig. 10*a*.

The predicted cyclic axial response is compared with two Type I loading cycles in Fig. 10*b*. Accumulated axial strains associated with shear coupling and creep have not been included in the experimental data so that only cyclic axial loading is compared in the figure. DUCM predicts a continuously degraded axial stiffness as well as permanent axial strain. The experimental axial stiffness for each cycle was less than that predicted by DUCM due to debonding that occurred in previous loading cycles. As for the [0] composite, DUCM predicted a stress drop at high applied strain levels that was not observed experimentally and under-predicted the amount of permanent deformation.

Type IA Loading—Dominant Tensile Load with Hold Times—Type IA load cycles were designed to distinguish time-dependent (creep) strains from time-independent strains. This was motivated by the experimental results shown in Fig. 10*a* that suggest the flat inelastic response may contain a time-dependent deformation component. Accumulated axial strains are plotted versus time elapsed from the point at which the axial stress level was held constant in Fig. 11. In Cycles 1 through 6 (Fig. 11*a*), where the maximum axial stress is 248 MPa (36 ksi), the accumulated strains were due primarily to shear coupling. Positive axial strains were induced by both shear loading and unloading. Conversely, the axial strains accrued in Cycles 7 through 12 (Fig. 11*b*) were principally due to creep. These creep strains increased dramatically as the maximum applied axial stress level was increased. Since the current version of DUCM relies upon displacement loading, no creep behavior was predicted with the model, even though a viscoplastic model has been used for the matrix material. However, short-term creep tests conducted on bulk Ti-15-3 at room temperature indicated that creep behavior can initiate after the proportional limit has been exceeded [*39*]. Optical microscopy revealed extensive plastic slip, elongated grains, and some grain boundary voids near fracture [*36*]. The specimens crept to failure, exhibiting a very ductile fracture surface; no embrittlement or cracking was evident. Significantly, creep response initiated at an applied axial stress of approximately 290 MPa (42 ksi) for $[\pm 45]_S$ SiC/Ti, while it does not occur until 690 MPa (100 ksi) for a specimen of bulk Ti-15-3.

Type II Loading—Dominant Shear Load—Type II loading was limited by the torsional capacity of the load stand, 2200 N-m (20 000 lbf-in.). Figure 9 indicates that torsion of a $[\pm 45]_S$ tube causes tensile transverse stresses, σ_{22}, in the third (and second) ply and compressive transverse stresses in the fourth (and first) ply. No shear stresses, τ_{12}, are present in any ply. Thus, torsional loading is expected to have a relatively minimal effect on stiffness degradation, because only two of the four plies experience stresses that cause interfacial debonding. For the small range of shear stresses that could be applied experimentally, only a small amount of stiffness degradation was observed. The predictions from DUCM agreed well with the experimental results.

Conclusions

A micromechanics-based model has been developed to predict the inelastic response of laminated MMCs in the presence of interfacial debonding. Predictions from the model were compared with experimental results from multiaxial tests on SiC/Ti tubes. Several conclusions can be drawn from studies of unidirectional and angle-ply composites.

Unidirectional [0] SiC/Ti

1. The axial shear response exhibits a catastrophic loss of stiffness due to sudden interfacial debonding just prior to initial matrix yielding.

2. Transverse tensile loading results in interfacial debonding prior to initial matrix yielding.
3. Fiber/matrix interfacial debonding is nonuniform and a major factor in the transverse and axial shear responses.
4. The axial tensile response is independent of the condition of the interface.

Angle-Ply $[\pm 45]_S$ SiC/Ti

1. Interfacial debonding has a major influence on the tensile response.

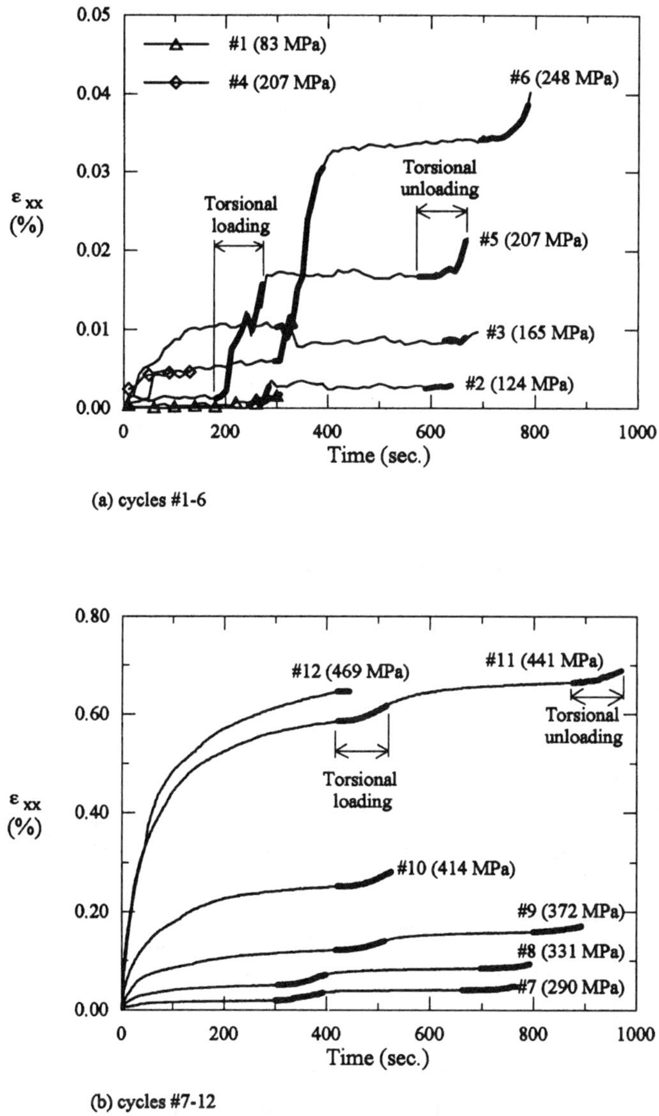

FIG. 11—*$[\pm 45]_S$ creep response—Type IA loading.*

2. At room temperature, longitudinal creep strains occur at applied stresses less than 50% of the stress level required to initiate creep response in bulk Ti-15-3.
3. Under tensile loading, interfacial debonding precedes and accelerates matrix yielding.
4. The axial modulus exhibits a damage-induced degradation of as much as 50% in cyclic tension tests to sequentially higher stress levels.
5. Imperfect bonding exists in the as-fabricated material as a result of thermal cool-down during processing.
6. Inelastic coupling between axial and shear strains occurs for shear loading in the presence of constant axial stress.
7. Acoustic emissions follow the Kaiser effect and provide a qualitative indication of when damage is occurring.

Acknowledgments

This work was supported by NASA Langley Research Center (NASA Grant NAG-1-745), the National Science Foundation (NSF Grant MSS-9115328), and the Center for Light Thermal Structures at the University of Virginia.

References

[*1*] Johnson, W. S., Lubowinski, S. J., and Highsmith, A. L., "Mechanical Characterization of Unnotched SCS_6/Ti-15-3 Metal Matrix Composites at Room Temperature," *Thermal and Mechanical Behavior of Metal Matrix and Ceramic Matrix Composites, ASTM STP 1080,* J. M. Kennedy, H. H. Moeller, and W. S. Johnson, Eds., American Society for Testing and Materials, Philadelphia, 1990, pp. 193–218.

[*2*] Nimmer, R. P., Bankert, R. J., Russel, E. S., Smith, G. A., and Wright, P. K., "Micromechanical Modeling of Fiber/Matrix Interface Effects in Transversely Loaded SiC/Ti-6-4 Metal Matrix Composites," *Journal of Composites Technology and Research,* Vol. 13, 1990, pp. 3–13.

[*3*] Lerch, B. A. and Saltsman, J. F., "Tensile Deformation in SiC/Ti-15-3 Laminates," *Composite Materials: Fatigue and Fracture, Fourth Volume, ASTM STP 1156,* W. W. Stinchcomb and N. E. Ashbaugh, Eds., American Society for Testing and Materials, Philadelphia, 1993, pp. 161–175.

[*4*] Majumdar, B. S. and Newaz, G. M., "Inelastic Deformation of Metal Matrix Composites: Plasticity and Damage Mechanisms," *Philosophical Magazine,* Vol. A66, 1992, pp. 187–212.

[*5*] Majumdar, B. S., Newaz, G. M., and Ellis, J. R., "Evolution of Damage and Plasticity in Titanium-Based, Fiber-Reinforced Composites," *Metallurgical Transactions A,* Vol. 24A, 1993, pp. 1597–1610.

[*6*] Aboudi, J., "Constitutive Equations for Elastoplastic Composites with Imperfect Bonding," *International Journal of Plasticity,* Vol. 4, 1988, pp. 103–125.

[*7*] Hashin, Z., "Thermoelastic Properties of Fiber Composites with Imperfect Interface," *Mechanics of Materials,* Vol. 8, 1990, pp. 333–348.

[*8*] Benveniste, Y., Dvorak, G. J., and Chen, T., "Stress Fields in Composites with Coated Inclusions," *Mechanics of Materials,* Vol. 7, 1989, pp. 305–317.

[*9*] Chen, T., Dvorak, G. J., and Benveniste, Y., "Stress Fields in Composites Reinforced by Coated Cylindrically Orthotropic Fibers," *Mechanics of Materials,* Vol. 9, 1990, pp. 17–32.

[*10*] Benveniste, Y., Dvorak, G. J., and Chen, T., "On Effective Properties of Composites with Coated Cylindrically Orthotropic Fibers," *Mechanics of Materials,* Vol. 12, 1991, pp. 289–297.

[*11*] Herakovich, C. T. and Hidde, J. S., "Response of Metal Matrix Composites with Imperfect Bonding," *Ultramicroscopy,* Vol. 40, 1992, pp. 215–228.

[*12*] Adams, D. F., "A Micromechanical Analysis of Crack Propagation in an Elastoplastic Composite Material," *Fiber Science and Technology,* Vol. 7, 1974, pp. 237–256.

[*13*] Aboudi, J., "Micromechanical Analysis of Fibrous Composites with Coulomb Frictional Slippage Between the Phases," *Mechanics of Materials,* Vol. 8, 1989, pp. 103–115.

[*14*] Wisnom, M. R., "Factors Affecting the Transverse Tensile Strength of Unidirectional Continuous Silicon Carbide Fiber Reinforced 6061 Aluminum," *Journal of Composite Materials,* Vol. 24, 1990, pp. 707–726.

[*15*] Achenbach, J. D. and Zhu, H., "Effect of Interphases on Micro and Macromechanical Behavior of Hexagonal-Array Fiber Composites," *Journal of Applied Mechanics,* Vol. 57, 1990, pp. 956–963.

[16] Zhu, H. and Achenbach, J. D., "Radial Matrix Cracking and Interphase Failure in Transversely Loaded Fiber Composites," *Mechanics of Materials,* Vol. 11, 1991, pp. 347–356.
[17] Nimmer, R. P., "Fiber-Matrix Interface Effects in the Presence of Thermally Induced Residual Stresses," *Journal of Composites Technology & Research,* Vol. 12, 1990, pp. 65–75.
[18] Brust, F. W., Newaz, G. M., and Majumdar, B. S., "Constitutive Response Analysis of Metal Matrix Composites via the Unit Cell Model," *Composite Materials: Testing and Design, Eleventh Volume, ASTM STP 1206,* American Society for Testing and Materials, Philadelphia, 1993.
[19] Needleman, A., "A Continuum Model for Void Nucleation by Inclusion Debonding," *Journal of Applied Mechanics,* Vol. 54, 1987, pp. 525–531.
[20] McGee, J. D. and Herakovich, C. T., "Micromechanics of Fiber/Matrix Debonding," Report AM-92-01, University of Virginia, Charlottesville, VA, 1992.
[21] Lissenden, C. J. and Herakovich, C. T., "Numerical Modeling of Damage Development and Viscoplasticity in Metal Matrix Composites," *Computer Methods in Applied Mechanics and Engineering,* Vol. 126, 1995, pp. 289–303.
[22] Eggleston, M. R., "Testing, Modeling and Analysis of Transverse Creep in SCS-6/Ti-6Al-4V Metal Matrix Composites at 482°C," Report 93CRD163, Physical Metallurgy Laboratory, General Electric Research and Development Center, Schenectady, NY, 1993.
[23] Allen, D. H., Jones, R. H., and Boyd, J. G., "Micromechanical Analysis of a Continuous Fiber Metal Matrix Composite Including the Effects of Matrix Viscoplasticity and Evolving Damage," *Journal of the Mechanics and Physics of Solids,* Vol. 42, 1994, pp. 505–520.
[24] Lissenden, C. J., Herakovich, C. T., and Pindera, M.-J., "Inelastic Deformation of Metal Matrix Composites," NASA Contractor Report 191522, Langley Research Center, Hampton, VA, 1993.
[25] Boyd, J. G., Constanzo, F., and Allen, D. H., "A Micromechanics Approach for Constructing Locally Averaged Damage Dependent Constitutive Equations in Inelastic Composites," *International Journal of Damage Mechanics,* Vol. 2, 1993, pp. 209–228.
[26] Bodner, S. R. and Partom, Y., "Constitutive Equations for Elastic-Viscoplastic Strain-Hardening Materials," *Journal of Applied Mechanics,* Vol. 42, 1975, pp. 385–389.
[27] Bodner, S. R., "Review of a Unified Elastic-Viscoplastic Theory," *Unified Constitutive Equations for Plastic Deformation and Creep of Engineering Alloys,* A. K. Miller, Ed., Elsevier, Amsterdam, 1987, pp. 273–301.
[28] Tvergaard, V., "Effect of Fiber Debonding in a Whisker-Reinforced Metal," *Materials Science and Engineering,* Vol. A125, 1990, pp. 203–213.
[29] Hidde, J. S. and Herakovich, C. T., "Inelastic Response of Hybrid Composite Laminates," *Journal of Composite Materials,* Vol. 26, 1992, pp. 2–19.
[30] Jones, R. M., *Mechanics of Composite Materials,* McGraw-Hill, New York, 1967.
[31] Lekhnitskii, S. G., *Theory of Elasticity of an Anisotropic Body,* Mir Publishers, Moscow, 1981 (translated from the revised 1977 Russian edition).
[32] Lissenden, C. J., Herakovich, C. T., and Pindera, M.-J., "Response of SiC/Ti Under Combined Loading—Part I: Theory and Experiment for Imperfect Bonding," *Journal of Composite Materials,* Vol. 29, 1995, pp. 130–155.
[33] Povirk, G. L., Needleman, A., and Nutt, S. R., "An Analysis of the Effect of Residual Stresses on Deformation and Damage Mechanisms in Al-SiC Composites," *Materials Science and Engineering,* Vol. A132, 1991, pp. 31–38.
[34] Lissenden, C. J., Herakovich, C. T., and Pindera, M.-J., "Stiffness Degradation of SiC/Ti Tubes Subjected to Biaxial Loading," *Composites Science and Technology,* Vol. 50, 1994, pp. 23–36.
[35] Derstine, M. S. and Pindera, M.-J., "Nonlinear Response of Composite Tubes under Combined Thermomechanical Loading," *Composites and Other New Materials for PVP: Design and Analysis Considerations,* ASME PVP-Vol. 174, D. Hui, T. J. Kozik, G. E. O. Widera, and M. Shiratori, Eds., American Society of Mechanical Engineers, New York, 1989, pp. 19–28.
[36] Lerch, B. A., personal communication, 1993.
[37] Kaiser, J., "Untersuchungen uber das Auftreten Gerauschen Biem Zugversuch," PhD thesis, Technisch Hochschule, Munich, 1950.
[38] Lissenden, C. J., Herakovich, C. T., and Pindera, M.-J., "Response of SiC/Ti Tubes Under Biaxial Loading in the Presence of Damage," *Damage Mechanics in Composites,* ASME-AMD–Vol. 150/AD–Vol. 32, D. H. Allen and D. C. Lagoudas, Eds., American Society of Mechanical Engineers, New York, 1992, pp. 73–90.
[39] Lissenden, C. J., Herakovich, C. T., and Pindera, M.-J., "Response of SiC/Ti Under Combined Loading—Part II: Room Temperature Creep Effects," *Journal of Composite Materials,* Vol. 29, 1995, pp. 1403–1417.

Catherine A. Bigelow,[1] *Yahei A. Bahei-El-Din,*[2] *and Massoud Mirdamadi*[3]

Time-Dependent Deformation of Titanium Metal Matrix Composites

REFERENCE: Bigelow, C. A., Bahei-El-Din, Y. A., and Mirdamadi, M., **"Time-Dependent Deformation of Titanium Metal Matrix Composites,"** *Life Prediction Methodology for Titanium Matrix Composites, ASTM STP 1253,* W. S. Johnson, J. M. Larsen, and B. N. Cox, Eds., American Society for Testing and Materials, 1996, pp. 278–296.

ABSTRACT: A three-dimensional finite element program called VISCOPAC was developed and used to conduct a micromechanics analysis of titanium metal matrix composites. The VISCOPAC program uses a modified Eisenberg-Yen thermo-viscoplastic constitutive model to predict matrix behavior under thermomechanical fatigue loading. The analysis incorporated temperature-dependent elastic properties in the fiber and temperature-dependent viscoplastic properties in the matrix. The material model was described and the necessary material constants were determined experimentally. Fiber-matrix interfacial behavior was analyzed using a discrete fiber-matrix model. The thermal residual stresses due to the fabrication cycle were predicted with a failed interface. The failed interface resulted in lower thermal residual stresses in the matrix and fiber. Stresses due to a uniform transverse load were calculated at two temperatures, room temperature and an elevated temperature of 650°C. At both temperatures, a large stress concentration was calculated when the interface had failed. The results indicate the importance of accurately accounting for fiber-matrix interface failure and the need for a micromechanics-based analytical technique to understand and predict the behavior of titanium metal matrix composites.

KEYWORDS: thermal residual stresses, interface, finite element analysis, viscoplasticity, discrete fiber-matrix model, titanium, titanium matrix composites, life prediction, titanium alloys, fatigue (materials), modeling

Fiber-matrix interfaces can play a key role in the mechanical behavior of continuous fiber-reinforced metal matrix composites (MMC) [*1*]. Interfaces govern the mode and extent of load transfer between the fiber and matrix. To predict composite behavior, it is necessary to understand interfacial behavior, including fiber-matrix debonding. Because of the large differences that can occur in the coefficients of thermal expansion of the fiber and the matrix, large thermal residual stresses can develop in the composite during cool-down from the fabrication temperature. Due to the high operating temperatures of MMC, the time-dependent behavior of the matrix should also be accounted for. Thus, it is important to model the thermal residual stresses that may be present as well as the time- and temperature-dependent behavior of the titanium matrix to understand and predict the composite behavior.

The fiber-matrix interface plays a particularly significant role in MMC with a matrix having a high yield strength. To study the stress state governing fiber-matrix debonding, a micromechanics analysis is required. A three-dimensional finite element (3D FE) program

[1]Senior scientist, FAA Technical Center, Atlantic City International Airport, Atlantic City, NJ 08405.

[2]Professor, University of Cairo, Cairo, Egypt.

[3]Research scientist, Analytical Services and Material, Hampton, VA 23666.

called VISCOPAC was used to conduct a micromechanics analysis. The VISCOPAC program uses a modified Eisenberg-Yen thermo-viscoplastic constitutive model to predict matrix behavior under thermomechanical fatigue (TMF) loading.

In the present paper, the capabilities of the VISCOPAC program are described, including the viscoplastic constitutive theory. To demonstrate the VISCOPAC program, a 3D FE unit cell micromechanics model is used to analyze the fiber-matrix interfacial behavior. The thermal residual stresses due to the fabrication cycle are predicted. The effects of the fiber-matrix interface failure and the time-dependent behavior of the matrix on the stress state in the unit cell are examined.

Analytical Method

Analytical predictions of the overall response of the composite material were performed with the VISCOPAC computer program developed by Bahei-El-Din for viscoplastic analysis of homogeneous and orthotropic materials subjected to thermomechanical loading. For homogeneous materials, the inelastic strain is predicted from the unified theory described in Appendix I. The viscoplastic constitutive theory implemented into the VISCOPAC program was developed by Bahei-El-Din, Shah, and Dvorak [*2*] for high temperature applications and is based on the viscoplastic theory of Eisenberg and Yen [*3*]. This constitutive theory is a unified theory, where creep and plasticity are combined with a yield surface. The theory assumes the existence of an equilibrium stress-strain response that corresponds to the theoretical lower bound of the dynamic response. State variables include the equilibrium stress and an isotropic hardening function. Evolution laws of the state variables account for thermal recovery. A two-surface plasticity theory, where the two surfaces are the yield surface and the bounding surface, describe the state variables under nonproportional loading. In addition, this constitutive viscoplastic model assumes that the elastic response is rate-independent and that inelastic rate-dependent deformation takes place if the current stress state is greater than the equilibrium stress. The theory requires three elastic constants and 20 viscoplastic constants as a function of temperature. The viscoplastic constants will also be dependent upon the loading rate. These constants are described in more detail in Appendix I. By assuming the effects of thermal recovery and isotropic hardening were negligible for this material, the number of required constants is reduced to three elastic and six viscoplastic constants. These material constants include the elastic modulus (E), Poisson's ratio (ν), the coefficient of thermal expansion (α), the yield stress (σ_{ys}), the yield stress of the bounding surface ($\bar{\sigma}_{ys}$), and the viscoplastic parameters: H_0, h, k, and p.

The overall strain of unidirectionally reinforced composite materials is evaluated from the inelastic strain of the homogeneous phases by means of elastic strain transformation factors that depend on the elastic properties of the fiber and matrix, the fiber volume fraction, and the micromechanical interaction of the phases. In the VISCOPAC program, the user has a choice of one of three averaging models: the self-consistent model [*4*], the Mori-Tanaka model [*5*], or the vanishing-fiber-diameter material (VFD) model [*6–8*]. Each model is briefly described in the following paragraphs.

The self-consistent model [*4*] is centered on the solution to the problem of a single fiber embedded in an unbounded homogeneous medium that is macroscopically indistinguishable from the composite. Under any uniform loading, the fiber strain is uniform. According to the basic idea of the self-consistent model, the fiber strain is assumed to be the average of the fibers in the composite.

Having noted the important role of image stress in work hardening of dispersion hardened materials, Mori and Tanaka [*5*] developed a method of calculating the average internal stress in the matrix of a material containing inclusions. They showed that the average stress in the matrix is uniform throughout the material and independent of the position within the domain. The actual stress in the matrix is the average stress plus the locally fluctuating stress, the average of which vanishes in the matrix. The model also considers the average elastic energy by accounting for the effects of the interaction among the inclusions and the presence of the free boundary.

The VFD model [6–8] consists of an elastic-viscoplastic matrix unidirectionally reinforced by continuous elastic fibers. Both constituents are assumed to be homogeneous and isotropic. The fibers are assumed to have a very small diameter, so that although the fibers occupy a finite volume fraction of the composite, they do not interfere with matrix deformation in the two transverse directions, but only in the axial (fiber) direction.

Materials and Test Specimens

The material modeled in this study was SCS-6/TIMETAL-21S that is fabricated by hot isostatic pressing (HIPing) Ti-15Mo-3Nb-3Al-0.2Si foils between tapes of unidirectional SCS-6 silicon-carbide fibers. A fiber volume fraction of 38.5% with a fiber diameter of 0.14 mm was used. The matrix material constants needed for the constitutive model used in VISCOPAC were determined experimentally by testing neat TIMETAL-21S material that had been subjected to the same processing cycle as the composite material. Prior to testing, all test specimens were heat treated (aged) at 621°C for 8 h in vacuum to stabilize the matrix material.

Experimental Techniques

All experiments were conducted using a 100-kN, closed-loop, servohydraulic test frame equipped with water-cooled hydraulic grips. The test specimens were heated using a 5-kW induction generator with three independent coils. The induction generator was controlled by a temperature profiler capable of running predetermined temperature profiles. K-type thermocouples were used to monitor and control the temperature along a 50-mm gage length section in the center of the specimen. In addition, an infrared thermovision camera was used to ensure that the temperature distribution was uniform along the gage section of the specimen.

Matrix Material Constants

The matrix material constants used in the VISCOPAC program are given in Table 1. These material constants include the elastic modulus (E), Poisson's ratio (ν), the coefficient of thermal expansion (α), the yield stress (σ_{ys}), the yield stress of the bounding surface ($\bar{\sigma}_{ys}$), and the viscoplastic parameters: H_0, h, k, and p. The coefficient of thermal expansion was determined from a load-controlled experiment. A test specimen, under zero load, was subjected to temperature increments ranging from -130 to 816°C. At each temperature increment, the temperature was held constant (about 1 min) to allow for stabilization. The temperatures and the corresponding thermal strains were fit to a third-degree polynomial using the method of least squares. The resulting polynomial function was differentiated with respect to temperature to determine the coefficient of thermal expansion as a function of temperature. The elastic modulus and the yield stress were determined from the linear portion and the onset of nonlinearity of the stress-strain curve, respectively, at each temperature.

Strain-controlled tests were used to define the equilibrium stress-strain curves, following the procedures defined by Mirdamadi et al. [9]. As schematically illustrated in Fig. 1, to allow stress relaxation, 5-min hold periods were imposed at predetermined strain levels. At

TABLE 1—*Material properties for neat TIMETAL-21S; stress rate of 2.56 MPa/s.*

	Temperature, °C			
	21.1	115	482	650
E, MPa	116 000	116 000	95 851	64 393
ν	0.34	0.34	0.34	0.34
α, mm/mm/°C	7.52e-6	7.52e-6	10.6e-6	11.8e-6
σ_{ys}, MPa	910	910	472	43.52
$\bar{\sigma}_{ys}$, MPa	969	969	561	69
H_0, MPa	829	829	1135	300
h, MPa	17 111	17 111	26 168	7552
k	2.1e-21	2.1e-21	5.4e-18	2.2e-12
p	8.59	8.59	6.35	3.46

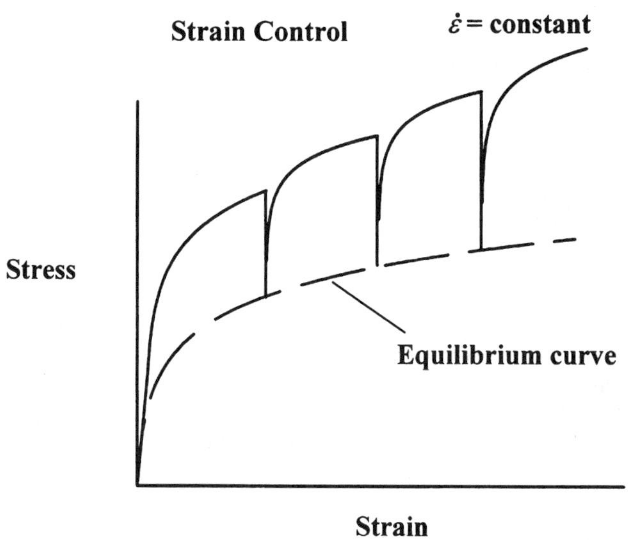

FIG. 1—*Determine equilibrium curve at each temperature of interest.*

each temperature, a minimum of five strain levels were selected to construct the equilibrium stress-strain curve. The strain rate during the loading was 1×10^{-4} mm/mm/s. The equilibrium curve was then approximated by fitting to a power law equation as described by Mirdamadi and Johnson [*10*]. Once the equilibrium curve is known, the constants, k and p, are determined by conducting a uniaxial tension test under load control. As shown schematically in Fig. 2, the overstress, R, is defined as the difference between the equilibrium curve and the loading curve. The inelastic strain rate, $\dot{\epsilon}_{in}$, can be written as a function of the overstress, R, and the constants, k and p, as $\dot{\epsilon}_{in} = kR^p$. By plotting the inelastic strain rate, $\dot{\epsilon}_{in}$, and R on a log-log plot, the constant, k, is found as the y-axis intercept and the constant, p, is found as the slope of the plot.

The remaining viscoplastic constants (H_0, h, σ_{ys}, and $\bar{\sigma}_{ys}$) are determined from the equilibrium curve as shown in Fig. 3. H_0 is the slope of the bounding curve, defined by the asymptotic value of the equilibrium curve. The yield stress of the bounding curve, $\bar{\sigma}_{ys}$, is

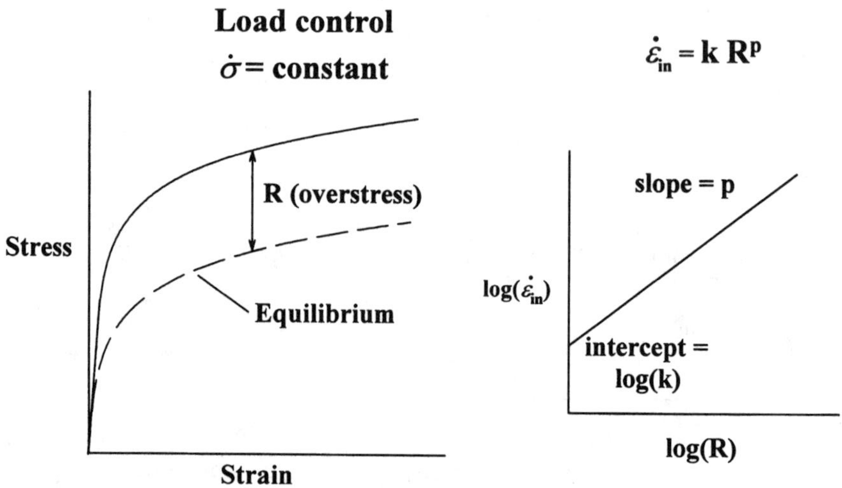

FIG. 2—*Determine constants* k *and* p *at each temperature and loading rate of interest.*

defined as the *y*-intercept of the line defining the slope, H_0. The constant, *h*, is determined from the equation of the instantaneous slope of the equilibrium curve, $H = H_0 + h[\delta/(\delta_{in} - \delta)]$. Figure 4 shows the predicted stress-strain curves at three temperatures, compared to the test data. The agreement between the viscoplastic theory and experiment is very good.

Analytical Modeling

The VISCOPAC program was used to analyze a discrete fiber-matrix (DFM) model assuming an infinitely repeating, rectangular array of fibers, rather than use one of the averaging models contained in the program. The VISCOPAC program uses three-dimensional, eight-noded, hexahedral elements and calculates the stresses at the element centroids. The ply thickness (0.104 mm), the fiber volume fraction (38.5%), and the fiber radius (0.070 mm) were used to calculate the dimensions of the model, as shown in Fig. 5. The ply

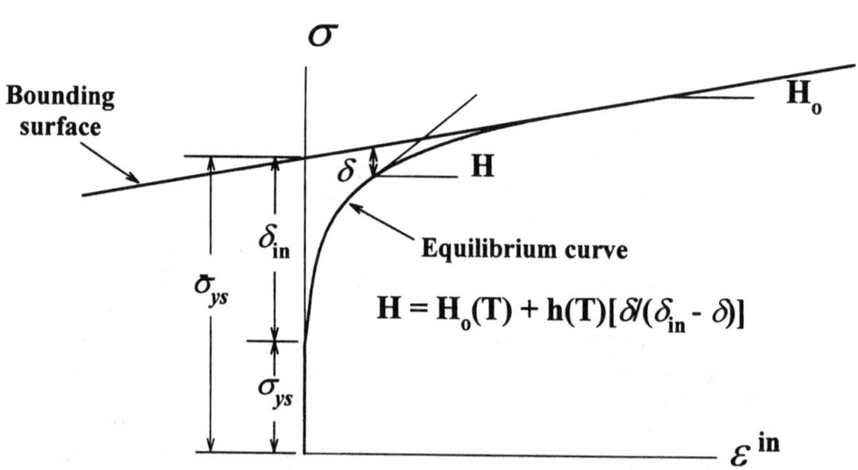

FIG. 3—*Determine* H_0, h, σ_{ys}, *and* $\bar{\sigma}_{ys}$ *from equilibrium curve.*

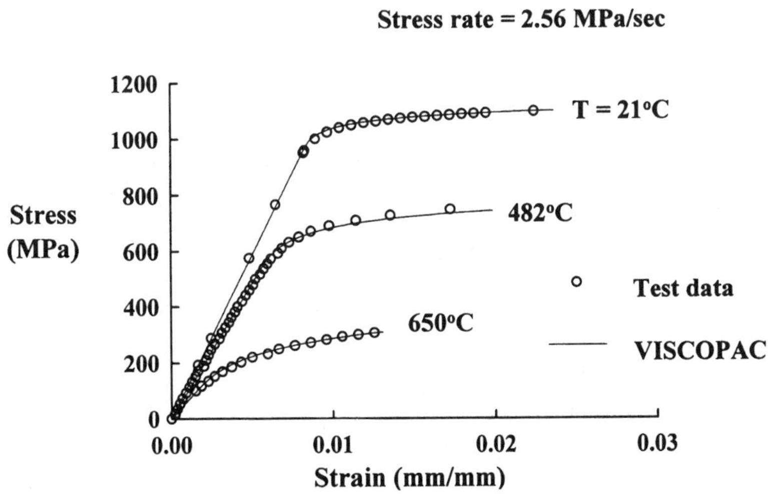

FIG. 4—*Predicted matrix stress-strain curves for TIMETAL-21S as a function of temperature.*

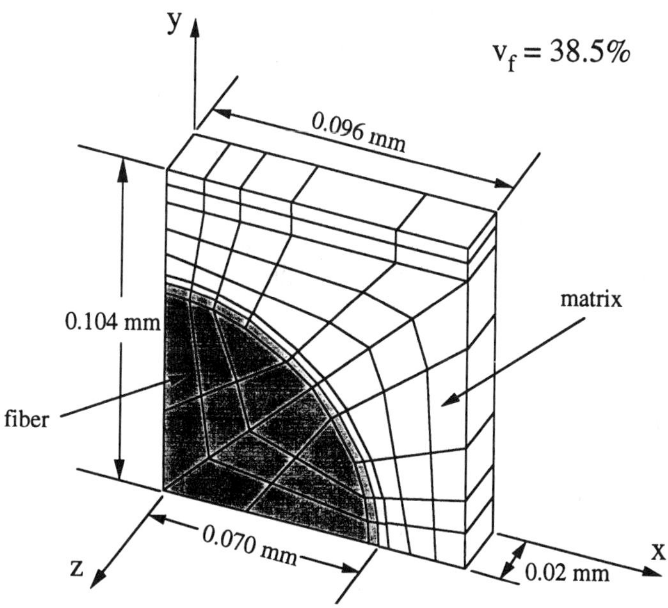

FIG. 5—*Finite element mesh used to model the discrete fiber-matrix unit cell.*

thickness, the fiber volume fraction, and the fiber diameter are typical for silicon-carbide/titanium matrix composites. Figure 5 also shows the finite element mesh refinement that was used. The model shown in Fig. 5 was used with the appropriate boundary conditions to represent a single fiber in an infinite array of fibers. For the single fiber in an infinite array, compatibility with adjacent unit cells was enforced on each face of the model by constraining all normal displacements to be equal. On the $x = 0$ face, the x-displacements were set to zero; on the $y = 0$ face, the y-displacements were set to zero; and on the $z = 0$ face, the z-displacements were set to zero. On the $x = 0.096$ mm face, the x-displacements were constrained to be equal to each other such that the plane remained plane. That is, all nodes in the plane were free to move, but all nodes in that plane moved the same amount in the x-direction. Likewise for the $y = 0.104$-mm plane and the $z = 0.02$-mm plane. A convergence study was done on the mesh shown in Fig. 5. For the thermal loading used here, the mesh shown in Fig. 5 predicted stresses that differed by less than 5% from those predicted by a mesh with twice the refinement.

Two loading conditions were analyzed. The thermal residual stresses due to the processing cycle and the mechanical stresses due to a uniform transverse loading were calculated. Stresses were calculated assuming an intact interface and a failed interface. A perfect bond between the fiber and matrix was assumed for the intact cases. The failed interface was modeled by modifying the element properties of the layer of matrix material next to the fiber. The interface was assumed to be failed from the start of the loading, that is, progressive damage was not modeled. A very thin layer of elements was introduced next to the fiber; the thickness of the interface was modeled as 0.0001 mm, compared to the fiber radius of 0.07 mm. To model the failed interface, the modulus of this interface layer was set to 50 MPa, compared to the room temperature modulus of the matrix of 116 GPa. To calculate the thermal residual stresses, the temperature was assumed to be uniform throughout the laminate and only the thermal cycle occurring during the fabrication process is analyzed in the present work. Thermal residual stresses were calculated assuming a temperature change of −629°C; that is, a temperature change from 650°C to room temperature 21.1°C. This temperature change is approximately one half of the melting point of the Ti-15-3 matrix. It was assumed that any residual stresses that developed during fabrication of the composite would be relieved due to relaxation at temperatures greater than one half the melting point of the matrix [*11*]. The fiber was assumed to remain elastic with temperature dependent properties. The elastic properties of the fiber are given in Table 2.

TABLE 2—*Elastic properties for SCS-6 fiber.*

Temperature, °C	E, Pa	ν	α, mm/mm/°C
21.11	3.93E11	0.25	3.564E-6
93.33	3.90E11	0.25	3.564E-6
204.44	3.86E11	0.25	3.618E-6
315.56	3.82E11	0.25	3.726E-6
426.67	3.78E11	0.25	3.906E-6
537.78	3.74E11	0.25	4.068E-6
648.89	3.70E11	0.25	4.266E-6
760.00	3.65E11	0.25	4.410E-6
871.11	3.61E11	0.25	4.572E-6
1093.30	3.54E11	0.25	...

Results and Discussion

Stress contours, based on element centroid values, are plotted for both loading conditions with an intact and a failed interface.

Thermal Residual Stresses

A time period of 1 h was used to simulate the fabrication process, although, since no viscoplastic behavior was predicted, the time used to simulate the cool-down cycle was immaterial. Figures 6 and 7 show the σ_{xx} stresses produced by the simulated cool-down cycle with an intact and failed interface, respectively. The dashed line shown in the figures represents the fiber-matrix boundary. For comparison purposes, the same scale for the stress contours is used in both figures. This results in some detail being lost in Fig. 6, where the stresses are greater due to the constraint provided by the intact interface. For the intact interface, Fig. 6, the σ_{xx} stresses range from approximately -400 to 475 MPa. The maximum tensile and compressive stresses are not shown in Fig. 6 due to the stress contour levels chosen. The maximum tensile stress occurred in the upper-left corner of the model, and the maximum compressive stresses occured in the lower-right-hand corner of the model. For the failed interface, Fig. 7, the σ_{xx} stresses range from approximately -200 to 295 MPa. The intact interface results in a greater constraint between the fiber and matrix, producing larger residual stresses. Although not shown, the calculated σ_{yy} stresses were nearly identical to the σ_{xx} stresses. If the DFM model had been symmetric, the σ_{xx} and σ_{yy} would have been identical. The transverse stresses, σ_{xx} and σ_{yy}, govern the radial and circumferential cracking in the matrix, fiber, and interface region. The lower stresses in the case of the failed interface indicate a lesser propensity for matrix cracking once the interface has failed.

Figures 8 and 9 show the σ_{zz} stresses produced by the simulated cool-down with an intact and a failed interface, respectively. In Fig. 8, the σ_{zz} stresses are nearly uniform in both the fiber and matrix, with a maximum compressive stress of -835 MPa in the fiber and a maximum tensile stress of 558 MPa in the matrix. In the interface region, next to the fiber, a

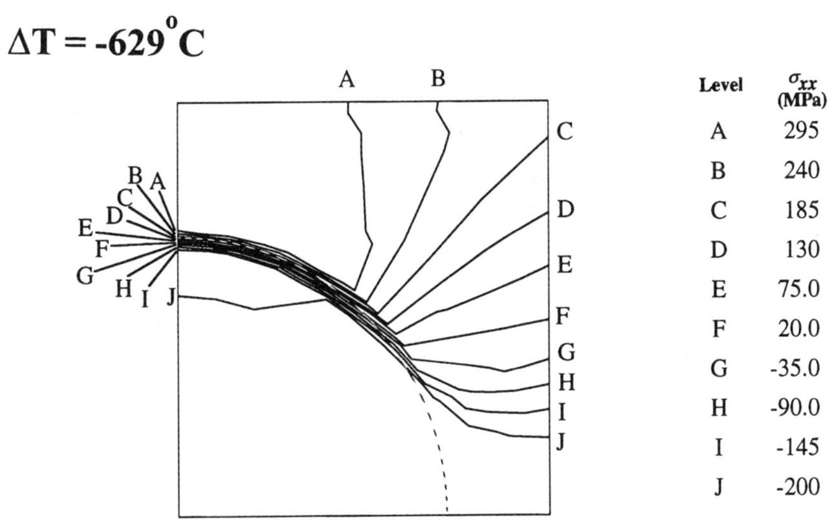

FIG. 6—*Thermal residual stresses, σ_{xx}, with an intact interface due to simulated cool-down,* $\Delta T = -629°C$, *[90] SCS-6/TIMETAL-21S.*

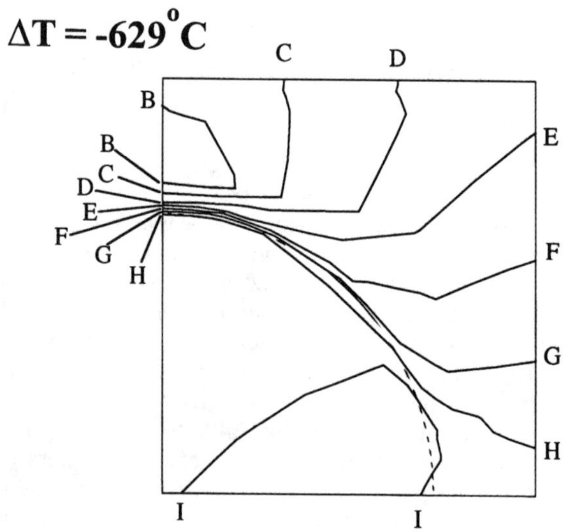

Level	σ_{xx} (MPa)
A	295
B	240
C	185
D	130
E	75.0
F	20.0
G	-35.0
H	-90.0
I	-145
J	-200

FIG. 7—*Thermal residual stresses, σ_{xx}, with a failed interface due to simulated cool-down,* $\Delta T = -629°C$, *[90] SCS-6/TIMETAL-21S.*

steep stress gradient is found due to the intact interface. In Fig. 9, the σ_{zz} stresses are still nearly uniform in the fiber, but with a maximum value of -768 MPa, compared to the case of the intact interface. The region of stress gradient next to the fiber is larger in the case of the failed interface and the maximum value of tensile σ_{zz} stress in the matrix is smaller for the failed interface, 487 MPa compared to the value of 558 MPa for the intact interface.

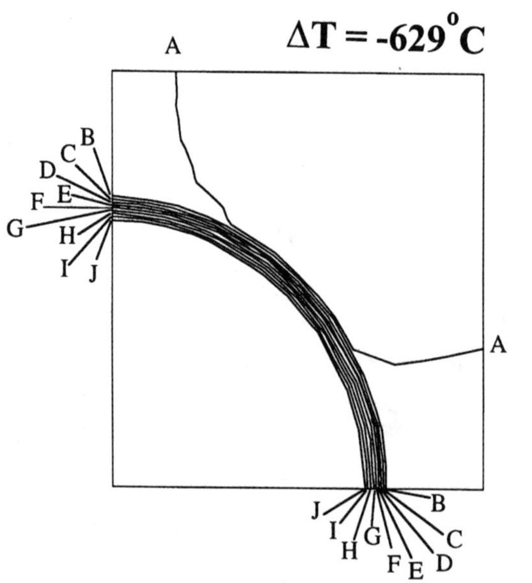

Level	σ_{zz} (MPa)
A	505
B	360
C	215
D	70.0
E	-75.0
F	-220
G	-365
H	-510
I	-655
J	-800

FIG. 8—*Thermal residual stresses, σ_{zz}, with an intact interface due to simulated cool-down,* $\Delta T = -629°C$, *[90] SCS-6/TIMETAL-21S.*

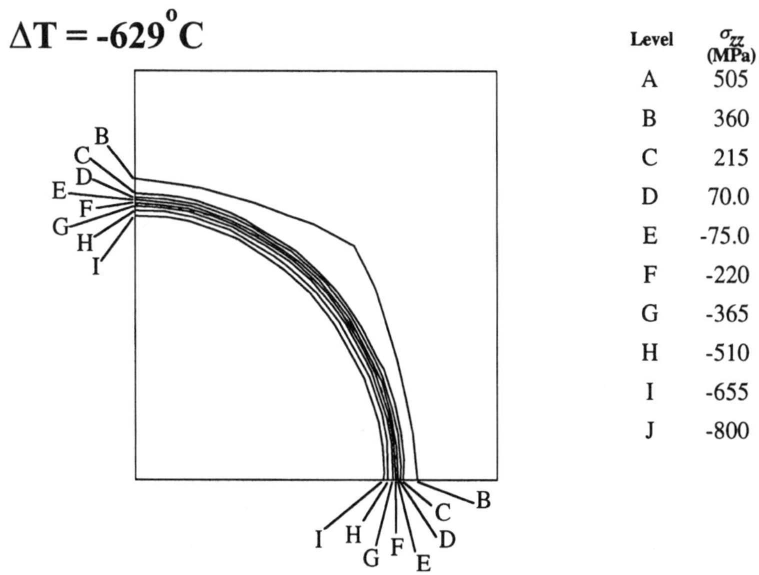

FIG. 9—*Thermal residual stresses, σ_{zz}, with a failed interface due to simulated cool-down,* $\Delta T = -629°C$, *[90] SCS-6/TIMETAL-21S.*

Again, the intact interface produces a greater constraint between the fiber and matrix, resulting in larger thermal residual stresses. The difference between the intact and failed cases is not as great for σ_{zz}, the axial stress, as for the transverse stresses. The axial stress governs axial cracking in both the fiber and the matrix. The thermal residual stresses would act as a prestress that could affect the composite properties and subsequent mechanical behavior. The axial stress calculations indicate that a smaller mechanical axial load would be necessary for the failed interface compared to the intact interface to overcome the thermal residual stresses.

Mechanical Loading

A transverse loading was applied to the model shown in Fig. 5. A uniform stress in the y-direction was applied to the $y = 0.104$-mm face with a loading rate of 2.56 MPa/s. Two isothermal conditions were analyzed, $T = 21.1$ and 650°C. Both the intact and failed interface were modeled. The loading was applied and the stress contours were plotted at time $t = 200$ s. As shown in Fig. 10 for the case of intact interfaces, at this time ($t = 200$ s), for $T = 21.1$°C, the overall behavior of the composite is still within the elastic regime, while for $T = 650$°C, the overall behavior is well into the inelastic regime. Inelastic behavior of the matrix was determined by comparing the von Mises equivalent stress calculated at appropriate temperature to the yield stress shown in Table 2. When the von Mises equivalent stress was greater than or equal to the yield stress, inelastic behavior of the matrix was assumed. The von Mises equivalent stress, σ_{vm}, is defined as follows

$$\sigma_{vm} = \sqrt{\sigma_x^2 + \sigma_y^2 + \sigma_z^2 - \sigma_x\sigma_y - \sigma_y\sigma_z - \sigma_z\sigma_x + 3(\tau_{xy}^2 + \tau_{yz}^2 + \tau_{zx}^2)}$$

Figures 11 and 12 show the calculated von Mises stress contours at $T = 21.1$°C, $t = 200$ s, for the intact and failed interface cases, respectively. The matrix yield stress at $T =$

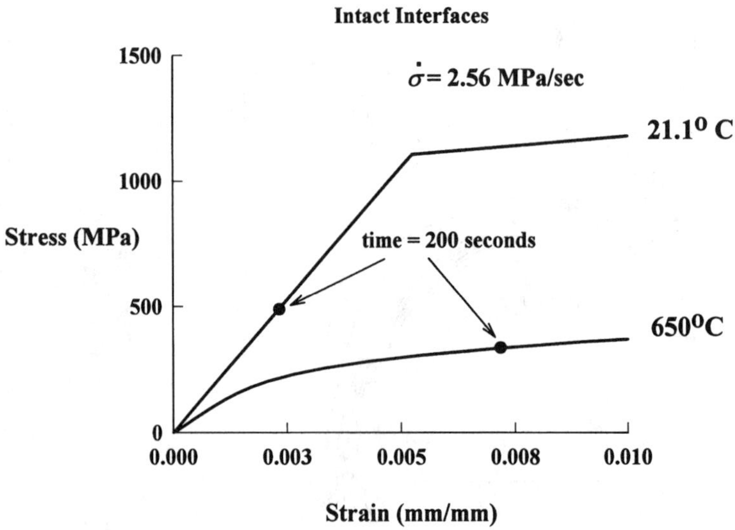

FIG. 10—*Predicted stress-strain curves for [90] SCS-6/TIMETAL-21S with intact interfaces.*

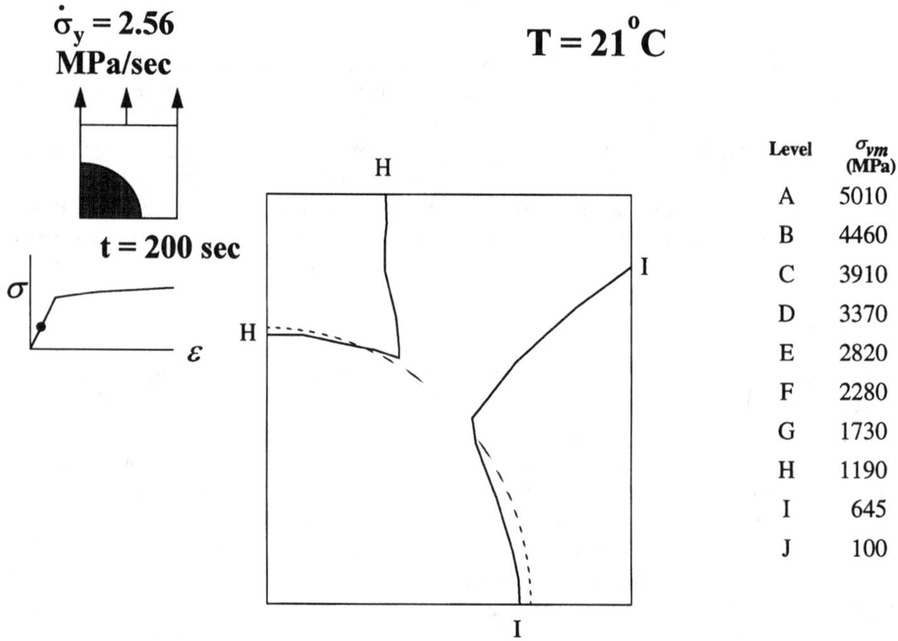

FIG. 11—*von Mises stress contours for intact interfaces,* T = *21.1°C,* t = *200 s, [90] SCS-6/TIMETAL-21S.*

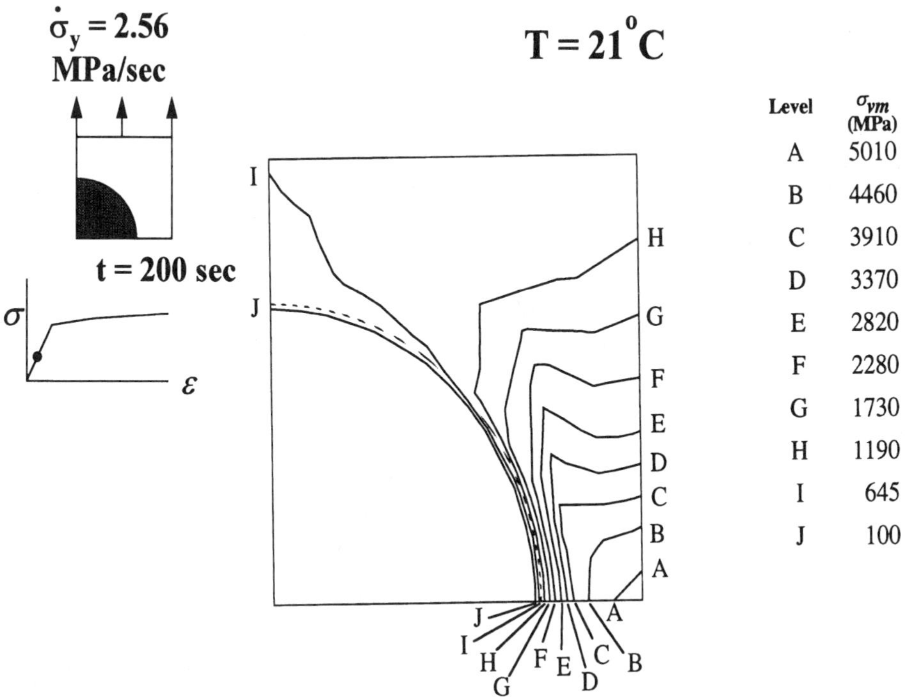

FIG. 12—*von Mises stress contours for failed interfaces,* T = *21.1°C,* t = *200 s, [90] SCS-6/TIMETAL-21S.*

21.1°C is 910 MPa. As shown in Fig. 11, even though the overall stress-strain behavior is still within the elastic region, for a small portion of the matrix (the upper-left-hand corner of the model), the von Mises stress is greater than the matrix yield stress, indicating inelastic behavior. For the case of the failed interface, as shown in Fig. 12, the matrix is well into the inelastic regime. A large stress concentration is calculated in the lower-right-hand corner of the model. The maximum stress calculated for the case of the failed interface is nearly five times greater than the maximum stress in the case of the intact interface.

Figures 13 and 14 show the calculated von Mises stress contours at T = 650°C, t = 200 s, for the intact and failed interface cases, respectively. The matrix yield stress at T = 650°C is 43.5 MPa. At this temperature, the overall stress-strain behavior is exhibiting considerable inelastic behavior, even with an intact interface. As shown in Fig. 13, all of the matrix is at a stress level greater than the yield stress. A stress concentration above the fiber is calculated for the intact interface case. For the case of the failed interface (Fig. 14), a stress concentration is calculated in the lower-right-hand corner of the model. Here, the maximum stress is about 10% greater than for the intact interface case and the region of maximum stress is considerably larger than for the intact interface case. The stress in the fiber is reduced by about 20% for the failed interface, compared to the intact interface case.

Comparing the stress states with a failed interface at the two temperatures, Figs. 12 and 14, shows the two stress states have very similar trends. The magnitude of the stress is more than twice as large at room temperature than at the elevated temperature. The stress states at the two temperatures with the intact interfaces, Figs. 11 and 13, however, are considerably different. As shown in Fig. 11, the stress state at room temperature is more uniform than at

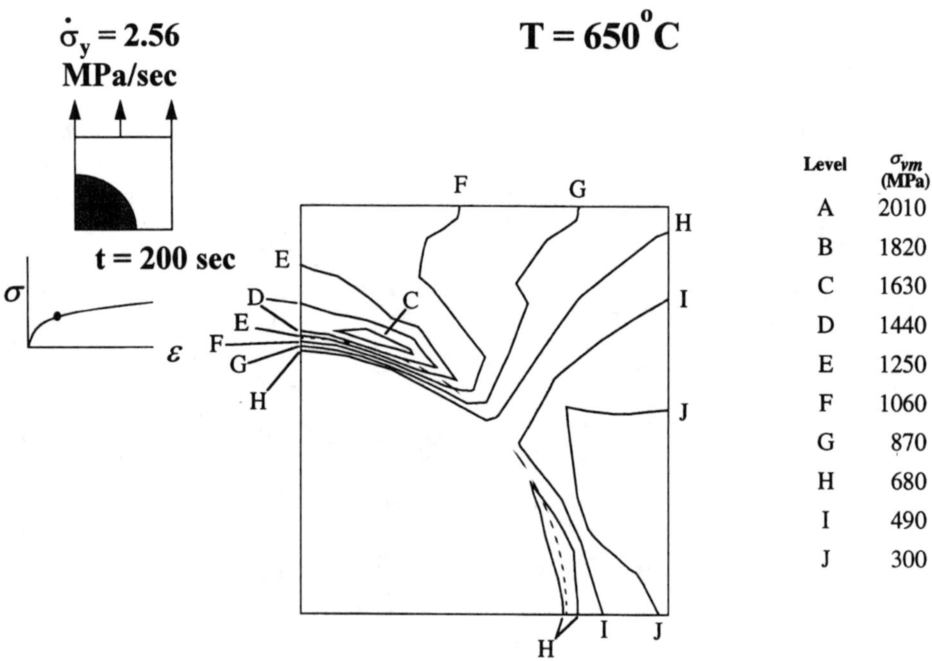

FIG. 13—*von Mises stress contours for intact interfaces,* T = *650°C,* t = *200 s, [90] SCS-6/TIMETAL-21S.*

the elevated temperature, even though the magnitudes of the stresses in both cases are nearly the same.

Concluding Remarks

A three-dimensional finite element program called VISCOPAC was developed and used to conduct a micromechanics analysis of titanium metal matrix composites. The VISCOPAC program uses a modified Eisenberg-Yen thermo-viscoplastic constitutive model to predict matrix behavior under thermomechanical fatigue loadings. The analysis incorporated temperature-dependent elastic properties in the fiber and temperature-dependent viscoplastic properties in the matrix. The material model was described and the necessary material constants were determined experimentally. The predictions of matrix behavior were accurate at a range of temperatures.

A micromechanics model was used to analyze the fiber-matrix interfacial behavior. The thermal residual stresses due to the fabrication cycle were calculated. The stresses due to a transverse mechanical loading under isothermal loading conditions were also calculated. The effects of the fiber-matrix interface failure and the time-dependent behavior of the matrix on the stress state in the unit cell were examined.

The thermal residual stresses produced during the fabrication of the composite were simulated by applying a thermal load of −629°C. Significant stresses were calculated due to the simulated cool-down. The failed interface resulted in lower thermal residual stresses in the matrix and fiber. There was less constraint between the fiber and matrix in this case, thus, lowering the stresses in the constituents.

Stresses due to a uniform transverse load were calculated at two temperatures, room

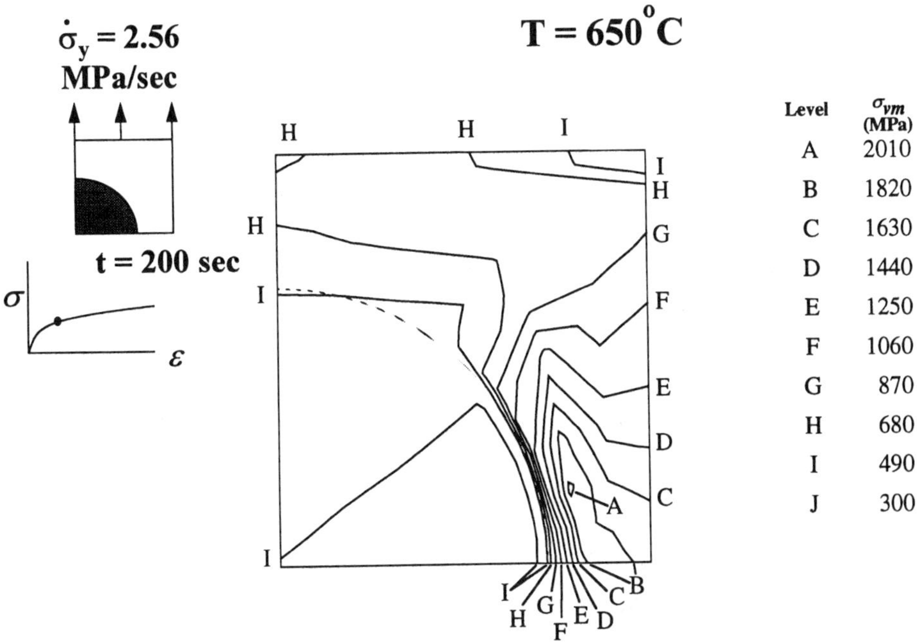

FIG. 14—*von Mises stress contours for failed interfaces,* T = *650°C,* t = *200 s, [90] SCS-6/TIMETAL-21S.*

temperature and an elevated temperature of 650°C. The extent of inelastic behavior was studied by comparing the calculated von Mises stress state to the yield stress of the matrix. At both temperatures, a large stress concentration was calculated when the interface had failed. At room temperature, the stress with the failed interface was over five times as large as the stress with the intact interface. The difference was not as great at the elevated temperature. Significant inelastic behavior was shown at both temperatures when the interface was failed, although the magnitude of the stresses was considerably lower at the elevated temperature.

The results indicate the importance of accurately accounting for fiber-matrix interface failure and the need for a micromechanics-based analytical technique to understand and predict the behavior of titanium metal matrix composites.

APPENDIX I

Viscoplastic Constitutive Material Model

The thermo-viscoplastic constitutive equations developed by Bahei et al. [2] and Shah [12] were used to describe the behavior of each homogenous phase of the composite material. At low homologous temperatures and isothermal conditions, the constitutive theory reduces to the formulation by Eisenberg and Yen [3] in a form more suitable for non-proportional loading.

The total strain rate is decomposed into elastic, thermal, and inelastic parts. The inelastic strain rate is found as a function of the overstress measured from an equilibrium yield surface that delineates the stress states that can be reached from the current state by purely elastic deformations. In the presence of kinematic and isotropic hardening, a von Mises form of the current equilibrium yield surface can be written as

$$f = \frac{3}{2}(s_{ij}^* - \beta_{ij})(s_{ij}^* - \beta_{ij}) - (Y + Q)^2 \tag{1}$$

where s_{ij}^* is the deviatoric equilibrium stress tensor, β_{ij} denotes the center of the yield surface, $Y = Y(T)$ is the temperature-dependent yield stress in tension that is independent of the loading rate, and Q is an isotropic hardening function. Figure 15 depicts the equilibrium yield surface, f, in the deviatoric stress space, where K_0 is the initial yield stress in shear.

Corresponding to a given stress tensor, s_{ij}, which lies outside the yield surface (Eq 1), there exists an equilibrium stress, s_{ij}^*, that satisfies Eq 1 and is determined as the intersection of the equilibrium surface and the line connecting the stress point and the center of the equilibrium surface, β_{ij}. Hence

$$s_{ij}^* = \left[\frac{2[Y(T) + Q(T)]^2}{3(s_{kl} - \beta_{kl})(s_{kl} - \beta_{kl})}\right]^{1/2}(s_{ij} - \beta_{ij}) + \beta_{ij} \tag{2}$$

The effective overstress, R, is a measure of the distance between the actual stress point, s_{ij}, and the equilibrium stress point, s_{ij}^*, such that R vanishes if the actual stress point lies on, or falls within, the yield surface. Thus

$$R = \frac{3}{2}[(s_{ij} - s_{ij}^*)(s_{ij} - s_{ij}^*)]^{1/2}(Y + Q)^2, \qquad \text{if} \quad f(s_{ij} - \beta_{ij}) > 0 \tag{3}$$

$$R = 0, \qquad \text{if} \quad f(s_{ij} - \beta_{ij}) \leq 0 \tag{4}$$

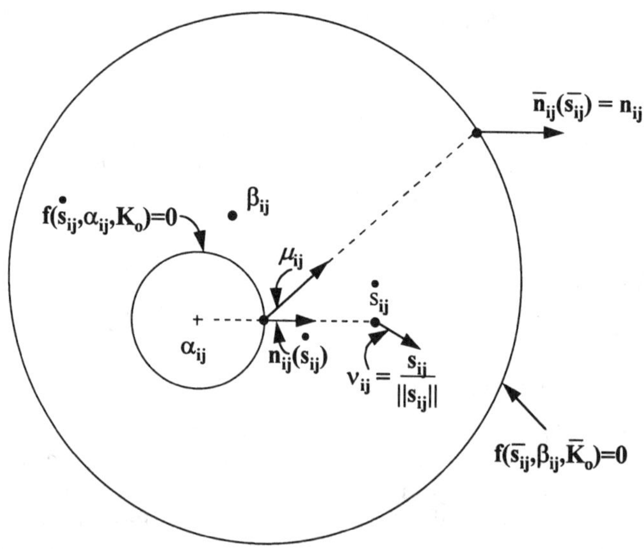

FIG. 15—*Equilibrium and bounding surfaces in deviatoric stress space.*

The inelastic strain rate is found from an associated flow rule in which the strain rate is normal to the equilibrium yield surface and its magnitude is assumed to be in the form of a power law of the overstress [3]

$$\dot{\epsilon}_{ij}^{in} = \sqrt{3/2}k(T)R^{p(T)}n_{ij}(s_{ij}^{*}) \tag{5}$$

$$n_{ij} = \sqrt{3/2}\left(\frac{s_{ij}^{*} - \beta_{ij}}{Y + Q}\right) \tag{6}$$

where the functions, $k(T)$ and $p(T)$, are material parameters and n_{ij} is the unit normal to the yield surface (Eq 1) at the current equilibrium stress point.

The evolution equation for Q, which includes the effect of inelastic deformation and thermal recovery on the yield stress, is given by

$$\dot{Q} = q(T)[Q_a(T) - q]\dot{\bar{\epsilon}}^{in} - b_r(T)|Q - Q_r(T)|^{n_r(T)-1}[Q - Q_r(T)] \tag{7}$$

The functions, $Q_a(T)$, $q(T)$, $b_r(T)$, $Q_r(T)$, and $n_r(T)$, are temperature-dependent material parameters, and $\dot{\bar{\epsilon}}^{in}$ is the effective inelastic strain rate

$$\dot{\bar{\epsilon}}^{in} = \left[\frac{2}{3}\dot{\epsilon}_{ij}^{in}\dot{\epsilon}_{ij}^{in}\right]^{1/2} = k(T)R^{p(T)}, \qquad \dot{\epsilon}_{kk}^{in} = 0 \tag{8}$$

Total ($Q_r(T) = 0$) or partial ($Q_r(T) \neq 0$) thermal recovery is represented by the second term in Eq 7.

In analogy with Eq 7 and permitting complete thermal recovery of kinematic hardening, the evolution equation for the center of the yield surface, β_{ij}, can be written as

$$\dot{\alpha}_{ij} = \dot{\mu}\nu_{ij} - c_r(T)\bar{\alpha}^{(m_r(T)-1)}\alpha_{ij}, \qquad \bar{\alpha} = (\alpha_{kl}\alpha_{kl})^{1/2} \tag{9}$$

where $c_r(T)$ and $m_r(T)$ are material parameters. The unit tensor, ν_{ij}, defines the direction of translation of the yield surface in the deviatoric stress space and can be specified according to the hardening rules applied in rate-dependent plasticity theories. If the Phillips hardening rule is selected then

$$\nu_{ij} = \dot{s}_{ij}/(\dot{s}_{kl}\dot{s}_{kl})^{1/2} \qquad \text{if} \quad \dot{s}_{ij} \neq 0 \tag{10}$$

$$\nu_{ij} = n_{ij} \qquad \text{if} \quad \dot{s}_{ij} = 0 \tag{11}$$

The factor, $\dot{\mu}$, in Eq 9 is found from Praeger's consistency condition, $\dot{f} = 0$, when translation of the yield surface is specified by the first term in Eq 9. The result is

$$\dot{\mu} = \sqrt{2/3}k(T)R^{p(T)}[H(T) - q(T)[Q_a(T) - Q]]/n_{kl}\nu_{kl} \tag{12}$$

A two-surface plasticity theory [13] can be used to describe evolution of the instantaneous tangent modulus, H, in terms of the distance, in the deviatoric space, between the equilibrium yield surface and a bounding surface at which the instantaneous tangent modulus assumes an experimentally measured asymptotic value, H_0, as shown in Fig. 3. A von Mises form of the equilibrium bounding surface is written as

$$f = \frac{3}{2}(\bar{s}_{ij} - \bar{\beta}_{ij})(\bar{s}_{ij} - \bar{\beta}_{ij}) - (\bar{Y} + \bar{Q})^2 = 0 \tag{13}$$

where $\bar{s}_{ij}$ is the bounding stress tensor, $\bar{\beta}_{ij}$ denotes the center of the bounding surface, $\bar{Y} = \bar{Y}(T)$ is the tensile bounding stress that is independent of the loading rate, and $\bar{Q}$ is an isotropic hardening function. Figure 15 depicts the bounding surface in the deviatoric stress space, where $\bar{K}_0$ is the initial yield stress in shear.

The instantaneous tangent modulus for the quasi-static response, H, is then found as

$$H(T) = H_0(T) + h(T)[\delta/(\delta_{in} - \delta)]^{m(T)} \tag{14}$$

$$\delta = \left[\frac{3}{2}(\bar{s}_{ij} - s_{ij}^*)(\bar{s}_{ij} - s_{ij}^*)\right]^{1/2} \tag{15}$$

where δ_{in} is the distance between the yield surface and the bounding surface at the onset of inelastic deformation. When the equilibrium stress point lies on the bounding surface, the plastic tangent modulus assumes the asymptotic value, $H_0(T)$. The h_0, h, and m parameters need to be determined experimentally as functions of temperature.

In analogy with the equilibrium yield surface, thermal recovery of isotropic as well as kinematic hardening of the bounding surface are included in the model. The rate equations for $\bar{Q}$ and $\bar{\beta}_{ij}$ are

$$\dot{\bar{Q}} = \bar{q}(T)[\bar{Q}_a(T) - \bar{Q}]\dot{\bar{\epsilon}}^{in} - \bar{b}_r(T)|\bar{Q} - \bar{Q}_r(T)|^{(\bar{n}(T)-1)}[\bar{Q} - \bar{Q}_r(T)] \tag{16}$$

$$\dot{\bar{\beta}}_{ij} - \dot{\beta}_{ij} = -\dot{\bar{\mu}}u_{ij} - \bar{c}_r(T)\bar{\beta}^{(\bar{m}_r(T)-1)}(\bar{\beta}_{ij} - \beta_{ij}) \tag{17}$$

Functions $\bar{Q}_a(T)$, $\bar{q}(T)$, $\bar{b}_r(T)$, $\bar{Q}_r(T)$, $\bar{n}_r(T)$, $\bar{c}_r(T)$, and $\bar{m}_r(T)$ are material parameters that must be estimated based on experimental results. The unit tensor, u_{ij}, defined the direction of translation of the bounding surface relative to the equilibrium surface in the deviatoric stress space, as shown in Fig. 15.

APPENDIX II

Failure Criteria

The form of the failure criteria specified in the VISCOPAC program depends on the type of material. For isotropic materials, the following failure envelope is specified in the stress space

$$f = \left(\frac{\bar{\sigma}_n}{\sigma^u}\right)^2 - \left(\frac{\bar{\sigma}_s}{\tau^u}\right)^2 - 1.0 = 0 \tag{18}$$

where

$$\bar{\sigma}_n = \sigma_{11}^2 + \sigma_{22}^2 + \sigma_{33}^2 - \sigma_{11}\sigma_{22} - \sigma_{22}\sigma_{33} - \sigma_{33}\sigma_{11} \tag{19}$$

$$\bar{\sigma}_s = 3(\sigma_{12}^2 + \sigma_{13}^2 + \sigma_{23}^2) \tag{20}$$

and σ_{ij} is the average stress in an element and σ^u and τ^u are the strength of the material under normal and shear stress, respectively. The failure surface given by Eqs 18 through 20 is reminiscent of the von Mises yield surface but with different strength magnitudes under normal and shear stresses.

For fibrous composite materials, the user can choose among two failure criteria available in the VISCOPAC program; one based on the overall stress and one based on the fiber stress. In either case, only the axial normal stress and the longitudinal shear stress components appear in the failure criteria. Specifically, the failure envelope is given by

$$f = \left(\frac{\sigma_{33}}{\sigma^u}\right)^2 + \left(\frac{\sigma_{13}}{\tau^u}\right)^2 + \left(\frac{\sigma_{23}}{\tau^u}\right)^2 - 1.0 = 0 \tag{21}$$

or

$$f = \left(\frac{\sigma^f_{33}}{\sigma^{uf}}\right)^2 + \left(\frac{\sigma^f_{13}}{\tau^{uf}}\right)^2 + \left(\frac{\sigma^f_{23}}{\tau^{uf}}\right)^2 - 1.0 = 0 \tag{22}$$

Here, σ_{ij} is the overall stress and σ^f_{ij} is the fiber stress specified in a Cartesian coordinate system $\overline{x_k}$, $k = 1, 2, 3$, such that the $\overline{x_1}$-axis coincides with the fiber direction and $\overline{x_2 x_3}$ coincides with the transverse plane. The symbols, σ^u and τ^u, denote the overall strength of the fibrous composite under axial normal stress and longitudinal shear stress, respectively. Similarly, σ^{uf} and τ^{uf} denote the fiber strength under axial normal stress and longitudinal shear stress, respectively.

References

[1] Johnson, W. S., Lubowinski, S. J., and Highsmith, A. L., "Mechanical Characterization of Unnotched SCS-6/Ti-15-3 Metal Matrix Composites at Room Temperature," *Thermal and Mechanical Behavior of Metal Matrix and Ceramic Matrix Composites, ASTM STP 1080,* J. M. Kennedy, H. H. Moeller, and W. S. Johnson, Eds., American Society for Testing and Materials, Philadelphia, 1990, pp. 193–218.

[2] Bahei-El-Din, Y. A., Shah, R. S., and Dvorak, G. J., "Numerical Analysis of the Rate-Dependent Behavior of High Temperature Fibrous Composites," Applied Mechanics Division, Vol. 18, *Mechanics of Composites at Elevated and Cryogenic Temperatures,* S. N. Singhal, W. F. Jones, C. T. Herakovich, and T. Cruse, Eds., Book No. G00618, American Society of Mechanical Engineers, 1991, pp. 67–78.

[3] Eisenberg, M. A. and Yen, C. F., "A Theory of Multiaxial Anisotropic Viscoplasticity, *ASME Journal of Applied Mechanics,* Vol. 48, 1981, pp. 276–284.

[4] Hill, R., "Theory of Mechanical Properties of Fiber-Strengthened Materials—III. Self-Consistent Model," *Journal of the Mechanics and Physics of Solids,* Vol. 13, 1965, pp. 189–198.

[5] Mori, T. and Tanaka, K., "Average Stress in Matrix and Average Elastic Energy of Materials with Misfitting Inclusions," *Acta Metallurgica,* Vol. 21, May 1973, pp. 571–574.

[6] Bahei-El-Din, Y. A., "Plastic Analysis of Metal-Matrix Composite Laminates," Ph.D. thesis, Duke University, Durham, NC, 1979.

[7] Bahei-El-Din, Y. A. and Dvorak, G. J., "A Review of Plasticity Theory of Fibrous Composite Materials," *Metal Matrix Composites: Testing, Analysis, and Failure Modes, ASTM STP 1032,* W. S. Johnson, Ed., American Society for Testing and Materials, Philadelphia, 1989, pp. 103–129.

[8] Dvorak, G. J. and Bahei-El-Din, Y. A., "Plasticity Analysis of Fibrous Composites," *Journal of Applied Mechanics,* Vol. 49, 1982, pp. 327–335.

[9] Mirdamadi, M., Johnson, W. S., Bahei-El-Din, Y. A., and Castelli, M. G., "Analysis of Thermomechanical Fatigue of Unidirectional Metal Matrix Composite," *Composite Materials: Fatigue and Fracture, Fourth Volume, ASTM 1156,* W. W. Stinchcomb and N. E. Ashbaugh, Eds., American Society for Testing and Materials, Philadelphia, 1993.

[10] Mirdamadi, M. and Johnson, W. S., "Prediction of Stress-Strain Response of SCS-6/TIMETAL-21S Subjected to the Hypersonic Flight Profile," NASA TM-109026, NASA Langley Research Center, Hampton, VA, 1994.

[11] Dieter, G. E., *Mechanical Metallurgy,* 2nd ed., McGraw-Hill, New York, 1976, pp. 451–489.
[12] Shah, R. S., "Modeling and Analysis of High Temperature Inelastic Deformation in Metal Matrix Composites," Ph.D. thesis, Rensaleer Polytechnic Institute, Troy, NY, 1991.
[13] Dafalias, Y. F. and Popov, E. P., "Plastic Internal Variable Formalism of Cyclic Plasticity," *ASME Journal of Applied Mechanics,* Vol. 43, 1976, pp. 645–651.

Joseph L. Kroupa,[1] Richard W. Neu,[2] Theodore Nicholas,[2] Demirkan Coker,[1] D. Dave Robertson,[3] and Shankar Mall[3]

A Comparison of Analysis Tools for Predicting the Inelastic Cyclic Response of Cross-Ply Titanium Matrix Composites

REFERENCE: Kroupa, J. L., Neu, R. W., Nicholas, T., Coker, D., Robertson, D. D., and Mall, S., **"A Comparison of Analysis Tools for Predicting the Inelastic Cyclic Response of Cross-Ply Titanium Matrix Composites,"** *Life Prediction Methodology for Titanium Matrix Composites, ASTM STP 1253,* W. S. Johnson, J. M. Larsen, and B. N. Cox, Eds., American Society for Testing and Materials, 1996, pp. 297–327.

ABSTRACT: Several micromechanical models that are currently being used for predicting the thermal and mechanical behavior of a cross-ply, [0/90], titanium matrix composite are evaluated. Six computer programs or methods are compared: (1) VISCOPLY; (2) METCAN; (3) FIDEP, an enhanced concentric cylinder model; (4) LISOL, a modified method of cells approach; (5) an elementary approach where the [90] ply is assumed to have the same properties as the matrix; and (6) a finite element method. Comparisons are made for the thermal residual stresses at room temperature resulting from processing, as well as for stresses and strains in two isothermal and two thermomechanical fatigue test cases. For each case, the laminate response of the models is compared to experimental behavior, while the responses of the constituents are compared among the models. The capability of each model to predict frequency effects, inelastic cyclic strain (hysteresis) behavior, and strain ratchetting with cycling is shown. The basis of formulation for the micromechanical models, the constitutive relationships used for the matrix and fiber, and the modeling technique of the [90] ply are all found to be important factors for determining the accurate behavior of the [0/90] composite.

KEYWORDS: micromechanical model, metal matrix composite, titanium, fatigue (materials), thermomechanical fatigue, laminates, cross-ply composites, titanium matrix composites, life prediction, titanium alloys, modeling

Titanium matrix composites (TMCs) are being considered for use in a number of aerospace structural and gas turbine engine applications where light-weight, stiff, and high strength materials are required to perform at elevated temperatures. Predictive models are required to characterize the mechanical behavior of these new materials under typical service conditions where both thermal and mechanical loads may be applied. In the case of TMCs, the internal stress-strain behavior can be quite complex due to the differences in the coefficient of thermal expansion (CTE) and modulus between the fiber and matrix. Furthermore, the material properties of the constituents are both time- and temperature-dependent.

[1]Associate research engineer and associate research engineer, respectively, University of Dayton Research Institute, Dayton, OH 45969.

[2]NRC associate (formerly) and senior scientist, respectively, Wright Laboratory Materials Directorate, Wright-Patterson Air Force Base, OH 45433. Mr. Neu is presently, Woodroff School of Mechanical Engineering, Georgia Institute of Technology, Atlanta, GA, 30332.

[3]Assistant professor and chairman, Department of Aeronautics and Astronautics, respectively, Air Force Institute of Technology, Wright-Patterson Air Force Base, OH 45433.

If these TMCs are to be used in high temperature applications with confidence, their behavior must be well understood and accurately predicted under typical operating environments.

One critical aspect of characterizing advanced TMC composites is the ability to predict the fatigue life, especially under combined thermal and mechanical loadings. Tools and procedures capable of modeling the composite behavior are required to make confident life predictions under these conditions. Furthermore, since thermomechanical fatigue (TMF) life has been shown to be dependent on the behavior of the fiber and matrix as well as the [90] ply [*1–6*], accurate micromechanics analysis methods are required. For example, in-phase TMF life is dependent on the stress in the fiber [*1,6*], whereas out-of-phase TMF life is dependent on the behavior of the [90] ply where the crack initiates and grows [*2–5*]. Therefore, not only is the laminate behavior necessary, but also the behavior of the constituents must be determined in order to develop a fatigue life prediction methodology.

Modeling the fatigue behavior of cross-ply TMCs is more challenging than that of unidirectional TMCs. In many TMCs, the bonding between the fiber and matrix is weak [*7–10*]. The interfacial strength is dependent on the strength of the chemical bond as well as the amount of clamping due to the thermal residual stresses that arise from the difference between the higher CTE of the matrix compared to that of the fiber. For the particular TMC under investigation, SCS-6/TIMETAL 21S,[4] chemical bonding is rather weak. This is attributed primarily to the carbon coating on the fiber. Because of the weak bonding, the fiber and matrix in the [90] ply slide and separate during loading, and this "damage" significantly influences the response of the cross-ply TMC [*6–10*]. Such response of the [90] ply has to be accounted for in analysis in order to accurately characterize the cross-ply composite.

Another feature of the behavior of TMCs is the inelastic response of the matrix material at high temperatures [*11*]. At elevated temperatures, the matrix behaves in a viscoplastic manner that produces time-dependent inelastic behavior of the composite under monotonic loading and strain ratchetting under stress-controlled fatigue [*2,3,12*]. During fatigue at elevated temperature, most of the inelastic behavior generally occurs during a transient period within the first ten cycles or so, where the composite strain increases under a positive stress ratio. Eventually, a near steady-state behavior is reached. The scope of the present work includes modeling the composite behavior up to the point of steady-state response.

In an earlier investigation, Bigelow et al. [*13*] compared four micromechanics models for metal matrix composites: FEM, AGLPLY, method of cells, and METCAN. These models are discussed later. In that study, they modeled the behavior of [0] and [0/90] boron/aluminum (B/Al) and SCS-6/Ti-15-3 composites. They used temperature-dependent, but generally time-independent, elastic-plastic matrix properties for the B/Al material, but only considered the elastic properties of the titanium composite because of its high yield strength. They also considered only the case of a perfect (no fiber/matrix separation or sliding) interface. Computations compared (1) elastic properties of the laminate, (2) laminate and constituent stresses under monotonic loading at room temperature, and (3) thermal residual stresses in a representative volume element (RVE). They concluded that all four methods predict the laminate response of the [0] and [0/90] composite of B/Al, which has a strong fiber/matrix interface at room temperature, very well. They also compared thermal residual stresses in a cross section. The major difference among the models was the type of inputs required rather than the accuracy. Since elastic constituent properties are involved in most of the solutions, comparisons can only be made regarding the formulation of the micromechanics models, not about constituent constitutive models or interfacial separation.

[4]Registered trademark of Temet Corporation, Henderson, NV.

This investigation extends the scope of the previous comparisons by Bigelow et al. [*13*] by focusing on the modeling of cyclic behavior, by analyzing a cross-ply $[0/90]_S$ TMC with a weak fiber/matrix interface, and by considering thermoviscoplastic matrix behavior. As a representative TMC system, this investigation considers the response of a $[0/90]_S$ SCS-6/TIMETAL 21S, a silicon-carbide-reinforced, metastable beta titanium alloy composite. This study addresses the cyclic behavior of the composite under isothermal fatigue at high temperature at two different frequencies, and under two cases of TMF. The computational predictions are compared to experimental data as well as among the models.

In all, six computational methods are examined, and the solutions from each are compared in this paper. Four of these methods are enhanced versions of the methods examined in the comparisons by Bigelow et al. [*13*]. A concentric cylinder model and a modified method of cells model are also included in this study. Each method is referred to by its respective software package name or acronym throughout the paper. The six methods considered are: (1) VISCOPLY, a laminate analysis that incorporates the vanishing fiber diameter model [*14,15*]; (2) METCAN, a multi-cell model in a laminate-based code [*16–19*]; (3) FIDEP, a concentric cylinder model with a parallel [90] element [*20,21*]; (4) LISOL, a modification of Aboudi's method of cells [*22–25*] approach [*26–28*]; (5) ELAM (ELementary Analysis Method), a simplified analysis where the [90] ply is assumed to have the properties of the matrix; and (6) FEM, a finite element method applied to a representative volume element (RVE). In all of the methods except METCAN, the fiber is treated as thermoelastic using the identical temperature-dependent modulus in each. The matrix material within FIDEP, LISOL, and FEM is represented by the unified constitutive model of Bodner and Partom [*29–30*]. The matrix response in VISCOPLY and ELAM is characterized by the "over-stress" model of Eisenberg and Yen [*31*]. In METCAN, a multifactor interaction relationship (MFIR) is used to describe the time- and temperature-dependence of the constituents [*18*]. The methods of representing the fiber/matrix debonding, as well as the micromechanics formulations, are different in each model. The calculated material response, therefore, reflects all of these differences.

Review of the Analysis Tools

VISCOPLY

The computer program, VISCOPLY, was designed to evaluate the thermoviscoplastic response of composites with any symmetric layup. The underlying theory of the treatment for each laminate emerged from the vanishing fiber diameter model (VFD), as proposed by Dvorak and Bahei-El-Din [*14*] in the late 1970s. In this model, the fibers are assumed not to affect matrix deformation in the transverse or thickness directions within a ply, but only to affect the behavior in the longitudinal direction. The unit cell for the [0/90] composite is represented in each ply by parallel fiber and matrix plates with axial coupling as illustrated in Fig. 1. The VFD essentially treats the transverse properties as if the fiber and matrix are in series with each other. To account for additional strains due to fiber/matrix separation, the transverse modulus of the fiber can be reduced. Mirdamadi and Johnson [*32*] used this procedure to obtain a best fit to experimental data. Typically, a transverse fiber modulus approximately 10% or less of the longitudinal modulus is required. This same procedure is followed in this investigation.

The methodology is embodied in VISCOPLY, a two-dimensional laminate computer code, that uses temperature-dependent elastic properties for the fiber and elastic-plastic properties for the matrix [*15*]. The matrix model was later enhanced by using the

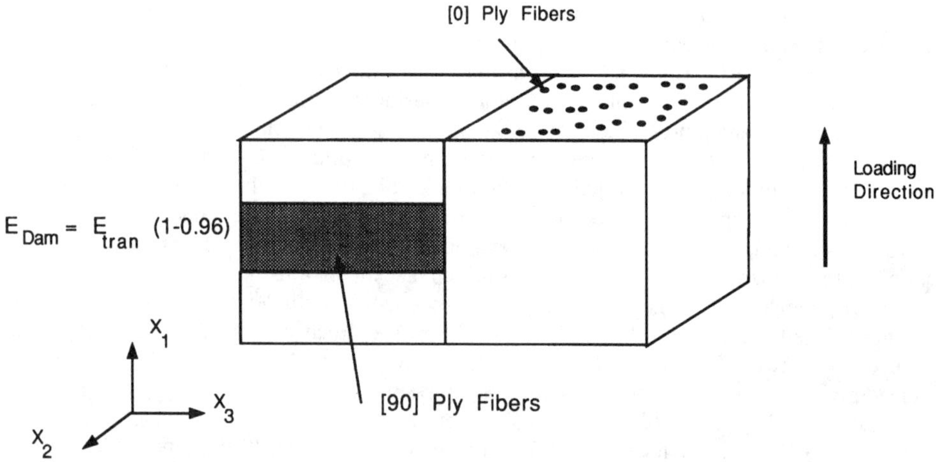

FIG. 1—*Vanishing fiber diameter representation of the [0/90] composite used by VISCOPLY.*

thermoviscoplastic model of Eisenberg and Yen [*31*] that is incorporated into the code used in this investigation [*33*].

METCAN

METCAN (METal Matrix Composite ANalyzer) was developed by Hopkins and Chamis [*16–19*] at NASA Lewis Research Center in the mid-1980s. The code is designed to obtain the global response of a composite laminate based on its constituent properties. A composite laminate of any orientation and number of plies and stacking sequence can be analyzed. In addition, for each ply, different materials and temperatures can be specified and any general loading typically employed in classical laminated plate theory can be applied. METCAN can simulate any laminate with nonlinear constituent properties, which was a notable advance over the elastic analysis given by classical laminated plate theory. The response obtained from the code can be used in structural codes. The constituent response can also be obtained from METCAN, although the model (especially, the material model) is not optimized specifically to obtain the most accurate constituent response.

The RVE of a ply is described as a regular square array pattern that simulates round fibers (Fig. 2) [*16*]. The cell is partitioned into a fiber region, an interphase region, and a matrix region. The matrix is further divided into Regions A, B, and C. Within these regions, the state of stress is assumed to be uniform. The method uses a mechanics of materials approach with the following assumptions: (1) the fiber and matrix regions are subjected to the same strain in the fiber direction of a unidirectional fiber composite (displacement compatibility) and (2) the same transverse stress is applied to both the fiber and matrix regions in the loading direction transverse to the fiber (force equilibrium) [*16,18*]. Classical laminated plate theory assumptions for strain are used to integrate the ply level properties and obtain the laminate level behavior. A convergence criterion is based on the global/laminate level strains, while the global incremental strains are preserved and compared at each iteration.

Material nonlinearity in METCAN is described by a multifactor interaction relationship (MFIR) [*17*]. Each material property is described as a product of many power law equations with each describing how that property is affected by one parameter. Since the details of the general form are given in a number of METCAN references [*18*], a specific example

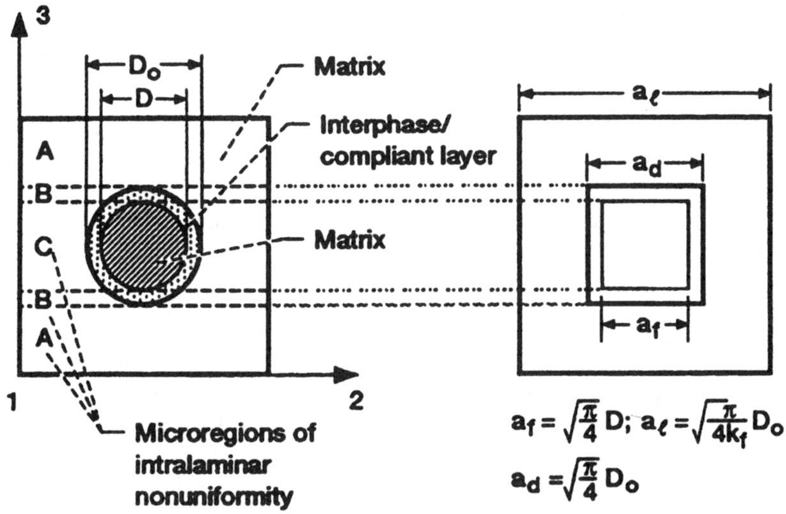

FIG. 2—*Micromechanical representation of the [90] ply found in METCAN.*

showing how it works is presented here. As an illustration, the modulus of a constituent is given by

$$\frac{E}{E_o} = \left[\frac{T_M - T}{T_M - T_o}\right]^n \left[\frac{S_F - \sigma}{S_F - \sigma_o}\right]^m \left[\frac{\dot{S}_F - \dot{\sigma}_o}{\dot{S}_F - \dot{\sigma}}\right]^l \tag{1}$$

where T_M is the melting temperature, S_F is the fracture stress at T, and $\dot{S}_F$ is an appropriately selected stress rate, which in METCAN is defined by the square root of the ratio of the modulus and the weight density multiplied by a constant. The subscript, o, represents the reference values. The reference values are typically defined, $T_o = 23°C$ (70.0°F) and $\sigma_o = \dot{\sigma}_o = 0$; E_o is the initial modulus at room temperature. Additional factors can be added to make the properties dependent on time and cycles [*19*] to simulate the degradation of the properties with these factors. These were left at the default settings in the present version of METCAN, since adjustments to these factors are not known a priori. The MFIR describes the decrease in E with increasing T through the first factor in Eq 1. The MFIR describes the inelastic behavior by reducing the modulus with stress. There is no separation of the elastic and plastic moduli, therefore, the MFIR cannot predict linear elastic unloading; instead, unloading generally follows the loading path. Time-dependent effects can be simulated through the stress rate factor. Since E is a function of a number of factors, sharp yield points in the matrix behavior cannot be modeled very well. The Poisson's ratio, CTE, and strength are also determined by similar MFIRs.

To model the weak fiber/matrix interface, an interphase is considered in the analysis. The thickness of the interphase is assumed to be 0.05 times the fiber diameter (140 μm). The modulus and strength values of the interphase are set to 25% of the matrix values. Although no complete separation of fiber/matrix is simulated, this method is shown to match transverse loading experiments in other published METCAN results [*19*].

FIDEP

FIDEP (FInite-Difference Code for Elastic-Plastic Analysis) was developed as a PC-compatible computer code for predicting micromechanical stresses in metal matrix composites under complex thermal and mechanical loading histories. The early version of FIDEP was a multiple concentric cylinder model used to represent the behavior of a unidirectional composite with elastic-plastic constituents [*34,35*]. The program was modified to include different constitutive models and loading histories [*20*]. Since it is a multiple concentric cylinder model, an interphase with different properties can also be included, though this is not done in this investigation. One distinguishing point about FIDEP compared to the laminate-based codes is that it considers the three-dimensional stresses in the fiber and matrix, especially their variation with distance from the fiber/matrix interface. Both bilinear elastic-plastic as well as unified viscoplastic models such as the Bodner-Partom model can be used.

To model a [0/90] cross-ply composite, a parallel element representing the [90] ply was recently added to the code as shown in Fig. 3 [*20,21*]. Strain compatibility and stress equilibrium are invoked in the loading direction, but lateral strains or Poisson's ratio effects between the [0] and [90] plies are not considered [*36*]. The [90] ply behavior is represented by a material constitutive model that incorporates the damage due to fiber/matrix separation. The [90] ply response is described as the average homogeneous response consisting of a bilinear elastic-plastic fit to the finite element results predicted with a unit cell with a weak (frictionless) interface, loaded transverse to the fiber.

LISOL

The micromechanical formulation present within computer program LISOL (Laminate Inelastic SOLver) was developed by Robertson and Mall [*26–28*] and consists of a three-dimensional analysis similar to Aboudi's method of cells [*22–25*]. In this method, a regular rectangular array of fibers is assumed and symmetry is invoked to isolate one quarter of a RVE for the analysis. The cell is then partitioned into various regions consisting a single fiber region, three matrix regions, and two infinitely thin interface regions as shown in Fig. 4. These regions are assumed to possess a state of uniform stress, and it is required that the normal stresses between adjacent regions be equal. Further, the stresses in any region through any cross-section of the analysis cell are in equilibrium with the lamina stresses. Additionally, the normal strain in the fiber direction is the same for each region, and the average strain along each of the external faces of the analysis cell is equal to the composite strain. The main assumption of this approach is that continuity of slip between adjacent

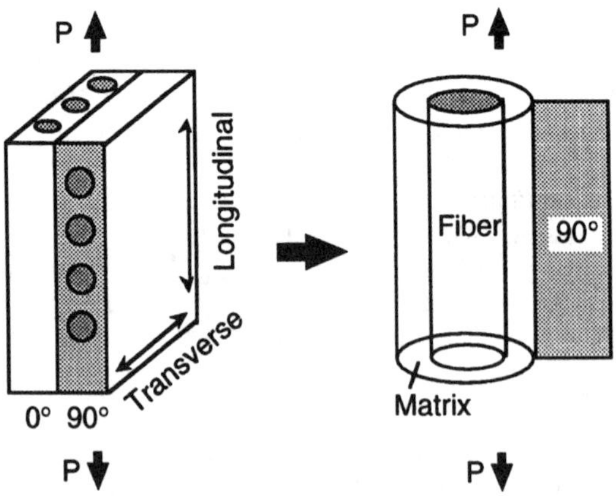

FIG. 3—*Concentric cylinder with parallel [90] element used in FIDEP.*

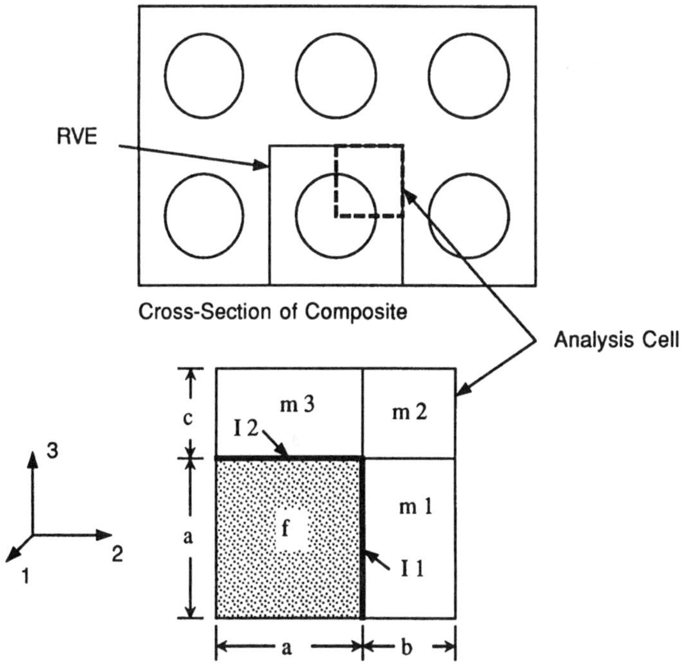

FIG. 4—*Micromechanics model embedded in LISOL.*

regions in the plane perpendicular to the fiber (2–3 plane in Fig. 4) is considered only along the boundary of the analysis cell. All other displacements are continuous at every point. The resulting constitutive relationships for normal stress and strain as well as transverse shear (shear applied in 2–3 plane) are identical to Aboudi's method of cells. The equations for longitudinal shear differ from the method of cells in that the LISOL model ensures complete continuity of displacements and satisfies equilibrium only in an average sense between regions. In contrast, the method of cells approach ensures complete equilibrium while satisfying continuity of displacements only in an average sense. Hence, if the interface and inelastic material characteristics are equivalent, the LISOL results are identical to the method of cells except when off-angle plies are considered. For off-angle plies, LISOL generally will produce a slightly stiffer response that the method of cells.

This method was formulated for both unidirectional and laminated composites. The classical laminated plate theory assumption for strain throughout the layup is employed to perform the laminate analysis. Therefore, similar to METCAN, a composite laminate of any orientation and number of plies and stacking sequence can be analyzed for any general loading typically employed in classical laminated plate theory. The micromechanics formulation presently possesses the capability to include various nonlinear material models such as either a thermoelastic-plastic model with a von Mises yield criterion or a thermoviscoplastic model with the unified viscoplastic theory of Bodner and Partom. In addition, the fiber/matrix interfacial bond may be modeled as either strong, weak, or with progressive failure during loading. The weak interface is accomplished by employing the infinitely thin interfacial regions as gap or contact elements whose stiffness is dependent on whether the interface is in tension or compression [*27*]. The solutions in this study were performed assuming a weak fiber/matrix interface and TIMETAL 21S material parameters for the Bodner-Partom model [*29,30*].

ELAM

The ELementary Analysis Method (ELAM) is a procedure for simple calculations of composite response with a minimum of sophistication. It was incorporated into this investigation as a baseline to evaluate the degree to which more sophisticated methods represent composite response. In an investigation comparing the behavior of three TMCs with different layups ($[0]_4$, $[0/90]_S$, and $[0/\pm45/90]_S$) [*3*], it was found that TMF strain accumulation could be correlated by using the fiber volume fraction of the fibers parallel to the loading. This assumes that the off-axis plies behave the same as the matrix. This assumption can simplify the analysis, since the complex behavior of the damage in the [90] ply is not involved in the problem. Thus, a concentric cylinder model or other unidirectional model can be employed, and by simply adjusting V_f, the behavior of other layups can be easily predicted. Thus, for this study, the [0/90] composite becomes a unidirectional composite with half the V_f of the [0/90] composite. The results reported in this paper for ELAM were obtained from calculations using VISCOPLY. Though not shown here, solutions were also obtained using FIDEP, which produces results that are generally stiffer (about 10%). The differences between the two methods are the matrix viscoplastic model and the three-dimensional stress state that is considered in FIDEP.

Finite Element Method

The finite element method (FEM) is able to calculate the micromechanical stresses throughout a RVE. This method can provide a more detailed description of stress and strain fields within the composite, though the numerical solution commonly takes an order of magnitude longer in time for computation than the other methods. Although the constituent behavior that cannot be measured directly with experiments, FEM is considered the most accurate representation of the micromechanical stresses and provides the best resolution of stress, strain, and strain-rate fields within all the micromechanical approaches considered in this investigation. To model the cross-ply behavior, two fiber and matrix cubes, each being a RVE for one-ply orientation, are arranged in such a way that a quarter fiber is parallel to the loading and one is transverse to it as shown in Fig. 5. This FEM mesh design is similar to the one used by Bigelow [*37*].

The computations are carried out using three-dimensional brick elements in the finite element code, ADINA [*38*]. For adequate resolution of stress and strain gradients in the radial direction, a 12-noded nonisoparametric element is employed with a quadratic formation in the radial direction. Linear strain fields are assumed for the hoop and longitudinal directions of the fibers. The boundary conditions for both the [0] and [90] RVE assume periodic packing, for which the geometry of the cells remains unchanged by allowing the faces of the element to move parallel to their original position.

The weak fiber/matrix interface, commonly referred to as imperfect bonding, is simulated by allowing frictionless slippage and separation between the matrix and the fiber. The interface is allowed to transfer compressive loads, but not tensile ones. A frictionless interface is assumed for simplicity because of a lack of any information on coefficient of friction. The matrix behavior is represented by the version of the Bodner-Partom model with directional hardening [*29,30*] that is incorporated into ADINA through user-defined subroutines [*11,38*].

Material Models of the Constituents

In this investigation, the SCS-6/TIMETAL 21S $[0/90]_s$ is composite system modeled. The matrix material, TIMETAL 21S (Ti-15Mo-3Nb-3Al-0.2Si), is a metastable beta titanium alloy while the SCS-6 silicon carbide fiber is a beta silicon-carbide monofilament having a

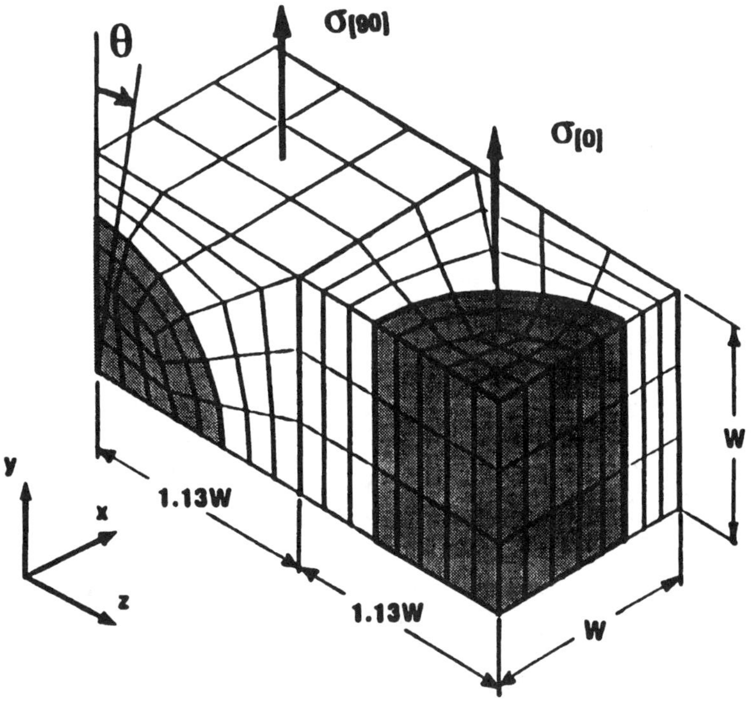

FIG. 5—*Representative volume element for the [0/90] composite employed by FEM.*

diameter of 140 μm. Details of the materials, composite fabrication, and experimental techniques are given in Ref *1*.

Except for METCAN, all models assume that the SCS-6 fiber behaves as an isotropic thermoelastic material for simplicity, with properties as given in Table 1. Using the MFIR in METCAN, the temperature-dependent properties for SCS-6 are those given in Table 2 and were obtained from the METCAN data bank (denoted as "SICA"). The modulus values given by METCAN are slightly higher than those that more closely represent current published values [*19*], but they were left as is since this is the standard database used to represent the SCS-6 fiber in METCAN.

For the matrix, three models are employed, each of which is both time- and temperature-dependent. For LISOL, FEM, and FIDEP, the Bodner-Partom model with directional hardening [*29*] is used and the material parameters for TIMETAL 21S are obtained from Ref *30* and are presented in Table 3. For VISCOPLY and ELAM, the Eisenberg-Yen model 31 is used and the parameters for TIMETAL 21S, which are provided with the code, are given in Table 4. Unfortunately, there are currently no material parameters specifically for TIMETAL 21S in the METCAN data bank. Since it was desired to include METCAN in this comparison study, the material parameters for Ti-15-3 ("TI15" in the data bank), which are the closest to TIMETAL 21S, are used and are shown as a function of temperature in Table 5. Since the MFIR does not completely capture the details of the viscoplasticity, the recalibration of material parameters for Ti-15-3 into those for TIMETAL 21S is not expected to change the predicted response significantly. The modulus at room temperature is smaller, but at 650°C the modulus of TIMETAL 21S and the METCAN parameters for Ti-15-3 are about the same.

The geometry of the micromechanics model as well as the fiber and matrix material

TABLE 1—*Thermoelastic properties of the SCS-6 fiber used by VISCOPLY, FIDEP, LISOL, ELAM, and FEM.*

Temperature Independent Material Parameters

$\nu = 0.25$

Temperature Dependent Material Parameters

Temperature, °C	Elastic Modulus, GPa	Thermal Expansion Coefficient, 10^{-6}/°C[a]
21.	393.	3.564
93.	390.	3.660
204.	386.	3.618
316.	382.	3.638
427.	378.	3.687
538.	374.	3.752
649.	370.	3.826
760.	365.	3.903
871.	361.	3.980
1093.	354.	4.103

[a]Secant CTE with $T_{ref} = 25$°C, from Ref *39*.

TABLE 2—*Temperature-dependent properties of the SCS-6 fiber used by the MFIR in METCAN.*

T, °C	*E*, GPa	ν	CTE,[a] 10^{-6}/°C	S_F, MPa
21	427.5	0.300	4.90	3447.5
93	424.6	0.298	4.93	3424.0
204	420.0	0.295	4.98	3386.8
316	415.1	0.291	5.04	3348.0
427	410.2	0.288	5.10	3308.1
538	405.1	0.284	5.17	3266.7
649	399.7	0.281	5.24	3223.7
760	394.2	0.277	5.31	3178.9
871	388.4	0.273	5.39	3132.1
1093	375.9	0.264	5.57	3031.7

[a]Tangent CTE.

models should be important in explaining the differences among the computations. Except for the matrix material for METCAN, all fiber and matrix material parameters are for SCS-6 and TIMETAL 21S, respectively. However, constants are not generally determined from the same experimental data base, so some differences arise due to this. Also, the models do not generally represent the exact experimental behavior, since each model tends to fit data somewhat differently and the fit is generally dependent on the user. To illustrate differences in the matrix models, simulations of the matrix behavior were conducted at a temperature not used to obtain the constants to reduce any bias on the results. Computational results are compared to the experimental behavior in Fig. 6 where it is seen that the behavior is very sensitive to strain rate at 566°C. This provides one condition where the capability of these models to represent the viscoplastic behavior can be evaluated critically.

Of the three models shown, the Bodner-Partom model matches these particular experiments the best. The Eisenberg-Yen model predicts the strain-rate layering quite well; however, the saturation stresses are over-predicted at all three strain rates. For the MFIR in

TABLE 3—*Bodner-Partom material parameters for TIMETAL 21S used by FIDEP, LISOL, and FEM.*

Temperature-Independent Constants

m_1, MPa^{-1}	r_1	r_2	Z_1, MPa	D_0, s^{-1}	ν
0	3	3	1600	1×10^4	0.34

Temperature-Dependent Constants

Temperature, °C	E, MPa	α,[a] 10^{-6}/°C	n	$Z_0 = Z_2$, MPa	Z_3, MPa	m_2, MPa^{-1}	$A_1 = A_2$, s^{-1}
23	**112 000**	**6.31**	**4.8**	**1550**	**100**	**0.35**	**0**
260	**108 000**	**7.26**	**3.5**	**1300**	**300**	**0.35**	**0**
315	*106 000*	*7.48*	◇[b]	◇	*390*	◇	*0*
365	*104 000*	*7.68*	◇	◇	*500*	◇	*0.00 03*
415	*101 000*	*7.88*	◇	◇	*660*	◇	*0.00 13*
465	*99 100*	*8.08*	◇	◇	*960*	◇	*0.00 50*
482	**98 100**	**8.15**	**1.7**	**1100**	**1100**	**5**	**0.00 764**
500	*97 000*	*8.22*	*1.5*	◇	*1300*	◇	*0.01 16*
525	*95 500*	*8.32*	*1.28*	◇	*1670*	◇	*0.02 03*
550	*93 900*	*8.43*	*1.1*	◇	*2100*	◇	*0.03 42*
575	*92 200*	*8.53*	*0.97*	◇	*2600*	◇	*0.05 59*
600	*72 000*	*8.63*	**0.82**	◇	**3700**	**10**	*0.08 87*
650	**86 600**	**8.83**	**0.74**	**1000**	**3800**	**10**	**0.20 8**
760	**77 200**	**9.27**	**0.58**	**600**	**4000**	**15**	**1.01**
815	**72 000**	**9.49**	**0.55**	**300**	**4100**	**30**	**1.97**

[a]Secant CTE with $T_{ref} = 25$°C.
[b]◇ = linear interpolation between values given in table (No experimental data was available).
NOTE—Bold are values determined based on experiments.
Italics are values that describe smooth and continuous parameter changes with temperature.

TABLE 4—*Eisenberg-Yen material parameters for TIMETAL 21S used by VISCOPLY and ELAM.*

Temperature Independent Material Properties

$\nu = 0.34$

Temperature Dependent Material Properties[a]

Temperature, °C	Elastic Modulus, GPa	CTE,[a] 10^{-6}/°C	Yield Strength, MPa	k,[b,c] MPa^{-p}/s	p[c]	A^*,[d] MPa	n^*[d]
21	116	8.62	910	2.10×10^{-21}	8.59	946	0.0145
150	116	8.62	910	2.10×10^{-21}	8.59	946	0.0145
316	104	9.23	640	2.92×10^{-16}	6.52	679	0.0350
427	103	9.89	600	1.66×10^{-18}	5.62	666	0.0350
482	96	10.26	472	5.38×10^{-13}	6.35	531	0.321
566	85.8	10.97	190	8.10×10^{-17}	5.1	231	0.117
621	80.14	11.5	85	6.05×10^{-13}	3.58	105	0.136
650	64.39	11.79	50	2.20×10^{-12}	3.46	61.8	0.0661
704	53.0	12.20	20	2.76×10^{-10}	3.00	29.8	0.140
760	49.9	12.48	7.5	1.00×10^{-8}	2.4	11.1	0.132

[a]Tangent CTE.
[b]Determined at $\dot{\sigma} = 2.56$ MPa/s.
[c]Constitutive equations, $\epsilon^{in} = k(H)^p$, where H is "over-stress."
[d]Equilibrium stress-inelastic strain curve, $\sigma^* = A^* (\epsilon^{in})^{n^*}$.

TABLE 5—*Temperature-dependent properties of Ti-15-3 used by the MFIR in METCAN.*

T, °C	E, GPa	ν	CTE,[a] 10^{-6}/°C	S_F, MPa
21	84.8	0.320	8.10	896.4
150	78.9	0.298	8.70	834.1
316	70.6	0.266	9.73	746.3
427	64.5	0.243	10.66	681.3
482	61.2	0.231	11.23	646.7
566	55.8	0.211	12.31	589.9
621	52.0	0.196	13.21	549.5
650	49.9	0.188	13.78	527.0
704	45.6	0.172	15.05	482.3
760	40.8	0.154	16.85	431.0
815	35.4	0.133	19.42	373.9

[a]Tangent CTE.

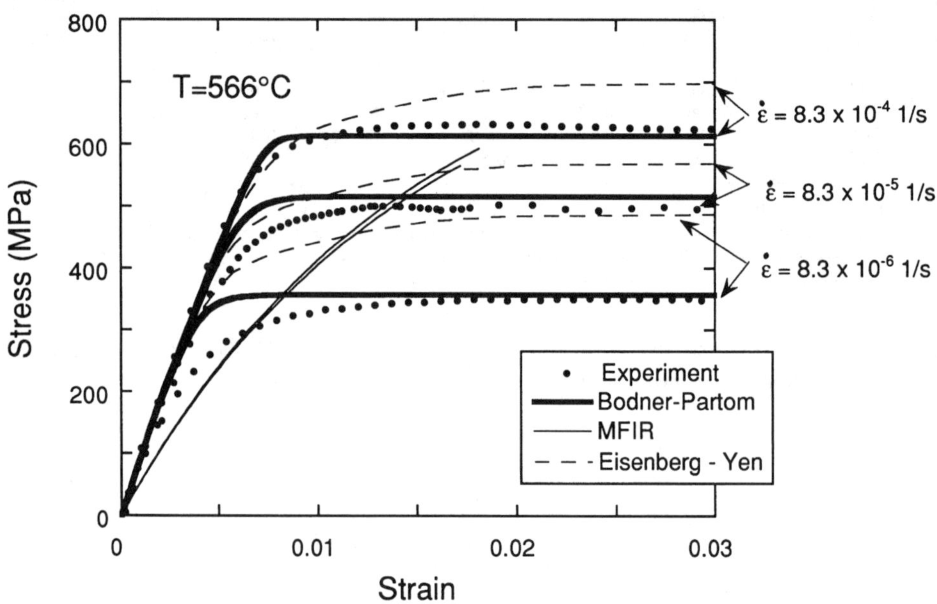

FIG. 6—*Strain-rate sensitivity of TIMETAL 21S at 566°C.*

METCAN, two stress rates were run: one that results in a strain rate near the minimum and one near the maximum. The MFIR does not show a sharp yield point and displays very little time-dependent behavior compared to the other models. The MFIR matches the experiments less accurately as deformation increases and significant inelastic deformation occurs.

Results

Each micromechanical approach described earlier was configured into a [0/90] composite model with a fiber-volume content of 0.35. The latest version of each code was run without modifications. In the cases of FIDEP, LISOL, and FEM, the code is used by the developer who is expected to be most familiar with all of the features and intracacies of each code.

ELAM is a simplified procedure in which results are obtained using VISCOPLY. In the cases of VISCOPLY and METCAN, the code was run based on available documentation and some input from the developers. Most of these codes can be modified to improve the results in various ways, such as incorporating some type of damage or degradation with time or cycling, so that others may be able to obtain slightly different results than shown here, particularly for VISCOPLY and METCAN.

Comparisons among the codes are limited to a number of uniaxial loading conditions of a [0/90] composite. For each loading condition, the predicted stress-strain response of each code is compared to experimental data. In addition, the average values for the [0] fiber stress, the [0] matrix stress, and the [90] stress in the direction of the applied load are examined for each model. Even though off-axis behavior is also of interest, this topic is not considered because it is beyond the scope of this paper.

Thermal Residual Stress

Before any of the cyclic analyses were conducted, the processing history was applied to generate the initial residual stress state in the composite. The processing history includes the cooling from an assumed consolidation temperature (900°C) as well as a hold at an intermediate temperature. The resulting room temperature residual stresses from each method are presented in Fig. 7. Since the CTE of the fiber is smaller than that of the matrix, and assuming a stress-free condition at the consolidation temperature, the fiber stress at room temperature is compressive while the matrix and [90] ply stresses are tensile. Four of the models predict a state of residual stress within the [0/90] composite within a narrow range of values, while METCAN and ELAM produce somewhat different values. The ELAM model shows that by representing the [90] ply as a matrix material, it tends to predict higher [90] ply stresses, thus, reducing the associated fiber and matrix stress within the ELAM model. FEM and the modified method of cells (LISOL) produce results that are

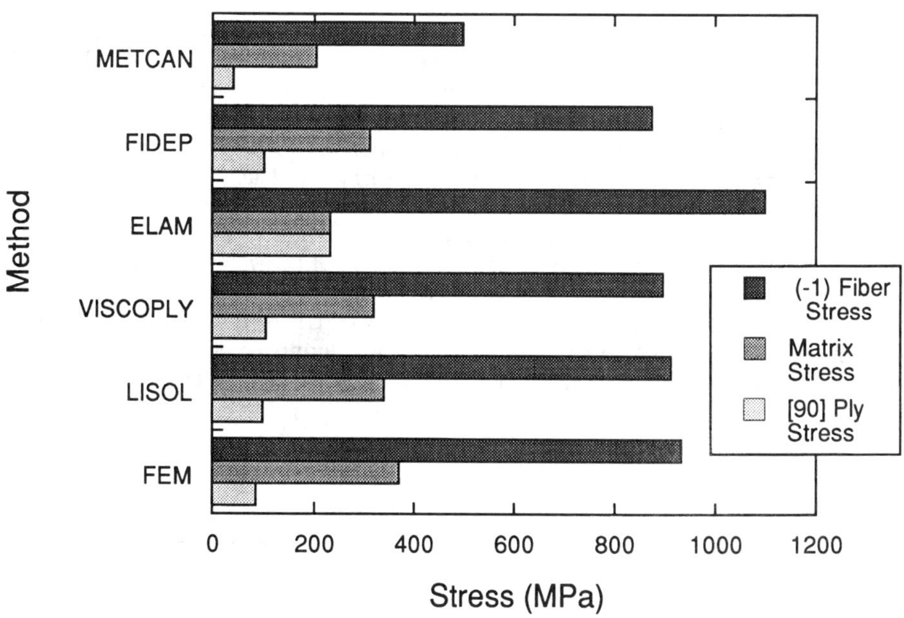

FIG. 7—*Residual stresses in the [0/90] composite at 25°C.*

fairly close to those provided by VISCOPLY. The results from FIDEP, which considers three-dimensional stresses, are nearly identical while METCAN predicts somewhat lower ply stresses and, consequently, lower fiber and matrix residual stresses. These trends are similar to those observed by Bigelow et al. [*13*].

Cyclic Behavior

To examine how well the models predict different types of cyclic behavior, simulations of four tests were performed with each model. A summary of the four cases is given in Table 6. All these represent loading in the longitudinal direction under stress-controlled mode. In all cases, the processing history is applied to simulate the initial composite residual stress state. Each case illustrates various capabilities of the models. The two isothermal fatigue cases compare how the models predict the effect of frequency on the viscoplastic response of the composite. The applied stress chosen is sufficient to produce inelastic strains in the composite, making comparisons between the models easier to discern. Since the stress ratio is positive, some strain ratchetting occurs with continued cycling, thus, ten cycles are simulated with each method in order to compare the capabilities of the codes to predict the ratchetting. Two thermomechanical fatigue loading conditions are used to examine the behavior when both mechanical and thermal strains are generated during a cycle. Two distinct phasings between load and temperature are examined: in-phase (maximum stress occurs at maximum temperature) and out-of-phase (maximum stress occurs at minimum temperature). Under in-phase TMF, ratchetting generally occurs, whereas under out-of-phase TMF, the response is nearly stable after the first cycle. The models are examined to see if they can correctly predict the TMF behavior under these two distinct phasings.

Isothermal Fatigue–Low Frequency—The first cycle of isothermal fatigue at 650°C and 0.01 Hz is predicted for each method and the results are shown in Fig. 8. First, the stress-strain behavior of the composite during initial loading and, in particular, the magnitude of inelastic strain in the first cycle is examined. The basic shape of the initial loading curve of the first cycle shows generally good agreement among all the solutions. All the methods predict the initial modulus quite well. The unloading portion of the curve depicts the permanent strain determined from each method. The METCAN approach displays the least amount of permanent strain, although it displays the greatest nonlinearity during loading that matches the experimental data quite well. On the contrary, LISOL, ELAM and FEM predict the most permanent strain, consistent with the experimental data. VISCOPLY and FIDEP predict progressively smaller amounts of permanent strain.

The ability of the models to predict strain ratchetting can be evaluated by comparing the response at Cycle 10 (Fig. 9). Of all the methods, FIDEP displays the least amount of strain ratchetting, while METCAN produces no ratchetting. LISOL is able to match the experiments the best while FEM, ELAM, and VISCOPLY produce successively smaller amounts of permanent strain after ten complete cycles. During the cycling that produces the strain ratchetting, there is a redistribution of stress within the composite such that the matrix

TABLE 6—*Summary of cyclic loadings.*

Case	Loading	S_{max}	Stress Ratio	Temperature	Frequency
1	isothermal	570	0.1	650	0.01
2	isothermal	570	0.1	650	1.0
3	in-phase TMF	500	0.1	150 to 650	0.005 56
4	out-of-phase TMF	500	0.1	650 to 150	0.005 56

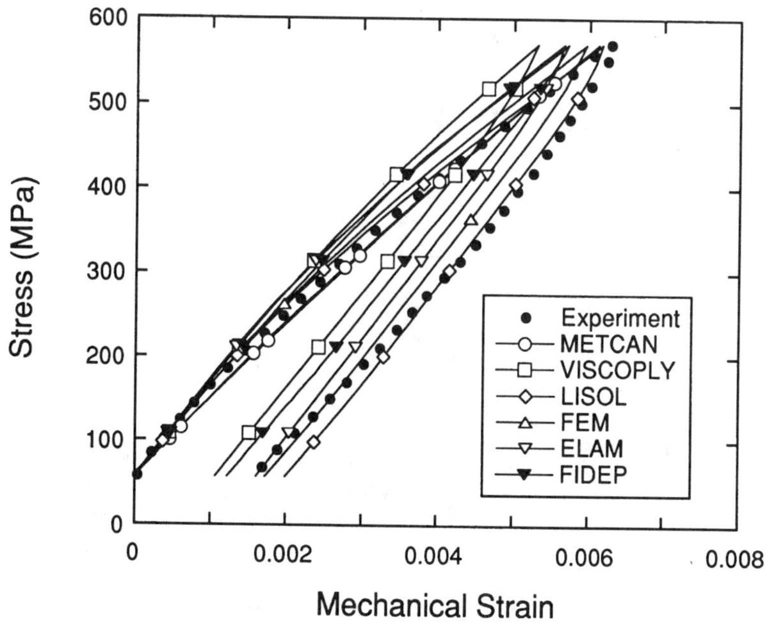

FIG. 8—*First cycle of isothermal fatigue at 650°C and 0.01 Hz.*

stress decreases while the fiber stress increases. This process of cyclic stress relaxation in the matrix can be illustrated by examining the maximum stress in the matrix and the [90] lamina at Cycle 10 as shown in Figs. 10 and 11, respectively. The corresponding increase in fiber stress is shown in Fig. 12. While four of the methods, FEM, LISOL, ELAM, and

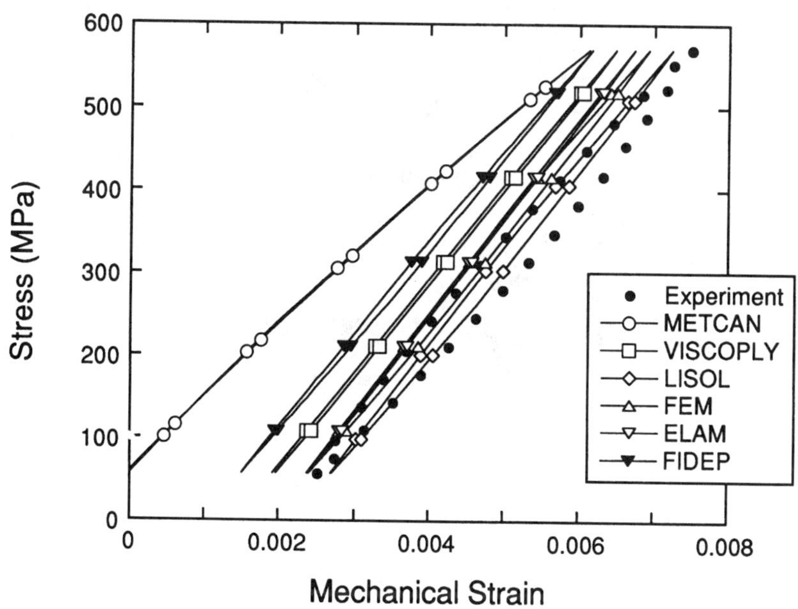

FIG. 9—*Tenth cycle of isothermal fatigue at 650°C and 0.01 Hz.*

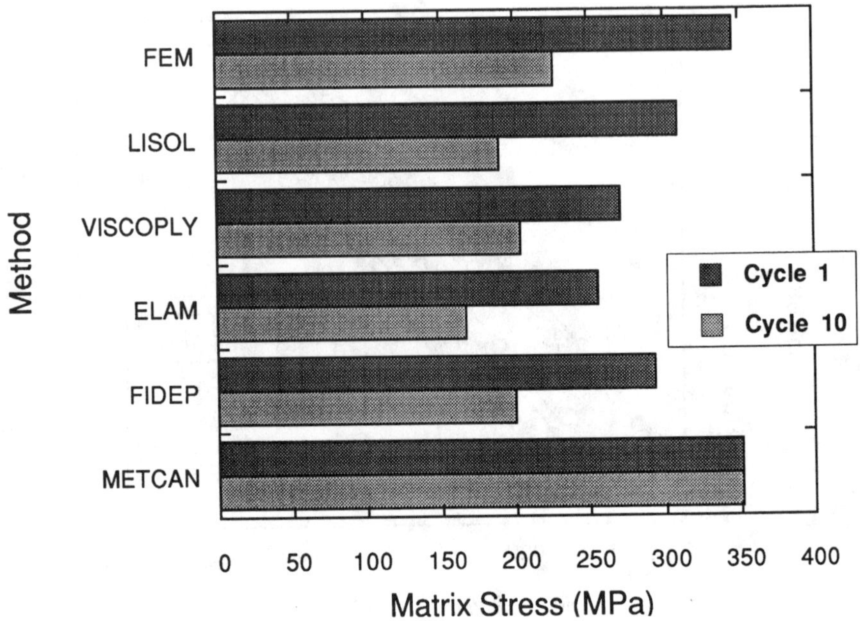

FIG. 10—*Peak matrix stress at Cycles 1 and 10 of isothermal fatigue at 650°C and 0.01 Hz.*

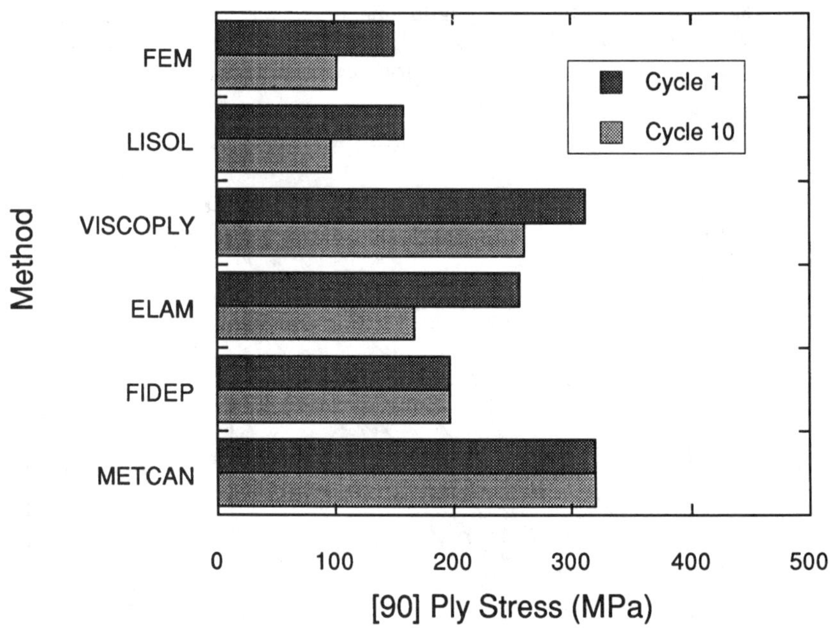

FIG. 11—*Peak [90] ply stress at Cycles 1 and 10 of isothermal fatigue at 650°C and 0.01 Hz.*

VISCOPLY produce comparable results for accumulated strain, these methods predict a significant amount of variation in calculated stresses among them, particularly with regard to [90] lamina stresses as shown in Fig. 11. The observations noted earlier regarding strain ratchetting are further demonstrated by examining the fiber stresses as shown in Fig. 12. Since the fiber is elastic, the fiber stress in a load-controlled test is a measure of the accumulated strain. Figure 12 shows that all the methods predict ratchetting with the exception of METCAN, while FIDEP produces only a small amount of permanent strain or change in fiber stress.

Isothermal Fatigue–High Frequency—The capability of the models to predict time-dependent or frequency effects can be seen by comparing the results from simulations of two isothermal fatigue tests conducted at different frequencies. In comparison to the previous results obtained at a frequency of 0.01 Hz (Fig. 8), the predictions of the first cycle of isothermal fatigue at a higher frequency of 1 Hz are shown in Fig. 13. At the higher frequency, there is less inelastic deformation, thus, all the models tend to predict the higher frequency response better than the 0.01-Hz response. As seen in Fig. 13, FIDEP matches the experimental data the best, while FEM and LISOL produce results that are almost identical to each other and predict permanent strains that are slightly less than the experimental values. It should be noted that the representation of the [90] ply in FIDEP is an elastic-plastic fit to experimental data for uniaxial response of a [90] layup, thus, the excellent fit is not surprising. The other three models, METCAN, VISCOPLY, and ELAM, do not capture any measurable inelastic strain on the first cycle at this frequency. Even though the 1-Hz response produces much less inelastic strain than the 0.01-Hz cycle, a significant amount of strain ratchetting is still observed after ten cycles, as shown in Fig. 14. The nature of the ratchetting is the same as that observed at the lower frequency, and the correlation with experimental data ranks approximately the same as in that case. FEM and LISOL predict the

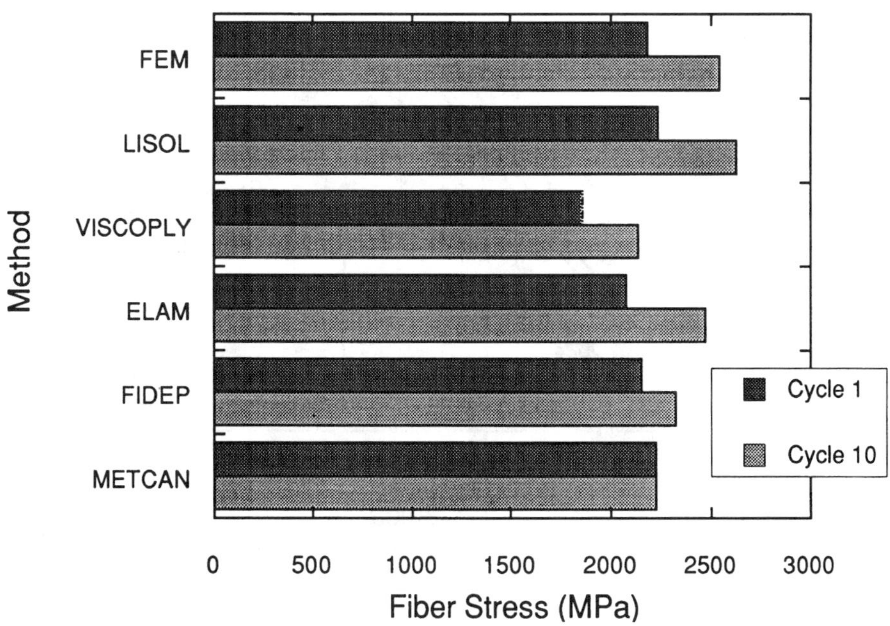

FIG. 12—*Peak fiber stress at Cycles 1 and 10 of isothermal fatigue at 650°C and 0.01 Hz.*

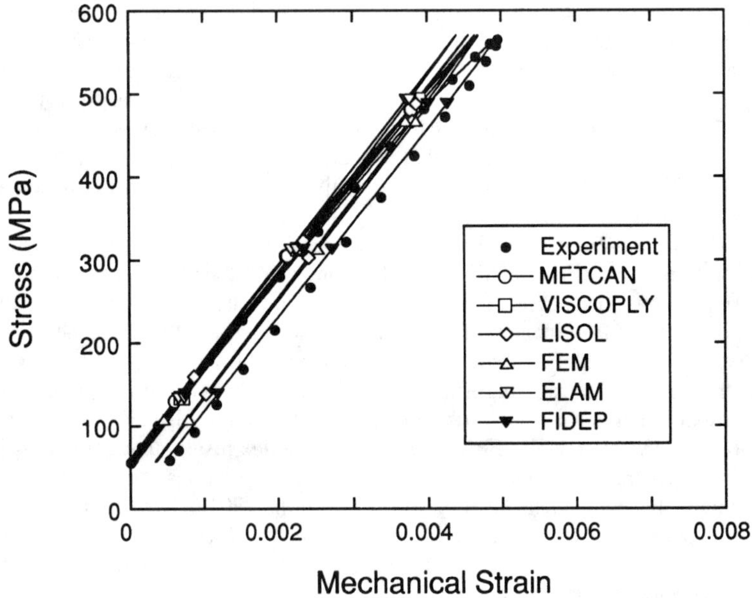

FIG. 13—*First cycle of isothermal fatigue at 650°C and 1 Hz.*

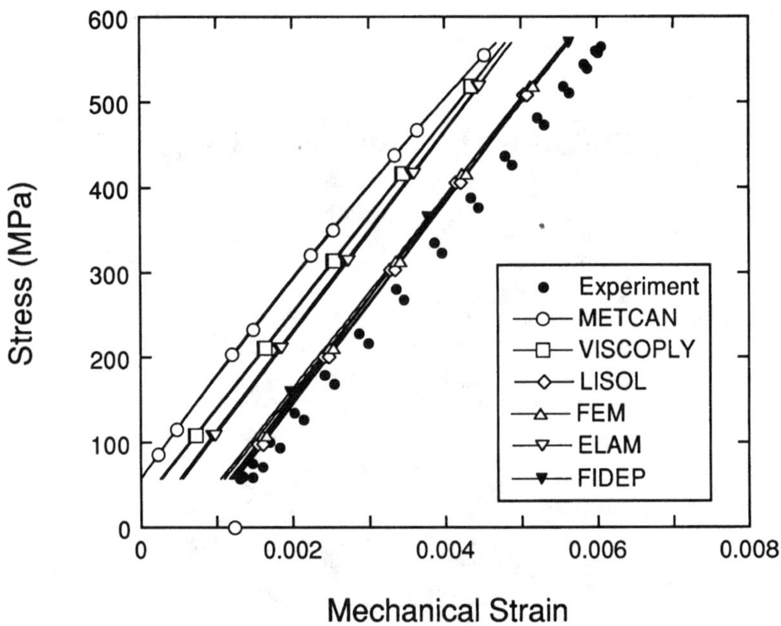

FIG. 14—*Tenth cycle of isothermal fatigue at 650°C and 1 Hz.*

most strain ratchetting, while METCAN does not display any permanent strain. Of the remaining approaches, VISCOPLY produces the least amount of permanent strain compared with the experimental data, although VISCOPLY, FIDEP, and ELAM predict responses that are very similar to each other.

The effect of frequency on the model responses can also be evaluated by comparing the stresses in the matrix, [90] ply, and fiber at Cycle 10, when a steady state has been nearly reached. The difference in the matrix stress between 1 and 0.01 Hz for the different models is quite significant (Fig. 15). The two methods that best match the experimental stress-strain behavior, FEM and LISOL, predict a reduction in the matrix stress by nearly a factor of two when reducing the frequency from 1.0 to 0.01 Hz. This demonstrates that the additional time in a low-frequency cycle produces more stress relaxation in the matrix. METCAN, in contrast, predicts that the matrix stress decreases slightly at the higher frequency. Likewise, going from 1 to 0.01 Hz, the stress in the [90] ply decreases (Fig. 16) and the fiber stress increases (Fig. 17) for all the models, except METCAN.

In-Phase TMF—The thermomechanical fatigue (TMF) cases were chosen to evaluate the nonisothermal capabilities of each method. The thermomechanical loading consists of triangular wave forms with temperature and mechanical load changing simultaneously. The predicted and experimental stress-strain responses during the first cycle of in-phase TMF are shown in Fig. 18. The composite mechanical strain, as shown, is defined as the composite total strain minus the composite thermal strain. The composite mechanical strain is plotted for convenience so that the effective composite stiffness can be seen. The figure shows that the effective stiffness is continuously changing because of temperature dependency of the material properties. The models all predict the effective stiffness fairly well, except for ELAM, which predicts a slightly higher stiffness than the others. As in the previous comparisons for isothermal fatigue, METCAN and FIDEP display a minimum amount of

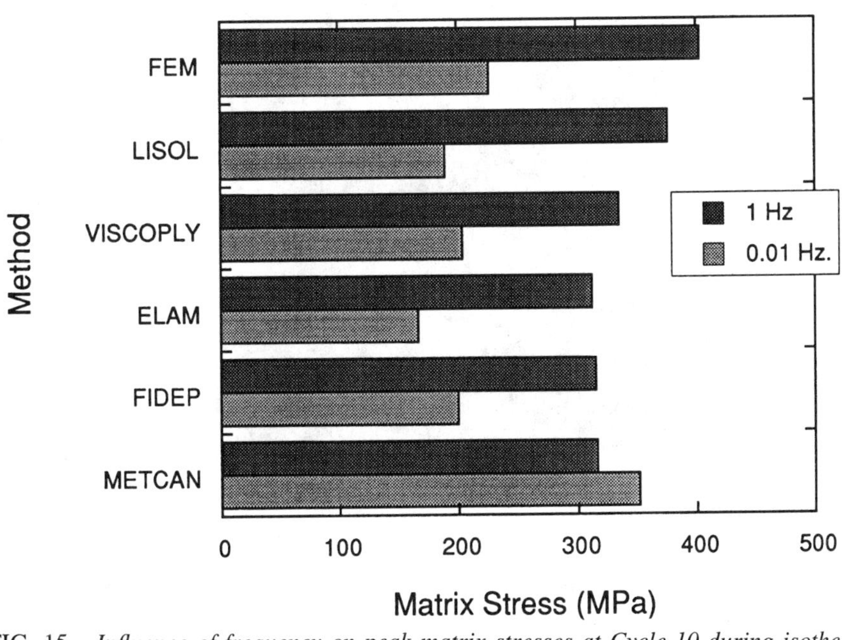

FIG. 15—*Influence of frequency on peak matrix stresses at Cycle 10 during isothermal fatigue at 650°C.*

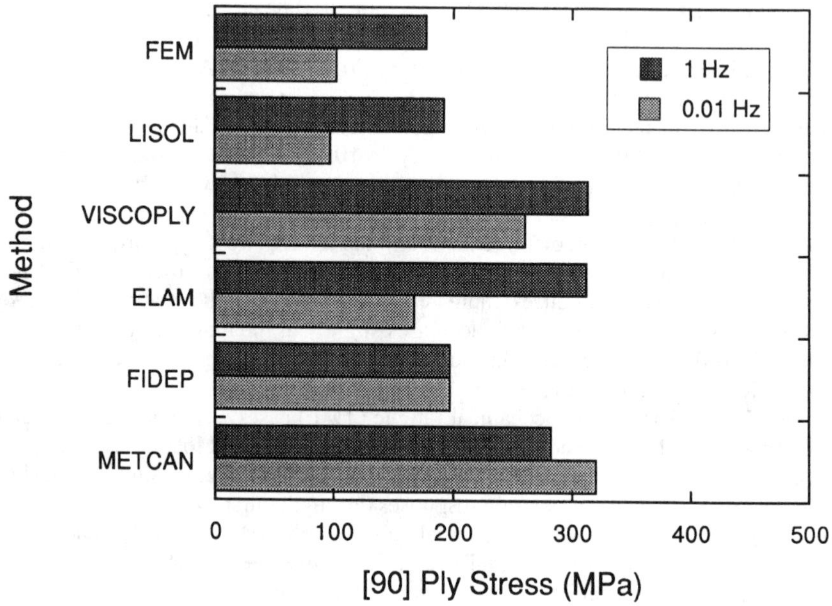

FIG. 16—*Influence of frequency on peak [90] ply stresses at Cycle 10 during isothermal fatigue at 650°C.*

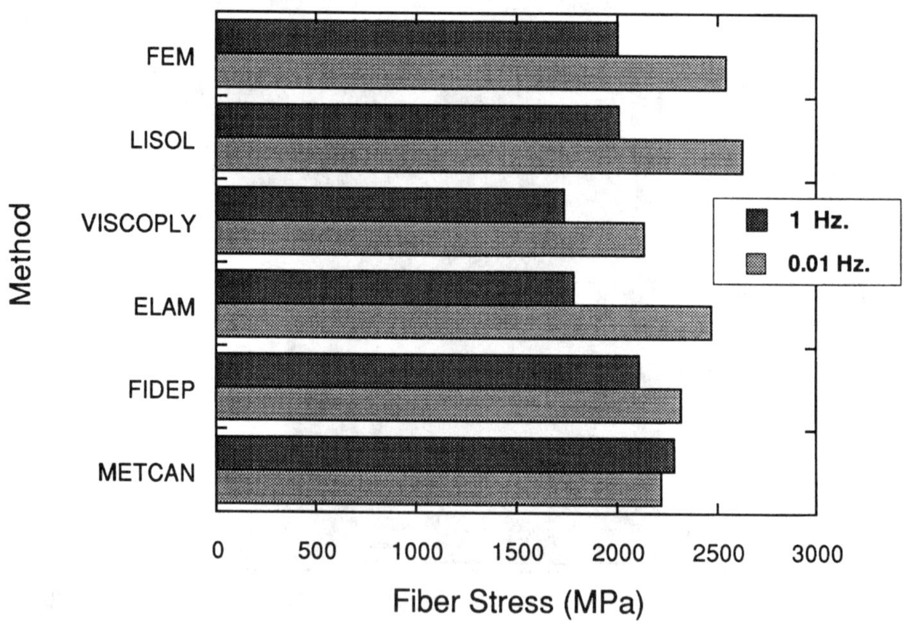

FIG. 17—*Influence of frequency on peak fiber stresses at Cycle 10 during isothermal fatigue at 650°C.*

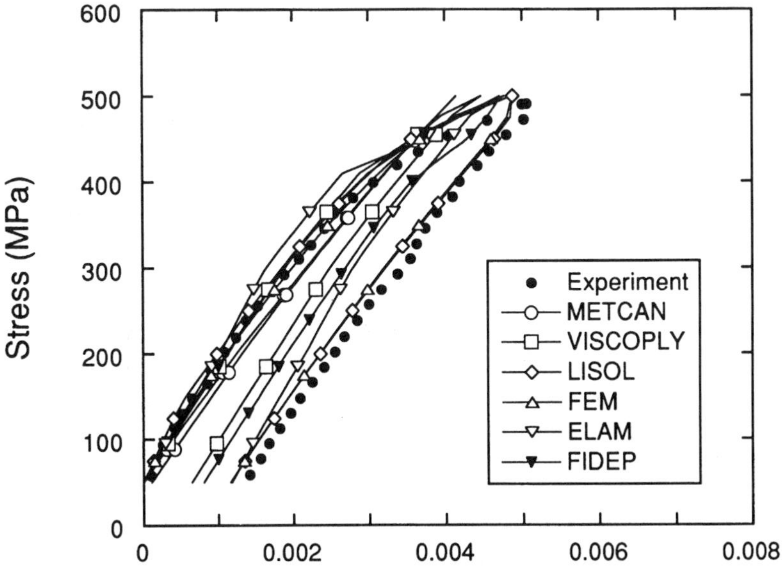

FIG. 18—*First cycle of in-phase TMF conducted from 150 to 650°C.*

permanent strain, while VISCOPLY predicts less permanent strain than the remaining models and the experimental results. LISOL, ELAM, and FEM match the experiment the closest, although ELAM, which does not incorporate damage in any fashion, produces the stiffest behavior upon unloading.

Under in-phase TMF, significant strain ratchetting is observed in the experiments where the composite accumulates a permanent mechanical strain of approximately 0.003 by Cycle 10 (Fig. 19). The prediction of the accumulated strain or ratchetting is much lower than experimentally observed for models that do not predict sufficient inelastic strain on the first cycle. The prediction of the ratchetting among the models follows the same trends observed in isothermal fatigue cases, with FEM and LISOL providing the best correlation with experimental data. ELAM predicts more strain than VISCOPLY, but both produce strains that are much lower than experimentally observed. None of the models predict as much strain as is observed experimentally, although some of this additional strain may be attributed to fiber breakage [*40*], which is not considered in any of the models. In addition to stress-strain behavior, the trends among the models in stress relaxation of the matrix and [90] ply along with the increase in fiber stress are similar to that seen in isothermal fatigue. The fiber stress at the maximum applied stress for Cycles 1 and 10 is illustrated in Fig. 20. The fiber stress should increase with cycling as strain accumulates because the amount of fiber stress ratchetting is directly proportional to the amount of composite strain ratchetting. This is clearly seen in comparing Fig. 20 with the strain ratchetting results of Fig. 19, where FEM and LISOL predict the maximum effect of strain ratchetting.

The prediction of matrix stresses and [90] ply stresses is more complex in thermomechanical fatigue than for the isothermal case since the applied load is changing simultaneously with temperature and, therefore, the properties of the constituents are changing continuously during cycling. For in-phase TMF, the matrix and fiber stresses increase with increasing load. At the same time, as temperature increases, the compressive thermal stresses in the fiber and the tensile matrix stresses are reduced. The combined effect of

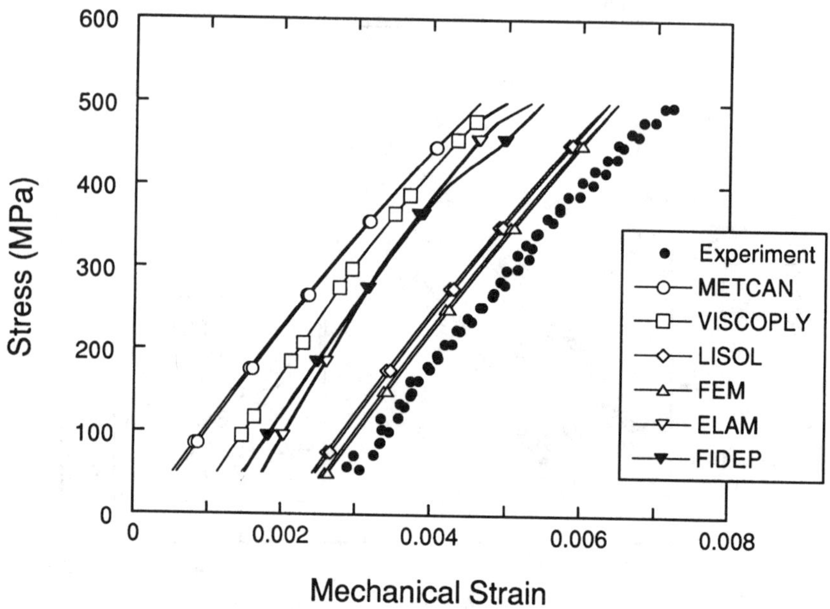

FIG. 19—*Tenth cycle of in-phase TMF conducted from 150 to 650°C.*

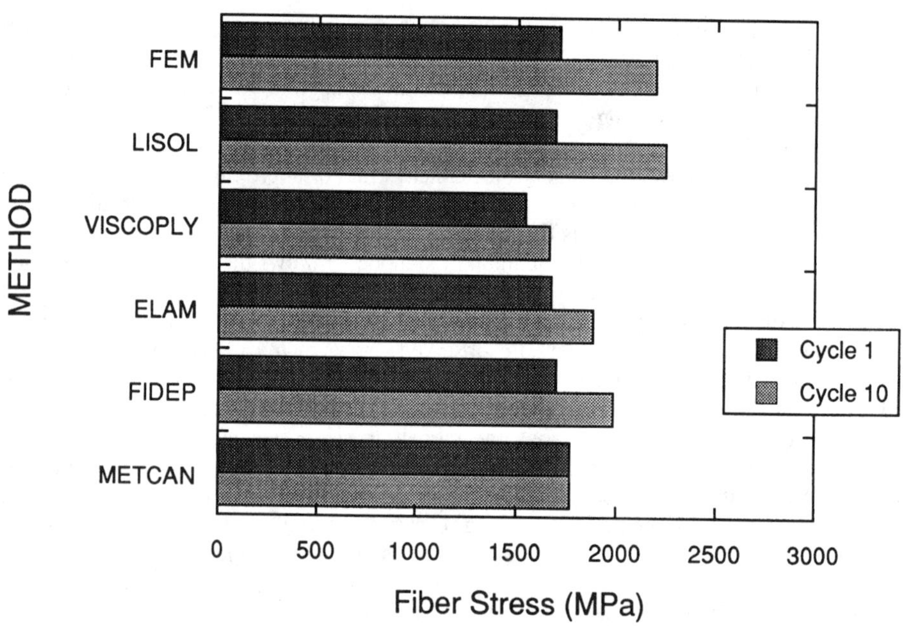

FIG. 20—*Peak fiber stress at Cycles 1 and 10 of in-phase TMF.*

thermal and mechanical loading on the matrix stress is shown for the first 180-s cycle in Fig. 21. The initial values at $t = 0$ are obtained from the cool-down stress calculations. Four of the models show an increase in stress during loading followed by a reduction in stress before the maximum load is reached at $t = 90$ s. This is attributed to the matrix reaching a stress level at which significant inelastic deformation occurs. This stress level decreases with increasing temperature. Only METCAN and FIDEP do not display a reduction in matrix stress on loading at higher temperatures. During the unloading portion of the cycle, the six models predict very similar trends in matrix stress with time. The major difference is in the starting point at maximum stress that varies considerably among the models. This, in turn, is influenced substantially by the starting values of matrix stress at $t = 0$ that result from the cool-down calculations.

The stress response of the [90] ply is a more sensitive measure of the effect of fiber/matrix separation than can be observed in the composite (see Fig. 22). During loading, METCAN and VISCOPLY exhibit no strong indications of yielding that the other four models apparently show. With the exception of ELAM, which starts with a much higher cool-down stress in the [90] ply, all of the models show similar trends during the initial loading. The deviations appear to involve the onset of yielding or damage or both in LISOL and FEM. During the unloading portion of the cycle, all of the models show an almost monotonic decrease in stress with time. The exception is FIDEP that displays a slight increase during the initial unloading, probably attributable to the time-independent representation of the [90] ply. Another difference is the slope of the curves for FEM and LISOL compared to the other four models during the unloading portion of the cycle. This lower slope may be due to a combination of the viscoplastic behavior of the matrix and the method of representing fiber/matrix separation in these two models.

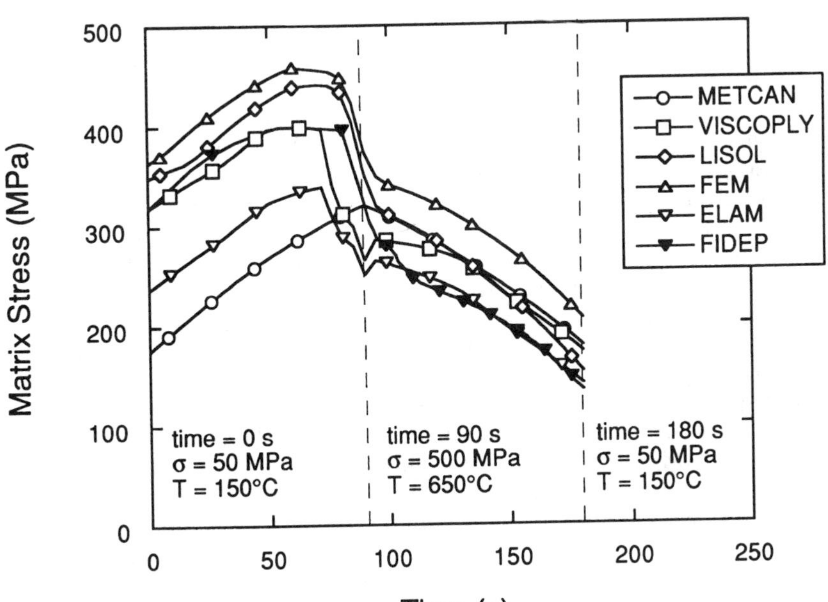

FIG. 21—*Matrix stress in first cycle of in-phase TMF.*

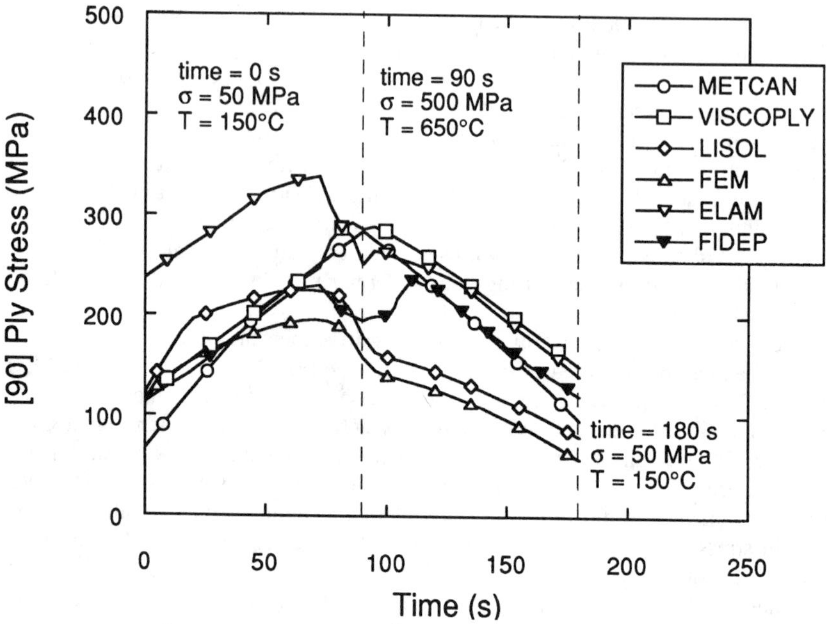

FIG. 22—*[90] ply stress during first cycle of in-phase TMF.*

Out-of-Phase TMF—Under out-of-phase TMF, the combination of mechanical and thermal effects tends to produce relatively low fiber stresses, high matrix, and [90] ply stresses. This generally produces cracks that initiate at the fiber/matrix interface in the [90] ply and grow in the matrix; thus, the mechanical response is significantly different than the other cases studied. The predicted laminate stress-mechanical strain response after ten cycles of out-of-phase TMF, along with experimental data, are shown in Fig. 23. None of the methods predict ratchetting, which is in agreement with experiments. Except for METCAN and ELAM, the predicted effective stiffnesses also match the experiments quite well. The METCAN prediction is too complaint, whereas the ELAM prediction is too stiff. However, the computed matrix stresses from these two models produce lower values at peak load than the other models during a single cycle as seen in Fig. 24. The lower stress in METCAN is consistent with its tendency to produce less stiff behavior in the composite in other simulations, however, the lower stress in ELAM is not consistent with its generally stiffer composite behavior. A more compliant composite has larger total strains, thus the fiber strains and stresses, which are elastic, should be larger. This requires that the combination of matrix stress and [90] ply stress should be lower. The [90] ply stresses over one complete TMF cycle are shown in Fig. 25. Except for ELAM, the ply stresses at maximum load are approximately equal. Thus, the low stiffness of the composite in METCAN, Fig. 23, should produce a higher matrix stress, as demonstrated in Fig. 24, since the ply stress is no different. For ELAM, however, a stiffer composite should produce higher stresses in the combined matrix and [90] ply. Since the matrix stress is low, Fig. 24, the [90] ply stress must be much higher, as shown in Fig. 25. It is clear that the representation of the [90] ply by pure matrix material in ELAM produces a [90] ply stiffness much higher than the other models under out-of-phase TMF loading. METCAN, on the other hand, appears to model the matrix material in the [0] ply as being stiffer than in the other models.

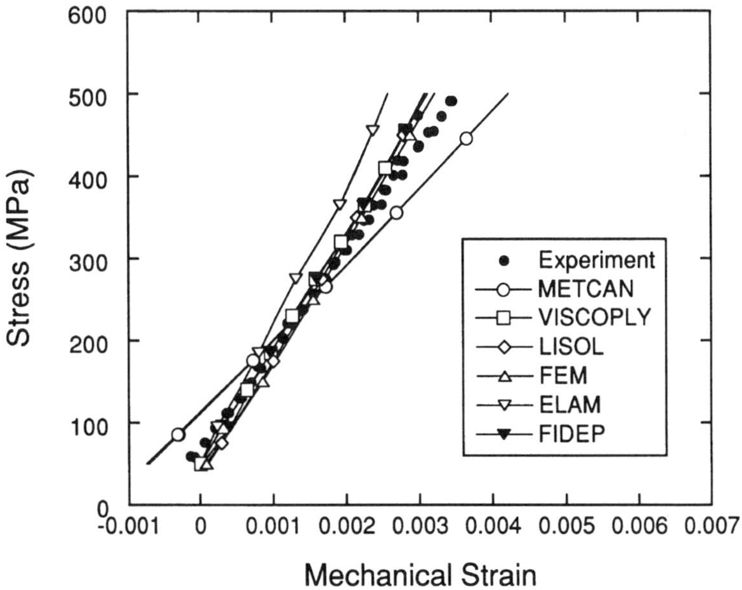

FIG. 23—*Tenth cycle of out-of-phase TMF conducted from 150 to 650°C.*

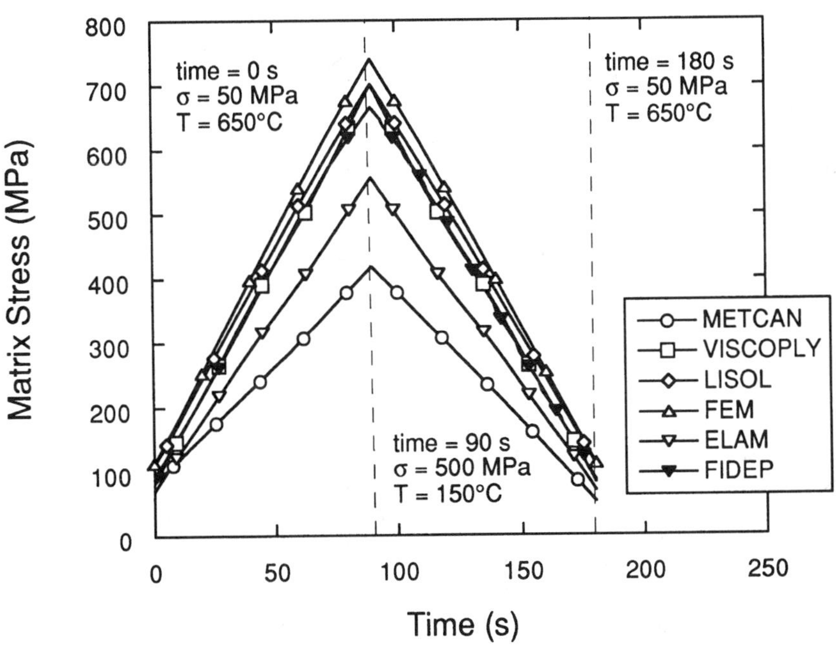

FIG. 24—*Matrix stress in first cycle of out-of-phase TMF.*

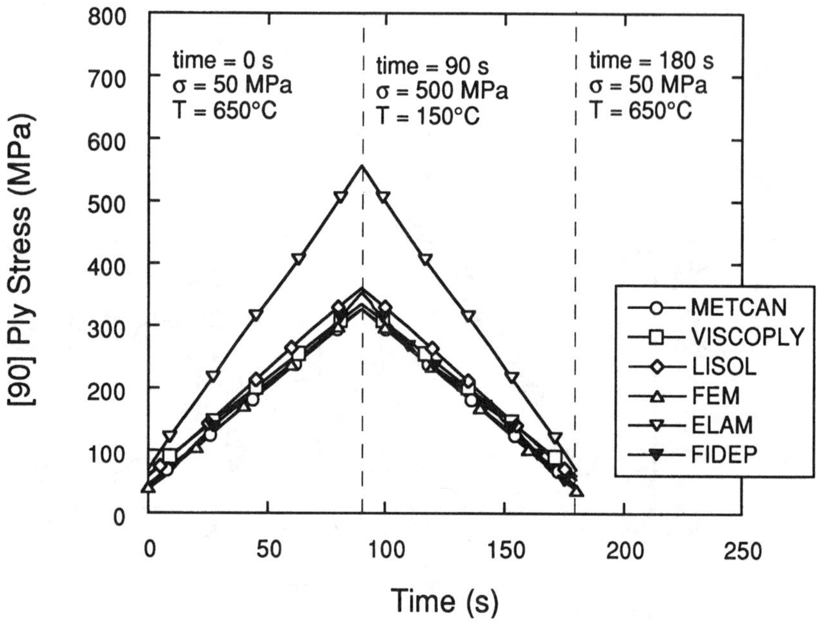

FIG. 25—*[90] ply stress in first cycle of out-of-phase TMF.*

Discussion

Three features of the models used to analyze a [0/90] composite are discussed. These are (1) the geometry of micromechanical models, (2) the constitutive models used for the matrix and fiber, and (3) the treatment of the [90] ply. In addition, the accuracy, efficiency, and availability of the models used in this paper are discussed. While the results herein pertain to a specific composite material, the comparisons and conclusions of this investigation are anticipated to be applicable to any high temperature metal matrix composite exhibiting thermoviscoplastic matrix behavior, elastic fiber response, and a weak fiber-matrix interface.

Model Geometry

One important aspect of a micromechanical model is the physical representation of the composite, which is typically based on a RVE consisting of fiber and matrix. The fiber array can be square, rectangular, hexagonal, or other, but for these comparisons a square array is considered sufficient and is used as the basis of all of the models. Of the various methods, FEM is considered to have the most accurate representation of an RVE with a nearly circular representation of the fiber. VISCOPLY does not require a specific representation of a fiber geometry. Rather, the behavior is represented by separate equations for the [0] and [90] plies. Both METCAN and LISOL use a square fiber in both plies, which does not appear to be a serious deficiency, considering that the LISOL results are typically very close to those of FEM. This indicates that the RVE for METCAN, which also uses a square fiber, is not the reason for the poor prediction of inelastic strains compared to experiments. The concentric cylinder model, which uses a circular RVE, has been shown to provide good correlation with results from FEM for axial loading in a unidirectional composite [*34*]. Therefore, the somewhat different results from FIDEP compared to FEM, should not be attributed to the circular cross section of the RVE for the [0] ply. Instead, the treatment of the [90] ply is significant in describing the differences among the models and is discussed later.

Matrix Constitutive Model

Another critical aspect is the modeling of the viscoplastic behavior of the matrix, especially in TMCs. A viscoplastic model is needed to capture the observed inelastic strain rate effects and ratchetting during fatigue. In this investigation, both the Bodner-Partom (B-P) and Eisenberg-Yen (E-Y) models appear to accomplish this task fairly well, as evidenced by the strain ratchetting and strain-rate sensitivity predictions of FEM and LISOL, which use B-P, and VISCOPLY, which uses E-Y. The MFIR of METCAN, as formulated for this investigation, does not include a capability to produce significant inelastic strains and viscoplastic behavior and therefore transient ratchetting and strain-rate sensitivity are not well predicted. METCAN can simulate ratchetting in another way through the use of degradation factors in the MFIR, which are dependent on time and cycles. However, the degradation behavior must be known a priori in order to determine the constants. This feature was not exercized in these model comparisons.

Treatment of the [90] Ply

Of the models considered here, the FEM is considered to provide the most realistic physical representation of the fiber/matrix interface. With this model, the interface is treated as frictionless and allows for progressive fiber/matrix separation. LISOL incorporates a hybrid contact element that can model nonlinear response of the fiber-matrix interface and, thus, can model progressive separation as an equivalent deformation process. A simpler method of treating the interface, employed by VISCOPLY and METCAN, involves reducing the fiber or interphase stiffness and strength properties. FIDEP uses a bilinear elastic-plastic fit to FEM results for a [90]. The time-independent representation for the [90] ply in FIDEP limits the capability of the model to capture all of the features of the other models and hampers its predictive capability.

The ELAM model is introduced into this study to evaluate how the predictions of the various models are influenced by the different representations of fiber/matrix separation. ELAM provides a simple baseline where the interfacial separation or sliding in the [90] ply is not accounted for. The results from ELAM generally produce stiffer composite behavior than the other models and predict less ratchetting compared to other models that also represent the matrix with a viscoplastic model. The amount of fiber ratchetting is very close to FEM, especially for the isothermal fatigue, and even greater than VISCOPLY, which was used to run the ELAM simulations. ELAM deviates for in-phase TMF where it predicts a [90] ply stress nearly twice that of all the other models (Fig. 22). For out-of-phase TMF, ELAM appears to underpredict the matrix stress in the [0] ply (Fig. 24), which results in a higher fiber stress and stiffer composite behavior (Fig. 23). The ability to represent fiber/matrix interfacial separation appears to be important in most of the simulations, especially for predicting the correct stiffness.

Accuracy

When the RVE, constituent behavior and [90] ply behavior are accurately modeled, the predictions match experimental data quite well. The best overall correlations with experimental data are provided by FEM and LISOL. Both FEM and LISOL have an adequate RVE, use an accurate viscoplastic matrix model, and realistically represent fiber/matrix separation in the [90] ply. Other models have limitations or differences in one or more of these features that produce predictions that differ from each other as well as from experimental results. METCAN uses an RVE similar to LISOL, but the matrix constitutive model appears to limit the capabilities of this model to reproduce all of the experimental observations. A limitation in VISCOPLY appears to be the method of treatment of the fiber/matrix separation in the [90] ply. The viscoplastic matrix model and RVE do not appear to limit the capability of this model. The RVE of FIDEP is also adequate, but the treatment of the [90] ply as an elastic-plastic material with no rate dependence, and no specific treatment of fiber/matrix separation, causes FIDEP to produce results that differ from the other models and the experimental data. ELAM treats the [90] ply as pure matrix and incorporates a very simple RVE, yet this method produces adequate results in some cases, primarily because of a realistic viscoplastic matrix model.

The viscoplastic model used in the matrix is extremely important in accurately predicting different cyclic behavior. The differences in the results among the various models seems to be less sensitive to the RVE used. In addition, FEM, LISOL, and to some extent ELAM best display the stress ratchetting (see Figs. 9 and 19) and strain-rate effects (see Fig. 18). The common feature of these models is a continuous load-carrying matrix ligament in the [90] ply. This is very important for cases where the fiber/matrix interface is weak. The results of this investigation indicate that accurate modeling of a [0/90] composite requires an accurate viscoplastic representation of the constituents and realistic physical representation of the [90] ply behavior.

Another key aspect in assessing the accuracy of various models is the sensitivity of the results to the intended use. One of the pieces of information obtained from the models, which cannot be measured experimentally, is the micromechanical stress state in the constituents. Of primary concern in predicting fatigue life is the maximum stress or stress range in the fiber. Consider the case of isothermal fatigue at the low frequency of 0.01 Hz, the results of which are shown in Fig. 12. Among all of the methods, the maximum fiber stress prediction at Cycle 10 ranged from 1925 to 2627 MPa, which is a difference of 702 MPa. It has been found that small differences in fiber stress, as little as 100 MPa, can greatly influence fatigue life prediction, especially in cases where the mode of failure is fiber-dominated such as in-phase TMF [*21*]. The predicted fiber stress is strongly dependent on the representation of the [90] ply behavior, thus emphasizing that a good representation of the thermoviscoplastic matrix behavior as well as the fiber/matrix separation in the [90] ply are both required to accurately predict the fiber stresses. The average stresses in the matrix are used for comparisons, since some codes provide only average values. Although local stresses, as determined in codes such as FEM or FIDEP, may influence failure more than average stresses in the matrix, no experimental data are available to support this hypothesis.

Efficiency

In general, there is a trade-off between the accuracy of the solution and the efficiency of obtaining the solution. In many analyses, a quick answer is desired and some accuracy must be sacrificed for efficiency. Therefore, the time it takes to learn how to use a particular model, to set up the specific analysis, to calibrate a material model, to run the analysis, and to examine the output for the desired results are all important. Since different computers were used in carrying out the analyses with the various models, only general observations can be made regarding efficiency. FEM was by far the most time-consuming computational procedure to exercise, whereas the other five methods were much easier to use and ran an order of magnitude or more quicker. As an illustration, LISOL was able to run ten in-phase TMF cycles plus processing cool-down in less than 10 min on a Sun workstation, whereas it took the FEM model tens of hours on a Cray supercomputer.

Availability of the Codes

Most of the codes are still considered to be in the developmental stage, so enhancements and modifications are continuing to be made to them. The versions of the codes available in 1993 were used in this investigation.

VISCOPLY was obtained from NASA Langley Research Center, Hampton, Virginia. No user's manual is available, since the program is user friendly and interactive.

METCAN and the source code was obtained from NASA Lewis Research Center, Cleveland, Ohio. Version 4.0 of the code was used. Since the source code is provided, the code can be modified. Presently, the time and cycle degradation parameters are "hardwired" into the code. The METCAN *User's Manual,* METCAN *Demonstration Manual,* and the METCAN *Theoretical Manual* are also available from NASA Lewis Research Center, Cleveland, Ohio. The standard features of METCAN are well documented in these manuals.

FIDEP is available from the University of Dayton Research Institute, Dayton, Ohio. Version 2.0 of the code was used. A manual on its basic operation is also available [*34*].

LISOL is available from Air Force Institute of Technology, Wright-Patterson Air Force Base, Ohio. A beta version (unfinalized first version) of the code was used, and no manual is currently available.

For FEM, the computations were carried out using ADINA. Most general purpose codes, which allow user-defined subroutines and incorporate contact elements, can handle this type of problem. In general, the user will have to obtain or write user-defined subroutines for the viscoplastic model.

Conclusions

This investigation utilizes six state-of-the-art micromechanical methods for predicing the [0/90] composite response under simulated processing conditions as well as under isothermal and thermomechanical loading conditions that are used in fatigue life predictions.

The three features essential to accurately predicting strain ratchetting and frequency effects are (1) geometry of the representative volume element describing the constituents, (2) the material models used for the matrix and fiber, and (3) the treatment of the weak fiber/matrix interface under transverse loading in the [90] ply. FEM and the modified method of cells approach (LISOL) handle these three features well, with LISOL running orders of magnitude faster. The remaining models provide various degrees of scatter in the results. For cases where the inelastic deformation is large, such as isothermal fatigue at low frequency and in-phase TMF, the differences among the solutions are quite large. The models using a good viscoplastic model predict these conditions the best. When inelastic deformation is smaller and less time-dependent, such as under isothermal fatigue at high frequency and out-of-phase TMF, the differences among the solutions are minimal and, in general, all the models give similar results. The best mix of numerical efficiency and accuracy is obtained by the modified method of cells approach (LISOL) using the unified viscoplastic theory of Bodner and Partom to represent the matrix material.

References

[*1*] Nicholas, T., Russ, S. M., Neu, R. W., and Schehl, N., "Life Prediction of a [0/90] Metal Matrix Composite Under Isothermal and Thermomechanical Fatigue," in this volume.

[*2*] Castelli, M. G., "Thermomechanical Fatigue Damage/Failure Mechanisms in SCS-6/TIMETAL 21S $[0/90]_s$ Composite," NASA CR-191115, NASA Lewis Research Center, Cleveland, OH, 1993.

[*3*] Neu, R. W. and Nicholas, T., "Effect of Laminate Orientation on the Thermomechanical Fatigue

Behavior of a Titanium Matrix Composite,'' *Journal of Composites Technology & Research,* Vol. 16, No. 3, 1994, pp. 214–224.

[*4*] Mirdamadi, M., Johnson, W. S., Bahei-El-Din, Y. A., and Castelli, M. G., ''Analysis of Thermomechanical Fatigue of Unidirectional Titanium Metal Matrix Composites,'' NASA Technical Memorandum 104105, NASA Langley Research Center, Hampton, VA, July 1991.

[*5*] Russ, S. M., Nicholas, T., Hanson, D. G., and Mall, S., ''Isothermal and Thermomechanical Fatigue of Cross-Ply SCS-6/β21-S,'' *Science and Engineering of Composite Materials,* Vol. 3, 1994, pp. 177–189.

[*6*] Mall, S. and Schubbe, J. J., ''Thermomechanical Fatigue Behavior of a Cross-Ply SCS-6/Ti-15-3 Metal Matrix Composite,'' *Composite Science and Technology,* Vol. 50, 1994, pp. 49–57.

[*7*] Johnson, W. S., Lubowinski, and Highsmith, A. L., ''Mechanical Characterization of Unnotched SCS-6/Tl-15-3 Metal Matrix Composites at Room Temperatures,'' *Thermal and Mechanical Behavior of Metal Matrix and Ceramic Matrix Composites, ASTM STP 1080,* J. M. Kennedy, H. H. Moeller, and W. S. Johnson, Eds., American Society for Testing and Materials, Philadelphia, 1990, pp. 193–218.

[*8*] Nimmer, R. P., ''Fiber-Matrix Interface Effects in the Presence of Thermally Induced Residual Stress,'' *Journal of Composites Technology and Research,* Vol. 12, No. 2, 1990, pp. 65–75.

[*9*] Nimmer, R. P., Bankert, R. J., Russell, E. S., Smith, G. A., and Wright, P. K., ''Micromechanical Modeling of Fiber/Matrix Interface Effects in Transversely Loaded SiC/Ti-6-4 Metal Matrix Composites,'' *Journal of Composites Technology and Research,* Vol. 13, No. 1, 1991, pp. 3–13.

[*10*] Mall, S., Hanson, D. G., Nicholas, T., and Russ, S. M., ''Thermomechanical Fatigue Behavior of a Cross-Ply SCS-6/β21-S Metal Matrix Composite,'' *Constitutive Behavior of High Temperature Composites,* MD-Vol. 40, B. S. Majumdar, G. M. Newaz, and S. Mall, Eds., American Society of Mechanical Engineers, New York, 1992, pp. 91–106.

[*11*] Sherwood, J. A. and Boyle, M. J., ''Investigation of the Thermomechanical Response of a Titanium-Aluminide/Silicon-Carbide Composite Using a Unified State Variable Model in ADINA,'' *Computers and Structures,* Vol. 40, 1991, pp. 243–255.

[*12*] Russ, S. M., Nicholas, T., Bates, M., and Mall, S., ''Thermomechanical Fatigue of SCS-6/Ti-24Al-11Nb Metal Matrix Composite,'' *Failure Mechanisms in High Temperature Composite Materials,* AD-Vol. 22/AMD-Vol. 122, G. K. Haritos, G. Newaz, and S. Mall, Eds., American Society of Mechanical Engineers, New York, 1991, pp. 37–43.

[*13*] Bigelow, C. A., Johnson, W. S., and Naik, R. A., ''A Comparison of Various Micromechanics Models for Metal Matrix Composites,'' *Mechanics of Composite Materials and Structures,* J. N. Reddy and J. L. Teply, Eds., American Society of Mechanical Engineers, New York, 1989, pp. 21–31.

[*14*] Dvorak, G. J. and Bahei-El-Din, Y. A., ''Plasticity Analysis of Fibrous Composites,'' *Journal of Applied Mechanics,* Vol. 49, 1982, pp. 327–335.

[*15*] Bahei-El-Din, Y. A., ''Plasticity Analysis of Fibrous Composite Laminates Under Thermomechanical Loads,'' *Thermal and Mechanical Behavior of Metal Matrix and Ceramic Matrix Composites, ASTM STP 1080,* J. M. Kennedy, H. H. Moeller, and W. S. Johnson, Eds., American Society for Testing and Materials, Philadelphia, 1990, pp. 20–39.

[*16*] Murthy, P. L. N. and Chamis, C. C., ''METal Matrix Composite ANalyzer (METCAN): Theoretical Manual,'' NASA Technical Memorandum 106025, NASA Lewis Research Center, Cleveland, OH, Feb. 1993.

[*17*] Chamis, C. C. and Hopkins, D. A., ''Thermoviscoplastic Nonlinear Constitutive Relationships for Structural Analysis of High-Temperature Metal Matrix Composites,'' *Testing Technology of Metal Matrix Composites, ASTM STP 964,* P. R. DiGiovanni and N. R. Adsit, Eds., American Society for Testing and Materials, Philadelphia, 1988, pp. 177–196.

[*18*] Hopkins, D. A. and Chamis, C. C., ''A Unique Set of Micromechanics Equations for High-Temperature Metal Matrix Composites,'' *Testing Technology of Metal Matrix Composites, ASTM STP 964,* P. R. DiGiovanni and N. R. Adsit, Eds., American Society for Testing and Materials, Philadelphia, 1988, pp. 159–176.

[*19*] Murthy, P. L. N., Hopkins, D. A., and Chamis, C. C., ''Metal Matrix Composite Micromechanics: In-situ Behavior Influence on Composite Properties,'' *Mechanics of Composite Materials and Structures,* J. N. Reddy and J. L. Teply, Eds., American Society of Mechanical Engineers, New York, 1989, pp. 41–48.

[*20*] Coker, D., Ashbaugh, N. E., and Nicholas, T., ''Analysis of the Thermomechanical Fatigue Behavior of [0] and [0/90] SCS-6/TIMETAL®21S Composites,'' *Thermomechanical Behavior of Advanced Structural Materials,* AD-Vol. 34, AMD-Vol. 173, W. F. Jones, Ed., American Society of Mechanical Engineers, New York, 1993, pp. 1–16.

[21] Neu, R. W., Coker, D., and Nicholas, T., "Cyclic Behavior of Unidirectional and Cross-Ply Titanium Matrix Composites," *International Journal of Plasticity*, submitted for publication.

[22] Aboudi, J., "A Continuum Theory for Fiber-Reinforced Elastic-Viscoplastic Composites," *International Journal of Engineering and Science*, Vol. 20, No. 5, 1982, pp. 605–621.

[23] Aboudi, J., "Closed Form Constitutive Equations for Metal Matrix Composites," *International Journal of Engineering and Science*, Vol. 25, No. 9, 1987, pp. 1229–1240.

[24] Aboudi, J., "Micromechanical Analysis of Composites by the Method of Cells," *Applied Mechanics Review*, Vol. 42, No. 7, July 1989, pp. 193–221.

[25] Aboudi, J., Hidde, J. S., and Herakovich, C. T., "Thermo-mechanical Response Predictions for Metal Matrix Composite Laminates," *Mechanics of Composites at Elevated and Cryogenic Temperatures*, AMD-Vol. 118, American Society of Mechanical Engineers, New York, 1991, pp. 1–8.

[26] Robertson, D. D. and Mall, S., "Micromechanical Relations for Fiber-Reinforced Composites Using the Free Transverse Shear Approach," *Journal of Composites Technology & Research*, Vol. 15, No. 3, 1993, pp. 181–192.

[27] Robertson, D. and Mall, S., "Micromechanical Analysis for Thermoviscoplastic Behavior of Unidirectional Fibrous Composites," *Composites Science and Technology*, accepted for publication.

[28] Robertson, D. D., "A Nonlinear Three-Dimensional Micromechanics Model for Fiber-Reinforced Laminated Composites," Ph.D. thesis, AFIT/DS/AA/93-3, Air Force Institute of Technology, Wright-Patterson AFB, OH, 1993.

[29] Chan, K. S., Bodner, S. R., and Lindholm, U. S., "Phenomenological Modeling of Hardening and Thermal Recovery in Metals," *Journal of Engineering Materials and Technology*, Vol. 110, Jan. 1988, pp. 1–8.

[30] Neu, R. W., "Nonisothermal Material Parameters for the Bodner-Partom Model," *Material Parameter Estimation for Modern Constitutive Equations*, MD-Vol. 43/AMD-Vol. 168, L. A. Bertram, S. B. Brown, and A. D. Freed, Eds., American Society of Mechanical Engineers, New York, 1993, pp. 211–226.

[31] Eisenberg, M. A. and Yen, C. F., "A Theory of Multiaxial Anisotropic Viscoplastic," *Journal of Applied Mechanics*, Vol. 48, 1981, pp. 276–284.

[32] Mirdamadi, M. and Johnson, W. S., "Stress-strain Analysis of a [0/90]2s Titanium Matrix Laminate Subjected to a Generic Hypersonic Flight Profile," NASA TM 107584, NASA Langley Research Center, Hampton, VA, March 1992.

[33] Bahei-El-Din, Y. A., Shah, R. S., and Dvorak, G. J., "Numerical Analysis of the Rate-Dependent Behavior of High Temperature Fibrous Composites," *Proceedings*, Symposium on Mechanics of Composites at Elevated and Cryogenic Temperatures, ASME Applied Mechanics Division Meeting, Columbus, OH, 16–19 June 1991, pp. 67–78.

[34] Coker, D. and Ashbaugh, N. E., "Elastic-Plastic Finite-Difference Analysis of Unidirectional Composites Subjected to Thermomechanical Cyclic Loading," WL-TR-93-4043, Wright Laboratory, Wright-Patterson AFB, OH, 1992.

[35] Coker, D., Ashbaugh, N. E., and Nicholas, T., "Analysis of Thermomechanical Cyclic Behavior of Unidirectional Metal Matrix Composites," *Thermomechanical Fatigue Behavior of Materials, ASTM STP 1186*, H. Sehitoglu, Ed., American Society for Testing and Materials, Philadelphia, 1993, pp. 50–69.

[36] Nicholas, T., Kroupa, J. L., and Neu, R. W., "Analysis of a [0/90] Metal Matrix Composite Under Thermomechanical Fatigue Loading," *Composites Engineering*, Vol. 3, No. 7–8, 1993, pp. 675–689.

[37] Bigelow, C. A., "Thermal Residual Stresses in a Silicon-Carbide/Titanium [0/90] Laminate," *Journal of Composites Technology and Research*, Vol. 15, 1993, pp. 304–310.

[38] Kroupa, J. K. and Neu, R. W., "Implementation of a Nonisothermal Unified Inelastic-Strain Theory into ADINA6.0 for a Titanium Alloy—Users Guild," WL-TR-93-4005, Wright Laboratory, Wright-Patterson AFB, OH, 1993.

[39] Hillmer, N. J., "Thermal Expansion of High-Modulus Fibers," *International Journal of Thermophysics*, Vol. 12, 1991, pp. 741–750.

[40] Nicholas, T. and Ahmad, J., "Modeling Fiber Breakage in a Metal Matrix Composite," *Composites Science and Technology*, Vol. 52, 1994, pp. 29–38.

Yehia A. Bahei-El-Din,[1] *George J. Dvorak,*[2] *Himanshu Nigam,*[3] *and Amr M. Wafa*[4]

Modeling of Thermomechanical Fatigue in $(0/\pm 45/90)_s$ Sigma/TIMETAL 21S Laminates

REFERENCE: Bahei-El-Din, Y. A., Dvorak, G. J., Nigam, H., and Wafa, A. M., "**Modeling of Thermomechanical Fatigue in $(0/\pm 45/90)_s$ Sigma/TIMETAL 21S Laminates,**" *Life Prediction Methodology for Titanium Matrix Composites, ASTM STP 1253,* W. S. Johnson, J. M. Larsen, and B. N. Cox, Eds., American Society for Testing and Materials, 1996, pp. 328–356.

ABSTRACT: Micromechanical modeling is employed to evaluate the stresses and strains in laminated composite plates reinforced by elastic fibers in a thermo–viscoplastic matrix. Loading is limited to uniform in–plane stresses and to uniform changes in temperature. The local fields and macroscopic response of the plies are estimated with the transformation field analysis method that utilizes overall elastic properties and transformation influence coefficients derived from the Mori–Tanaka model. Fiber debonding and sliding are modeled by modifying certain fiber elastic moduli. This methodology, incorporated in a finite element scheme, is applied in simulations of hot isostatic pressing and subsequent loading of a $(0/\pm 45/90)_s$ Sigma/TIMETAL 21S laminate. Axial tension/tension stress cycles are applied at constant temperature and in–phase cyclic temperature changes. The results show that the inelastic deformation of the matrix, assisted by extensive fiber debonding in the off–axis plies, promotes stress transfer to the fibers in the 0° plies. Under repeated loads, the fiber stress increases gradually at a certain rate and the laminate fails when the fiber ultimate strength is exceeded.

KEYWORDS: titanium matrix composites, laminates, fabrication, fatigue (materials), micromechanics, viscoplasticity, damage (materials), titanium, life prediction, titanium alloys, modeling

The response of metal matrix fibrous laminates to fabrication and processing conditions, and to cyclic thermomechanical loading in service, produces complex internal stress states in the plies and in the fiber and matrix phases. Moreover, damage processes such as interface debonding by sliding or separation, matrix fatigue cracking possibly assisted by corrosion, and fiber breaks, change the microstructural geometry and local stresses. Under such circumstances, intuitive guesses of the local states offer no reliable insight, however, considerable progress can be achieved with micromechanical analysis of the deformation and damage processes that take place during manufacture and in service.

The purpose of the present paper is to outline a micromechanical modeling procedure that

[1]Professor, Structural Engineering Department, Cairo University, Giza, Egypt.

[2]William Howard Hart Professor of Mechanics, Center for Composite Materials and Structures and Department of Civil Engineering, Rensselaer Polytechnic Institute, Troy, NY 12180–3590.

[3]Formerly assistant professor, Department of Civil Engineering, Rensselaer Polytechnic Institute, Troy, NY 12180–3590.

[4]Research scientist, General Electric Corporate Research and Development, Schenectady, NY 12301.

can be useful in obtaining a quantitative description of the overall response and of the local fields in thermomechanically loaded laminated plates. Results of isothermal and in–phase thermomechanical fatigue (TMF) experiments are also interpreted. Matrix behavior is described by a unified thermo–viscoplasticity theory summarized in the next section. In the subsequent section, the transformation field analysis approach to analysis of inelastic composites is described [*1*]. The $(0/\pm45/90)_s$ Sigma/TIMETAL 21S laminate is then analyzed under cyclic tension applied at 650°C, and under in–phase thermomechanical load cycles in the temperature range of 150 to 650°C. Comparisons of the results with experimental data suggest that fiber overloading may be responsible for fatigue failure under constant temperature and in–phase TMF.

Constitutive Equations of the Fiber and Matrix

Both constituents are regarded as homogeneous solids with a prescribed elastic symmetry, and with temperature–dependent properties. The fiber remains elastic until failure. The matrix is elastic if loaded within a certain equilibrium surface in the stress space, but becomes viscoplastic if loaded beyond this surface. The inelastic response of the matrix is derived here from a unified viscoplastic theory described by Bahei-El-Din et al. [*2*] and Shah [*3*]. In either phase, the total strain rate, $\dot{\epsilon}_{ij}$, is additively decomposed into elastic, thermal, and inelastic components

$$\dot{\epsilon}_{ij} = \dot{\epsilon}^{e}_{ij} + \dot{\epsilon}^{t}_{ij} + \dot{\epsilon}^{in}_{ij} \tag{1}$$

The elastic and thermal rates are found from

$$\dot{\epsilon}^{e}_{ij} = M^{e}_{ijkl}(\theta)\dot{\sigma}_{kl} \tag{2}$$

$$\dot{\epsilon}^{t}_{ij} = [(dM^{e}_{ijkl}(\theta)/d\theta)\sigma_{kl} + \delta_{ij}\beta(\theta)]\dot{\theta} \tag{3}$$

where θ is the current temperature, $M^{e}_{ijkl}(\theta)$ is the elastic compliance, δ_{ij} is the Kronecker tensor, and $\beta(\theta)$ is the linear coefficient of thermal expansion. Equation 3 follows from Eq 27 in the Appendix.

The inelastic strain is found as a function of the overstress measured from an equilibrium yield surface that contains all stress states that can be reached from the current state by purely elastic deformation [*2,3*]. For a von Mises material, the equilibrium surface is given by

$$g = \frac{3}{2}(s^{*}_{ij} - a_{ij})(s^{*}_{ij} - a_{ij}) - (Y + Q)^2 = 0 \tag{4}$$

where s^{*}_{ij} is the deviatoric equilibrium stress tensor; a_{ij} denotes the center of the yield surface; $Y = Y(\theta)$ is the initial yield stress in tension, independent of the loading rate; and $Q = Q(\theta)$ is an isotropic hardening function.

For a given stress state, s_{ij}, outside the surface (Eq 4), there exists an equilibrium stress s^{*}_{ij} that satisfies Eq 4 such that the stress points, s_{ij} and s^{*}_{ij}, and the center of the equilibrium surface, a_{ij}, are collinear

$$s^{*}_{ij} - \left[\frac{2(Y + Q)^2}{3(s_{k\ell} - a_{k\ell})(s_{k\ell} - a_{k\ell})}\right]^{1/2}(s_{ij} - a_{ij}) + a_{ij} \tag{5}$$

The effective overstress, R, is a measure of the distance between the actual stress point, s_{ij}, and the equilibrium stress point, s^*_{ij}. It vanishes if the stress point lies on, or falls inside, the yield surface. In particular

$$\begin{aligned} R &= \left[\frac{3}{2}(s_{ij} - s^*_{ij})(s_{ij} - s^*_{ij})\right]^{1/2} \qquad \text{if} \quad g(s_{ij} - a_{ij}) > 0 \\ R &= 0 \qquad \text{if} \quad g(s_{ij} - a_{ij}) \leq 0 \end{aligned} \tag{6}$$

The inelastic strain rate is then written as

$$\dot{\epsilon}^{in}_{ij} = \sqrt{(3/2)}k(\theta)R^{p(\theta)}n_{ij} \tag{7}$$

where the functions, $k(\theta)$ and $p(\theta)$, are material parameters and n_{ij} is a unit normal to the yield surface (Eq 4) at the current equilibrium stress point. From Eq 4, one finds this as

$$n_{ij} = \frac{(s^*_{ij} - a_{ij})}{[(s^*_{k\ell} - a_{k\ell})(s^*_{k\ell} - a_{k\ell})]^{1/2}} = \sqrt{(3/2)}\,\frac{(s^*_{ij} - a_{ij})}{(Y + Q)} \tag{8}$$

The evolution equation for Q that includes the effect of inelastic deformation and thermal recovery on the yield stress is given as

$$\dot{Q} = q(\theta)[Q_a(\theta) - Q]\dot{\bar{\epsilon}}^{in} - b_r(\theta)|Q - Q_r(\theta)|^{(n_r(\theta)-1)}[Q - Q_r(\theta)] \tag{9}$$

where the functions $q(\theta)$, $Q_a(\theta)$, $b_r(\theta)$, $Q_r(\theta)$, and $n_r(\theta)$ are material parameters; total ($Q_r = 0$) or partial ($Q_r \neq 0$) thermal recovery is represented by the second term in Eq 9; and $\dot{\bar{\epsilon}}^{in}$ is the effective inelastic strain rate

$$\dot{\bar{\epsilon}}^{in} = \left[\frac{2}{3}\dot{\epsilon}^{in}_{ij}\dot{\epsilon}^{in}_{ij}\right]^{1/2} = k(\theta)R^{p(\theta)} \tag{10}$$

where $\dot{\epsilon}^{in}_{kk} = 0$.

In analogy to Eq 9 and permitting complete thermal recovery of kinematic hardening, the evolution equation for the center a_{ij} of the yield surface can be written as

$$\dot{a}_{ij} = \dot{\mu}\nu_{ij} - v_r(\theta)\bar{a}^{(w_r(\theta)-1)}a_{ij}, \qquad \bar{a} = (a_{k\ell}a_{k\ell})^{1/2} \tag{11}$$

where $v_r(\theta)$ and $w_r(\theta)$ are material parameters. The unit tensor, ν_{ij}, defines the direction of translation of the yield surface in the deviatoric stress space and can be specified according to the hardening rules applied in rate–independent plasticity theories. If the Phillips hardening rule is selected, then

$$\nu_{ij} = \dot{s}_{ij}/(\dot{s}_{k\ell}\dot{s}_{k\ell})^{1/2} \qquad \text{if} \quad \dot{s}_{ij} \neq 0 \tag{12}$$

$$\nu_{ij} = n_{ij} \qquad \text{if} \quad \dot{s}_{ij} = 0 \tag{13}$$

The factor, $\dot{\mu}$, in Eq 11 is found from Prager's consistency condition, $\dot{g} = 0$, with translation of the yield surface specified by the first term in Eq 11. The result is

$$\dot{\mu} = \sqrt{(2/3)}\,k(\theta)R^{p(\theta)}[H(\theta) - q(\theta)[Q_a(\theta) - Q]]/n_{k\ell}\nu_{k\ell} \tag{14}$$

A variant of the two–surface plasticity theory by Dafalias and Popov [*4*] is used to describe evolution of the instantaneous tangent modulus, H

$$H(\theta) = H_0(\theta) + h(\theta)[\delta/(\delta_0 - \delta)]^{m(\theta)} \tag{15}$$

$$\delta = \left[\frac{3}{2}(\bar{s}_{ij} - s^*_{ij})(\bar{s}_{ij} - s^*_{ij})\right]^{1/2} \tag{16}$$

where δ is the distance between the stress point, s^*_{ij}, on the equilibrium surface and a point, $\bar{s}_{ij}$, on the bounding surface with normal equals n_{ij}, and δ_0 is the value of δ at the onset of inelastic deformation. When the equilibrium stress point lies on the bounding surface, the plastic tangent modulus, $H(\theta)$, assumes the asymptotic value, $H_0(\theta)$, which together with $h(\theta)$ and $m(\theta)$ need to be determined experimentally. In analogy with the equilibrium yield surface, thermal recovery of isotropic as well as kinematic hardening of the bounding surface can be included in the model. This is omitted here for brevity. We only mention that the recovery terms for isotropic and kinematic hardening of the bounding surface assume a form similar to those suggested earlier for the yield surface, but with new material parameters that we denote by script letters. For example, the von Mises form of the bounding surface is written as

$$\mathcal{G} = \frac{3}{2}(\bar{s}_{ij} - \mathcal{A}_{ij})(\bar{s}_{ij} - \mathcal{A}_{ij}) - (\mathcal{Y} + \mathcal{Q})^2 = 0 \tag{17}$$

where $\mathcal{A}_{ij}$ is the center of the bounding surface in the deviatoric stress space, $\mathcal{Y} = \mathcal{Y}(\theta)$ is the bounding stress given by the intersection of the asymptotic part of the uniaxial stress–plastic strain curve and the stress axis, and $\mathcal{Q} = \mathcal{Q}(\theta)$ is an isotropic hardening function.

In the Sigma/TIMETAL 21S system, the fiber was regarded as elastic and the matrix as thermo–viscoplastic. Elastic moduli of the Sigma fiber were assumed to be the same as those of the SCS6 fiber, and the material parameters for the TIMETAL 21S were derived, in part, from experiments performed at NASA Langley Research Center [*5,6*]. Isotropic hardening of the equilibrium (Eq 4) and bounding (Eq 17) surfaces was neglected; thermal recovery in Eqs 9 and 11 was minor. The material parameters remaining after reductions appear in Tables 1 and 2.

TABLE 1—*Elastic moduli and coefficients of thermal expansion.*

Sigma Fiber				TIMETAL 21S Matrix			
θ, °C	E, GPa	ν	α, 10^{-6}/°C	θ, °C	E, GPa	ν	α, 10^{-6}/°C
21	393	0.25	3.56	21	116	0.338	8.62
93	390	0.25	3.56	150	116	0.347	8.72
204	386	0.25	3.62	316	104	0.358	9.23
426	378	0.25	3.91	427	103	0.363	9.87
538	374	0.25	4.07	482	96	0.365	10.30
649	370	0.25	4.27	565	86	0.365	11.10
760	365	0.25	4.41	621	80	0.365	11.50
871	361	0.25	4.57	650	64	0.365	11.80
				704	53	0.374	12.00
				760	50	0.382	12.50

TABLE 2—*Viscoplastic parameters for TIMETAL 21S.*

θ, °C	Y, MPa	$\mathcal{Y}$, MPa	H_0, MPa	h, MPa	m	k, $(\text{MPa})^{-p}$/s	p
21	925	1000	72	4024	1.365	2.10E−21	8.59
150	925	1000	72	4024	1.365	2.10E−21	8.59
316	645	785	146	5432	1.355	2.92E−16	6.52
427	630	770	143	5432	1.365	1.66E−14	5.62
482	505	615	103	3165	1.478	5.38E−13	6.35
565	190	365	256	6311	1.211	6.07E−12	5.52
621	85	185	149	2618	1.200	6.05E−10	3.58
650	30	87	42	1043	1.000	1.76E−08	2.00
704	24	53	45	829	1.000	1.50E−08	2.00
760	9	20	15	250	1.000	1.00E−08	2.00

Micromechanical Models of Composite Laminates

Model Selection

Among the available approaches to modeling of inelastic composite materials, we use here the transformation field analysis (TFA) of the inelastic lamina [*1*] in conjunction with the laminated plate theory for symmetric laminates subjected to in–plane uniform overall stresses and changes in temperature. When used with a selected micromechanical model, the TFA method provides piecewise uniform approximations of the overall instantaneous stiffness, compliance, and the thermal stress and strain tensors of the plies that are needed in the evaluation of ply stress and strain averages in the analysis of laminate response, as well as averages of local fields in the fiber and matrix of each ply. The governing equations of the TFA method are described in some detail subsequently. These equations were implemented into the laminated plate analysis with the finite element method in the VISCOPAC program [*7*] for a class of averaging micromechanical models. Specifically, the vanishing fiber diameter [*8,9*], the self–consistent [*10*], and the Mori–Tanaka [*11,12*] models are included. Work in progress uses the Periodic Hexagonal Array (PHA) model [*13,14*] that provides piecewise uniform approximations of the local fields in the matrix, fiber, and at their interfaces, and can thus follow in more closely their evolution along the applied loading path.

Transformation Field Analysis

A representative volume of a fibrous composite ply is considered, under uniform overall stress or strain, and a uniform change in temperature. The goal is to estimate the stress and (small) strain averages caused in the elastic fiber and inelastic matrix by the applied loads. In the transformation field analysis, these averages are found by superposition of two load systems, one consisting of the applied mechanical loads, in the form of uniform overall stress or strain, and one associated with internal eigenstrains that include thermal strains, any inelastic strains, and also equivalent eigenstrains designed to simulate damage development, for example, by fiber sliding and debonding. This identification of thermal, inelastic, and damage–induced deformations with eigenstrains implies that both load systems act on an elastic, undamaged composite. Of course, except for the thermal strains, the eigenstrain load system is not known a priori; it is evaluated incrementally from certain governing equations provided by the transformation analysis method.

The procedure has been described in several references, for example, in Refs *1* and *15*. An exposition of these results that is particularly easy to implement uses symbolic notation where boldface lowercase letters represent symmetric second–order tensors written as (6 × 1) matrices, and boldface capital letters denote symmetric fourth–order tensors written as (6 × 6) matrices. Connections with the notation used earlier are easily established. For example, the strain tensor, ϵ_{ij}, with the symmetry, $\epsilon_{ij} = \epsilon_{ji}$, is written as $\boldsymbol{\epsilon} = \{\epsilon_{11}\ \epsilon_{22}\ \epsilon_{33}\ 2\epsilon_{23}\ 2\epsilon_{31}\ 2\epsilon_{12}\}^T$; fourth–order tensors having at least the symmetries $A_{ijkl} = A_{jikl} = A_{ijlk}$ are reduced to (6 × 6) matrices, $\mathbf{A}$, such that $\mathbf{A}\mathbf{A}^{-1} = \mathbf{A}^{-1}\mathbf{A} = \mathbf{I}$, the identity matrix.

The selected representative volume, V, of a composite ply is assumed to contain many fibers and to be loaded by an overall stress, $\boldsymbol{\sigma}$, or strain, $\boldsymbol{\epsilon}$, both macroscopically uniform. A certain subdivision of V into subvolumes, V_r, V_s, is chosen ($r, s = 1, 2, \ldots N$). In what follows, $N = 2$ and V_r, V_s will be either fiber or matrix volumes V_f or V_m ($r, s = f, m$). The volume fractions of the phases will be denoted by $C_r = V_r/V$, $r = f, m$.

The elastic constitutive relationships for the fiber and matrix phases with initial strains or eigenstrains, $\boldsymbol{\mu}_r$, and eigenstresses, $\boldsymbol{\lambda}_r = -\mathbf{L}_r\boldsymbol{\mu}_r$, are written as

$$\boldsymbol{\sigma}_r = \mathbf{L}_r\boldsymbol{\epsilon}_r + \boldsymbol{\lambda}_r \qquad \boldsymbol{\epsilon}_r = \mathbf{M}_r\boldsymbol{\sigma}_r + \boldsymbol{\mu}_r \tag{18}$$

The $\mathbf{L}_r$ and $\mathbf{M}_r = \mathbf{L}_r^{-1}$ are phase elastic stiffness and compliance matrices, discussed in the Appendix. The $\boldsymbol{\mu}_r$ is the total strain that remains in the phase volume V_r after complete local unloading to zero stress, whereas the eigenstress, $\boldsymbol{\lambda}_r$, is seen to represent the total stress caused in a fully constrained volume, V_r, by the eigenstrain, $\boldsymbol{\mu}_r$. It then follows that $\boldsymbol{\lambda}_r = -\mathbf{L}_r\boldsymbol{\mu}_r$. Since the $\mathbf{L}_r$ and $\mathbf{M}_r$ depend only on elastic moduli, assumed to be constant, Eq 18 can be readily converted into a rate form

$$\dot{\boldsymbol{\sigma}}_r = \mathbf{L}_r\dot{\boldsymbol{\epsilon}}_r + \dot{\boldsymbol{\lambda}}_r \qquad \dot{\boldsymbol{\epsilon}}_r = \mathbf{M}_r\dot{\boldsymbol{\sigma}}_r + \dot{\boldsymbol{\mu}}_r \tag{19}$$

With reference to Eqs 1, 3, and 7, the eigenstrain rate, $\dot{\boldsymbol{\mu}}_r$, that represents the thermal and inelastic deformation may be written as

$$\dot{\boldsymbol{\epsilon}} = \dot{\boldsymbol{\epsilon}}^e + \dot{\boldsymbol{\mu}} \qquad \dot{\boldsymbol{\mu}} = \dot{\boldsymbol{\epsilon}}^t + \dot{\boldsymbol{\epsilon}}^{in} \tag{20}$$

The superposition of the two load states employed in the transformation field method provides the following form of the local fields

$$\dot{\boldsymbol{\epsilon}}_r = \mathbf{A}_r\dot{\boldsymbol{\epsilon}} + \sum_s \mathbf{D}_{rs}\dot{\boldsymbol{\mu}}_s \tag{21a}$$

$$r, s = f, m$$

$$\dot{\boldsymbol{\sigma}}_r = \mathbf{B}_r\dot{\boldsymbol{\sigma}} + \sum_s \mathbf{F}_{rs}\dot{\boldsymbol{\lambda}}_s \tag{21b}$$

The $\dot{\boldsymbol{\epsilon}}$ and $\dot{\boldsymbol{\sigma}}$ denote the uniform overall strain and stress rates, respectively, applied to the representative volume, and $\mathbf{A}_r$ and $\mathbf{B}_r$ are mechanical strain and stress concentration factor matrices. The $\mathbf{D}_{rs}$ and $\mathbf{F}_{rs}$ are transformation concentration factors that evaluate the contribution to the local field in V_r caused by a uniform transformation field of unit magnitude present in V_s. Both the mechanical and transformation concentration factors depend only on elastic moduli and phase geometry, and thus remain constant, except under large changes in geometry or temperature that affect the elastic moduli. In the latter case, the concentration factors may need to be recalculated to reflect such changes.

Of course, Eq 19 can be applied to any number of subdivisions of the representative volume. However, we limit our attention to phase average fields in two–phase solids. Then, the concentration factors can be found in a very simple form [*1,16,17*], involving only overall elastic moduli, **L**, or compliances, **M**, and the known phase elastic moduli and volume fractions

$$\begin{aligned} \mathbf{A}_r &= (\mathbf{L}_r - \mathbf{L}_s)^{-1}(\mathbf{L} - \mathbf{L}_s)/c_r \\ & \qquad\qquad\qquad\qquad r, s = f, m \\ \mathbf{B}_r &= (\mathbf{M}_r - \mathbf{M}_s)^{-1}(\mathbf{M} - \mathbf{M}_s)/c_r \end{aligned} \tag{22}$$

$$\begin{aligned} \mathbf{D}_{rf} &= (\mathbf{I} - \mathbf{A}_r)(\mathbf{L}_f - \mathbf{L}_m)^{-1}\mathbf{L}_f \qquad & \mathbf{D}_{rm} &= -(\mathbf{I} - \mathbf{A}_r)(\mathbf{L}_f - \mathbf{L}_m)^{-1}\mathbf{L}_m \\ \mathbf{F}_{rf} &= (\mathbf{I} - \mathbf{B}_r)(\mathbf{M}_f - \mathbf{M}_m)^{-1}\mathbf{M}_f \qquad & \mathbf{F}_{rm} &= -(\mathbf{I} - \mathbf{B}_r)(\mathbf{M}_f - \mathbf{M}_m)^{-1}\mathbf{M}_m \end{aligned} \tag{23}$$

The overall elastic stiffness can be either measured, bracketed by the Hashin–Rosen or other exact bounds, or estimated by several approximate methods. In this work we use the Mori–Tanaka estimates discussed in the Appendix.

The governing equations for the unknown local eigenstrain rates can be now completed by first substituting for the concentration factors from Eqs 22 and 23 and for the eigenstrains from Eq 20. In particular, the $\dot{\boldsymbol{\mu}}$ in Eq 20 is obtained from Eqs 3 and 7, converted into an eigenstress using the $\dot{\boldsymbol{\lambda}} = -\mathbf{L}\dot{\boldsymbol{\mu}}$ relationship that follows from Eqs 18 or 19, and substituted into Eq 21*b*. Therefore, the eigenstrain rates are replaced by expressions involving increments of temperature and of local stress rates, and Eq 21*b* is converted into a system of differential equations for evaluation of the local stresses. In a similar way, Eq 21*a* can be converted into a system of equations for the local strains. Solution techniques for either system are readily available [*15*].

Once the averages of the local stress or strain rates are known at a particular point of the loading path, the inelastic strains and other local eigenstrains can be found from the phase constitutive relationships. Overall averages of the total local stress and strain field rates follow from the connections

$$\dot{\boldsymbol{\sigma}} = \sum_r c_r \dot{\boldsymbol{\sigma}}_r \qquad \dot{\boldsymbol{\epsilon}} = \sum_r c_r \dot{\boldsymbol{\epsilon}}_r \tag{24}$$

while the overall transformation fields are found from

$$\dot{\boldsymbol{\lambda}} = \sum_r c_r \mathbf{A}_r^T \dot{\boldsymbol{\lambda}}_r \qquad \dot{\boldsymbol{\mu}} = \sum_r c_r \mathbf{B}_r^T \dot{\boldsymbol{\mu}}_r \tag{25}$$

The overall constitutive relationships of the ply are then written in a form similar to Eqs 18 and 19

$$\begin{aligned} \boldsymbol{\sigma} &= \mathbf{L}\boldsymbol{\epsilon}^0 + \boldsymbol{\lambda} \qquad & \boldsymbol{\epsilon} &= \mathbf{M}\boldsymbol{\sigma}^0 + \boldsymbol{\mu} \\ \dot{\boldsymbol{\sigma}} &= \mathbf{L}\dot{\boldsymbol{\epsilon}}^0 + \dot{\boldsymbol{\lambda}} \qquad & \dot{\boldsymbol{\epsilon}} &= \mathbf{M}\dot{\boldsymbol{\sigma}}^0 + \dot{\boldsymbol{\mu}} \end{aligned} \tag{26}$$

where $\dot{\boldsymbol{\sigma}}^0$ or $\dot{\boldsymbol{\epsilon}}^0$ are the overall applied rates, the overall elastic stiffness, **L**, and compliance, **M**, are obtained as in Eq 22, from experimental measurements, bounds, or estimates.

The transformation analysis procedure can be used with any physically acceptable phase constitutive relationships and micromechanical technique for elastic composites, including the unit–cell models [*13,14*], and also in laminate analysis. The only difference resulting from the various model selection is in the number of subdivisions of the representative volume ($r = 1, 2, \ldots N$), and in the evaluation of the concentration factor tensors for each subvolume, V_r. Of course, if more local volumes are used, the piecewise uniform approximation of the local fields found from Eq 21 should be more responsive to the inelastic constitutive relationships of the phases and thus more accurate. However, for given constitutive relationships, the structure of the governing equations and the solution procedures are identical for all models.

This approach differs from that where the inclusion problem is solved for an elastic matrix whose compliance is identified with the instantaneous plastic compliance derived from the average matrix stress history [*18*]. The latter calls for repetitive solutions of the inelastic inclusion problem in a generally anisotropic matrix, not needed in our approach; it also may violate Eq 25. On the other hand, if the concentration factors in Eq 21 are found using a finite element analysis of a subdivided unit cell model, such as the periodic hexagonal array (PHA) model [*13,14*], then the solution of Eq 21 coincides with an inelastic finite element solution [*15*].

Modeling of Interface Damage

Fiber debonding or sliding or both are usually the initial damage states in fibrous composite plies subjected to off–axis loading of sufficient magnitude. The interface strengths are typically quite small. For example, Clyne and Withers [*19*] give interface shear strength magnitudes in the range of 50 to 100 MPa; normal strength values do not appear to be available, but are expected to be lower. Cyclic shear sliding tends to reduce the number of adhering asperities and thus the apparent interface shear strength. Therefore, transverse tension and shear, and longitudinal shear stresses in plies can initiate interface damage. Of course, processing and heat treatment may affect the actual magnitudes; for example, growth of brittle interfacial reaction layers may promote early debonding and matrix crack growth, whereas treatments that promote favorable residual stress states may increase the apparent strength.

However, as long as some adhering asperities are left at the interface after debonding, they should provide for shear stress transfer that may prevent fiber pullout. For example, the average surface shear (τ) and normal (σ) stresses in the fiber are connected by the force equilibrium relationship, $2\pi r L \tau = \pi r^2 \sigma$, over a length, L, of a fiber with radius, r, that is, $\sigma/\tau = 2L/r$. If $L = 25$ mm is selected as a transfer length, and the radius of the Sigma fiber, $r = 50$ μm, then one finds that $\sigma = 1000\ \tau$. This suggests that large normal stress may be transferred into a fiber even if the interface shear stress is very low. Therefore, fiber pullout is not expected, and the integrity of the fibrous plies in the axial direction should not be impaired by fiber sliding.

In the present analysis, we represented the interface failure under normal or shear stresses by reducing certain fiber stress components to zero through abrupt changes in the fiber elastic moduli. In particular, the fiber was regarded as transversely isotropic, with moduli equal to isotropic material magnitudes (see the Appendix). If the interface normal stress component reached the selected strength magnitude of 50 MPa, then the transverse shear modulus, G_T^f, and Young's elastic moduli, E_T^f, were reduced to zero. Similarly, if the interface longitudinal shear stress exceeded the selected 80-MPa shear strength, the longitudinal shear modulus, G_L^f, was reduced to zero. This approach has been implemented in the transformation analysis method. The ply elastic moduli in Eqs 22 and 23 have been found by the Mori–Tanaka method. After fiber moduli reductions, the ply stiffnesses and compliances, as well as the concentration factors, were reevaluated and introduced in the governing equations (Eq 21). The laminate analysis was performed by the finite element method [*7*].

More advanced stages of damage, such as matrix cracking and fiber breaks, were not considered in the present analysis.

Modeling of Fatigue Behavior

Isothermal Fatigue

Here we examine the internal stresses generated in the $(0/\pm45/90)_s$ Sigma/TIMETAL 21S laminate ($c_b = 0.409$) by hot isostatic pressing (HIP), subsequent cooling to room temperature, aging at 621°C for 8 h, reheating to 650°C, and by cyclic axial tensile stress applied at that temperature. The loading conditions are $R = S_{min}/S_{max} = 0.1$, where $S_{max} =$ 300 MPa at 0.001 Hz. Figure 1 shows the temperature and load history applied during fabrication and testing of the laminate, with eleven load cycles at 0.001 Hz that were sustained by an actual test sample prior to failure [*20*]. We note that the aging period of 8 h, during which the titanium matrix remained elastic, has been reduced in Fig. 1 and the subsequent figures to 2 h.

Under the selected values of 50 MPa for the interface normal bond strength, σ_b, and 80 MPa for the shear bond strength, τ_b, failure of the fiber interface took place in the various plies of the $(0/\pm45/90)_s$ laminate during mechanical loading to the mean stress at 1.375 MPa/s. The overall stress levels at the onset of interface failure, and the corresponding local stress averages in the individual plies are indicated in Table 3. The local stress averages in the fiber and matrix are shown in Tables 4 and 5. The local stresses in the ply, fiber, and matrix are defined in the coordinates of each ply, such that σ_{11} is the normal stress parallel to

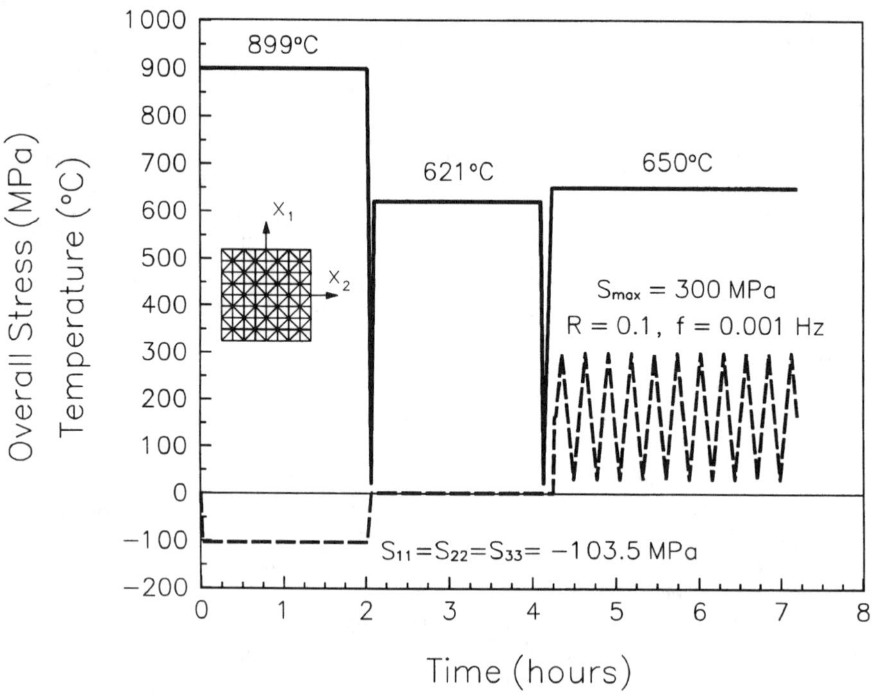

FIG. 1—*Thermomechanical loading history applied to the $(0/\pm45/90)_s$ laminate during HIP and isothermal cyclic loading.*

TABLE 3—*Local stress averages (MPa) in plies of a Sigma/TIMETAL 21S (0/±45/90)$_s$ laminate caused by hot isostatic pressing and subsequent static loading at 650°C.*

Ply	Ply[e] Stress	HIP Residual Stress		Under Overall Axial Stress, MPa			
		21°C	650°C	18[a]	26[b]	59[c]	65[d]
0°	σ_{11}	−157	−55	−5	39	210	217
	σ_{22}	157	55	54	59	85	40
45°	σ_{11}	−157	−55	−33	−10	53	77
	σ_{22}	157	55	69	84	47	50
	σ_{12}	0	0	15	26	54	54
90°	σ_{11}	−157	−55	−61	−82	−77	−58
	σ_{22}	157	55	84	43	61	65

[a]Corresponds to onset of fiber debonding in 90° ply at σ_b = 50 MPa.
[b]Corresponds to onset of fiber debonding in ±45° plies at σ_b = 50 MPa.
[c]Corresponds to onset of fiber debonding in 0° ply at σ_b = 50 MPa.
[d]Corresponds to onset of fiber sliding in ±45° plies at τ_b = 80 MPa.
[e]In local coordinates of each ply.

TABLE 4—*Fiber stress averages (MPa) in plies of a Sigma/TIMETAL 21S (0/±45/90)$_s$ laminate caused by hot isostatic pressing and subsequent static loading at 650°C.*

Ply	Fiber[e] Stress	HIP Residual Stress		Under Overall Axial Stress, MPa			
		21°C	650°C	18[a]	26[b]	59[c]	65[d]
0°	σ_{11}	−1095	−293	−195	−109	256	320
	σ_{22}	−52	15	15	20	50	0
	σ_{33}	−232	−53	−53	−55	−65	0
45°	σ_{11}	−1095	−293	−255	−214	−30	21
	σ_{22}	−52	15	32	50	0	0
	σ_{33}	−232	−53	−54	−56	0	0
	σ_{12}	0	0	20	34	78	80
90°	σ_{11}	−1095	−293	−314	−325	−319	−281
	σ_{22}	−52	15	50	0	0	0
	σ_{33}	−232	−53	−54	0	0	0

[a]Corresponds to onset of fiber debonding in 90° ply at σ_b = 50 MPa.
[b]Corresponds to onset of fiber debonding in ±45° plies at σ_b = 50 MPa.
[c]Corresponds to onset of fiber debonding in 0° ply at σ_b = 50 MPa.
[d]Corresponds to onset of fiber sliding in ±45° plies at τ_b = 80 MPa.
[e]In local coordinates of each ply.

the fiber longitudinal axis, σ_{22} and σ_{33} are the normal stresses in the transverse plane, and σ_{12} is the longitudinal shear stress.

Also shown in Tables 3 through 5 are the residual stresses caused by fabrication at room temperature and at the test temperature (650°C). As expected, in the quasi–isotropic laminate, the residual stresses after HIP are independent of the fiber orientation. Moreover, it is evident that the residual stresses caused by processing do contribute to the subsequent stress magnitudes. This contribution need not be detrimental; in the present case, the residual axial fiber stress, σ_{11}, is compressive and thus helps to reduce the tensile stress magnitude under subsequent tensile mechanical loading. However, a tensile residual stress, σ_{22}, of 15 MPa is

TABLE 5—*Matrix stress averages (MPa) in plies of a Sigma/TIMETAL 21S (0/±45/90)$_s$ laminate caused by hot isostatic pressing and subsequent static loading at 650°C.*

Ply	Matrix[e] Stress	HIP Residual Stress		Under Overall Axial Stress, MPa			
		21°C	650°C	18[a]	26[b]	59[c]	65[d]
0°	σ_{11}	492	110	127	141	177	145
	σ_{22}	302	82	82	86	110	67
	σ_{33}	161	36	37	38	45	0
45°	σ_{11}	492	110	121	131	110	115
	σ_{22}	302	82	95	108	80	85
	σ_{33}	161	36	37	39	0	0
	σ_{12}	0	0	11	19	37	37
90°	σ_{11}	492	110	115	85	90	97
	σ_{22}	302	82	107	72	103	110
	σ_{33}	161	36	38	0	0	0

[a]Corresponds to onset of fiber debonding in 90° ply at σ_b = 50 MPa.
[b]Corresponds to onset of fiber debonding in ±45° plies at σ_b = 50 MPa.
[c]Corresponds to onset of fiber debonding in 0° ply at σ_b = 50 MPa.
[d]Corresponds to onset of fiber sliding in ±45° plies at τ_b = 80 MPa.
[e]In local coordinates of each ply.

present in the fiber at 650°C and is likely to promote debonding under subsequent axial tension loading. Indeed, the transverse tensile stress in the 90° fiber reached the selected interface strength of 50 MPa at an overall axial stress of only 18 MPa. Subsequent loading to 26 MPa caused fiber debonding in the ±45° plies, followed by debonding of the 0° fiber at 59 MPa. Finally, the magnitude of longitudinal shear stress in the fiber of the ±45° plies reached the selected shear bond strength of 80 MPa at overall stress of 65 MPa.

Under subsequent loading, the debonded fiber in the off–axis plies as well as the fiber in the axial direction sustain load only in the axial direction. Consequently, the stiffness of the off–axis plies in the loading direction is reduced substantially, and the stress concentration factor, given by the ratio of the local stress increment and the overall stress increment, in the 0° lamina in the axial direction increases. For example, this ratio, computed from Table 3, increased from 2.8 at the overall stress of 18 MPa, where all fibers are perfectly bonded, to 4.2 at 65 MPa, where the fibers in all plies suffered interfacial debonding. The corresponding ratios for the fiber stress in the 0° ply, Table 4, are 5.4 and 9.4 indicating a substantial increase in the fiber stress in the axial direction due to interface damage of the off–axis plies.

Evolution of the local stresses during processing and subsequent isothermal cyclic load history is given in Figs. 2 through 10 for the individual plies. Figure 2 shows the average stresses generated in the 0° ply, and Figs. 3 and 4 provide the local stress averages in the fiber and matrix. The ply and local stresses in the ±45° plies are shown in Figs. 5 through 7, and the stresses in the 90° ply are presented in Figs. 8 through 10. The ratcheting behavior found in the stress histories is caused by the viscoplastic response of the titanium matrix modeled by the constitutive equations given earlier. In general, the matrix stresses are relaxed and the fiber axial stress in all plies is enhanced under the applied tension/tension load cycles. Under slow load cycles applied at 0.001 Hz, a nearly steady-state response is reached after only a few cycles.

The cyclic stress maxima and minima computed in the first cycle and in the failure Cycle 11 are shown in Tables 6 through 8 under the cyclic loading conditions indicated in Fig. 1

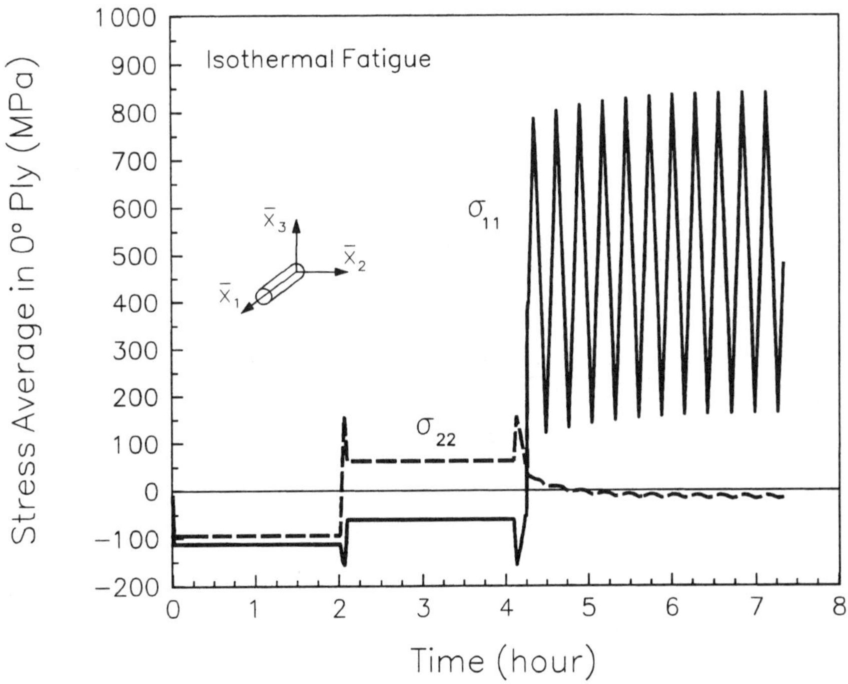

FIG. 2—*Stress average computed in the 0° ply of a Sigma/TIMETAL 21S (0/±45/90)$_s$ laminate subjected to HIP and isothermal cyclic loading.*

for the slow fatigue cycles at 0.001 Hz (Figs. 2 through 10). The stresses computed for a fast cycle at 0.1 Hz are also indicated. Actual test sample subjected to load cycles at 0.1 Hz sustained 207 cycles [*20*]. In this case, the computed local stresses reached a steady state after 90 cycles. The local stresses were then extrapolated to the failure cycle using a logarithmic function of the cycle number.

The rate effect on the local stresses is evident in the results given in Tables 6 through 8 for the first cycle. For example, the matrix stress, σ_{11}, computed in the 0° ply at S_{max} = 300 MPa is 119 MPa under the slow load cycle and 206 MPa under the fast cycle, Table 8. In contrast, the axial stress in the 0° fiber is enhanced by the slow loading rate. The maximum fiber stress found at failure, however, appears to be independent of the loading frequency. Specifically, the axial stress computed in the 0° fiber at failure is 1981 MPa under 0.001 Hz and 1988 MPa under 0.1 Hz. This suggests that failure of the laminate is controlled by the strength of the fiber in the 0° ply, and that the fatigue life is dependent on the rate at which the fiber tensile strength is reached.

The average transverse stresses, as well as the longitudinal shear stress, in the matrix are low in the plies with debonded fibers and decrease with cycling. Also, the maximum axial stress, σ_{11}, in the ±45° fiber remains low in the range of 687 MPa at 0.001 Hz and 613 MPa at 0.1 Hz. The corresponding fiber stress in the 90° ply is compressive at 609 and 756 MPa.

Collectively, these results suggest the existence of internal stresses that cause early failure of the interface bond. The axial fiber stress is very high in the 0° ply, and is apparently responsible for laminate failure. On the other hand, the average matrix stresses induced by the cyclic loading are low in all plies, and probably insufficient to cause significant matrix cracking early in the loading process.

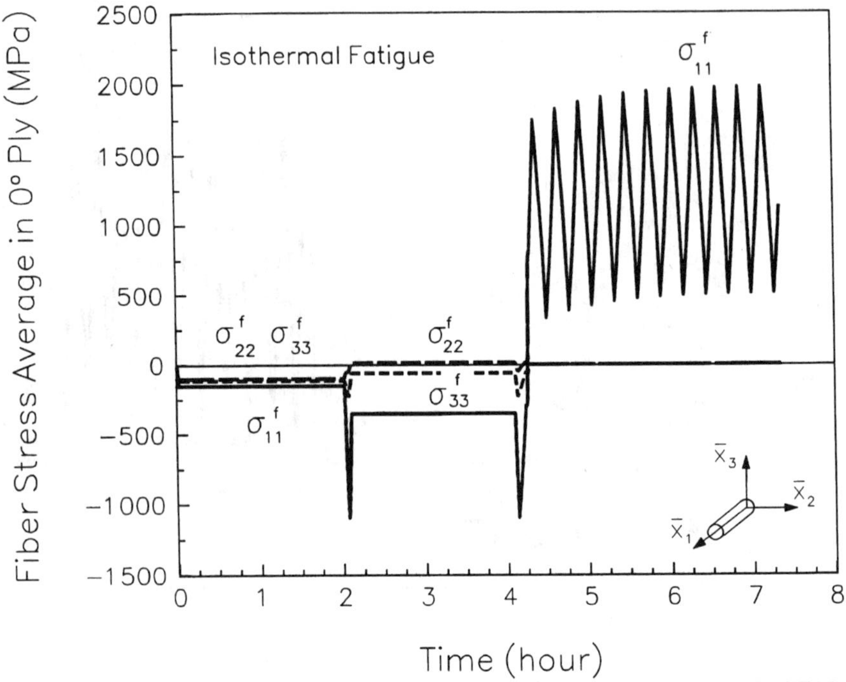

FIG. 3—*Fiber stress average computed in the 0° ply of a Sigma/TIMETAL 21S* $(0/\pm45/90)_s$ *laminate subjected to HIP and isothermal cyclic loading.*

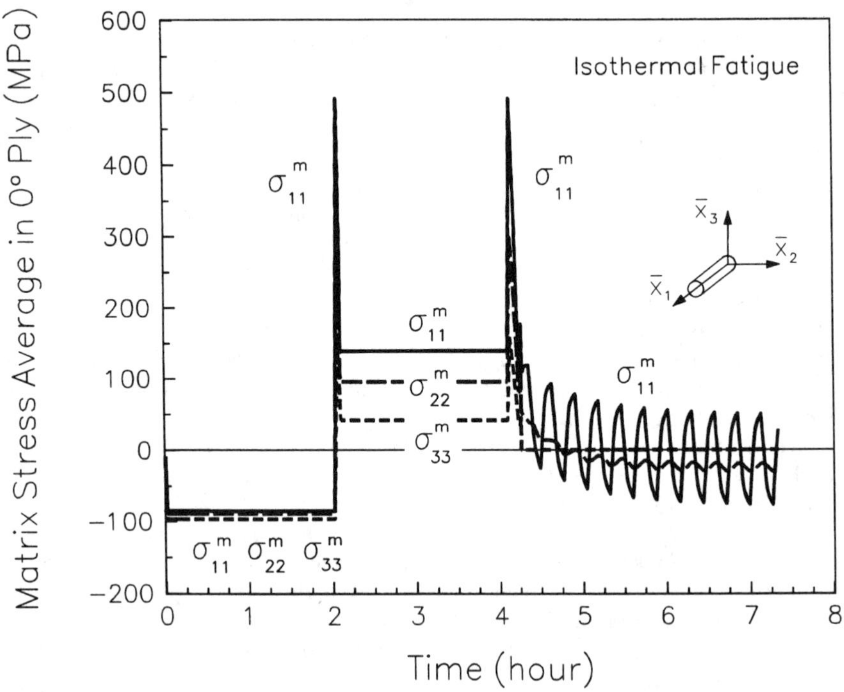

FIG. 4—*Matrix stress average computed in the 0° ply of a Sigma/TIMETAL 21S* $(0/\pm45/90)_s$ *laminate subjected to HIP and isothermal cyclic loading.*

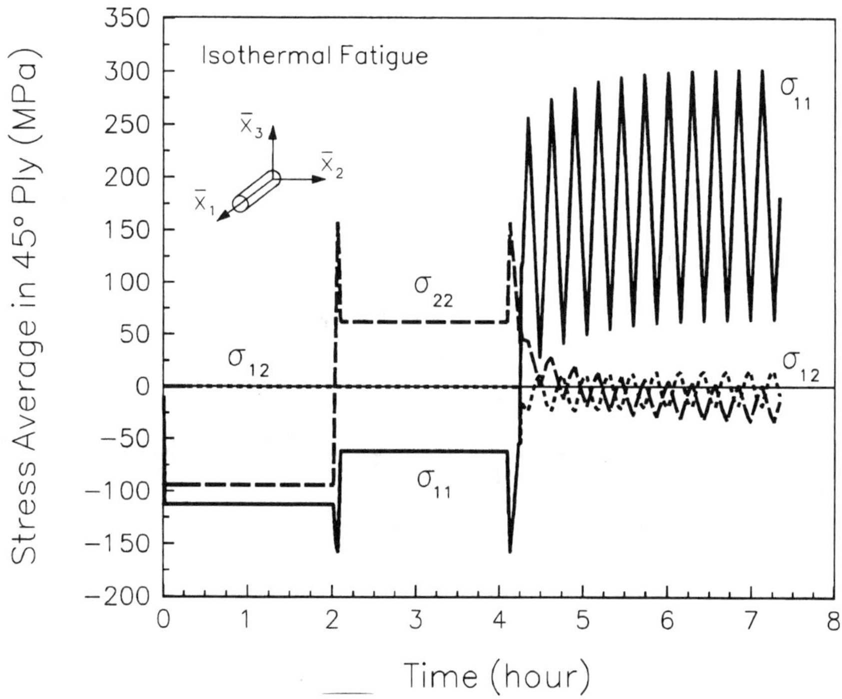

FIG. 5—*Stress average computed in the 45° ply of a Sigma/TIMETAL 21S* $(0/\pm 45/90)_s$ *laminate subjected to HIP and isothermal cyclic loading.*

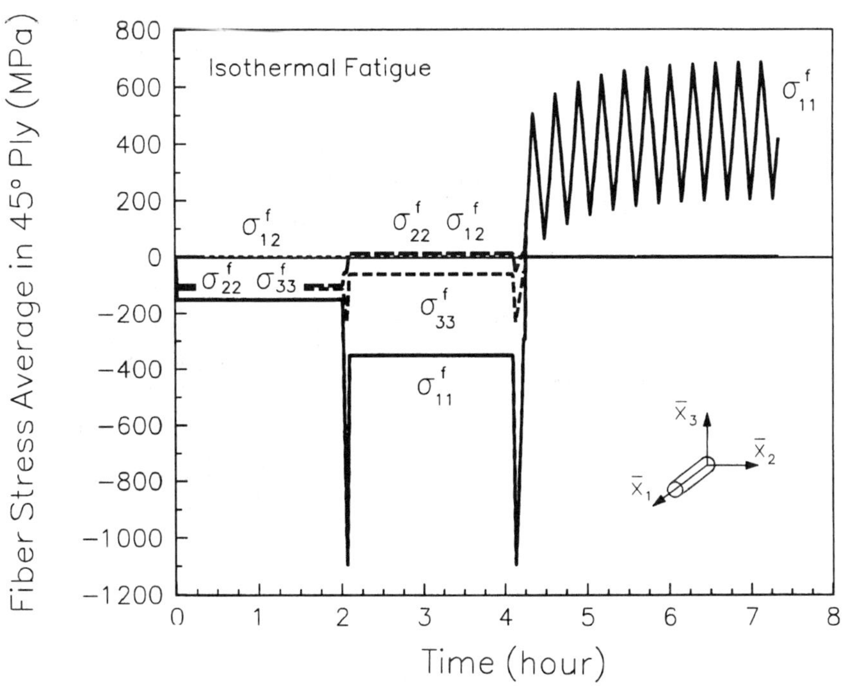

FIG. 6—*Fiber stress average computed in the 45° ply of a Sigma/TIMETAL 21S* $(0/\pm 45/90)_s$ *laminate subjected to HIP and isothermal cyclic loading.*

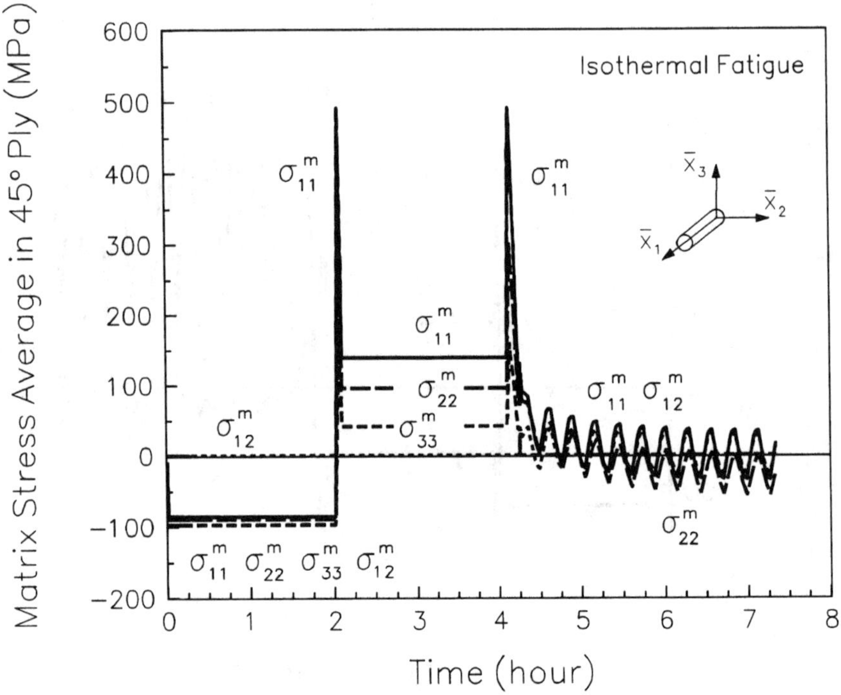

FIG. 7—*Matrix stress average computed in the 45° ply of a Sigma/TIMETAL 21S $(0/\pm45/90)_s$ laminate subjected to HIP and isothermal cyclic loading.*

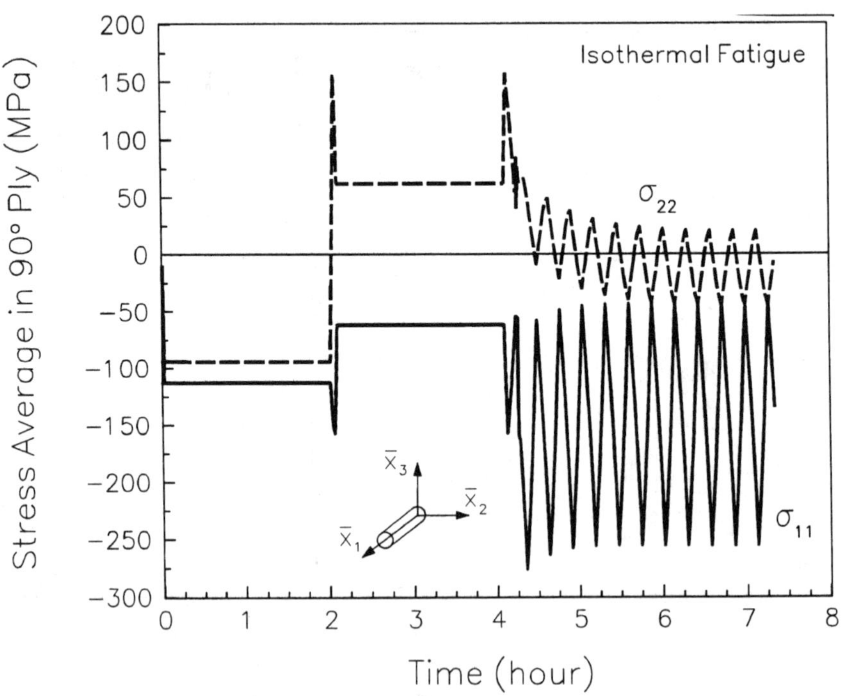

FIG. 8—*Stress average computed in the 90° ply of a Sigma/TIMETAL 21S $(0/\pm45/90)_s$ laminate subjected to HIP and isothermal cyclic loading.*

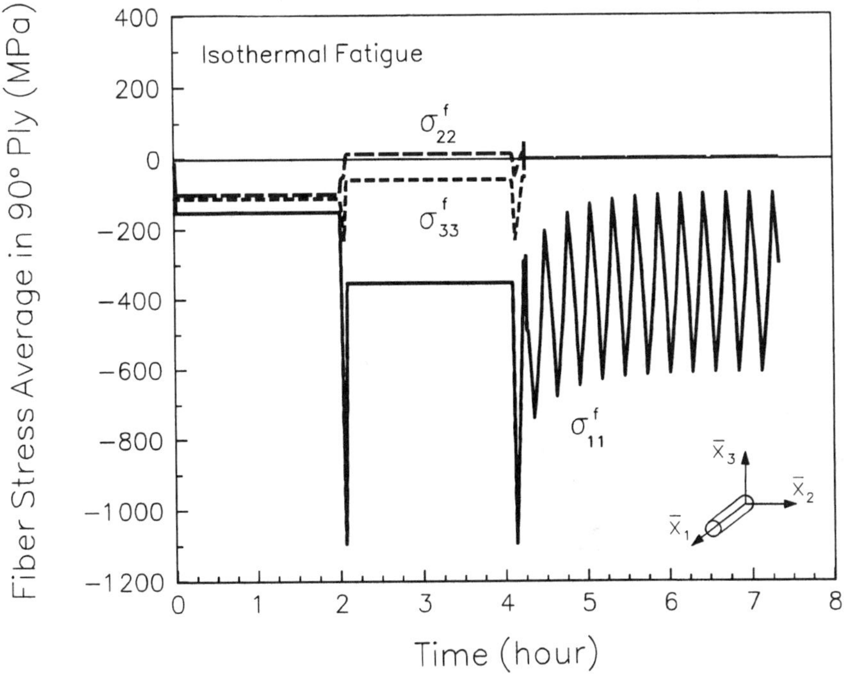

FIG. 9—*Fiber stress average computed in the 90° ply of a Sigma/TIMETAL 21S $(0/\pm45/90)_s$ laminate subjected to HIP and isothermal cyclic loading.*

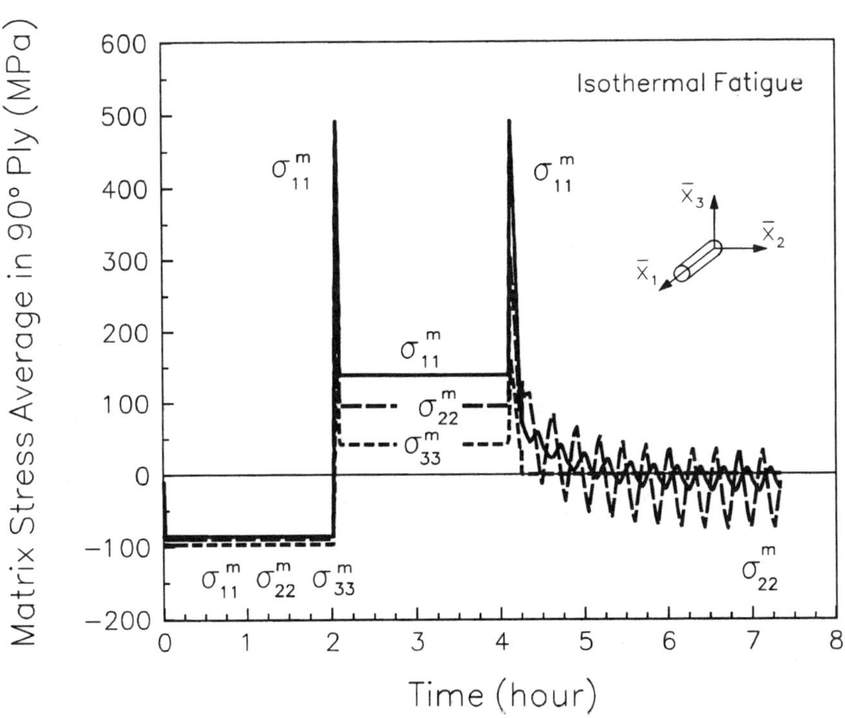

FIG. 10—*Matrix stress average computed in the 90° ply of a Sigma/TIMETAL 21S $(0/\pm45/90)_s$ laminate subjected to HIP and isothermal cyclic loading.*

TABLE 6—*Local stress averages (MPa) in plies of a Sigma/TIMETAL 21S (0/±45/90)$_s$ laminate caused by axial load cycles (S_{max} = 300 MPa, R = 0.1, and θ = 650°C).*

		f = 0.001 Hz				f = 0.1 Hz			
		Cycle 1		Cycle 11[a]		Cycle 1		Cycle 207[a]	
Ply	Ply[b] Stress	max	min	max	min	max	min	max	min
0°	σ_{11}	786	121	840	165	732	86	882	231
	σ_{22}	23	7	−11	−18	31	27	14	10
45°	σ_{11}	257	29	302	65	239	8	270	38
	σ_{22}	43	0	1	−33	62	24	15	−24
	σ_{12}	24	−11	18	−16	36	−10	4	−41
90°	σ_{11}	−276	−58	−256	−44	−261	−78	−289	−106
	σ_{22}	67	−8	21	−45	95	22	17	−56

[a]Number of cycles at failure found experimentally [*20*].
[b]In local coordinates of each ply.

TABLE 7—*Fiber stress averages (MPa) in plies of a Sigma/TIMETAL 21S (0/±45/90)$_s$ laminate caused by axial load cycles (S_{max} = 300 MPa, R = 0.1, and θ = 650°C).*

		f = 0.001 Hz				f = 0.1 Hz			
		Cycle 1		Cycle 11[a]		Cycle 1		Cycle 207[a]	
Ply	Fiber[b] Stress	max	min	max	min	max	min	max	min
0°	σ_{11}	1749	333	1981	513	1492	243	1988	713
45°	σ_{11}	506	65	687	208	389	−32	613	185
90°	σ_{11}	−739	−205	−609	−100	−717	−309	−756	−343

[a]Number of cycles at failure found experimentally [*20*].
[b]In local coordinates of each ply.

TABLE 8—*Matrix stress averages (MPa) in plies of a Sigma/TIMETAL 21S (0/±45/90)$_s$ laminate caused by axial load cycles (S_{max} = 300 MPa, R = 0.1, and θ = 650°C).*

		f = 0.001 Hz				f = 0.1 Hz			
		Cycle 1		Cycle 11[a]		Cycle 1		Cycle 207[a]	
Ply	Matrix[b] Stress	max	min	max	min	max	min	max	min
0°	σ_{11}	119	−25	51	−77	206	−22	117	−103
	σ_{22}	40	13	−19	−31	53	47	23	18
45°	σ_{11}	84	3	35	−34	135	35	34	−63
	σ_{22}	73	1	2	−56	106	40	23	−41
	σ_{12}	40	−18	31	−26	60	−17	7	−69
90°	σ_{11}	44	44	−11	−5	54	81	32	59
	σ_{22}	113	−14	35	−76	162	37	28	−94

[a]Number of cycles at failure found experimentally [*20*].
[b]In local coordinates of each ply.

In–Phase Thermomechanical Fatigue

The internal stresses generated in the Sigma/TIMETAL 21S $(0/\pm45/90)_s$ laminate under in–phase thermomechanical load cycles were evaluated analytically. As before, the fatigue loading was preceded by hot isostatic pressing, subsequent cooling to room temperature, and aging at 621°C for 8 h. The loading conditions are $S_{max} = 270$ MPa, $R = 0.1$, $\theta_{max} = 650$°C, and $\theta_{min} = 150$°C at 3 min per cycle ($f = 0.00555$ Hz, Fig. 11). Actual test samples reinforced in the SCS6 fiber and subjected to this loading condition sustained approximately 3000 cycles [*21*]. Apart from the differences in diameter, the sigma fiber and the SCS6 fiber are known to have similar mechanical properties. The laminate was taken to the mean temperature of 400°C, followed by loading to the mean stress of 148.5 MPa in 120 s.

Tables 9 through 11 show the local residual stress averages computed in the plies and in the fiber and matrix at the mean temperature, and at the overall axial stress levels that correspond to fiber interface damage. As in the previous tables, the local stresses are defined in the coordinate system of each ply, such that σ_{11} is the normal stress parallel to the fiber longitudinal axis, σ_{22} and σ_{33} are the normal stresses in the transverse plane, and σ_{12} is the longitudinal shear stress. In contrast to the test temperature of 650°C examined earlier, the transverse stress, σ_{22}, in the fiber at 400°C is compressive. This delayed the onset of fiber debonding somewhat, but did not prevent it. All off–axis plies suffered fiber interface damage either by debonding or sliding or both well before application of the thermomechanical load cycles.

Figures 12 through 20 show the evolution of the local stresses during the first ten thermomechanical load cycles. The average stresses generated in the 0° ply are given in Fig.

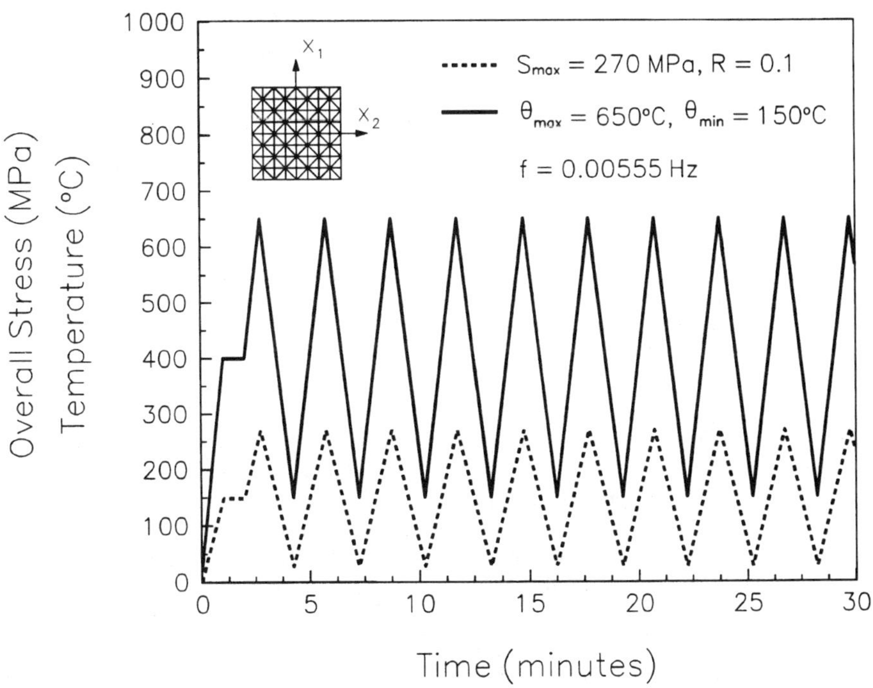

FIG. 11—*In–phase thermomechanical loading history applied to the* $(0/\pm45/90)_s$ *laminate.*

TABLE 9—*Local stress averages (MPa) in plies of a Sigma/TIMETAL 21S (0/±45/90)$_s$ laminate caused by hot isostatic pressing and subsequent static loading at 400°C.*

Ply	Ply[d] Stress	HIP Residual Stress		Under Overall Axial Stress, MPa		
		21°C	400°C	28[a]	41[b]	57[c]
0°	σ_{11}	−157	−103	−34	23	122
	σ_{22}	157	103	103	113	153
45°	σ_{11}	−157	−103	−71	−38	2
	σ_{22}	157	103	128	156	122
	σ_{12}	0	0	26	43	62
90°	σ_{11}	−157	−103	−108	−144	−118
	σ_{22}	157	103	154	101	117

[a]Corresponds to onset of fiber debonding in 90° ply at σ_b = 50 MPa.
[b]Corresponds to onset of fiber debonding in ±45° plies at σ_b = 50 MPa.
[c]Corresponds to onset of fiber sliding in ±45° plies at τ_b = 80 MPa.
[d]In local coordinates of each ply.

TABLE 10—*Fiber stress averages (MPa) in plies of a Sigma/TIMETAL 21S (0/±45/90)$_s$ laminate caused by hot isostatic pressing and subsequent static loading at 400°C.*

Ply	Fiber[d] Stress	HIP Residual Stress		Under Overall Axial Stress, MPa		
		21°C	400°C	28[a]	41[b]	57[c]
0°	σ_{11}	−1095	−663	−543	−446	−284
	σ_{22}	−52	−9	−11	0	43
	σ_{33}	−232	−128	−129	−131	−135
45°	σ_{11}	−1095	−663	−615	−564	−471
	σ_{22}	−52	−9	20	50	0
	σ_{33}	−232	−128	−130	−133	0
	σ_{12}	0	0	33	55	80
90°	σ_{11}	−1095	−663	−686	−695	−659
	σ_{22}	−52	−9	50	0	0
	σ_{33}	−232	−128	−131	0	0

[a]Corresponds to onset of fiber debonding in 90° ply at σ_b = 50 MPa.
[b]Corresponds to onset of fiber debonding in ±45° plies at σ_b = 50 MPa.
[c]Corresponds to onset of fiber sliding in ±45° plies at τ_b = 80 MPa.
[d]In local coordinates of each ply.

12. The corresponding local stress averages in the fiber and matrix are given in Figs. 13 and 14. The ply and local stresses in the ±45° plies are shown in Figs. 15 through 17, and the stresses in the 90° ply are presented in Figs. 18 through 20. The nonzero stresses found in Figs. 12 through 20 at the beginning of the thermomechanical loading represent estimates of the residual stress averages at room temperature caused by processing. The numerical values of these stresses are given in the third column of Tables 9 through 11.

The maxima and minima of the local stresses in the first and in the nearly steady-state tenth cycle are given in Tables 12 through 14. The stress averages extrapolated to the failure Cycle 3000 are also shown. Compared to the isothermal case at 650°C, less ratcheting is found in the titanium matrix under in–phase thermomechanical loading. This causes the internal stresses to stabilize in only a few cycles. We note that relatively large stresses are

TABLE 11—*Matrix stress averages (MPa) in plies of a Sigma/TIMETAL 21S $(0/\pm 45/90)_s$ laminate caused by hot isostatic pressing and subsequent static loading at 400°C.*

Ply	Matrix[d] Stress	HIP Residual Stress		Under Overall Axial Stress, MPa		
		21°C	400°C	28[a]	41[b]	57[c]
0°	σ_{11}	492	285	318	348	404
	σ_{22}	302	180	181	191	228
	σ_{33}	161	89	90	91	94
45°	σ_{11}	492	285	305	326	300
	σ_{22}	302	180	204	229	178
	σ_{33}	161	89	90	92	0
	σ_{12}	0	0	21	35	50
90°	σ_{11}	492	285	292	237	256
	σ_{22}	302	180	227	170	199
	σ_{33}	161	89	91	0	0

[a]Corresponds to onset of fiber debonding in 90° ply at $\sigma_b = 50$ MPa.
[b]Corresponds to onset of fiber debonding in ±45° plies at $\sigma_b = 50$ MPa.
[c]Corresponds to onset of fiber sliding in ±45° plies at $\tau_b = 80$ MPa.
[d]In local coordinates of each ply.

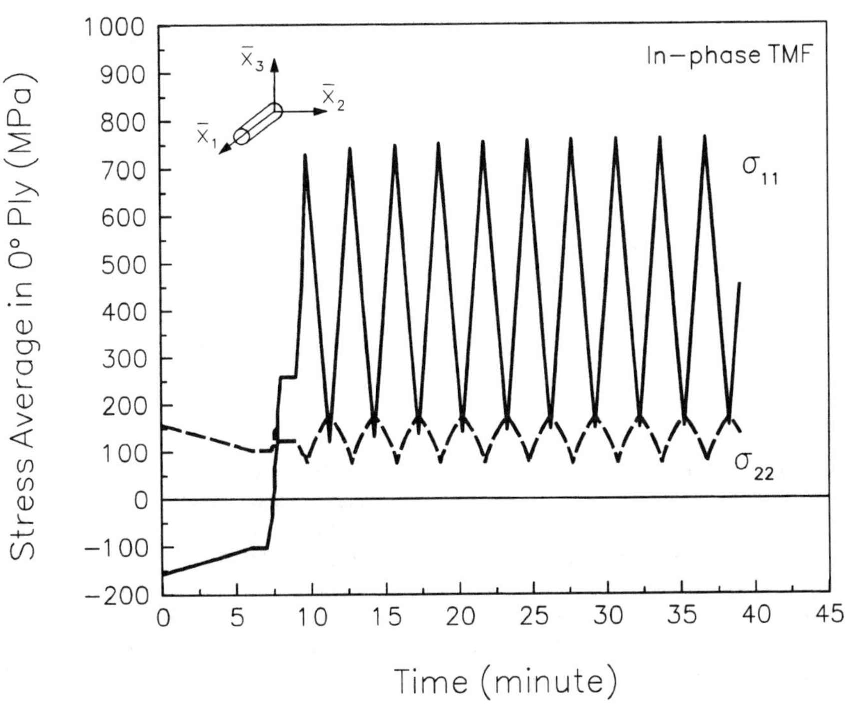

FIG. 12—*Stress average computed in the 0° ply of a Sigma/TIMETAL 21S $(0/\pm 45/90)_s$ laminate subjected to HIP and in-phase thermomechanical cyclic loading.*

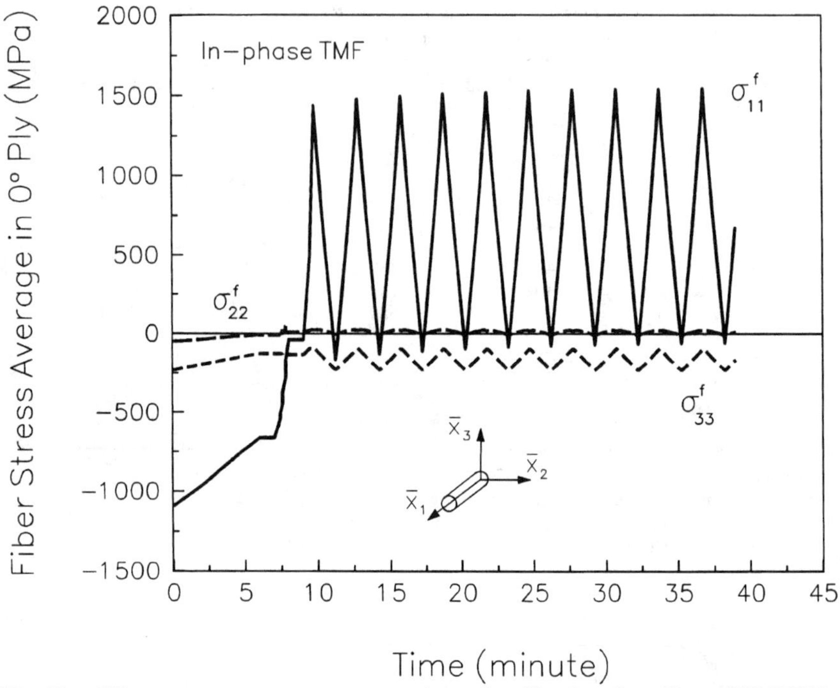

FIG. 13—*Fiber stress average computed in the 0° ply of a Sigma/TIMETAL 21S $(0/\pm45/90)_s$ laminate subjected to HIP and in–phase thermomechanical cyclic loading.*

supported by the matrix in the in–phase TMF case, Table 14. This may cause internal cracks in the matrix and elevate the fiber stress in the 0° ply. Suppressing matrix damage in modeling the local stresses, the axial stress in the 0° fiber was estimated at 1674 MPa in the failure cycle, Table 13; that is, a 16% reduction from the fiber ultimate strength computed from the isothermal case. A simple calculation in which the maximum axial stress of 814 MPa in the 0° ply, Table 12, is supported by the fiber alone provides fiber ultimate strength of 1990 MPa. This magnitude is consistent with the fiber maximum stress computed in the range of 1980 MPa for the two isothermal fatigue tests discussed earlier.

Conclusions

Although limited in scope, the computed internal stress fields point to the possible causes of fatigue failure in the $(0/\pm45/90)_s$ Sigma/TIMETAL 21S laminate. Under constant temperature, the fiber stresses in the 0° plies increase with the number of cycles, until a certain overloading situation is reached. The rate of change of the fiber stress with cycles is reflected by the ratcheting rate of the viscoplastic titanium matrix, particularly at high temperature, and is a function of the loading frequency. The axial stresses estimated in the 0° fiber under identical isothermal loading conditions, but different frequencies are equal. Substantial interface damage is expected to take place in the fiber of the off–axis plies, but the matrix stresses are small and may not cause significant matrix cracking.

Failure under in–phase thermomechanical load cycles is also caused by overloading the fibers in the 0° plies. In this case, however, relatively large stresses may build up in the matrix and eventually cause internal cracks. Consideration of this damage mode in estimating the local stresses under in–phase TMF provides fiber stress in the 0° ply at failure in

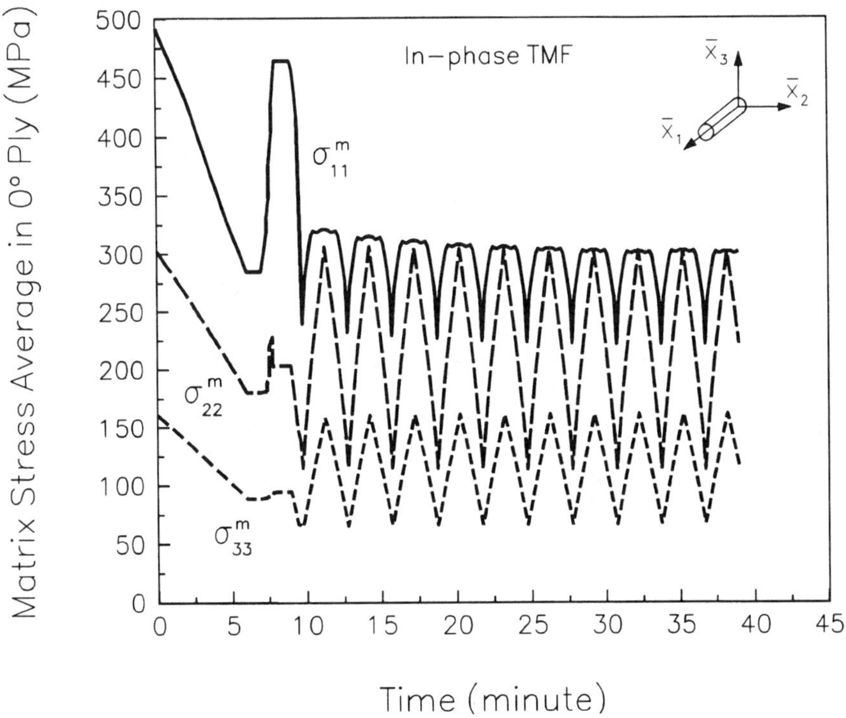

FIG. 14—*Matrix stress average computed in the 0° ply of a Sigma/TIMETAL 21S $(0/\pm 45/90)_s$ laminate subjected to HIP and in–phase thermomechanical cyclic loading.*

the range of 1990 MPa, which is consistent with the fiber ultimate strength found in isothermal fatigue tests.

Acknowledgments

This work was supported by the Air Force Office of Scientific Research and was performed in cooperation with the National Institute for the Mechanics and Life Prediction of High Temperature Composites (NIC) at Wright Laboratories. Drs. Walter Jones, Ted Nicolas, and Joseph Zuiker served as program monitors and advisors.

APPENDIX

Suppose that the fiber and matrix moduli and coefficients of thermal expansion are represented by continuous functions of temperature, θ, derived from experimental data. After a uniform change, $\Delta\theta$, from a reference temperature, θ_0, applied under a uniform stress, σ_r, an unconstrained phase deforms uniformly

$$\epsilon_r = \mathbf{M}_r(\theta)\sigma_r + \int_{\theta_0}^{\theta} \boldsymbol{\beta}_r(\theta)\, d\theta \tag{27}$$

If the temperature changes under an applied uniform strain, the stress is

$$\sigma_r = \mathbf{L}_r(\theta)\left[\epsilon_r - \int_{\theta_0}^{\theta} \boldsymbol{\beta}_r(\theta)\, d\theta\right] \tag{28}$$

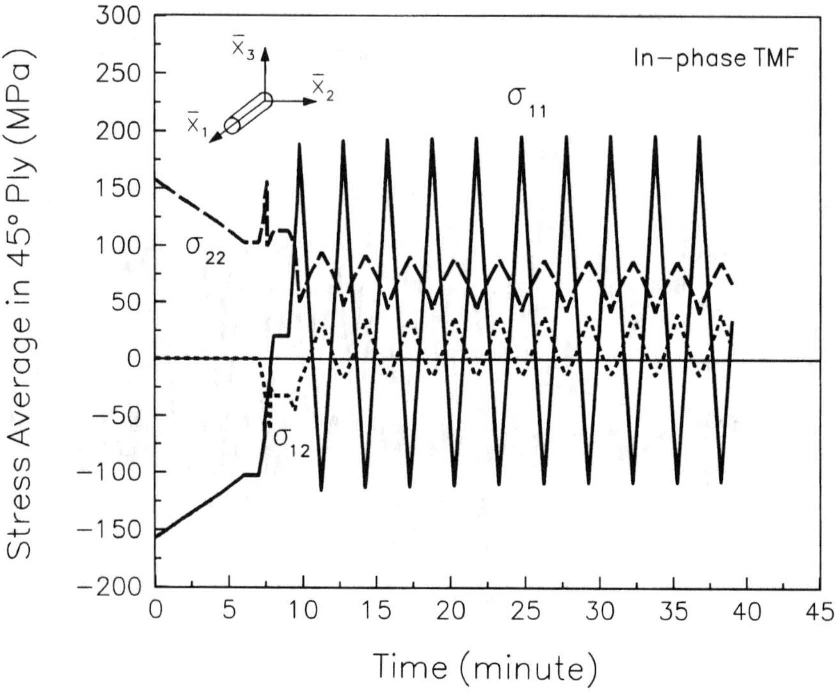

FIG. 15—*Stress average computed in the 45° ply of a Sigma/TIMETAL 21S (0/±45/90)$_s$ laminate subjected to HIP and in–phase thermomechanical cyclic loading.*

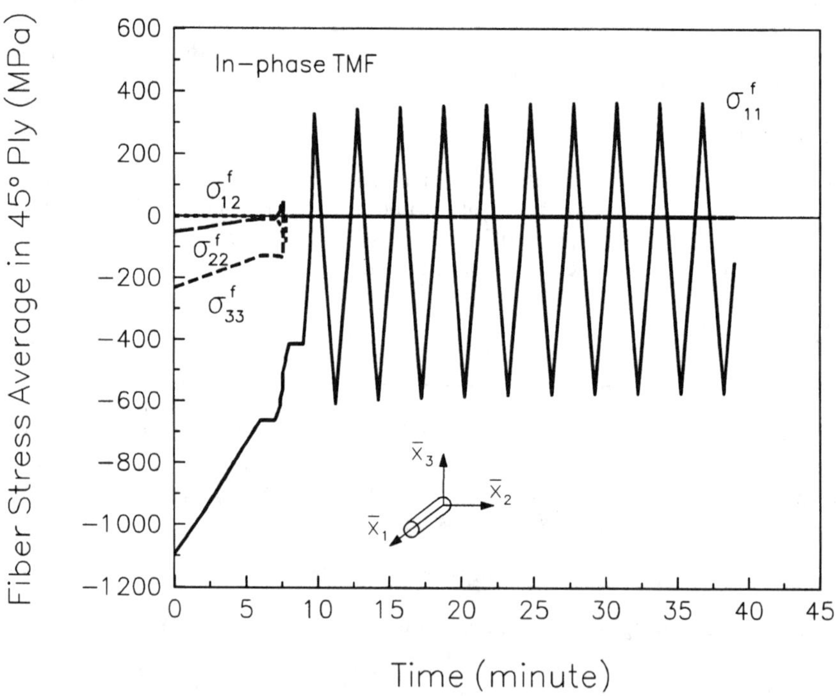

FIG. 16—*Fiber stress average computed in the 45° ply of a Sigma/TIMETAL 21S (0/±45/90)$_s$ laminate subjected to HIP and in–phase thermomechanical cyclic loading.*

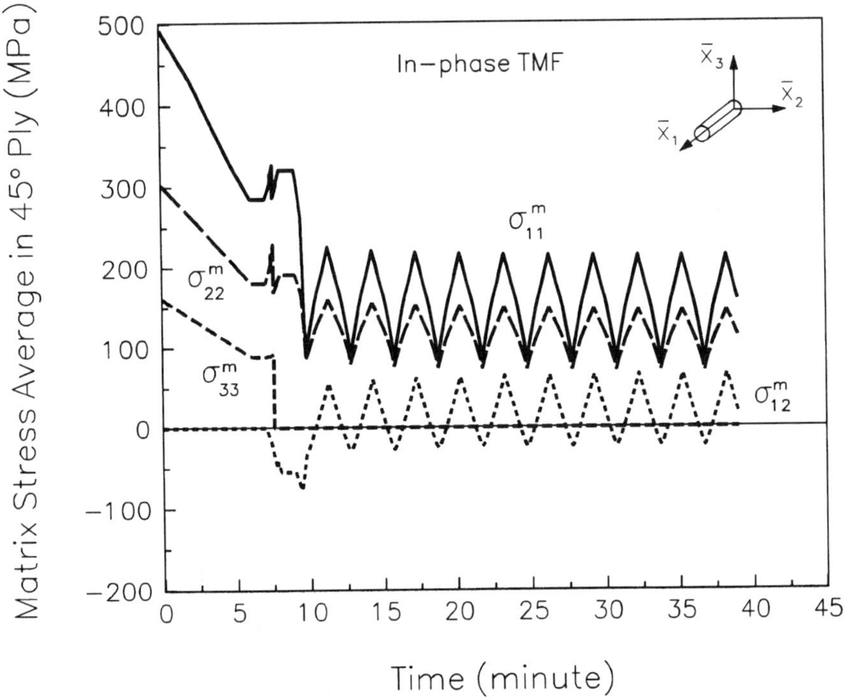

FIG. 17—*Matrix stress average computed in the 45° ply of a Sigma/TIMETAL 21S $(0/\pm45/90)_s$ laminate subjected to HIP and in–phase thermomechanical cyclic loading.*

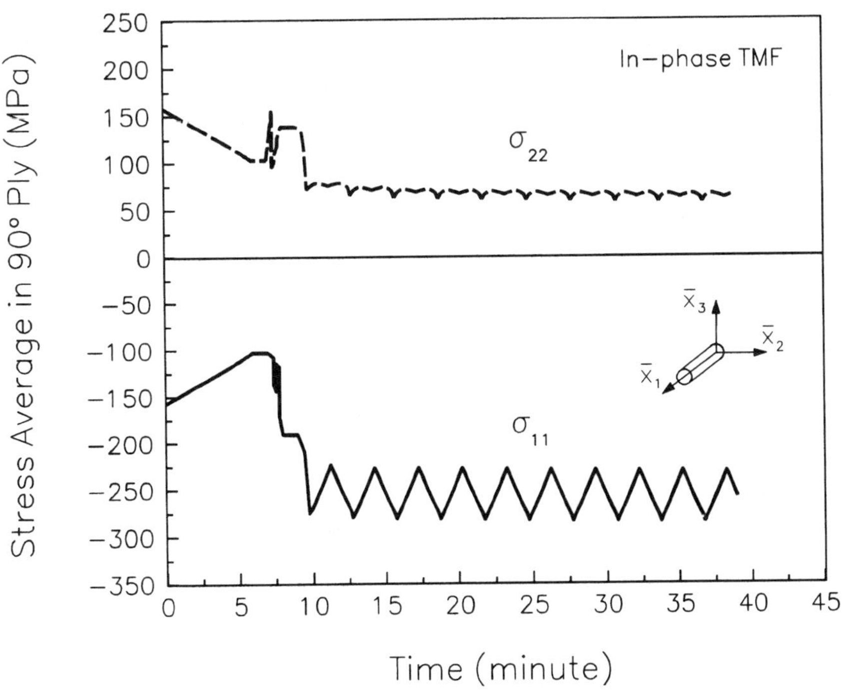

FIG. 18—*Stress average computed in the 90° ply of a Sigma/TIMETAL 21S $(0/\pm45/90)_s$ laminate subjected to HIP and in–phase thermomechanical cyclic loading.*

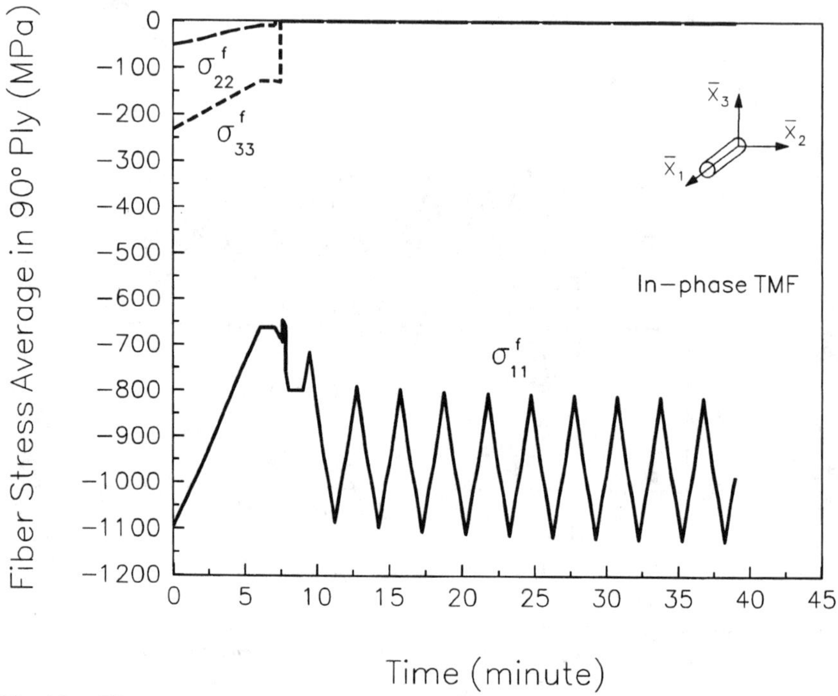

FIG. 19—*Fiber stress average computed in the 90° ply of a Sigma/TIMETAL 21S $(0/\pm 45/90)_s$ laminate subjected to HIP and in–phase thermomechanical cyclic loading.*

where $r = f, m$ denotes the phase, $\mathbf{L}_r(\theta)$ and $\mathbf{M}_r(\theta) = [\mathbf{L}_r(\theta)]^{-1}$ are temperature dependent, (6×6) phase stiffness and compliance matrices, $\boldsymbol{\beta}_r(\theta)$ is the (6×1) thermal strain vector of expansion coefficients, their particular forms depend on the elastic symmetry of the phase.

For a transversely isotropic solid, with x_1 as the axis of rotational symmetry, and compliance matrix, $\mathbf{M}$, the constitutive relationship, $\epsilon = \mathbf{M}\sigma$, is written using the engineering moduli

$$\begin{Bmatrix} \epsilon_{11} \\ \epsilon_{22} \\ \epsilon_{33} \\ 2\epsilon_{23} \\ 2\epsilon_{31} \\ 2\epsilon_{12} \end{Bmatrix} = \begin{bmatrix} 1/E_{11} & -\nu_{21}/E_{22} & -\nu_{21}/E_{22} & 0 & 0 & 0 \\ & 1/E_{22} & -\nu_{32}/E_{22} & 0 & 0 & 0 \\ & & 1/E_{22} & 0 & 0 & 0 \\ & & & 1/G_{23} & 0 & 0 \\ & & & & 1/G_{12} & 0 \\ & \text{SYM} & & & & 1/G_{12} \end{bmatrix} \begin{Bmatrix} \sigma_{11} \\ \sigma_{22} \\ \sigma_{33} \\ \sigma_{23} \\ \sigma_{31} \\ \sigma_{12} \end{Bmatrix} \tag{29}$$

The stiffness, $\mathbf{L} = \mathbf{M}^{-1}$, is best written in terms of Hill's [22] moduli

$$k = -[1/G_{23} - 4/E_{22} + 4\nu_{21}^2/E_{11}]^{-1}$$
$$n = E_{11} + 4k\,\nu_{21}^2 = E_{11} + \ell^2/k$$
$$\ell = 2k\,\nu_{21}$$
$$m = G_{23}$$
$$p = G_{12}$$

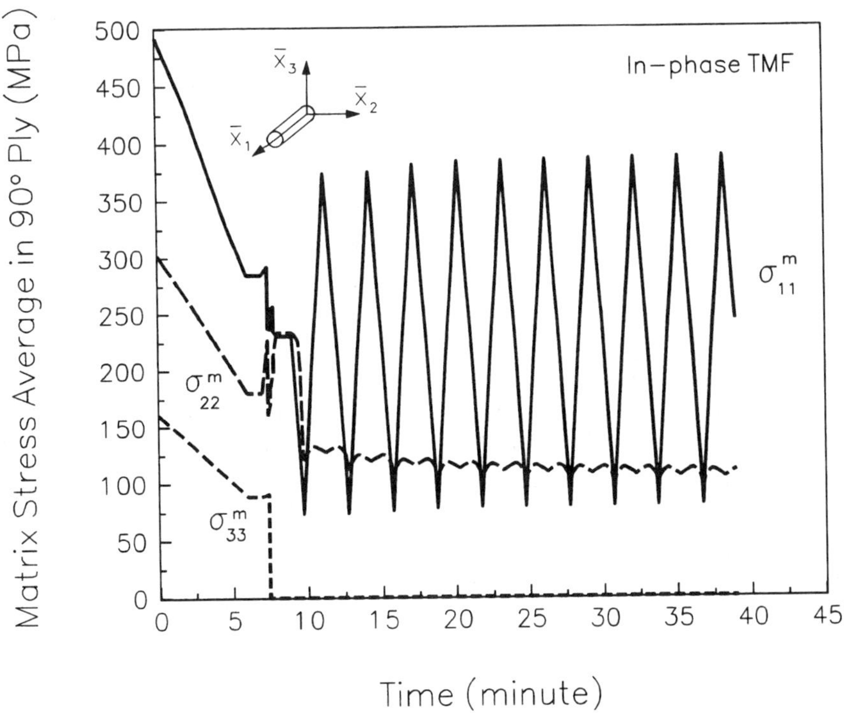

FIG. 20—*Matrix stress average computed in the 90° ply of a Sigma/TIMETAL 21S $(0/\pm45/90)_s$ laminate subjected to HIP and in-phase thermomechanical cyclic loading.*

TABLE 12—*Local stress averages (MPa) in plies of a Sigma/TIMETAL 21S $(0/\pm45/90)_s$ laminate caused by in-phase thermomechanical loading (S_{max} = 270 MPa, R = 0.1, θ_{max} = 650°C, θ_{min} = 150°C, and f = 0.00555).*

Ply	Ply[b] Stress	Cycle 1		Cycle 10		Cycle 3000[a]	
		max	min	max	min	max	min
0°	σ_{11}	730	121	763	154	814	207
	σ_{22}	76	178	74	173	72	165
45°	σ_{11}	189	−115	197	−107	205	−95
	σ_{22}	50	95	41	86	28	74
	σ_{12}	20	−33	14	−40	3	−49
90°	σ_{11}	−275	−224	−285	−231	−293	−249
	σ_{22}	71	75	57	62	31	30

[a]Number of cycles at failure found experimentally [21].
[b]In local coordinates of each ply.

TABLE 13—*Fiber stress averages (MPa) in plies of a Sigma/TIMETAL 21S (0/±45/90)$_s$ laminate caused by in-phase thermomechanical loading* (S_{max} = *270 MPa*, R = *0.1*, θ_{max} = *650°C*, θ_{min} = *150°C, and* f = *0.00555).*

Ply	Fiber[b] Stress	Cycle 1 max	Cycle 1 min	Cycle 10 max	Cycle 10 min	Cycle 3000[a] max	Cycle 3000[a] min
0°	σ_{11}	1439	−167	1547	−59	1674	142
45°	σ_{11}	328	−608	366	−572	396	−514
90°	σ_{11}	−780	−1086	−814	−1125	−362	−1209

[a]Number of cycles at failure found experimentally [*21*].
[b]In local coordinates of each ply.

TABLE 14—*Matrix stress averages (MPa) in plies of a Sigma/TIMETAL 21S (0/±45/90)$_s$ laminate caused by in-phase thermomechanical loading* (S_{max} = *270 MPa*, R = *0.1*, θ_{max} = *650°C*, θ_{min} = *150°C, and* f = *0.00555).*

Ply	Matrix[b] Stress	Cycle 1 max	Cycle 1 min	Cycle 10 max	Cycle 10 min	Cycle 3000[a] max	Cycle 3000[a] min
0°	σ_{11}	239	321	221	301	219	300
	σ_{22}	115	306	113	302	113	292
45°	σ_{11}	92	226	80	214	73	204
	σ_{22}	85	160	70	145	47	117
	σ_{12}	33	−57	23	−67	1	−92
90°	σ_{11}	74	374	81	388	101	416
	σ_{22}	121	128	96	104	53	51

[a]Number of cycles at failure found experimentally [*21*].
[b]In local coordinates of each ply.

The constitutive relationship, $\boldsymbol{\sigma} = \mathbf{L}\boldsymbol{\epsilon}$, then is

$$\begin{Bmatrix} \sigma_{11} \\ \sigma_{22} \\ \sigma_{33} \\ \sigma_{23} \\ \sigma_{31} \\ \sigma_{12} \end{Bmatrix} = \begin{bmatrix} n & \ell & \ell & 0 & 0 & 0 \\ & (k+m) & (k-m) & 0 & 0 & 0 \\ & & (k+m) & 0 & 0 & 0 \\ & & & m & 0 & 0 \\ & & & & p & 0 \\ & \text{SYM} & & & & p \end{bmatrix} \begin{Bmatrix} \epsilon_{11} \\ \epsilon_{22} \\ \epsilon_{33} \\ 2\epsilon_{23} \\ 2\epsilon_{31} \\ 2\epsilon_{12} \end{Bmatrix} \tag{30}$$

Estimates of the overall moduli, **L**, of heterogeneous solids by the Mori–Tanaka method are given in Refs *11* and *12*. Also, the following formulas for the overall Hill's moduli of a fibrous composite in terms of the local moduli and the fiber volume fraction, c_f, were derived by Chen et al. [*23*] for the Mori–Tanaka method

$$p = \frac{2c_f p_m p_f + c_m(p_m p_f + p_m^2)}{2c_f p_m + c_m(p_f + p_m)} \tag{31}$$

$$m = \frac{m_m m_f(k_m + 2m_m) + k_m m_m(c_f m_f + c_m m_m)}{k_m m_m + (k_m + 2m_m)(c_f m_m + c_m m_f)} \tag{32}$$

$$k = \frac{c_f k_f (k_m + m_m) + c_m k_m (k_f + m_m)}{c_f (k_m + m_m) + c_m (k_f + m_m)} \tag{33}$$

$$\ell = \frac{c_f \ell_f (k_m + m_m) + c_m \ell_m (k_f + m_m)}{c_f (k_m + m_m) + c_m (k_f + m_m)} \tag{34}$$

$$n = c_m n_m + c_f n_f + (\ell - c_f \ell_f - c_m \ell_m) \frac{\ell_f - \ell_m}{k_f - k_m} \tag{35}$$

References

[1] Dvorak, G. J., ''Transformation Field Analysis of Inelastic Composite Materials,'' *Proceedings,* Royal Society of London, Vol. A437, 1992, pp. 311–327.

[2] Bahei–El–Din, Y. A., Shah, R. S., and Dvorak, G. J., ''Numerical Analysis of the Rate–Dependent Behavior of High Temperature Fibrous Composites,'' *Mechanics of Composites at Elevated and Cryogenic Temperatures,* S. N. Singhal, W. F. Jones, T. Cruse, and C. T. Herakovich, Eds., AMD–Vol. 188, American Society of Mechanical Engineers, New York, 1991, pp. 67–78.

[3] Shah, R. S., ''Modeling and Analysis of High Temperature Inelastic Deformation in Metal Matrix Composites,'' Ph.D. thesis, Rensselaer Polytechnic Institute, Troy, NY, 1991.

[4] Dafalias, Y. F. and Popov, E. P., ''Plastic Internal Variables Formalism of Cyclic Plasticity,'' *Journal of Applied Mechanics,* Vol. 43, 1976, pp. 645–651.

[5] Johnson, W. S., Mirdamadi, M., and Bahei–El–Din, Y. A., ''Stress–Strain Analysis of a $[0/90]_{2s}$ Titanium Matrix Laminate Subjected to a Generic Hypersonic Flight Profile,'' *Journal of Composites Technology and Research,* Vol. 15, 1993, pp. 297–303.

[6] Johnson, W. S. and Mirdamadi, M., private communication, 1993.

[7] Bahei–El–Din, Y. A., ''VISCOPAC Finite Element Program for Viscoplastic Analysis of Composites, User's Manual,'' Structural Engineering Department, Cairo University, Giza, Egypt, 1993.

[8] Dvorak, G. J. and Bahei–El–Din, Y. A., ''Plasticity Analysis of Fibrous Composites,'' *Journal of Applied Mechanics,* Vol. 104, 1982, pp. 327–335.

[9] Bahei–El–Din, Y. A., ''Plasticity Analysis of Fibrous Composite Laminates Under Thermomechanical Loads,'' *Thermal and Mechanical Behavior of Metal Matrix and Ceramic Matrix Composites, ASTM STP 1080,* J. M. Kennedy, H. H. Moeller, and W. S. Johnson, Eds., American Society for Testing and Materials, Philadelphia, 1990, pp. 20–39.

[10] Hill, R., ''Theory of Mechanical Properties of Fiber–Strengthened Materials; III. Self–Consistent Model,'' *Journal of the Mechanics and Physics of Solids,* Vol. 13, 1965, pp. 189–198.

[11] Mori, T. and Tanaka, K., ''Average Stress in Matrix and Average Elastic Energy of Materials with Misfitting Inclusions,'' *Acta Metallurgica,* Vol. 21, 1973, pp. 571–574.

[12] Benveniste, Y., ''A New Approach to the Application of Mori–Tanaka's Theory in Composite Materials,'' *Mechanics of Materials,* Vol. 6, 1987, pp. 147–157.

[13] Dvorak, G. J. and Teply, J. L., ''Periodic Hexagonal Array Models for Plasticity Analysis of Composite Materials,'' *Plasticity Today: Modeling, Methods and Applications, W. Olsazak Memorial Volume,* A. Sawczuk and V. Bianchi, Eds., Elsevier Scientific Publishing Co., Amsterdam, 1985, pp. 623–642.

[14] Teply, J. L. and Dvorak, G. J., ''Bounds on Overall Instantaneous Properties of Elastic–Plastic Composites,'' *Journal of the Mechanics and Physics of Solids,* Vol. 36, 1988, pp. 29–58.

[15] Dvorak, G. J., Bahei–El–Din, Y. A., and Wafa, A. M., ''Implementation of the Transformation Field Analysis for Inelastic Composite Materials,'' *Computational Mechanics,* Vol. 14, 1994, pp. 201–228.

[16] Dvorak, G. J., ''Plasticity Theories for Fibrous Composite Materials,'' *Metal Matrix Composites, Mechanisms and Properties,* R. K. Everett and R. J. Arsenault, Eds., Vol. 2, Academic Press, Boston, 1991, pp. 1–77.

[17] Dvorak, G. J. and Benveniste, Y., ''On Transformation Strains and Uniform Fields in Multiphase Elastic Media,'' *Proceedings,* Royal Society of London, Vol. A437, 1992, pp. 291–310.

[18] Hill, R., ''A Self–Consistent Mechanics of Composite Materials,'' *Journal of the Mechanics and Physics of Solids,* Vol. 13, 1965, pp. 213–222.

[19] Clyne, T. W. and Withers, P. J., *An Introduction to Metal Matrix Composites,* Cambridge University Press, Cambridge, 1993.

[20] Dvorak, G. J., Bahei–El–Din, Y. A., and Nigam, H., "Static and Fatigue Damage in High Temperature Composites," Annual Progress Report, Contract Number F–4962092–J–0391, Air Force Office of Scientific Research, Washington, DC, 1994.

[21] Neu, R. W. and Nicholas, T., "Effect of Laminate Orientation on the Behavior of a Titanium Matrix Composite," *Journal of Composites Technology and Research,* Vol. 16, 1994, pp. 214–224.

[22] Hill, R., "Elastic Properties of Reinforced Solids: Some Theoretical Principles," *Journal of the Mechanics and Physics of Solids,* Vol. 11, 1963, pp. 357–372.

[23] Chen, T., Dvorak, G. J., and Benveniste, Y., "Mori–Tanaka Estimates of the Overall Elastic Moduli of Certain Composite Materials," *Journal of Applied Mechanics,* Vol. 59, 1992, pp. 539–546.

Fatigue

Douglas J. Herrmann,[1] *Geoffrey T. Ward,*[1] *Eric J. Lawson,*[1] *and Ben M. Hillberry*[1]

Prediction of Matrix Fatigue Crack Initiation from Notches in Titanium Matrix Composites

REFERENCE: Herrmann, D. J., Ward, G. T., Lawson, E. J., and Hillberry, B. M., **"Prediction of Matrix Fatigue Crack Initiation from Notches in Titanium Matrix Composites,"** *Life Prediction Methodology for Titanium Matrix Composites, ASTM STP 1253,* W. S. Johnson, J. M. Larsen, and B. N. Cox, Eds., American Society for Testing and Materials, 1996, pp. 359–376.

ABSTRACT: Fatigue crack initiation from notches in titanium matrix composites (TMCs) is of interest in the design of TMC structures. A model is presented for predicting the fatigue crack initiation life of longitudinal, transverse, and symmetric cross-ply laminates using the strain-life behavior of the matrix material. The effective strain parameter, $\Delta\epsilon_{eff}$, proposed by Hillberry and Johnson is used to account for the residual stress state imposed during composite processing. Since the effective strain parameter requires knowledge of the local residual and applied strain in the matrix, methods for calculating these strains are developed. Model predictions were verified with crack initiation tests of $[0]_4$, $[90]_4$, and $[0/90]_s$ SCS-6/TIMETAL 21S center-hole specimens at room temperature and $[0_4/\overline{90}]_s$ SCS-6/Ti-15-3 center-hole specimens at 300 and 550°C.

KEYWORDS: fatigue (materials), titanium matrix composites, silicon-carbide fibers, notches, crack initiation, titanium, life prediction, titanium alloys, modeling

Continuous fiber titanium matrix composites (TMCs) are being considered for a number of advanced aerospace applications because of their good elevated temperature mechanical properties and relatively low density. Potential applications include structures for hypersonic aircraft and gas turbine structures. Early fatigue crack initiation from notches is of interest when designing TMC structures. Crack initiation can be a significant portion of the total fatigue life of a notched TMC component subject to high applied stresses. At lower applied loads, crack initiation may be a small portion of the total fatigue life compared to fatigue crack propagation [*1*], but knowledge of the crack initiation lives for various notch locations in a structure can assist in determining the fatigue critical locations.

Fatigue tests of center-hole and double-edge-notch SCS-6/Ti-15-3 TMC specimens show that fatigue cracks initiate in the Ti-15-3 matrix material and continue to grow leaving the SCS-6 fibers intact [*1–3*]. Hillberry and Johnson [*3*] have developed a local effective strain approach incorporating the Smith, Watson, Topper parameter [*4*] to predict fatigue crack initiation in the matrix at a notch in longitudinal plies of SCS-6/Ti-15-3. This method is based on strain-life fatigue analysis and uses the fatigue behavior of the matrix material to predict the cycles to crack initiation at a notch in the composite. While the model developed by Hillberry and Johnson predicts the crack initiation life in longitudinal plies of [0] and $[0/90]_s$ laminates, their model does not predict the cycles to crack initiation in the transverse

[1]Graduate student, graduate student, graduate student, and professor, respectively, School of Mechanical Engineering, Purdue University, West Lafayette, IN 47907–1288.

plies. Due to the lower strength and stiffness of the transverse plies, cracks may initiate in the transverse plies first and then propagate to the longitudinal plies. This was observed by Bakuckas et al. [*1*] in fatigue tests of $[0/90]_{2s}$ SCS-6/Ti-15-3. Since the surface ply is typically a longitudinal ply, ignoring the transverse plies could result in an unconservative life prediction. Also, previous studies of crack initiation from notches in TMCs have been limited to room temperature even though these materials are being considered for high temperature environments.

This paper presents a model for predicting the fatigue crack initiation life of longitudinal and transverse plies of TMCs at room and elevated temperatures based on the strain-life behavior of the matrix. The residual stress state imposed during composite processing is accounted for using the effective strain parameter, $\Delta\epsilon_{eff}$, proposed by Hillberry and Johnson [*3*]. Methods for calculating the local residual mechanical strain, the applied strain, and the applied strain range in the matrix required to determine $\Delta\epsilon_{eff}$ are also presented. Model predictions are compared with crack initiation test results for SCS-6/TIMETAL 21S[2] $[0]_4$, $[90]_4$, and $[0/90]_s$ center-hole specimens at room temperature and SCS-6/Ti-15-3 $[0_4/\overline{90}]_s$ center-hole specimens at 300 and 550°C.

Model Development

In continuous fiber titanium matrix composites such as SCS-6/Ti-15-3 and SCS-6/TIMETAL 21S, the elastic modulus of the silicon-carbide fiber is about four times the elastic modulus of the titanium matrix. Since the high stiffness fibers remain elastic during deformation, the matrix is subjected to a strain-control loading by the fibers in the loading direction. In TMCs, fatigue cracks tend to initiate in the matrix material. Thus, Hillberry and Johnson [*3*] proposed that the cycles to matrix fatigue crack initiation in the composite can be predicted using strain-life fatigue analysis for the matrix material under the restriction that the matrix strain state at the notch is dependent on the composite properties as well as the notch geometry.

Residual Mechanical Strain

Since TMCs are manufactured at high temperatures, heat treated, and cooled to room temperature, residual stresses are induced in the fiber and the matrix due to the coefficient of thermal expansion mismatch between the two materials. This residual stress state can be significant and must be accounted for when determining the matrix strain in the composite. These residual stresses lead directly to a residual matrix mechanical strain, ϵ_{res}. This residual strain causes a mean strain effect during fatigue cycling that has the effect of changing the strain ratio, R_ϵ, in the matrix. To account for this R_ϵ effect in the matrix material, Hillberry and Johnson [*3*] modified the Smith, Watson, and Topper effective parameter [*4*] to include the residual mechanical strain. The Hillberry and Johnson relationship for the effective strain parameter, $\Delta\epsilon_{eff}$, is

$$\Delta\epsilon_{eff} = \sqrt{\epsilon_{max}\frac{\Delta\epsilon}{2}} = \epsilon_{max}\sqrt{\frac{(1 - R_\epsilon)}{2}} \tag{1}$$

and

$$\epsilon_{max} = \epsilon_{max}^{app} + \epsilon_{res} \qquad \text{and} \qquad \Delta\epsilon = \epsilon_{max} - \epsilon_{min} \tag{2}$$

[2]Registered trademark of the Timet Corporation, Henderson, NV.

where

ϵ_{max} = maximum matrix strain,
ϵ_{min} = minimum matrix strain,
$\Delta\epsilon$ = matrix strain range,
ϵ_{max}^{app} = maximum applied strain, and
ϵ_{res} = matrix residual mechanical strain.

Once $\Delta\epsilon_{eff}$ is determined, the fatigue crack initiation life, N_i, can be calculated

$$N_i = N_i(\Delta\epsilon_{eff}) \tag{3}$$

where Eq 3 is constructed from smooth specimen, strain-controlled fatigue test results for the matrix material.

Longitudinal Ply Strain

The axial (fiber direction) strain can be assumed to be uniform in a longitudinal ply, thus the axial strain applied to the matrix is equal to the strain applied to the composite. The strain at the notch is accounted for using the strain concentration factor, K_ϵ. The value of K_ϵ can be determined by performing a finite element analysis of the component using the appropriate orthotropic material constants. Equations 1 and 2 can be written in terms of K_ϵ and the remote applied composite strains as

$$\Delta\epsilon_{eff} = \sqrt{(K_\epsilon e_{max} + \epsilon_{res})\frac{K_\epsilon \Delta e}{2}} \tag{4}$$

where

e_{max} = remote applied composite strain in the axial direction and
Δe = remote applied composite strain range in the axial direction.

Equation 4 is the same as the expression used by Hillberry and Johnson [*3*] if it is assumed that $\epsilon_{res} = \sigma_{zz}^{res}/E_m$ where σ_{zz}^{res} is the matrix residual stress in the axial (fiber) direction and E_m is the elastic modulus of the matrix.

Therefore, knowing e_{max}, Δe, K_ϵ, and ϵ_{res}, Eq 4 can be used to find $\Delta\epsilon_{eff}$ and then Eq 3 to determine the initiation life. Hillberry and Johnson showed that this model was in good agreement with matrix fatigue crack initiation data for $[0]_8$ and $[0/90]_{2s}$ SCS-6/Ti-15-3 center-hole and edge-notch specimens [*3*].

Transverse Ply Strain

The model of Hillberry and Johnson requires knowledge of the applied local matrix strains, ϵ_{max} and $\Delta\epsilon$, and residual mechanical strain, ϵ_{res}, to calculate $\Delta\epsilon_{eff}$ using Eq 1. The matrix fatigue crack initiation life is then predicted from Eq 3. Therefore, this model should be applicable to off-axis plies as long as the local applied and residual mechanical strains can be determined. For off-axis plies, this is difficult since the composite and matrix strains cannot be directly related as is the case for longitudinal plies.

For a transverse ply, the stress-strain state in the matrix can be calculated by applying the strain in the notch to a unit cell of the composite (Fig. 1). The applied notch strain can be

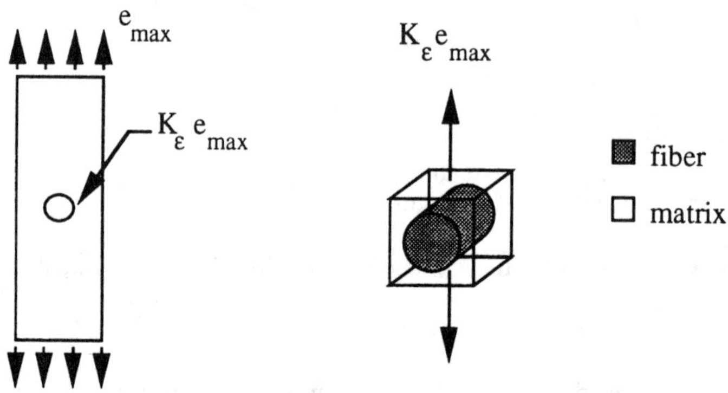

FIG. 1—*Relationship between the component and the fiber/matrix unit cell.*

easily determined from finite element analysis of the component using the appropriate orthotropic material constants. The stresses and strains in the unit cell can be calculated by a number of methods including finite element analysis, boundary element analysis, or the closed-form solution of Naik and Crews [*5*]. The maximum value of $\Delta\epsilon_{eff}$ calculated from the unit cell results can be used to predict the cycles to crack initiation. Only elastic deformation is considered here.

Determination of the matrix strains in the transverse ply is complicated by the nonlinear stress-strain behavior exhibited at low applied stresses by a number of TMCs including SCS-6/TIMETAL 21S and SCS-6/Ti-15-3. Johnson et al. [*6*] showed that the fiber/matrix bond for SCS-6/Ti-15-3 is very weak and can be exceeded at low load levels. After this bond is broken on the initial loading cycle, only the residual stresses due to processing keep the fiber and matrix in contact. After the fiber and matrix separate, the composite continues to deform elastically but the fiber no longer contributes to the transverse stiffness of the unidirectional composite. This significantly reduces stiffness and load carrying capability of the ply (Fig. 2). Therefore, the effect of fiber/matrix debonding must be included in calculating the strain in the matrix when the applied composite stress is large enough to overcome the residual stresses.

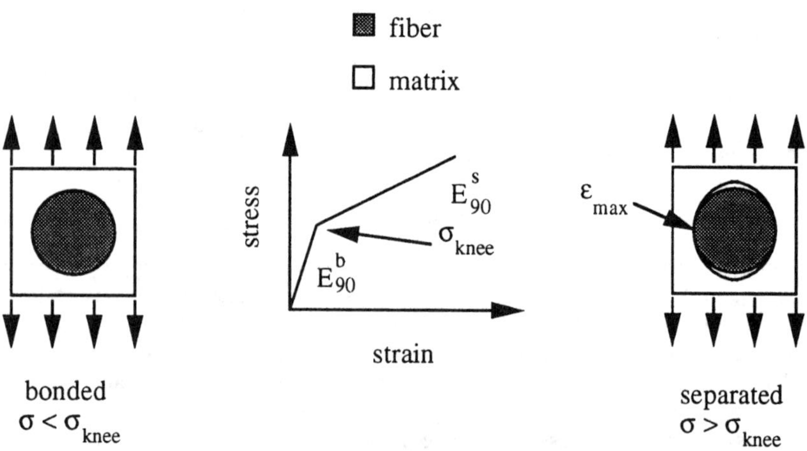

FIG. 2—*Stress-strain curve for fiber/matrix separation.*

Modeling Fiber/Matrix Separation—The stress-strain state in the matrix during fiber/matrix separation could be calculated by finite element analysis including the interface condition between the fiber and matrix. This is complicated by the fact that the fiber/matrix interface zone in TMCs is not well understood. For these reasons, an alternate method for approximating the matrix stress-strain state when fiber/matrix separation occurs was used in this research.

When the stress applied to the transverse ply is less than the stress at which the fiber and matrix separate, σ_{knee} (Fig. 2), the results from the unit cell model can be used directly. If the stress applied to the transverse ply is greater than σ_{knee}, fiber/matrix separation occurs and must be modeled. Mirdamadi and Johnson [7] approximated the stress-strain behavior after debonding by reducing the fiber transverse elastic modulus in their micro-mechanics model. Mirdamadi and Johnson chose the reduced value of the fiber transverse elastic modulus by fitting the results of their micromechanics model to the experimental stress-strain behavior of the composite. From the SCS-6/Ti-15-3 stress-strain data in Ref 7, the best fit for room temperature was determined to be

$$E_f^{\mathrm{T}} = 0.03E_f^{\mathrm{L}} \tag{5a}$$

where

E_f^{L} = fiber longitudinal modulus and
E_f^{T} = fiber transverse modulus.

This ratio was used in the unit cell model to account for the fiber/matrix separation at room temperature. At elevated temperatures

$$E_f^{\mathrm{T}} = 0.1E_f^{\mathrm{L}} \tag{5b}$$

was used as suggested by Mirdamadi and Johnson [7]. From the unit cell model for the debonded case, it is readily seen that the location in the matrix of maximum strain (critical location) is at the fiber/matrix interface perpendicular to the loading direction (Fig. 3).

Matrix Strain Calculation—For this study, the closed-form unit cell model of Naik and Crews [5] was used to determine the matrix strains for the transverse plies. As mentioned earlier, a finite element model could be used instead. However, the closed-form solution of

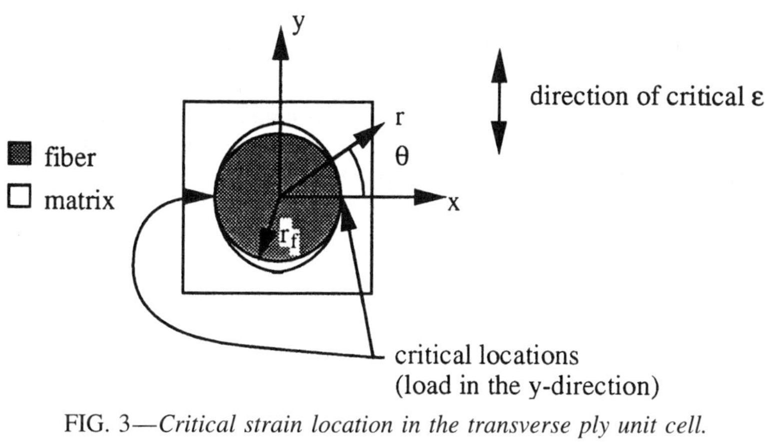

FIG. 3—*Critical strain location in the transverse ply unit cell.*

Naik and Crews is advantageous because of the savings in computational time and effort. The unit cell stress-strain analysis must be performed for the residual stress conditions and the transversely loaded composite for both the bonded and the debonded conditions. From these results, the value of ϵ_{max} and ϵ_{min} must be determined at the critical location, $r = r_f$ and $\theta = 0$ (Fig. 3), so that the matrix crack initiation life can be calculated using Eqs 1 through 3. The total mechanical strain is determined by superimposing the strains calculated from the unit cell model for the thermal residual stresses, the bonded composite with an applied stress, and the debonded composite when the applied stress exceeds σ_{knee}. This superposition can be used to determine the strain in the matrix because only elastic, reversible deformation is being considered.

Prior to fiber/matrix separation ($K_\epsilon S_{90} < \sigma_{knee}$), the strain at the critical location in the transverse direction is

$$\epsilon = \frac{K_\epsilon S_{90}}{E_m}[a^b_{\theta\theta} - \nu_m(a^b_{rr} + a^b_{zz})] + \epsilon_{res} \qquad \text{for} \quad K_\epsilon S_{90} < \sigma_{knee} \tag{6}$$

where

a^b_{rr} = influence coefficient in the radial direction for the bonded case,
$a^b_{\theta\theta}$ = influence coefficients in the tangential direction for the bonded case,
a^b_{zz} = influence coefficients in the axial direction for the bonded case,
E_m = elastic modulus of the matrix,
ν_m = Poisson's ratio of the matrix, and
S_{90} = remote stress applied to the transverse ply.

The influence coefficients and σ_{knee} can be determined from the bonded and residual stress unit cell results (Appendix I). Since only elastic deformation is being considered, K_ϵ is the same as the elastic stress concentration factor, K_t.

After the fiber and matrix separate ($K_\epsilon S_{90} > \sigma_{knee}$), the matrix strain in the transverse direction can be written as

$$\epsilon = \frac{K_\epsilon S_{90} - \sigma_{knee}}{E_m}[a^s_{\theta\theta} - \nu_m(a^s_{rr} + a^s_{zz}] + \frac{\sigma_{knee}}{E_m}[a^b_{\theta\theta} - \nu_m(a^b_{rr} + a^b_{zz})] + \epsilon_{res} \tag{7}$$

where

a^s_{rr} = influence coefficient in the radial direction for the case of fiber/matrix separation,
$a^s_{\theta\theta}$ = influence coefficient in the tangential direction for the case of fiber/matrix separation, and
a^s_{zz} = influence coefficient in the axial direction for the case of fiber/matrix separation.

The influence coefficients for the case of fiber/matrix separation can be determined from unit cell results (Appendix I). Using Eqs 6 and 7, the minimum and maximum local matrix strains can be calculated knowing the minimum and maximum remote stresses applied in the transverse ply. Knowing ϵ_{max} and ϵ_{min}, $\Delta\epsilon_{eff}$ can be calculated and the cycles to matrix fatigue crack initiation predicted (Eqs 1 through 3).

Cross-Ply Laminates

For cross-ply laminates, matrix fatigue crack initiation predictions can be made for the longitudinal and transverse plies separately. To perform these predictions, the portion of the load carried by each ply must be determined. Assuming that the strain through the thickness of the laminate is constant and is elastic deformation, the stresses in the 0° and 90° plies of a

$[0_m/90_n]_s$ laminate can be calculated readily using a rule of mixtures approach (Appendix II). The resulting applied remote strain in the longitudinal plies is

$$e = \frac{S}{E^b_{0/90}} \qquad \text{for} \quad K_\epsilon S < \sigma^c_{\text{knee}} \tag{8}$$

$$e = \frac{1}{K_\epsilon}\left[\frac{\sigma^c_{\text{knee}}}{E^b_{0/90}} + \frac{K_\epsilon S - \sigma^c_{\text{knee}}}{E^s_{0/90}}\right] \qquad \text{for} \quad K_\epsilon S > \sigma^c_{\text{knee}}$$

where

e = remote applied strain,
S = the remote applied composite stress,
$E^b_{0/90}$ = elastic modulus of the $[0_m/90_n]_s$ laminate prior to fiber/matrix separation,
$E^s_{0/90}$ = elastic modulus of the $[0_m/90_n]_s$ laminate prior to fiber/matrix separation, and
σ^c_{knee} = stress applied to the $[0_m/90_n]_s$ laminate at which fiber/matrix separation occurs.

The strain (stress) concentration factor, K_ϵ, is for the $[0_m/90_n]_s$ laminate. The results of Eq 8 can be substituted into Eq 4 to determine $\Delta\epsilon_{\text{eff}}$ for the longitudinal plies and thus the cycles to crack initiation in the longitudinal ply by Eq 3. In the same way the stress at the notch of the transverse plies can be written as

$$K_\epsilon S_{90} = K_\epsilon S \frac{E^b_{90}}{E^b_{0/90}} \qquad \text{for} \quad K_\epsilon S < \sigma^c_{\text{knee}} \tag{9}$$

$$K_\epsilon S_{90} = \sigma_{\text{knee}} + (K_\epsilon S - \sigma^c_{\text{knee}}) \frac{E^s_{90}}{E^s_{0/90}} \qquad \text{for} \quad K_\epsilon S > \sigma^c_{\text{knee}}$$

where

E^h_{90} = transverse elastic modulus prior to fiber/matrix separation and
E^s_{90} = transverse elastic modulus after fiber/matrix separation

The results of Eq 9 can be substituted into Eqs 6, 7, and 1 to determine $\Delta\epsilon_{\text{eff}}$. The cycles to crack initiation in the transverse ply are then obtained using Eq 3.

Materials and Methods

Materials

The materials tested as part of this research were SCS-6/TIMETAL 21S and SCS-6/Ti-15-3 metal matrix composites. SCS-6/TIMETAL 21S was tested at room temperature and SCS-6/Ti-15-3 was tested at elevated temperature. TIMETAL 21S (previously called Ti-β21S) is a metastable beta titanium alloy with the composition Ti-15Mo-3Al-2.7Nb-0.2Si (percent by weight). Ti-15-3 is a metastable beta titanium alloy with the composition Ti-15V-3Cr-3Al-3Sn (percent by weight). The matrix is reinforced with continuous SCS-6 fibers. The 0.14-mm-diameter fibers are comprised of silicon around a carbon core and are manufactured by Textron Specialty Materials Division.[3]

The SCS-6/TIMETAL 21S laminates were fabricated with no weave material by hot-pressing foils of TIMETAL 21S between unidirectional tapes of the SCS-6 fibers at a temperature near 1000°C. The plates were then cooled to 621°C and aged for 8 h to stabilize the matrix material, and finally cooled to room temperature. The SCS-6/Ti-15-3 material

[3]Textron Specialty Materials Division, Lowell, MA.

was manufactured similarly but was aged at 593°C for 24 h. Textron manufactured the materials used in this study.

The residual stress state for the SCS-6/TIMETAL 21S laminates was determined based on a stress-free temperature of 621°C cooled to the test temperature, 21°C. For the SCS-6/Ti-15-3 composite, a temperature change of 572°C was assumed for the cooling from the stress-free state to room temperature. These stress-free states assume that all residual stresses from hot pressing the fibers and matrix are relieved during the age heat treatment.

Test Specimens

Center hole specimens with a width of 19 mm, length of 150 mm, and hole diameter of 6.35 mm were tested in this study. The SCS-6/TIMETAL 21S $[0]_4$ and $[90]_4$ specimens were cut from the same plate. The fiber volume fraction was 36.3% and the mean thickness was 0.88 mm. The SCS-6/TIMETAL 21S $[0/90]_s$ specimens had fiber volume fraction of 34.8% and a mean thickness of 0.92 mm. The SCS-6/Ti-15-3 $[0_4/\overline{90}]_s$ specimens had an average thickness of 2.02 mm and a fiber volume fraction of 33.5%. All specimens tested were polished to a mirror finish.

The elastic strain (stress) concentration factor was calculated for each composite layup from a plane stress finite element model (Table 1). The required elastic constants were determined using the ANGLPLY laminate program [*8*] from the fiber and matrix properties (Table 2).

Mechanical Testing

Mechanical testing was performed using a 44.5 kN (10 000 lb) closed-loop, servohydraulic test system, controlled with a software program developed for a Sun 3 workstation [*9*]. The elevated temperature tests were performed using a free convection

TABLE 1—K_ϵ *for the center-hole specimens.*

Material	Temperature, °C	Layup	V_f, %	$K_\epsilon = K_t$
SCS-6/TIMETAL 21S	21	$[0]_4$	36.3	3.95
SCS-6/TIMETAL 21S	21	$[90]_4$	36.3	3.46
SCS-6/TIMETAL 21S	21	$[0/\overline{90}]_s$	34.8	3.72
SCS-6/Ti-15-3	300	$[0_4/\overline{90}]_s$	33.4	3.84
SCS-6/Ti-15-3	550	$[0_4/\overline{90}]_s$	33.4	3.92

TABLE 2—*Mechanical properties of the fiber and matrices.*

Material	Temperature, °C	E, GPa	ν	α, mm/mm/°C
SCS-6 [7]	23	393	0.25	3.79×10^{-6} [b]
SCS-6 [7]	300	393	0.25	3.92×10^{-6} [c]
SCS-6 [7]	550	393	0.25	4.11×10^{-6} [d]
TIMETAL 21S[a]	23	112	0.35	9.48×10^{-6} [b]
Ti-15-3 [7]	300	81	0.36	9.52×10^{-6} [c]
Ti-15-3 [7]	550	66	0.36	9.99×10^{-6} [d]

[a] Data from Dr. Massoud Mirdamadi, NASA Langley Research Center.
[b] Average α for ΔT from 621 to 21°C.
[c] Average α for ΔT from 576 to 300°C.
[d] Average α for ΔT from 576 to 550°C.

furnace [10]. The temperature distribution was controlled over a 12.5 mm gage section within ±5°C. All tests were performed in load control with a stress ratio of $R = 0.1$. The fatigue test frequencies were 10 Hz at room temperature, 1 Hz at 300°C, and 0.167 Hz at 550°C. These frequencies are consistent with those used in strain control fatigue tests of the matrix materials [11,12].

Crack initiation was defined as a crack on the specimen surface of 0.3 to 0.5 mm. The specimen surface was monitored for crack initiation using a video imaging system with magnification capability up to ×800 and surface replicas. To eliminate any contamination or etching of the titanium due to acetone impurities, the highest available grade acetone, 99.85%, was used for making the acetate tape replicas.

Model Predictions

$\Delta\epsilon_{eff}$ was calculated for the longitudinal and transverse plies of the SCS-6/TIMETAL 21S and SCS-6/Ti-15-3 laminates that were tested (Fig. 4). Predictions of the crack initiation life of SCS-6/TIMETAL 21S were made using the model presented here for both the 36.3% fiber volume fraction $[0]_4$ and $[90]_4$ specimens as well as the 34.8% fiber volume fraction $[0/90]_s$ specimens. A bilinear curve-fit of the TIMETAL 21S strain-life data presented by Ward et al. [13] was used to determine N_i from $\Delta\epsilon_{eff}$ (Appendix III). The matrix strain-life data can be transformed to $\Delta\epsilon_{eff}$ basis using Eq 1 and $\epsilon_{res} = 0$

$$\Delta\epsilon_{eff} = \frac{\Delta\epsilon}{\sqrt{2(1 - R_\epsilon)}} = \epsilon_{max}\sqrt{\frac{(1 - R_\epsilon)}{2}} \tag{10}$$

Elevated temperature Ti-15-3 strain control fatigue data has been generated by Gayda et al. [12]. These results are limited (seven data points) and exhibit significant scatter. Therefore, comparing the matrix and composite results on a $\Delta\epsilon_{eff}$ basis is more appropriate in evaluating the model presented here than attempting a life prediction.

Results and Discussion

The room temperature crack initiation life of the $[90]_4$ SCS-6/TIMETAL 21S specimens was significantly shorter than that of the $[0]_4$ specimens (Fig. 5, Table 3). This is expected since the longitudinal direction is significantly stiffer and stronger than the transverse direction. The $[0/90]_s$ specimens had crack initiation lives falling between those of the $[0]_4$ and $[90]_4$ specimens. Fatigue cracks in the $[0]_4$ and $[0/90]_s$ laminates initiated 60° to 75° from the loading axis that is consistent with previous results for SCS-6/Ti-15-3 [1,3,14]. Cracks in the $[90]_4$ specimens initiated 90° from the loading axis, as is the case in isotropic materials.

The predictions for the $[0]_4$ and $[90]_4$ specimens are in agreement with the experimental results (Fig. 6). This indicates that for the range of applied stresses considered here that the model of Hillberry and Johnson [3] for longitudinal plies can be applied to SCS-6/TIMETAL 21S. Also the model presented here for calculating the crack initiation life for transverse plies agrees well with the experimental values. The matrix fatigue crack initiation life for the $[90]_4$ specimens is very sensitive to the applied stress level. This is because after the fiber and matrix separate, the fibers no longer carry any load and large strains develop in the matrix.

The cycles to crack initiation for the $[0/90]_s$ specimens fall between the predictions for the transverse plies and the longitudinal plies of the cross-ply laminate (Fig. 7). This suggests that cracks may initiate in the transverse plies and grow to the surface of the specimen that is

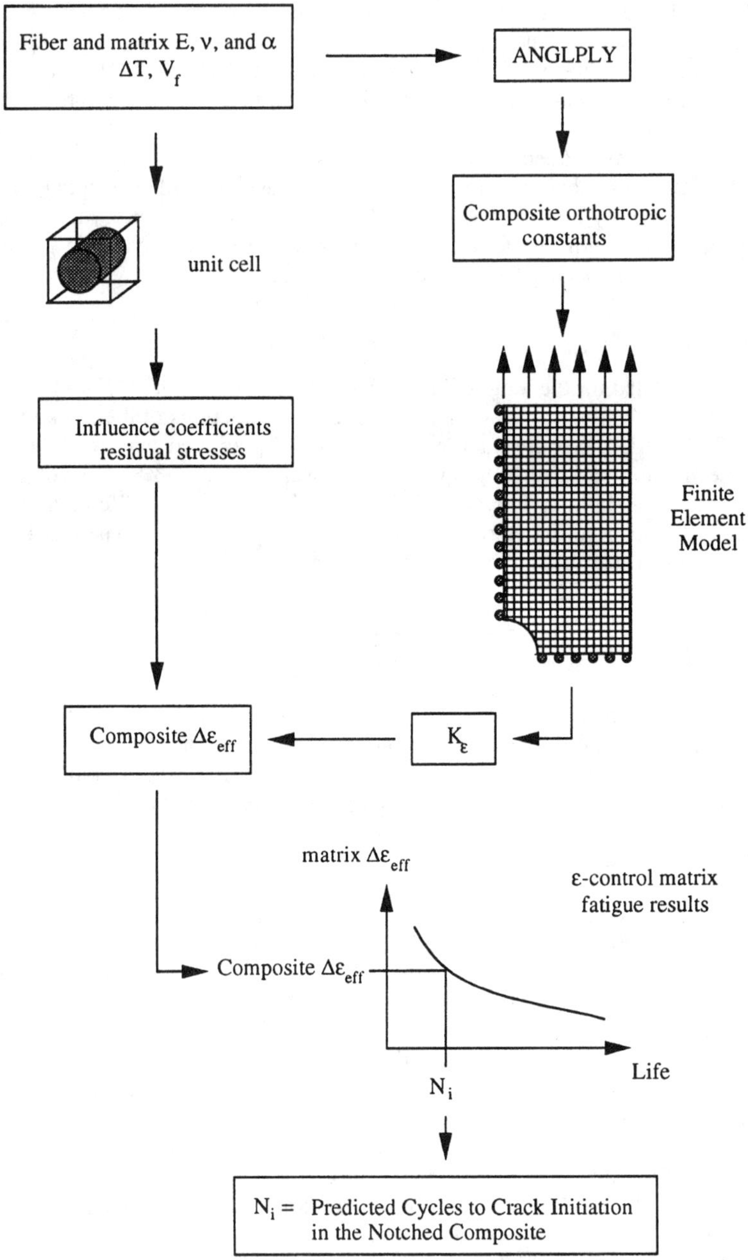

FIG. 4—*Life prediction model flowchart.*

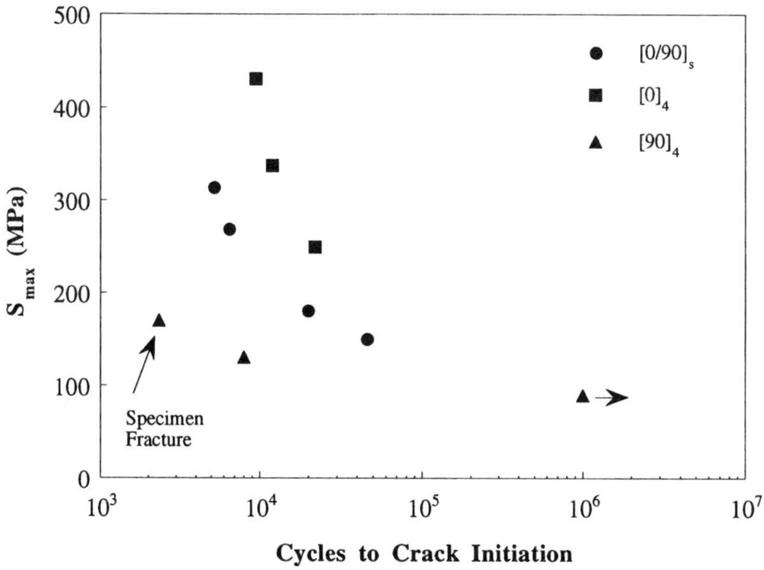

FIG. 5—*Results of the room temperature* R = *0.1 SCS-6/TIMETAL 21S fatigue crack initiation tests.*

TABLE 3—*Fatigue crack initiation data for the 21°C SCS-6/TIMETAL 21S center-hole specimens.*

Layup	S_{max}, MPa	Cycles to Initiation
$[0]_4$	431	9 500
$[0]_4$	337	12 000
$[0]_4$	249	22 000
$[90]_4$	170	2 332[a]
$[90]_4$	130	8 000
$[90]_4$	89	1 000 000[b]
$[0/90]_s$	314	5 200
$[0/90]_s$	268	6 500
$[0/90]_s$	180	20 000
$[0/90]_s$	150	46 100

[a]Specimen fractured.
[b]Test stopped at 1 000 000 cycles.

a longitudinal ply. This idea is supported by the fact that the cracks in all four of the $[0/90]_s$ specimens were through the thickness of the specimen. Also, Bakuckas et al. [*1*] identified SCS-6/Ti-15-3 matrix fatigue cracks initiating in transverse plies and growing into the longitudinal plies from edge replicas of center-hole fatigue specimens. Ward et al. [*13*] identified similar behavior in SCS-6/TIMETAL 21S from edge replicas of smooth fatigue specimens. Predictions for notched $[0/90]_s$ SCS-6/Ti-15-3 specimens yield similar results to these for SCS-6/TIMETAL 21S [*15*].

For the elevated temperature results, $\Delta\epsilon_{eff}$ was calculated for a longitudinal ply of the $[0_4/\overline{90}]_s$ laminate using the cross-ply analysis. Only the longitudinal-ply $\Delta\epsilon_{eff}$ is used

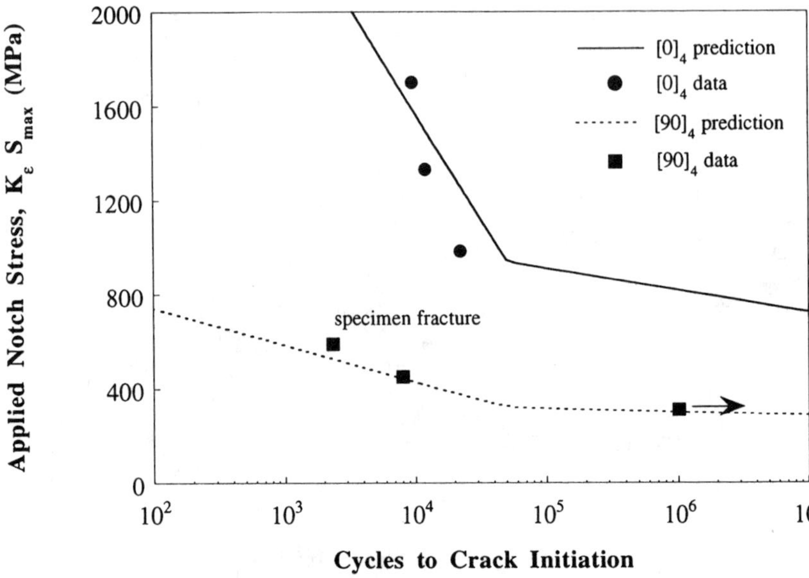

FIG. 6—*Unidirectional fatigue crack initiation life compared to unidirectional ply predictions for SCS-6/TIMETAL 21S with* V_f *= 36.3% and* R *= 0.1.*

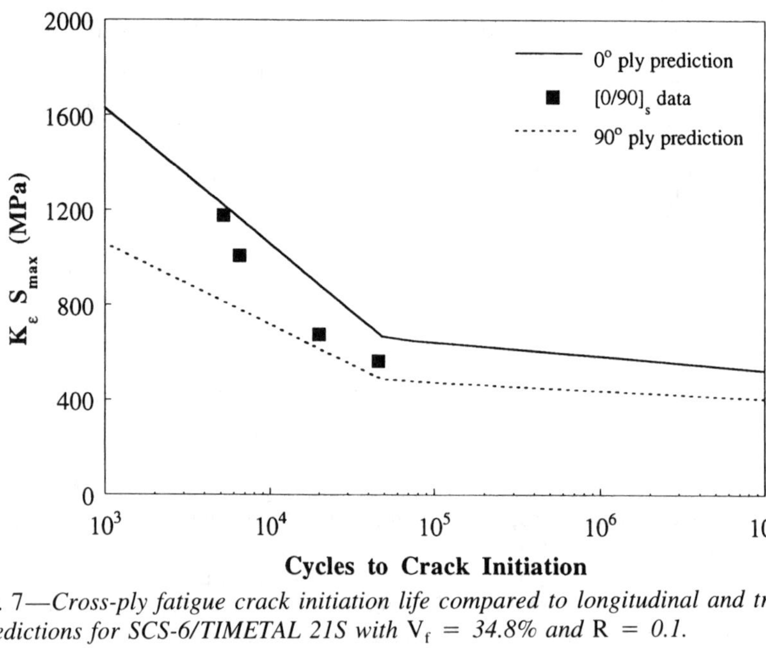

FIG. 7—*Cross-ply fatigue crack initiation life compared to longitudinal and transverse ply predictions for SCS-6/TIMETAL 21S with* V_f *= 34.8% and* R *= 0.1.*

because the longitudinal plies will dominate the behavior since there is only one transverse ply in the center of the laminate compared to four longitudinal plies on each side. $\Delta\epsilon_{eff}$ for the Ti-15-3 matrix was calculated for the smooth specimens tested by Gayda et al. [*12*]. The resulting values of $\Delta\epsilon_{eff}$ for the composite and matrix are plotted versus the number of cycles to crack initiation (Fig. 8, Table 4). The model would be in perfect agreement with the experimental data if the composite results fell on the matrix $\Delta\epsilon_{eff}$ versus life results. The results show very good agreement suggesting that $\Delta\epsilon_{eff}$ is an appropriate fatigue crack initiation parameter at elevated temperatures as well as room temperature.

Gayda et al. [*12*] note that the Ti-15-3 matrix undergoes a significant amount of stress relaxation at 550°C causing the mean stress to approach zero at a rapid rate. Even though the $\Delta\epsilon_{eff}$ model only considers elastic deformation in the calculation of $\Delta\epsilon_{eff}$, the effect of mean stress relaxation on life is accounted for by the use of strain-controlled matrix data. Similar to room temperature results [*1,3,13*], cracks in the 300 and 550°C $[0_4/\overline{90}]_s$ SCS-6/Ti-15-3 center-hole specimens initiated 60° to 75° from the loading axis.

Summary

The cycles to matrix fatigue crack initiation from notches in longitudinal and transverse plies of TMC laminates can be predicted using the effective strain parameter. Predictions for $[0]_4$, $[90]_4$, and $[0/90]_s$ laminates of SCS-6/TIMETAL 21S agree with experimental results. Fatigue crack initiation tests of notched $[0_4/\overline{90}]_s$ SCS-6/Ti-15-3 at elevated temperatures demonstrate the applicability of the effective strain model at elevated temperatures.

Acknowledgments

Support and SCS-6/TIMETAL 21S material was provided by NASA Langley Research Center under Grant NAG-1-1316 directed by Dr. W. Steven Johnson. The SCS-6/Ti-15-3 material was provided by Pratt and Whitney GESP.

APPENDIX I

Influence Coefficients

The matrix strain in a transverse ply at the critical location, Eqs 6 and 7, requires the results of a composite unit cell model subject to several loading conditions to determine the residual stress-strain state and the influence coefficients. Since in most finite element models and in the closed-form model [*5*] stresses are output as a default option, the residual strain and influence coefficients are written in terms of stresses resulting from the unit cell analysis. The residual stresses due to processing are calculated by applying the appropriate temperature change, ΔT, to the unit cell. The matrix residual strain, ϵ_{res}, in the tangential direction at $r = r_f$ and $\theta = 0$ (Fig. 3) is

$$\epsilon_{res} = \frac{1}{E_m}[\sigma_{\theta\theta}^{res} - \nu_m(\sigma_{rr}^{res} + \sigma_{zz}^{res})] \tag{11}$$

where

$\sigma_{\theta\theta}^{res}$ = matrix residual stress in the tangential direction at $r = r_f$ and $\theta = 0, \pi/2$;
σ_{rr}^{res} = matrix residual stress in the radial direction at $r = r_f$ and $\theta = 0, \pi/2$; and
σ_{zz}^{res} = matrix residual stress in the axial direction at $r = r_f$ and $\theta = 0, \pi/2$.

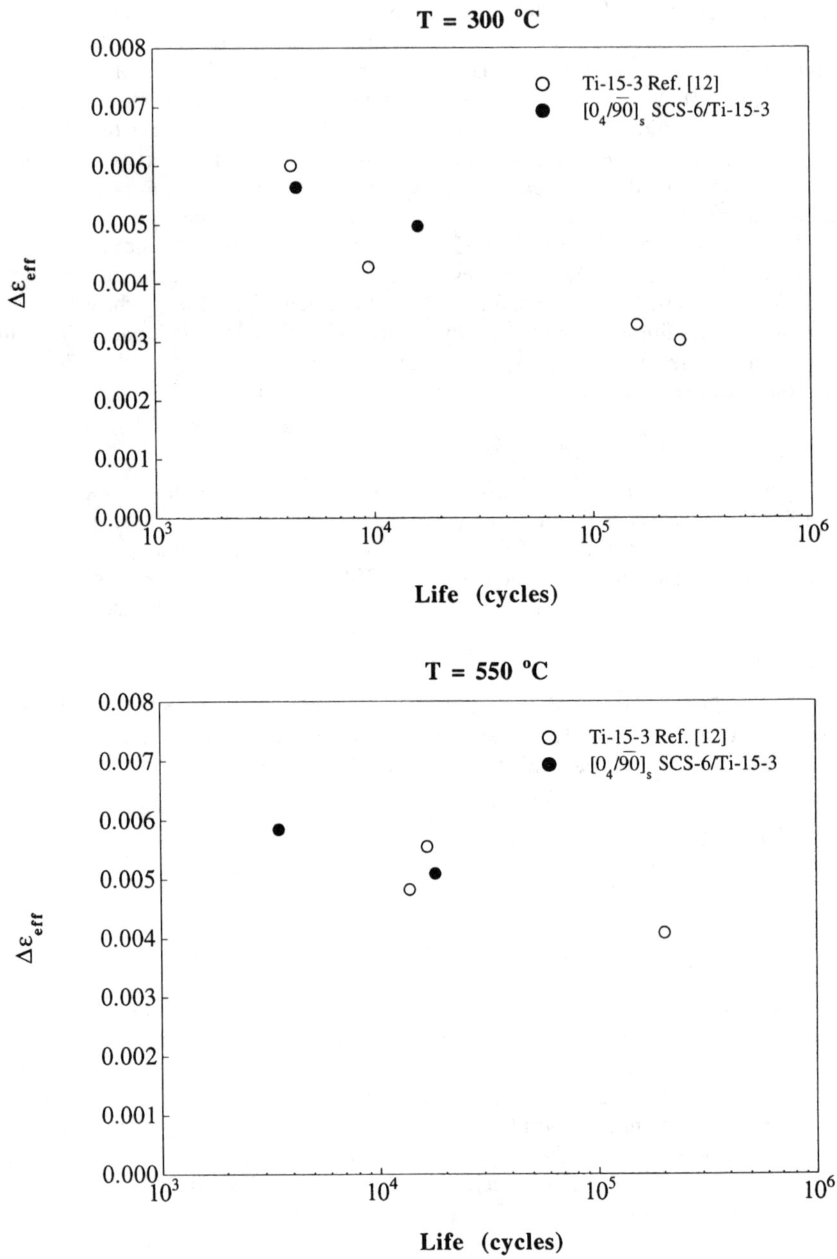

FIG. 8—*Comparison of Ti-15-3 and notched SCS-6/Ti-15-3 fatigue at* (a) *300°C and* (b) *550°C based on the effective strain parameter,* $\Delta\epsilon_{eff}$.

TABLE 4—*Fatigue crack initiation data for the $[0_4/\overline{90}]_s$ SCS-6/Ti-15-3 center-hole specimens.*

Temperature, °C	S_{max}, MPa	Cycles to Initiation
300	315	4 500
300	361	16 000
550	315	3 500
550	361	18 000

The stress-strain state before the fiber and matrix separate is calculated by applying a uniform displacement to the unit cell in the transverse direction and enforcing the symmetry boundary conditions. The results can be used to determine the bonded case influence coefficients a^b_{rr}, $a^b_{\theta\theta}$, and a^b_{zz}

$$a^b_{rr} = \frac{\sigma_{rr}(r_f,0)}{\bar{\sigma}_{\text{app}}}$$
$$a^b_{\theta\theta} = \frac{\sigma_{\theta\theta}(r_f,0)}{\bar{\sigma}_{\text{app}}} \tag{12}$$
$$a^b_{zz} = \frac{\sigma_{zz}(r_f,0)}{\bar{\sigma}_{\text{app}}}$$

where

$\bar{\sigma}_{app}$ = average stress applied to the unit cell resulting from the uniform applied displacement,
$\sigma_{rr}(r,\theta)$ = radial stress in the matrix,
$\sigma_{\theta\theta}(r,\theta)$ = tangential stress in the matrix, and
$\sigma_{zz}(r,\theta)$ = axial stress in the matrix.

The σ_{knee} can be determined from the bonded and debonded unit cell results. Fiber/matrix separation is considered to occur when

$$\sigma_{rr}(r_f,\pi/2) = 0 = (a^b_{rr})_{\pi/2}K_\epsilon S_{90} + \sigma^{\text{res}}_{rr} \quad \text{at} \quad K_\epsilon S_{90} = \sigma_{\text{knee}} \tag{13}$$

where

$$(a^b_{rr})_{\pi/2} = \frac{\sigma_{rr}(r_f,\pi/2)}{\bar{\sigma}_{\text{app}}} \tag{14}$$

Therefore, fiber/matrix separation occurs if

$$K_\epsilon S_{90} > \sigma_{\text{knee}} = \frac{-\sigma^{\text{res}}_{rr}}{(a^b_{rr})_{\pi/2}} \tag{15}$$

The additional strain that occurs after the fiber and matrix separate is calculated by applying a uniform displacement to the unit cell in the transverse direction, enforcing the symmetry boundary conditions, and using the debond condition, Eq 5. The results can be used to determine the influence coefficients for the case of fiber/matrix separation a^s_{rr}, $a^s_{\theta\theta}$, and a^s_{zz}

$$a^s_{rr} = \frac{\sigma_{rr}(r_f,0)}{\bar{\sigma}_{\text{app}}}$$

$$a^s_{\theta\theta} = \frac{\sigma_{\theta\theta}(r_f,0)}{\bar{\sigma}_{\text{app}}} \tag{16}$$

$$a^s_{zz} = \frac{\sigma_{zz}(r_f,0)}{\bar{\sigma}_{\text{app}}}$$

APPENDIX II

$[0_m/90_n]_s$ Elastic Moduli

Assuming that the strain through the thickness of the laminate is constant and elastic deformation, the stresses in the 0° and 90° plies of a $[0_m/90_n]_s$ laminate can be calculated. Referring to Fig. 9 and using a rule of mixtures approach, the elastic modulus of the $[0_m/90_n]_s$ laminate before fiber/matrix separation in the transverse plies is

$$E^b_{0/90} = \frac{1}{n + m}[mE_0 + nE^b_{90}] \tag{17}$$

where

E_0 = elastic modulus of the longitudinal ply,
E^b_{90} = elastic modulus of the transverse ply given by the unit cell analysis with the fiber and matrix bonded,

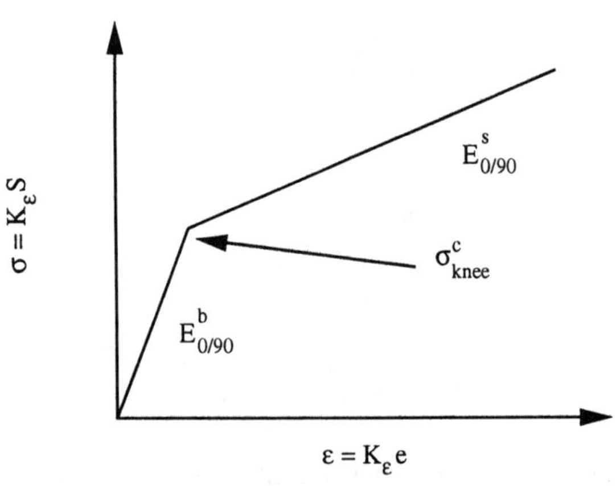

FIG. 9—*Elastic stress-strain relationship for a $[0_m/90_n]_s$ laminate.*

$E^b_{0/90}$ = elastic modulus of the cross-ply laminate while the fiber and matrix are bonded,
m = number of longitudinal plies, and
n = number of transverse plies.

After fiber/matrix separation, the slope of the cross-ply stress-strain relation becomes

$$E^s_{0/90} = \frac{1}{n+m}[mE_0 + nE^s_{90}] \tag{18}$$

where

E^s_{90} = elastic modulus of the transverse ply given by the unit cell analysis with the fiber and matrix separated and
$E^s_{0/90}$ = elastic modulus of the cross-ply laminate while the fiber and matrix are separated.

The stress level at which the slope in the stress-strain curve changes due to fiber/matrix separation for the $[0_m/90_n]_s$ laminate, σ^c_{knee}, can be written as

$$\sigma^c_{knee} = \sigma_{knee}\frac{E^b_{0/90}}{E^b_{90}} \tag{19}$$

APPENDIX III

$\Delta\epsilon_{eff} - N_i$ Relationship for TIMETAL 21S

The results of the strain controlled TIMETAL 21S fatigue tests presented in Ref *13* were converted to effective strain versus life ($\Delta\epsilon_{eff}$ versus N_i) using Eq 10. The data were fit with a bilinear curve-fit. The resulting relationship is

$$\log_{10} N_i = \frac{0.0177 - \Delta\epsilon_{eff}}{0.00301} \qquad \Delta\epsilon_{eff} > 0.00347 \tag{20}$$

$$\log_{10} N_i = \frac{0.0379 - \Delta\epsilon_{eff}}{0.0000671} \qquad \Delta\epsilon_{eff} < 0.00347$$

References

[*1*] Bakuckas, J. G., Jr., Johnson, W. S., and Bigelow, C. A., "Fatigue Damage in Cross-Ply Titanium Metal Matrix Composites Containing Center Holes," NASA TM 104197, National Aeronautics and Space Administration, Washington, DC, Jan. 1992.

[*2*] Naik, R. A. and Johnson, W. S., in "Fatigue of Continuous Fiber Reinforced Titanium Metal Matrix Composites," *Composite Materials: Fatigue and Fracture, 3rd Volume, ASTM STP 1110,* T. K. O'Brien, Ed., American Society for Testing and Materials, Philadelphia, 1991.

[*3*] Hillberry, B. M. and Johnson, W. S., "Prediction of Matrix Fatigue Crack Initiation in Notched SCS-6/Ti-15-3 Metal Matrix Composites," *Journal of Composites Technology & Research,* Vol. 14, No. 4, Winter 1992, pp. 221–224.

[*4*] Smith, K. N., Watson, P., and Topper, T. H., "A Stress-Strain Function for the Fatigue of Metals," *Journal of Materials,* Vol. 5, No. 4, Dec. 1970, pp. 767–778.

[5] Naik, R. A. and Crews, J. H., Jr., "Closed-Form Analysis of Fiber-Matrix Interface Stresses Under Thermo-Mechanical Loadings," NASA TM 107575, National Aeronautics and Space Administration, Washington, DC, Oct. 1991.

[6] Johnson, W. S., Lubowinski, S. J., and Highsmith, A. L., "Mechanical Characterization of Unnotched SCS-6/Ti-15-3 Metal Matrix Composites at Room Temperature," *Thermal and Mechanical Behavior of Ceramic and Metal Matrix Composites, ASTM STP 1080,* J. M. Kennedy, H. H. Moeller, and W. S. Johnson, Eds., American Society for Testing and Materials, Philadelphia, 1990, pp. 193–218.

[7] Mirdamadi, M. and Johnson, W. S., "Stress-Strain Analysis of a $[0/90]_{2s}$ Titanium Matrix Laminate Subjected to a Generic Hypersonic Flight Profile," NASA TM-107584, National Aeronautics and Space Administration, Washington, DC, March 1992.

[8] Bahei-El-Din, Y. A. and Dvorak, G. J., "Plasticity Analysis of Laminated Composite Plates," *Journal of Applied Mechanics,* Vol. 49, No. 4, 1982, pp. 740–746.

[9] McKeighan, P. C., Evans, R. D., and Hillberry, B. M., "Fatigue and Fracture Testing Using a Multitasking Minicomputer Workstation," *Applications of Automation Technology to Fatigue and Fracture Testing, ASTM STP 1092,* A. A. Braun, N. E. Ashbaugh, and F. M. Smith, Eds., American Society for Testing and Materials, Philadelphia, 1990, pp. 52–67.

[10] Lawson, E. J., "Design and Evaluation of a Free Convection Furnace for Thermal and Thermomechanical Testing of Materials," Master's thesis, Purdue University, West Lafayette, IN, Aug. 1994.

[11] Ward, G. T., Herrmann, D. J., and Hillberry, B. M., "Stress-Life Behavior of Unnotched SCS-6/Ti-β21S Cross-Ply Metal Matrix Composites at Room Temperature," *Proceedings,* Fatigue 93, Montreal, May 1993, Engineering Materials Advisory Services, Ltd., Warley, West Midlands, UK, pp. 1067–1072.

[12] Gayda, J., Jr., Gabb, T. P., and Freed, A. D., "The Isothermal Fatigue Behavior of a Unidirectional SiC/Ti Composite and the Ti Alloy Matrix," NASA TM-101984, National Aeronautics and Space Administration, Washington, DC, April 1989.

[13] Ward, G. T., Herrmann, D. J., and Hillberry, B. M., "Fatigue-Life Behavior and Matrix Fatigue Crack Spacing in Unnotched SCS-6/TIMETAL® 21S Metal Matrix Composites," *Journal of Composites Technology and Research,* Vol. 17, No. 3, July 1995, pp. 205–211.

[14] Newaz, G. M. and Majumdar, B. S., "Crack Initiation Around Holes in a Unidirectional MMC Under Fatigue Loading," *Engineering Fracture Mechanics,* Vol. 42, No. 4, 1992, pp. 699–711.

[15] Herrmann, D. J., Ward, G. T., and Hillberry, B. M., "Prediction of Matrix Fatigue Crack Initiation in Longitudinal and Transverse Plies of Notched Metal Matrix Composites," *Proceedings,* Fatigue 93, Montreal, May 1993, Engineering Materials Advisory Services, Ltd., Warley, West Midlands, UK, pp. 1097–1102.

S. M. Jeng,[1] *P. C. Wang,*[1] *and J.-M. Yang*[1]

Fatigue Damage Evolution and Degradation of Mechanical Properties in Silicon-Carbide (SiC) Fiber-Reinforced Titanium Matrix Composites

REFERENCE: Jeng, S. M., Wang, P. C., and Yang, J.-M., "**Fatigue Damage Evolution and Degradation of Mechanical Properties in Silicon-Carbide (SiC) Fiber-Reinforced Titanium Matrix Composites,**" *Life Prediction Methodology for Titanium Matrix Composites, ASTM STP 1253,* W. S. Johnson, J. M. Larsen, and B. N. Cox, Eds., American Society for Testing and Materials, 1996, pp. 377–394.

ABSTRACT: Fatigue damage accumulation and failure modes of several unidirectional and angle-ply SCS-6 fiber-reinforced titanium matrix composites are reviewed. The fatigue damage characteristics of the notched and unnotched composites were classified using several schematic diagrams based upon the constituent's properties and loading conditions. Quantitative relationships between the microstructural damage and stiffness reduction were also established. A shear-lag model incorporating a damage function (which involved the cracking density and length, fiber/matrix frictional stress, and debonding length) was also used to predict residual stiffness of the composites.

KEYWORDS: titanium, titanium matrix composites, life prediction, titanium alloys, fatigue (materials), modeling, degradation, damage evolution

Continuous fiber-reinforced titanium matrix composites are one of the most promising structural materials for advanced aerospace structural applications. Owing to their highly anisotropic behavior, unidirectional fiber-reinforced titanium can only be used in very specific applications such as hoop-wound turbine engine rings and nozzles [*1*]. For components subjected to multiaxial loading, such as the hat-channel-stiffened channels in hypersonic airframes, composites with a balance of in-plane properties need to be developed. This can be achieved through the proper construction of multilayer, angle-ply laminates. However, before these materials can be used, a comprehensive understanding of their damage mechanisms and fracture behavior under different thermal or mechanical loading conditions or both must be established. Meanwhile, the effect of these damage mechanisms on the degradation of the mechanical properties of the composites also needs to be developed. In general, fatigue damage in these composites is a complex process that depends on many variables including constituent properties, lay-up sequences, loading conditions, and specimen geometries. The constituent properties that will affect the mechanical degradation of the composite during the fatigue loading include the statistical characteristic of the fiber strength, strength of the interfacial reaction layer, interfacial bonding strength, frictional stress, and matrix toughness [*2,3*].

In this paper, the fatigue properties and fatigue damage characteristics of several SCS-6

[1]Post doctoral fellow, graduate student, and associate professor, respectively, Department of Materials Science and Engineering, University of California, Los Angeles, CA 90024–1595.

fiber-reinforced titanium matrix composites are examined. Emphasis will be placed on correlating the microstructural damage with property degradation. A modified theoretical model based upon shear-lag analysis was also adopted to predict the effect of microstructural damage on degradation of mechanical properties.

Mechanical Properties of the Constituents and Composites

A wide range of SCS-6 fiber-reinforced titanium composites have been developed in recent years. The matrix alloys include Ti-6-4 ($\alpha+\beta$), Ti-15-3 (metastable β), Ti-6242 (near α), β-21S (metastable β), Ti-24-11 ($\alpha_2+\beta$), Ti-25-10 ($\alpha_2+\beta$), and Ti-22-23 (orthorhombic Ti_3Al). Typical mechanical properties of the matrix alloys and composites are shown in Table 1 [*4–10*]. The fracture toughness and ductility of conventional titanium alloys can be as high as 100 to 120 MPa·m$^{1/2}$ and 10%, respectively. However, the toughness and ductility of brittle titanium aluminide (Ti-25-10 and Ti-24-11) are only around 15 to 20 MPa·m$^{1/2}$ and 2 to 3%, respectively.

A SCS-6 fiber is a chemical vapor deposition (CVD)-silicone-carbide (SiC) fiber with a carbon core and a 3-μm thick carbon-rich surface coating. The strength, elastic modulus, and fracture strain of the as-fabricated fiber are 4000 MPa, 400 GPa, and 1%, respectively. However, after high temperature consolidation of the composite, a reaction zone was formed at the fiber/matrix interface that may adversely affect the fiber strength. The reaction products have been identified as titanium carbide and silicide in the conventional titanium matrix composites; because complex carbides and silicides were identified in the titanium aluminide matrix composites. Meanwhile, a significant number of microvoids were also found in the reaction layer. Since the carbide and silicides are very brittle, the strength of the interfacial reaction layer is usually low and shows a thickness dependence as shown in Fig. 1 [*10*]. Meanwhile, the Weibull modulus of the interfacial reaction layer was found to be around 2. Due to the low strength of the reaction layer, early cracking of this layer would induce a notch effect on the fiber under mechanical loading. As a result, the fiber would certainly suffer strength degradation depending on the reaction layer thickness. A previous

TABLE 1—*Typical mechanical properties of the SCS-6 fiber-reinforced titanium matrix composites and alloys.*

	Matrix				Composite[a]		
	Modulus, GPa	Ultimate Tensile Strength, MPa	Fracture Strain, %	Toughness, MPa·m$^{1/2}$	Modulus, GPa	Ultimate Tensile Strength, MPa	Fracture Strain, %
Ti-6-4[b]	114	896	10	115	213	1634	1.1
Ti-15-3[b]	90	772	8 to 12	100	210	1572	0.7
Ti-6242[c]	103	972	15 to 17	NA[d]	217	1655	0.87
β-21S[c]	78	890	6 to 21	NA	205	1964	NA
Ti-25-10[c]	125	880	2	14 to 17	189	1300	0.76
Ti-24-11[c]	93	724	2	18 to 21	166	1271	0.78
Ti-22-23[c]	117	1159	~4	NA	NA	~1780	NA

[a]The fiber volume fraction for all the composites are around 33 to 40%.

[b]The mechanical data for Ti-6-4 and Ti-15-3 are from bulk specimen.

[c]The mechanical data for Ti-6242, β-21S, Ti-25-10, Ti-24-11, and Ti-22-23 matrix are from foil specimen. Meanwhile, the properties for Ti-6242 and β-21S are strongly dependent on heat treatment.

[d]NA = not available.

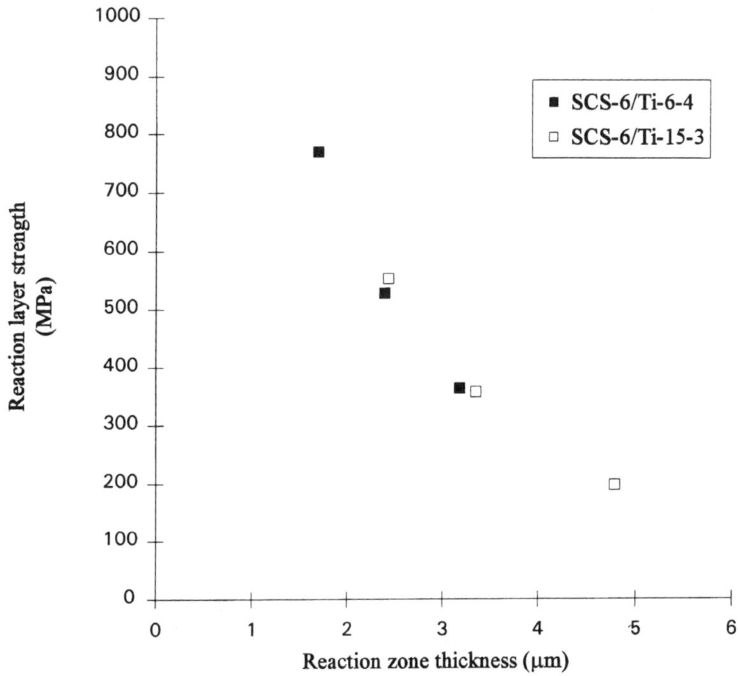

FIG. 1—*Variation of tensile strength of interfacial reaction layer as a function of reaction layer thickness of SCS-6/Ti alloy matrix composites.*

study showed that the relationship between the fiber strength and thickness of the reaction layer followed the equation

$$\sigma_f = 1/(\lambda c^{1/2})$$

where λ is a material constant and c is the reaction layer thickness, as shown in Fig. 2 [*11*]. The results also showed that the fiber can retain its strength if the interfacial reaction layer can be controlled under 1 μm. Meanwhile, the interfacial reaction layer increased the roughness of the SCS-6 fiber surface that, in turn, increased the scattering of fiber strength. A previous study indicated that the Weibull modulus of the SCS-6 fiber decreased from $m \approx 10$ in the as-fabricated state to $m \approx 3$ after consolidation [*11*].

The interfacial shear strength (debonding strength) and post-debonded frictional stress between the fiber and matrix are two important factors that affect the mechanical properties of these composites. The interfacial shear strength of the SCS-6/Ti matrix composites were measured using the fiber push-out method and ranged from 115 to 165 MPa (Table 2), depending on processing conditions, matrix alloy composition, and heat-treatment history [*11*]. The widely scattered interfacial shear strength can be roughly classified into three categories: weak interface (<110 MPa), medium strong interface (~125 MPa), and strong interface (>155 MPa). However, the weak or strong interface is dependent on relative fiber strength, matrix toughness, and loading condition. The frictional stresses were also measured by pushing-back the debonded fiber and were found to range from 58 to 90 MPa (Table 2). Meanwhile, it has been shown that the frictional stress is sensitive to loading history. The frictional stress in SCS-6/Ti-15-3 composites have also been studied after being subjected to different loading conditions. The fiber push-out-back method was used to study

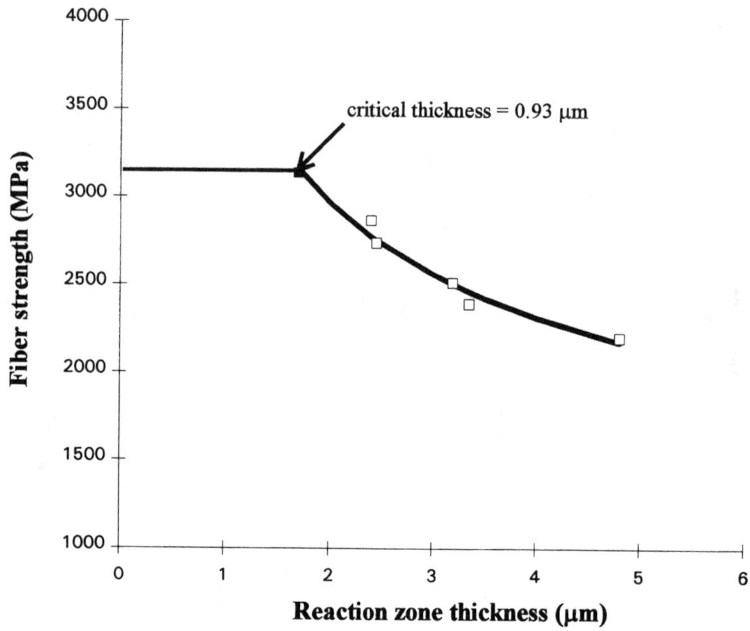

FIG. 2—*Variation of the SCS-6 fiber strength as a function of the reaction layer thickness in titanium alloy matrix composites.*

the reverse loading effect. The results showed that the frictional stresses gradually decreased as the reverse loading cycles increased. The frictional stress was around half of the original stress (40 MPa) for the SCS-6/Ti-15-3 composite after being pushed out and back 5 to 6 times [*12*]. The frictional stresses in the SCS-6/Ti-15-3 composite measured from the fatigue matrix cracking region and single-edge fatigue crack propagation specimens were found to decrease to around 15 to 35 MPa [*13,14*].

Fatigue Damage Evolution in Unnotched Composites

The fatigue properties and behavior of unnotched composites discussed later are based upon tests conducted under load-controlled, tension-tension fatigue with $f = 10$ Hz and $R = 0.1$.

Unidirectional Composites

The relationship between fatigue life, the maximum applied stresses in the fiber, and dominant microstructural damage mechanisms in SCS-6 fiber-reinforced titanium matrix

TABLE 2—*Interfacial mechanical properties of the SCS-6/Ti alloy matrix composites.*

	Interfacial Shear Strength,[a] MPa	Frictional Stress,[a] MPa
SCS-6/Ti-6-4	156 to 167	87 to 90
SCS-6/Ti-15-3	120 to 130	~80
SCS-6/Ti-25-10	96 to 115	58 to 62

[a]The interfacial shear strength and frictional stress were measured for the composite in as-fabricated state and heat treated at 800°C for 50 to 100 h.

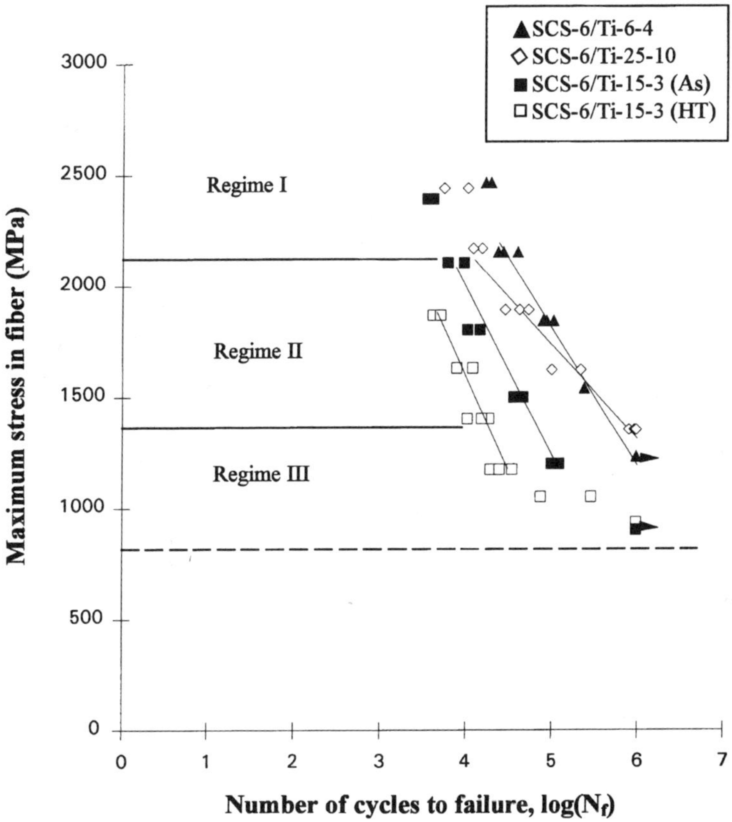

FIG. 3—*Fatigue damage diagram for the SCS-6/Ti alloy matrix composites (AS = as-fabricated and HT = 800°C/50 h).*

composites can be described by a fatigue damage life diagram as shown in Fig. 3 [*2*]. The stresses in the fibers were calculated from the maximum applied stresses by using the initial composite elastic modulus and rule-of-mixtures. The thermal residual stress was not taken into account in the calculation, since it may be released through matrix cracking or other fatigue damage mechanisms or both. Depending on the microstructural damage mechanisms, three regimes can be roughly classified. In the high applied stress regime (Regime I), the fatigue life was less than 10^3 cycles and the composite showed a catastrophic failure behavior. For the intermediate stress regime (Regime II), the fatigue life was around 10^3 to 10^5 cycles and exhibited a linear relationship with the applied stresses. However, when the applied stresses were further decreased (Regime III), the fatigue life increased dramatically. The fatigue life is usually higher than 10^6 cycles. However, the boundary between Regimes II and III can not be clearly defined. Meanwhile, the scattering of fatigue life in each regime formed a scatter band for all the composites. This is due to the statistical distribution of the fatigue data.

For Regime I, the lower bound of applied stress in the fibers were found to be around 2000 to 2500 MPa in all the composites. This stress level was around 80% of the SCS-6 fiber strength after high temperature consolidation, and the probability of the fiber failure was found to be around 40%. Microstructural observation indicated that fiber breakage was the dominant mechanism with no evidence of fatigue damage in the matrix being found.

This is due to the accumulation of fiber breakage that controlled the failure of the composites in this applied stress range. This suggests that the fatigue life in this regime is controlled by fiber strength and its statistical distribution.

In Regime II (progressive damage regime), microstructural observation indicated that matrix cracking, cracking of the interfacial reaction layer, and fiber breakage accumulated progressively during fatigue. First, cracks initiated from the specimen edge (from damaged fibers and interfaces due to machining) and the interfacial reaction layer away from the surface. As these cracks propagated, bridging fibers were left in the wake of the matrix cracks. The bridging fibers imposed a significant closing pressure on the cracked matrix. This significantly reduced the driving force (stress intensity factor) at the matrix crack tip (discussed in detail in the crack propagation section), which in turn, reduced the crack propagation rate. However, as the fatigue loading continued, random fiber breakage occurred near the cracked matrix. This is due to the statistical fiber strength distribution and increasing of stresses in the bridging fibers near the cracked matrix. Furthermore, the population of the random fiber breakage decreased as the applied stresses decreased. Meanwhile, significant interfacial debonding was also observed near the cracked matrix region. The interfacial debonding lengths were quite scattered and ranged from 20 to 100 μm. Furthermore, in this regime, the crack initiation locations were very sensitive to the finish of the specimen surface (especially the specimen edge). For the roughly polished[2] specimens, the dominating cracks were initiated from the damaged fiber and interfacial reaction layer. These defects are created from mechanical machining during the specimen preparation. Very few cracks that initiated from the interfacial reaction layers away from the specimen edges were observed. As a result, the damaged specimen contained significant numbers of long matrix cracks with almost equal spacing. However, for the finely polished[3] specimens, cracking of the interior interfacial reaction layer became the dominant crack initiation location. This lead to a significant amount of short matrix cracks randomly distributed across the specimen. As a result, the finely polished specimens also had a higher fatigue life ($\sim$10 times at σ_{max} = 800 MPa) than the roughly polished ones [*2*].

For Regime III, matrix cracking initiated primarily from surface defects and propagated across the whole specimen section. Cracking of the interior interfacial reaction layer was not observed. Meanwhile, the fiber breakage that was observed in Regime II was not seen in this regime. After 10^6 fatigue cycles in this applied stress regime, the residual strength and elastic modulus of the unbroken composites were close to those calculated by the rule-of-mixtures using fiber properties only ($\sigma_c = \sigma_f V_f$, $Ec = E_f V_f$) [*2*]. Meanwhile, the stresses in the fibers that were calculated using the residual elastic modulus were found to be around 1500 to 1600 MPa for all of the SCS-6/Ti composites. This implied that the fibers may exhibit a threshold stress below which the composite will not fail during fatigue loading.

In order to understand the effect of damage accumulation on the degradation of mechanical properties, a series of elastic modulus measurements was conducted using an extensometer at applied stresses near the boundary of the Regimes II and III on the finely-polished SCS-6/Ti-15-3 composite specimens. Figure 4 shows the degradation of elastic modulus at different applied maximum stress levels [*15*]. Meanwhile, residual stiffness of composites after heat-treatment at 800°C for 50 h (interface thickness was around 5 μm) were also conducted to investigate the effect of the reaction layer thickness on the stiffness reduction.

[2]Roughly polished refers to machining followed by grinding using 180, 240, 400, and 600 SiC papers.

[3]Finely polished refers to machining followed by grinding using a 30-μm diamond disk to remove a 75-μm thickness from edge and fine polishing by 180, 240, 400, and 600 SiC paper and 1-μm diamond paste.

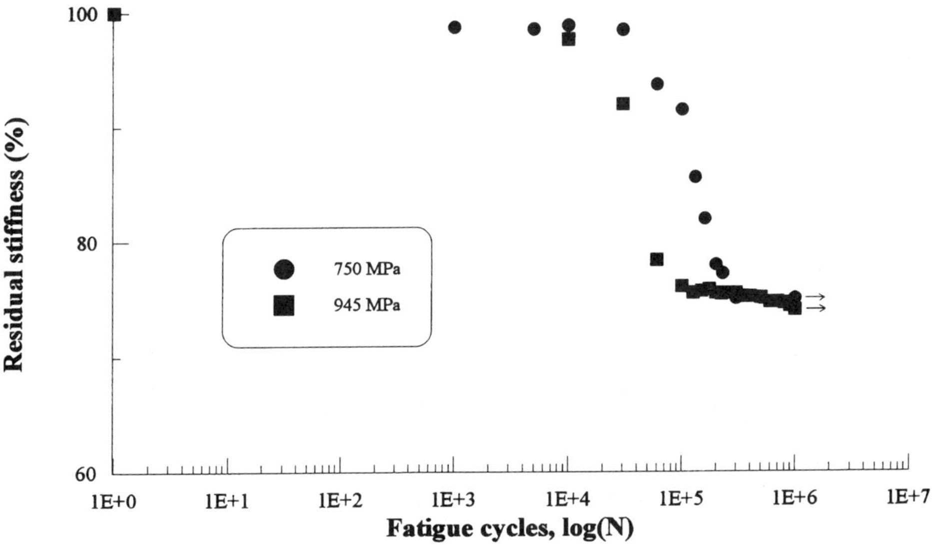

FIG. 4—*Residual stiffness as a function of fatigue cycles at both 750 and 945 MPa maximum applied stress levels of unidirectional SCS-6/Ti-15-3 composite.*

The stiffness reduction curves of the as-fabricated and heat-treated composites at an applied maximum stress of 750 MPa are shown in Fig. 5. All the curves exhibited an initial slow stiffness degradation followed by a rapid stiffness drop and then a saturation stage with minimal change in stiffness. The transition from slow reduction regime to rapid stiffness drop regime occurred around 10^4 cycles, where the residual stiffness was around 98%. The rapid stiffness drop occurred between $\sim 10^4$ to $\sim 10^5$ cycles, and the residual stiffness

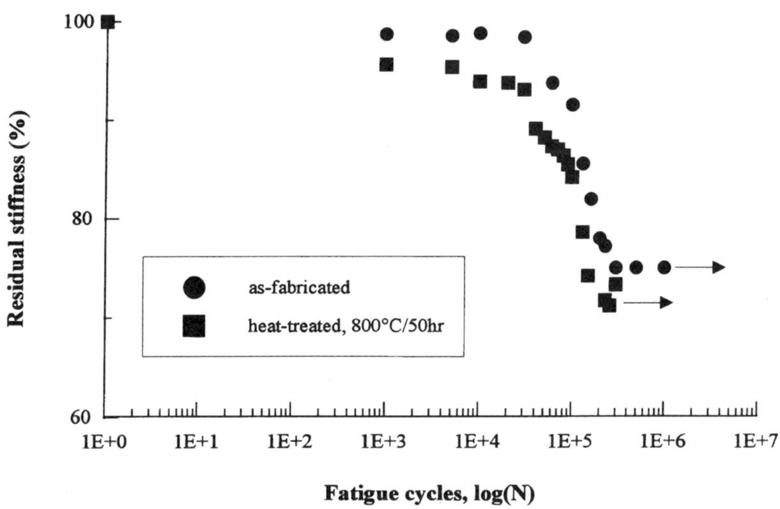

FIG. 5—*Residual stiffness curves for the as-fabricated and heat-treated SCS-6/Ti-15-3 composites tested at σ_{max} = 750 MPa.*

decreased from 98% to around 70%. In the saturation stage, the composite exhibited almost a near constant residual elastic modulus. Furthermore, the heat-treated composite showed a more severe stiffness reduction than did the as-fabricated composite in the slow reduction regime. However, the stiffness in the rapid drop and saturation regimes were almost the same for these two composites.

The crack length and density in the interfacial reaction layer and matrix were measured in composites tested at different fatigue cycles as shown in Table 3. Figure 6 shows the cracking density as a function of fatigue cycles for the as-fabricated and heat-treated composites tested at σ_{max} = 750 MPa. Several important characteristics of damage accumulation were observed from these results. The crack density in the interfacial reaction layer is always higher than that of matrix. Meanwhile, the crack density in the interfacial reaction layer and matrix increased dramatically between 10^4 to 10^5 cycles. These fatigue cycles corresponded to the transition between Regime I and Regime II. However, in this stage, the average matrix cracking length increased only slightly. This implied that the new initiated matrix cracks propagated very fast. The rapid nucleation and growth of the matrix cracks were the primary damage mechanisms in the rapid stiffness reduction regime. Meanwhile, the heat-treated composite exhibited a more severe stiffness reduction than the as-fabricated one. This was due to a thicker and weaker interfacial reaction layer in the heat-treated composite than that in the as-fabricated one. The rapid stiffness drop in the unidirectional composites was due to rapid accumulation of cracks in the interfacial reaction layer and the matrix. Since the cracking of the interface (especially long matrix cracks) induced a severe stress concentration along the fiber/matrix interface, a certain amount of interfacial debonding would also occur as the fatigue loading progressed. This interfacial debonding would reduce the load transfer efficiency between the fiber and matrix that, in turn, would reduce the effective elastic modulus of the composite. The increase of interface cracking and the matrix cracking population suggest that the effective interfacial debonding area would increase significantly. As a result, the residual stiffness of the composite would drop rapidly. Furthermore, the final failure of the composite was caused by the connection of several matrix fatigue cracks. This implied that none of the matrix cracks propagate through the specimen during the fatigue loading life. For the saturated regime, the average matrix cracking length for different applied stresses were very close, ranging from 35 to 45 μm. Meanwhile, the cracking density of the reaction layer was also very similar for all the testing

TABLE 3—*Matrix cracking spacing and crack length of SCS-6/Ti-15-3 composite at different fatigue loading stage.*

Residual Stiffness, %	Average Crack Spacing, μm	Average Crack Length, μm
	σ_{max} = 945 MPa	
97.3	615.7	32.75
89.1	364.8	24.75
85	251.2	30.99
74	146.4	33.63
	σ_{max} = 845 MPa	
96.2	912.6	28.71
81.4	307.2	38.39
67.8	169.9	38.36
	σ_{max} = 750 MPa	
77	531.2	41.1
75	413.2	43.3
71.9	302.3	42.2

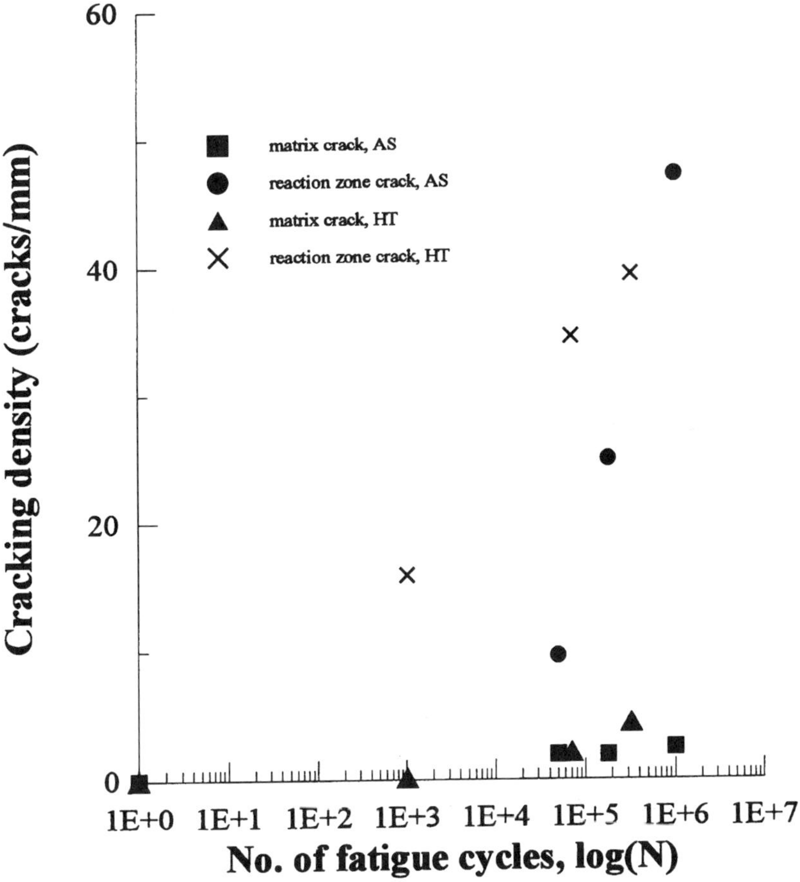

FIG. 6—*Cracking density as a function of fatigue cycles for the as-fabricated (AS) and heat-treated (HT, 800°C/50 h) composites tested at* σ_{max} = *750 MPa.*

conditions, ranging from 40 to 47 mm^{-1}. These results suggest that the unidirectional composite may exhibit a damage saturation stage with minimal change in the cracking density of the interfacial reaction layer and matrix.

Angle-Ply Composites

The fatigue behavior of several angle-ply SCS-6/Ti-15-3 laminates have also been investigated. Typical fatigue life curves for three different SCS-6/Ti-15-3 laminates are shown in Fig. 7 [*16*]. The initial applied stress was also converted to the stress in the 0° fiber for comparison. The results indicated that all of the composites exhibited a linear relationship between the initial stresses in fibers and fatigue life in the range of 10^3 to 10^5 cycles. Meanwhile, the fatigue life of the angle-ply composite showed more stress sensitivity than did the unidirectional composites. When the fatigue life exceeded 10^6 cycles, the corresponding initial stresses in the 0° fibers were found to be around 800 to 1000 MPa for these three laminates.

Microstructural observation showed that fiber breakage was still the major damage mech-

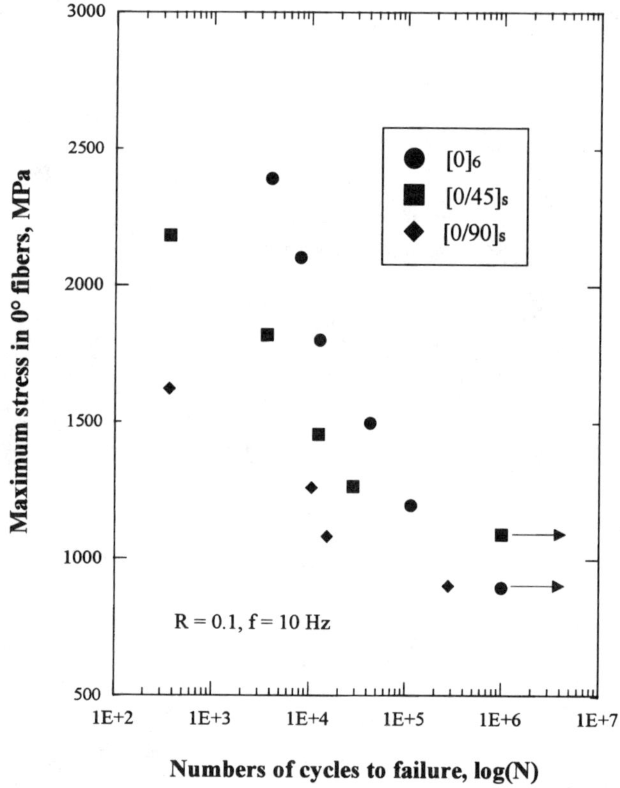

FIG. 7—*Fatigue life for three different SCS-6/Ti-15-3 laminates.*

anism at high stress levels as discussed in unidirectional composites. In the progressive damage regime, several damage mechanisms different from those in unidirectional composites were observed. These include: fiber bridging by angle-ply fibers, breakage of 0° fibers, and irregular matrix cracking. Unlike cracking of the interfacial reaction layer in the unidirectional composites, cracks initiated primarily from the separation of the fiber/matrix interface in the off-axis (±45° or 90°) layers. This was due to the fact that the interfacial opening stresses in the angle-ply layers (212 and 144 MPa for $[0/\pm45]_s$ and $[0/90]_s$, respectively) were the weakest locations in the composites. However, as fatigue loading proceeded, these debonded cracks would be deflected and would propagate into the matrix perpendicular to the fiber in the ±45° layers (Fig. 8*b*). Meanwhile, these cracks also extended into the neighboring 0° layer, leading to cracking in the 0° layer along the same direction as that in the ±45° layers as shown in Fig. 8*a*. These crack propagation characteristics resulted in an irregular cracking pattern in the fatigued specimen. Furthermore, the 0° fibers were found randomly broken across the specimen even though the initial maximum applied stress was as low as 425 MPa ($>10^6$ fatigue cycles). However, the angle-ply fibers were still intact and bridged the cracked matrix. This may be caused by the debonding of the interface that reduced the stress concentration in the off-axis fibers and prevented the fiber breakage. For the $[0/90]_s$ composites, the separation cracks in the 90° layers would propagate along the interface and extend into the 0° layers. Meanwhile, no cracks initiated from the interfacial reaction layer nor extended into matrix in the 90° layers. As a result, only

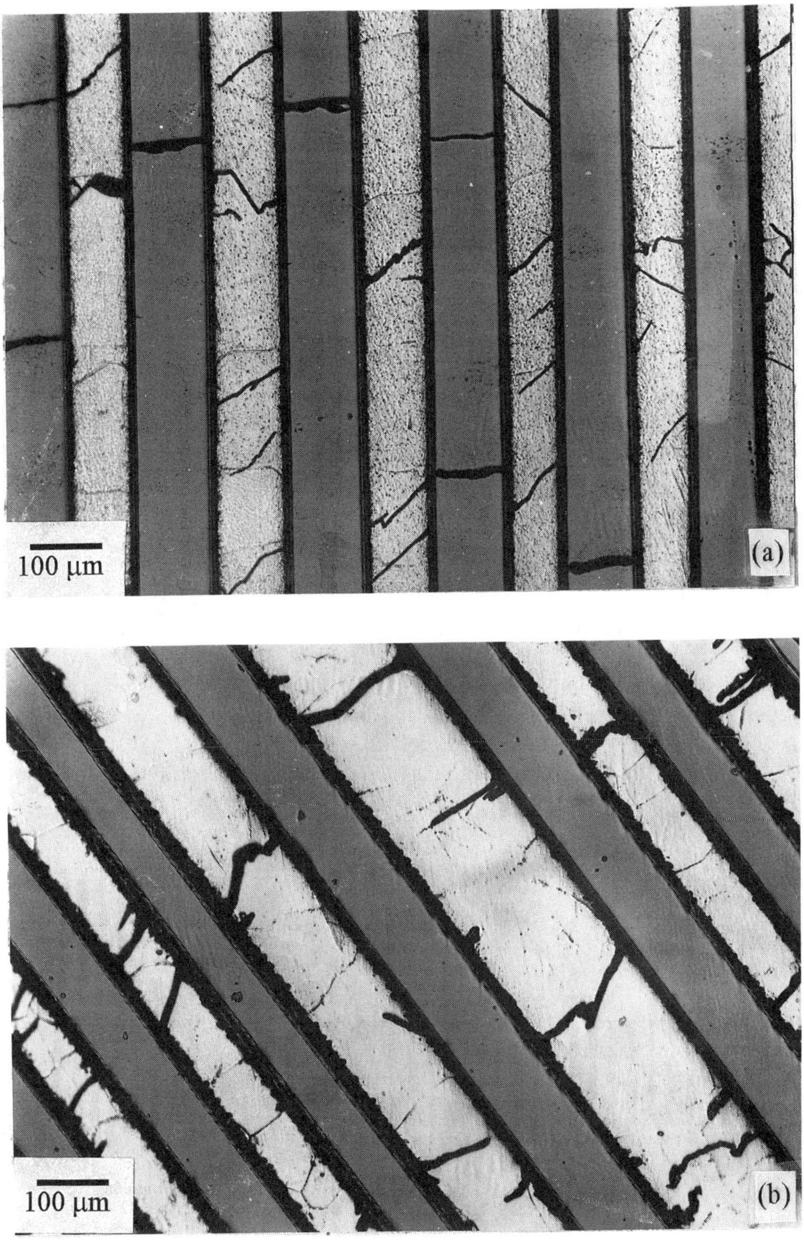

FIG. 8—*Fatigue crack pattern in* (a) *0° layer and* (b) *45° layer for the* $[0/\pm 45]_s$ *SCS-6/Ti-15-3 composite.*

several long matrix cracks across the specimen were observed in the fatigued composite. A detailed study will be conducted to understand the effects of these damage mechanisms on the degradation of mechanical properties.

Fatigue Crack Propagation in Notched Composite

Fatigue crack propagation tests were conducted using single-edge notch specimens under load-controlled, tension-tension fatigue loading with $R = 0.1$ and $f = 20$ Hz. Crack propagation was monitored by interrupting the test and measuring the crack length using an optical microscope at a magnification of $\times 250$.

Unidirectional Composites

The fatigue crack propagation mechanisms of SCS-6 fiber-reinforced titanium matrix composites can be schematically described by a diagram (Fig. 9) based upon the ratio of fiber strength to interfacial bonding strength and initial applied stress intensity factor to matrix toughness [*3*]. The crack propagation modes can be classified into Mode I catastrophic, Mode I non-catastrophic, multiple Mode I, and mixed-mode cracking. The corresponding fatigue crack growth curves are shown in Fig. 10 [*3*]. In the composite with a strong interface (SCS-6/Ti-6-4), a single Mode I crack was observed. However, depending on the initial applied stress intensity factor, the composite exhibited three different fatigue crack propagation characteristics: Mode I catastrophic failure, crack growth deceleration followed by acceleration until catastrophic failure and Mode I non-catastrophic (crack arrest). For the catastrophic failure mode, fiber breakage was found to occur at the crack tip. This suggests that the fiber would break and matrix would crack when the crack tip encounters them. As a result, the crack growth behavior was just like that of a typical metallic alloy where the crack propagation rate accelerated as the fatigue cycles increased.

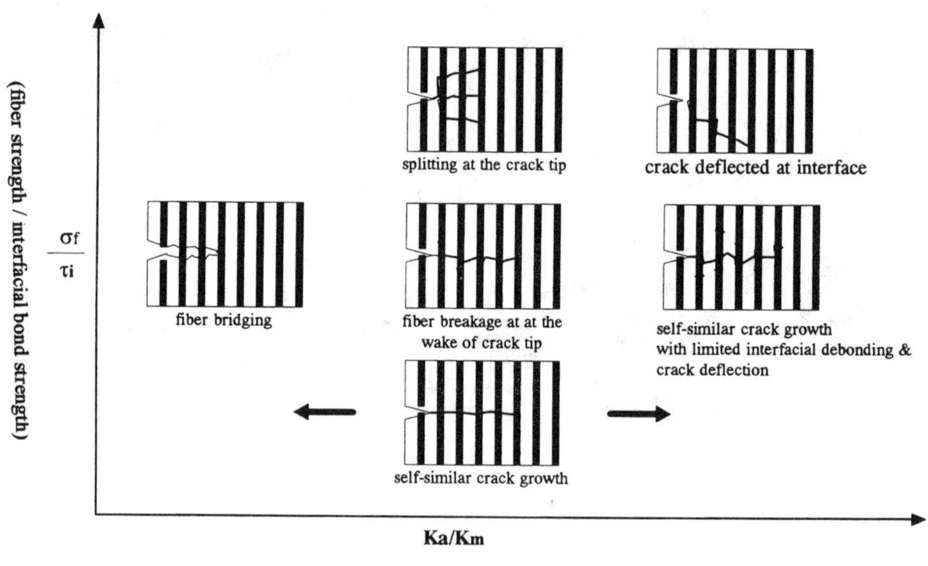

FIG. 9—*Fatigue crack propagation mechanisms of the SCS-6/Ti alloy matrix composites.*

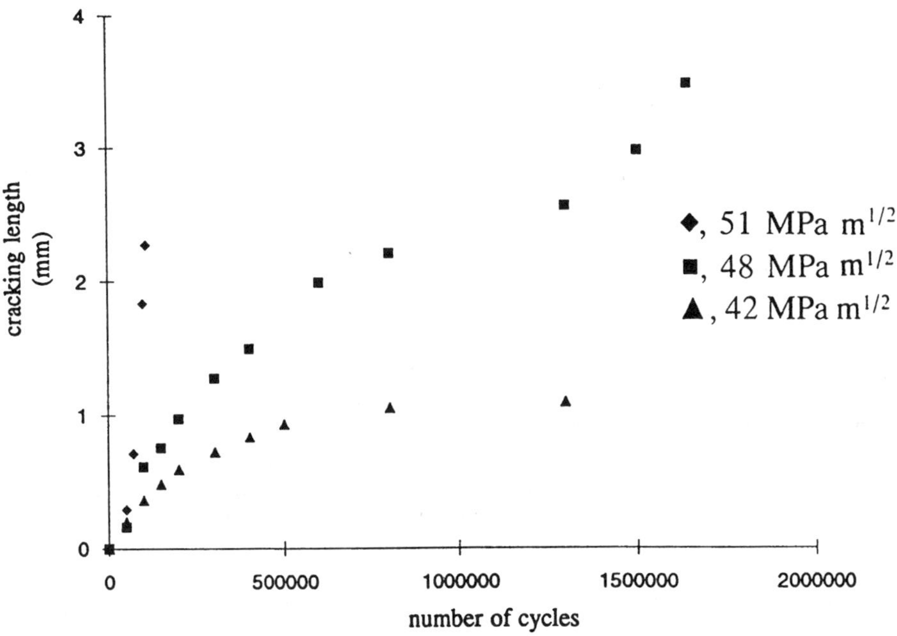

FIG. 10—*Typical crack propagation behavior of the SCS-6/Ti alloy matrix composite at various initial applied* K_{max}.

However, when the initial applied *K* was reduced to a certain level, the crack would slow down for a certain number of fatigue cycles. This was due to the stress intensity at the crack tip not being high enough to cause fiber breakage. However, the matrix crack would propagate around the fiber and leave bridging fiber in the wake of the crack tip. This would reduce the effective crack-tip stress intensity factor as the crack growth. As the crack continues to grow, the stresses in the bridging fiber would increase and two possible events may occur. If the stresses in the bridging fibers are higher than the strengths of some of the fibers, random fiber breakage will occur in the wake of the crack tip and crack growth would accelerate until the composite failed. However, if the initial applied *K* (or applied stress) is further reduced to a threshold value, the bridging fiber could survive and arrest the crack.

Multiple Mode I cracks growing from the notch tip were observed in the composites with medium interfacial strengths (SCS-6/Ti-15-3). A single crack initiating from the notch tip occurred at an early stage of cyclic loading. However, the growth of this crack would gradually slow down and arrest. Meanwhile, crack splitting occurred along the fiber/matrix interface (Mode II) near the notch tip. As the fatigue loading proceeded further, multiple Mode I cracks parallel to the original crack would continuously initiate from the Mode II crack and grow. These multiple Mode I cracks exhibited similar crack growth behavior to the original crack. Also, these cracks were arrested almost at the same distance from the notch tip. Microstructural observation indicated that the fibers in the wake of these Mode I cracks were unbroken and bridged the cracked matrix. Furthermore, the spacing between each Mode I crack was almost constant and equal to the matrix crack spacing that was observed in the unnotched composite.

Mixed-mode cracking was observed in composites with weak interfaces and brittle matrices (SCS-6/Ti-25-10). In general, a single Mode I crack was initiated from the notch tip.

However, this Mode I crack would be deflected and turn into a Mode II crack after encountering the fiber/matrix interface. Instead of multiple Mode I cracks initiating from the Mode II crack, this Mode II crack would be deflected again and become a Mode I crack again. As the cyclic fatigue loading proceeded, the Mode I and II cracking cycle would repeat. Meanwhile, the fibers in the wake of the cracked matrix were still intact at low applied stress levels. However, fiber breakage near the cracked matrix region was observed at the high applied stress level. In this failure mode, the crack propagation rate is very difficult to track and define.

Angle-Ply Composites

The fatigue crack propagation behavior of notched angle-ply ($[0/\pm45]_s$) and cross-ply ($[0/90]_s$) SCS-6/Ti-15-3 composites were also studied [*17*]. Crack propagation in both of the angle-ply and cross-ply composites is more complicated than that observed in the unidirectional composites. A common characteristic of the angle-ply and cross-ply laminates was the multiplication of damage in the form of many microcracks in front of the crack tip during fatigue cycling. Initially, a primary crack emanated from the notch tip. However, the growth rate of this primary crack decelerated and arrested just like the bridging case in the unidirectional composites. When the fatigue loading further proceeded, a huge damage zone containing many microcracks was gradually formed in front of the crack tip. With further cycling, one of the microcracks interacted with the primary crack, leading to an explosively accelerating "large" crack. Usually, this "coalesced" crack would also decelerate and arrest. The damage accumulation cycle then repeated itself. Microstructural observation indicated that the primary cracks in the 0° and ±45° layers were not concurrent, and crack lengths were also different [*17*]. Meanwhile, all cracks in the 0° and ±45° layers propagated perpendicular to the fiber direction. At the initial applied stress intensities (20 to 25 MPa·m$^{1/2}$), the ±45° fibers were still intact. However, the 0° fibers were broken randomly near the matrix cracking region. Furthermore, secondary cracks that formed in the damage zone were initiated from the separation of fiber/matrix interface in the off-axis layers and subsequently deflected into the matrix for the angle-ply composite. However, the secondary cracks found in the cross-ply composite were caused by the fiber/matrix interfacial opening in the 90° layers.

Modeling Degradation of Mechanical Properties

The preceding discussion reveals that a significant amount of microstructural damage including interfacial reaction layer cracking, matrix cracking, and interfacial debonding accumulated during fatigue loading, leading to significant stiffness reduction. In order to understand the effect of microstructural damage on stiffness degradation, a shear-lag model developed by Kuo et al. [*18,19*] was modified and is schematically shown in Fig. 11. In this model, a cylinder of given length containing a single fiber and matrix was assumed. The fiber volume fraction was considered by calculating the radius of the fiber and matrix. The fiber/matrix interface was assumed to be partially debonded and the matrix partially cracked. The interfacial frictional stress is assumed to be a constant in the debonded area and the stress in the fiber builds up linearly in this debonded region. Also, the stress in the fiber and the displacement in the bonded region are linearly correlated. Since there was no physical contact between the fiber and matrix, the strain in both fiber and matrix were assumed to be the same on the matrix cracking plane. Furthermore, the stress concentration at the tip of the partial matrix crack was neglected for simplification. The resulting stress distributions along the fiber were obtained as follows

1. in the bonded region: $x < S/2 - S_d$

$$\sigma_f = \left(\frac{E_f r^2 \sigma_c}{E_f V_f r^2 + E_m r^2 - E_m V_f r_0^2}\right) - \frac{2\tau_i}{r}\left(\frac{S}{2} - x\right) \tag{1}$$

2. for the debonded region: $S/2 - S_d < x < S/2$

$$\sigma_f = \frac{E_f \sigma_c}{E_c} + \frac{\cosh \beta x}{\cosh \beta\left(\frac{S}{2} - S_d\right)}\left[\left(\frac{E_f \sigma_c}{E_f V_f + E_m\left(1 - \frac{V_f r_0^2}{r^2}\right)}\right) - \frac{2\tau_i}{r} S_d\right] \tag{2}$$

where

E_f and E_m = the Young's modulus of the fiber and matrix, respectively;
V_f and V_m = the volume fractions of fiber and matrix, respectively;
$E_c = E_f V_f + E_m V_m$ = the Young's modulus of the composite; and

$$\beta = \left[\frac{2E_c}{E_f(1 + \nu_m)\left(\ln \frac{1}{V_f} + V_f - 1\right)}\right]^{1/2} \frac{1}{r}$$

where

σ_c = the stress applied to the composite,
ν_m = the Poisson's ratio of the matrix,
S = the matrix crack spacing,
S_d = the debonding length,
τ_i = the interfacial frictional stress,
r = the fiber radius, and
r_0 = the matrix crack length.

The average stress in the fiber can be calculated from Eqs 1 and 2

$$\tilde{\sigma}_f = \frac{2}{S}\int_0^{S/2} \sigma_f(x)\, dx \tag{3}$$

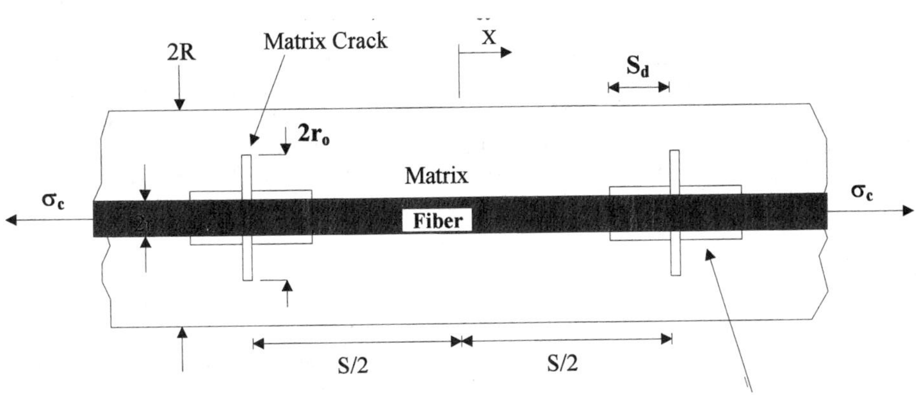

FIG. 11—*The unit cell for modeling the residual stiffness of the composite.*

The average strain in the composite, $\tilde{\epsilon}_c$, which is the same as the average strain in the fiber, can be calculated by dividing the average stress in the fiber by the fiber modulus

$$\tilde{\epsilon}_c = \frac{\tilde{\sigma}_f}{E_f} \tag{4}$$

The residual stiffness is then given by taking the ratio of the applied stress and the average strain of the composite. The resulting residual stiffness of a cracked composite as a function of matrix crack length, crack spacing, interfacial debonding length, and interfacial frictional stress is given by

$$E_c = \left\{ \frac{1}{E_c}\left(1 - \frac{2S_d}{S}\right) + \frac{2S_d}{S\left[E_fV_f + E_m\left(1 - V_f\left(\frac{r_0}{r}\right)^2\right)\right]} - \frac{2\tau_i S_d^2}{SrE_f\sigma_c} \right. \tag{5}$$
$$\left. + \frac{2}{S\beta}\tanh\beta\left(\frac{S}{2} - S_d\right)\left[\frac{1}{E_fV_f + E_m\left(1 - V_f\left(\frac{r_0}{r}\right)^2\right)} - \frac{1}{E_c} - \frac{2\tau_i S_d}{rE_f\sigma_c}\right]\right\}^{-1}$$

In order to study the effects of the different damage mechanisms on stiffness reduction, the residual stiffness were calculated using different interfacial debonding and matrix cracking lengths. The results are plotted as a function of crack spacing as shown in Figs. 12*a* and *b*. The experimental results were also plotted for comparison. In these results, the interfacial frictional stress was assumed to be 25 MPa, which is the average stress from reported data. Meanwhile, the lower bound crack spacing used in the calculation was the measured experimental saturated crack spacing. Since the thickness of the interfacial reaction layer is around 2 μm, the calculated result using a cracking length of 2 μm and limited amount of interfacial debonding (20 μm) is considered to be the degradation caused by interface cracking only. The results showed that this damage mechanism has limited influence on the stiffness reduction as shown in Fig. 12*a*. However, as the matrix cracks grew, the whole stiffness reduction trend is similar to that of the experimental results, that is, a slow stiffness reduction followed by a rapid drop of the residual stiffness. This implied that only the fast nucleation of the interfacial reaction layer cracks could not contribute to the degration of the rapid stiffness reduction. It should be accompanied by the fast growth of the newly initiated cracks. Meanwhile, the calculated and experimental results match quite well at the applied stress of 945 MPa using an interfacial debonding length of 20 μm as shown in Fig. 12*a*. However, for the composite tested at 750 MPa, the experimental data for residual stiffness of 75 to 77% is much lower than that of the calculated results using an interfacial debonding length of 20 μm. When using a debonding length of 100 μm, the results agree well with the experimental values as shown in Fig. 12*b*. This implies that the debonding length is a function of fatigue loading conditions. A detailed analysis is needed to understand the relationship between interfacial debonding length and fatigue loading history.

Concluding Remarks

1. Depending on the interfacial shear strength, matrix toughness, fiber orientation, and fatigue loading conditions, a significant difference in the damage mechanisms (or modes) and its accumulation were observed in the SCS-6 fiber-reinforced titanium matrix composites. A criteria for transition between these damage modes has to be established.

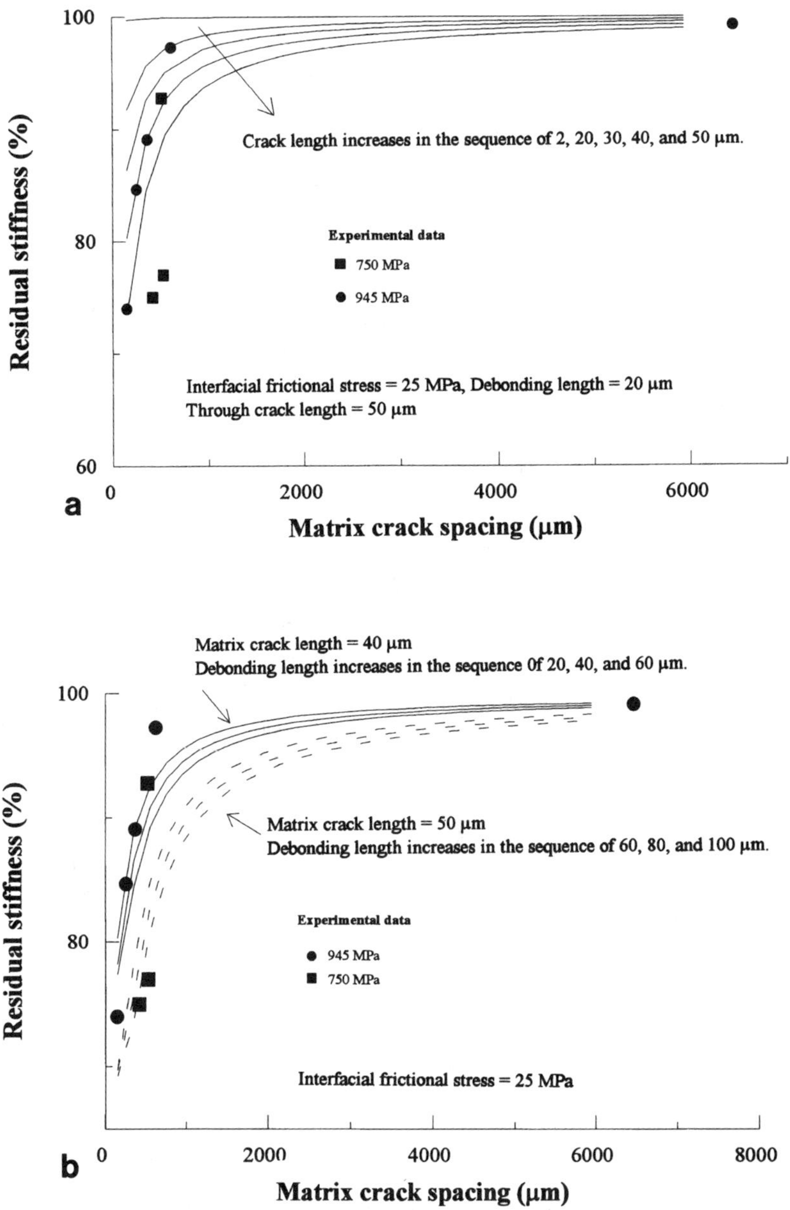

FIG. 12*a* and *b*—*The calculated residual stiffness as a function of damage parameters for the SCS-6/Ti-15-3 composite.*

2. Three regimes of stiffness reduction were observed in the unnotched composite under fatigue loading. These include a slow stiffness reduction followed by a rapid drop and a minimal stiffness change in a saturation regime. The controlling mechanisms were interfacial debonding and cracking of the interface and matrix.

3. The theoretical model used to predict the residual stiffness was quite consistent with experimental results. However, a detailed understanding of the mechanics of debonding to determine the length as a function of loading history is needed. Meanwhile, a quantitative analysis of damage accumulation as a function of fatigue loading and fatigue cycles has to be established.

Acknowledgments

This work was partially supported by the National Science Foundation (DDM-9057030) and the Air Force Office of Scientific Research (F49620-93-1-0320, Dr. Walter Jones, program monitor). We also thank Mr. Cyrus Liu for reviewing the manuscript.

References

[*1*] Larsen, J. M., Revelos, W. C., and Gambone, M. L., *Intermetallic Matrix Composites II,* D. B. Miracle, D. L. Anton, and J. A. Graves, Eds., Materials Research Society, Pittsburgh, PA, Vol. 273, 1992, pp. 3–16.

[*2*] Jeng, S. M., Alassoeur, P., and Yang, J.-M., *Materials Science Engineering,* Vol. A148, 1991, pp. 67–77.

[*3*] Jeng, S. M., Alassoeur, P., and Yang, J.-M., *Materials Science Engineering,* Vol. A154, 1992, pp. 11–19.

[*4*] Technical data from Texas Instruments, 1992.

[*5*] Bania, P. J., Lenning, G. A., and Hall, J. A., in *Proceedings,* Beta Titanium Alloys in the 80's, R. R. Boyer and H. W. Rosenberg, Eds., Atlanta, GA, 1983, pp. 209–229.

[*6*] Donachie, M. J., in *Ti and Ti Alloy,* American Society for Metals, Metals Park, OH, 1982.

[*7*] Larsen, J. M., Russ, S. M., and Jones, J. W., *Proceedings,* NATO AGARD Conference on Characterization of Fiber Reinforced Titanium Metal Matrix Composites, Bordeaux, France, Sept. 1993.

[*8*] Sukonnik, I. M., Smith, P. R., Graves, J. A., Rhodes, C. G., Shaw, M. C., Krishnamurthy, S., and Pandey, A., in *Titanium Matrix Composite,* WL-TR-92-4035, P. R. Smith and W. C. Revelos, Eds., Wright Laboratory, Cleveland, OH, 1991, pp. 98–114.

[*9*] Smith, P. R., Graves, J. A., Rhodes, C. G., James, M. R., and Porter, J. R., in *Titanium Matrix Composite,* WL-TR-92-4035, P. R. Smith and W. C. Revelos, Eds., Wright Laboratory, Cleveland, OH, 1991, pp. 115–144.

[*10*] Jeng, S. M., Yang, J.-M., and Yang, C. J., *Materials Science and Engineering,* Vol. A138, 1991, pp. 169–180.

[*11*] Yang, J.-M., Jeng, S. M., and Yang, C. J., *Materials Science and Engineering,* Vol. A138, 1991, pp. 155–167.

[*12*] Jeng, S. M. and Yang, C. J., unpublished work.

[*13*] Kantzos, P., Eldridge, J., Koss, D. A., and Ghosn, L. J., *Proceedings,* Materials Research Society Symposium, Vol. 273, 1992, pp. 135–142.

[*14*] Walls, D. P., Bao, G., and Zok, F. W., *Acta Metallurgica & Materials,* Vol. 41, No. 7, 1993, p. 2061.

[*15*] Wang, P. C., Jeng, S. M., and Yang, J.-M., in *Science of Advanced Materials and Processing, Engineering Series,* Vol. 39, K. Drake, J. Bauer, T. Serafini, and P. Cheng, Eds., Society for the Advancement of Materials and Process Engineering, (SAMPE), Covina, CA, 1994, pp. 3304–3316.

[*16*] Wang, P. C., Jeng, S. M., Chiu, H.-P., and Yang, J.-M., *Journal of Materials Science,* Vol. 30, 1995, pp. 1818–1826.

[*17*] Bruce Nguyen, T.-H., Jeng, S. M., and Yang, J.-M., *Materials Science and Engineering,* Vol. A138, 1994, pp. 1–9.

[*18*] Kuo, W.-S. and Chou, T. W., in *Metal & Ceramic Matrix Composites: Processing, Modeling & Mechanical Behavior,* by R. B. Bhagat et al., The Metals Society, Warrendale, PA, 1990, pp. 311–318.

[*19*] Karandikar, P. and Chou, T. W., *Composites Science & Technology,* Vol. 46, 1993, pp. 253–263.

Timothy P. Gabb[1] and John Gayda[1]

Matrix Fatigue Cracking in α_2 Titanium Matrix Composites for Hypersonic Applications

REFERENCE: Gabb, T. P. and Gayda, J., "**Matrix Fatigue Cracking in α_2 Titanium Matrix Composites for Hypersonic Applications,**" *Life Prediction Methodology for Titanium Matrix Composites, ASTM STP 1253,* W. S. Johnson, J. M. Larsen, and B. N. Cox, Eds., American Society for Testing and Materials, 1996, pp. 395–411.

ABSTRACT: The objective of this work was to understand matrix cracking mechanisms in a unidirectional α_2 titanium matrix composite (TMC) in possible hypersonic applications. A $[0]_8$ SCS-6/Ti-24Al-11Nb (atomic %) TMC was first subjected to a variety of simple isothermal and nonisothermal fatigue cycles to evaluate the damage mechanisms in simple conditions. A modified ascent mission cycle test was then performed to evaluate the combined effects of loading modes. This cycle mixes mechanical cycling at 150 and 483°C, sustained loads, and a slow thermal cycle to 815°C. At low cyclic stresses and strains more common in hypersonic applications, environment-assisted surface cracking limited fatigue resistance. This damage mechanism was most acute for out-of-phase nonisothermal cycles having extended cycle periods and the ascent mission cycle. A simple linear fraction damage model was employed to help understand this damage mechanism. Time-dependent environmental damage was found to strongly influence out-of-phase and mission life, with mechanical cycling damage due to the combination of external loading and CTE mismatch stresses playing a smaller role. The mechanical cycling and sustained loads in the mission cycle also had a smaller role.

KEYWORDS: titanium matrix composites, fatigue (materials), thermomechanical fatigue, hypersonic, damage mechanisms, titanium, life prediction, titanium alloys, modeling

Titanium alloy matrix composites (TMCs) have potential in aerospace structural applications requiring high strength and stiffness combined with low density. In gas turbine engines, these composites are being considered for highly stressed rotating components such as compressor disks and blades, and a variety of static components [*1*]. In projected hypersonic applications, composites have been considered for airframe panels consisting of skins with stiffeners and joints [*2*].

Investigations of the fatigue response of TMCs have provided some insights of TMC fatigue failure mechanisms possible in such applications. Fatigue loading of unidirectional 0° fiber reinforced composites foreseen in rotating components at high cyclic stresses ($\sigma_{max} > 0.6\sigma_{uts}$) such as disks can induce stress relaxation in the matrix, where some of the load is shifted to the fibers. In load-controlled fatigue conditions, this can cause fiber-dominated failures [*3,4*]. This failure mode is accentuated in cycles where maximum loads are concentrated at maximum temperatures, such as isothermal fatigue tests with maximum load dwells [*5*], and in-phase thermomechanical fatigue (TMF) tests [*6,7*]. However, in fatigue loading of unidirectional 0° fiber reinforced composites at the lower cyclic stresses ($\sigma_{max} < 0.5\sigma_{uts}$)

[1]Research metallurgists, NASA Lewis Research Center, Cleveland, OH 44135.

foreseen in hypersonic applications, the fatigue durability is often limited by matrix cracking mechanisms [*3–8*] for isothermal fatigue and thermomechanical fatigue. Composites of other laminate orientations, such as $[90]_8$, $[0/90]_s$, and $[0/\pm45/90]_s$ having off-axis fibers, also suffer fiber-matrix debonding in isothermal [*9,10*] and thermomechanical [*11–13*] fatigue, and matrix cracks often initiate at the debonded interfaces. Comparative tests in vacuum [*14,15*] indicate the environment has a role in many of these matrix cracking mechanisms. α_2 titanium aluminide alloys generally have better high temperature strength and environmental resistance (as measured by static weight gain) than β titanium alloys [*16*] and, therefore, could have potential in hypersonic applications where temperatures approach 815°C.

The objective of this study was to investigate the matrix cracking mechanisms in a unidirectional 0° fiber-reinforced α_2 TMC at low applied stresses to determine the primary factors driving these mechanisms in TMCs and the governing factor/damage relationships. Isothermal fatigue and nonisothermal fatigue tests were first employed to understand the failure mechanisms induced in simple fatigue cycles that were relevent to hypersonic load-temperature-time service cycles. A more complex modified hypersonic ascent mission cycle was then employed to evaluate the damage modes in more realistic conditions where several loading modes are superimposed at variable temperatures. A linear life fraction damage model was applied to aid in understanding the relative contributions of the different loading and damage modes.

Experimental Procedure

The composite test material was made of Ti-24Al-11Nb (atomic %) matrix, reinforced by 35% by volume of SCS-6 SiC (silicon-carbide) fibers oriented at 0° to the specimen loading axis, Fig. 1. This composite was fabricated by Textron Specialty Materials through foil-fiber processing. The fibers were first woven with titanium-niobium (Ti-Nb) wire crossweave into unidirectional mats. Eight alternating layers of matrix foil and 0° fiber mat were then arranged and consolidated by hot isostatic pressing to form $[0]_8$ panels. Rectangular and reduced midsection specimens each having gage sections 12.5 mm long, 6.3 mm wide, and 1.6 mm thick were waterjet machined from the panels. Load-controlled fatigue tests were performed in air on a servohydraulic test system utilizing direct induction heating and an axial extensometer. As shown in the load and temperature waveforms of Fig. 2, standard

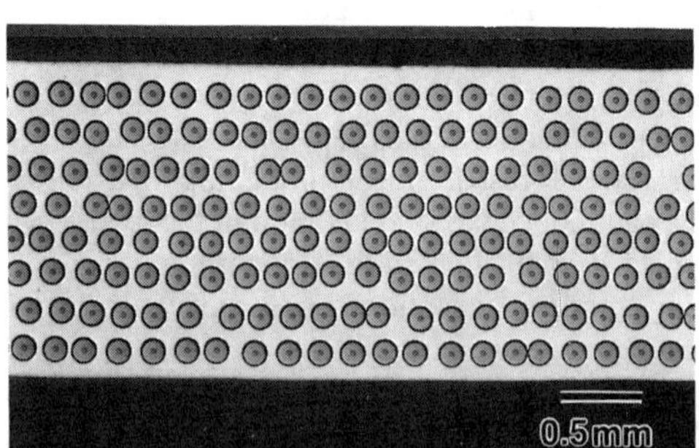

FIG. 1—*SCS-6/Ti-24Al-11Nb* $[0]_8$ *composite cross section.*

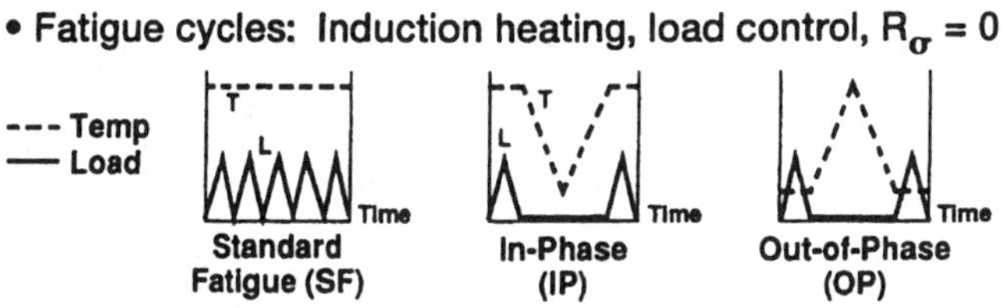

FIG. 2—*Initial isothermal and nonisothermal fatigue test cycles.*

fatigue (SF) tests were first performed at 0.33 Hz and $R_\sigma = 0$ (minimum/maximum stress ratio) at constant temperature. Simple nonisothermal tests, where load and temperature are sequentially cycled, were then performed to simplify initial damage analyses. Load was applied at minimum temperature for the out-of-phase (OP) tests and at maximum temperature for several comparative in-phase (IP) tests. More complex nonisothermal and hypersonic mission cycles were subsequently performed, as will be described later. Failure for all cyclic types is defined as complete specimen fracture into two pieces. The damage mechanisms induced in interrupted and failed test specimens were subsequently characterized with optical and scanning electron microscopy.

Results and Discussion

Isothermal and Nonisothermal Fatigue

In all test types, cyclic life increased with decreasing stress range as usual [5]. Figure 3 compares cyclic life at an equivalent stress range of about 690 MPa and mechanical strain range of about 0.40%. Standard isothermal fatigue life decreased with increasing temperature, however OP cycling was life-limiting at these low stress ranges. The operative damage mechanisms are evident from the fracture surfaces shown in Fig. 4. Standard isothermal fatigue at 150°C induced relatively few surface-initiated fatigue cracks that propagated long distances of 1 to 2 mm. Standard isothermal fatigue at 650°C induced relatively few fatigue cracks initiating principally at damaged fibers on the specimen edges. These cracks grew

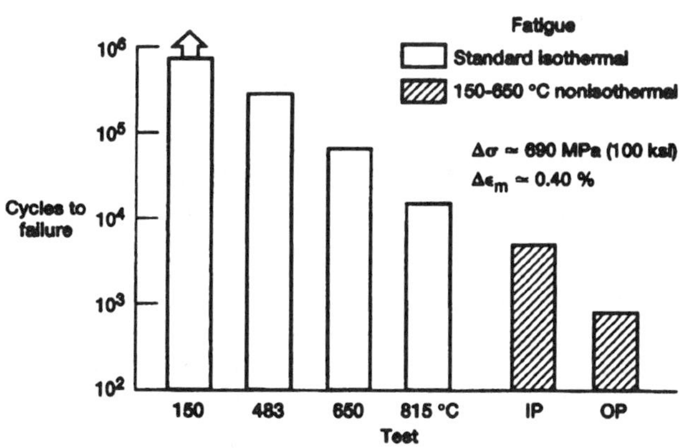

FIG. 3—*Comparison of fatigue life at an approximate stress range, $\Delta\sigma$, of 690 MPa and mechanical strain range, $\Delta\epsilon$, of 0.40%.*

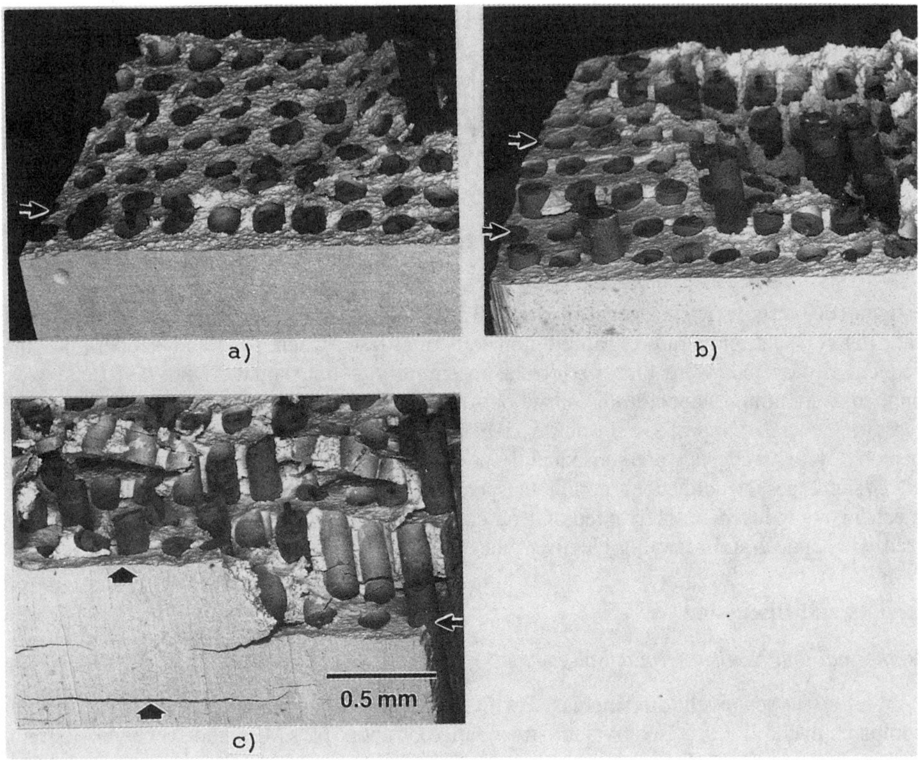

FIG. 4—*Fracture surfaces of specimens tested in* (a) *150°C standard isothermal fatigue,* (b) *650°C standard isothermal fatigue, and* (c) *150 to 650°C out-of-phase nonisothermal fatigue. Load oriented vertically.*

moderate distances of 0.1 to 1.5 mm. IP cycling from 150 to 650°C induced edge cracking similar to 650°C standard fatigue. OP cycling from 150 to 650°C induced a different damage mechanism of enhanced surface crack initiation in the matrix. The growth of the out-of-phase cracks subsequently allowed considerable fiber-matrix interface oxidation and damage. Examination of metallographic sections of the test specimens indicated that in each of these cases, cracks preferentially grew in the matrix while the fibers often remained uncracked near the crack tip [5].

The factors driving these matrix cracking mechanisms were then explored by performing a series of isothermal and nonisothermal tests varying temperature cycle range (ΔT) and maximum temperature (T_{max}) at a constant stress range. OP rather than IP nonisothermal cycling was selected for this evaluation, due to the lower life and increased matrix cracking of the OP cycle at low cyclic stresses. Resulting fatigue life is shown as a function of ΔT in Fig. 5*a*. Isothermal fatigue life decreased consistently when going from 150 to 815°C. But OP life was always much lower than the isothermal life at the same loading temperature. As clearly shown for a loading temperature of 150°C, fatigue life consistently decreased with increasing ΔT for a given loading temperature. However, life varied with loading temperature. Life is expressed as a function of T_{max} in Fig. 5*b*. OP life was related more closely to T_{max}. OP damage correspondingly increased with increasing T_{max}, as shown by comparing the specimen surfaces in Fig. 6. Standard isothermal fatigue also produced considerable

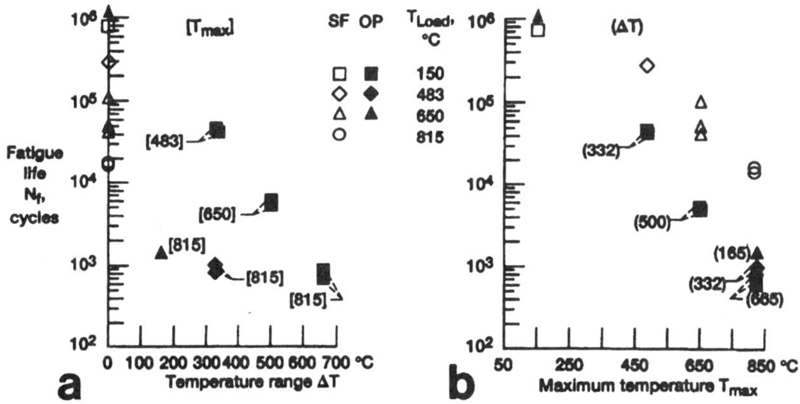

FIG. 5—*Fatigue life at* $\Delta\sigma \approx$ *690 MPa,* $\Delta\epsilon \approx$ *0.40% as a function of* (a) *temperature range of* ΔT *and* (b) *maximum temperature of* T_{max}.

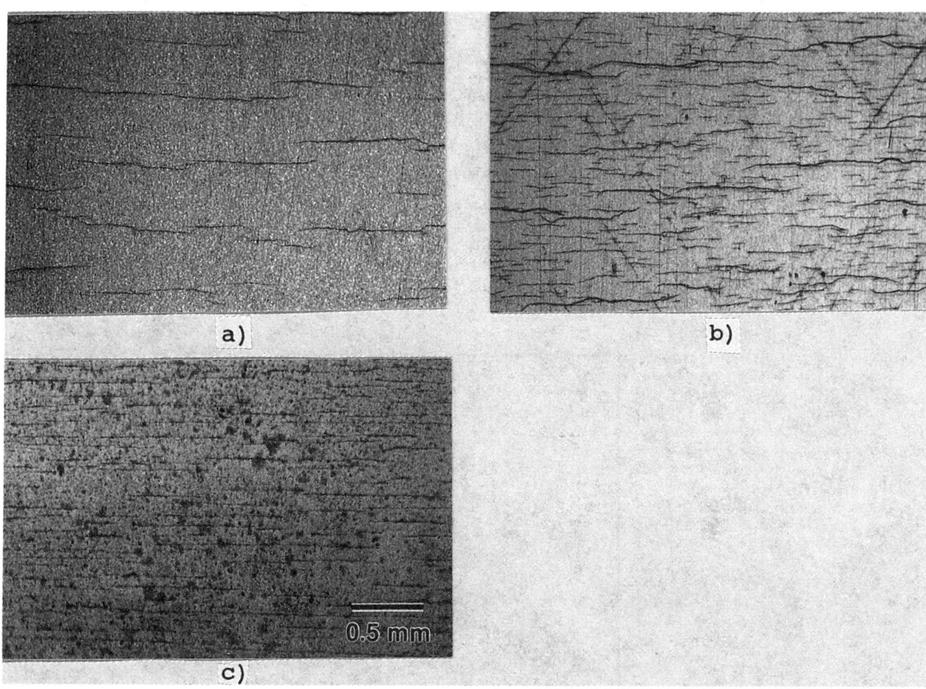

FIG. 6—*Specimen gage sides after* (a) *150 to 483°C out-of-phase nonisothermal fatigue,* (b) *150 to 815°C out-of-phase nonisothermal fatigue, and* (c) *815°C standard isothermal fatigue. Load oriented vertically.*

surface cracking at 815°C, as shown in Fig. 6. This suggested time and temperature-dependent environmental effects were dominating damage and life in both isothermal and nonisothermal tests at high temperatures, and that a single damage relationship might be able to describe the damage processes.

Linear Damage Formulation

Damage relationships were evaluated quantitatively to aid understanding, using an approach similar to that of Nicholas and Russ [*17*]. The purpose of this modeling was only to provide some understanding of the factor/damage relationships, rather than to provide a generalized TMC life prediction model. A simple linear summation of mechanical damage, $\Sigma\, N/N_m$, and time-dependent environmental damage, $\Sigma\, N\, P/t_E$, was applied, where N is the number of cycles; N_m is the mechanical life at a strain range, $\Delta\epsilon_m$, and temperature, T; and t_E is the accumulated time-to-failure at cycle period, P, and temperature, T. For isothermal and nonisothermal OP cyclic life, this simplifies [*18*] to

$$1/N_m + \Sigma\, P/t_E = 1/N_f$$

where the time-temperature waveform of the OP cycle is approximated with isothermal segments of 1-s duration to determine $\Sigma\, P/t_E$. Environmental damage relationships were characterized with isothermal fatigue cycles with $\sigma_{max} = 690$ MPa and dwells at zero load to vary cycle period. This would incorporate fatigue influences. However, such an approach precluded application to monotonic creep tests that were not a subject of this investigation. Figure 7 depicts the accumulated time-to-failure versus cycle period in tests at several temperatures. Accumulated time-to-failure, t_E, did not vary appreciably with cycle period over this range. This suggested that an environmental damage relationship based on simple summation of time at a temperature could be applicable under these test conditions. The accumulated time-to-failure decreased with increasing temperature, reflecting the increasing

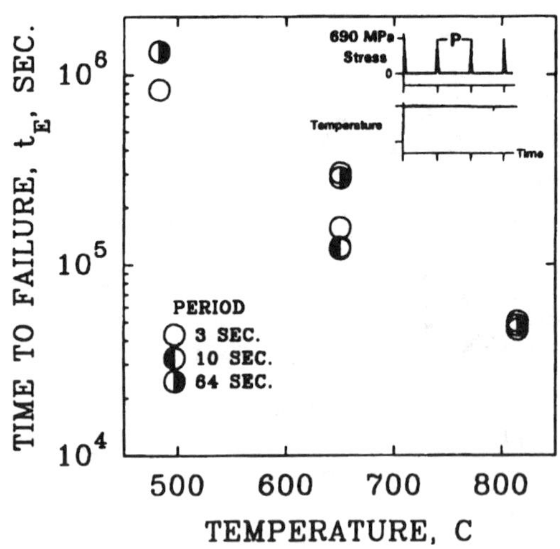

FIG. 7—*Accumulated time to failure as a function of temperature in isothermal 0 load dwell tests at* σ_{max} = *690 MPa.*

damage severity with increasing T_{max}. This gave the environmental damage term, $\Sigma P/t_E$, as a function of temperature. Mechanical damage relationships were characterized with isothermal fatigue tests run at $\sigma_{max} = 690$ MPa in argon at a high frequency of 5 Hz to minimize time-dependent environment effects. In comparison to air tests, cyclic life increased in the 5 Hz, argon tests. However, these tests failed at lower accumulated times than tests in air. Therefore, it was assumed that mechanical damage was dominant, and that environmental damage was minimized. These tests gave mechanical fatigue life and the mechanical damage term, $1/N_m$, as a function of temperature, Fig. 8.

OP cycling produces higher mechanical cyclic strains in the matrix than isothermal tests, due to thermal expansion mismatch strains generated by the ΔT. The mechanical damage in OP tests was therefore adjusted for the additional thermal mismatch strains generated in the matrix along the axial direction, normal to observed surface cracking. The axial thermal mismatch strain was approximated as $(\alpha_{matrix} - \alpha_{comp.})\ \Delta T$ and added to the applied mechanical strain for OP cycling to give the total matrix strain. A strain range–fatigue life line was not available from the argon environment tests. The mechanical damage life at the loading temperature was instead multiplied by the ratio of lives observed in tests at the applied and total matrix strain ranges performed at 150°C in air

$$1/N_{m,\,OP} = 1/N_m(N_{\Delta\epsilon_a}/N_{\Delta\epsilon_t})_{150°C}$$

This provided some measure of the magnitude of the out-of-phase cycles' coefficient of thermal expansion (CTE) effects on mechanical damage. The damage relationships generated from isothermal tests could now be utilized to handle both isothermal and nonisothermal OP life.

The simple damage formulation predicted nonisothermal OP life fairly well using the isothermally generated relationships. Calculated fatigue lives are compared to observed fatigue lives for standard fatigue, isothermal dwell, and OP tests in Fig. 9. Calculated

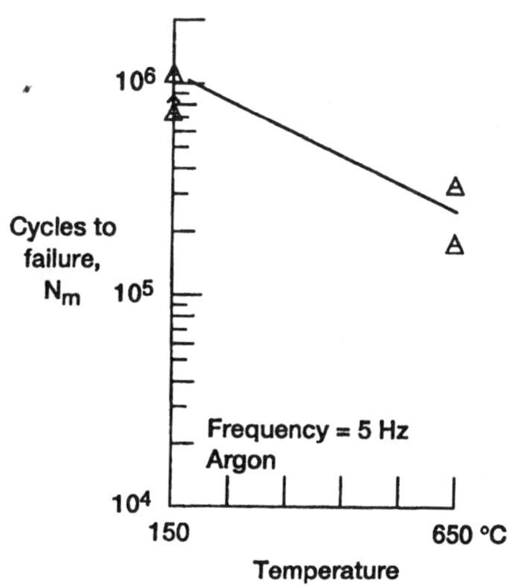

FIG. 8—*Mechanical cycling life in isothermal 5-Hz tests at σ_{max} = 690 MPa in argon as a function of temperature.*

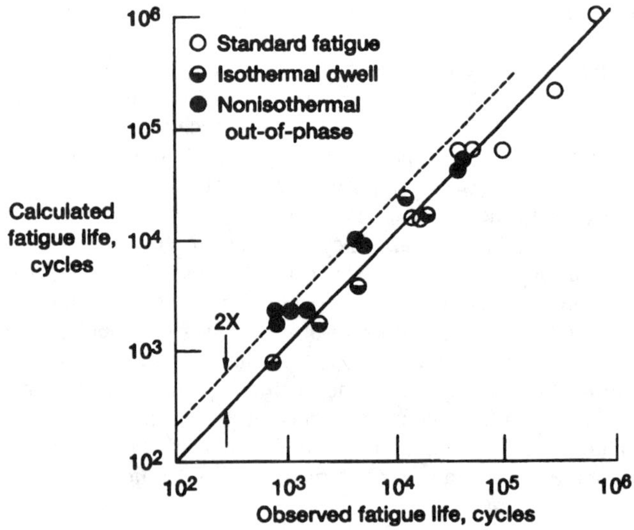

FIG. 9—*Observed fatigue lives versus calculated lives.*

isothermal lives agree well with observed lives, indicating the model properly reproduces the data employed in deriving the relationships. OP life is predicted within about two times (2×) of observed life. The environmental damage term contributed most of the damage in tests at high T_{max} and low life, while the mechanical damage term began to contribute substantial damage in low T_{max}, long life tests. This is consistent with the propensity of surface cracking observed experimentally. OP life predictions at high T_{max} and low life were consistently higher than experimental data. This may indicate the presence of additional environmental damage attributable to TMF-environment interactions, not included in the simple damage formulation. But the simple damage formulation still provided a useful tool for comparing mechanical and environmental sources of damage in different cycles. Additional linear fraction damage interaction terms have been successfully developed by Neu and Nicholas [*19*].

Hypersonic Service Cycles

Actual hypersonic structural applications will produce complex service cycles mixing cyclic loads, sustained loads, and multiple temperature changes. It is important to understand the effects of such cycles on durability and failure mechanisms. Therefore, additional tests were performed to evaluate these effects. An ascent mission cycle [*20*] was chosen and modified to represent an example service cycle, shown in Fig. 10. This cycle has subcycling at 150 and 425°C, intermediate sustained loads, and a slow thermal ramp from 425 to 815°C. The maximum applied stress was again 690 MPa, but the cycle period was 26 min. This mission cycle does have OP cycle traits, with maximum applied stresses attained at low-intermediate temperatures within the cycle. The average mission cycle life of duplicate tests was only 101 cycles. SEM micrographs of mission cycle fracture surfaces and gage sections indicated the dominant damage mechanism was again environment-assisted surface cracking.

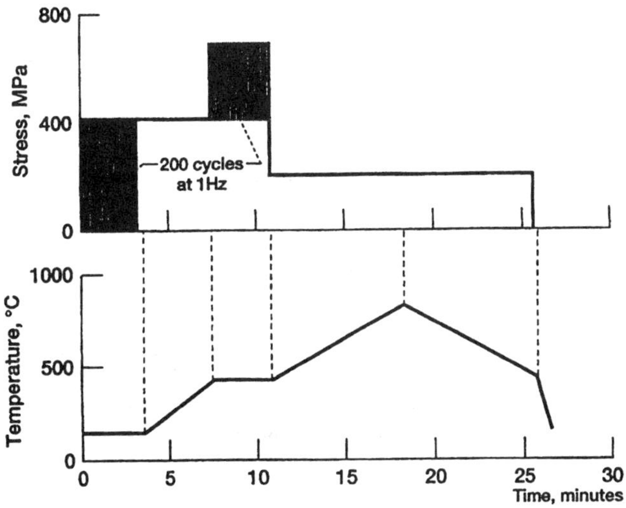

FIG. 10—*Modified ascent mission cycle* [20].

The mission cycle life of 101 cycles was much lower than the OP life of 764 cycles at the same maximum stress of 690 MPa and ΔT of 150 to 815°C. Several characteristics unique to the mission cycle could explain the lower life of the mission cycle. Sustained loads were present throughout the cycle. Subcycling was applied at low and intermediate temperatures. The cycle time was much longer than the standard OP cycle period. A series of tests were performed to understand the effects of each of these service characteristics. These tests were all performed with a maximum temperature of 815°C, and a maximum stress of 690 MPa.

Several tests were performed to evaluate the effects of sustained load. A creep test was performed at maximum stress and maximum temperature. A specimen was thermally cycled from 150 to 815°C at a sustained stress of 690 MPa. The resulting lives are compared to standard OP and mission cycle lives in Fig. 11. The creep test was interrupted at 526 h with no signs of matrix surface cracking. The thermal cycling life was comparable to the standard OP life on both a time-to-failure and cycles-to-failure basis, and had comparable surface cracking damage. However, the thermal cycling life did not approach the low mission cycle life. The simple damage formulation was not specifically designed to handle sustained loads. However, a thermal cycling life prediction using the model's environmental damage term P/t_E alone was within 2X, Fig. 11. These sustained load tests indicated that sustained loads applied at maximum temperature and sustained load thermal cycling were no more damaging than the OP test.

Subcycling effects were evaluated in tests using the standard OP thermal cycle, with a block of n subcycles applied at 1 Hz at the minimum temperature of 150°C between each thermal cycle at zero load to 815°C. The resulting lives in blocks to failure are shown as a function of subcycles per block in Fig. 12. Subcycling did not strongly influence OP life and could not explain the reduced mission cycle life. The additional mechanical damage produced by even 200 subcycles between each thermal cycle was predicted to remain small relative to the environmental damage and have little effect on OP life, consistent with the experimentally observed trend.

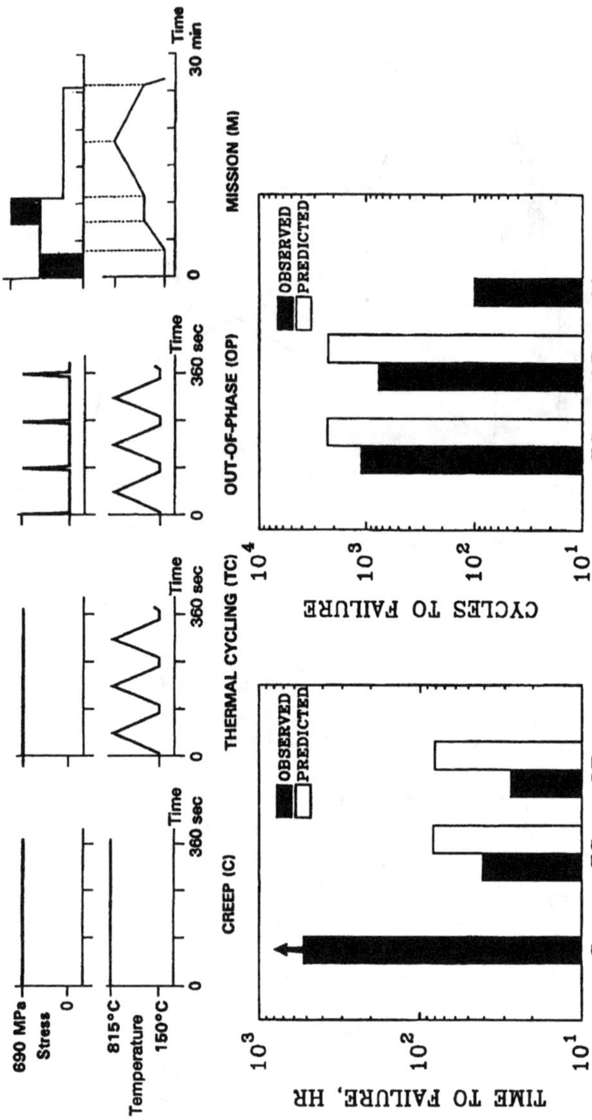

FIG. 11—*Sustained load effects on life.*

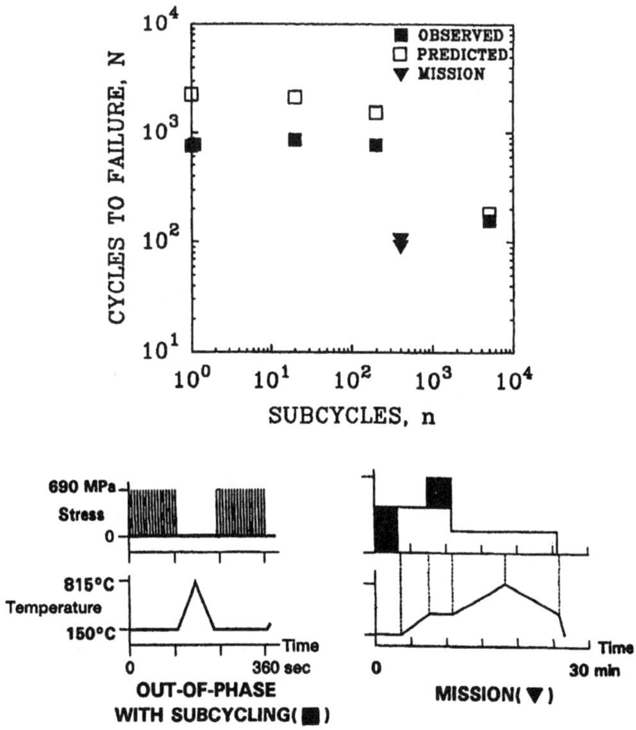

FIG. 12—*Effects of subcycling on out-of-phase cyclic life* (■).

The preceding tests have shown severe sustained applied stresses were no more damaging than the same cyclic stress applied at minimum temperatures, and did not strongly influence OP life. Subcycling did not strongly influence OP life. These results suggested the low mission cycle life was largely due to the additional environmental effects incurred by the extended mission cycle period. This indicated that a simple standard OP cycle with an extended cycle period might largely reproduce the mission cycle damage and life.

An OP test was performed with an extended cycle period of 26 min. Life as a function of cycle period is compared for OP and mission cycle tests in Fig. 13. An extended cycle period reduced OP life to the mission cycle levels. Comparison of the specimen gage surfaces in Fig. 14 show severe environment-assisted surface cracking in both extended cycle period and mission tests. The slow nonisothermal OP test appears to be a useful simple screening test for resistance to environment-assisted surface cracking in hypersonic applications, as both the slow OP and mission test lives are dominated by this time-dependent environment-driven damage. The linear fraction model was applied to the slow OP cycle as before. Based on the results of previous tests of mission cycle effects, an attempt could be made to also employ the model to predict mission cycle life. Sustained loads were handled as previously described. Subcycling mechanical damage was computed as before, with a simple stress ratio correction using a simplified Goodman approach. Both predictions were within 2X, with time-dependent environmental damage terms strongly dominating in both cases. The additional mechanical damage contributed by the mission's subcycling was negligible, so that the slow OP and mission cycle life predictions were essentially the same.

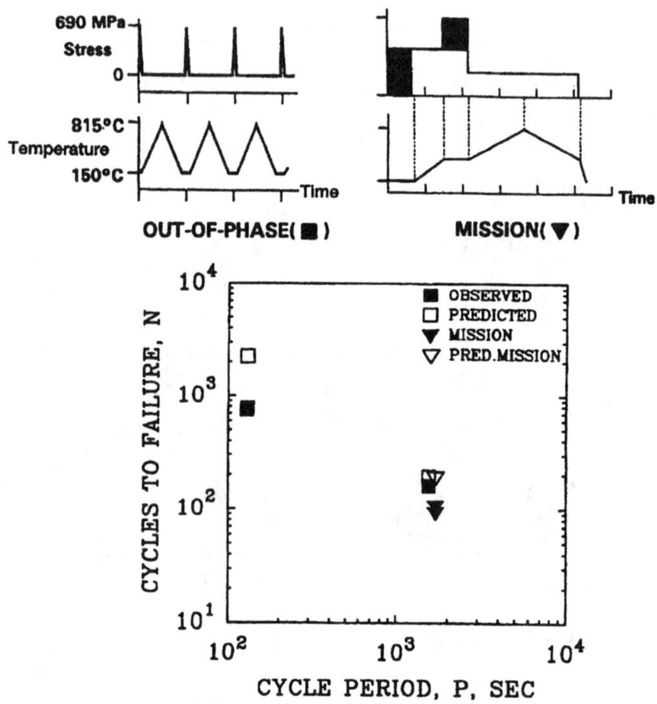

FIG. 13—*Life as a function of cycle period in OP (■) and mission (▼) tests.*

Observed fatigue lives are compared to lives calculated using the simple linear fraction damage formulation for all tests in Fig. 15. This simple formulation did a reasonable job predicting the nonisothermal lives that span several orders of magnitude. This appears to be due in part to the very strong effects of time-dependent environment-driven damage in elevated temperature tests, including the ascent mission cycle.

Out-of-Phase Environment-Assisted Surface Cracking

The results indicate that a most serious, life-limiting cracking mechanism in hypersonic applications could be associated with OP environment-assisted cracking. Environment-assisted surface cracking was enhanced in these OP tests over that in isothermal and isothermal dwell tests having comparable times at high temperatures. Further, a creep test at 815°C and 690 MPa did not produce significant surface cracking after 526 h. This indicated environment-assisted surface cracking was not proportional to only time at high temperatures. These facts suggested that the low OP and mission lives were due to surface cracking enhanced by TMF-environment interactions. In order to experimentally explore this, the damage of the thermal and loading cycles of the OP test were evaluated separately. A specimen was first subjected to 750 thermal cycles from 150 to 815°C at zero load and then examined. Although the surface was oxidized, only a very few minor surface cracks up to 0.4 mm long were observed, Fig. 16*a*. This specimen was then fatigue tested at 150°C and 690 MPa. This test was interrupted after 738 930 cycles and produced relatively few large

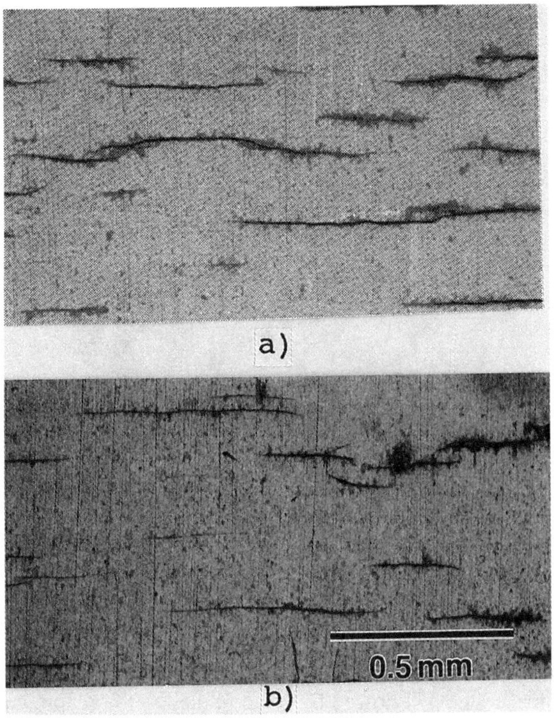

FIG. 14—*Gage sides after fatigue tests with* $\sigma_{max} \approx$ *690 MPa, 26-min cycle period:* (a) *OP and* (b) *mission.*

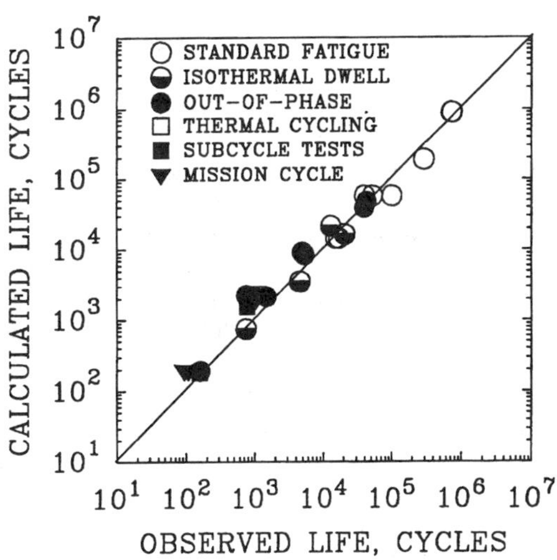

FIG. 15—*Observed fatigue lives versus calculated lives.*

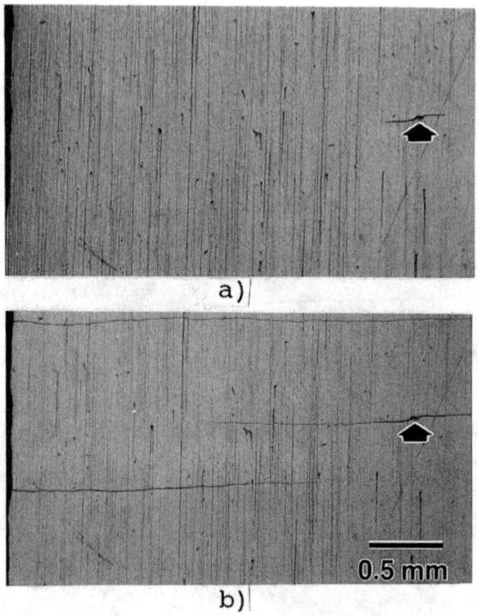

FIG. 16—*Gage surfaces after* (a) *750 thermal cycles at 0 load from 150 to 815°C and* (b) *above plus 739 000 fatigue cycles, 150°C, $\Delta\sigma$ = 690 MPa.*

surface cracks as in standard 150°C isothermal fatigue tests, Fig. 16*b*. A thermal cycle to high temperatures immediately followed by or combined with an applied tensile load at low temperatures apparently interacts to produce much more environment-assisted surface cracking damage than that produced separately by thermal cycling following by low temperature fatigue cycling. It should be noted that the simple linear damage formulation would predict the same damage and life in this damage separation test as in the conventional OP test, as it has no cycle or damage sequence dependencies. This illustrates that such a simple formulation artificially separates the damage of an inherently combined damage mechanism. While such a formulation is a useful tool, it does not completely describe or predict such complex damage mechanisms.

An oxygen-affected surface layer forms in the titanium aluminide matrix during high temperature excursions. This region has been shown [*21*] to be made up of a nonprotective brittle oxide rich in TiO_2 over matrix embrittled by additional oxygen in solid solution. An applied cyclic load could preferentially crack this brittle region. As the kinetics of titanium oxide formation and oxygen diffusion in titanium are rapid, the embrittling effect of oxygen could advance beyond a crack tip during each thermal cycle. The crack could then advance through this embrittled region in subsequent loading, especially during low temperature loading where the ductility of titanium alloys such Ti-24Al-11Nb generally decreases with decreasing temperature.

The environment-assisted surface cracking mechanism may be difficult to prevent in titanium matrix composites. Titanium alloys (greater than 50 atomic % titanium) generally form TiO_2-rich oxide layers at temperatures approaching 815°C and have high oxygen diffusivities. The γ TiAl alloys have a more protective Al_2O_3 oxide at temperatures up to 815°C and could offer improved resistance to this cracking. TMC coatings or claddings of more oxidation-resistant materials may offer benefits, provided crack initiation problems of even ductile coatings can be surmounted [*20*]. The effects of environment-assisted surface cracking could be reduced at the low partial pressures of oxygen foreseen during high elevation, high temperature segments of the hypersonic service cycles. OP testing of SCS-6/Ti-24Al-11Nb in flowing argon containing 10^{-6} ppm of oxygen increased life by more than 100 times over tests in air [*15*]. However, surface crack initiations in the matrix still limited life even at this low oxygen level, and therefore still could do so in many hypersonic applications.

Summary and Conclusions

TMC fatigue damage mechanisms involving matrix cracking were evaluated at low applied stresses in a $[0]_8$ SCS-6/Ti-24Al-11Nb TMC. Simple isothermal and nonisothermal fatigue tests were first performed to understand the basic factor damage relationships. More complex cycles that mixed cyclic loads, sustained loads, and multiple temperature changes were then employed to extend this understanding to complex service cycles. Matrix cracking occurred in both isothermal and nonisothermal tests. Time-dependent environment-assisted surface cracking was activated in tests to high temperatures. This environment-assisted surface cracking was enhanced to produce lowest lives in out-of-phase fatigue tests having extended cycle periods. It can be concluded from this work that:

1. Matrix cracking mechanisms can often limit the fatigue durability of unidirectional 0° fiber TMCs such as SCS-6/Ti-24Al-11Nb.
2. Environment-assisted surface cracking in the matrix can occur in out-of-phase service cycles that combine low temperature tensile loads with high temperature excursions to most seriously limit TMC durability. The slow out-of-phase characteristics of the hypersonic ascent mission cycle largely governed life.
3. The governing factor-damage relationships could be modeled reasonably well with a linear damage formulation for these test conditions, however the enhanced out-of-phase cracking mechanism has an inherent out-of-phase TMF-environment interaction that is not fully described in this formulation.
4. This enhanced environment-assisted matrix surface cracking may have reduced effects in hypersonic applications with high temperature exposures at low oxygen partial pressures, but the cracking may well still occur and limit life.

Acknowledgments

The authors wish to gratefully acknowledge the support of the National Institute for the Mechanics and Life Prediction of High Temperature Composites.

References

[*1*] Larsen, J. M., Russ, S. M., and Jones, J. W., "Possibilities and Pitfalls in Aerospace Applications of Titanium Matrix Composites," *Proceedings,* AGARD Conference on the Characterization of Fiber Reinforced Titanium Metal Matrix Composites, Bordeaux, France, Sept. 1993.

[2] Sorensen, J., "Titanium Matrix Composites—NASP Materials and Structures Augmentation Program," AIAA Report AIAA-90-5207, American Institute of Aeronautics and Astronautics, Washington, DC, 1990.

[3] Gayda, J., Gabb, T. P., and Freed, A. D., ''The Isothermal Fatigue Behavior of a Unidirectional SiC/Ti Composite and the Ti Alloy Matrix,'' *Fundamental Relationships Between Microstructure and Mechanical Properties of Metal Matrix Composites,* P. K. Liaw and M. N. Gungor, Eds., The Metallurgical Society of the American Institute of Mining, Metallurgical, and Petroleum Engineers, Warrendale, PA, 1990, pp. 497–514.
[4] Majumdar, B. S. and Newaz, G. M., ''Isothermal Fatigue Failure Mechanisms in Ti-Based Metal Matrix Composites,'' NASA Contract Report CR-191181, NASA Lewis Research Center, Cleveland, OH, 1993.
[5] Gabb, T. P. and Gayda, J., ''Isothermal and Nonisothermal Fatigue Damage/Failure Mechanisms in SiC/Ti-14Al-21Nb,'' *Titanium Matrix Composites,* Workshop Proceedings, WL-TR-92-4035, P. R. Smith and W. C. Revelos, Eds., Wright Laboratory, Wright-Patterson AFB, OH, 1992, pp. 292–305.
[6] Castelli, M. G., Bartolotta, P. A., and Ellis, J. R., ''Thermomechanical Testing of High-Temperature Composites: Thermomechanical Fatigue (TMF) Behavior of SiC(SCS-6)/Ti-15-3,'' *Composite Materials: Testing and Design, 10th Volume, ASTM STP 1120,* G. C. Grimes, Ed., American Society for Testing and Materials, Philadelphia, 1992, pp. 70–86.
[7] Russ, S. M., Nicholas, T., Bates, M., and Mall, S., ''Thermomechanical Fatigue of SCS-6/Ti-24Al-11Nb Metal Matrix Composite,'' *Proceedings,* Symposium on Failure Mechanisms in High Temperature Composites, American Society for Mechanical Engineers, New York, 1991.
[8] Jeng, S. M., Yang, J.-M., and Aksoy, S., ''Damage Mechanisms of SCS-6/Ti-6Al-4V Composites Under Thermal-Mechanical Fatigue,'' *Materials Science and Engineering,* Vol. A156, 1992, pp. 117–124.
[9] Johnson, W. S., Lubowsinki, S. J., and Highsmith, A. L., ''Mechanical Characterization of Unnotched SCS-6/Ti-15-3 Metal Matrix Composites at Room Temperature,'' *Thermal and Mechanical Behavior of Metal Matrix and Ceramic Matrix Composites, ASTM STP 1080,* J. M. Kennedy, H. H. Moeller, and W. S. Johnson, Eds., American Society for Testing and Materials, Philadelphia, 1990.
[10] Gayda, J. and Gabb, T. P., ''Isothermal Fatigue Behaviour of a $[90°]_8$ SiC/Ti-15-3 Composite at 426°C,'' *International Journal of Fatigue,* Vol. 14, Jan. 1992, pp. 14–20.
[11] Castelli, M. G., ''Thermomechanical and Isothermal Fatigue Behavior of a $[90]_8$ Titanium Matrix Composite,'' *Proceedings,* American Society for Composites, Technomic Publishing, Lancaster, PA, 1993, pp. 884–892.
[12] Schubbe, J. J. and Mall, S., ''Damage Mechanisms in a Cross-Ply Metal Matrix Composite Under Thermal-Mechanical Cycling,'' *Composites, Proceedings,* 8th International Conference on Composite Materials (ICCM/8), Society for the Advancement of Material and Process Engineering, Covina, CA, 1991, pp. 20–B–1 to 20–B–9.
[13] Castelli, M. G. and Ellis, J. R., ''Isothermal and Thermomechanical Fatigue Damage/Failure Mechanisms in SCS-6/TIMETAL 21S $[0/90]_s$ Composite,'' *Titanium Metal Matrix Composites II,* Workshop Proceedings, WL-TR-93-4105, P. R. Smith and W. C. Revelos, Eds., Wright Laboratory, Wright-Patterson AFB, OH, 1993, pp. 442–456.
[14] Gayda, J., Gabb, T. P., and Lerch, B. A., ''Fatigue-Environment Interactions in a SiC/Ti-15-3 Composite,'' *International Journal of Fatigue,* Vol. 15, Jan. 1993, pp. 41–45.
[15] Bartolotta, P. A. and Verrilli, M. J., ''Thermomechanical Fatigue Behavior of SiC/Ti-24Al-11Nb,'' NASA Report TM-105723, NASA Lewis Research Center, Cleveland, OH, 1992.
[16] Khobaib, M. and Vahldiek, F. W., ''High Temperature Oxidation Behavior of Ti_3Al Alloys,'' *Space Age Metal Technology, Proceedings,* 2nd International SAMPE Metals Conference, Society for the Advancement of Material and Process Engineering, Covina, CA, 1988, pp. 262–270.
[17] Nicholas, T. and Russ, S. M., ''Elevated Temperature Fatigue Behavior of SCS-6/Ti-24Al-11Nb,'' *Materials Science and Engineering,* Vol. A153, 1992, pp. 514–519.
[18] Gabb, T. P. and Gayda, J., ''Nonisothermal Fatigue Damage of SiC/Ti_3Al+Nb,'' NASA Report CP-10104, NASA Lewis Research Center, Cleveland, OH, 1992, pp. 48–1 to 48–12.
[19] Neu, R. W. and Nicholas, T., ''Thermomechanical Fatigue of SCS-6/TIMETAL 21S Under Out-of-Phase Loading,'' *Proceedings,* Symposium on Thermomechanical Behavior of Advanced Structural Materials, ASME Winter Annual Meeting, American Society of Mechanical Engineers, New York, 1993.
[20] Niemann, J. T., Bowden, D. M., and Thies, D. J., ''NASP Materials and Structures Augmentation Program Final Report, Titanium Matrix Composites, Vol. II—Materials and Process

Development,'' NASP Report CR-1145, Vol. II, NASA Langley Research Center, Hampton, VA, 1993.

[*21*] Brindley, W. J. and Smialek, J. L., ''Environmental Effects on Orthorhombic Ti-22Al-23Nb,'' NASA Report CP-19117, NASA Lewis Research Center, Cleveland, OH, 1993, pp. 34–1 to 34–12.

Michael G. Castelli[1]

Characterization of Damage Progression in SCS-6/TIMETAL 21S $[0]_4$ Under Thermomechanical Fatigue Loading

REFERENCE: Castelli, M. G., "**Characterization of Damage Progression in SCS-6/TIMETAL 21S $[0]_4$ Under Thermomechanical Fatigue Loading,**" *Life Prediction Methodology for Titanium Matrix Composites, ASTM STP 1253,* W. S. Johnson, J. M. Larsen, and B. N. Cox, Eds., American Society for Testing and Materials, 1996, pp. 412–431.

ABSTRACT: A detailed experimental investigation was performed at a single maximum cyclic stress (σ_{max}) level to physically characterize the progression of thermomechanical fatigue (TMF) damage in continuously reinforced [0°] SCS-6/TIMETAL 21S, a titanium matrix composite. In-phase (IP) and out-of-phase (OP) loadings were investigated at σ_{max} = 1000 MPa with a temperature cycle from 150 to 650°C. Damage progression, in terms of macroscopic property degradation, was experimentally quantified through an advanced TMF test methodology that incorporates explicit measurements of the isothermal static moduli at the TMF temperature extremes and the coefficient of thermal expansion (CTE) as functions of the TMF cycles. Detailed characterization of the physical damage progression at the microstructural level was performed by interrupting multiple TMF tests at various stages of mechanical property degradation and analyzing the microstructure through extensive destructive metallography. Further, the extent of damage was also quantified through residual static strength measurements. Results indicated that damage initiation occurred very early in cyclic life ($N < 0.1N_f$) for both the IP and OP TMF loadings. IP TMF damage was found to be dominated by fiber breakage with a physical damage progression in the microstructure that was difficult to quantify. OP TMF loadings produced matrix cracking exclusively associated with surface initiations. Here, damage progression was easily distinguished in terms of both the number of cracks and their relative inward progressions toward the outer fiber rows with increased cycling. The point at which the leading cracks reached the outer fiber rows (when localized fiber/matrix debonding and matrix crack bridging occurred) appeared to be reflected in the macroscopic property degradation curves.

KEYWORDS: metal matrix composites, damage progression, thermomechanical fatigue, mechanical properties, stiffness, coefficient of thermal expansion, titanium, titanium matrix composites, life prediction, titanium alloys, fatigue (materials), modeling

Silicon-carbon fiber-reinforced titanium matrix composites (TMCs) have recently received considerable attention in the research community for potential use in advanced high-temperature airframe and propulsion system applications. The obvious attractions of such materials are their relatively high stiffness and strength to weight ratios at elevated temperatures, when compared to monolithic materials. Studies aimed at experimentally characterizing the elevated temperature behavior of TMCs have revealed many of the complexities introduced by incorporating constituents with vastly differing mechanical properties. Many of these complexities, such as coefficient of thermal expansion (CTE) mismatch and matrix load

[1]Research engineer, NYMA, Inc., Engineering Services Division, NASA Lewis Research Center Group, Cleveland, OH 44135.

shedding, are particularly highlighted under thermomechanical fatigue (TMF) conditions [*1–10*] where dominant damage mechanisms have been clearly shown to be functions of not only the load-temperature phasing, but also the multitude of additional environmental and mechanical test parameters.

If damage modeling for this class of materials is to be entirely successful under general TMF loading conditions, the fundamental mechanisms detailed in the cited studies must be represented. Experimental studies often discuss the TMF mechanisms in terms of the various observations made through fractography and metallography on specimens cycled to failure. Though this information is extremely useful, it is often times only qualitative and lacks specific details concerning the physical progression of damage. To supplement this information, a recent advance in TMF testing was introduced to experimentally quantify TMF damage progression at the macroscopic property level [*11*]. This technique incorporated explicit measurements to track the degradations of the isothermal stiffness (E) and coefficient of thermal expansion (CTE) as functions of accumulated cycles, and has been found to provide unique insights to the nature and progression of TMF damage in TMCs [*5–7,11*]. By combining this information with that obtained from the microscopy, a further characterization is obtained; however, questions still persist concerning the physical damage progression at the microstructural level and its relative impact on the macroscopic mechanical properties. The study presented here was undertaken to address this concern.

The objective of this study was to mechanically and physically characterize the progression of damage in SCS-6/TIMETAL 21S $[0]_4$ under idealized thermomechanical fatigue (TMF) loading conditions at a single maximum cyclic stress (S_{max}) level. The general TMF deformation and life trends of this TMC are briefly reviewed from a previous study [*5*] with particular reference to the damage progression as gaged by the E and CTE degradation measurements. With this information, a series of specimens were cycled under identical TMF conditions at a S_{max} level of 1000 MPa to 10, 50, and 90% of the expected cyclic life under OP conditions and to an initial damage state and a mid-to-late damage state under IP conditions. At these various states of degradation, the specimens were subjected to destructive metallography to examine the extent of physical damage in the microstructure. Optical and scanning electron microscopy (SEM) were employed. Further, a second set of specimens was cycled in identical fashion for the OP TMF condition and tension tested at room temperature to investigate residual strength properties. Results from these various progressive damage tests are discussed and interpreted in view of the macroscopic property degradation data.

Material and Test Details

The TMC used in this study was SCS-6/TIMETAL 21S[2] $[0]_4$. The SCS-6 is a silicon carbide (SiC) fiber (manufactured by Textron Specialty Materials) with a nominal diameter of 140 μm. TIMETAL 21S is a beta-phase titanium alloy with a nominal composition of Ti-15Mo-3Nb-3Al-0.2Si (percent by weight). The TMC panel was fabricated by hot isostatic pressing alternate layers of TIMETAL 21S foil and SCS-6 fiber mat; TiNb ribbon was used for crossweave material in the fiber mats. The composite had a nominal fiber volume fraction of 0.33 and a nominal thickness of 1.0 mm. Coupon test specimens were cut from the panel using abrasive waterjet machining with no additional surface or edge preparation. All specimens were subjected to a pre-test heat treatment consisting of an 8-h soak in vacuum at 621°C to stabilized the alpha plus beta titanium microstructure of the TIMETAL 21S. The coupon specimen design is a "dogbone" geometry featuring a length of 15.2 cm, a

[2]TIMETAL 21S is a registered trademark of Timet, Titanium Metals Corporation, Toronto, OH.

radius cut of 36.8 cm, and a central parallel section with a length and width of 26 and 10 mm, respectively. This design proved to be very successful at ensuring specimen failures within the stress and temperature controlled parallel section. Metal tabs were attached to the ends of the specimen to alleviate stress concentrations induced at the gripping locations.

All fatigue tests were conducted in air on a computer-controlled servohydraulic test system equipped with water-cooled grips. A triangular waveform was used for the load under load-controlled, zero-tension (min/max stress ratio = 0) conditions. In-phase (IP) and out-of-phase (OP) TMF loadings with S_{max} = 1000 MPa were examined with a cyclic frequency of 5.56×10^{-3} Hz (3-min period) and a temperature cycle using a triangular waveform from 150 to 650°C. IP and OP loadings were defined as 0° and 180°, respectively, time phase shifts between the load and temperature waveforms. Axial strain measurements were made over a centrally located 12.7 mm gage length with a water-cooled extensometer mounted on the edge of the specimen. Intrinsic K-type thermocouples were used for temperature monitoring and closed-loop control. The thermocouple locations did not exhibit premature cracking, and thus were considered to have had a negligible influence on the specimen fatigue life. Specimen fatigue life is defined as the number of cycles to failure (N_f) where failure represents complete separation of the specimen into two pieces. Tensile residual strength tests were conducted at room temperature in strain control with a tensile strain rate of 1.0×10^{-4} s^{-1}.

The TMF test technique used in this study [*11*] to experimentally quantify both the stiffness and the CTE degradations during TMF testing is shown graphically in Fig. 1. After each TMF cycle designated for data collection, the specimen was subjected to a thermal

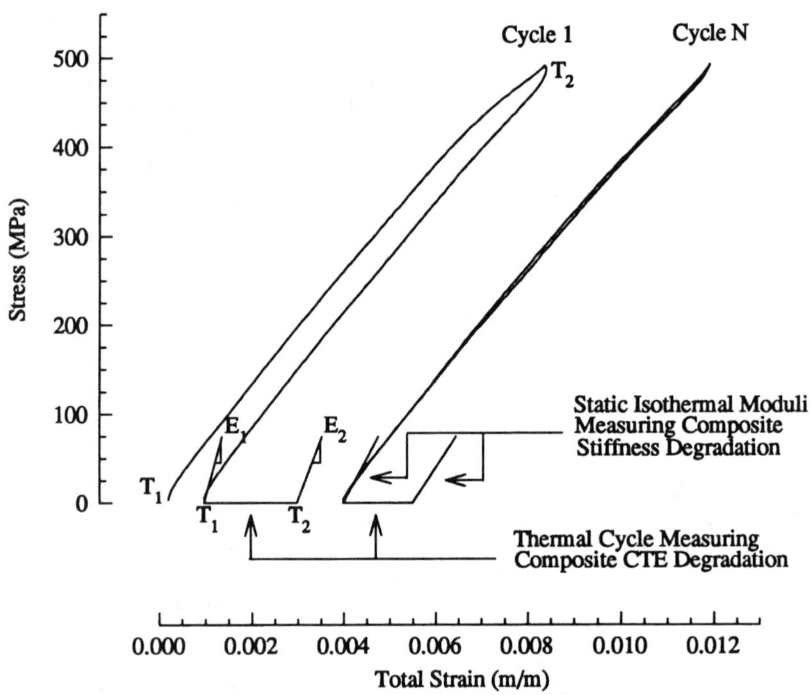

FIG. 1—*Advanced test technique enabling explicit measurement of the isothermal static moduli and CTE during TMF testing.*

cycle under zero load, represented by the horizontal line from T_1 to T_2 at zero load. This thermal cycle allowed the real-time or "current" thermal strain properties of the composite to be explicitly measured with the extensometer. At the extremes of the thermal cycle (that is, 150 and 650°C), the temperatures were briefly stabilized and small "elastic" loads were applied; these loadings are represented by the stress-strain lines in Fig. 1 with slopes marked E_1 and E_2. This allowed explicit measurement of the "current" isothermal material stiffness at both temperature extremes. Note that the E measurements are essentially a Young's modulus measurement (see ASTM Test Method for Young's Modulus, Tangent Modulus, and Chord Modulus (E 111-82)) from zero macroscopic load. The maximum stress levels used to obtain E were kept relatively small (80 MPa) to facilitate a nondamaging, linear-elastic response.

In comparison to an uninterrupted "conventional" TMF test, this technique introduced several additional thermal cycles at zero load and several relatively small elastic loads. These additional loadings were found to be inconsequential with respect to the ongoing material damage and resulting TMF life at the stress levels investigated. This was verified throughout the development and implementation of the test technique [*11*].

One advantage of this test methodology was the ability to record the "real-time" thermal strain information, which was "essential" for accurate post-test assessment of the mechanical strain response (total minus thermal strain) [*5–7,11*]. Also, and most critical to the present work, the CTE and isothermal E measurements provided definitive, quantitative means for tracking the TMF damage progression in terms of macroscopic property degradation.

Results and Discussion

Deformation and Life

Prior to discussing the TMF damage progression, it is worthwhile to briefly review the general TMF deformation and life exhibited by this TMC [*5*]. Representative deformation behaviors for OP and IP TMF with S_{max} = 1000 MPa are shown in Fig. 2, where maximum cyclic mechanical strain values are plotted. As indicated, OP TMF loadings did not promote excessive strain ratchetting, as the OP S_{max} occurs at T_{min} (150°C) where matrix load shedding due to time-dependent deformation and tensile-strain ratchetting was minor. Further, the maximum strain increases illustrated in Fig. 2 that occurred later in TMF life will be shown to have resulted primarily from increases in composite compliance (stiffness degradation). In contrast, IP load-controlled TMF conditions were highly conducive to strain ratchetting, indicated in Fig. 2 by the large maximum mechanical strain increases early in cyclic life. Here, S_{max} and T_{max} loadings coincide, promoting matrix load shedding due to time-dependent deformation, as seen in previous TMF experimental studies on [0] TMCs [*1–5*] and modeled through various time-dependent and viscoplastic analyses [*12–15*]. Subsequent discussions of the stiffness degradations under IP conditions will show that the majority of mechanical strain increase is associated with strain ratchetting rather than an increase in composite compliance.

TMF life results for SCS-6/TIMETAL 21S $[0]_4$ are plotted in Fig. 3. In general, IP TMF loading conditions were found to be life-limiting under relatively high S_{max} conditions where the strain ratchetting mechanisms mentioned earlier led to excessive strains. The IP and OP TMF life trends crossed at a S_{max} of approximately 1000 MPa, below which OP conditions resulted in shorter cyclic lives. Based on these TMF life trends, a S_{max} of 1000 MPa was chosen for the present study, so as to investigate the damage progression under IP and OP

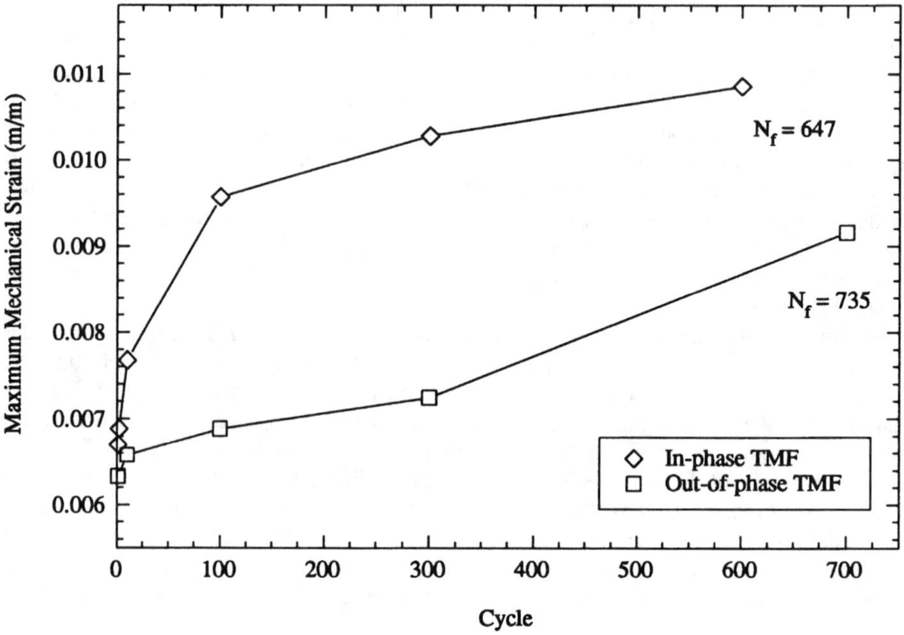

FIG. 2—*Mechanical strain deformation behavior of SCS-6/TIMETAL 21S $[0]_4$ under IP and OP loadings from 150 to 650°C with* S_{max} = *1000 MPa.*

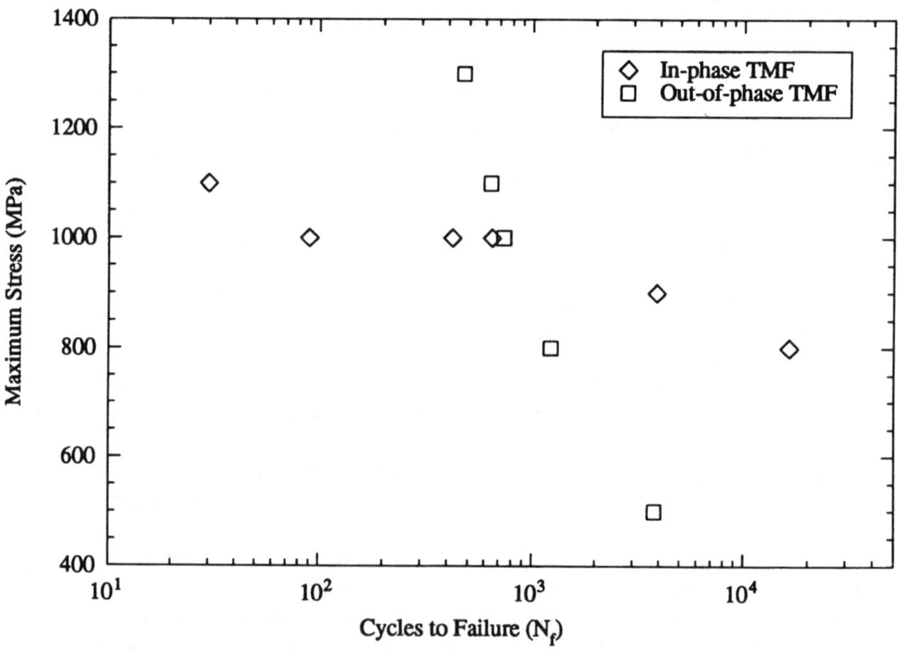

FIG. 3—*TMF life behavior of SCS-6/TIMETAL 21S $[0]_4$ under IP and OP loadings from 150 to 650°C.*

conditions resulting in comparable lives. This selection also served to equalize temperature exposure times and thus, the potential for environmental degradation.

Macroscopic Property Degradation

Out-Of-Phase TMF

As previously stated, the TMF test technique used in this study explicitly measured isothermal *E* and CTE as functions of accumulated TMF cycles; these data are shown in Figs. 4*a* and *b*, respectively, for the OP TMF test with S_{max} = 1000 MPa. Note that the isothermal *E* measurements for this one test are represented in Fig. 4*a* by two sets of data consisting of the isothermal *E* values measured at the temperature extremes of the TMF cycle, that is, E_{150} and E_{650}. The dashed vertical lines on the graphs represent points at which progressive damage tests were interrupted and destructively examined; subsequently discussed.

Several noteworthy trends were evident in Fig. 4*a* concerning the *E* degradations. First, the *E* values measured from the OP test did not degrade equally, but rather, the E_{150} ultimately degraded much more severely, even to the point where the E_{150} and E_{650} were essentially equal just prior to failure (that is, in a highly damaged state). Previous studies [*1–5*] have shown that the dominant damage mechanism in [0] TMCs under OP TMF loadings was matrix damage in association with minimal fiber cracking. Using a simple rule of mixtures approach to composite stiffness, one notes that the contribution of the matrix is much greater at 150°C than at 650°C. This comes from the fact that the stiffness of the TIMETAL 21S matrix is strongly dependent upon temperature in this temperature range [*16*]; see Table 1. Given this fact, and considering a continual degradation of the effective matrix contribution in the presence of undamaged fibers, the relative degradation of the composite E_{150} values will be greater than that experienced by the composite E_{650} values. Further, in a case where the matrix approaches a "completely" damaged state, the composite *E* values at both temperature extremes will degrade to similar quantities, given that the SCS-6 fiber stiffness is relatively insensitive to temperature over this temperature range [*17*]; see Table 1. For example, using the values provided in Table 1, the composite stiffnesses as calculated based solely upon the undamaged fiber contribution are 128 and 122 GPa at 150 and 650°C, respectively.

It is important to note, however, that other factors, such as constituent stresses existing at zero macroscopic load, can influence this composite stiffness measurement and complicate a rule-of-mixtures interpretation. For example, subsequent to the matrix stress relaxation promoted during elevated temperature fatigue cycling, longitudinal matrix stresses induced by the fiber/matrix CTE mismatch can be compressive at relatively high temperatures and tensile at lower temperatures, all at zero macroscopic load. Such behavior can allow for closure of the matrix cracks at higher temperatures and crack opening at lower temperatures. Given the definition used for the composite stiffness, that is, the initial stress-strain slope from zero macroscopic load, it becomes apparent that a low-temperature static stiffness measurement can actually fall below that measured at a higher temperature (by inducing matrix crack closure at the higher temperature); see Ref *5* for a case in point.

Degradation of the CTE of the TMC during OP TMF with S_{max} = 1000 MPa is shown in Fig. 4*b*. This macroscopic property can be tracked either in terms of an instantaneous value at a given temperature, or a mean value between two temperatures. The latter approach is taken here for ease of discussion. The mean CTE was calculated by taking the thermal strain range, as measured under zero macroscopic load, from 150 to 650°C and dividing it by ΔT, that is, 500°C. Thus, it represents the average CTE for the TMF temperature cycle.

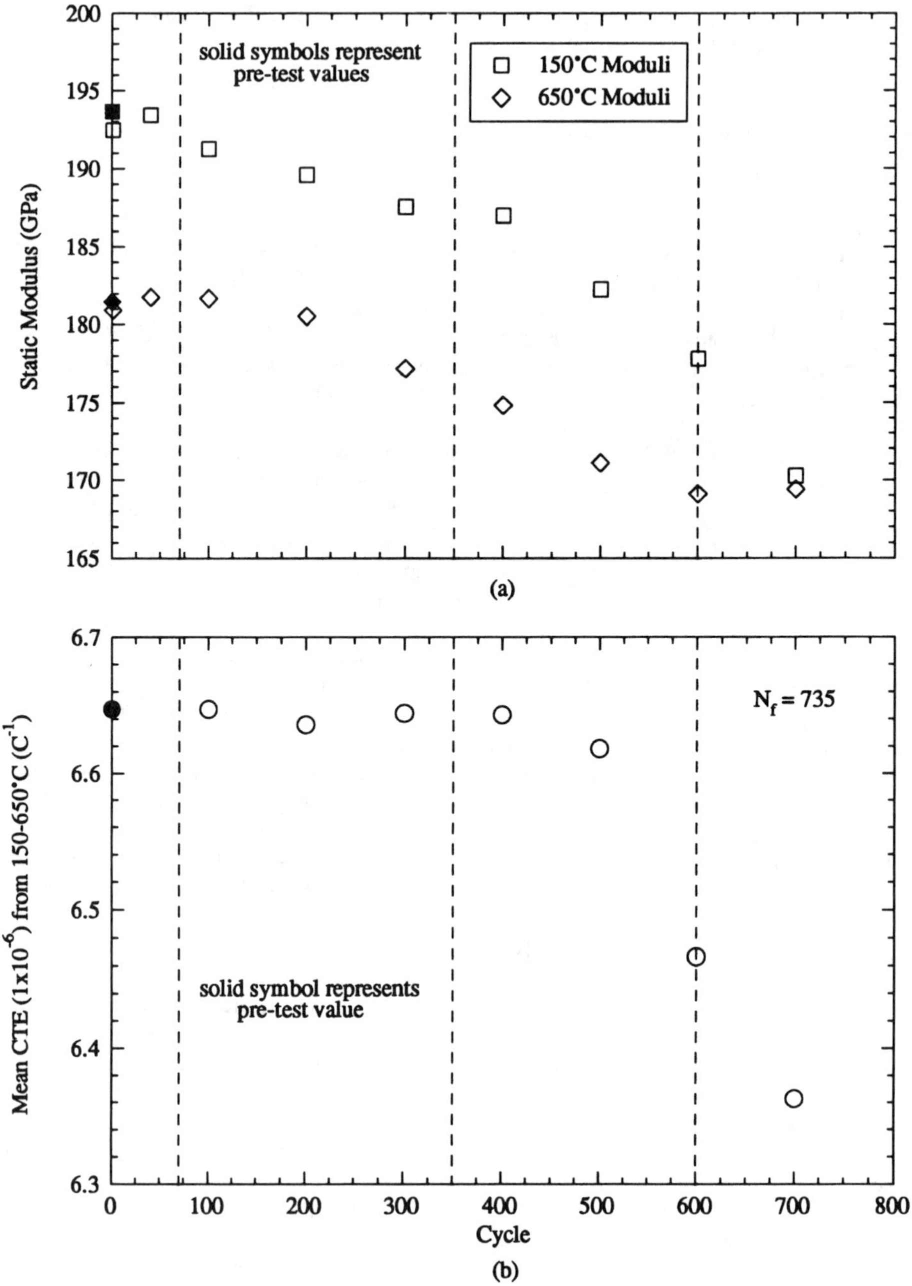

FIG. 4—*Thermomechanical fatigue degradation of the* (a) *isothermal moduli and* (b) *mean CTE under OP conditions with* S_{max} = *1000 MPa.*

TABLE 1—*Static stiffness values for the SCS-6 fiber* [17] *and TIMETAL 21S foil matrix* [16].

Temperature, °C	SCS-6 Fiber, GPa	TIMETAL 21S, GPa
150	388	110
650	370	60 to 75[a]
...	ΔE = 18	ΔE = 35 to 50

[a]This property exhibits a positive strain rate dependence.

As shown in Fig. 4*b*, the macroscopic mean CTE of the TMC is essentially constant up to approximately ½N_f. Subsequently, the property degrades a relatively small degree to a total degradation of approximately 5% prior to failure. Previous experimental studies where CTE degradation was monitored has shown this measurement to be a good indicator of matrix-dominated damage, particularly at relatively low S_{max} levels where matrix crack growth can mature and dominate the cross-section prior to overload failure [*5–7*]. In such cases, the degradation of this property has been observed to be as high as 25%. It is important to point out that this macroscopic property not only changes with damage, but has also been shown in other metal matrix composites to change with inelasticity of the matrix, and constituent stress redistributions [*18*]. However, the stiffness measurements discussed earlier clearly suggest the progression of damage.

In-Phase TMF

In similar fashion to that of Figs. 4*a* and *b*, Figs. 5*a* and *b* contain the isothermal *E* and CTE degradation curves, respectively, for an IP TMF test with S_{max} = 1000 MPa. Figure 5*a* reveals that the IP TMF damage promoted a relatively small, but detectable stiffness degradation. In contrast to the OP TMF degradation, the magnitude of degradation was comparable at both temperature extremes. This change is often not obvious from the deformation data, as a clear assessment of stiffness/compliance is somewhat convoluted because of the strain ratchetting, continually changing temperature, and potentially changing thermal strain component [*5,11*]. Again, using a simple rule of mixtures interpretation and recalling from previous studies that typical IP TMF damage was dominated by fiber cracking [*1–5*], it is easily shown that a reduction of the effective SCS-6 contribution to the composite stiffness will promote reductions of comparable magnitudes in this measurement at 150 and 650°C because the SCS-6 fiber stiffness is relatively insensitive to temperature over this range. This may not suggest that the fiber stiffness is actually decreasing, but rather that the "effective fiber contribution" to the composite stiffness is decreasing. This effect may be a result of the fiber cracking and highly localized debonding associated with the crack locations. The data also reveals that the stiffness degradation occurred very early in cyclic life where strain ratchetting is most severe, and tended to stabilize after approximately 100 cycles.

Degradation of the TMC CTE during IP TMF is shown in Fig. 5*b*. As seen in this figure, the TMC experiences only a small decrease in mean CTE with S_{max} = 1000 MPa IP TMF cycling, as the total change represents a change in thermal strain range of approximately 100 $\mu m\sqrt{m}$. Note, however, the similarity in trends with the stiffness degradation and strain ratchetting, as the majority of change occurs early in cyclic life.

Evolution of Microstructural Damage Under OP Conditions

10% N_f—The first stage of microstructural damage investigated was that at 10% N_f for the OP TMF condition discussed in Figs. 2 and 4. Shown in Fig. 6 is a metallographic

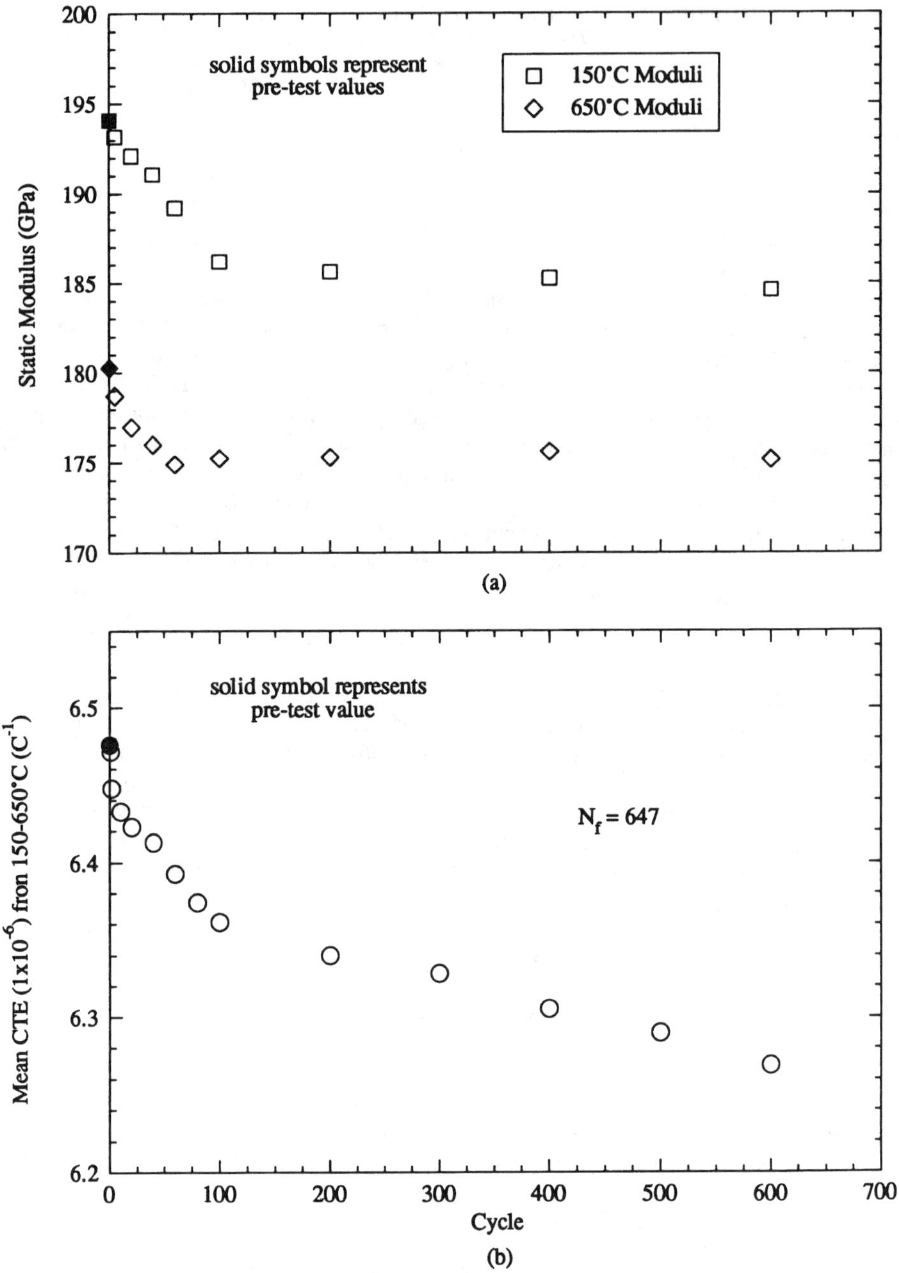

FIG. 5—*Thermomechanical fatigue degradation of the* (a) *isothermal moduli and* (b) *mean CTE under IP conditions with* S_{max} = *1000 MPa.*

section taken from a specimen interrupted at this level; the cut is a longitudinal section through the thickness (LST) with the micrograph revealing the surface edge of the specimen. As seen here, OP TMF conditions produced very fine surface initiated matrix cracks at this early stage in TMF life. Individual cracks were relatively numerous throughout the gage section and exhibited an average length into the specimen thickness of 10 to 20 μm. As shown in Fig. 7, this early stage of matrix cracking was readily detectible on the specimen surface through the use of scanning electron microscopy (SEM). Here, numerous cracks were found predominantly transverse to the loading direction with lengths across the specimen width as great as 80 μm. It is noteworthy to state that the majority of initiation points detected were found to be associated with "rough" surface features. (Recall that specimens were tested with as-received surface conditions, as typical of most TMC testing to date.) As would be expected with most metallic materials, this suggests that surface polishing may extend the time-to-crack initiation. It is apparent that the slight degradation detected in Fig. 4*a* by the E_{150} measurements are indications of these very early surface initiations. Again, the fact that the E_{150} measurements detected this damage and the E_{650} measurements did not, is likely related to the respective matrix stress states at the two temperatures (discussed earlier).

Additional longitudinal sections through the width (LSW) and transverse sections (TS) did not reveal any further degradation in either of the constituents with one exception; highly localized fiber/matrix (F/M) debonds were evident at the titanium-niobium (TiNb)

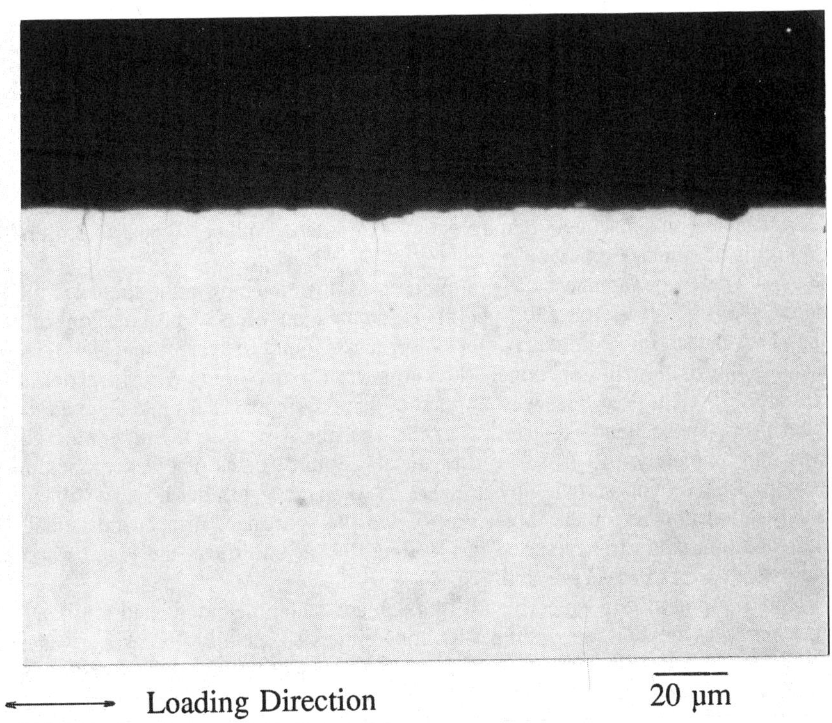

FIG. 6—*OP TMF damage progression at 10%* N_f *revealing surface initiated matrix cracks (longitudinal section into specimen thickness).*

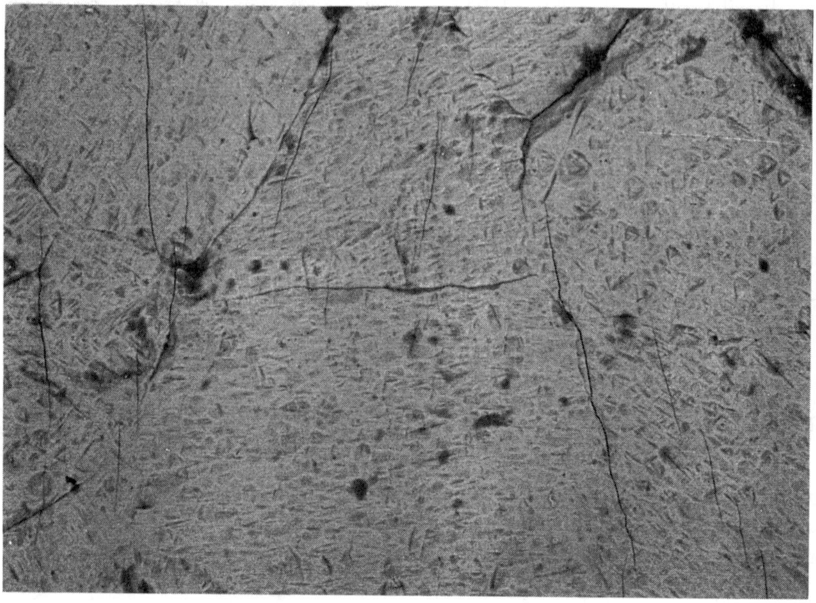

FIG. 7—*SEM of specimen surface revealing matrix cracks under OP TMF conditions at 10%* N_f.

crossweave locations, such as that pictured in Fig. 8 where the weave appears as a white "band" in the matrix, subsequent to a lactic acid etch. Although this effect was present at this 10% stage and all subsequent damage states examined, it did not appear to preferentially initiate or enhance matrix cracking.

50% N_f—The next damage state examined was that existing after the specimen was cycled to 50% N_f. Here, the OP specimen revealed an orderly progression of matrix cracking. The surface-initiated matrix cracks were now found to have propagated an average distance of 60 μm into the specimen thickness with a maximum visible penetration of approximately 100 μm. Nowhere was it apparent that the major cracks had reached the outer most fiber rows (subsequently referred to as the first fiber rows), as the outer surface of the fiber was approximately 130 μm from the surface. Initial stages of crack branching were visible in the LST mount (shown later in the $0.9N_f$ damage state), but it was visible on only a relatively small number of the larger cracks. On the specimen surface, individual cracks were still predominantly transverse to the loading, but at this stage, the lengths across the specimen width were on average 200 μm.

LSW and TS mounts confirmed that at this damage state, the cracks had not reached the first fiber rows, as no F/M debonding was apparent. Also, no fiber cracking was visible. Note from the E_{150} (Fig. 4*a*) and the CTE (Fig. 4*b*) degradation curves that slope decreases (increased degradation rates) occurred after approximately cycle 400, which was slightly after this interrupt point. From the microstructural evidence available, this was likely the point at which the major cracks reached the first fiber rows, as this would introduce an additional damage component, namely, F/M debonding. Prior to this event, however, it is

FIG. 8—*Highly localized debonds evident at the 10%* N_f *OP TMF condition (shown) and the initial damage state IP TMF condition (longitudinal section into specimen width).*

not clear as to why the CTE did not reveal some degradation associated with the matrix cracking.

90% N_f—The last OP specimen examined was interrupted at the 90% N_f level, as indicated again by the dashed line on Fig. 4. The microstructural damage was now quite extensive, with the most obvious progression being that the cracks had reached and propagated just beyond the first fiber rows. Many of the larger cracks exhibited extensive branching such as that illustrated in the LST mount shown in Fig. 9. Also shown in this figure was the debonding at the F/M interface that was, at this stage, common to the vast majority of fibers in the first fiber rows. This effect was most evident in the LSW mounts revealing one of the outer most fiber rows, such as that illustrated in Fig. 10. Here, substantial matrix cracking, F/M debonding, and crack bridging were all apparent. Even at this extensive state of damage, no fiber cracking was identified. A surface view of the matrix cracks revealed transverse lengths (across the specimen width) of up to 400 μm. At this 90% N_f level, both the E and CTE degradations were distinct, indicating the severe damage revealed in the microstructure. One effect that was not quantified in this study or mentioned thus far, but was clearly found to play a role in the failure mechanisms was environmental degradation. Failure surfaces from the TMF life study [*5*] revealed strong environmental interaction, as all matrix cracks were found to be heavily oxidized. A previous study designed to isolate the environmental degradation effects under TMF conditions revealed a dramatic impact on resulting OP TMF lives [*3*]. Thus, it is sufficient to state that all of the OP damage mechanisms discussed were enhanced by the oxidizing environment.

FIG. 9—*OP TMF damage progression at 90%* N_f *revealing extensive matrix crack branching and F/M debonding at first fiber row (longitudinal section into specimen thickness).*

Residual Strength under OP Conditions

In addition to the characterization of microstructural damage progression, room temperature tensile residual strengths were also examined to assist in quantifying the damage states. These tests were conducted after the specimens were subjected to the various percentages of OP TMF life with S_{max} = 1000 MPa. Shown in Fig. 11 are the results from the tensile residual strength measurements. At 10% of the expected N_f, reductions of 7 and 9% were realized in the ultimate tensile strength (S_{ult}) and strain to failure (e_f), respectively. Little or no reduction in room temperature E was detectable, as consistent with the E degradation curves given in Fig. 4*a*. After 350 cycles of OP TMF (50% N_f), the tensile response revealed clear signs of the matrix damage discussed earlier. With only a nominal reduction evident in the room temperature E, significant reductions of 18 and 22% were found to occur in the S_{ult} and e_f, respectively. Though the e_f remains moderate for this TMC, one can see that the onset of nonlinearity occurs much earlier as a result of the matrix cracks that were relatively close to the first fiber rows at this stage of TMF damage.

The final stage of damage investigated, 90% N_f, reveals a dramatic change in tensile behavior, as reductions of 40 and 45% were realized in the S_{ult} and e_f, respectively. Note the essentially linear behavior to complete failure. Recall that at this damage state, the matrix cracks had propagated from the surface to just beyond the first fiber rows, but the fibers appeared to be intact, beyond obvious damage to the outer carbon coatings. If a rule-of-mixtures calculation of S_{ult} is made (which is reasonably accurate for the unfatigued sample

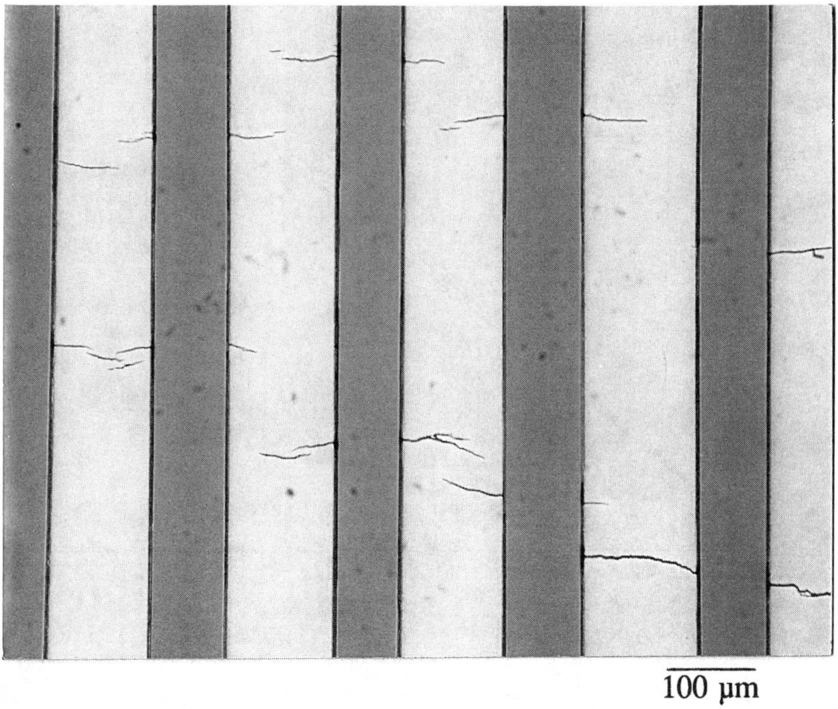

FIG. 10—*OP TMF damage progression at 90%* N_f *revealing considerable debonding and crack bridging evident in first fiber row (longitudinal section into specimen width).*

with S_{ult} of the fiber ≃ 4000 MPa and S_{ult} of the TIMETAL 21S ≃ 1100 MPa) assuming no matrix contribution, the result is 1320 MPa. This implies that the composite strength at this stage of damage is less than the "fiber only" strength and suggests that some degree of fiber strength degradation is present, especially in light of the fact that the micrographs from above do not show the matrix to be "fully damaged." This loss in fiber strength could result from a number of effects, including environmental degradation, the carbon coating damage cited earlier, and appropriate stress concentrations resulting from the matrix cracks.

To summarize the degradation of the room-temperature residual strength tensile properties, the stiffness, ultimate strength, and strain to failure are plotted in Fig. 12 as normalized values with respect to the properties obtained from the unfatigued sample. It is interesting to note that a straight line approximates the residual strength property data with reasonable accuracy.

Evolution of Microstructural Damage under IP Conditions

An investigation similar to that discussed previously for the OP condition was attempted for S_{max} = 1000 MPa under IP loadings. Unfortunately, the IP TMF damage progression and lives proved to be much less deterministic in nature. Thus, in the process of obtaining the 50 and 90% N_f specimens based upon the S_{max} = 1000 MPa TMF life illustrated in Figs. 2 and 5, "premature" specimen failures were realized. These life points are also plotted in Fig. 3. This was the case, even though the property degradation patterns were similar for each of the specimens. Note again that the vast majority of property degradation occurred

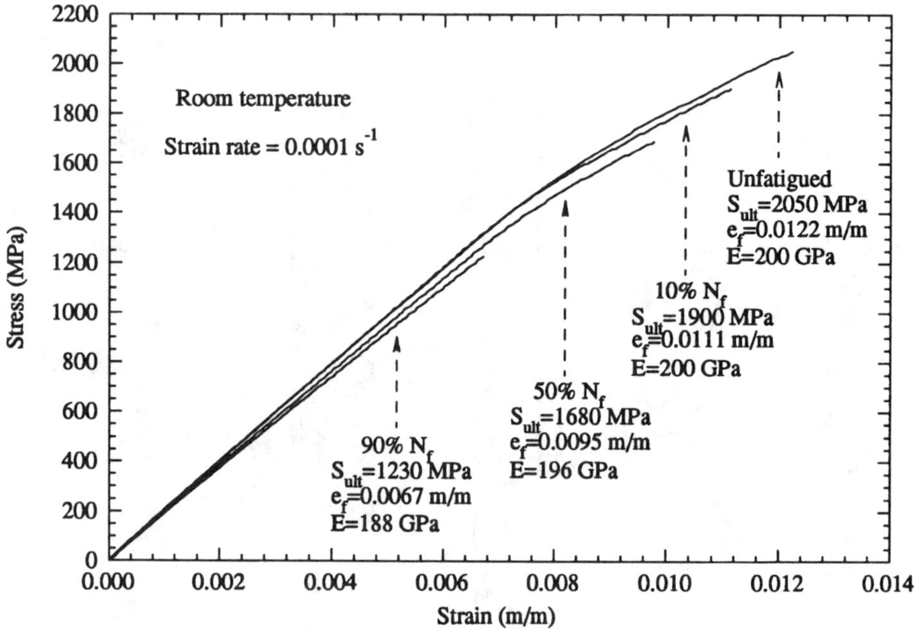

FIG. 11—*Room temperature tensile residual strength measurements of SCS-6/TIMETAL 21S* $[0]_4$ *subjected to various percentages of OP TMF life with* S_{max} *= 1000 MPa.*

very early in the TMF life. As a result of the nondeterministic failure experienced under the IP condition, it was not possible to state with sufficient confidence that a particular specimen was at a given percentage of expected life (given that N_f was found to vary from 90 to 647 cycles). However, it was worthwhile to consider the evolution of microstructural damage at both a relatively early stage in the TMF life and at a later stage, even if this state could not be explicitly quantified in terms of cyclic life. Thus, the following microstructural examination will address these two states.

Initial Damage State—Similar to the OP specimen that revealed microstructural damage after experiencing only 70 cycles (10% N_f) with $S_{\max}$ = 1000 MPa, the specimen subjected to 65 cycles of IP loading also revealed physical damage at this early stage. This damage was fiber cracking. Fiber cracking, such as that shown by the LSW mount in Fig. 13, was evident in multiple fibers; however, many fibers did not exhibit cracking. The cracks were relatively fine and most often found only in the silicon rich "outer" portion of the fiber and did not appear to continue through the fiber's carbon core. Although some degree of highly localized F/M debonding was likely associated with the fiber crack locations, it was not clearly visible on the metallographic mounts. This physical identification of fiber cracking very early in cyclic life is consistent with a similar TMF study by Neu and Roman [*19*] where acoustic emission (AE) indications suggested numerous fiber breakage events on the first IP TMF cycle at this $S_{\max}$ with a slightly higher fiber volume ratio.

As found in the OP specimen examined at 10% N_f, the initial damage state under IP conditions also revealed highly localized fiber/matrix (F/M) debonds at the TiNb cross-weave locations, similar to that illustrated in Fig. 8. Again, although this effect was present at this early stage of damage and in the latter stage examined, it did not appear to enhance or preferentially initiate the fiber cracks. Additional LSW, LST, and TS mounts did not reveal any matrix cracking or further degradation in either of the constituents.

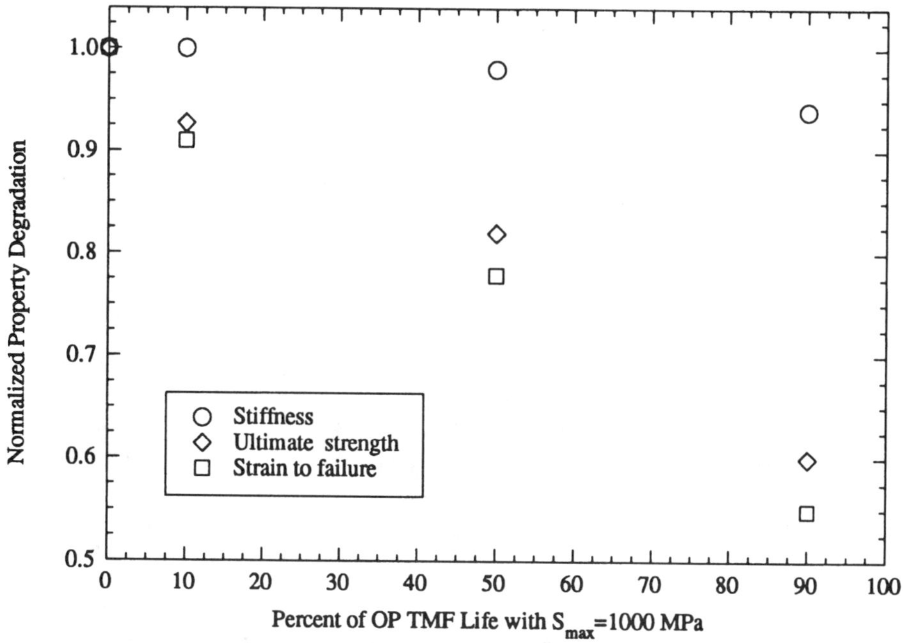

FIG. 12—*Normalized degradations of room temperature residual strength properties at various percentages of OP TMF life with* S_{max} = *1000 MPa.*

Mid-to-Late Damage State—In contrast to the OP damage progression, the IP TMF test interrupted at the mid-to-late damage state did not reveal a clear progression of damage from that evident at the earlier damage state. Subsequent to 325 IP TMF cycles, only a slight increase in fiber cracking was apparent. Transverse cracks were again evident in many of the fibers; however, each of the sections also revealed fibers that appeared uncracked. One feature that potentially did increase was the amount of visible "V-wedge" cracks, termed so because of their appearance in the plain exposed by a LSW mount, as shown in Fig. 14 on the four right-most fibers. This two-dimensional view may be suggesting a three-dimensional spiral-type crack path about the carbon core. F/M interface debonding at the fiber crack locations remains difficult to detect. Further, many cracks in the silicon-rich region of the SCS-6 fiber appear to stop (or begin) at the innermost carbon coating on the outside diameter of the fiber, with the outermost carbon coating appearing to be intact. Note, however, that this observation is subject to the various two-dimensional views enabled by the metallographic mounts. (See Ref *20* for a complete description of the SCS-6 fiber and its various microstructural features.)

In comparison to the observations made from the initial damage state, it can be suggested with only a small degree of confidence that at this mid-to-late damage state (1) the fiber cracks are more numerous, (2) the unbroken fiber lengths between cracks are on average smaller, and (3) the axial fiber crack openings are slightly larger. For these observations to be confirmed with even modest quantitative significance, it is apparent that a more comprehensive study involving multiple specimens and mounts at each damage state would likely be necessary. It can be said, however, that even at this mid-to-late stage of damage

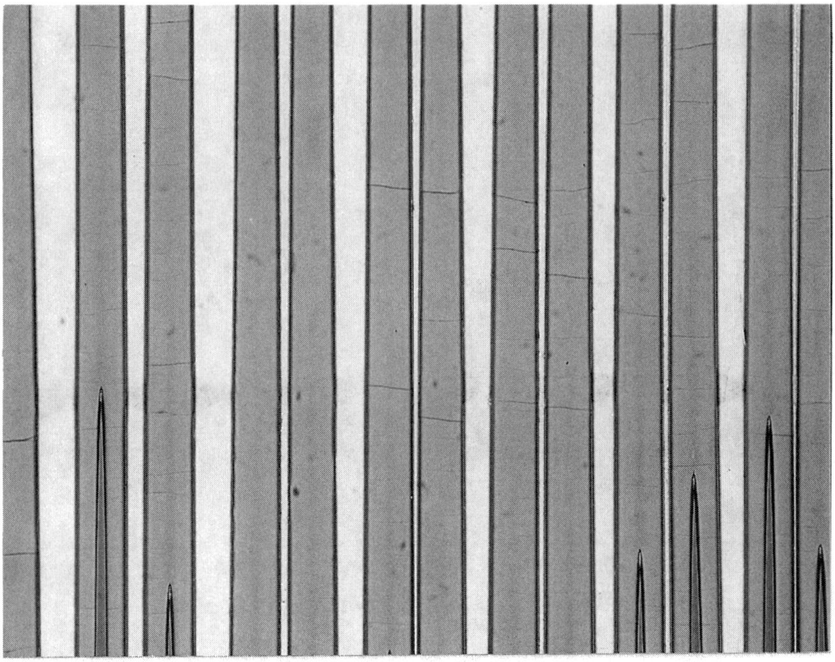

FIG. 13—*Initial damage state under IP TMF revealing fiber cracks (longitudinal section into specimen width).*

progression no matrix cracking could be identified on any of the mounts. Recalling the macroscopic property trends established in Fig. 5, one notes that very little additional stiffness degradation was detected beyond that which existed early in cyclic life, and only a relatively small CTE degradation could be measured. In contrast to what the data appeared to suggest here, it is noteworthy to state that the previously cited study [*19*] that utilized AE to detect fiber breaks, strongly suggested that additional fiber breaks occurred with continued IP TMF cycling in this life regime.

Residual strength measurements of specimens subjected to IP TMF cycling were not available for this report.

Summary/Conclusions

A detailed experimental investigation was conducted at a single maximum cyclic stress level (S_{max} = 1000 MPa) to characterize the thermomechanical fatigue (TMF) damage progression in SCS-6/TIMETAL 21S $[0]_4$ under in-phase (IP) and out-of-phase (OP) loading conditions with a temperature cycle from 150 to 650°C. The progression of damage was characterized at the macroscopic level through explicit measurements of the composite isothermal static moduli (E) and coefficient of thermal expansion (CTE). The progression of damage at the microstructural level was characterized through a systematic series of tests interrupted at various percentages of expected cyclic life (N_f) and then destructively prepared for examination using optical and scanning electron microscopy. In addition, OP TMF

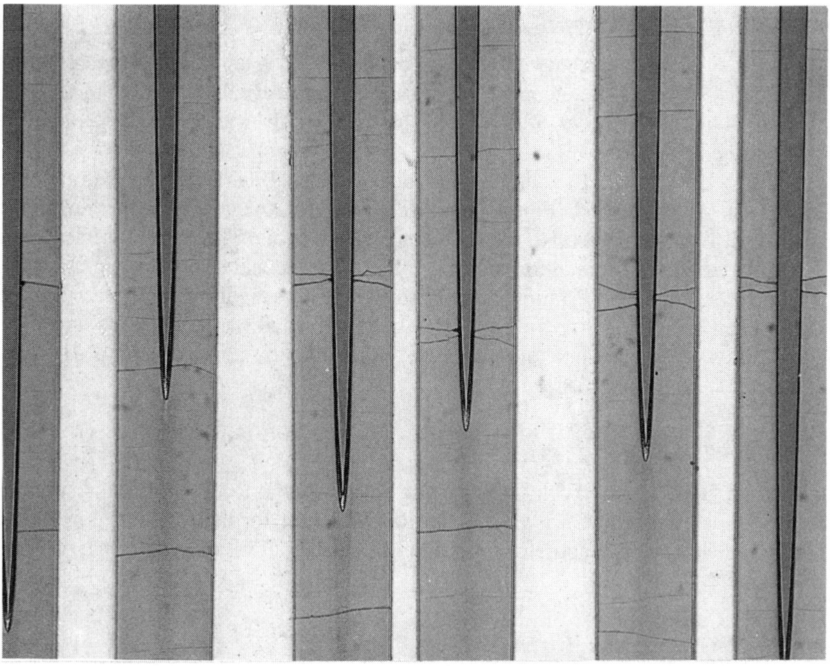

FIG. 14—*IP TMF damage progression at mid-to-late state revealing "V-wedge" type fiber cracks as seen in the four right-most fibers (longitudinal section into specimen width).*

damage was further quantified through residual static strength measurements on specimens cycled to various percentages of expected life. This study has indicated the following results.

Out-of-Phase Damage Progression

1. Initiation of damage occurs very early in cyclic life, as multiple-surface-initiated matrix cracks were found approximately 20 to 40 μm into the specimen thickness at $0.1N_f$. A modest drop of less than 10% was realized in the room temperature ultimate tensile strength (S_{ult}) and strain to failure (e_f).
2. At $0.5N_f$, surface initiated cracks were observed to have propagated at most a length of 100 μm into the specimen thickness, approximately 30 μm away from the outermost fibers. Initial crack branching was evident in the major cracks. No fiber cracking was apparent. Measurements of E at 150°C clearly indicated the presence of damage, but macroscopic CTE measurements revealed no change. Room temperature residual strength properties dropped by approximately 20%.
3. At $0.9N_f$, numerous matrix cracks had reached and propagated slightly beyond the first rows of fibers causing considerable fiber/matrix (F/M) debonding. Extensive matrix crack branching was evident, but fiber cracking remained undetectable. The E_{150} values had degraded by 13% and were approximately equal in magnitude to those measured at 650°C. CTE measurements also revealed a modest degradation. Room temperature residual strength properties were degraded by more than 40%, with S_{ult} = 1230 MPa and e_f = 0.0067 m/m.

In-Phase Damage Progression

1. Initiation of damage occurs very early in cyclic life, as relatively fine fiber cracks were identified. Macroscopic *E* and CTE measurements revealed small, but measurable degradations on the order of 3 to 5% with trends coincidental to those of the strain ratchetting.
2. At mid-to-late stages of cyclic life, the microstructural damage progression could not be clearly distinguished. Fiber/matrix debonding at the cracked fiber locations remained difficult to detect. No matrix damage was visible. The *E* measurements did not reveal any further degradation and the CTE values revealed only a slight decrease.
3. The nondeterministic IP failures observed in this investigation suggested that a more comprehensive study involving multiple specimens and mounts at each damage state would likely be necessary to appropriately quantify the microstructural damage progression.

Acknowledgments

Thanks be to God. The author also wishes to acknowledge Rod Ellis for his support and input to this research and to Chris Burke and Ralph Corner for their technical assistance in the Fatigue and Structures Laboratory and the Metallographic Laboratory at NASA Lewis Research Center.

References

[*1*] Castelli, M. G., Bartolotta, P. A., and Ellis, J. R., "Thermomechanical Testing of High-Temperature Composites: Thermomechanical Fatigue Behavior of SiC(SCS-6)/Ti-15-3," *Composite Materials: Testing and Design, Tenth Volume, ASTM STP 1120,* G. C. Grimes, Ed., American Society for Testing and Materials, Philadelphia, 1992, pp. 70–86.

[*2*] Russ, S. M., Nicholas, T., Bates, M., and Mall, S., "Thermomechanical Fatigue of SiC/Ti-24Al-11Nb Metal Matrix Composites," *Failure Mechanisms in High Temperature Composite Materials,* AD-Vol. 22 and AMD-Vol. 122, G. K. Haritos, G. Newaz, and S. Mall, Eds., American Society of Mechanical Engineers, New York, 1991, pp. 37–43.

[*3*] Bartolotta, P. A., Kantzos, P., Verrilli, M. J., and Dickerson, R. M., "Environmental Degradation of an Intermetallic Composite During Thermomechanical Fatigue," *Fatigue 93,* Vol. 2, J. P. Bailon and J. I. Dickson, Eds., 1993, pp. 1001–1006.

[*4*] Neu, R. W. and Nicholas, T., "Effect of Laminate Orientation on the Thermomechanical Fatigue Behavior of Titanium Matrix Composites," *Journal of Composites Technology and Research,* Vol. 16, No. 3, July 1994, pp. 214–224.

[*5*] Castelli, M. G., "Thermomechanical Fatigue Deformation, Damage and Life of SCS-6/TIMETAL 21S $[0]_4$," *International Journal of Fatigue,* submitted for publication.

[*6*] Castelli, M. G., "Thermomechanical Fatigue Damage/Failure Mechanisms in SCS-6/TIMETAL 21S $[0/90]_s$ Composite," *Composites Engineering,* Vol. 4, No. 9, 1994, pp. 931–946.

[*7*] Castelli, M. G., "Thermomechanical and Isothermal Fatigue Behavior of a $[90]_8$ Titanium Matrix Composite," *Proceedings,* The American Society for Composites 8th Technical Conference on Composite Materials, Mechanics, and Processing, Technomic Publications, Lancaster, PA, Oct. 1993, pp. 884–892.

[*8*] Majumdar, B. S. and Newaz, G. M., "Thermomechanical Fatigue of a Quasi-Isotropic Metal Matrix Composite," *Composite Materials: Fatigue and Fracture, Third Volume, ASTM STP 1110,* T. K. O'Brien, Ed., American Society for Testing and Materials, Philadelphia, 1991, pp. 732–752.

[*9*] Johnson, W. S., Mirdamadi, M., and Bahei-El-Din, Y. A., "Stress-Strain Analysis of a $[0/90]_{2s}$ Titanium Matrix Laminate Subjected to a Generic Hypersonic Flight Profile," *Journal of Composites Technology and Research,* Vol. 15, No. 4, Winter 1993, pp. 297–303.

[*10*] Mall, S., Hanson, D. G., Nicholas, T., and Russ, S. M., "Thermomechanical Fatigue Behavior of a Cross-Ply SCS-6/β21-S Metal Matrix Composite," *Constitutive Behavior of High Temperature Composites,* MD-Vol. 40, B. S. Majumdar, G. M. Newaz, and S. Mall, Eds., American Society of Mechanical Engineers, New York, 1992, pp. 91–106.

[11] Castelli, M. G., "An Advanced Test Technique to Quantify Thermomechanical Fatigue Damage Accumulation in Composite Materials," *Journal of Composites Technology and Research,* Vol. 16, No. 4, Oct. 1994.

[12] Mirdamadi, M., Johnson, W. S., Bahei-El-Din, Y. A., and Castelli, M. G., "Thermal Mechanical Fatigue of Titanium Metal Matrix Composites Literature Review and Analysis," *Composite Materials: Fatigue and Fracture, 4th Volume, ASTM STP 1156,* W. W. Stinchcomb and N. E. Ashbaugh, Eds., American Society for Testing and Materials, Philadelphia, 1993, pp. 591–607.

[13] Neu, R. W. and Nicholas, T., "Thermomechanical Fatigue of SCS-6/TIMETAL 21S Under Out-of-Phase Loading," *Thermomechanical Behavior of Advanced Structural Materials,* W. F. Jones, Ed., AD-Vol. 34/AMD-Vol. 173, American Society of Mechanical Engineers, New York, 1993, pp. 97–111.

[14] Neu, R. W., "A Mechanistic-Based Thermomechanical Fatigue Life Prediction Model for Metal Matrix Composites," *Fatigue and Fracture of Engineering Materials and Structures,* Vol. 16, No. 8, 1993, pp. 811–828.

[15] Chamis, C. C., Tong, M. T., and Murthy, P. L. N., "Computational Simulation of Time-Dependent Behavior in Metal Matrix Composites Using METCAN," *HITEMP Review 1992: Advanced High Temperature Engine Materials Technology Program,* Vol. II, NASA Report CP-10104, NASA Lewis Research Center, Cleveland, OH, Oct. 1992, pp. 43:1–10.

[16] Castelli, M. G., Arnold, S. M., and Saleeb, A. F., "Specialized Deformation Tests for the Characterization of a Viscoplastic Model: Application to a Titanium Alloy," NASA Report TM-106268, NASA Lewis Research Center, Cleveland, OH, 1996.

[17] Textron Specialty Materials Data Sheet, Silicon Carbide Composite Materials, Lowell, MA.

[18] Urguhart, E., Arnold, S. M., Pindera, M. J., and Aikin, B. J., "Simulation of Experimentally Observed Thermal Expansion Behavior of FeCrAlY-Based Composites," *HITEMP Review 1993: Advanced High Temperature Engine Materials Technology Program,* Vol. II, NASA Report CP-19117, NASA Lewis Research Center, Cleveland, OH, Oct. 1993, pp. 54:1–12.

[19] Neu, R. W. and Roman, I., "Acoustic Emission Monitoring of Damage in Metal Matrix Composites Subjected to Thermomechanical Fatigue," *Composites Science and Technology,* Vol. 52, No. 1, 1994, pp. 1–8.

[20] Ning, X. J. and Pirouz, P., "The Microstructure of SCS-6 SiC Fiber," *Journal of Materials Research,* Vol. 6, No. 10, Oct. 1991, pp. 2234–2248.

Frank W. Zok,[1] *Steven J. Connell,*[1] *and Zhen-Zhang Du*[1]

Fatigue Maps for Titanium Matrix Composites

REFERENCE: Zok, F. W., Connell, S. J., and Du, Z.-Z., **"Fatigue Maps for Titanium Matrix Composites,"** *Life Prediction Methodology for Titanium Matrix Composites, ASTM STP 1253,* W. S. Johnson, J. M. Larsen, and B. N. Cox, Eds., American Society for Testing and Materials, 1996, 432–460.

ABSTRACT: The present article attempts to bring together many of the observations and measurements pertaining to fatigue cracking and fracture of unidirectional titanium matrix composites at ambient temperature into a fatigue fracture map. The map identifies the regimes in which various mechanisms operate, allowing fatigue life predictions to be made. The map is constructed using micromechanical models that incorporate crack bridging. The article also demonstrates the utility of the models in describing the experimental data for one specific composite system.

KEYWORDS: titanium matrix composites, notched strength, fatigue (materials), fracture (materials), crack bridging, titanium, life prediction, titanium alloys, modeling

Nomenclature

a	Crack length
$2a_0$	Notch width or hole diameter
a_*	Bridging length scale, $E\Gamma/\bar{\sigma}_0^2$
Δa	Crack extension
A	Orthotropy factor
d	Fiber diameter
D	Crack opening displacement (COD) at notch tip
E	Longitudinal Young's modulus of composite
E_f	Fiber Young's modulus
E_m	Matrix Young's modulus
$\bar{E}$	Effective composite modulus
f	Fiber volume fraction
f_i	Geometric parameters, Fig. 9
h_p	Plastic zone width
K	Crack-tip stress intensity factor
ΔK_t	Stress intensity factor amplitude at crack tip
ΔK_a	Stress intensity factor due to applied stress
ΔK_b	Stress intensity factor due to bridging fibers
n	Paris law exponent, Eq 25
N	Number of cycles
N_f	Number of cycles to failure

[1]Associate professor, graduate student, and post-graduate researcher, respectively, Materials Department, University of California, Santa Barbara, CA 93106-5050.

P Load
$\mathcal{R}$ Stress ratio (minimum/maximum)
S_f Fiber strength
t Panel thickness
u Load point displacement
u_c Critical value of u (at fracture)
Δu Crack opening displacement range
Δu_a Δu due to applied stress
Δu_b Δu due to bridging fibers
$2w$ Specimen width
x Distance from crack (or notch) center
Y Geometric function, Eq 24
α Normalized notch length, $a_0\bar{\sigma}_0^2/\Gamma E$
β Paris law parameter, Eq 25
δ Crack opening displacement, COD
δ_f COD needed for fiber fracture
δ_c COD needed for matrix fracture
ϵ_f Fiber failure strain
γ Normalized ultimate tensile strength, Eq 30*c*
Γ Fracture energy
Γ_1, Γ_2 Components of fracture energies from bridging
Γ_b Fracture energy derived from bridging
Γ_0 Intrinsic toughness
λ Toughness ratio, $\lambda = \Gamma_0/\Gamma_b$
ν Poisson's ratio
Ω Bridging law parameter for fatigue, Eq 21
ϕ Normalized fiber strength, Eq 30*b*
σ Remote tensile stress
$\Delta\sigma$ Stress amplitude (maximum-minimum)
$\Delta\sigma_{th}$ Threshold stress amplitude
σ_b Bridging stress
$\bar{\sigma}$ Notched tensile strength
$\bar{\sigma}_0$ Unnotched tensile strength
$\bar{\sigma}_1$ Critical stress in bridging law
σ_m^y Matrix yield stress
τ Interfacial sliding stress
Σ Stress ratio, $\Delta\sigma/(1 - \mathcal{R})\bar{\sigma}_0$
ξ Elastic parameter, Eq 27*b*

There has been considerable work devoted towards understanding the fatigue behavior of unidirectionally reinforced titanium matrix composites (TMCs) [*1–12*]. The present understanding is as follows. Under tensile loading parallel to the fibers, matrix fatigue cracks initiate from defects such as broken fibers, brittle reaction layers at the fiber-matrix interfaces, and holes or notches machined into the composite. When the peak stress in the loading spectrum is sufficiently small, the cracks grow past the fibers, leaving the fibers intact in the crack wake. This process is accommodated by debonding and sliding along the fiber/matrix interfaces. The bridging fibers exert closing tractions on the crack faces, reducing the crack-tip stress intensity range and the rate of fatigue crack growth. There exists a fatigue threshold below which the stresses borne by the bridging fibers reach a steady-state value that is "less" than the fiber strength. As a result, the fibers never break and the fatigue crack growth rate reaches a steady-state value, independent of crack length. Above the threshold, fibers begin to break when the crack length reaches a critical value, dependent on the fiber strength and the size of the notch or hole. The subsequent fatigue crack growth rate accelerates rapidly, ultimately leading to fracture. At yet higher stresses, fibers fail "ahead" of the matrix crack tip, with the crack growth behavior being dominated by the fatigue properties of the matrix. The upper limit on the stress at which these processes can occur is the ultimate tensile strength of the composite, as influenced by the presence of any holes or notches.

The purpose of the present article is to bring together many of the observations pertaining to fatigue cracking and fracture of TMCs at ambient temperature into a "fatigue fracture map." In its present form, the map provides information pertaining to the "mechanisms" governing fatigue and fracture. Contours corresponding to a prescribed fatigue life can be superimposed onto the map, though such developments are, as yet, incomplete. The map is constructed using micromechanical models that describe the processes occurring near the crack tip. In most cases, the processes can be described using concepts of crack bridging. The present article includes select experiments on a Ti/SiC composite. Some of these results have been presented elsewhere [*13–15*], whereas others are new. The experimental results are compared with the predictions of the models in order to assess the utility of the models. The models are then combined to construct a fatigue map for the composite of interest.

The maps are described in terms of two nondimensional parameters. The first, Σ, is the ratio of the peak stress in the loading spectrum, σ_p, to the ultimate tensile strength of the composite, $\bar{\sigma}_0$. For tests conducted at a fixed stress range, $\Delta\sigma$, and fixed stress ratio, $\mathcal{R}$, this ratio becomes

$$\Sigma = \Delta\sigma/(1 - \mathcal{R})\bar{\sigma}_0 \tag{1a}$$

The second parameter, α, is the ratio of the notch or hole size, a_0, to a characteristic bridging length scale, a_*, given by [*14,16*]

$$a_* = E\Gamma/\bar{\sigma}_0^2 \tag{1b}$$

where E is the longitudinal composite modulus and Γ is the (steady-state) fracture energy. When α is large compared with unity, the problem becomes one of small-scale bridging (SSB). In this case, the mechanics associated with bridging are straightforward [*16,17*]. However, experience with a variety of composites suggests that such conditions are rarely achieved. In most cases, $\alpha \lesssim 1$. The problem is then inherently one of large-scale bridging (LSB) and thus requires a more sophisticated analysis.

In addition to the two nondimensional parameters, a number of material properties are required to construct the map. Some of these can be obtained readily: the fiber modulus, E_f, the matrix modulus, E_m, the fiber volume fraction, f, and the fiber diameter, d. Others are somewhat problematic. For example, the interfacial sliding stress, τ, changes dramatically during cycling, a result of wear of the fiber coatings [*6,18,19*]. Because of this phenomenon, the value of τ measured in pristine material cannot be used in simulating crack growth. Moreover, it raises concerns regarding the assumption in the bridging models [*7,8*] that τ remains constant. Despite these problems, it has been found that a reasonably consistent description of fatigue cracking can be obtained over a range of loading conditions, using a constant, "reduced" value of τ in the simulations [*6*]. It will be shown that this reduced value is consistent with measurements made using fiber pushout tests "following cyclic loading." The fiber strength, S_f, is also somewhat ill-defined. Notably, the wear process occurring in the fiber coating may lead to a degradation in fiber strength [30]. Presently, the available data regarding the fiber strength distribution and its degradation are sparse. Notwithstanding such difficulties, models based on the assumption of a deterministic fiber strength have been found to be useful in describing the transition from a fully bridged crack to one in which the fibers are breaking [6,8]. In the subsequent development, the characteristic fiber strength is assumed to be constant, though in general it too may be reduced relative to that of the pristine fibers. Presently, no models are available to account for a progressive degradation in strength with cycling.

The paper is organized in the following way. The next section considers the problem of the notched tensile strength of TMCs. In the context of fatigue, the notched strength represents the stress corresponding to $N_f = \frac{1}{2}$, with N_f being the number of loading cycles to fracture. It will be shown that fracture is preceded by the development of a stable plastic strip ahead of the notch. The plastic strip can be viewed as a Mode I crack with bridging stresses acting across it, a concept initially proposed by Cottrell and co-workers some 30 years ago [20,21]. The third section illustrates the three main types of fatigue cracking in TMCs: (1) fully bridged matrix cracks; (2) partially bridged cracks, with fiber fracture occurring in the crack wake; and (3) unbridged cracks, with fiber fracture occurring ahead of the crack tip. This section also contains comparisons between crack growth measurements and model predictions. Finally, in the fourth section, the key results are combined into a fatigue fracture map. The map shows the transition between the three types of fatigue behavior and the locus corresponding to the (monotonic) notched strength. All of the experimental results presented here are based on a Ti-6Al-4V alloy reinforced with 32% of unidirectional silicon-carbide, SiC(σ), fibers. The panel thickness is $t = 1.0$ mm. In all the mechanical tests, the load was applied parallel to the fiber direction. A complete list of terms is included in the nomenclature. Additional details of the material can be found elsewhere [13,14].

Notched Strength

Experimental Techniques

Uniaxial tension tests were conducted on specimens containing either notches or circular holes, located at the specimen center. The holes ranged in diameter from 0.6 to 6 mm. The notches were ~400 μm wide and ranged in length from 1.5 to 6 mm. The ratio relevant of notch or hole size, $2a_0$, to specimen width, $2w$, was fixed at 0.2. A summary of the specimen dimensions and the corresponding strengths is presented in Table 1. Prior to testing, one face of each specimen was polished to a 1 μm finish. Beveled stainless steel tabs were bonded to the specimen ends with an epoxy adhesive. In some instances, a 0.8-mm strain gage was attached immediately ahead of the notch or hole. The tests were conducted in a servohydraulic testing machine, using hydraulic wedge grips to load the specimen. For comparison, uniaxial tests were also conducted on straight (unnotched) tensile specimens, 6 mm wide, with axial strains measured using a contacting extensometer.

During the tests, the region immediately ahead of the notch or hole was monitored using a traveling stereomicroscope and recorded using a video camera. These observations were used to establish the extent of plasticity. In some cases, the tests were interrupted following the development of an extensive plastic zone (~1 to 3 mm), but prior to fracture. The extent of fiber fracture within this zone was determined by dissolving the matrix in the vicinity of the notch with a 49% hydrogen fluoride (HF) solution and examining the underlying fibers in a scanning electron microscope (SEM).

The composite toughness, Γ, was evaluated using an edge-notched four-point work-of-

TABLE 1—*Summary of tension test results.*

	Hole or Notch Size, $2a_0$ (mm)	Specimen Width, $2w$ (mm)	Ultimate Tensile Strength, $\bar{\sigma}_0$ (MPa)
Unnotched	...	6.0	1590 ± 100[a]
Hole	0.6	3.0	1180
Hole	1.5	7.5	980
Hole	3.0	15.0	910
Hole	6.0	30.0	860
Notch	1.5	7.5	910
Notch	3.0	15.0	850
Notch	6.0	30.0	810

[a]Average and standard deviation from eight tests.

rupture specimen [*22,23*] (shown later in the inset of Fig. 5). (For reasons described below, the notched tension tests can not be used to obtain Γ.) The test was conducted at a displacement rate of 0.5 mm/min. The toughness (or fracture energy) was evaluated using the relationship

$$\Gamma = \frac{\int_0^{u_c} P\,du}{t(w - a_0)} \tag{2}$$

where P is the load, u is the load point displacement, u_c is the displacement at fracture, and t is the specimen thickness.

As detailed in a subsequent section, the effects of holes and notches on the tensile strength can be rationalized in terms of a crack-bridging model. One of the important parameters involved in the bridging traction law is the crack opening displacement (COD) at fracture, δ_c. This parameter was determined in the notched specimens from measurements of the notch width both before and after fracture. The post-fracture notch width was taken to be the sum of the normal distances from the notch surfaces to the tips of the fracture surface (as shown later in the inset of Fig. 6*b*). Similar measurements were made on specimens with small holes (0.6 and 1.5 mm diameter).

Another key parameter in the traction law is the displacement δ_f at the onset of fiber failure. This displacement can be estimated using the relationship

$$\delta_f \approx h_p \epsilon_f \tag{3}$$

where h_p is the width of the plastic zone measured parallel to the loading direction and ϵ_f is the failure strain of the fibers ($\epsilon_f \approx 1\%$). The plastic zone size was measured off of micrographs of both sides of the fractured specimen, taken using Nomarski interference microscopy. At each notch tip, ~20 to 30 such measurements were made, starting immediately ahead of the notch tip and proceeding at intervals of ~0.2 mm.

Yet another parameter in the traction law is the strength of the composite following fiber fracture. This strength was measured using a two-step procedure. First, a specimen with a 6-mm notch was loaded in tension until plastic strips ~3 mm long had developed on both sides of the notch. SEM examinations of similar specimens indicated that all the fibers had indeed fractured within this strip. Narrow, longitudinal bars (~2 mm wide), passing through the plastic strips, were then cut from the tested specimen. These bars were tested in tension, with a 10-mm clip gage placed across the plastic strip to measure the local displacements.

Experimental Results

Figures 1 and 2 show a typical stress-local strain response of a notched panel and a corresponding series of optical micrographs taken during the test. Initially, the response was linear, with nonlinearity occurring at a stress of ~450 MPa. At a slightly higher stress (590 MPa), a plastic zone was observed at the notch tip (Fig. 2*a*). Upon further loading, the length of the plastic strip increased stably, reaching ~3 mm at a stress of ~750 MPa. The specimen was subsequently unloaded for further examination. The local strain at the notch tip was ~5%: considerably higher than the fracture strain measured in the unnotched tensile specimens (~1%). The ultimate tensile strength of a similar specimen was ~850 MPa.

Examinations of specimens interrupted prior to fracture showed that "all" of the fibers contained within the plastic strips had broken. Figure 3 shows one such example, from a specimen with a notch length, $2a_0 = 3$ mm. In this case, both the plastic strip and the "plane" of fractured fibers follow the same, somewhat nonplanar, path. The specimens containing circular holes exhibited similar behavior.

The trends in tensile strength, $\bar{\sigma}$, with notch or hole size, $2a_0$, are summarized in Fig. 4. Evidently, the strength drops quickly with a_0 in the regime $0 < 2a_0 < 1.5$ mm. For larger values, $2a_0 \gtrsim 1.5$ mm, the strength continues to decrease, though at a much slower rate. The specimens containing holes exhibited similar trends, though the strengths were slightly higher than those corresponding to notches of the same size, by ~6 to 7%.

The results of the work of rupture test are shown in Fig. 5. Combining these measurements with Eq 2 yields a toughness, $\Gamma = 72$ kJ m^{-2}. In this configuration, a similar process of plastic yielding was observed ahead of the notch tip. Once a crack had formed, it propagated stably across the specimen along a Mode I path, resulting in a relatively smooth, continuous load-displacement curve.

The measurements of plastic zone width, h_p, and local displacement at fracture, δ_c, are summarized in Fig. 6. The values of both parameters are independent of the specimen size

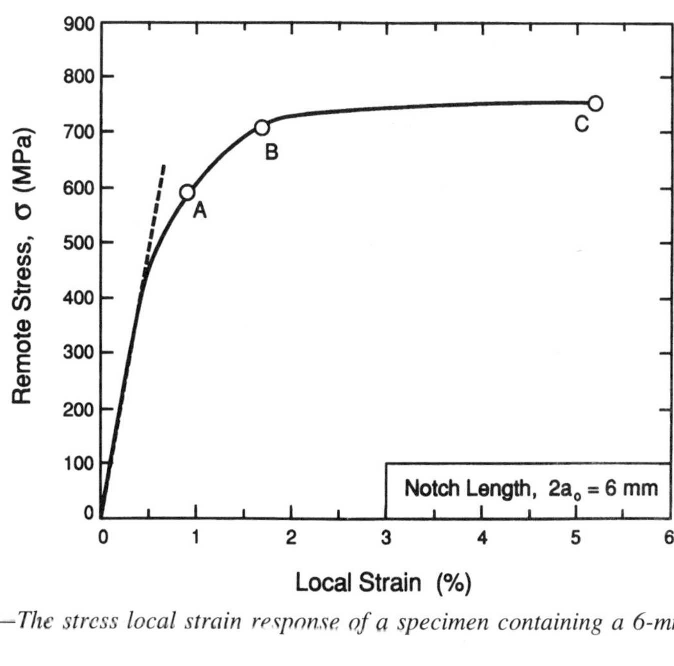

FIG. 1—*The stress local strain response of a specimen containing a 6-mm notch.*

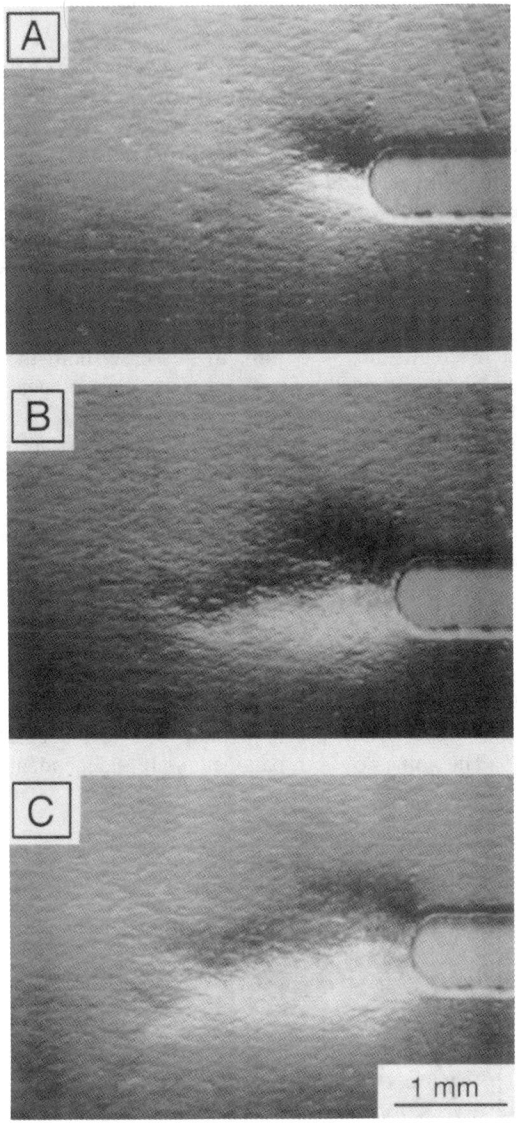

FIG. 2—*Optical micrographs showing the progression of plastic strip development. The labels A, B, and C correspond to the stress levels shown in Fig. 2.*

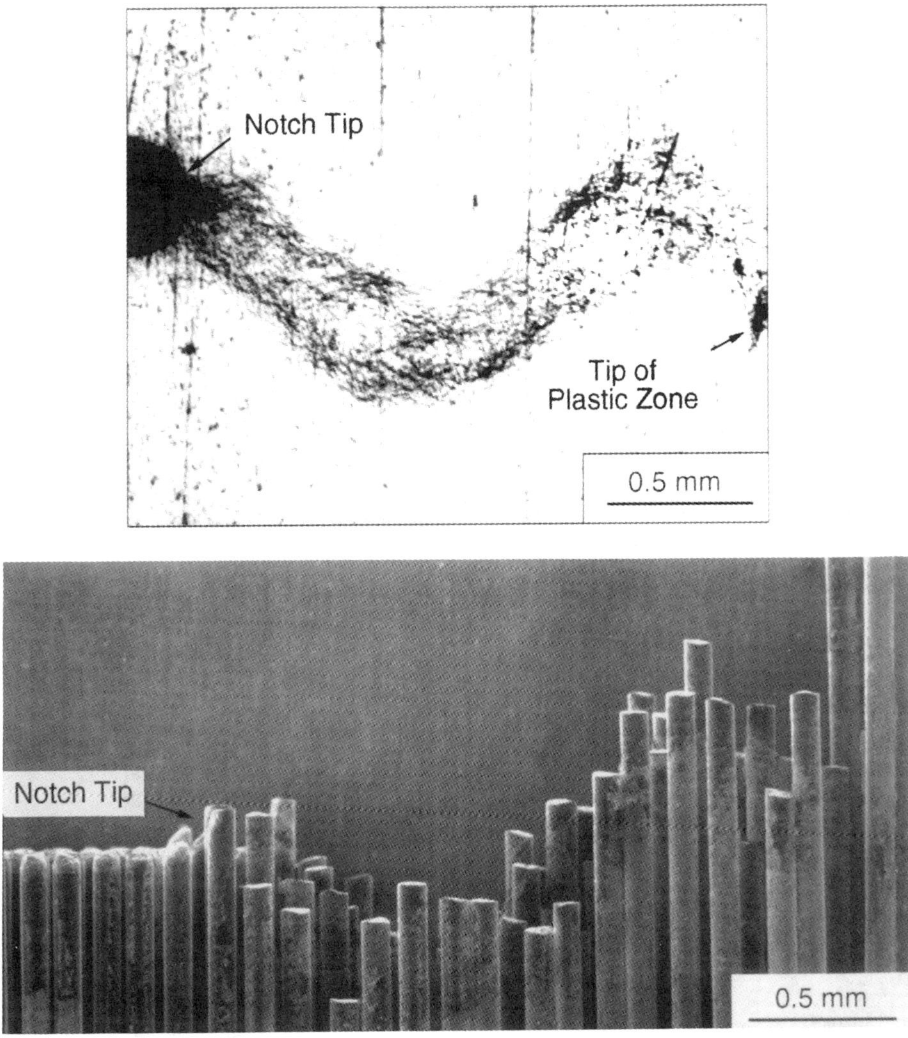

FIG. 3—*Comparison of* (a) *the plastic strip ahead of a notch tip, with* (b) *the pattern of fiber fracture following matrix dissolution. The micrographs show identical regions of the specimen.*

and the specimen configuration (notch versus hole), with average values, $h_p \approx 0.9$ mm (Fig. 6*a*) and $\delta_c \approx 80$ μm (Fig. 6*b*).

Figure 7 shows the results of a tension test conducted on a specimen containing a plastic strip within which all fibers had been broken. The results are presented in the form of stress versus displacement (not strain), since virtually all the inelastic strain was localized within the previously yielded strip. In this test, the response was essentially elastic-perfectly plastic, with a "yield stress" of ~800 to 850 MPa (approximately one half of the unnotched tensile strength). The contribution expected from the matrix alone is

$$\bar{\sigma}_1 \approx (1 - f)\sigma_m^y \tag{4}$$

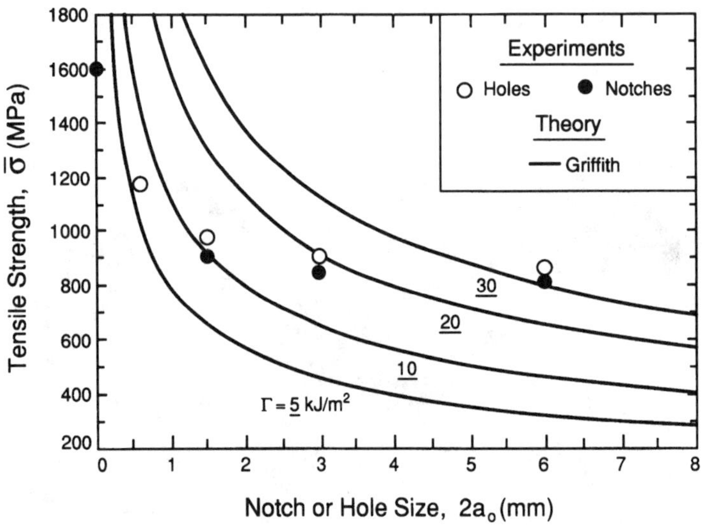

FIG. 4—*Trends in tensile strength with hole or notch size.*

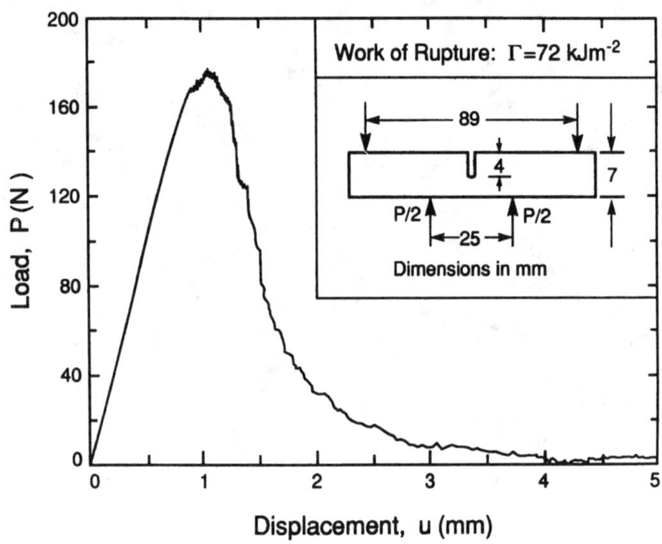

FIG. 5—*Load-displacement curve measured on work-of-rupture specimen.*

Taking $\sigma_m^y \sim 1000$ MPa [*13*] yields the estimate, $\bar{\sigma}_1 \approx 680$ MPa, somewhat lower than the measured value. The difference is attributable to the contribution from the broken fibers, analogous to the strengthening derived from discontinuous fiber reinforcements.

Analysis and Modeling

A preliminary attempt at rationalizing the trends in notched strength was made using the Griffith relationship. Figure 4 shows the predicted trends in strength with notch size for values of Γ ranging from 5 to 30 kJ m^{-2}. It is apparent that this approach does not provide a good representation of the measurements. The disparity between experiment and theory

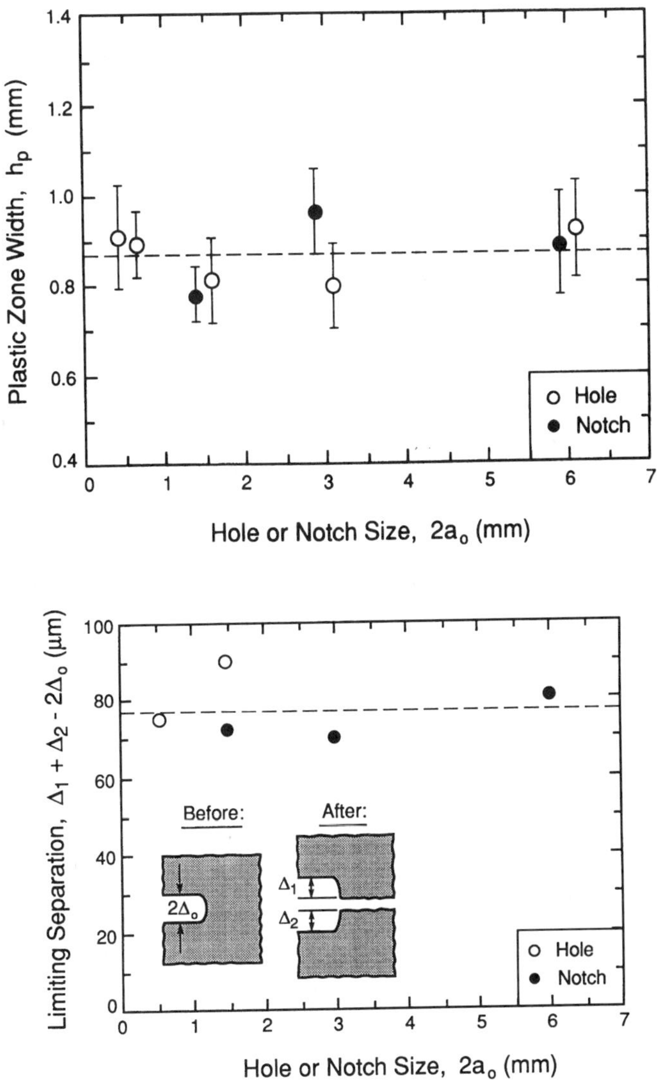

FIG. 6—*Trends in* (a) *plastic zone width,* h_p, *and* (b) *local displacement at fracture,* δ_c, *with hole or notch size. The error bars in* (a) *represent the standard deviation from ~30 measurements made on each specimen.*

provides the motivation for the subsequent modeling effort. Moreover, it demonstrates that the trends in tensile strength with notch length cannot be used to infer the composite toughness.

The present measurements and observations suggest that the notch strength characteristics of the Ti/SiC composite can be rationalized in terms of a bridging traction law having the features shown by the solid line in Fig. 8. In this law, the stress initially increases with displacement, analogous to the stress-strain response measured in a uniaxial tension test. At a critical stress, taken to be the unnotched tensile strength of the composite, $\bar{\sigma}_0$, the fiber bundle fails, causing the stress to drop precipitously with increasing COD. The correspond-

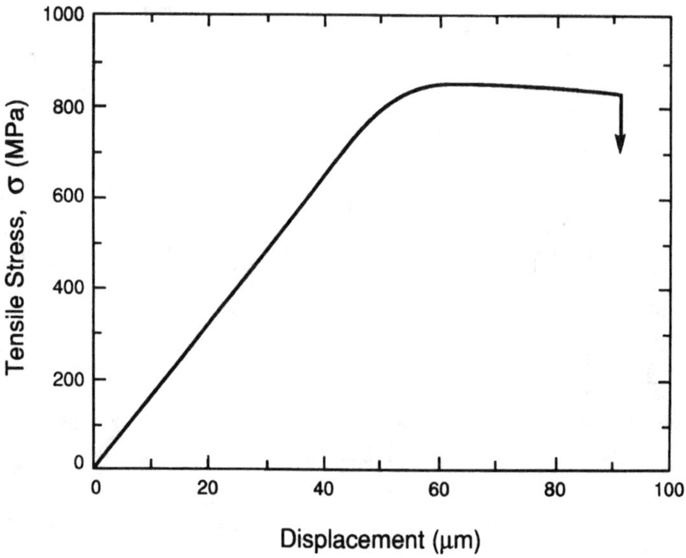

FIG. 7—*Tension test of specimen containing plastic strip with broken fibers.*

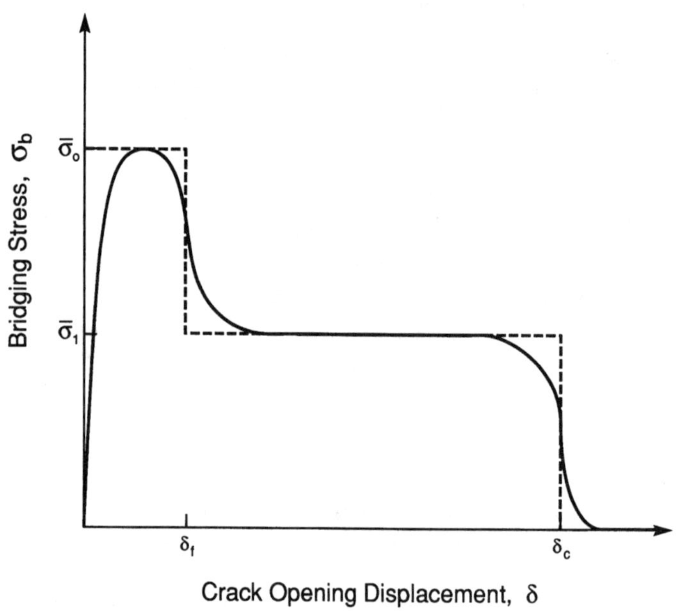

FIG. 8—*Schematic diagram of the proposed traction law governing fracture in the Ti/SiC composite.*

ing critical displacement, δ_f, depends on the width of the plastic strip (being the effective gage length) and the fiber failure strain, in accordance with Eq 3. Upon further crack opening, the bridging stress reaches a saturation level, dictated by the yield stress of the metal. At a yet larger COD, the local strain reaches the failure strain of the metal, leading to the formation of a matrix crack and a loss in bridging.

For modeling purposes, it is convenient to represent this behavior by a two-level bridging traction law, shown schematically by the dashed lines in Fig. 8. Each of the two parts of the traction law is characterized by a strength ($\bar{\sigma}_0$ or $\bar{\sigma}_1$) and a critical displacement (δ_f or δ_c). The total (or steady-state) fracture energy, obtained using the *J*-integral, is [24]

$$\Gamma = \int_0^{\delta_c} \sigma_b \, d\delta = \Gamma_1 + \Gamma_2 \qquad (5a)$$

where Γ_1 and Γ_2 are the areas contained within the top and bottom parts of the traction law, given by

$$\Gamma_1 = (\bar{\sigma}_0 - \bar{\sigma}_1)\delta_f \qquad (5b)$$

and

$$\Gamma_2 = \bar{\sigma}_1 \delta_c \qquad (5c)$$

A preliminary assessment of the proposed traction law can be made by comparing the total fracture energy, computed using Eq 5 along with the measured values of the traction law parameters, with the value obtained through the work-of-rupture test. The experimental measurements indicate that $\bar{\sigma}_0 = 1590$ MPa, $\bar{\sigma}_1 = 850$ MPa (Fig. 7), and $\delta_c = 80$ μm. Moreover, combining the plastic zone width, $h_p = 0.9$ mm, and the unnotched tensile fracture strain, $\epsilon_f = 1\%$, with Eq 3 yields $\delta_f \approx 9$ μm. Equation 5 thus predicts a fracture energy, $\Gamma = 71$ kJ m^{-2}, essentially identical to the value obtained experimentally ($\Gamma = 72$ kJ m^{-2}). This correlation provides some initial confidence in the proposed traction law.

For the purpose of simulating the notched strength, the two-level rectilinear law was incorporated into a fracture mechanics model. In the model, the composite panel contains a sharp, through-thickness notch of length, $2a_0$, with a specimen width, $2w = 10\ a_0$. The panel is loaded remotely with a uniform tensile stress, σ. The material is assumed to be linearly elastic, except within the plastic strips formed ahead of the notch tips. The total crack length (notch and plastic strip) is designated $2a$, and the COD at the notch tip is designated D.

The problem can be simplified in the following way. First, it is noted that the energy dissipated in the top portion of the traction law is small in relation to the total (that is, $\Gamma_1/\Gamma_2 \sim 0.1$). Moreover, the critical displacement, δ_f, is small in relation to δ_c. Consequently, the energy dissipated in the top portion can be lumped into an *intrinsic* fracture energy, Γ_0, whereupon

$$\Gamma_0 = \Gamma_1 = (\bar{\sigma}_0 - \bar{\sigma}_1)\delta_f \qquad (6)$$

The fracture resistance can thus be characterized by Γ_0 in combination with a single-level rectilinear bridging traction law, with a characteristic strength, $\bar{\sigma}_1$, and a critical displacement, δ_c. The toughness derived from bridging is thus

$$\Gamma_b \equiv \Gamma_2 = \bar{\sigma}_1 \delta_c \qquad (7)$$

The ratio of the two contributions is characterized by the parameter

$$\lambda \equiv \Gamma_0/\Gamma_b \qquad (8)$$

The notched strength of such a system can be evaluated from the stress-crack length relationships, obtained from handbook solutions of stress intensity factors and CODs [25]. In nondimensional form, the relevant solutions can be expressed as

$$\frac{DE}{a_0\sigma_1} = f_1 \frac{\sigma}{\bar{\sigma}_1} - f_2 \qquad (9a)$$

and

$$\frac{K}{\bar{\sigma}_1\sqrt{\pi a_0}} = f_3\frac{\sigma}{\bar{\sigma}_1} - f_4 \tag{9b}$$

where K is the crack-tip stress intensity factor and f_i are functions of a_0/w and a_0/a. The trends in f_i with a_0/a for $a_0/w = 0.2$ are shown in Fig. 9 [*26*]. Setting K equal to the critical value

$$K = K_c = \sqrt{E\Gamma_0} \tag{10}$$

and using the definition of λ (Eq 8) allows Eq 9 to be rewritten as

$$\left(\frac{\bar{\sigma}_0}{\bar{\sigma}_1}\right)^2 \frac{1}{\alpha(1 + \lambda)} = f_1\frac{\sigma}{\bar{\sigma}_1} - f_2 \tag{11}$$

and

$$\frac{\bar{\sigma}_0}{\bar{\sigma}_1}\sqrt{\frac{\lambda}{\pi\alpha(1 + \lambda)}} = f_3\frac{\sigma}{\bar{\sigma}_1} - f_4 \tag{12}$$

where α is the normalized notch size

$$\alpha \equiv \frac{a_0\bar{\sigma}_0^2}{\Gamma E} \tag{13}$$

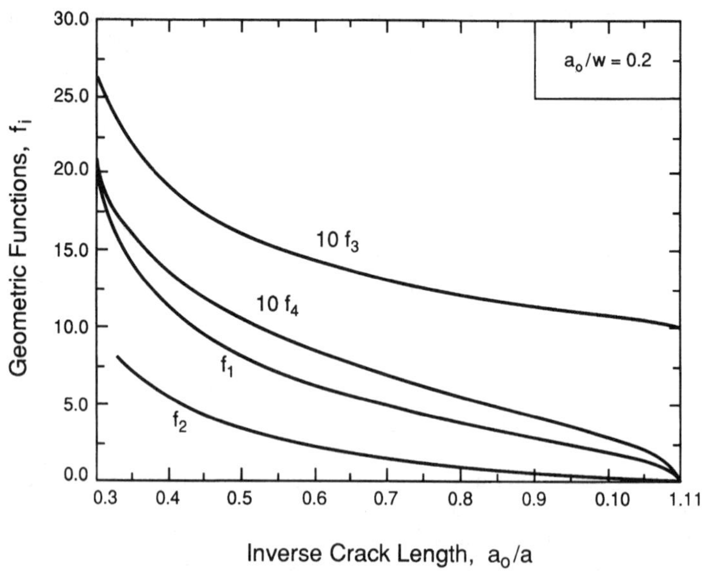

FIG. 9—*Geometric functions,* f_i, *for* $a_0/w = 0.2$.

The stress-crack opening relationship is obtained by incrementally stepping through values of a/a_0, starting from zero. At each point, the stress, σ, is evaluated from Eq 11 and the result combined with Eq 12 to obtain D. The relationship between strength, $\bar{\sigma}$, and notch length is then obtained from the maximum points in the σ versus D curves. In the limit wherein $\lambda = 0$ and $a_0/w = 0$, Eqs 11 and 12 reduce to a result similar to the one developed by Cottrell and coworkers [*20,21*], namely

$$\frac{\bar{\sigma}}{\bar{\sigma}_0} = \frac{2\bar{\sigma}_1}{\pi\bar{\sigma}_0}\cos^{-1}\left(\exp\frac{-\pi\bar{\sigma}_0^2}{8\alpha\bar{\sigma}_1^2}\right) \tag{14}$$

Figure 10 shows the predicted trends in strength with notch length for values of λ ranging from 0 to 0.2. The results are plotted in terms of the two nondimensional parameters: Σ and α. Also shown are the values obtained experimentally. The comparisons show that the predictions are in close agreement with the measurements for $\lambda \approx 0.1$. The inferred value of λ is comparable to the one calculated using the values of $\bar{\sigma}_0$, $\bar{\sigma}_1$, δ_f, and δ_c quoted earlier: $\lambda \sim 0.09$. For smaller values of α, the model predicts a strength-notch length relationship of the form

$$\frac{\bar{\sigma}}{\bar{\sigma}_0} = \sqrt{\frac{E\Gamma_0}{\pi a_0\bar{\sigma}_0^2}} \tag{15a}$$

or, equivalently

$$\Sigma = 1/\sqrt{\pi\alpha} \tag{15b}$$

in accordance with the Griffith equation. In this regime, the model overestimates the notched strength. Consequently, an alternate approach is required to capture the relevant trends.

Because of the similarities in the strength characteristics of the specimen containing holes and notches, no calculations were conducted for holes. The similarities are consistent with

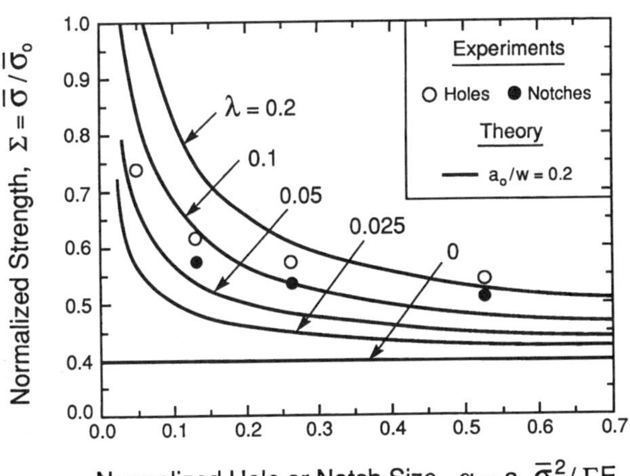

FIG. 10—*Comparisons between experimental measurements and predictions of the model based on the simple rectilinear law, with an intrinsic toughness,* Γ_0.

previous calculations [16] that show that the strength is insensitive to the shape of the discontinuity, provided the bridging length scale, a_*, is sufficiently large compared with a_0.

In summary, the correlation between experiment and theory demonstrates that the notched strength characteristics of fiber-reinforced titanium matrix composites can be described using a relatively simple crack-bridging model. The parameters controlling the bridging traction law have been measured and found to provide a consistent description of both the notch sensitivity and the steady-state composite toughness.

Fatigue Crack Growth

Experimental Techniques

Fatigue tests were conducted in the 0° orientation using center-notched tensile specimens. In all cases, notch length was $2a_0 = 3$ mm and the specimen width was $2w = 15$ mm. One face of each specimen was diamond polished to a 1-μm finish. Tests were conducted on a servohydraulic mechanical test system at fixed stress range, $\Delta\sigma$. In all cases, the stress ratio (minimum/maximum), $\mathcal{R}$, was maintained at 0.1. Crack extension was monitored using one of two techniques: (1) indirect potential drop (with thin foil crack gages mounted at the notch tips), and (2) a traveling stereomicroscope. In some instances, the crack-tip regions were recorded using a video camera. The values of $\Delta\sigma$ ranged from 300 to 800 MPa.

Observations of fiber bridging and fiber fracture were also made. For this purpose, tested specimens were sectioned along a plane ~3 mm above the matrix crack plane, and the matrix subsequently dissolved down to a depth of ~6 mm. During matrix dissolution, the fractured fibers were removed, whereas the intact fibers continued to span the entire length of the specimen. A comparison of the spatial distribution of fractured fibers with the matrix crack prior to dissolution provide a direct measure of the length over which intact fibers had bridged the matrix crack.

Experimental Results

A summary of the crack growth measurement for all tests is presented in Fig. 11. The data are plotted as crack extension, Δa, versus number of loading cycles, N. Logarithmic coordinates are used here because of the wide range of values of N. Additionally, the measurements for each test are plotted separately in linear coordinates in Figs. 12(*a* through *e*) and compared with model predictions, detailed later. For convenience, the results are categorized according to the level of stress: low, intermediate, and high. This categorization is motivated by the observed changes in mechanisms.

At low stresses ($\Delta\sigma$ = 300 and 400 MPa, Fig. 12*a*), the crack growth rates decrease monotonically with increasing crack extension. This deceleration is a manifestation of crack bridging by intact fibers. The presence of such fibers was confirmed through SEM examination, shown in Fig. 13.

At intermediate stress levels ($\Delta\sigma$ = 675 MPa, Fig. 12*b*), the crack growth rate initially decreases with crack extension. At a critical length, a transition occurs in which the growth rate accelerates rapidly. This transition has previously been correlated with the onset of fiber fracture. Observations of the crack-tip region (Fig. 14) indicate that, shortly after the onset of acceleration, a zone of intense plasticity is developed ahead of the crack tip. Based on observations of specimens used for the earlier notched strength study, it can be inferred that all the fibers in the crack wake and some of the fibers ahead of the crack tip have broken. Beyond this point, the growth rate reaches values as high as ~5 μm/cycle: a factor of 40 larger than the minimum rate established within the bridging regime. Moreover, the length of the plastic zone increases rapidly, reaching a value of ~1 mm at fracture (Fig. 14*c*).

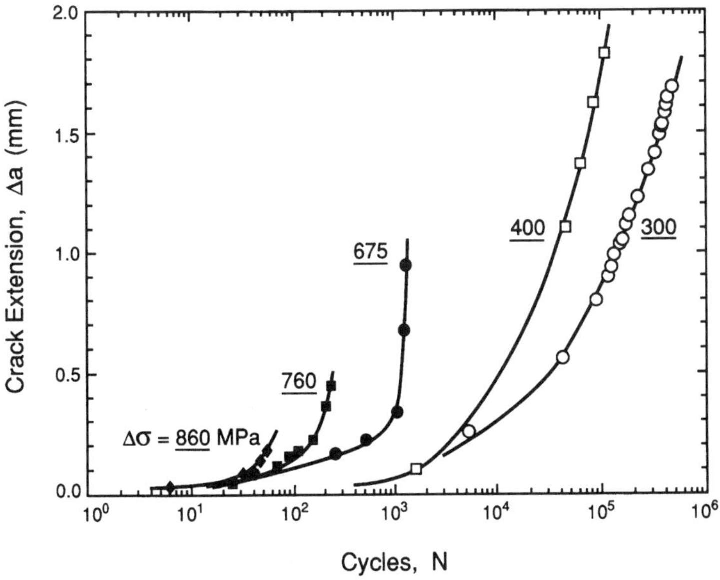

FIG. 11—*Summary of fatigue crack growth measurements.*

At high stresses ($\Delta\sigma$ = 800 MPa, Fig. 12*d*), there is no evidence of crack bridging. Intense plastic zones are developed ahead of the notches on the *first* loading cycle (Fig. 15), leading to fiber fracture *prior to the initiation of a matrix crack.* The strain measurements at the notch root clearly indicate the development of plasticity at stresses $\gtrsim$800 MPa (Fig. 16). The peak strain obtained in the first loading cycle is ~1.4%, exceeding the fiber failure strain by ~0.4%. Once a matrix crack forms, it grows at a rapid rate: ~5 to 10 μm/cycle after ~40 loading cycles.

The tests conducted at $\Delta\sigma$ = 760 MPa (Fig. 12*c*) exhibited a behavior intermediate to those of the *intermediate* and *high* stress regimes described earlier. On the first loading cycle, there was some evidence of plasticity from the strain gage measurements, though the peak strain (~0.75%) did not exceed the fiber failure strain (Fig. 16). At this point, there was no observable plasticity. However, upon further cycling, a fatigue crack was formed, followed by the development of a large plastic zone and a slight acceleration in crack growth rate.

Modeling and Analysis

The mechanics of crack bridging by frictionally constrained fibers in notch matrix composites under cyclic loading has been well established [*7–10*]. A fundamental assumption in the analysis is that the driving force for crack extension is the crack-tip stress intensity factor amplitude, ΔK_t, given by

$$\Delta K_t = \Delta K_a + \Delta K_b \tag{16a}$$

where ΔK_a is the component due to the applied stress amplitude ($\Delta\sigma$), and ΔK_b is the component due to the bridging tractions ($\Delta\sigma_b$) exerted by intact fibers in the crack wake. For an "infinite" center-cracked tensile panel, these components are given by [*25*]

$$\Delta K_a = \Delta\sigma\sqrt{\pi a} \tag{16b}$$

and

$$\Delta K_b = -2\sqrt{\frac{a}{\pi}}\int_{a_0}^{a}\frac{\Delta\sigma_b(x)}{\sqrt{a^2 - x^1}}\,dx \tag{16c}$$

where $2a_0$ is the initial notch length, $2a$ is the current crack length, and x is the distance from the crack center.

To evaluate the distribution of bridging tractions, $\Delta\sigma_b(x)$, it is first necessary to specify the contributions to the change in COD, Δu, due to the applied stress, Δu_a, and that due to the bridging fibers, Δu_b [25]

$$\Delta u_a = \frac{4}{\bar{E}}\Delta\sigma_a\sqrt{a^2 - x^2} \tag{17a}$$

and

$$\Delta u_b = \frac{-4}{\bar{E}}\int_{a_0}^{a}\Delta\sigma_b(t)\,H(t,x,a)\,dt \tag{17b}$$

where $\bar{E}$ is an effective composite modulus (taking account of material orthotropy) and the function, H, is [25]

$$H(t,x,a) = \frac{1}{\pi}\log\left|\frac{\sqrt{a^2 - x^2} + \sqrt{a^2 - t^2}}{\sqrt{a^2 - x^2} - \sqrt{a^2 - t^2}}\right| \tag{18}$$

The sum of these components

$$\Delta u = \Delta u_a + \Delta u_b \tag{19}$$

is required to be consistent with the cyclic traction law (taking into account reverse slip during unloading) [7]

$$\Delta u = \frac{1}{2}\Omega\,\Delta\sigma_b^2 \tag{20}$$

Here, Ω is a material parameter defined by

$$\Omega = \frac{d(1 - f)^2E_m^2}{4E^2E_f\tau f^2} \tag{21}$$

where d is the fiber diameter; f is the fiber volume fraction; E_m and E_f are the matrix and fiber Young's moduli, respectively; and E is the rule-of-mixtures composite modulus ($= fE_f + (1 - f)E_m$). Combining Eqs 17 through 20 gives an integral equation of the form

$$\frac{1}{2}\Omega\Delta\sigma_b^2 = \frac{4}{\bar{E}}\Delta\sigma_a\sqrt{a^2 - x^2} - \frac{4}{\bar{E}}\int_{a_0}^{a}\Delta\sigma_b(t)H(t,x,a)\,dt \tag{22}$$

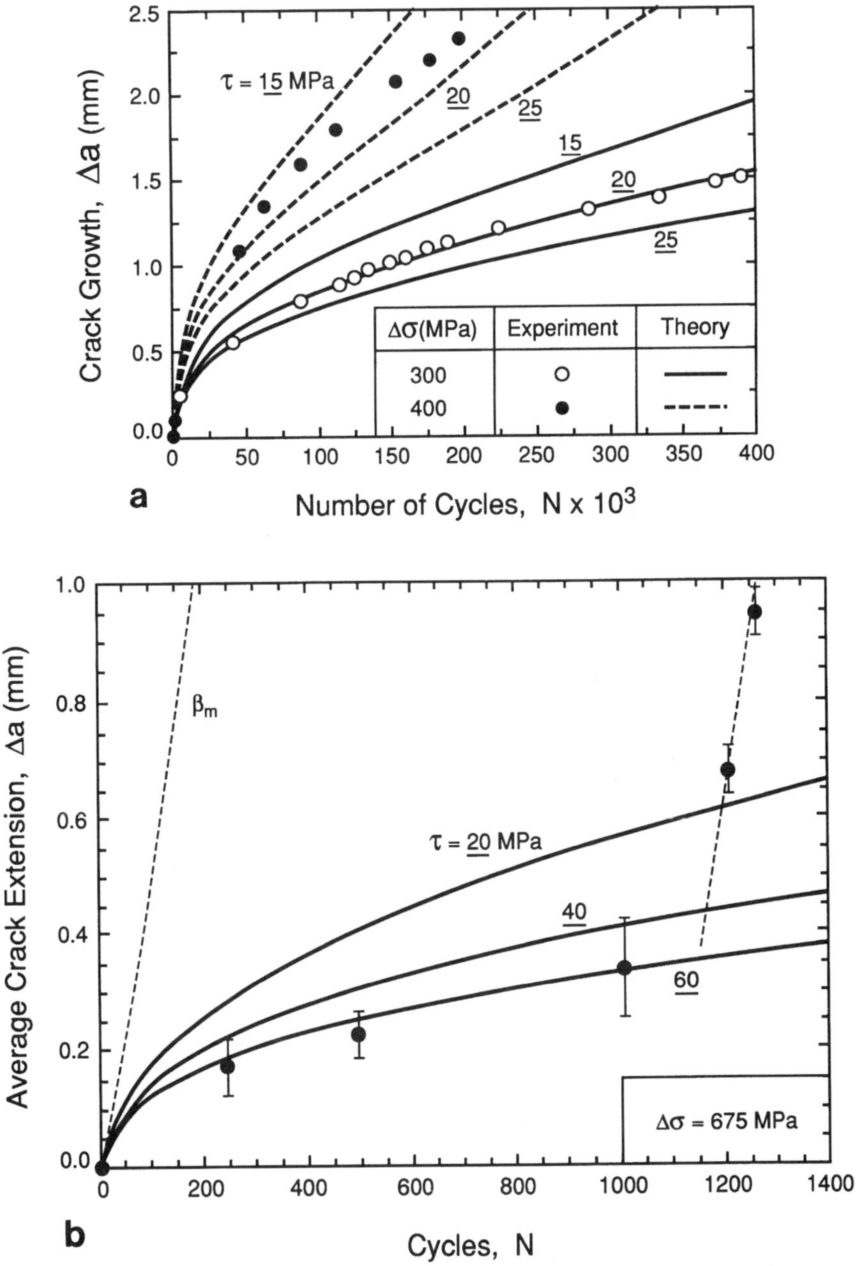

FIG. 12—*Comparisons of experimental crack growth curves with model predictions for* $\Delta\sigma$ *values of* (a) *300 and 400 MPa,* (b) *675 MPa,* (c) *760 MPa, and* (d) *800 MPa. The errors bars represent the range of values measured at the two notch tips. The dashed line on the right side of* (b) *represents matrix behavior, shifted to show the similarity with the fatigue crack growth rate in the composite following fiber fracture.*

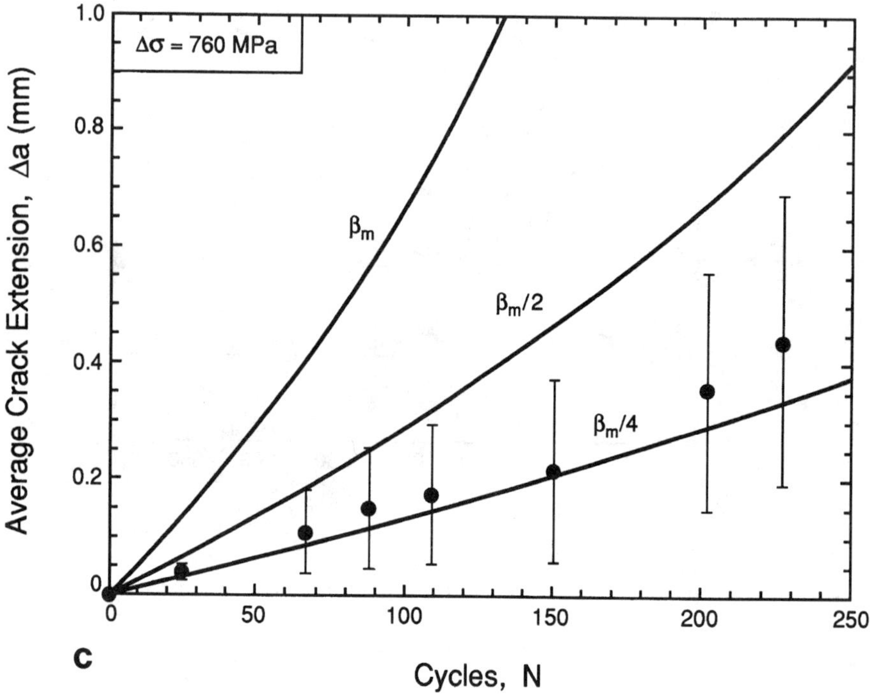

c

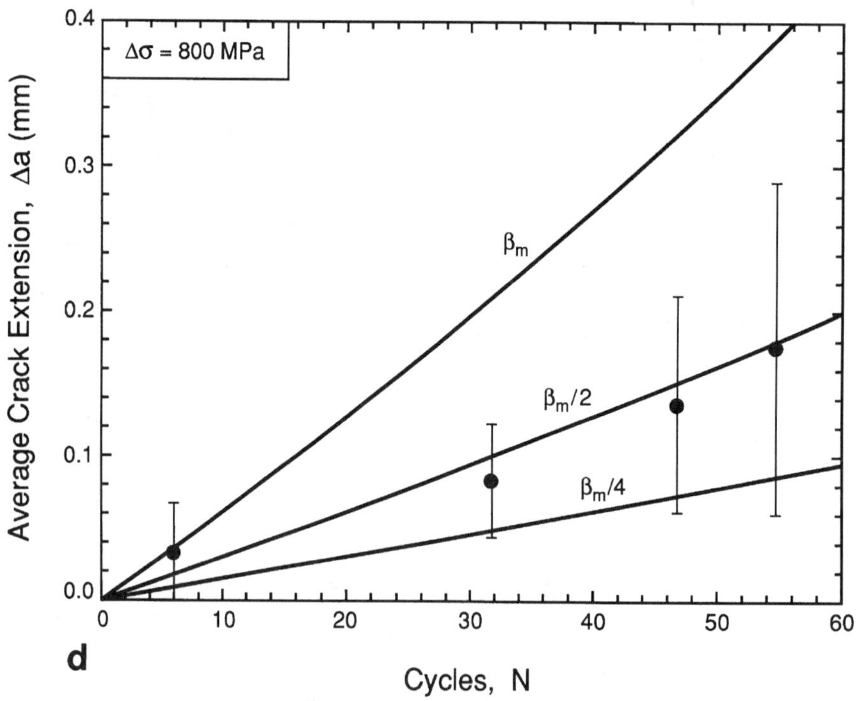

d

FIG. 12—*(Continued).*

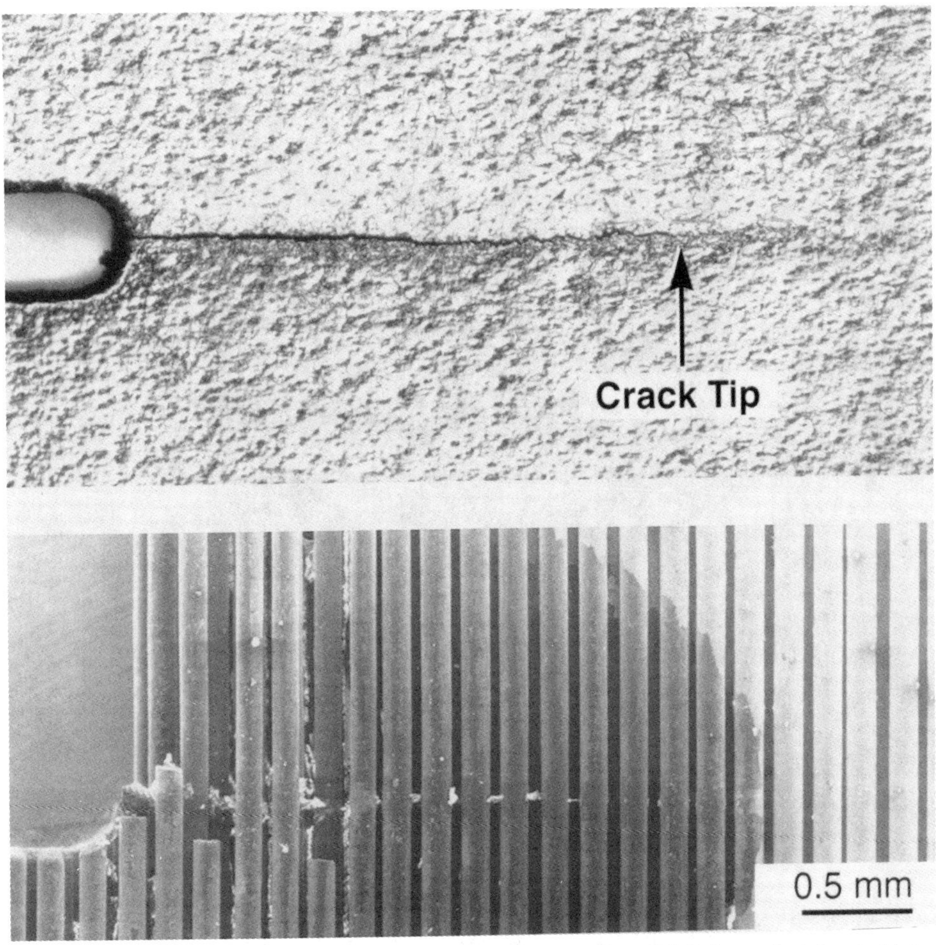

FIG. 13—*Comparison between* (a) *matrix fatigue crack and* (b) *the underlying bridging fibers, following matrix dissolution. The fatigue test was conducted at* $\Delta\sigma = 400$ *MPa.*

This equation is solved numerically for $\Delta\sigma_b$ using an iterative scheme, and the result combined with Eqs 16*a* through *c* to evaluate ΔK_t.

The effects of finite specimen width, $2w$, have also been studied through calculations based on finite element methods [8]. For specimens with a normalized notch size of $a_0/w = 0.2$ (the value used in the present experiments) and crack lengths in the range of $a/w \lesssim 0.5$, the effects of finite width on the crack-tip stress intensity amplitude can be approximated by the relationship

$$\Delta K_t(a/w,\Delta\sigma) \approx Y(a/w)\,\Delta K_t(a/\infty,\Delta\sigma) \tag{23}$$

where $\Delta K_t(a/w,\Delta\sigma)$ and $\Delta K_t(a/\infty,\Delta\sigma)$ represent values for the finite and infinite specimens, respectively, and

$$Y(a/w) = \sqrt{\sec \pi a/2w} \tag{24}$$

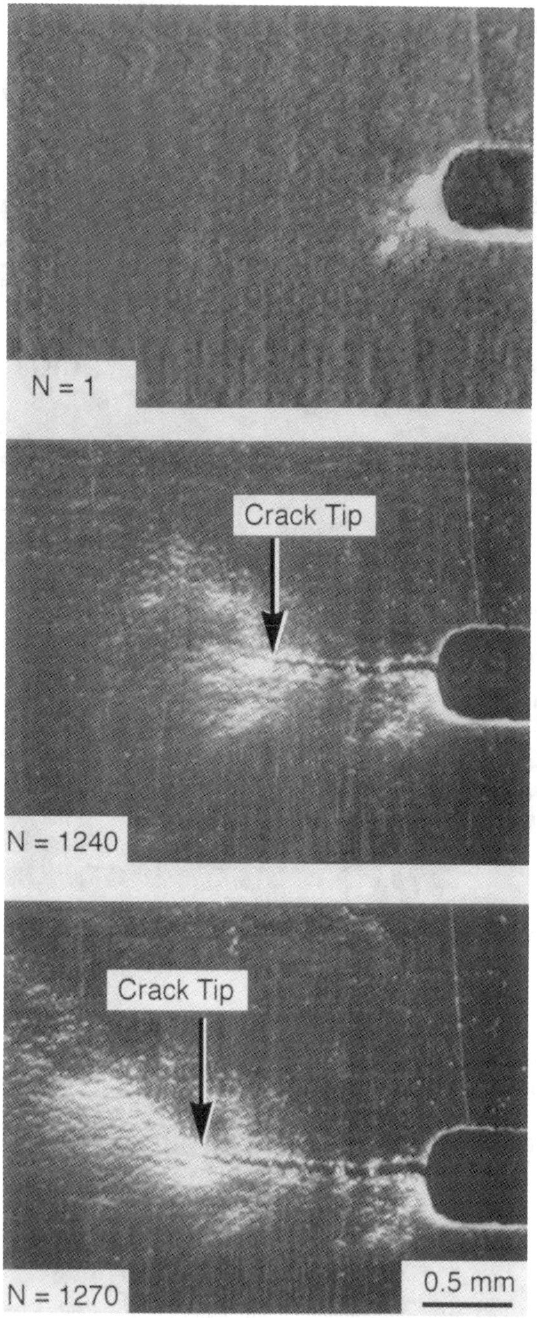

FIG. 14—*Micrographs showing the development of the fatigue crack and the plastic zone for the test conducted at* $\Delta\sigma$ = *675 MPa.*

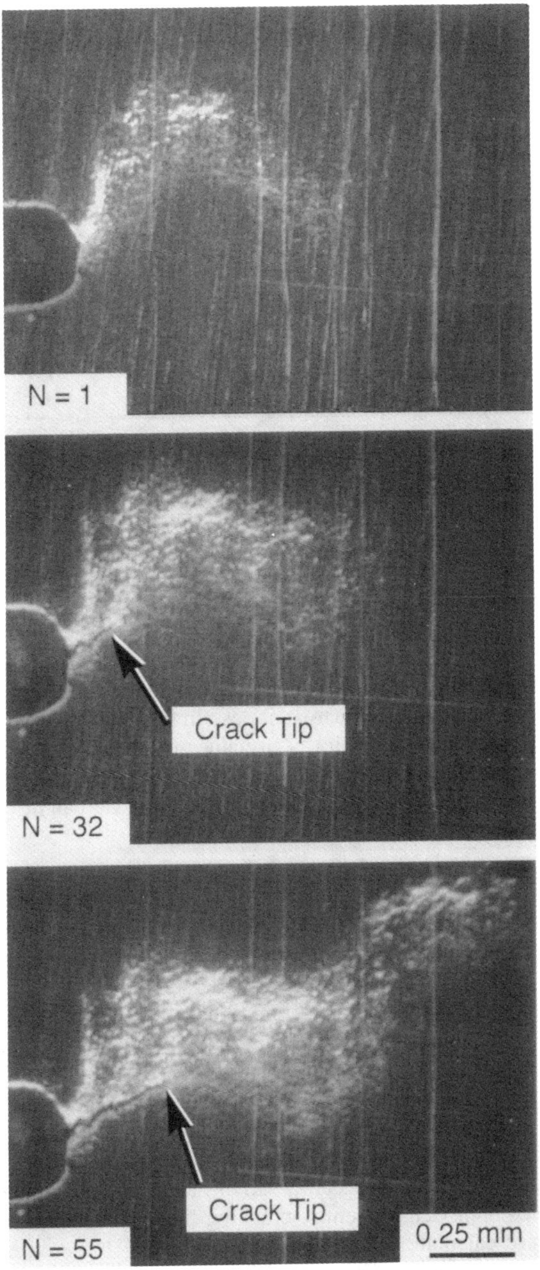

FIG. 15—*Micrographs showing the development of the plastic zone followed by a fatigue crack for the test conducted at* $\Delta\sigma = 800$ *MPa.*

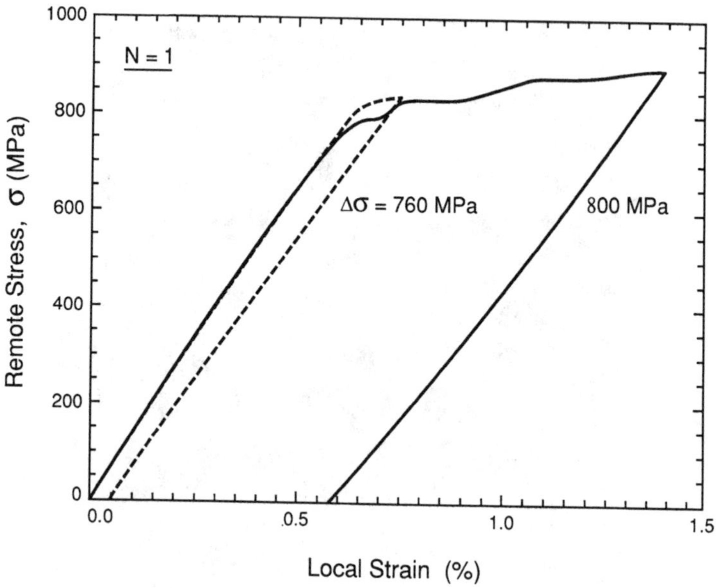

FIG. 16—*The development of local strain for fatigue tests conducted at* $\Delta\sigma = 760$ *and* 800 *MPa.*

(This is the usual finite width correction used in calculating the *applied* stress intensity [*12*]). The error introduced by this approximation is less than ~3%. As seen later, this range of crack lengths is consistent with the majority of values measured experimentally, making the approximate width correction (Eqs 23 and 24) suitable for subsequent calculations.

Once ΔK_t has been obtained, the rate of fatigue crack growth can be described through an empirical relationship of the form

$$da/dN = \beta(\Delta K_t)^n \tag{25}$$

where N is the number of loading cycles. The parameters, β and n, represent the behavior of a matrix crack propagating through an array of elastic fibers and are thus properties of the "composite." However, in view of the lack of understanding of the effects of the fibers on the processes occurring at the crack tip, it seems adequate to select values of β and n that are representative of the monolithic matrix alloy (β_m and n_m). With this approach, the effects of the fibers at the crack tip are neglected.

Once the fibers begin to fail, their contribution to crack-tip shielding is reduced substantially. To incorporate such effects in the model, a deterministic criterion for fiber failure has been used [*8*]. The calculations are conducted by continuously adjusting the unbridged portion of the crack to maintain a stress at the tip of the unbridged segment equal to the fiber strength. Through this approach, the entire cracking history (a versus N) can be simulated.

The results of these calculations can also be used to develop a criterion for a *threshold* stress amplitude, $\Delta\sigma_{th}$, below which fiber failure does not occur for *any* crack length [*6,8,27*]. Within this regime, the crack growth rate approaches a steady-state value, with all fibers in the crack wake remaining intact. The variation in the *threshold* stress amplitude with fiber strength, S_f, is plotted in Fig. 17. The maximum value of $\Delta\sigma_{th}$ occurs when there is no notch, that is, $a_0 = 0$, whereupon

$$\Delta\sigma_{th}/fS_f(1 - \mathcal{R}) = 1 \tag{26}$$

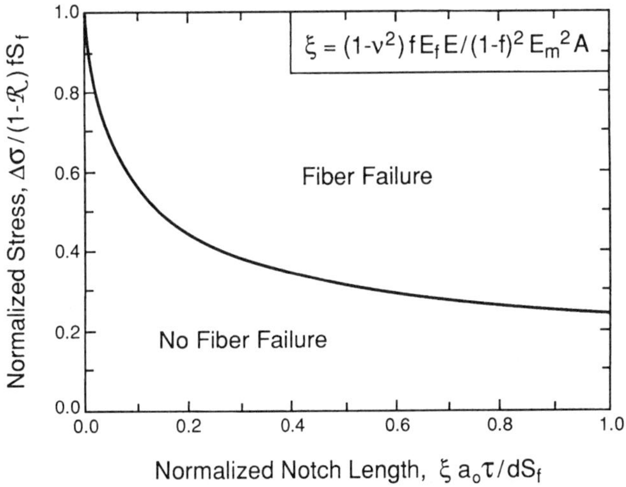

FIG. 17—*Influence of notch length and material properties on the fatigue threshold.*

For finite notch lengths, the threshold stress can be accurately described by the relationship [*27*]

$$\frac{6\pi\xi a_0\tau}{DS_f} = \left\{ \left(\frac{\Delta\sigma}{fS_f(1 - \mathcal{R})}\right)^{-4/3} - \left(\frac{\Delta\sigma}{fS_f(1 - \mathcal{R})}\right)^{2/3} \right\}^{3/2} \quad (27a)$$

where ξ is a nondimensional parameter defined by

$$\xi = fE_fE(1 - \nu^2)/(1 - f)^2E_m^2A \quad (27b)$$

with ν being Poisson's ratio and A being an orthotropy factor of order unity.

Comparison Between Experiment and Theory

The measured crack growth curves at *low* stress levels have been compared with the predictions of the model described in the preceding section. The parameters, β and n, in Eq 25 are taken to be those for the matrix alloy [*28*] given in Table 2. The various elastic moduli [*29*] are also given in Table 2. The material parameter that is subject to the most uncertainty is the interface sliding stress, τ. Consequently, the approach adopted here was to compare the experimental data with the model predictions for *a range of values* of τ and then assess whether consistency is achieved over the entire range of measurements. A similar approach is used to model the behavior at *intermediate* stress levels, though the predictions also account for fiber failure, assuming a deterministic fiber strength, S_f. In the regime following fiber fracture, the calculations are based on a fixed value of τ (chosen to be consistent with the data in the regime prior to fiber failure) and comparisons made for a range of values of S_f. The inferred value of S_f is then compared with values reported elsewhere. At *high* stress levels, the crack growth curves are compared with predictions of the Paris law, assuming $\Delta K_t = \Delta K_a$ (that is, $\Delta K_b = 0$). The exponent, n, is again taken to be the same as that for the matrix and the parameter, β, varied from β_m (the value for the matrix) to $\beta_m/4$.

TABLE 2—*Summary of constituent and composite properties used in constructing the fatigue map.*

	MATRIX
Elastic modulus [*29*]	E_m = 115 GPa
Coefficient in Paris law [*28*]	$\beta_m = 3.6 \times 10^{-29}m^{11/2}/N^3cycle$
Exponent in Paris law [*28*]	n_m = 3
	FIBERS
Elastic modulus	E_f = 400 GPa
Mean strength [*15*]	$S_f \approx$ 4 GPa
Diameter	d = 100 μm
Volume fraction	f = 0.32
	INTERFACE
Sliding stress	τ = 20 MPa
	COMPOSITE
Longitudinal modulus [*13*]	E = 200 GPa
Effective modulus [*8*]	$\bar{E}$ = 193 GPa

Figure 12 shows the comparisons between experiment and theory. At low stresses, the experimental data are consistent with the model for $\tau \approx 20$ MPa. The value of τ inferred in this manner has been compared with measurements on both pristine and *fatigued* fibers, using single fiber pushout tests [*19*] (Fig. 18). Specimens with fatigued fibers are prepared by cutting composite sections ~600 μm thick, adjacent to a matrix fatigue crack. The sections are then ground and polished to a final section thickness of ~400 μm. The pristine specimens are prepared in a similar fashion, using undeformed material. The pushout tests show that the sliding resistance of the pristine fibers is ~120 to 130 MPa. In contrast, the sliding stress for the fatigued fibers is initially only ~30 to 50 MPa. These values are of the

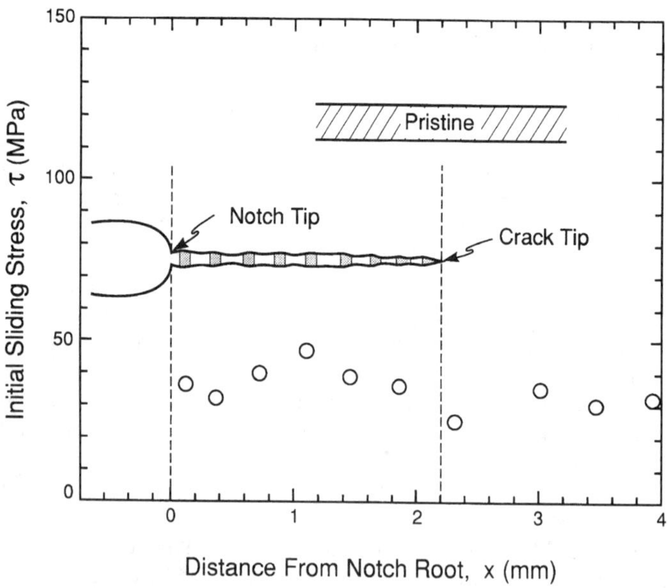

FIG. 18—*Trends in interfacial sliding stress with distance from the notch tip.*

same order as the one inferred from the fatigue crack growth experiments (within a factor of ~2).

At the intermediate stress level, the initial growth behavior is consistent with the model predictions for τ in the range of ~50 to 70 MPa. Again, these values are similar to those obtained using pushout measurements. In the regime following the onset of fiber fracture, the crack growth rate is comparable to that in the matrix alloy at the same level of crack extension (~5 μm/cycle at Δa = 0.8 mm). This correlation suggests that the fibers no longer provide a contribution to fatigue resistance.

At the high stress levels ($\Delta\sigma$ = 760 and 800 MPa), the crack growth behavior appears to follow the Paris law, with $\Delta K_b = 0$. The values of β inferred from the comparisons are in the range of $\beta_m/2$ to $\beta_m/4$, indicating that the fatigue resistance of the composite is comparable to that of the matrix alone.

Fatigue Fracture Maps

The preceding results for fatigue and fracture can be combined to form a fatigue fracture map, shown in Fig. 19. The purpose of the map is to graphically represent the combinations of stress, notch size, and material properties over which the various types of mechanisms operate. As noted earlier, the axes of the maps are two nondimensional parameters: Σ and α. For simplicity, the a_0/w ratio is taken to be zero, though the results can be generalized for finite values. The material properties are selected to coincide with those of the Ti-6Al-4V/SiC(σ) composite studied here. The properties of the constituents are summarized in Table 2.

The maps are constructed in the following way. First, the locus of notched strength is plotted, following the approach outlined in the section on experimental results. This locus corresponds to N_f = ½. The portion of the curve at very small values of α is disregarded, since this predicts a strength in excess of the unnotched tensile strength.[2]

Immediately below the fracture locus, fatigue is dominated by the matrix, since the fibers break *ahead* of the growing crack and do not contribute substantially to the fatigue resistance. As the stress is decreased, a transition occurs in which the fibers initially bridge the crack and reduce the rate of matrix cracking. However, the fibers eventually break, leading to a finite fatigue life. The transition between these two regimes can be modeled in an approximate manner through the two-level bridging law used to describe the notched strength. Here, the criterion for the onset of fiber failure ahead of the crack tip is

$$D \geq \delta_f \tag{28}$$

where, as before, D is the notch-tip opening displacement and δ_f is the critical opening for fiber fracture, shown in Fig. 8. This criterion is satisfied when the energy in the first portion of the bridging law ($0 \leq \delta \leq \delta_f$) is expended. The stress needed to satisfy this condition is evaluated by re-interpreting the rectilinear bridging results of Eq 14, whereupon

$$\Sigma = \frac{2}{\pi}\cos^{-1}\left(\exp\frac{-\pi}{8\alpha g}\right) \tag{29a}$$

with g given by

$$g = \left(1 + \frac{1}{\lambda}\right)\left(1 - \frac{\bar{\sigma}_1}{\bar{\sigma}_0}\right) \tag{29b}$$

[2]See Ref 14 for an alternate approach to this problem.

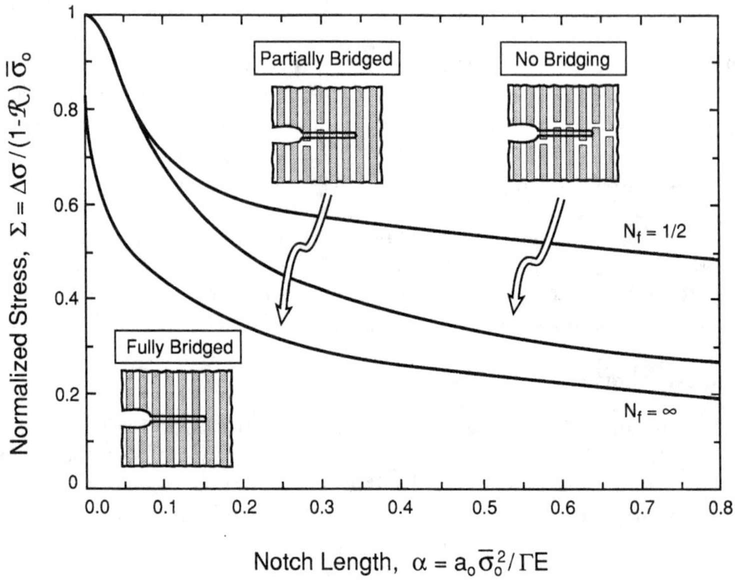

FIG. 19—*Fatigue fracture map for the Ti-6Al-4V/SiC composite.*

Note that the prediction of Eq 29 makes a smooth transition to $\Sigma = 1$ at $\alpha = 0$, as required. Indeed, this curve represents the asymptotic limit of the fracture locus as $\alpha \rightarrow 0$.

At yet lower stresses, another transition occurs. The transition represents the fatigue threshold: the stress below which the bridging fibers *never* break, regardless of crack length. This condition, described by Eq 27, can be re-expressed in terms of the coordinates of the map

$$\alpha = \phi\{(\gamma\Sigma)^{-4/3} - (\gamma\Sigma)^{2/3}\}^{3/2} \tag{30a}$$

where ϕ and γ are dimensionless parameters defined by

$$\phi = \frac{AdS_f(1 - f)^2E_m^2\bar{\sigma}_0^2}{6\pi(1 - \nu^2)\tau f E_f\Gamma E^2} \tag{30b}$$

and

$$\gamma = \bar{\sigma}_0/fS_f \tag{30c}$$

Using the values in Table 2 gives $\phi \approx 0.05$ and $\gamma \approx 1.25$. This line is appropriately labeled $N_f = \infty$.

The fatigue map can be used in the following way. Below the threshold stress (corresponding to $N_f = \infty$), the fatigue crack growth behavior can be simulated using the bridging models, neglecting fiber fracture. Though composite fracture never occurs, matrix cracking may lead to a local degradation in properties (including stiffness and thermal conductivity) that exceeds allowable limits. Consequently, a criterion for failure in this regime may be the attainment of a critical crack length. Above the threshold, similar simulations of matrix cracking can be conducted, incorporating fiber fracture. The experimental measurements presented here and elsewhere [*6*] suggest that the growth rate accelerates extremely rapidly following the onset of fiber failure, with only minimal retained fatigue life. A conservative estimate of fatigue life can thus be obtained by computing the number of loading cycles required to *initiate* fiber failure. In the high stress regime, the fatigue life is obtained by integrating the Paris law in the usual manner, using the values of β and n representative of the composite. The limiting crack length in the integration is simply the difference between the initial crack (or notch) length and the crack length corresponding to $N_f = ½$. This difference can be read directly off the map.

Concluding Remarks

The present article demonstrates how fatigue fracture maps can be constructed for fiber-reinforced TMCs. The map identifies the relevant mechanisms controlling fatigue cracking and fracture. Having established the regimes in which the various mechanisms operate, appropriate models can be used to predict crack growth and fatigue life. These results can then be plotted onto the map as contours of fixed N_f or Δa_f, where Δa_f represents the crack extension at the onset of composite fracture. Such developments are the focus of current work.

Acknowledgments

Funding for this work was supplied by the Advanced Research Project Agency (ARPA) University Research Initiative Program at University of California at Santa Barbara (UCSB) under Office of Naval Research (ONR) Contract N0014-92-J-1808. The authors gratefully acknowledge fruitful discussions with R. M. McMeeking.

References

[1] Cox, B. N. and Marshall, D. B., *Fatigue and Fracture of Engineering Materials and Structures*, Vol. 14, 1991, p. 847.
[2] Sensmeier, M. D. and Wright, P. K. in *Fundamental Relationships Between Microstructure and Mechanical Properties of Metal Matrix Composites*, P. K. Liaw and M. N. Gungor, Eds., The Minerals Metals and Materials Society, Warrendale, PA, 1990, pp. 441–457.
[3] Wright, P. K., Nimmer, R., Smith, G., Sensmeier, M., and Brun, M. in *Interfaces in Metal-Ceramic Composites*, R. Y. Lin, R. J. Arsenault, G. P. Martins, and S. G. Fishman, Eds., The Minerals, Metals and Materials Society, Warrendale, PA, 1989, pp. 559–581.
[4] Harmon, D. M. and Saff, C. R. in *Metal Matrix Composites: Testing, Analysis, and Failure Modes, ASTM STP 1032*, W. S. Johnson, Ed., American Society for Testing and Materials, Philadelphia, 1989.
[5] Kantzos, P., Telesman, J., and Ghosn, L., NASA TM-103095, Cleveland, OH, 1989.
[6] Walls, D. P., Bao, G., and Zok, F. W., *Acta Metallurgica Materials*, Vol. 41, 1993, pp. 2061–2071.
[7] McMeeking, R. M. and Evans, A. G., *Mechanics and Materials*, Vol. 9, 1990, p. 217.
[8] Bao, G. and McMeeking, R. M., *Acta Metallurgica Materials*, Vol. 42, 1994, pp. 2415–2425.
[9] Cox, B. N. and Lo, C. S., *Acta Metallurgica Materials*, Vol. 40, 1992, p. 69.
[10] Cox, B. N. and Marshall, D. B., *Acta Metallurgica Materials*, Vol. 42, 1994, pp. 341–363.
[11] Bowen, P., Ibbotson, A. R., and Beevers, C. J. in *Fatigue of Advanced Materials*, R. O. Ritchie, R. H. Dauskardt, and B. N. Cox, Eds., Materials and Component Engineering Publishing Ltd., Birmingham, UK, 1991, pp. 379–393.
[12] Bakuckas, J. G., Jr. and Johnson, W. S., *Journal of Composite Technology and Research*, Vol. 15, 1993, pp. 242–255.
[13] Weber, C. H., Chen, X., Connell, S. J., and Zok, F. W., *Acta Metallurgica Materials*, Vol. 42, 1994, pp. 3443–3450.
[14] Connell, S. J., Zok, F. W., Du, Z. Z., and Suo, Z., *Acta Metallurgica Materials*, Vol. 42, 1994, pp. 3451–3461.
[15] Zok, F. W., Chen, X., and Weber, C. H., *Journal*, American Ceramics Society, Vol. 78, 1995, pp. 1965–1968.
[16] Suo, Z., Ho, S., and Gong, X., *Journal of Engineering Materials and Technology*, Vol. 115, 1993, pp. 319–326.
[17] Bao, G. and Zok, F., *Acta Metallurgica Materials*, Vol. 41, 1993, pp. 3515–3524.

[18] Walls, D. P and Zok, F. W., *Acta Metallurgica Materials,* Vol. 42, 1994, pp. 2675–2681.
[19] Warren, P. D., Mackin, T. J., and Evans, A. G., *Acta Metallurgica Materials,* Vol. 40, 1992, pp. 1243–1249.
[20] Bilby, B. A., Cottrell, A. H., and Swindon, K. H., *Proceedings,* Royal Society, London, Vol. A272, 1963, pp. 304–314.
[21] Cottrell, A. H., *Mechanics of Fracture,* Tewksbury Symposium on Fracture, University of Melbourne Press, Jan. 1963.
[22] Tattersall, H. G. and Tappin, G., *Journal of Materials Science,* Vol. 1, 1966, pp. 296–301.
[23] Bluhm, J. I., *Engineering Fracture Mechanics,* Vol. 7, 1975, pp. 593–604.
[24] Rice, J. R., *Journal of Applied Mechanics,* June 1968, pp. 379–386.
[25] Tada, H., Paris, P. C., and Irwin, G. R., *The Stress Analysis of Cracks Handbook,* Del Research, St. Louis, 1985.
[26] Du, Z. Z. and Zok, F. W., to be published.
[27] Budiansky, B. and Cui, L., *Journal of the Mechanics and Physics of Solids,* Vol. 42, 1994, pp. 1–19.
[28] *Damage Tolerant Design Handbook, Part 1,* Metals and Ceramics Information Center, Batelle's Columbus Laboratories, Columbus, OH, 1975.
[29] "Properties and Selection: Stainless Steels, Tool Materials and Special Purpose Metals," *Metals Handbook, Ninth Ed. (1990), Vol. 3,* American Society of Metals, Metals Park, OH, pp. 388–391.
[30] Kantzos, P., Eldridge, J., Koss, D. A., and Ghosn, L. J., *Proceedings,* Materials Research Society, Vol. 273, 1992, pp. 135–142.

Paul Bowen[1]

Characterization of Crack Growth Resistance Under Cyclic Loading in the Presence of an Unbridged Defect in Fiber-Reinforced Titanium Metal Matrix Composites

REFERENCE: Bowen, P., "**Characterization of Crack Growth Resistance Under Cyclic Loading in the Presence of an Unbridged Defect in Fiber-Reinforced Titanium Metal Matrix Composites,**" *Life Prediction Methodology for Titanium Matrix Composites, ASTM STP 1253,* W. S. Johnson, J. M. Larsen, and B. N. Cox, Eds., American Society for Testing and Materials, 1996, pp. 461–479.

ABSTRACT: This paper considers the experimental characterization of crack growth from unbridged defects in fiber-reinforced titanium metal matrix composites subjected to cyclic loading by the use of fracture mechanics parameters. The conditions under which parameters such as the nominal applied stress intensity range, ΔK_{app}, the nominal maximum stress intensity factor, K_{max}, and the effective stress intensity range, ΔK_{eff}, are of use, and their experimental measurement are considered. Effects of fiber fracture, stress intensity factor range, mean stress, loading configuration (bending versus tension), test temperature, crack size, crack shape, and fiber-matrix interfacial strength on fatigue crack growth resistance are highlighted. The experimental determination of crack arrest in such composites is outlined.

KEYWORDS: crack growth, defects (materials), titanium, titanium matrix composites, life prediction, titanium alloys, fatigue (materials), modeling

This paper reviews the experimental assessment of the fatigue crack growth resistance of fiber-reinforced titanium metal matrix composites (MMCs) in the presence of an unbridged defect. Under cyclic loading, such defects result in localized dominant matrix cracks that are bridged by fibers. During sub-critical crack growth for all studies to date, fibers do not fail ahead of the advancing matrix crack, but rather they fail only when they have been breached, that is, they fail behind the crack-tip and in its wake. The single most important feature in controlling the fatigue crack growth resistance of titanium fiber-reinforced MMCs is therefore the number of intact fibers bridging in the crack wake. Once "large" numbers of fibers are bridging a crack, then it becomes relatively harmless, and the composite displays excellent damage tolerance. To define the number of fibers required, and to predict whether this state will be reached is one principal aim of current studies. In this context, the behavior of partly or fully bridged cracks is central to the modeling and life prediction of titanium fiber-reinforced composites.

At present, there is considerable interest in silicon-carbide fiber-reinforced titanium MMCs for use in aerospace applications at elevated temperatures. At ambient temperature,

[1]Professor of Mechanical Metallurgy, School of Metallurgy and Materials/IRC in Materials for High Performance Applications, The University of Birmingham, Edgbaston, B15 2TT, UK.

commercial unidirectionally reinforced materials can possess longitudinal tensile strengths in excess of 1600 MPa, and they can exhibit fatigue lifetimes of better than 10^4 cycles when stressed at up to 75% of their longitudinal tensile strength. To justify their use, it is likely that they will be stressed highly in service.

Under such envisaged loading conditions, it is imperative to establish their "fitness for purpose" and reliability. Indeed, to utilize their potential fully, it is necessary to develop realistic methods of assessment of damage and validated lifing procedures. It is natural, in view of their envisaged substitution for monolithic alloys in many cases, to consider the use of fracture mechanics parameters including the fracture toughness, stress intensity factor range, and integrated crack growth rates for the prediction of fatigue life. Such parameters are not usually applied to characterize damage in fiber-reinforced composites, but it is important to recognize that many studies have established in these titanium MMCs that localized dominant cracks develop under cyclic loading [*1–7*]. Examples are shown in Fig. 1. Such observations give confidence that a fracture mechanics-based approach may be of use provided that effective crack-tip stress intensities can be determined, and such evaluation is the subject of many numerical micromodels [*8–14*]. In such models, the influence of "bridging length scale" must be considered carefully [*6,15–17*].

Unfortunately, because there can be no unique relationship between a nominal applied stress intensity factor range, ΔK_{app}, and the crack increment per cycle, da/dN, which is independent of the number and position of bridging fibers, this introduces complexity into any experimental program. Indeed, care is required to avoid a large experimental data base that will still not identify the factors that control crack growth resistance. Even with care, an extensive program is required to produce a quantitative description. Some important factors that need to be identified with respect to fiber fracture and hence to controlling fatigue crack growth resistance include:

1. the role of matrix-fiber interfacial strength (for interfacial strengths that allow fiber debonding ahead of the advancing matrix crack);
2. the influence of matrix microstructure;

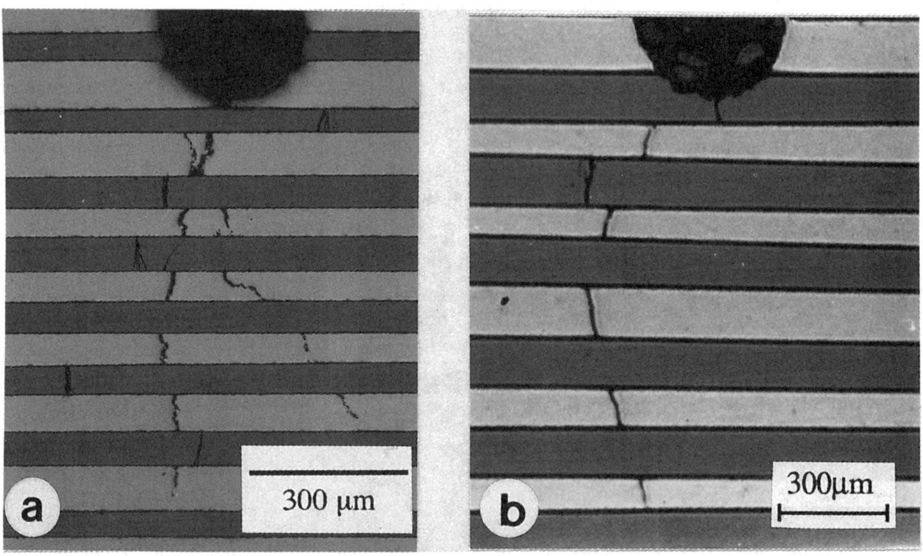

FIG. 1—*Optical micrographs of sections through interrupted fatigue crack growth:* (a) *Ti-6A1-4V/SCS6 and* (b) *Ti-15-3/SCS6.*

3. the effect of unbridged crack shape and crack size;
4. the influence of test temperature and environment (air versus vacuum);
5. the importance of load range, mean load, testpiece geometry, and testpiece size; and
6. conditions that promote crack arrest and hence ensure component integrity.

The influence of many of these parameters on fiber fracture and hence on fatigue crack growth resistance is now considered. (Other factors that may also influence the propensity of fibers to fracture such as: the material processing quality, for example, fiber bunching, fiber contact, and incomplete consolidation; the mode of loading, for example, load versus strain control, creep/environmental interactions; and thermo-mechanical fatigue; are outside the scope of this present paper). The purpose of this paper is thus to address some of the issues that relate to the characterization of crack growth resistance from an initial unbridged defect subjected to cyclic loading.

Experimental Studies

Materials

Commercially available titanium alloys/silicon-carbide fiber composite systems: Ti-15-3/SCS6; Ti-6Al-4V/SCS6; Ti-6Al-4V/SM1240; and Ti-β21S/SCS6 have all been studied. It is noted that both types of fiber considered here have similar elastic moduli and tensile strengths, SCS6 fibers are of a nominal diameter of 140 μm, and are grown by chemical vapor deposition onto a carbon core, while SM1240 fibers are of a nominal diameter of 100 μm and are grown by chemical vapor deposition onto a tungsten core. In addition, different external coatings are applied to these fibers. Details of all systems are given elsewhere [*3–7,17–21*]. All composites reinforced with SCS6 fibers have a nominal fiber volume fraction of 0.35, and are eight-ply composites while those reinforced with SM1240 fibers have a nominal volume fraction of fibers of 0.31, and six-ply composites have been considered. In the metastable β systems, aging treatments have been found to increase the interfacial strength. In most cases, unidirectionally reinforced materials only have been considered, but tests on cross-ply laminates are now in progress and results obtained on the Ti-6Al-4V/SCS6 cross-ply laminate system reinforced to a nominal fiber volume fraction of 0.35 are also reported.

Fatigue Crack Growth Resistance Curves

Initial tests were confined to single-edged through-thickness notch testpieces in bending, but comparisons for some systems have now been made with single-edged notch testpieces in tension-tension loading. Initial normalized unbridged notch depths of between 0.05 and 0.25 have been considered, where the notch depth (a_0) is normalized to the testpiece width (W). But for ease of comparison, most tests have been performed for an initial normalized unbridged notch (crack) depth, $a_0/W = 0.25$. All testpieces, machined from composites reinforced with SCS6 fibers are of width, $W = 4$ mm, and thickness, $B = 2$ mm. For testpieces machined from composites with SM1240 fibers, they are all of width, $W = 4.5$ mm, and thickness $B = 1.05$ mm. Under bending, an overall span (S) to width (W) value of $S/W \sim 15$ has been employed, although it is noted that similar crack growth resistance curves can be obtained for the range $5 \leq S/W \leq 15$ [*19*]. Crack growth was monitored using the direct-current potential difference technique.

In many cases, care is required to establish appropriate calibration curves of normalized potential to normalized crack depth because of the small size of the testpiece under consideration [*5*]. Changes in potential difference relate to matrix crack growth alone, and the technique gives the crack depth averaged across the testpiece thickness. In some tests, fractography has also allowed local crack growth rates to be measured by the observation of fatigue striations [*3,22,23*].

In addition, the assessment of part-through-thickness unbridged defects is now under study [*24,25*]. Initial work has addressed both unbridged corner cracks and unbridged semi-elliptical cracks [*24,25*].

Tests have almost exclusively been performed using a constant cyclic load range, ΔP. Therefore, the nominal applied stress intensity range, ΔK_{app}, increases with increases in fatigue crack length. Crack growth resistance curves can be usefully represented under such testing conditions by plotting crack growth rate per cycle, da/dN, versus either ΔK_{app} or total crack length. Servo-hydraulic testing machines, operating at test frequencies of up to 10 Hz, have been used throughout these programs. Since these tests in bending require very low loads, the largest load-cell employed was rated to 10 kN. (Under tensile–tensile loading a 100-kN load cell has been utilized. Note that hydraulic grips are required to perform tests under cyclic tensile loading on these materials.) Most test programs to date have utilized load ratios of $R = 0.5$ and $R = 0.1$ (where $R = P_{min}/P_{max}$ and P_{min} and P_{max} are the minimum and maximum loads applied over the fatigue cycle, respectively), but some have now considered load ratios of $R = 0.7$ (under bending). Tests have also been carried out routinely at temperatures of up to 550°C both in air (bending and tension loading) and in vacuum (bending only) [*19,25,26*].

A methodology of evaluating crack arrest for a given geometry of a specimen and initial unbridged defect size has also been developed [*7,17,24*]. It involves, for a given mean stress (load ratio) and a single composite system, testing a range of initial ΔK_{app} values until crack arrest (defined as $da/dN \leq 10^{-8}$ mm/cycle) has been achieved under a constant cyclic load range. It is important to note that multiple tests are required, because once the crack is bridged by fibers then comparisons between testpieces are valid only if the testpiece geometry (including unbridged to bridged scaling length) is similar, and even then great care is necessary. In this context, it is critical to note that useful comparisons between tests on different composites or different microstructural conditions or both can be made easily, only if the initial normalized defect size (a_0/W) is held constant during crack arrest determination.

In-situ Determination of Fiber Fracture during Cyclic Loading

Early studies deduced the onset of fiber fracture from audible emissions and rapid voltage excursions observed during crack growth [*4,27*]. Such events have now been quantified by the in-situ use of both acoustic emission and direct optical observation of the failure of (surface) fibers during cyclic loading. A commercially available PAC LOCAN AE analyzer has been used, and under the conditions of use, fiber fracture is characterized most easily by amplitude events of ≥90 dB. Details of this system and analysis are given elsewhere [*22,28*]. It has now also been employed during tests at elevated test temperatures. Although this technique has proved invaluable in combination with potential difference measurements for determining the number of intact fibers bridging a given matrix crack, it cannot identify the exact spatial position of fiber failure. A second method has now been developed [*25*], and it consists of measuring the electrical resistance of fiber cores with a fine probe. It is most easily used on cores of tungsten (for example as found in SM1240), and further details can be found elsewhere [*25*].

Evaluation of Fiber-Matrix Interfacial Strength

Early studies employed micro-hardness push-out tests [*5,20*], but recently a continuous load-displacement fiber push-out test has been developed at the University of Birmingham.

Here, a remote high-resolution optical telescope and video recording system have been used to monitor the behavior of the individual fibers during the push-out tests. Details are given elsewhere [*21,29*], but it is sufficient to note that such a system allows both the debonding shear stress and the frictional shear stress of the (debonded) interface to be evaluated.

In general, for the systems studied to date, the value of the frictional shear stress appears to be higher than that of the debonding shear stress (provided that the slice is of sufficient thickness).

The wide range of assessment techniques employed during these studies have allowed micromechanisms of crack growth in fiber-reinforced composites to be elucidated, and they have also highlighted many important factors for controlling crack growth resistance. A selection of some of these results is now presented.

Results

Fatigue Crack Growth Resistance Curves

In monolithic Ti-6Al-4V material (manufactured by a foil-foil processing route similar to that used for many of the composites under study here), a characteristic fatigue crack growth resistance curve is obtained, see Fig. 2. For $\Delta K \geq 4.5$ PMa$\sqrt{m}$, the curve can be represented well by a Paris-Erdogan relationship to the form $da/dN = C\Delta K^m$, where C and m are numerical constants, equal to 9.2 10^{-9} mm/cycle and 3.12, respectively (for ΔK in MPa$\sqrt{m}$) [*12*].

In sharp contrast for the Ti-6Al-4V/SM1240 composite system, the dependence of crack growth rates observed experimentally, *da/dN*, is shown versus total crack length, Fig. 3, for an initial ΔK_{app} of 7.5 MPa$\sqrt{m}$. At this stress intensity range, crack growth rates decrease with increases in fatigue crack length until crack arrest is obtained, despite the increase in nominal stress intensity range with fatigue crack length increase (because these tests are performed at constant cyclic load range) [*12*].

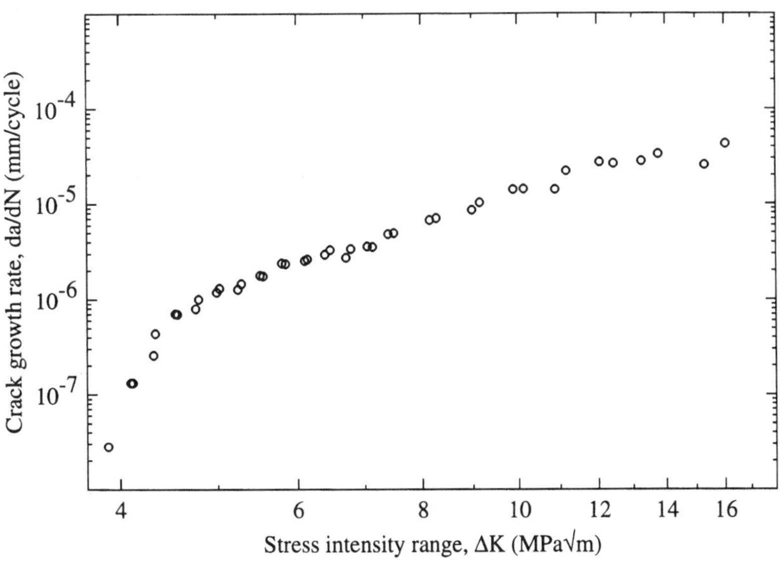

FIG. 2—*Crack growth resistance curve:* da/dN *versus* ΔK, R = *0.1, and* υ = *0.5 Hz. Monolithic Ti-6Al-4V alloy, ambient temperature.*

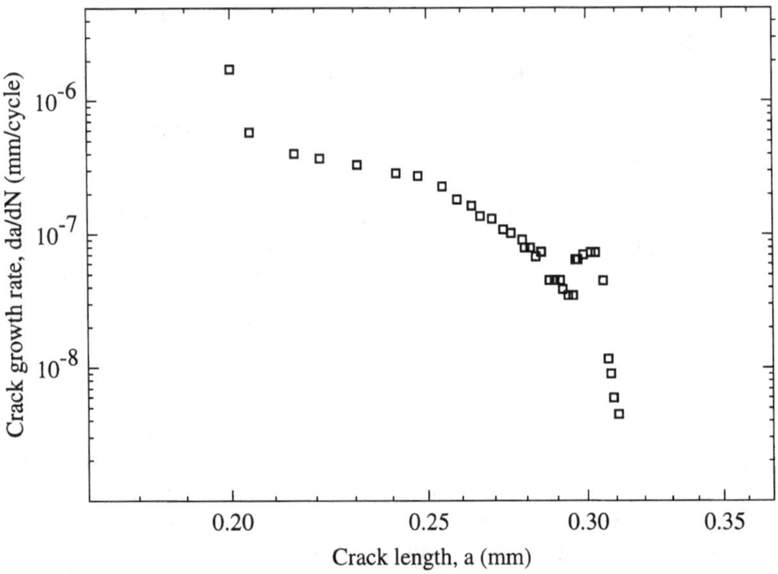

FIG. 3—*Crack growth resistance curve:* da/dN *versus crack length for initial* $\Delta K_{app} \approx$ *7.5* $MPa\sqrt{m}$, R = *0.5, and* υ = *10 Hz. Ti-6Al-4V/SM1240[*0_6*] composite, ambient temperature.*

Such observations are quite general in all systems studied to date, where crack growth increments per cycle, *da/dN*, usually decrease initially with crack length increase ahead of the initial unbridged defect. Examples are given in Figs. 3 through 6. These decreases in crack growth rates are despite an increase in nominal applied stress intensity range, ΔK_{app}, with crack length increase. An explanation for this unusual crack growth behavior is given by considering optical sections through interrupted fatigue crack growth tests, see Fig. 1. The micromechanism of crack growth is well characterized by initial crack growth through the matrix that breaches (debonded) fibers in its wake. In studies to date, no fibers have been observed to fail ahead of the growing fatigue crack tip. The bridging fibers then reduce the effective stress intensity at the growing crack tip, ΔK_{eff}, and if no (or very few) fibers fail then crack growth rates decrease eventually to crack arrest defined here by $da/dN \leq 10^{-8}$ mm/cycle. Fiber failure in the crack wake often results in the sharp increases in crack growth rates seen in Figs. 4 and 6 (and this is discussed further in the next section).

The balance between sub-critical crack growth and catastrophic failure will be governed primarily by the number of fibers remaining intact within the crack wake, see Fig. 1 and Table 1. In all tests performed to date, crack growth either arrests eventually or catastrophic failure occurs (defined by the symbol ★ in all figures).

The micromechanisms of matrix fatigue crack growth have been further elucidated by fractographic observations of striation spacings, such as shown in Fig. 7. In general, crack growth rates from these striation spacings [*3,22,23*] are much greater than those measured by direct-current potential difference techniques, see Table 2, and they can be deduced to reflect local fiber volume fraction differences. In all cases, striation spacings define local crack growth directions around fibers; indicate discontinuous matrix crack growth (deduced to be due to both periodic arrest and crack growth excursions following fiber fracture); and allow an assessment of effective matrix crack-tip stress intensity factor ranges, ΔK_{eff}, to be

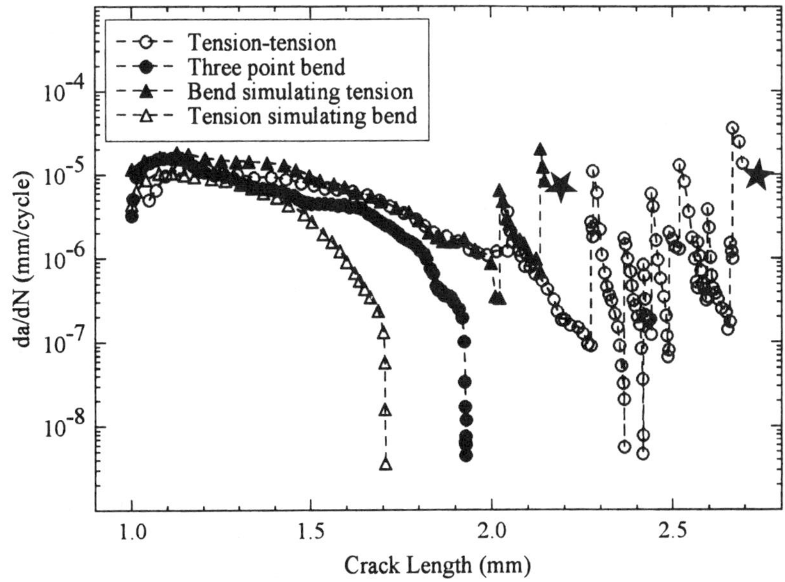

FIG. 4—*Crack growth resistance curves:* da/dN *versus total crack length for initial* ΔK_{app} *17* $MPa\sqrt{m}$, R = *0.5, and* υ = *10 Hz. Ti-15-3/SCS6[0*$_8$*] composite, solution-treated, and tested at ambient temperature under cyclic bending and cyclic tension.*

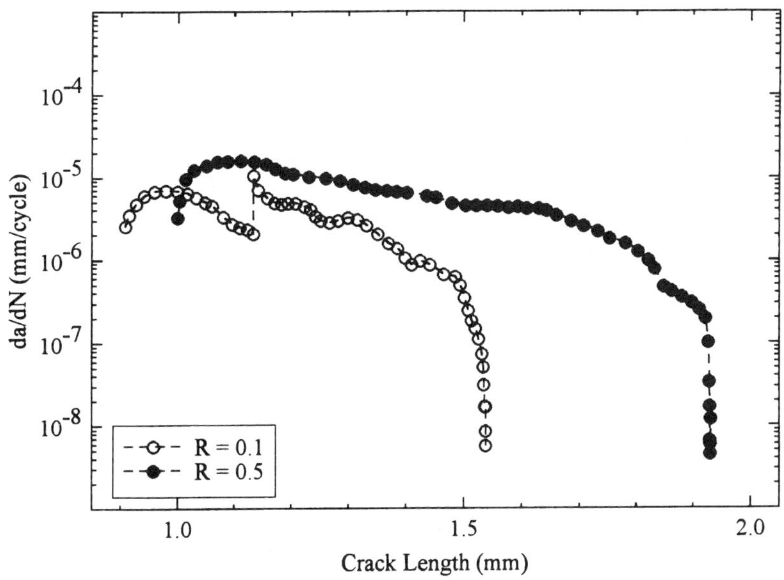

FIG. 5—*Effect of load ratio on crack growth resistance curves:* da/dN *versus crack length for initial* $\Delta K_{app} \approx$ *16* $MPa\sqrt{m}$, υ = *10 Hz. Ti-15-3/SCS6 [0*$_8$*] composite, solution-treated condition ambient temperature.*

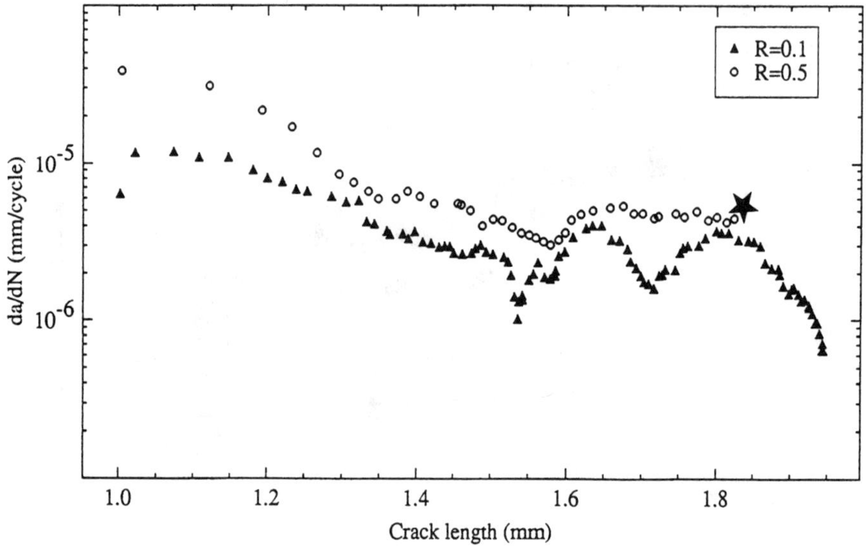

FIG. 6—*Effect of load ratio on crack growth resistance curves:* da/dN *versus crack length for initial* $\Delta K_{app} \approx 18\ MPa\sqrt{m}$, $\upsilon = 0.5\ Hz$. *Ti-6Al-4V/SCS6 [0_8] composite, ambient temperature.*

TABLE 1—*The importance of the number of bridging fibers intact in the crack wake on stable/unstable crack growth transitions (TI-6Al-4V/SCS6[0_8])* [4,27].

Number of Bridging Fibers Remaining Intact	K_{max} (nominal), $MPa\sqrt{m}$	Type of Event
8	39	catastrophic failure
12	64	catastrophic failure
26	66	no growth
21	80	no growth
16	51	no growth
7	64	catastrophic failure

made by comparison with growth rates obtained in the unreinforced monolithic alloy (Fig. 2 and Table 2).

Effects of Fiber Fracture

A second general feature of all tests to date is that the failure of discrete fibers often sharply increases local crack growth rates, see Fig. 8, where the failure of fibers has been identified unequivocally by acoustic emission techniques. (For reference, a complete amplitude distribution for events detected during this test is given in Fig. 9). Note, to measure such local increases in crack growth rate, it is essential to monitor crack growth rates by continuous potential difference techniques. Further examples are given in Fig. 10 (following monotonic overloads to fracture individual fibers). The sharpest increases in local fatigue crack growth rates are observed at near crack arrest growth rates, and the magnitude of these growth rate excursions is also in part determined by the precise secant method employed to

FIG. 7—*Striations visible in matrix regions after failure Ti-6Al-4V/SCS6 $[0_8]$. Note the discontinuous change in striation spacings. The direction of macroscopic crack growth is from top to bottom of the fractograph* [23].

TABLE 2—*A comparison of local crack growth rates (striation spacings) and average crack growth rates (direct-current potential difference) for Ti-6Al-4V/SCS6 unidirectional composites* [23] *and cross-ply laminates* [22]. *Effective stress intensity factor ranges, ΔK_{eff}, have been deduced from monolithic alloy data, Fig. 2* [13].

Material	$\Delta K_{nominal}$, $MPa\sqrt{m}$	da/dN, mm/cycle (DCPD)	ΔK_{eff}, $MPa\sqrt{m}$ (DCPD)	da/dN, mm/cycle (striations)	ΔK_{eff}, $MPa\sqrt{m}$ (striations)
$[0_8]$	23	$6.6.10^{-6}$	8.2	$8.6.10^{-5}$	18.7
	26	$4.5.10^{-6}$	7.3	$8.8.10^{-5}$	18.9
	30	$1.5.10^{-6}$	5.1	$2.4.10^{-4}$	26.0
	31	$9.4.10^{-7}$	4.4	$6.9.10^{-5}$	17.5
	34	$1.4.10^{-6}$	5.0	$1.3.10^{-4}$	21.4
$[0/90]_{2s}$	13	$1.2.10^{-5}$	10.0	$2.6.10^{-4}$	26.7
	13	$1.2.10^{-5}$	10.0	$1.5.10^{-4}$	22.4
	18	$6.5.10^{-6}$	8.2	$1.7.10^{-4}$	23.3
	18	$6.5.10^{-6}$	8.2	$1.1.10^{-4}$	20.3
$[90/0]_{2s}$	23	$2.7.10^{-4}$	27.0	$2.9.10^{-4}$	27.7
	24	$2.7.10^{-4}$	27.0	$9.5.10^{-5}$	19.3

determine crack growth rates. These observations indicate that to model these systems the fibers must be considered as discrete entities [*12–14*]. Moreover, many tests have illustrated that fibers may fail after extended cycling in near crack arrest regions (for example, see Fig. 4).

Effects of Test Temperature

In tests performed to date at elevated temperatures, crack growth rates usually increase relative to those measured at room temperature [*18,19*]. However, a full picture of the influence of test temperature on testpiece life or crack arrest or both has yet to emerge, and it is expected in part to depend upon the influence of test temperature on the fiber-matrix interfacial strength. Certainly, in systems that can age during testing, pronounced effects of

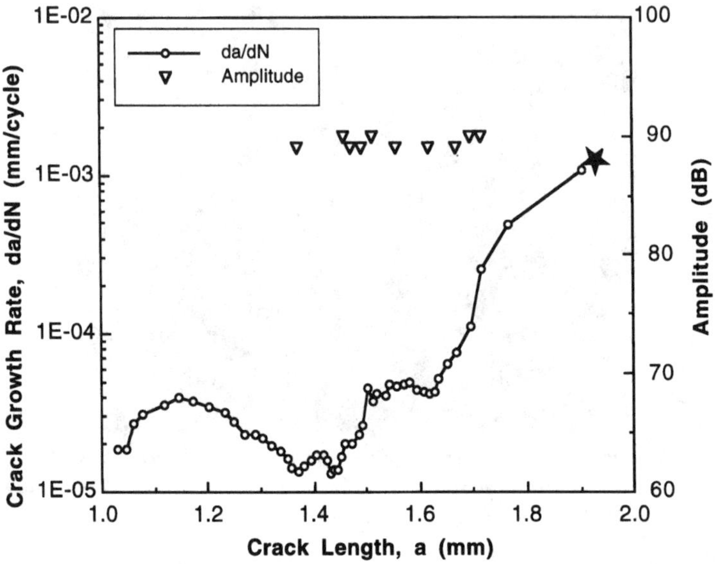

FIG. 8—*Crack growth resistance curve:* da/dN *versus nominal* ΔK_{app} *for initial* ΔK_{app} *18* $MPa\sqrt{m}$, R = *0.1, and* υ = *0.5 Hz. Ti-6Al-4V/SCS6 [0/90] cross-ply laminate, tested at ambient temperature. Individual fiber failures measured using acoustic emission events are arrowed.*

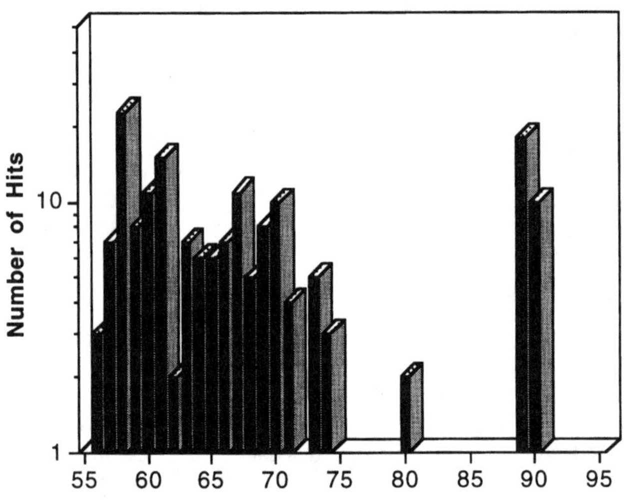

FIG. 9—*Acoustic emission amplitude spectrum measured after failure for the [0/90] cross-ply laminate.*

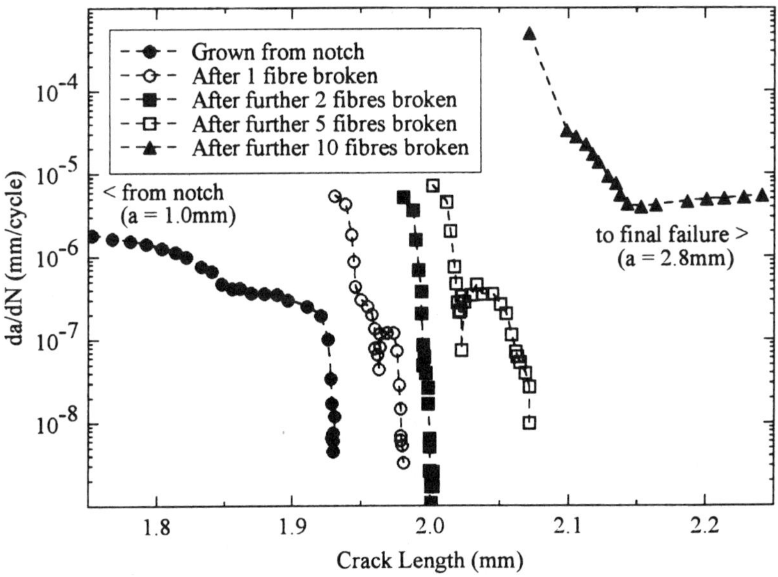

FIG. 10—*Effect of monotonic overloads on crack growth resistance curves:* da/dN *versus crack length,* R = *0.5, and* υ = *10 Hz. Ti-15-3/SCS6 [0_8] composite, ambient temperature, solution-treated condition.*

test temperature and test frequency are observed, see Table 3. In general, the poorer crack growth resistance of these particular composites that is displayed at elevated temperatures requires both faster matrix crack growth rates and easier fiber fracture (consistent with an increased fiber-matrix interfacial strength on aging).

Recent results [*30*] for Ti-6Al-4V/SCS6 $[0/90]_{2s}$ laminates have suggested that the influence of elevated test temperature on fatigue crack growth resistance is insignificant at a temperature of 300°C, but decreased crack growth resistance is obtained at the higher test temperature of 450°C in air. It is of interest to note that no reduction in fatigue crack growth resistance was obtained at a temperature of 450°C for tests carried out under vacuum [*30*].

Effects of Fiber-Matrix Interfacial Strength

After peak aging, the interfacial strength of the β21S/SCS6 composite system increases significantly, compared with solution-treated or as-processed conditions. Values of the debonding shear stress increase from approximately 80 to 120 MPa, and values of the

TABLE 3—*Comparison of the numbers of cycles to specimen failure for tests conducted at different temperatures, load ratios, and frequencies for an initial $\Delta K_{app} \sim 16$ MPa$\sqrt{m}$ (Ti-15-3/SCS6[0_8])* [18].

		25°C	200°C	350°C	500°C
R = 0.1	10.0 Hz	crack arrest		1 354 700	
R = 0.5	10.0 Hz	crack arrest	697 800	323 600	38 400
				201 900	
				167 200	
	2.0 Hz				12 000
	0.5 Hz			147 000	16 200

frictional shear stress (after debonding) increase from 100 to 180 MPa. Moreover, the number of cycles to failure decreases sharply from a minimum of 1 800 000 for as-processed and solution-treated conditions to 88 000 for the peak-aged condition. These studies are reported elsewhere [*21*], but here it is noted that the initial stress intensity factor range was sufficient to promote fiber failure in all conditions. Under these circumstances an increase in interfacial strength would be expected to increase the incidence of fiber failure and to promote earlier fiber failure, because the fibers will be more highly stressed as a result of more efficient load transfer. Recent studies have confirmed that earlier fiber failure does occur for the peak-aged condition [*31*].

Effects of Loading Geometry

Figure 4 suggests that crack arrest can occur more readily under cyclic bending than under cyclic tensile loading, for tests performed at identical initial ΔK_{app} conditions. However, this can be shown to be a function primarily of the rate of increase in stress intensity factor range with crack increment by matching the rate of nominal ΔK_{app} increase for each geometry. (This is achieved simply by load-range shedding for the tensile geometry and load-range increases for the bending geometry after appropriate crack depth increments). Under such conditions, although closely similar crack growth behavior can be produced for each testpiece geometry, this illustrates the care required to interpret nominal stress intensity parameters in comparing crack growth even in "geometrically similar testpieces" [*32*]. In this context, the influence of testpiece size is expected to introduce further significant effects and is the subject of continuing studies.

Effects of (Load Ratio) Mean Stress

In general, two different regimes of behavior can be distinguished under equivalent initial ΔK_{app} conditions, and they appear to depend upon whether increased load ratio (mean stress) promotes major instances of fiber failure.

In situations where crack arrest will eventually occur at both mean stresses, Fig. 5, the influence of mean stress can be seen only to promote crack growth to longer crack lengths (that is, an increased number of load cycles occur prior to arrest). This may be the result of minor instances of fiber failure but may also be a consequence of decreased threshold values, ΔK_{th}, that often result in monolithic alloys at increased mean stress [*18*]. In other examples, cracks can grow towards arrest at low mean stress conditions, $R = 0.1$, but can grow to catastrophic failure under equivalent ΔK_{app} conditions at high mean stress, $R = 0.5$ (see Fig. 6). In such situations, little effect of the higher mean stress is observed on crack growth rates prior to catastrophic failure. In general, the failure of bridging fibers is deduced to occur more readily at high mean stress because they will be required to resist greater local crack opening displacements during the load cycle. However, an important feature is that as each fiber is stressed more highly it will exert a higher clampback stress across the crack surfaces in the wake of the crack provided it does not fail, that is, it is both more potent at shielding the crack tip and closer to reaching its fiber fracture stress in tension. Such considerations are inherent to numerical and analytical micromodeling studies and are discussed elsewhere [*6,12–14*]. In summary, care is required to assess the effects of mean stress in these fiber-reinforced composites, and particularly in the definition of crack arrest (see next section).

Effects of Initial Applied Stress Intensity Range

For a given depth of initial unbridged crack (notch), an example of the influence of ΔK_{app} on crack growth resistance is shown in Fig. 11. Crack growth rates are similar initially for both levels of ΔK_{app}, but then crack arrest is observed for the lower applied stress intensity range only. Such observations indicate the delicate balance than can sometimes exist between promoting crack arrest and catastrophic failure in these composites [*4,18*].

Recently, further systematic studies have considered both effects of mean stress and initial applied stress intensity range for a constant initial unbridged defect size and a single composite system [*7,17,33*]. These studies have resulted in a simple model for the prediction of fiber failure for testpieces of a specific geometry and for the given composite system (Ti-6Al-4V/SM1240). This is reported elsewhere in more detail [*17,24,34*], but some of the results in which the model is based are also summarized in Table 4. Provided that K_{max} values (where K_{max} is the maximum nominal stress intensity factor applied over the fatigue cycle) calculated for a crack that has just reached the first row of fiber are left below 27.5 MPa$\sqrt{m}$, then few fibers fail and crack arrest is observed experimentally. Such calculations of K_{max} values may thus be of potential use in predicting conditions for crack arrest. From Table 4, the influence of mean stress in defining crack arrest conditions in terms of K_{max} after the first row of breached fibers can also be observed. As the mean stress increases, then it would appear that increased values of K_{max} can be tolerated after the first row of fibers. This observation needs careful discussion, and it is considered in detail elsewhere [*34*]. It is of interest to note that such values of K_{max} that can be tolerated vary with composite system for testpieces of identical geometry and are greatly reduced for a 0/90 cross-ply laminate (which has only half the number of 0° bridging fibers present for an equivalent increment of crack depth), see Table 5.

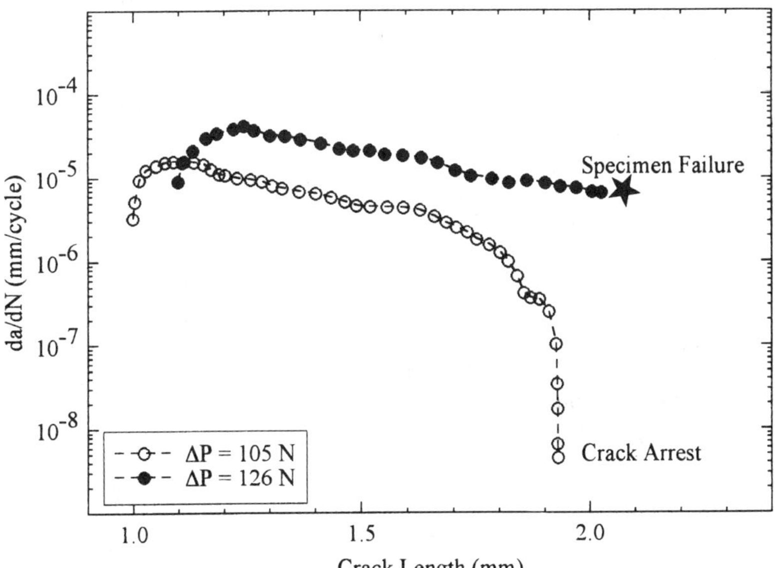

FIG. 11—*Effect of initial* ΔK_{app} *on crack growth resistance curves:* da/dN *versus crack length for* R = *0.5,* υ = *10 Hz. Ti-15-3/SCS6 [0_8] composite, solution-treated condition, ambient temperature. The curves correspond to initial* $\Delta K_{app} \approx$ *16 and 19 MPa$\sqrt{m}$, respectively.*

TABLE 4—*Effects of mean stress; initial* ΔK_{app}, K_{max}*;* K_{max} *after the first row of fibers; on crack growth resistance (Ti-6Al-4V/SM1240[0$_6$])* [17].

Stress Ratio	Initial ΔK, $MPa\sqrt{m}$	Initial K_{max}, $MPa\sqrt{m}$	K_{max} after 1st Row Fibers, $MPa\sqrt{m}$	Test Result
0.5	12.1	24.2	25.8	crack arrest
0.5	13.8	27.6	29.8	catastrophic failure
0.1	22.3	24.8	27.5	crack arrest
0.1	25.0	27.8	30.2	catastrophic failure
0.7	8.5	28.3	31.0	crack arrest
0.7	9.0	30.1	33.0	catastrophic failure

TABLE 5—*Comparison of experimental observations of crack growth behavior related to* K_{max} *after the first row of breached fibers* [7,17,33].

Material	K_{max} after First Row of Fibers, $MPa\sqrt{m}$	Test Result
Ti-6Al-4V/SCS6	22.5	crack arrest
$[0_8]$	28.8	catastrophic failure
Ti-β21S/SCS6	39.6	crack arrest
$[0_8]$	40.6	catastrophic failure
	45.6	catastrophic failure
Ti-15-3/SCS6	35.0	crack arrest
$[0_8]$	42.2	catastrophic failure
Ti-6Al-4V/SCS6	15.0	crack arrest
$[0/90]_{2s}$	19.8	catastrophic failure

Effects of Crack Shape and Crack Size

It is important to recognize that an approach based on K_{max} values after a row of breached fibers implicitly includes a bridging scale length argument because the crack depth increment to breach the first row of fibers will be constant for a given fiber architecture. Thus, this method of analysis can account easily for the experimental observations, Fig. 12, that a smaller initial unbridged normalized crack depth, a_0/W, can be more damaging than a larger unbridged defect when tested under an identical initial stress intensity range ΔK_{app} ($\approx$18 $MPa\sqrt{m}$). This is simply because the fixed absolute increment of crack depth required to breach the first row of fibers will produce a larger increase in K_{max} for the smaller initial crack. The approach has clear potential to produce crack arrest maps [*7,17,24,34*].

In addition, the dependence of crack arrest on the absolute number of bridging fibers has profound implications for the assessment of part-through thickness defects, since the earlier crack arrest criterion is specific to the discrete number of intact fibers bridging in the crack wake and acting over the crack front [*34*]. Because, in the general case, a part-through thickness crack may interact with fewer fibers than an equivalent through-thickness crack, then lower values of K_{max} can be tolerated. Experimental evidence in support of this concept is considered in detail elsewhere [*24,25*], but preliminary observations are given in Table 6.

Discussion

Crack Growth Resistance Curves

Several factors that may control crack growth resistance have been demonstrated experimentally. Although no unique relationship between an applied nominal stress intensity

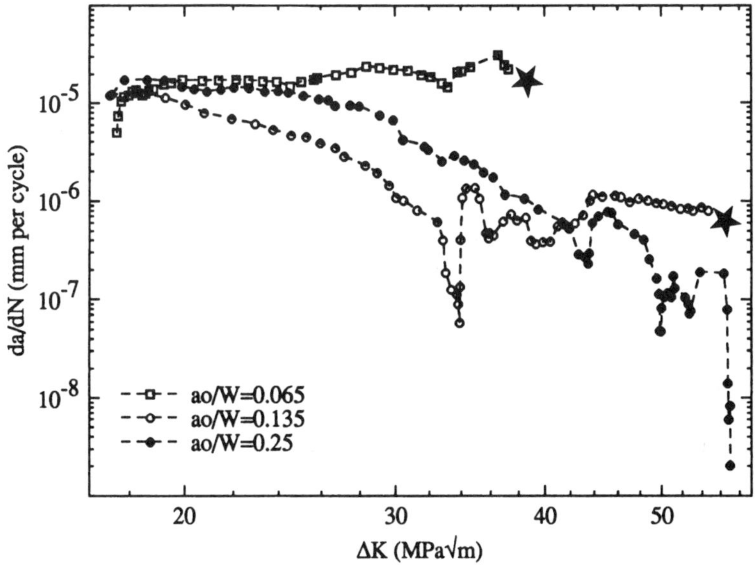

FIG. 12—*Effect of initial unbridged crack length on crack growth resistance curves:* da/dN *versus* ΔK_{app}, υ = *10 Hz, and* R = *0.5. Ti-β21S/SCS6 [0_8] composite, ambient temperature.*

factor range and crack growth rate can be defined, it is hoped that the use of parameters such as ΔK_{app} and K_{max} to characterize crack growth resistance for specific testpiece geometries has been illustrated. In particular, such nominal parameters appear to be of use to explain the influence of loading condition (bending versus tension), Fig. 4, and unbridged crack depth, Fig. 12, on fatigue crack growth resistance for specific testpiece geometrics, crack shapes, and fiber volume fractions. Indeed, it is essential that any parameter used to describe crack growth resistance must include a crack scaling dimension, and hence a stress intensity factor is a natural choice. Several models attempt to evaluate ΔK_{eff} values in the presence of fiber bridging, and the importance of this approach is clear from the measured crack growth rates of Figs. 3 through 6, 8, 10 through 12, and Table 2. Attention is now addressed to further critical factors that must be understood and incorporated into any quantitative characterization of fatigue crack growth resistance.

TABLE 6—*A comparison of crack growth stability as a function of initial stress intensity factor range and crack shape for Ti-6Al-4V/SCS6* [25] *and Ti-6Al-4V/SM1240* [24] *composites.*

Material	ΔK_{app}, $MPa\sqrt{m}$	*R* ratio	Crack Shape	Observation
Ti-6Al-4V/SCS6	18.0	0.1	through thickness	crack arrest
[0_6]	23.0	0.1	through thickness	catastrophic failure
	9.8	0.1	semi-elliptical	catastrophic failure
	6.2	0.1	semi-elliptical	crack arrest
Ti-6Al-4V/SM1240	12.1	0.5	through thickness	crack arrest
[0_6]	13.8	0.5	through thickness	catastrophic failure
	10.3	0.5	corner crack	catastrophic failure
	6.9	0.5	corner crack	crack arrest

Effects of Fiber Fracture on Crack Growth Resistance

The number of intact bridging fibers is the most important single factor in controlling crack growth resistance and crack stability in these composites. Hence, it is essential to use monitoring techniques to determine fiber fracture in situ, and fibers must be modeled as discrete entities. Experimentally, it is clear that once a fiber has broken it becomes insignificant in providing crack-tip shielding (despite average pull-out lengths for Ti-6Al-4V/SCS6 composites in bending of approximately 120 μm [*25*]), compared with the surrounding intact fibers. This, presumably occurs because a fiber fails relatively close to the crack plane (and under the local crack-tip opening displacements that generate sufficient tensile stress in the fiber to cause failure). Experimental evidence has been obtained to indicate that fibers fail at the peak of the load cycle [*32*], and it is therefore natural to consider parameters such as the local maximum crack-tip opening displacement to describe fiber failure. It is important to recognize that fiber failure is promoted by the presence of the crack and that fibers fail only within the crack wake. Thus, any quantitative description of local fiber failure must require measurements of local crack-tip displacements under load combined with detailed numerical micromodeling (because the crack is embedded in regions of local plasticity). However, for the limited number of geometrics presented here, a good approximation observed experimentally, to the onset of fiber failure (for a given number of fibers and a given testpiece size) appears to be estimated from the nominal value of K_{max} as the crack tip passes the first row of fibers, see Tables 4 and 5. (A better estimate would be given by evaluating the value of $K_{max\ eff}$ for a given situation [*35,36*], but to date such measurements have not been made). In the present study, this deduction arises from the onset of crack arrest (see Table 5) combined with in-situ monitoring of little or no fiber failure for a given situation. Thus, it may provide the beginning of a predictive model to address crack arrest.

It should be noted that the crack arrest condition is specific to the actual number of fibers bridging in the crack wake over the entire crack front. Therefore, care is required if the growing crack front interacts with a reduced number of fibers, for example, because of nonuniform crack growth through-thickness. (The crack arrest conditions given in Tables 4 and 5 apply to through-thickness cracks that grow uniformly through thickness and hence, to a first approximation, they interact with the total number of reinforcing fibers simultaneously). Such caution is exemplified by the reduced K_{max} values that are required to allow crack arrest for testpieces containing corner cracks and semi-elliptical cracks, Table 6. These arrest conditions can be rationalized if the total number of bridging fibers interacting with the (reduced) crack front is considered [*34*]. These experimental observations have clear implications for the accurate prediction of life expectancy and the use of components in service [*34*].

This method of analysis to predict the onset of crack arrest (based on the failure of the first row of breached fibers for one system and testpiece geometry), in principle, can then be applied to a wide range of fiber architectures, different crack sizes and crack shapes, and other geometries of a testpiece (component). It avoids many of the problems inherent to the modeling of actual crack growth rates prior to catastrophic failure, but further work is required to establish if such crack arrest maps are unduly pessimistic with respect to the prediction of life expectancy of actual components [*34*]. In any event, before this method of analysis can be applied with confidence to predict the influence of testpiece geometry, crack size, and crack shape on crack arrest conditions, further experimental verification is required, and this is the subject of continuing studies.

The observation of fiber fracture after extended cycling in near crack arrest regions is observed commonly. Since, in such cases, the crack growth increments are very small, it would appear to suggest that the fibers themselves undergo some form of fatigue damage.

The most likely mechanisms would relate to changes in fiber-matrix interfacial regions on cycling. Clearly, it is possible that wear damage during cycling will impair the outermost fiber-coating systems. If these coatings impart a beneficial residual stress distribution to the fiber surface, then a reduction in effective fiber fracture strength will result. It also has been shown using repeated push-out tests [29] that the effective frictional stress after debonding decreases progressively. Plausibly, if such results can be translated to cyclic fatigue, this would suggest that to maintain the same bridging traction force across the crack surfaces, the fiber must then suffer a greater displacement, and a new debonded length will be created. Thus, more volume of the fiber will be "sampled" and a lower effective fiber fracture stress may result. Further work is required to establish such possible mechanisms of fiber strength degradation.

Optimization of Fatigue Crack Growth Resistance

Potential routes for optimization of a composite of a given fiber and fiber architecture include manipulation of fiber-matrix interfacial properties and microstructural control of the matrix alloy. The merit of such approaches can perhaps be deduced from the crack arrest conditions given in Table 5 for similar fiber volume fractions and testpiece geometries that vary markedly with different matrix alloys. In this context, the experimental determination of crack arrest appears to provide a useful basis for comparison between different composite systems: such comparisons would be impossible to make from single tests performed on different composites.

Defining an optimized interfacial strength appears problematic, at present, given that debonding occurs ahead of the advancing matrix crack tip. In some tests on β21S/SCS6 composites, an increase in interfacial strength on peak aging has led to a reduction in the number of cycles to failure (that is, in a regime where fiber failure occurs for both conditions), but conditions for crack arrest are similar [29]. If a reduction in interfacial strength is to be beneficial, then this must be as a result of less efficient stress transfer in near-crack-plane regions. Thus, fibers are less highly stressed and are therefore less likely to fail. However, if this less efficient crack bridging promotes increased local crack-opening displacements, then it is plausible that the fibers could be more highly displaced and the net effect could be highly stressed fibers once again. Indeed, this argument may have merit when tests at elevated temperatures are performed because, in general, increased crack growth resistance is not obtained at such temperatures despite a reduced fiber-matrix interfacial strength. Possible complications that arise in such tests at elevated temperatures include environmental factors, load shedding from the matrix (for tests performed under load control) onto the fibers (as the matrix strength reduces), and decreased matrix crack growth resistance. Hence, further work is required to establish mechanisms of crack growth resistance at elevated temperatures, although initial studies suggest that test temperatures of 300°C can be tolerated easily in existing systems.

Increasing the fatigue crack growth resistance of the matrix alloy by microstructural manipulation should improve the crack growth resistance of the composite. Indeed, crack arrest ($\leq 10^{-8}$ mm/cycle) results from a reduction in stress intensity factor range below the long crack "matrix" threshold stress intensity factor range, ΔK_{th}, and this is sensitive to microstructure. Near threshold crack growth rates in the matrix alloy can also be reduced. To achieve the required degree of microstructural control, heat treatment subsequent to composite consolidation is likely to be required, and this will usually involve temperatures significantly higher than those used for consolidation. Stability of the fiber coating system at such temperatures is therefore required and this is the subject of ongoing studies.

In summary, the degree of complexity inherent to crack growth resistance in these materials should not be underestimated. Effects of interfacial strength, mean stress, testpiece geometry, and initial applied stress intensity factor range on crack growth resistance under cyclic loading will always be significant if they alter the balance of the number of bridging fibers that fail for any given crack depth increment. Great care is needed if fracture mechanics concepts are to be applied to the reliability of components containing a range of crack shapes and sizes, and the approach will require both predictive modeling and experimental validation.

Conclusions

Dominant localized fatigue cracks are produced from unbridged defects in fiber-reinforced titanium alloy composites subjected to cyclic loading. The growth of these cracks can be characterized by the use of fracture mechanics parameters, and their experimental measurement has been considered. Crack growth resistance is controlled by the number of intact fibers bridging behind the growing fatigue crack tip in its wake. Effects of crack shape, crack size, test temperature, interfacial strength, applied stress intensity factor range, mean stress, and testpiece geometry on crack growth resistance will be marked if the balance of the number of bridging fibers that fail for a given increment of matrix crack growth is altered. For specific testpiece geometries and fiber architecture, conditions for crack arrest can be predicted.

Acknowledgments

The author is indebted to many members of the fracture and fatigue group at the University of Birmingham. Particular thanks are due to D. C. Cardona (now at Rolls-Royce plc); F. Brisset, P. J. Cotterill, A. R. Ibbotson, J. F. Knott, M. Strangwood, and K. Takashima (Kumamoto University, Japan) together with C. Barney, K. M. Fox, and J. G. Pursell. All studies have been supported by EPSRC (Engineering and Physical Sciences Research Council), with additional funding from Rolls-Royce plc, Derby, and BP Research Limited, Sunbury. The support of M. V. Hartley at Rolls-Royce plc has been invaluable to many of the programs. The provision of facilities both within the Interdisciplinary Research Centre in Materials for High Performance Applications and the School of Metallurgy and Materials is also gratefully acknowledged.

References

[*1*] Johnson, W. S., *Mechanical Fatigue of Advanced Materials,* R. O. Ritchie et al., Eds., MCE Publications, Santa Barbara, Jan. 1991, pp. 357–377.
[*2*] Walls, D., Bao, G., and Zok, F., *Mechanical Fatigue of Advanced Materials,* R. O. Ritchie et al., Eds., Santa Barbara, Jan. 1991, pp. 343–356.
[*3*] Barney, C., Beevers, C. J., and Bowen, P., *Journal of Composites,* Vol. 24, 1993, pp. 229–234.
[*4*] Bowen, P., Ibbotson, A. R., and Beevers, C. J., *Mechanical Fatigue of Advanced Materials,* R. O. Ritchie et al., Eds., MCE Publications, Santa Barbara, Jan. 1991, pp. 379–394.
[*5*] Bowen, P., Cotterill, P. J., and Ibbotson, A. R., *Test Techniques for Metal Matrix Composites,* N. D. R. Goddard, Ed., Institute of Physics Publishing, London, 1990, pp. 82–96.
[*6*] Bowen, P., *The Processing, Properties and Applications of Metallic Materials,* M. H. Loretto and C. J. Beevers, Eds., MCE Publications, Birmingham, Sept. 1992, pp. 697–718.
[*7*] Bowen, P., "Characterisation of Fibre Reinforced Titanium Matrix Composites," AGARD Report 796, Paper 18.1–16, The 77th Meeting of the AGARD Structures and Materials Panel, Bordeaux, France, Sept. 1993.
[*8*] Budiansky, B., Hutchinson, J. W., and Evans, A. G., *Journal of the Mechanics and Physics of Solids,* Vol. 34, 1986, pp. 167–189.
[*9*] Marshall, D. B., Cox, B. N., and Evans, A. G., *Acta Metallurgica,* Vol. 33, 1985, pp. 2013–2021.
[*10*] Marshall, D. B. and Cox, B. N., *Acta Metallurgica,* Vol. 35, 1987, pp. 2607–2619.

[11] McMeeking, R. M. and Evans, A. G., *Mechanics of Materials,* Vol. 9, 1990, pp. 217–227.
[12] Cardona, D. C., Barney, C., and Bowen, P., *Journal of Composites,* Vol. 24, 1993, pp. 122–128.
[13] Bowen, P., Cardona, D. C., and Ibbotson, A. R., *Materials Science and Engineering A,* Vol. 153, 1992, pp. 628–634.
[14] Cardona, D. C. and Bowen, P., *Theoretical Concepts and Numerical Analysis of Fatigue,* A. F. Blom and C. J. Beevers, Eds., EMAS West Midlands, Birmingham, UK, May 1992, pp. 505–522.
[15] Cox, B. N., *Acta Metallurgica et Materials,* Vol. 39, 1991, pp. 1189–1201.
[16] Cox, B. N. and Lo, C. S., *Acta Metallurgica et Materials,* Vol. 40, 1992, pp. 69–80.
[17] Barney, C., Cardona, D. C., and Bowen, P., *Proceedings,* ICCM9, Madrid, Vol. 1, A. Miravete, Ed., University of Zaragoza/Woodhead Publishing Limited, July 1993, pp. 541–548.
[18] Cotterill, P. J. and Bowen, P., *Journal of Composites,* Vol. 24, 1993, pp. 241–247.
[19] Ibbotson, A. R., Beevers, C. J., and Bowen, P., *Journal of Composites,* Vol. 24, 1993, pp. 241–247.
[20] Fox, K. M., Cotterill, P. J., Strangwood, M., and Bowen, P., *Titanium '92, Proceedings,* 7th World Conference on Titanium, Vol. III, F. H. Froes and I. L. Caplan, Eds., The Metals Society, San Diego, July 1992, pp. 2649–2658.
[21] Fox, K. M. and Bowen, P., *Proceedings,* ICCM9, Madrid, Vol. 1, A. Miravete, Ed., University of Zaragoza/Woodhead Publishing Limited, July 1993, pp. 675–682.
[22] Pursell, J. G., Takashima, K., and Bowen, P., *Proceedings,* Third Japan International SAMPE Symposium, Vol. 2, Chiba, Japan. T. Kishi, N. Takeda, and Y. Kagawa, Eds., Japan Chapter of Sampe, Tokyo, Dec. 1993, pp. 1818–1825.
[23] Ibbotson, A. R., Beevers, C. J., and Bowen, P., *Fatigue '93,* Vol. II, J. P. Bailon and J. I. Dickson, Eds., Montreal, EMAS, West Midlands, UK, May 1993, pp. 1079–1084.
[24] Barney, C., Cardona, D. C., and Bowen, P., *Proceedings,* Third Japan International SAMPE Symposium, Vol. 2, Chiba, Japan, T. Kishi, N. Takeda, and Y. Kagawa, Eds., Japan Chapter of Sampe, Tokyo, Dec. 1993, pp. 1830–1835.
[25] Ibbotson, A. R., ''Fatigue Crack Growth in Continuous Fibre Reinforced Titanium Alloy Matrix Composites,'' PhD thesis, The University of Birmingham, March 1994.
[26] Barney, C., ''Fatigue Crack Growth Resistance of Continuous Fibre Reinforced Titanium Matrix Composites in the Presence of Unbridged Defects,'' PhD thesis, The University of Birmingham, Jan. 1995.
[27] Ibbotson, A. R., Beevers, C. J., and Bowen, P., *Scripta Metallurgica et Materials,* Vol. 25, 1991, pp. 1781–1786.
[28] Takashima, K., Fox, K. M., and Bowen, P., *Proceedings,* Third Japan International SAMPE Symposium, Vol. 2, Chiba, Japan, T. Kishi, N. Takeda, and Y. Kagawa, Eds., Japan Chapter of Sampe, Tokyo, pp. 2365–2370.
[29] Fox, K. M., Strangwood, M., and Bowen, P., *Journal of Composites,* accepted for publication.
[30] Brisset, F. and Bowen, P., *Proceedings,* ICM 8, The Hague, May 1995.
[31] Fox, K. M., ''Effects of Interfacial Properties on Fatigue Crack Growth Resistance in SiC Fibre Reinforced Titanium Metal Matrix Composites,'' Ph.D thesis, The University of Birmingham, Oct. 1994.
[32] Cotterill, P. J. and Bowen, P., *Materials Science and Technology,* submitted for publication.
[33] Bowen, P., *Proceedings,* Vol. 1, ICCM9, Madrid, A. Miravete, Ed., University of Zaragoza/Woodhead Publishing Limited, July 1993, pp. 585–593.
[34] Cardona, D. C., Barney, C., and Bowen, P., ''Modeling and Prediction of Crack Arrest and Fiber Reinforced Composites,'' in this volume.
[35] Ritchie, R. O., private communication, 1994.
[36] Ritchie, R. O., Yu, W., and Bucci, R. J., *Engineering Fracture Mechanics,* Vol. 32, No. 3, 1989, pp. 361–377.

P. Kennard Wright[1]

A Hot-Spot Thermal Fatigue Test of a Titanium Matrix Composite

REFERENCE: Wright, P. K., **"A Hot-Spot Thermal Fatigue Test of a Titanium Matrix Composite,"** *Life Prediction Methodology for Titanium Matrix Composites ASTM STP 1253,* W. S. Johnson, J. M. Larsen, and B. N. Cox, Eds., American Society for Testing and Materials, 1996, pp. 480–493.

ABSTRACT: A thermal fatigue test was designed and executed based on heating and cooling of a local spot in a thin titanium matrix composite (TMC) sheet by quartz lamps. This approach offers two features not offered by the load-controlled testing of standard coupons: (1) the stress state is biaxial instead of uniaxial, and (2) the mean stress is primarily compressive instead of tensile. A test rig was constructed to subject a small spot in the center of a 51- by 51-mm sheet of TMC to alternate heating and cooling between room temperature (RT) and 615°C. A specimen of unidirectionally reinforced SCS-6/Ti-6-2-4-2 was subjected to cycling with intermittent inspection for damage. Minimal cracking was observed in this test even after 10 000 cycles. Load-controlled low cyclic fatigue (LCF) and thermomechanical fatigue (TMF) tests were also performed on this material. The observed life in the hot-spot test was significantly greater than uniaxial transverse tensile fatigue load-controlled data would suggest based on a thermal stress analysis of the test panel. However, it was in reasonably good agreement with longitudinal, uniaxial load-controlled out-of-phase TMF data.

KEYWORDS: titanium matrix composites, fatigue (materials), thermal fatigue, titanium matrix, silicon-carbide fibers, finite element analysis, scanning acoustic microscopy, titanium, life prediction, titanium alloys, modeling

Because of its high specific strength and stiffness, silicon-carbide (SiC) reinforced Ti-6Al-2Sn-4Zr-2Mo (Ti-6-2-4-2) composite is being considered for structural application to components in aircraft engines. These applications tend to involve localized cyclic heating and cooling that may subject them to thermal fatigue cracking. Titanium matrix composites (TMCs) are known to be susceptible to thermal fatigue [*1*], partly because of the thermal expansion mismatch between the fibers and matrix and partly because of sensitivity of titanium matrices to oxygen pickup at high temperatures and subsequent embrittlement at low temperatures.

Up to this point, studies of thermal fatigue in TMCs have been conducted using uniaxial, straight-sided test coupons in closed-loop hydraulic machines with isothermal or cyclic temperatures (TMF) [2]. These kinds of tests show a strong dependence of life and failure mode on strain-temperature phasing and material orientation. Thus, proper simulation of the actual thermal fatigue conditions in an engine component will probably be required in order to achieve an accurate assessment of TMC thermal fatigue resistance.

To try to get a more realistic simulation of thermal fatigue in a locally heated and cooled TMC structure, a thermal fatigue test was designed based on spot heating of a thin TMC panel by quartz lamps. This approach offers two special features not offered by the thermomechanical fatigue (TMF) approach:

[1]Principal engineer, GE-Aircraft Engines, Cincinnati, OH 45215.

1. The stress state biaxial instead of uniaxial. Most components undergoing thermal fatigue do experience some degree of biaxiality since the hottest locations tend to be localized within stiffer, cooler material. This produces thermal stresses in all directions. Since TMCs are anisotropic, the effect of the biaxial stress field may be significant.
2. The stress state in the spot-heated test consists of a large amount of compression. Nearly all uniaxial test specimen work has been under tension loading, since it is difficult to load thin sheet material in uniaxial compression without buckling. However, compression static strengths of TMC are known to be much higher (about ×2) than tensile strengths [3] and it might be expected that the same would be true for fatigue. The locally heated panel test produces compression because as the center spot is heated it tries to expand and is compressed by the cooler surrounding material.

Experimental Approach

A rig was constructed at GE-Aircraft Engines to conduct thermal cyclic fatigue testing to evaluate cracking response under biaxial thermal stress. The rig used two 750-W quartz lamps with elliptical focusing reflectors to heat from both sides a 6-mm-diameter spot at the center of a 51- by 51-mm by 0.88-mm-thick TMC panel. A heating rate of 85°C/s was achieved (room temperature (RT) to 615°C in 7 s). A schematic of this rig is given in Fig. 1.

Upon reaching peak temperature in the hot spot, the panel was forced air cooled by a side fan to room temperature in about 45 s. Automatic controls were used to switch the air source

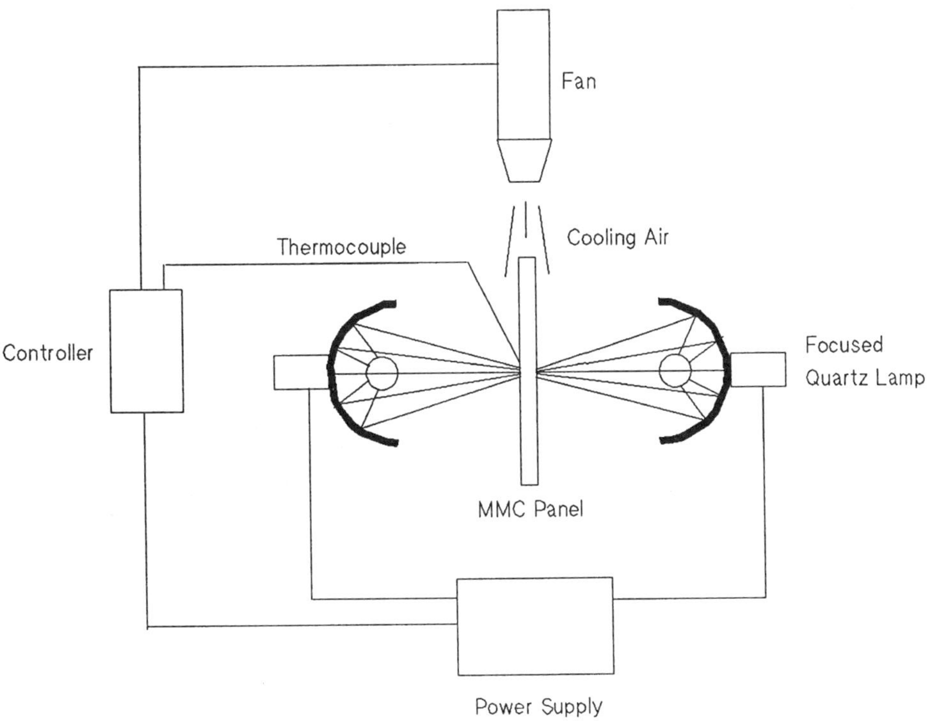

FIG. 1—*Schematic diagram of hot-spot thermal fatigue apparatus.*

and quartz lamps based on temperatures programmed into the controls. K-type thermocouples were used to control and monitor temperature levels in the panel and to protect against over-temperature.

A 51- by 51-mm square sample was cut from a unidirectionally reinforced 39% fiber fraction (V_f) SCS-6/Ti-6-2-4-2 panel and was instrumented with thermocouples in the panel center, 3 mm from the center, 12.5 mm from the center, and at the edge of the panel as shown in Fig. 2. Temperatures were measured as a function of time at each location to give the temperature distribution around the panel at various time points in the cycle. Heating and cooling parameters were adjusted to give the time-temperature profiles shown in Table 1 and Fig. 3. The panel was loosely supported at two corners with welded wires to minimize any external mechanical constraint on the panel.

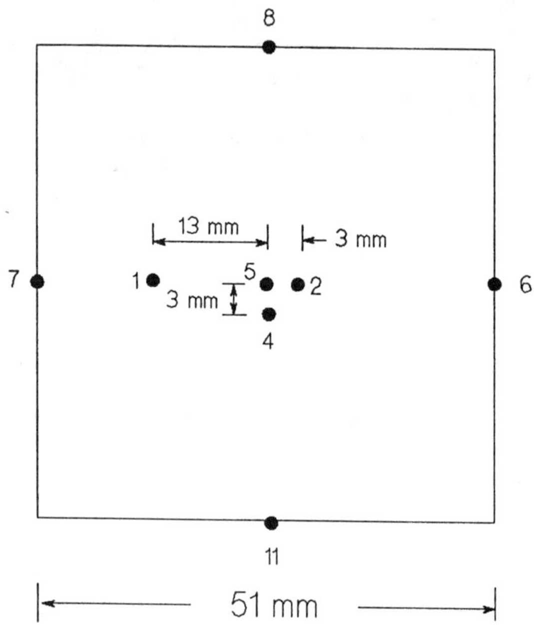

FIG. 2—*Locations of thermocouples on MMC panel.*

TABLE 1—*Time-temperature profiles for hot-spot test.*

Location (Fig. 2)	Time, s							
	0	7	9	11	15	19	27	45
1	21	268	263	221	132	85	46	21
2	21	527	477	399	229	121	46	21
4	21	560	499	413	213	127	46	21
5	21	613	488	435	302	166	66	21
6	21	168	154	115	52	32	25	21
7	21	171	174	138	110	85	46	21
8	21	216	207	182	127	85	43	21
11	21	213	204	182	121	82	46	21

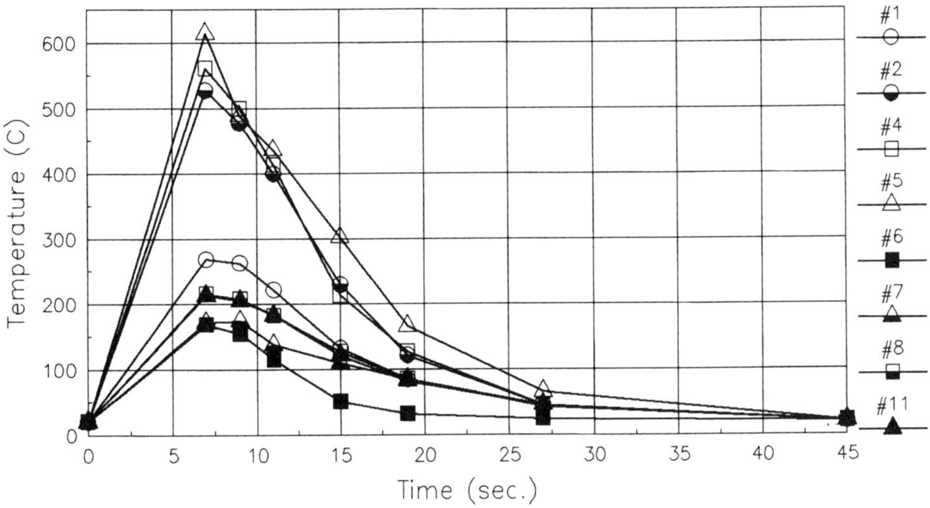

FIG. 3—*Time-temperature cycle at indicated thermocouple locations on metal-matrix composite (MMC) panel (see Fig. 2 for locations).*

Specimen Analysis

The thermal stresses were calculated using a two-dimensional ANSYS finite element analysis of the panel with quadrilateral elements. The temperature distribution in the panel was assumed to be constant through the panel thickness, and the temperature distribution from Table 1 was expanded as shown in Fig. 4 to cover the entire surface of the panel. Figure 4 makes more evident the slight directionality in temperature distribution that can be seen from inspection of Table 1. At a given distance from the panel center, the panel was hotter in the fiber or longitudinal direction than it was in the transverse direction. This is in qualitative agreement with measurements of thermal conductivity in SiC reinforced TMCs that show higher conductivity parallel to the fibers than perpendicular to them.

The finite element model was used with these temperature distributions to perform a two-dimensional elastic thermal plane stress analysis on the panel, with temperature-dependent homogeneous thermal expansion and elastic constants for SCS-6/Ti-6-2-4-2 as shown in Table 2. One corner of the panel was fixed with the other edges stress free and unconstrained. Since the analysis was two-dimensional, no out-of plane stresses were calculated. Out-of plane stresses would not be expected in a thin plate heated or cooled with no through-thickness temperature gradient. No observations of panel warping during testing were made. Plasticity effects were not incorporated because ANSYS does not support an anisotropic plasticity analysis. For the temperature distribution corresponding to the peak hot-spot temperature, the longitudinal and transverse normal stresses as well as the in-plane shear stresses were determined and are shown in Fig. 5. For this condition, the maximum compressive stresses in the longitudinal and transverse direction occur in the center of the panel at the hot spot, and are −230 MPa and −150 MPa, respectively. The maximum shear stress occurs away from the center and is quite small, 60 MPa. The maximum tensile stresses are at the edges of the panel and are about the same magnitude as the compressive stresses: +170 MPa longitudinal and +270 MPa transverse. These represent the maximum (absolute magnitude) stresses in the cycle, while the minimum stresses are at the coolest part of the cycle and are approximately zero.

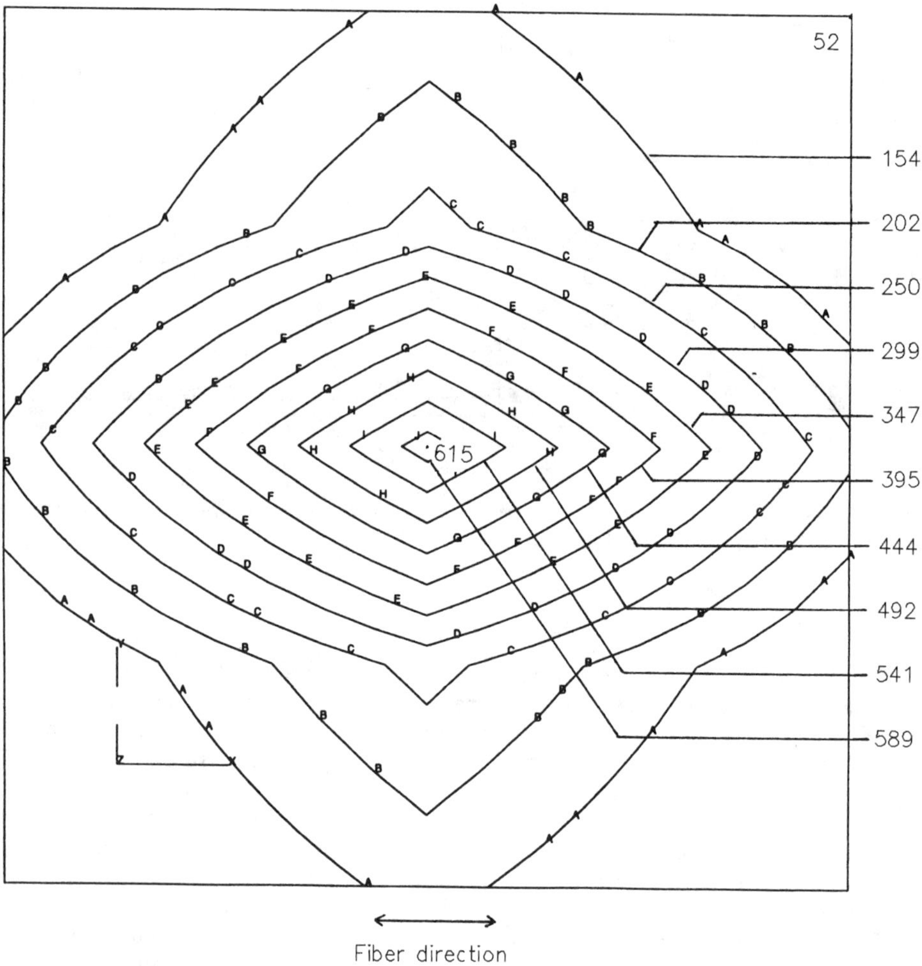

FIG. 4—*Temperature distribution at peak temperature used for finite element analysis.*

TABLE 2—*Thermal expansion and elastic moduli of SCS-6/Ti-6-2-4-2 as a function of temperature.*

Temperature, °C	α_L, 10^{-6}/°C	α_T, 10^{-6}/°C	E_L, GPa	ν_{LT}	E_T, GPa	G_{LT}, GPa
RT	5.9	7.9	188	0.26	138	52
204	5.9	7.9	183	0.27	138	50
427	6.5	8.3	172	0.29	124	42
640	6.8	9.0	152	0.31	94	28

This cycle is "out-of-phase" in TMF terminology: the highest (most tensile) stress occurs at minimum temperature and the lowest (most compressive) stress occurs at peak temperature. This is typical of many critical locations in thermal fatigue. Unlike the uniaxial specimen data available to date, the hot-spot cycle runs from zero to maximum compression rather than from maximum tension to zero. (In actuality, there may be some yielding at high temperature, especially in the transverse direction that would cause the stress at minimum

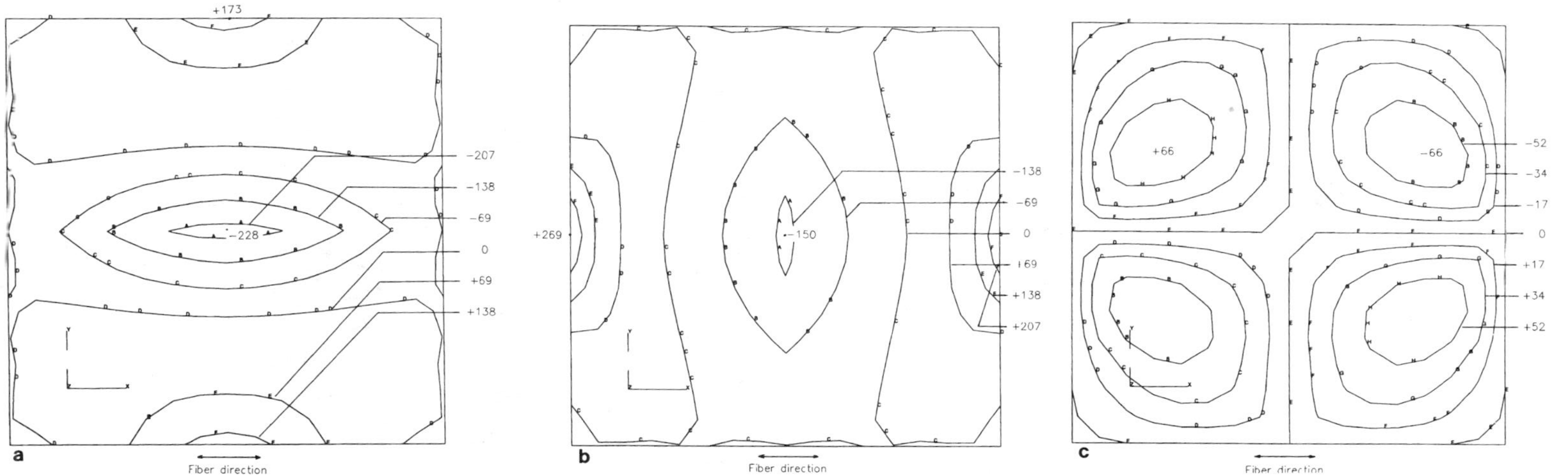

FIG. 5—*Stresses calculated in MMC panel by finite element analysis:* (a) *longitudinal stress (stress in MPa),* (b) ***transverse*** *stress (stress in MPa), and* (c) *in-plane shear stress (stress in MPa).*

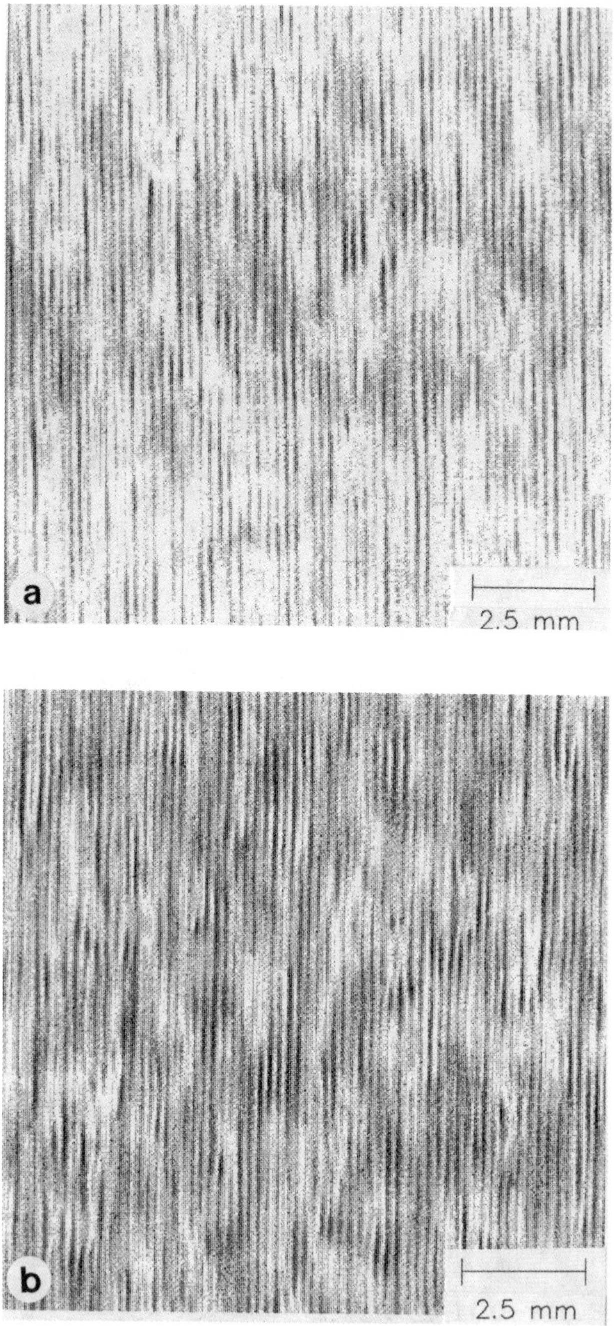

FIG. 6—*Scanning acoustic microscopy image of MMC panel prior to testing (longitu-nal wave scan):* (a) *outer ply, Side T and* (b) *outer ply, Side B.*

temperature to become positive. However, the stress ranges calculated should be unaffected.)

Results

The sample was inspected using scanning acoustic microscopy prior to testing, and uniform, undamaged material was seen, Fig. 6. The sample was subjected to 100 cycles, RT to 615°C, and re-inspected. These scans, shown in Fig. 7, showed the presence of the four thermocouples used to monitor and control temperature, but there are no clear indications of cracking away from these thermocouples. Visual inspection also showed no cracking. The sample was returned to the test rig and run an additional 9900 cycles (10 000 total). Optical examination at low magnification (×5 to ×25) showed a smooth, blue and straw colored spot on the panel center (Fig. 8) with a few small surface cracks extending away from the thermocouple welds on the B side (not visible in Fig. 6). Scanning acoustic microscopy, Fig. 9, showed indications at the welds, as well as several away from the welds when using a shear wave mode. The B side showed significantly more indications than the T side.

A section was cut from the center of the spot and examined by scanning electron microscopy and optical metallography. In confirmation of the acoustic microscopy inspections, several small cracks were seen at the thermocouple weld locations, Fig. 10, and a

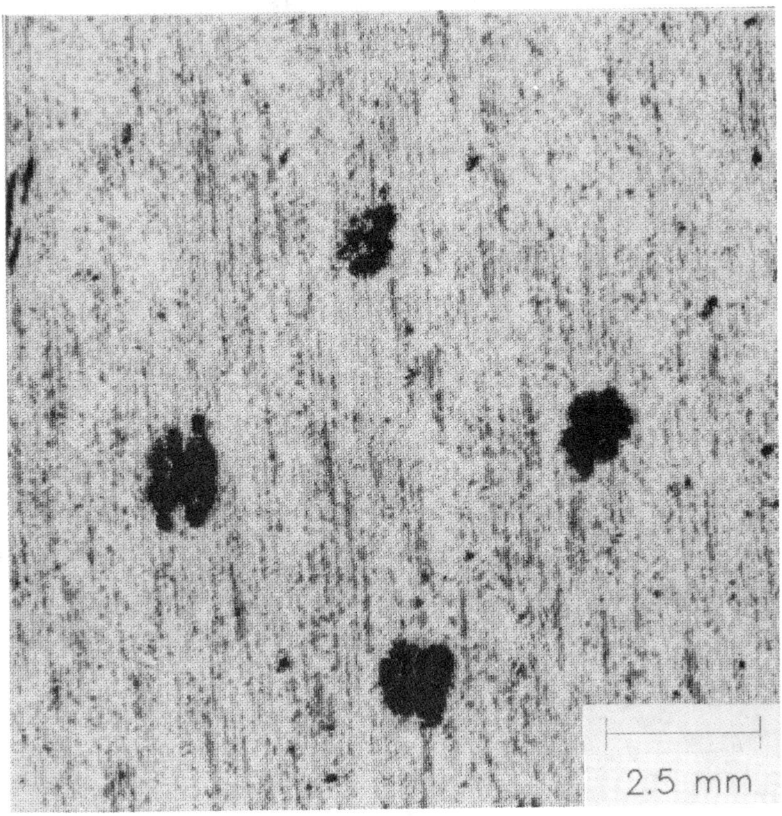

FIG. 7—*Scanning acoustic microscopy image of MMC panel after 100 cycles (shear wave scan perpendicular to fibers, outer ply, Side T).*

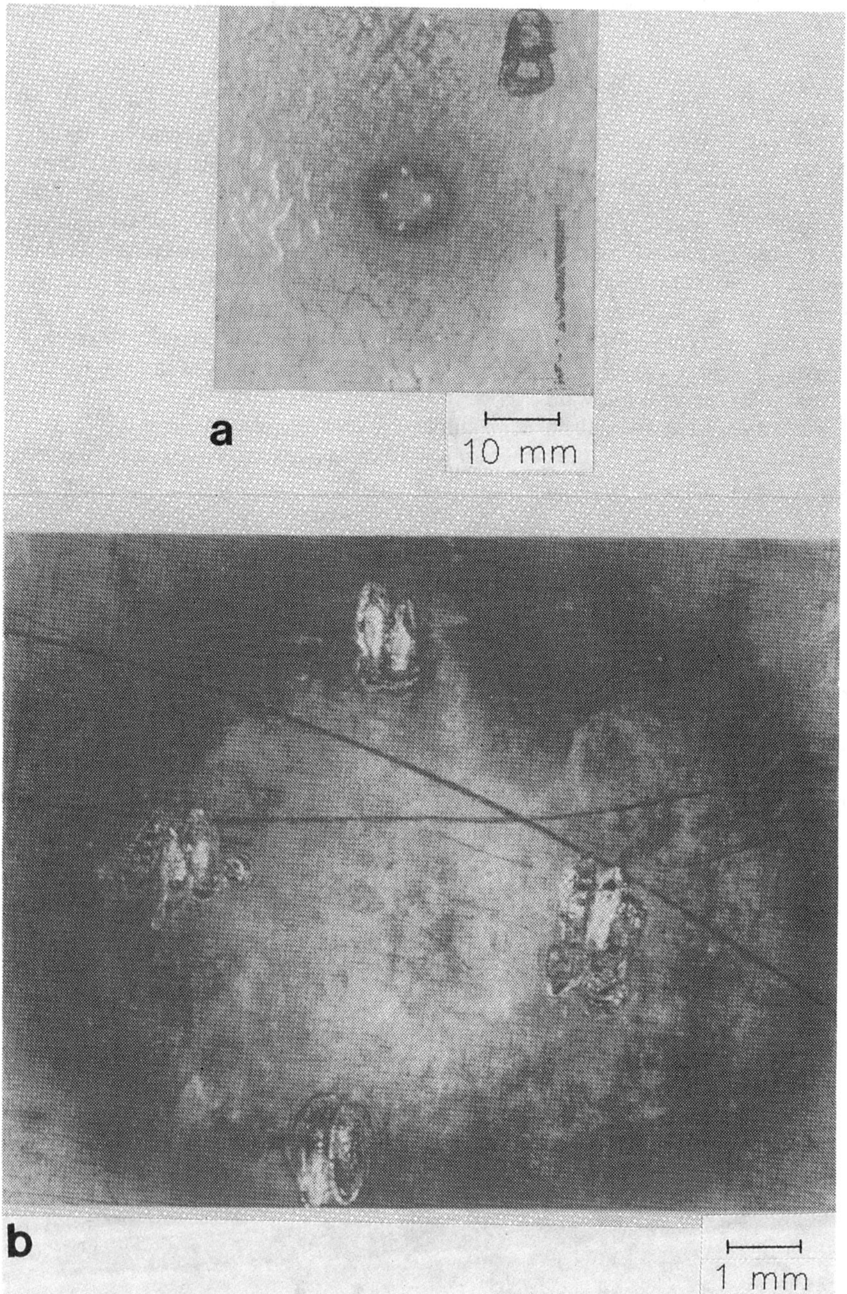

FIG. 8—*Visual condition of MMC panel after 10 000 cycles, Side B.*

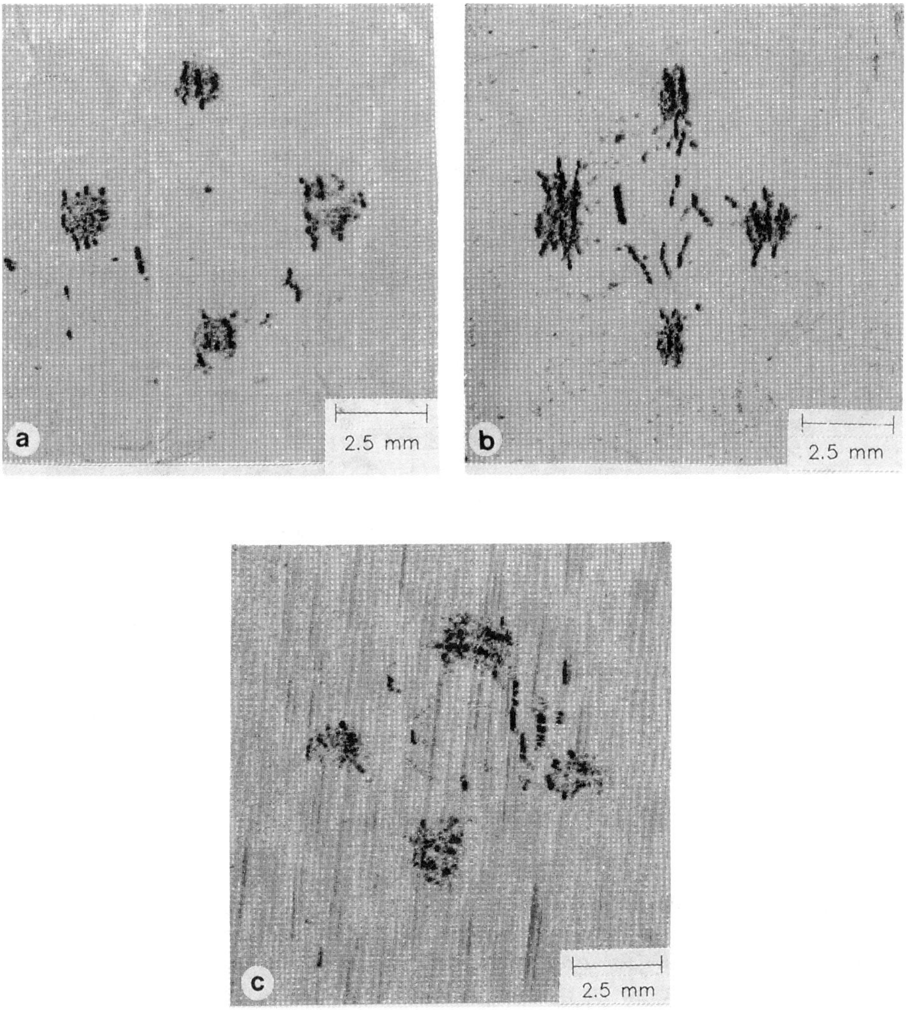

FIG. 9—*Scanning acoustic microscopy image of MMC panel after 10 000 cycles:* (a) *shear wave scan parallel to fibers, outer ply, Side T;* (b) *shear wave scan parallel to fibers, outer ply, Side B; and* (c) *shear wave scan perpendicular to fibers, outer ply, Side B.*

group of small cracks were found (on the B side only) in the center of the spot, Fig. 11. These cracks appeared to be principally but not entirely perpendicular to the fibers. Little or no evidence of oxidation of the cracks or the fiber-matrix interface was observed. In summary, the panel survived 10 000 cycles from RT to 615°C with only minor damage in the form of matrix cracks 1 mm long by 0.3 mm deep.

Uniaxial load-controlled fatigue data were acquired from a separate panel of unidirectionally reinforced 30% V_f SCS-6/Ti-6-2-4-2. These tests were run at $R = 0.05$ (near zero to maximum tension) at RT, 650°C, and in out-of phase TMF from 93 to 650°C (maximum load at minimum temperature). Tests were performed in the longitudinal and transverse panel directions. These tests results are presented in Table 3.

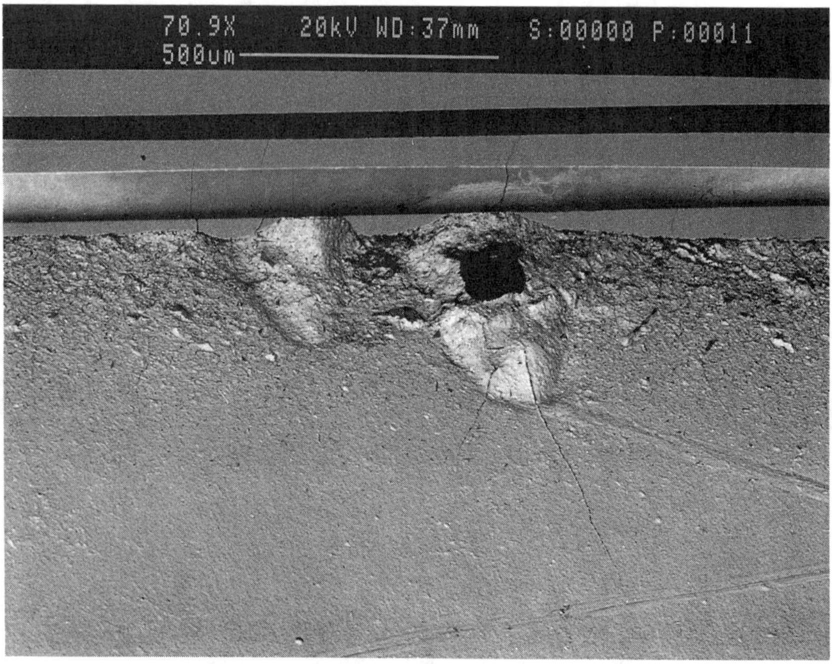

FIG. 10—*Scanning electron microscope image showing cracking emanating from thermocouple welds on surface.*

Discussion

The hot-spot behavior can be compared with various other measures of TMF capability obtained from uniaxial tests. Figure 12 shows the calculated longitudinal stress range of 230 MPa plotted on the isothermal and TMF data for longitudinal material from Table 3. The estimated stress places the coupon well below projected failure conditions based on longitudinal isothermal data at RT or 615°C. Extrapolation of the two longitudinal out-of-phase TMF points, however, suggests that failure in the longitudinal orientation (cracks perpendicular to the fibers) might be expected in as short as 2500 cycles. This prediction is in reasonable agreement with the observation of limited cracking perpendicular to the fibers in the hot-spot test. While this requires a long extrapolation of the two existing TMF test points, other out-of-phase TMF data on TMC [*4*] suggest that the fatigue curves for this condition do indeed proceed nearly linearly down in log-log coordinates. The similarity between the out-of-phase TMF and the hot-spot lives may be fortuitous, but it should be noted that the two test conditions have a high degree of similarity: they both have minimum cyclic stress at peak cycle temperature and maximum cyclic stress at minimum cycle temperature. They do differ in the mean stress applied; that is, the out-of-phase TMF test cycles between zero and tension and the hot-spot test cycles between compression and zero. Uniaxial out-of phase TMF tests should be performed using a compression-to-zero cycle to determine if the mean stress makes a significant difference.

Similarly, comparisons of fatigue life for the hot spot test can be made to the transverse uniaxial specimen tests. Figure 13 shows the hot-spot result (plotted at 150 MPa, the transverse stress range) along with transverse uniaxial data at RT, 650°C, and 93 to 650°C out-of-phase TMF. The 650°C low cycle fatigue (LCF) results fall far below the observed

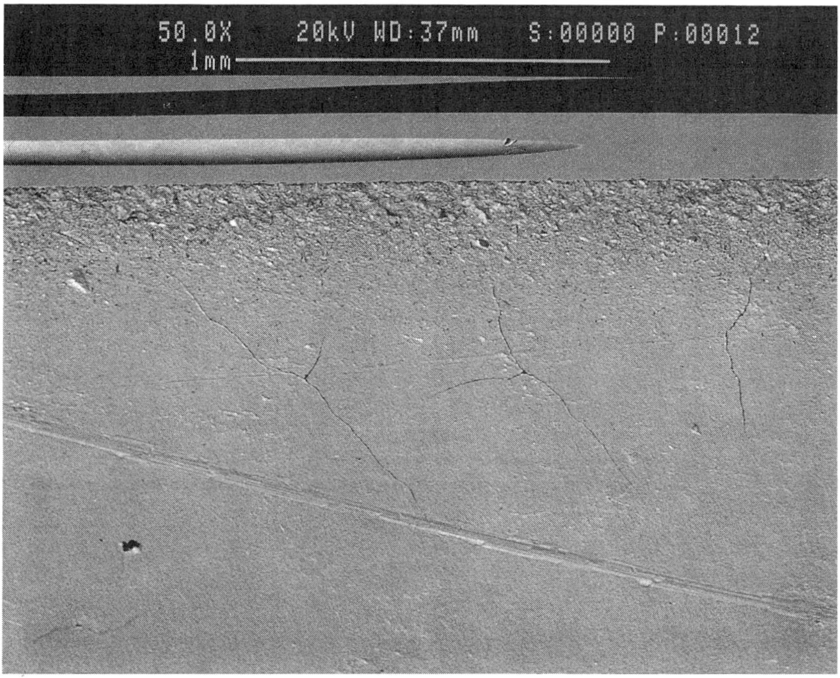

FIG. 11—*Scanning electron microscope image showing cracking in center of hot spot.*

TABLE 3—*Fatigue test results from uniaxial coupons of SCS-6/Ti-6-2-4-2 load controlled,* R = *0.05, 0.5 Hz.*

	Room Temperature		650°C		TMF[a]	
	Stress Range, MPa	Cycles to Fail	Stress Range, MPa	Cycles to Fail	Stress Range, MPa	Cycles to Fail
Longitudinal	1034	2 130	689	1 695	689	395
	689	4 377	483	25 641	483	741
	483	268 128	345	472 721		
Transverse	276	4 250	276	27	138	1181
	207	10 172	138	786		
	131	>1 230 824	69	4 359		

[a]93 to 650°C out-of-phase, 0.00167 Hz.

hot spot life. The single 93 to 650°C out-of-phase TMF data point suggests that there is little difference between it and the 650°C LCF behavior. The RT LCF curve falls closer, but above the hot-spot test result.

One significant difference between the uniaxial specimen results and the hot-spot test is the sign of the mean stress: compression in the hot spot and tension in the uniaxial specimen tests. In the transverse orientation, fatigue damage of SCS-6 reinforced TMCs under tension loading is observed to originate by separation at the fiber-matrix interface due to the weak

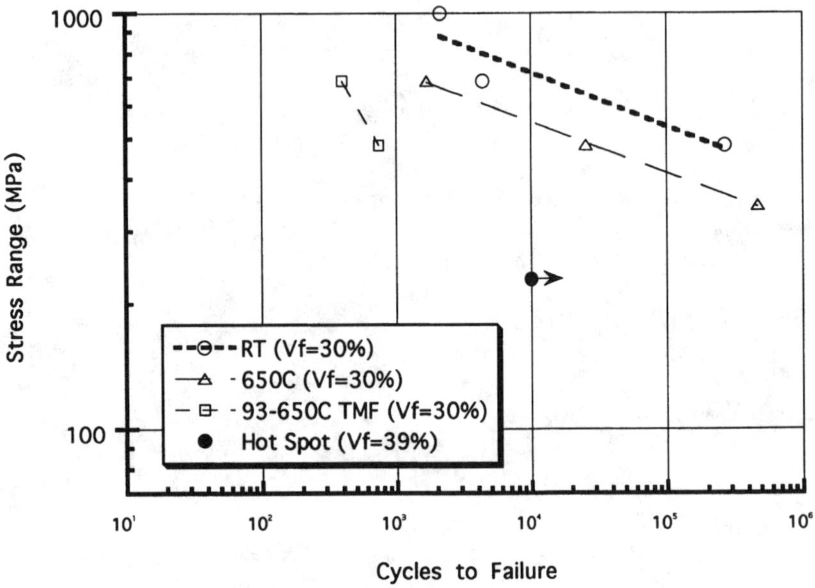

FIG. 12—*Longitudinal fatigue capability of SCS-6/Ti-6-2-4-2 in comparison to MMC hot-spot panel longitudinal stress.*

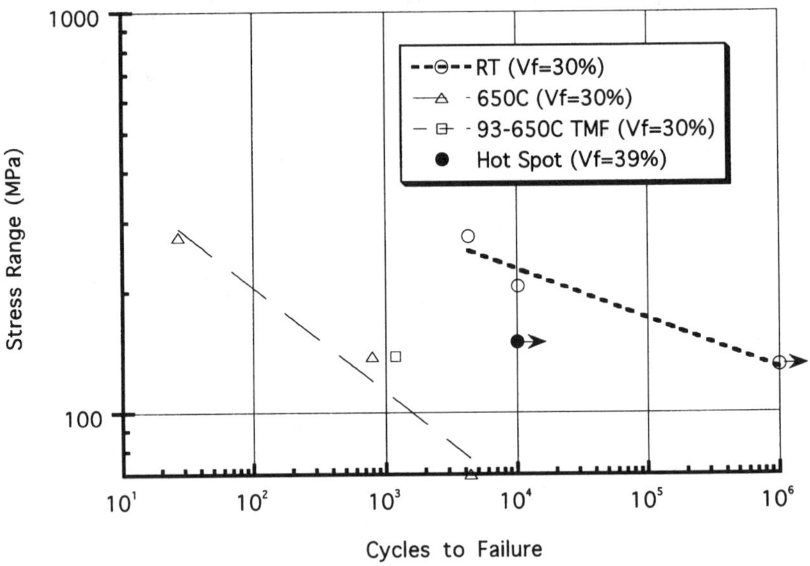

FIG. 13—*Transverse fatigue capability of SCS-6/Ti-6-2-4-2 in comparison to MMC hot-spot panel transverse stress.*

interfacial bonding [*5*]. Subsequently, cracks propagate through the remaining matrix ligaments between the fibers. No interfacial fiber-matrix debonding was observed in the hot-spot test. Under compressive loading, this failure mode is likely to be suppressed, and indeed, in static compression tests, SCS-6 reinforced TMCs do not debond at the interface and show approximately twice the strength in compression compared to tension [*3*]. It is likely that a significant improvement would also be observed in compression fatigue.

Besides the mean stress differences discussed earlier, other factors that need to be considered in attempting to apply the coupon fatigue data to predict the panel hot-spot cracking are: cycle frequency (relaxation and environmental effects), fiber volume fraction differences, the biaxial stress state, the control mode (stress versus strain), and the definition of failure (observation of matrix cracking versus specimen fracture).

Summary and Conclusions

A simple test rig was built to subject a TMC panel to cyclic spot heating to evaluate thermal fatigue resistance. One test was run on unidirectional SCS-6/Ti-6-2-4-2 material to 10 000 cycles with only relatively small cracks developing at the hot spot primarily perpendicular to the fibers. Despite the expectation from the uniaxial coupon test data, the longitudinal properties appear to be more limiting in the panel hot spot than the transverse properties. The observed life was significantly longer than uniaxial transverse tensile elevated temperature or TMF fatigue load-controlled data would suggest based on a thermal stress analysis of the test panel. However, it was in reasonably good agreement with expectations from longitudinal uniaxial load-controlled out-of-phase TMF data. Due to the complexity of TMC failure modes, considerable care should be used when applying laboratory coupon test results to component life prediction to ensure that comparable failure modes are being evaluated. Further uniaxial coupon tests at more simulative loading conditions should be performed to better assess the effects of the many variables involved in the life of a hot-spot test.

Application of these results to life prediction using various fatigue models is encouraged.

Acknowledgments

The author wishes to express his gratitude to David Veit for assistance in developing the test, to Mike Hartle for finite element analysis, to Tom Walrauch for testing, and to Rich Klaassen for scanning acoustic microscopy.

References

[*1*] Russ, S. M., "Thermal Fatigue of Ti-24Al-11Nb/SCS-6," *Metallurgical Transactions,* Vol. 21A, 1990, pp. 1595–1602.

[*2*] Russ, S. M., Nicolas, T., Bates, M., and Mall, S., "Thermomechanical Fatigue of SiC/Ti-24Al-11Nb Composites," *Failure Mechanisms in High Temperature Composite Materials,* AD-Vol. 22 and AMD-Vol. 122, G. K. Haritos, G. Newaz, and S. Mall, Eds., American Society of Mechanical Engineers, New York, 1991, pp. 37–43.

[*3*] Newaz, G. M. and Majumdar, B. S., *Inelastic Deformation Mechanisms in SCS-6/Ti 15-3 MMC Lamina Under Compression,* NASA Contractor Report 191170, Lewis Research Center, National Aeronautics and Space Administration, Cleveland, OH, Sept. 1993.

[*4*] Gambone, M. L., *Fatigue and Fracture of Titanium Aluminides,* WRDC-TR-89-4145, Vol. II, Wright Research and Development Center, Materials Laboratory, Wright-Patterson Air Force Base, OH, Feb. 1990.

[*5*] Castelli, M. G. and Gayda, J., *An Overview of Elevated Temperature Damage Mechanisms and Fatigue Behavior of a Unidirectional SCS-6/Ti-15-3 Composite,* NASA Technical Memorandum 106131, National Aeronautics and Space Administration, Washington, DC, April 1993.

Life Predictions

John G. Bakuckas, Jr.,[1] *and W. Stephen Johnson*[2]

A Methodology to Predict Damage Initiation, Damage Growth, and Residual Strength in Titanium Matrix Composites

REFERENCE: Bakuckas, J. G., Jr., and Johnson, W. S., **"A Methodology to Predict Damage Initiation, Damage Growth, and Residual Strength in Titanium Matrix Composites,"** *Life Prediction Methodology for Titanium Matrix Composites, ASTM STP 1253,* W. S. Johnson, J. M. Larsen, and B. N. Cox, Eds., American Society for Testing and Materials, 1996, pp. 497–519.

ABSTRACT: In this research, a methodology to predict damage initiation, damage growth, fatigue life, and residual strength in titanium matrix composites (TMC) is outlined. Emphasis was placed on micromechanics-based engineering approaches. Damage initiation was predicted using a local effective strain approach. A finite element analysis verified the prevailing assumptions made in the formulation of this model. Damage growth, namely, fiber-bridged matrix crack growth, was evaluated using a fiber bridging (FB) model that accounts for thermal residual stresses. This model combines continuum fracture mechanics and micromechanics analyses yielding stress-intensity factor solutions for fiber-bridged matrix cracks. In the FB model, fibers in the wake of the matrix crack are idealized as a closure pressure, and an unknown constant frictional shear stress is assumed to act along the debond length of the bridging fibers. This frictional shear stress was used as a curve-fitting parameter to the available experimental data. Figure life and post-fatigue residual strength were predicted based on the axial stress in the first intact 0° fiber calculated using the FB model and a three-dimensional finite element analysis.

KEYWORDS: fracture mechanics, fiber bridging model, matrix cracking, fiber breakage, thermal residual stresses, micromechanics

Nomenclature

a, a_o	Current crack length and unbridged initial crack length, m
d	Hole diameter, m
C, n, m	Crack growth coefficient constants
E_f, E_m	Fiber and matrix modulus, MPa
E_L, E_T, G_{LT}	Composite longitudinal, transverse, and shear modulus; MPa
F_{app}	Boundary correction factor for applied stress intensity factor
F_{par}	Correction factor for parallel cracks
G	Weight function
K_t^ϵ	Strain concentration factor
K_t	Orthotropic stress concentration factor
ΔK_m	Discrete stress-intensity factor range in matrix, $\text{MPa}\sqrt{\text{m}}$
ΔK_{Comp}	Continuum stress-intensity factor range in composite, $\text{MPa}\sqrt{\text{m}}$

[1]Senior research engineer, Galaxy Scientific Corporation, Egg Harbor Twp., NJ 08234.
[2]Professor, School of Material Science and Engineering, Georgia Institute of Technology, Atlanta, GA 30332-0245.

h	Distance between parallel cracks, m
Δp	Range in bridging pressure, MPa
r	Fiber radius, m
R	Hole radius, m
R_{Comp}	Composite stress ratio $= S^{\text{min}}/S^{\text{max}}$
R_m	Matrix stress ratio $= \sigma_m^{\text{min}}/\sigma_m^{\text{max}}$
ΔS	Applied stress range, MPa
ΔT	Change in temperature, °C
v_f, v_m	Fiber and matrix volume fractions
W	Width of specimen, m
$\tilde{x}$	Integration variable along crack from center, m
α_f, α_m	Coefficient of thermal expansion for fiber and matrix, mm/mm/°C
$\Delta\delta_{\text{Comp}}$	Continuum crack opening displacement range in composite, m
$\Delta\delta_m$	Discrete crack opening displacement range in matrix, m
$\Delta\epsilon_{\text{eff}}$	Effective strain parameter
$\Delta\epsilon$	Strain range
τ	Interfacial frictional shear stress, MPa
ν_{LT}	Composite Poisson's ratio
σ_f, σ_m	Axial stress in fiber and matrix, MPa
σ_m^{Therm}	Axial thermal residual stresses in the matrix, MPa

Titanium matrix composites (TMC) have excellent potential as structural materials for elevated temperature applications such as future hypersonic flight vehicles. However, a major problem facing TMC is the lack of a generalized life prediction methodology that can be applied in designs incorporating these advanced materials. The complex damage growth process that evolves in these materials often limits the use of conventional damage tolerance design approaches used for metals. On the composite microstructural level, damage consists of multiple failure mechanisms including fiber-matrix debonding, fiber breakage, and matrix cracking. The interaction and combination of these failure mechanisms can lead to extensive damage process zones having no single dominate crack. In such instances, classical fracture mechanics approaches are not suitable. An additional complexity includes thermal residual stresses that are induced in the fiber and matrix during the fabrication process that can have a profound effect on the fracture and fatigue behaviors of TMC. The heterogeneous nature of TMC poses challenges to the development of a damage tolerance design philosophy for this class of materials.

In an effort to address these challenges, a micromechanics-based life prediction methodology is proposed. The effects of applied stress, number of fatigue cycles, and specimen geometry (for example, stress concentrations and notches) on damage initiation, damage growth, and residual strength can be determined and could be used as a design limiter for notched TMC. Figure 1 is a flow chart describing this philosophy and how it can be used to predict the conditions causing the following fatigue scenarios: (1) initiation of cracks from notches using a local effective strain parameter; (2) growth of matrix cracks to a threshold condition where crack arrest occurs based on stress-intensity factor in the matrix calculated using a fiber bridging (FB) model; (3) continuous growth of matrix cracks based on stress-intensity factor in the matrix using the FB model; (4) specimen life based on the stress in the first intact fiber calculated using the FB model; and (5) post-fatigue residual strength based on the fiber stress calculated using either the FB model or a three-dimensional finite element analysis.

The objective of this research is to outline the framework for this generalized life

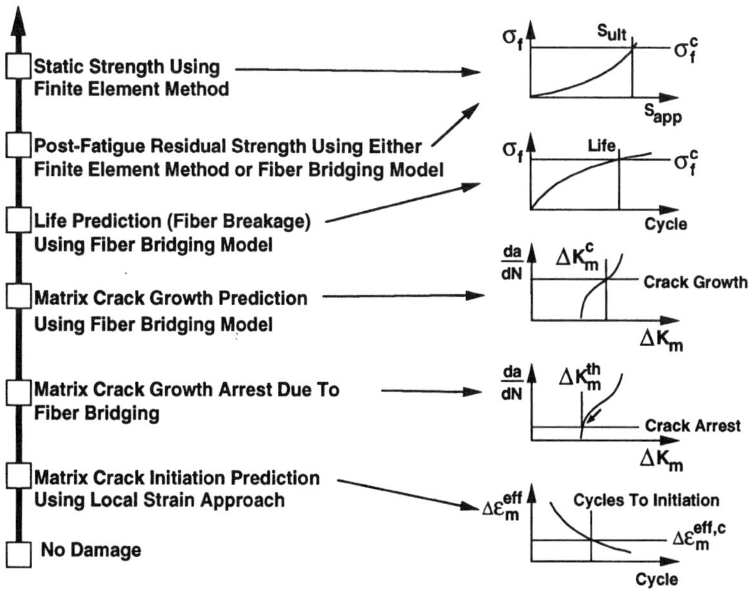

FIG. 1—*Micromechanics-based approaches used to predict and evaluate damage initiation, damage growth, and residual strength in titanium matrix composites.*

prediction methodology. This paper will review research addressing issues of damage initiation, damage growth, and post-fatigue residual strength. Emphasis was placed on micromechanics-based approaches. Damage initiation was predicted using a local effective strain approach developed by Hillberry and Johnson [*1*]. A finite element analysis verified the prevailing assumptions made in the formulation of this model. Next, damage growth and life prediction approaches for fiber-bridged matrix crack growth were developed based on a FB model. The FB model combines continuum fracture mechanics and micromechanics analyses to derive stress-intensity factor solutions of matrix cracks bridged by fibers. Thermal residual stresses were accounted for in this model by Bakuckas and Johnson [2]. The frictional shear stress that is assumed to act along the debond length of the bridging fibers was used as a curve-fitting parameter to the available experimental data. Finally, post-fatigue residual strengths were predicted based on the axial stress in the first intact 0° fiber calculated using the FB model developed in Ref *2* and using a three-dimensional finite element analysis of Bigelow et al. [*3–5*].

Experimental Procedures

Materials and Specimens

Two TMC systems were studied in this research, namely, SCS-6/Ti-15-3 and SCS-6/TIMETAL-21S laminates. Both material systems were fabricated by Textron by hot isostatic pressing (HIPing) titanium foils between unidirectional tapes of silicon-carbide (SCS-6) fibers having a diameter of 0.14 mm. In the SCS-6/Ti-15-3 and SCS-6/TIMETAL-21S laminates, fibers were held in place with molybdenum and titanium-niobium cross weave wires, respectively. The composition of matrix materials used in the SCS-6/Ti-15-3 and SCS-6/TIMETAL-21S laminates were Ti-15V-3Cr-3Al-3Sn and Ti-15Mo-3Al-2.7Nb-

TABLE 1—*Material Properties.*

Property	Material System SCS-6/Ti-15-3	SCS-6/TIMETAL-21S
	MATRIX	
E_m, GPa	92.4	116.0
ν_m	0.34	0.34
C	3.8E-11	5.5E-11
n	2.8	2.5
m	0.35	0.35
α_m, mm/mm/°C	9.7×10^{-6}	8.6×10^{-6}
	FIBER	
E_f, GPa	400.00	400.0
ν_f	0.25	0.25
α_f, mm/mm/°C	3.6×10^{-6}	3.6×10^{-6}
	COMPOSITE	
layup	$[0]_8$	$[0/90]_s$
v_f	0.33	0.36
E_L, GPa[a]	192.2	188.4
E_T, GPa[a]	130.1	188.4
G_{LT}, GPa[a]	45.9	58.1
ν_{LT}[a]	0.32	0.26
σ_m^{Therm}, MPa[a]	202.0[b]	221.9[c]
Layup	$[0/90]_s$	$[0/90]_{2s}$
v_f	0.35	0.38
E_L, GPa[a]	168.4	192.9
E_T, GPa[a]	168.4	192.9
G_{LT}, GPa[a]	47.4	59.3
ν_{LT}[a]	0.25	0.26
σ_m^{Therm}, MPa[a]	183.8[b]	230.5[c]

[a]Predicted using VISCOPLY [9] based on constituent properties.
[b]Axial thermal residual stress in matrix in 0° plies for cool-down from 555°C.
[c]Axial thermal residual stress in matrix in 0° plies for cool-down from 620°C.

0.2Si, respectively. For the SCS-6/Ti-15-3 laminates, the lay-ups studied were $[0]_8$ and $[0/90]_s$ with a range of measured fiber volume fractions (ν_f) from 0.33 to 0.36. For the SCS-6/TIMETAL-21S laminates, the lay-ups studied were $[0/90]_s$ and $[0/90]_{2s}$ with a range of measured fiber volume fractions (ν_f) from 0.36 to 0.38. The material properties are listed in Table 1.

Specimens were machined using a diamond wheel saw into straight-sided coupons with the 0° fibers in the loading direction. Each specimen was 152.5-mm in length. Center holes having diameters (d) of 6.3 ± 0.3 mm were drilled ultrasonically in specimens having a width of 19.1 mm. Center slits having lengths ($2a$) of 5.08, 7.62, and 8.89 mm were machined via electo-discharged machining (EDM) in specimens having a width (W) of 25.4 mm. The SCS-6/Ti-15-3 specimens were tested in the as-received condition while the SCS-6/TIMETAL-21S specimens were heat treated at a temperature of 620°C for 8 h in a vacuum and then furnace-cooled prior to testing. To enhance optical observations and surface replication, the surface of each specimen was polished to obtain a flat and lustrous finish. Thin sheet aluminum end tabs having dimensions 19.1 by 19.1 mm were bonded on the ends of all specimens in order to prevent grip failure. Information on the specimens tested is listed in Table 2.

TABLE 2—*Specimen loading history and fiber bridging analysis parameters.*

Material	Layup	Specimen Type	S_{max}, MPa	$2a/W$	R_m	h, mm	Temperature, °C	τ, MPa[a]	τ, MPa[b]
SCS-6/Ti-15-3	$[0]_8$	center notch	120	0.30	0.79	NA[c]	21	60	12 [6]
SCS-6/Ti-15-3	$[0]_8$	center notch	200	0.30	0.71	NA	21	60	20 [6]
SCS-6/Ti-15-3	$[0]_8$	center notch	300	0.30	0.62	NA	21	60	20 [6]
SCS-6/Ti-15-3	$[0]_8$	center notch	325	0.35	0.62	NA	21	60	40 [6]
SCS-6/Ti-15-3	$[0/90]_s$	center hole	150	0.35[d]	0.71	2.8	21	25	11 [7][e]
SCS-6/Ti-15-3	$[0/90]_s$	center hole	200	0.35[d]	0.65	2.1	21	25	11 [7][e]
SCS-6/Ti-15-3	$[0/90]_s$	center hole	250	0.35[d]	0.61	2.0	21	25	7 [7][e]
SCS-6/TIMETAL-21S	$[0/90]_s$	center hole	125	0.35[d]	0.77	1.5	21	100	30 [7][e]
SCS-6/TIMETAL-21S	$[0/90]_s$	center notch	150	0.20	0.75	NA	21	100	...
SCS-6/TIMETAL-21S	$[0/90]_s$	center notch	150	0.20	0.64	NA	200	32	...
SCS-6/TIMETAL-21S	$[0/90]_{2s}$	center hole	150	0.32[d]	0.74	1.9	21	100	35 [7][e]
SCS-6/TIMETAL-21S	$[0/90]_{2s}$	center hole	200	0.32[d]	0.69	2.0	21	100	35 [7][e]

[a] Frictional shear stress required to analyze data using ME relationships [*15*] when thermal residual stresses are considered.
[b] Frictional shear stress required to analyze data using ME relationships [*15*] when thermal residual stresses are neglected.
[c] Not applicable.
[d] Hole diameter to width ratio (d/W).
[e] Frictional shear stress required to analyze data using ME relationships [*15*] neglecting the effect of parallel cracks.

Test Procedures

Testing was conducted using a closed-loop servohydraulic test machine equipped with water cooled hydraulic grips. Tension-tension fatigue tests were done under load control with $R_{\text{Comp}} = 0.1$ ($S^{\min}/S^{\max}$) at a frequency of 10 Hz at both room temperature and an elevated temperature of 200°C. Results from these tests were generated in previous studies [*3,6,7*]. In these studies, damage initiation and progression were monitored and recorded using several techniques including a closed-circuit television system (CCTV), a long focal length microscopic system, surface and edge replicas, scanning electron microscopy (SEM), and optical microscopy. In addition, the acoustic emission (AE) technique was used where emphasis was placed on detecting the occurrence of fiber breakage. Details of the AE setup and the parameters used are reported in Ref *8*.

Analytical Procedures

Figure 1 is a flow chart outlining the micromechanics-based life prediction methodology developed. It was used in this study to determine the effects of applied stress, number of fatigue cycles, and specimen geometry (for example, stress concentrations and notches) on damage initiation, damage growth, fatigue life, and residual strength. The method consists of three analysis approaches: (1) local strain model used to predict damage initiation based on the value of the effective strain parameter; (2) FB model used to evaluated damage growth based on the stress-intensity factor in the matrix and used to evaluate fatigue life and residual strength based on the stress in the first intact fiber; and (3) a three-dimensional finite element model used to predict static and post-fatigue residual strength based on the stress in the first intact fiber. Each of the three approaches were developed in previous studies and were combined in this paper to make up the framework for a life prediction methodology. A brief description of these three approaches and the application to TMC to predict damage initiation, damage growth, fatigue life, and residual strength are discussed.

Matrix Crack Initiation

A procedure to predict the initiation of damage in the form of matrix cracks was introduced in Ref *1*. In that approach, an effective strain parameter was used that includes thermal residual stresses and stress concentrations

$$\Delta\epsilon_{\text{eff}} = \sqrt{\left(K_t\epsilon_{\max} + \frac{\sigma_m^{\text{Therm}}}{E_m}\right)\frac{K_t\Delta\epsilon}{2}} \tag{1}$$

In this equation, K_t, $\epsilon_{\max}$, σ_m^{Therm}, E_m, and $\Delta\epsilon$ are the orthotropic stress concentration factor, the maximum applied strain, the axial matrix thermal residual stress in the 0° plies, the matrix modulus, and the applied strain range, respectively. It is assumed in this model that the strain in the fiber, matrix, and composite are compatible and that the strain concentration factor in the matrix, K_t^ϵ, is equal to the orthotropic (composite) stress concentration factor, K_t.

The axial matrix thermal residual stress, σ_m^{Therm}, induced during the fabrication process was calculated using the VISCOPLY code [*9*], a micromechanics-based model. The thermo-viscoplastic properties of the constituents used in the calculation of σ_m^{Therm} are reported in Ref *9* for SCS-6/Ti-15-3 laminates and in Ref *10* for SCS-6/TIMETAL-21S laminates. In order to include thermal residual stresses in the calculations, the composite fabrication process was analytically simulated by modeling the cool-down cycle from an assumed stress-free temperature to room temperature. A stress-free temperature of 555°C was used for the SCS-6/Ti-15-3 laminates, and 620°C was used for the SCS-6/TIMETAL-21S laminates. For SCS-6/Ti-15-3 laminates, it was assumed that the stress-free condition occurs at one half the melting temperature of the titanium matrix, that is, at 555°C. For the SCS-6/TIMETAL-21S laminates, it was assumed that the residual stresses were removed during the heat treatment cycle at a temperature of 620°C. Values of σ_m^{Therm} induced during cool-down of the composite fabrication are listed in Table 1.

The effective strain parameter as a function of cycles to failure for the Ti-15-3 matrix material was determined in Ref *1* and is shown in Fig. 2. The solid line is a curve fit to the experimental data. For a notched composite with matrix thermal residual stresses, the number of cycles to matrix crack initiation can be predicted by comparing the modified effective strain parameter defined by Eq 1 with the solid line in Fig. 2.

Finite Element Analysis—A finite element analysis was conducted in this research to verify the prevailing assumptions made in the effective strain approach. The assumptions made in this model (Eq 1) are the following: (1) strain in the fiber, matrix, and composite are compatible; and (2) the orthotropic stress concentration factor is the same as the strain concentration factor for the fiber, matrix, and composite. Thus, the stress concentration calculated using a anisotropic continuum analysis is assumed to be equivalent to the strain concentration in the matrix. A concern with this assumption is that the local stress concentrations that occur near the ends of discrete machined fibers, particularly for hole configurations, are not accounted for. In some instances, cracks have been observed to initiate and grow as shown schematically in Fig. 3 where the initiation site is in the region of a machined fiber end. In order to determine the severity of the stress concentration factor around discrete fiber ends along the hole boundary, a two-dimensional finite element analysis was conducted for a unidirectional SCS-6/TIMETAL-21S monolayer having a fiber volume fraction, v_f = 0.36, and containing a center hole, d/W = 0.33.

The finite element model used is shown in Fig. 4 that consists of a global region and a local heterogeneous region. The global region was used to model the specimen geometry to which the far-field loads were applied and was treated as an orthotropic elastic continuum whose properties were calculated from the constituent properties using the rule-of-mixtures. The local heterogeneous region was used to conduct the micromechanics analysis near the center hole. In this region, the fiber and matrix was modeled as distinct zones using their

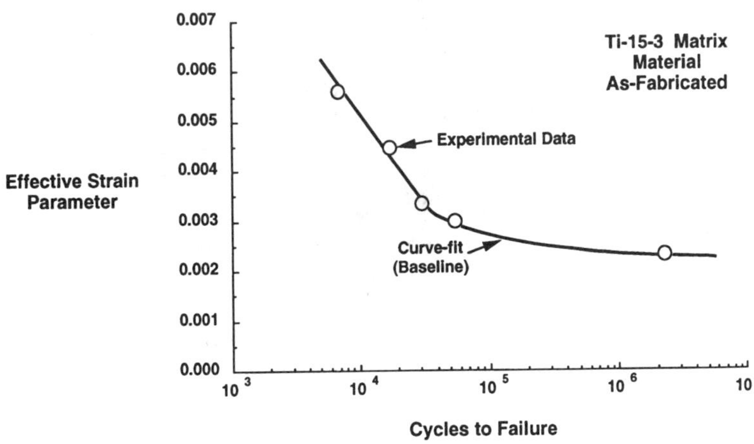

FIG. 2—*Effective strain parameter as a function of the number of cycles to failure for as-received Ti-15-3 matrix material* [1].

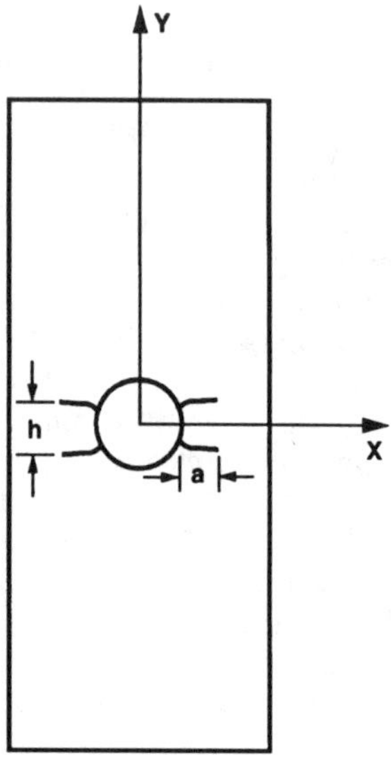

FIG. 3—*Schematics of cracking pattern commonly observed in TMC containing center hole.*

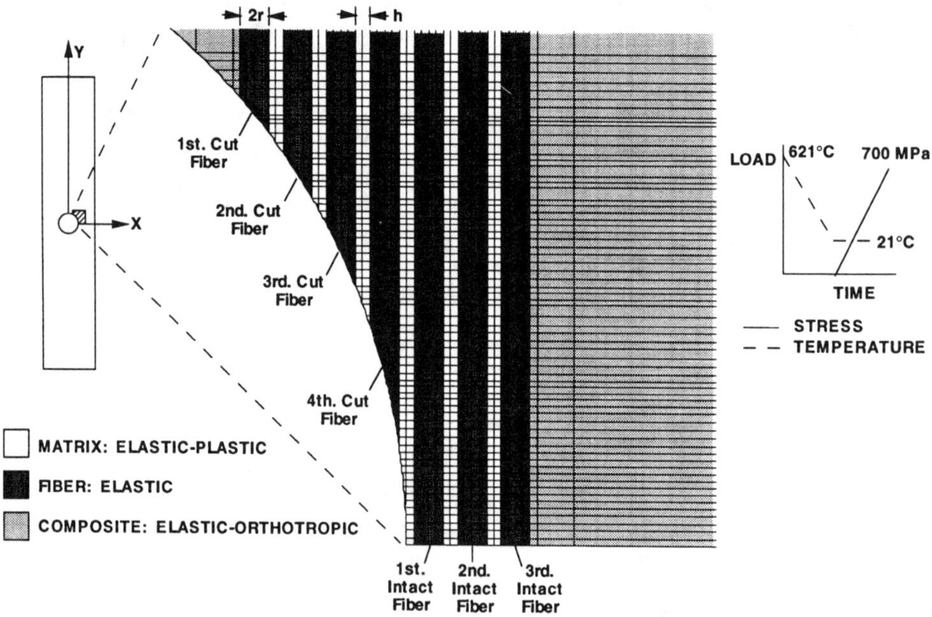

FIG. 4—*Finite element representation of SCS-6/TIMETAL-21S monolayer containing center hole and applied loads.*

respective neat properties. The constitutive properties used in the finite element analysis of the SCS-6/TIMETAL-21S monolayer are given in Table 1.

The entire finite element mesh consisted of 2588 elements and 10 581 nodes. Nine-noded isoparametric elements with 3 by 3 Gauss Numerical Integration was used. Thermal residual stresses were included in the analysis by simulating the cool-down cycle from an assumed stress-free temperature of 620°C to room temperature as described previously. Then, a mechanical load was applied incrementally up to 700 MPa along the top row of nodes.

The distribution of von Mises stresses in the matrix is shown in the stress contours in Fig. 5. As shown in this figure, stress concentrations do exist in the matrix in the vicinity of the machined fiber ends. A detailed stress contour in the matrix between the first and second cut fibers is shown in Fig. 6 and indicates that the stress builds up behind the second cut fiber. However, the maximum stress and strain concentration occurs as shown in Fig. 5. Thus, the center hole is the primary stress concentration while the machined fiber ends are secondary stress concentrations.

The local strain concentration factor in the matrix is defined as

$$K_t^{\epsilon} = \frac{\epsilon_{\text{axial}}^{\text{local-max}}}{\epsilon_{\text{axial}}^{\text{global}}} \tag{2}$$

where $\epsilon_{\text{axial}}^{\text{local-max}}$ is the maximum local axial strain and $\epsilon_{\text{axial}}^{\text{global}}$ is the remote global axial strain. For an elastic analysis from the initial load increment, $K_t^{\epsilon} = 3.81$. The global stress concentration factor for a hole in an orthotropic continuum is defined as [*11*]

$$K_t = \left[1 + \sqrt{\frac{E_L}{G} - 2\,\nu_{LT} + 2\sqrt{\frac{E_L}{E_T}}}\right]\alpha \tag{3}$$

The terms in this equation are the laminate properties as defined in the nomenclature. The term, α, is the finite width correction factor. In this study, the finite width correction factor for isotropic material was used [*12*]

$$\alpha = \frac{2 + \left(1 - \dfrac{d}{W}\right)^3}{3\left(1 - \dfrac{d}{W}\right)} \tag{4}$$

For the composite analyzed, $K_t = 3.83$ is in excellent agreement with the strain concentraction factor (K_t^{ϵ}) calculated from the discrete finite element analysis and, therefore, can be used in the definition of the local effective strain given by Eq 1.

Matrix Crack Growth

Methods to evaluate matrix crack growth were introduced in several studies using FB models [*13–15*]. The application of these FB models to analyze damage growth in unidirectional TMC was demonstrated in Ref *6*. The frictional shear stress term assumed in these models was used as a fitting parameter to crack growth data. It was shown Ref 6 that the FB models do provide an effective engineering tool for interpreting crack growth data in this class of materials. However, the values of the frictional shear stress term used to evaluate crack growth data varied with several parameters including the applied load and crack length. Further developments of FB analysis were realized when thermal residual stresses were considered [2]. It was shown in Ref *2* that a single value of frictional shear stress was sufficient to evaluate the crack growth data. While these studies are encouraging, the FB models still lack true predictive capabilities due to the unknown frictional shear stress term.

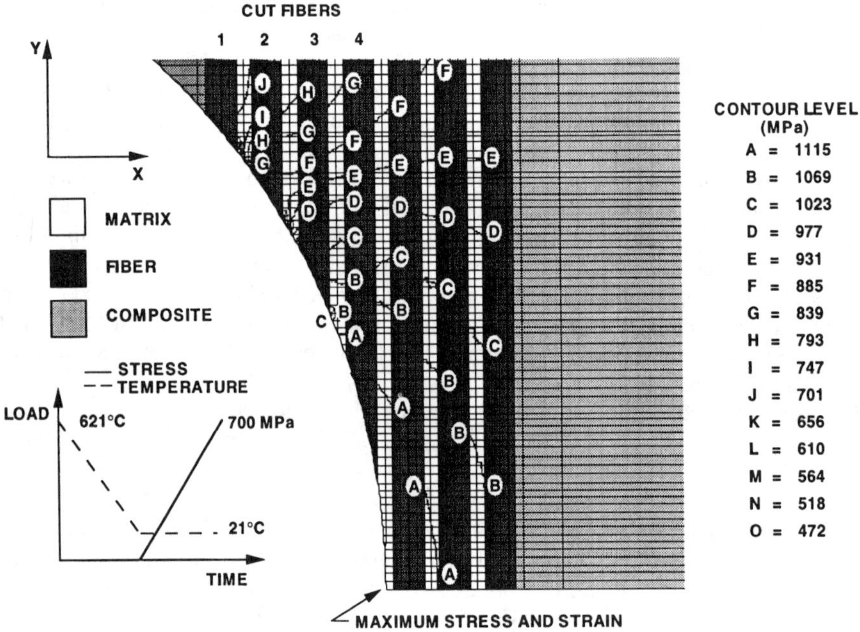

FIG. 5—*von Mises stress distribution in vicinity of machined fiber ends. Maximum stress and strain concentration is due to center hole. Machined fiber ends cause secondary stress and strain concentrations.*

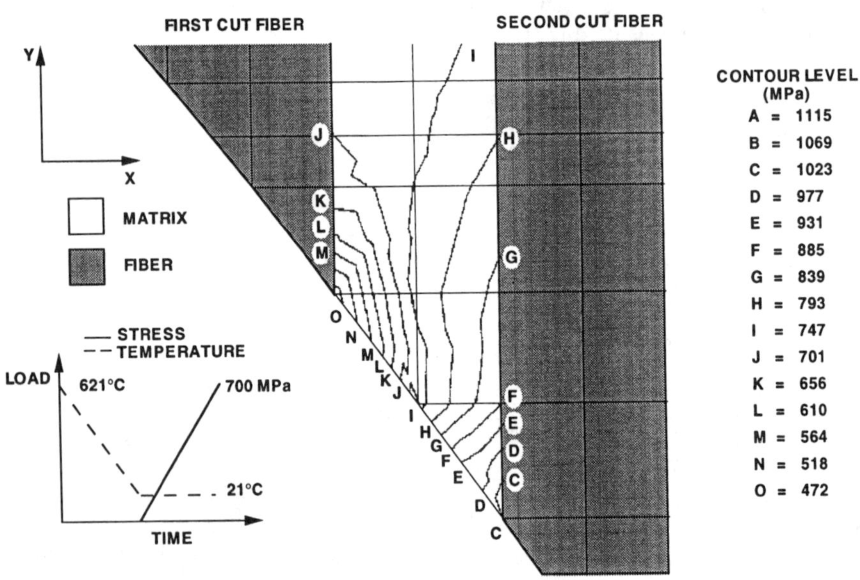

FIG. 6—*von Mises stress distribution between the first and second cut fibers. Stress builds up behind the second cut fiber; however, the maximum stress and strain concentration is due to center hole.*

Fiber-bridged matrix crack growth was analyzed using the approach outlined in Ref *2*. In that study, crack growth behavior was evaluated according to a version of the power law function introduced by Walker that includes *R*-ratio effects [*16*]

$$\frac{da}{dN} = C \cdot \left(\frac{\Delta K_m}{(1 - R_m)^{(1-m)}} \right)^n \tag{5}$$

where C, n, and m are material constants for the neat matrix material, R_m is the local stress ratio in the matrix ($R_m = \sigma_m^{\min}/\sigma_m^{\max}$), and ΔK_m is the Mode I stress intensity factor range in the matrix. By using Eq 5, thermal residual stresses can be included in the analysis through the quantities R_m and ΔK_m as described next.

Calculation of $\mathrm{R_m}$—The stress ratio in the matrix ($R_m = \sigma_m^{\min}/\sigma_m^{\max}$) was calculated using the VISCOPLY code [*9*]. After the composite fabrication simulation, a global mechanical load from $S^{\min}$ to $S^{\max}$ was applied, and the corresponding stresses in the matrix in the 0° plies were calculated. Using these stress values, the R_m value in the matrix in the 0° plies was calculated. The R_m values are listed in Table 2 for each specimen tested. As revealed by the results in Table 2, the R_m values were considerably higher than the value of the applied *R*-ratio ($R_{\mathrm{Comp}} = 0.1$). Thus, the R_{Comp} value does not represent the actual stress ratio in the matrix and the local R_m values listed in Table 2 should be used in the analysis.

Calculation of $\Delta\mathrm{K_m}$—The stress intensity factor range in the matrix, ΔK_m, was calculated using an FB model implemented with thermal residual stresses [2]. It is assumed in this model that fiber-matrix debonding takes place as a matrix crack progresses past the fibers. The intact fibers in the wake of the matrix crack are idealized as a crack-bridging pressure, Δp. An unknown but constant frictional shear stress, τ, is assumed to act along the debond length of the bridging fibers. A continuum fracture mechanics analysis and a micromechanical analysis are combined using discrete-continuum relationships to obtain stress-intensity factor solutions for FB matrix cracks of arbitrary size.

The procedure used to calculate ΔK_m involves a micromechanical analysis and a continuum fracture mechanics analysis. The crack opening displacement in terms of the bridging pressure is obtained from both analyses. Discrete-continuum relationships are used to combined the two analyses to yield a single nonlinear integral equation in terms of the bridging pressure. Solving for the bridging pressure, ΔK_m can be determined. Assuming the discrete-continuum relationships in Ref 15, the stress-intensity factor in the matrix is

$$\Delta K_m = \Delta S \sqrt{\pi a} F_{\mathrm{app}} - 2 \int_{a_o}^{a} \Delta p G(x,a,R,W) dx \tag{6}$$

In this equation, ΔS is the applied far-field stress, F_{app} is the applied stress boundary correction factor, and a_o is the initial unbridged crack length (hole radius or notch length). In addition, the term $G(x,a,R,W)$ is the weight function for the bridging stress-intensity factor range. Details on the calculation of can be found in Ref *2*.

Fatigue Life

A procedure to predict fatigue life is proposed in this study based on the axial stress in the first intact fiber calculated using the FB model. The axial stress in the fibers is approximated in terms of the bridging pressure (Δp) and the fiber volume fraction (v_f)

$$\Delta\sigma_f = \frac{\Delta p}{v_f} \tag{7}$$

This equation is valid for fibers closely spaced relative to the crack length. The bridging pressure is calculated using the aforementioned FB model. When the stress in the first intact fiber reaches a critical value, failure of that fiber occurs. During fatigue loading of TMC, catastrophic fracture of a specimen typically occurs shortly after the onset of fiber breakage. By determining the stress in the first intact fiber, the conditions under which a specimen will fail, namely, the loading conditions (applied load and number of cycles) and the specimen configuration (hole diameter and crack length), can be predicted.

For this purpose, the onset of fiber breakage was monitored and recorded in several cross-ply SCS-6/TIMETAL-21S specimens containing center holes using the AE technique and using the crack growth characteristics. A detailed description of the AE test setup and parameters used is provided in Ref *8*. Briefly, AE events having high amplitudes (that is, greater than or equal to 99 dB) were associated with fiber breakage.

The onset of fiber breakage was also determined from the crack growth characteristics. A schematic of the crack length as a function of number of cycles is shown in Fig. 7. For this example, the crack growth is characterized by three regions: (1) initiation and growth, (2) slow growth/crack arrest, and (3) rapid crack growth. After the initial region, the rate of crack growth decreases due to the FB phenomena. The number of fibers bridging the crack increases as the crack length increases thereby reducing the stress-intensity factor in the matrix. As the crack extends, the stress in the bridging fibers increases. Under certain conditions, the stresses in the fibers may become sufficiently high to cause fiber breakage resulting in the rapid crack growth in the third region. In this study, the onset of fiber breakage is assumed to occur at the "knee" of the crack growth curve between the second and third regions, Fig. 7.

Using the AE technique and the characteristics of the crack growth curve, the conditions (that is, specimen geometry, crack length, applied load, number of cycles, etc.) under which fiber breakage occurs was determined for several specimens. Using these conditions as input parameters, the failure stress range for the first intact fiber, $\Delta\sigma_f^{fail}$, was calculated using the FB model and was used as a failure criterion to predict fatigue life. These failure stresses are the range values ($\Delta\sigma_f^{fail} = \sigma_f^{max} - \sigma_f^{min}$) between S^{min} and S^{max} and are listed in Table 3.

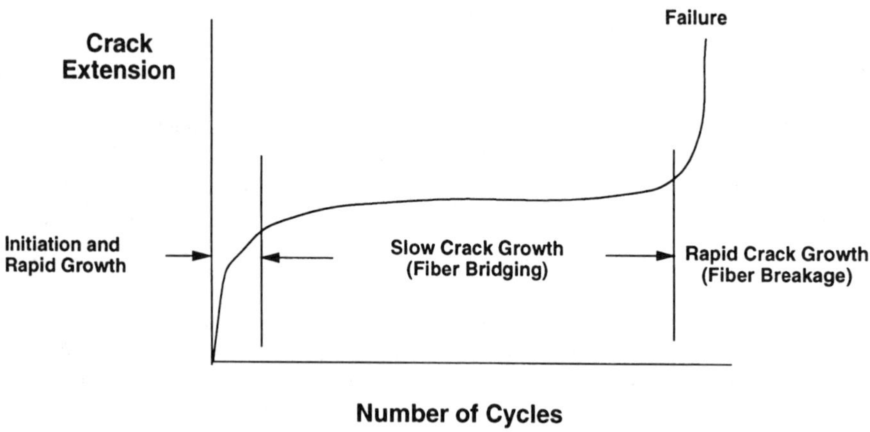

FIG. 7—*Schematics of typical crack growth curve observed in TMC. Characterized by three regions: (1) initiation and rapid crack growth, (2) slow crack growth, and (3) rapid crack growth up to failure.*

TABLE 3—*Calculated stress in first intact fiber at the onset of fiber breakage, SCS-6/TIMETAL-21S, MCE model* [13].

Layup	S_{max}, MPa	d/W	Δa_{avg}, mm	Detection of Breakage	Fiber Stress, τ = 50 MPa: Fatigue, $\Delta\sigma_f^{fail}$	Fiber Stress, τ = 50 MPa: Static, σ_f^{fail}
$[0/90]_s$, ν_f = 0.35	175	0.356	1.55	Crack Growth Characteristics	2430	1930
$[0/90]_{2s}$,	200	0.341	1.36	Acoustic Emission	2530	2030
ν_f = 0.37	250	0.314	0.78	Acoustic Emission	2650	2100
		Critical fiber stress ≈			2500	2000

Post-Fatigue Residual Strength

The post-fatigue residual strength was predicted using two approaches: (1) the FB model and (2) a three-dimensional finite element program, PAFAC [*17*]. The axial stress in the 0° fiber next to the notch was predicted as a function of applied load. The composite strength was assumed to be the applied load at which the axial stress in the 0° fiber next to the hole reached the fiber strength.

Selection of the approach used to predict the post-fatigue residual strength depends on the characteristics of the damage process zone near the notch. For damage classified as self-similar, Mode I matrix crack growth, the FB model is suitable. The FB models are applicable to the analysis of SCS-6/TIMETAL-21S laminates where the damage is primarily a single Mode I matrix crack. The FB analysis was previously described. The critical fiber strength, σ_f^{fail}, used to determine the onset of fiber breakage in SCS-6/TIMETAL-21S during static loading is listed in Table 3. These failure stresses are absolute values and are lower than the range values obtained under fatigue loading. Under fatigue loading, the fiber stresses at the minimum applied load, σ_f^{min}, is compressive due to the thermal residual stresses. Consequently, the fiber stress range under fatigue loading, $\Delta\sigma_f^{fail}$, is higher than the absolute fiber stress under static loading, σ_f^{fail}.

In TMC where extensive damage process zones develop, such as in SCS-6/Ti-15-3, a more robust approach must be used such as a three-dimensional finite element analysis. Details of this approach are outlined in Refs *3* through *5*. Briefly, a three-dimensional finite element code, PAFAC [*17*], is used to calculate the fiber stresses. The damage that evolves during fatigue loading, namely, matrix cracking and fiber-matrix debonding, are analytically simulated by using reduced stiffness properties of the matrix material.

Results and Discussion

The aforementioned fatigue crack growth methodology was used to determine the effects of applied stress, number of fatigue cycles, applied temperature, and specimen geometry (for example, stress concentrations and notches) on damage initiation, damage growth, fatigue life, and residual strength in TMC as discussed in the following sections.

Matrix Crack Initiation

Using Eq 1, the effective strain parameter versus the number of cycles to matrix crack initiation is plotted in Fig. 8 for the SCS-6/Ti-15-3 laminates tested in Refs *1* through *3*. In

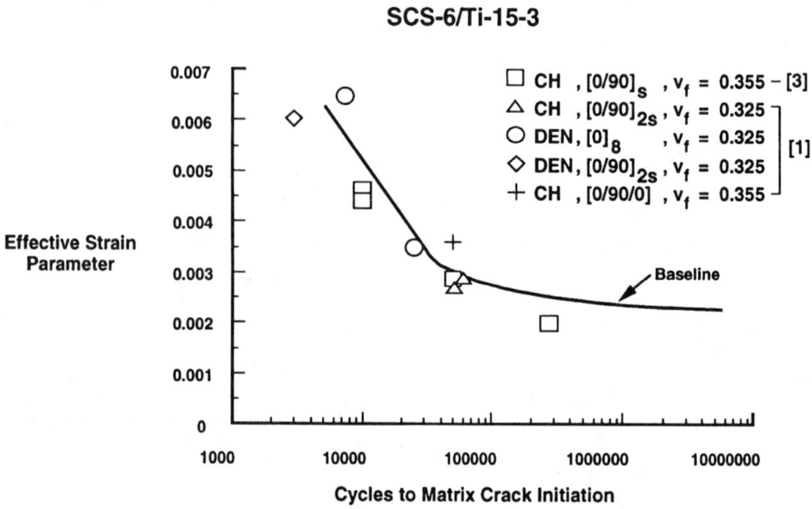

FIG. 8—*Effective strain parameter as a function of the number of cycles to matrix crack initiation. Good agreement is obtained between baseline curve and experiments.*

this figure, CH and DEN designate center hole and double edge notched specimens, respectively. The solid line is the baseline curve for the matrix material (same as that shown in Fig. 2). Experimentally, matrix crack initiation was assumed when the crack length reached between 0.3 and 0.5 mm. As shown in Fig. 8, by using Eq 1, the experimental data are consolidated along the baseline curve indicating that the effective strain parameter can be used to predict crack initiation in the matrix reasonably well.

Matrix Crack Growth

The matrix crack growth behavior observed in the two TMC systems was analyzed using the FB model implemented with thermal residual stresses [*2*]. The frictional shear stress term, τ, in the FB model was used as a curve-fitting parameter to matrix crack growth data reduced using the crack growth law defined by Eq 5. Results are presented in the subsequent sections. The effects of thermal residual stresses on the values of τ used in FB analysis are also discussed.

Calculated Crack Growth Curves—The crack length as a function of number of cycles was calculated and plotted in Figs. 9 through 12 using a value of τ that best-fit the experimental results for the two material systems having various external loads and geometric conditions. The discrete-continuum relationships developed by Marshall, Cox, and Evans (MCE) [*13*] were used in the calculations. For all cases, a good agreement was obtained between the experiments (open symbols) and calculations (filled symbols) using a single value of the frictional shear stress.

Of particular interest are the results presented in Fig. 12 showing the crack length as a function of the number of cycles at room temperature and at an elevated temperature of 200°C for two center-notched specimens of $[0/90]_s$ SCS-6/TIMETAL-21S. As shown in the figure, good agreement was obtained between experiments (open symbols) and the calculations (filled symbols) using the values of τ = 50 MPa and 15 MPa for the room temperature and elevated temperature test results, respectively. The difference in the calculated crack growth behavior at the two temperatures is due primarily to the differences in thermal residual stress states that affects the calculated values of R_m and ΔK_m.

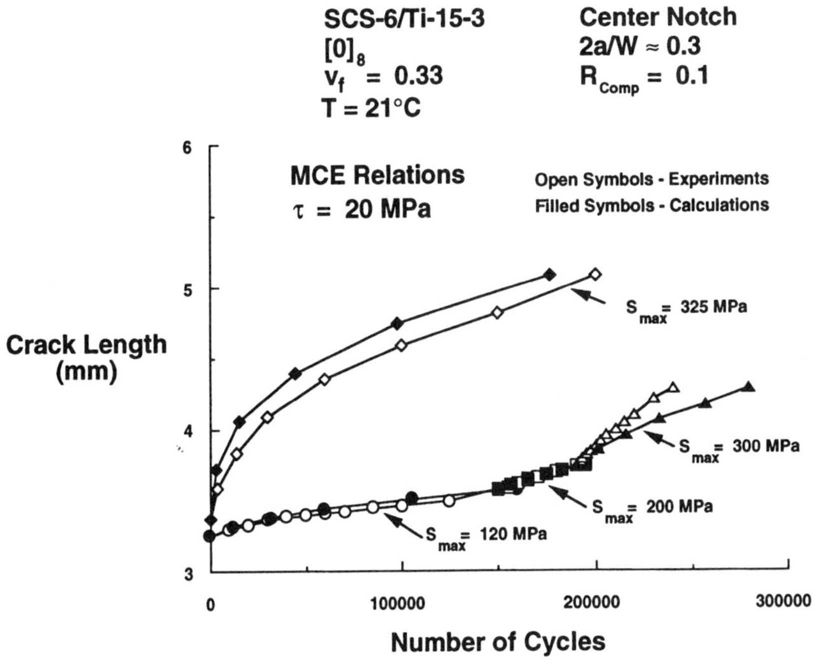

FIG. 9—*Crack growth data for center-notched* $[0]_8$ *SCS-6/Ti-15-3 laminates at several applied stress levels. Comparison between experiments and calculations made using FB model.*

If the mechanical properties of the matrix are assumed to be identical at both temperatures, only R_m and ΔK_m affect the crack growth rate defined by Eq 5. At elevated temperatures, the axial thermal residual stress in the matrix, σ_m^{Therm}, is lower than at room temperature that results in lower values of R_m. For the room and elevated temperature tests, the calculated local R-ratios in the matrix were $R_m = 0.75$ and 0.64, respectively. From Eq 5, for a given value of ΔK_m, the crack growth rate increases with an increase in the value of R_m. Based solely on the values of R_m for the two test conditions, the crack growth rate should have been higher for the room temperature test than the elevated temperature test. Results shown in Fig. 12, however, show the opposite trend; the crack growth rate data were higher for the elevated temperature test compared to the room temperature test.

Another factor affecting the crack growth rate is the stress-intensity factor range, ΔK_m. Thermal residual stresses affect the value of ΔK_m. The lower value of τ for the elevated temperature test translated into a higher value in the stress-intensity factor range as shown in Fig. 13. At the elevated temperature, the value of ΔK_m was higher than at room temperature. Using the crack growth law defined by Eq 5, for a given value of R_m, the crack growth rates increases with an increase in the value of ΔK_m that agrees with the trend in the results shown in Fig. 12. As a result of the lower thermal residual stresses at the elevated temperature test, the value of ΔK_m increased while the value of R_m decreased. The lower values of τ for the elevated temperature results suggest that the radial thermal residual stresses acting across the debonded fiber-matrix interfaces in the 0° plies were lower. These radial stresses "choke" the fibers that enhance the shear transfer between the fiber and matrix. By lowering the radial thermal stresses, the amount of shear transfer and displacement constraint between the fiber and matrix were lowered thereby increasing the stress-intensity factor. Thus, for the

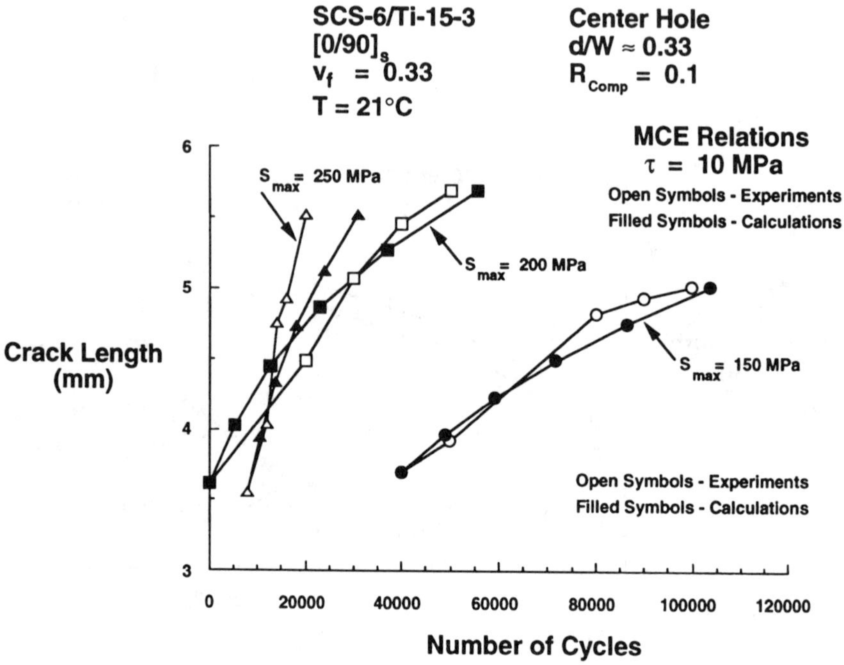

FIG. 10—*Crack growth data for center hole [0/90]$_s$ SCS-6/Ti-15-3 laminates at several applied stress levels. Comparison between experiments and calculations made using FB model.*

elevated temperature test, the increase in the value of ΔK_m had a more profound effect on the crack growth rate than the decrease in the value of R_m.

Effect of Thermal Residual Stress on the Frictional Shear Stress—The effect of thermal residual stresses on FB analysis is revealed in the values of the frictional shear stress term required to reduce the crack growth data that are listed in Table 2. The values of τ listed in this table were determined using the discrete-continuum relationships developed by McMeeking and Evans (ME) [*15*] with and without thermal residual stresses. For the unidirectional SCS-6/Ti-15-3 specimens tested in Ref *6*, the value of τ required to fit crack growth data varied from 12 to 41 MPa when thermal residual stresses are neglected (that is, original FB analysis). On the other hand, when thermal residual stresses were included in the analysis, a single value of frictional shear stress, τ = 60 MPa, was required to curve-fit the data. Thus, the scatter in the value of τ was reduced significantly by considering thermal residual stresses in the FB analysis. In fact, for a given material system, lay-up, and temperature, a single value of τ was sufficient to analyze the crack growth data. Thermal residual stresses are an important factor overlooked in the original formulations of the FB models.

Fatigue Life

Using the values of frictional shear stress used to evaluate the crack growth data presented in the previous section, the fatigue life was estimated based on the stress in the first intact fiber. The critical value of stress in the fiber at which fiber breakage occurred was first determined for several specimens of cross-ply SCS-6/TIMETAL-21S. For these specimens,

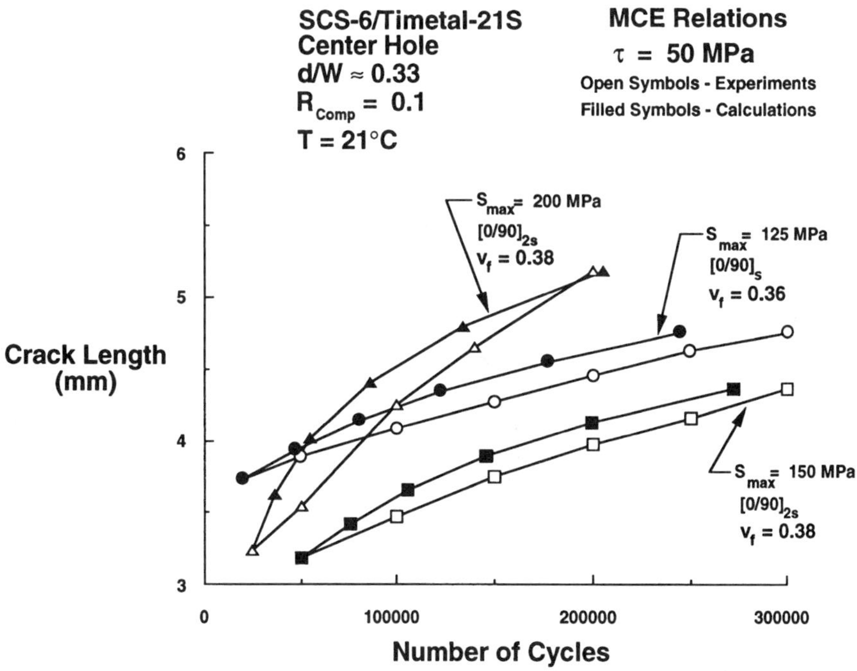

FIG. 11—*Crack growth data for center hole cross-ply SCS-6/TIMETAL-21S laminates at several applied stress levels. Comparison between experiments and calculations made using FB model.*

the AE technique and the characteristics of the crack growth curve were used to determine the conditions (that is, specimen geometry, crack length, applied load, number of cycles, etc.) under which fiber breakage occurred as described previously. Using these conditions as input parameters, the value of stress in the first intact fiber when fiber breakage occurs was calculated for both static and fatigue applied loadings using the FB model. Results are presented in Table 3. As shown, the value of the axial stress in the first intact fiber was nearly a constant value for the specimens listed. In this study, the average value of the stresses listed in Table 3 was used as a critical value to determine the onset of fiber breakage. Under fatigue loading conditions, the value of stress used was 2500 MPa and under static loading conditions, the value used was 2000 MPa.

Having the critical value of fiber stress, the life prediction methodology was applied to a $[0/90]_{2s}$ SCS-6/TIMETAL-21S laminate containing a center hole subjected to $S_{max} = 200$ MPa and $R_{Comp} = 0.1$. The crack length as a function of the number of cycles was monitored and is plotted in Fig. 11 along with the calculation made using the FB model and a value of $\tau = 50$ MPa. Crack growth was recorded up to 200 000 cycles. This specimen failed after 222 600 cycles. The stress in the first intact fiber as a function of the number of fatigue cycles was calculated and plotted in Fig. 14. As shown in the figure, the stress in the first intact fiber reaches the average critical value (2500 MPa) after 285 000 cycles.

Post-Fatigue Residual Strength

The post-fatigue residual strength was evaluated in two cross-ply TMC laminates containing center holes, namely, in a SCS-Ti-15-3 specimen and a SCS-6/TIMETAL-21S speci-

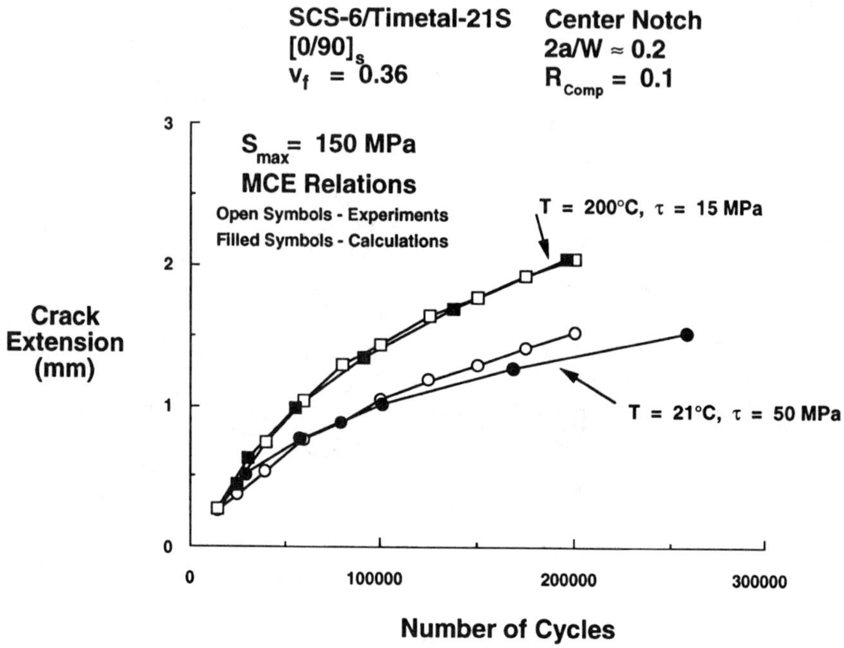

FIG. 12—*Crack growth data for center-notched [0/90]$_s$ SCS-6/TIMETAL-21S laminates at two temperature levels. Comparison between experiments and calculations made using FB model.*

men. The SCS-6/Ti-15-3 specimen had been subjected to S_{max} = 250 MPa for 200 000 cycles during which extensive matrix cracking occurred [3]. This specimen was loaded quasi-statically in tension to failure and a residual strength of 325 MPa was measured.

The SCS-6/TIMETAL-21S specimen had been subjected to S_{max} = 150 MPa for 200 000 cycles during which two symmetric cracks grew from the hole. This specimen was then loaded quasi-statically in tension to failure and a residual strength of 350 MPa was measured. The AE technique was used to determine the occurrence of fiber breakage. For this specimen, the number of high-amplitude events, that is, AE events having an amplitude greater than 98 dB, associated with fiber breakage, is plotted in Fig. 15 as a function of the applied load. As shown, fiber breakage occurs at an applied load of 180 MPa. The number of high-amplitude events increases up to failure at 350 MPa. This indicates that stable fiber breakage occurred prior to catastrophic fracture. From the onset of fiber breakage, an increase in applied load was required for subsequent fiber breakages.

Fiber Bridging (FB) Analysis—The FB model was used to evaluate the post-fatigue residual strength for the SCS-6/TIMETAL-21S specimen. Since the fatigue damage that occurred in the SCS-6/TIMETAL-21S specimen can be considered as Mode I self-similar crack growth, the application of the FB model in the analysis was suitable. The axial stress in the first intact fiber as a function of the applied load was calculated. The onset of fiber breakage was assumed to occur when the stress in the first intact fiber equaled the critical value of 2000 MPa. The critical value was calculated as described previously. Using the FB model, the stress in the first intact fiber reaches the critical value at an applied load of 235 MPa. Using the AE technique, the emission of high-amplitude events associated with fiber breakage occurred at 180 MPa. A comparison is shown in Fig. 15. While both these values are in agreement, this specimen failed catastrophically at 350 MPa. This indicates that the

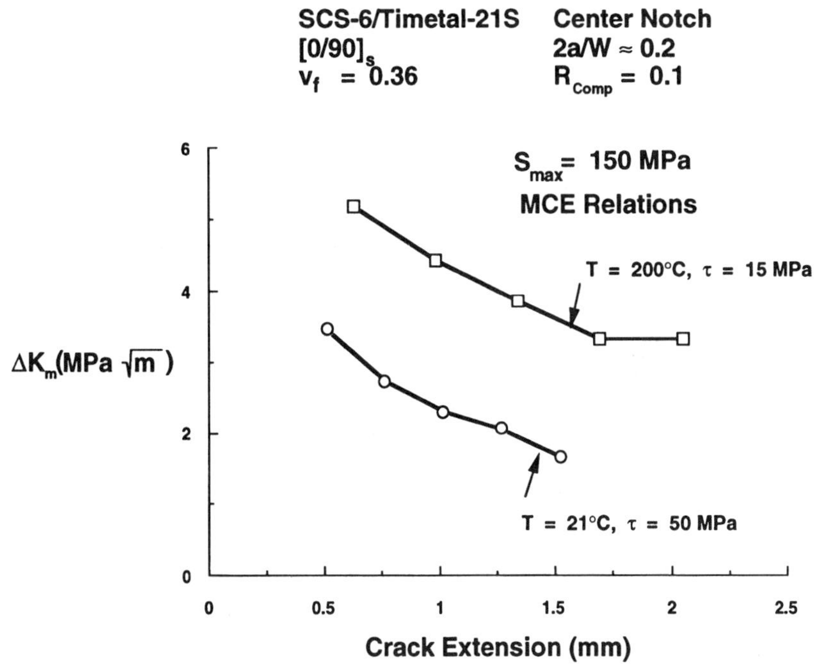

FIG. 13—*Variation of stress intensity with crack extension in center-notched $[0/90]_s$ SCS-6/TIMETAL-21S laminates at two temperature levels.*

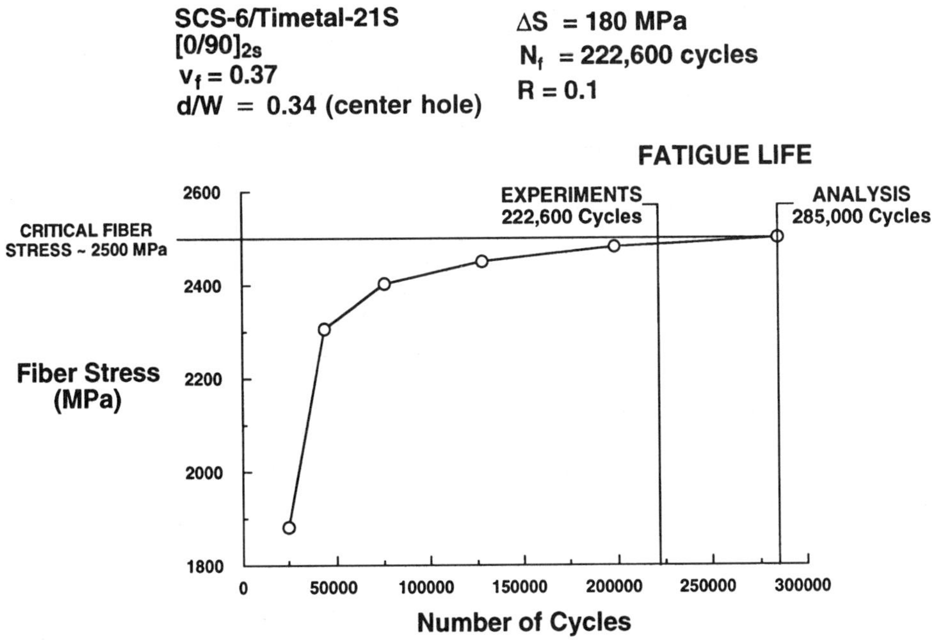

FIG. 14—*Fatigue life prediction based on stress in the first intact fiber calculated using the FD model.*

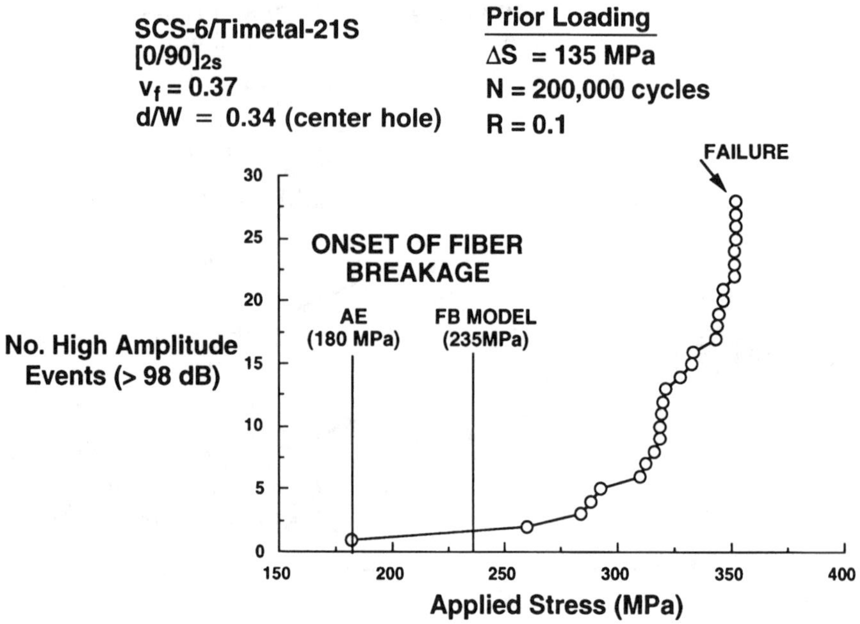

FIG. 15—*Determination of fiber breakage via AE technique and FB model.*

occurrence of first fiber breakage may not be a good indicator of post-fatigue residual strength for the SCS-6/TIMETAL-21S. The results shown in Fig. 15 suggest that a stable fiber breakage occurred prior to catastrophic fracture.

PAFAC Analysis—A three-dimensional finite element analysis (PAFAC) was used to predict the virgin static and post-fatigue residual strengths in cross-ply SCS-6/Ti-15-3 laminates containing center holes. For the virgin specimen, the constitutent properties shown in Table 1 were used. In addition, fiber-matrix debonding in the 90° plies was modeled as described in Ref *5*. The damage that occurred in the fatigued specimen was quite extensive and widespread consisting of multiple matrix cracks and fiber-matrix debonding. This damage was simulated using a reduced matrix modulus as described in Ref *3*.

The axial stress in the 0° fiber next to the center hole as a function of the applied stress is plotted in Fig. 16 for both cases. The horizontal line indicates the assumed fiber strength of 4400 MPa, and the two vertical lines show the experimental strengths of the two conditions. The fiber strength was calculated from the measured strain to failure of an $[0/90]_s$ unnotched specimen (ϵ_f^{ult} = 0.0011) multiplied by the fiber modulus (400 GPa). As shown in Fig. 16, the applied stress at which the 0° fiber stress equals the fiber strength corresponds closely to the measured strength for both conditions. Using the axial stress in the 0° fiber next to the center hole as the criterion for laminate failure, the predicted strengths for the virgin and post-fatigue conditions were 483 and 301 MPa, respectively. Both predictions are within 8% of the experimental values.

Summary

In this study, a micromechanics-based life prediction methodology was presented and used to determine the effects of applied stress, number of fatigue cycles and specimen geometry on damage initiation, damage growth, and residual strength in notched titanium

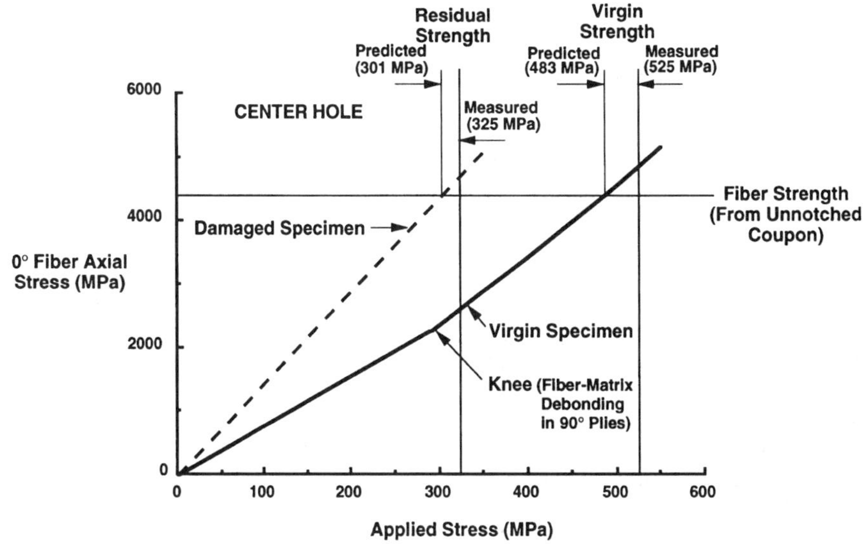

FIG. 16—*Axial stress in 0° fiber next to hole as a function of the applied stress for virgin and post-fatigue conditions in* $[0/90]_s$ *SCS-6/Ti-15-3 laminates,* $v_f = 0.355$. *Applied stress at which 0° fiber stress reaches the fiber strength corresponds to the notched strength for both conditions.*

matrix composite (TMC). This methodology was applied to two TMC systems, namely, SCS-6/Ti-15-3 and SCS-6/TIMETAL-21S.

Damage initiation was predicted using a local effective strain approach. A finite element analysis of a SCS-6/TIMETAL-21S monolayer containing a center hole verified the prevailing assumptions made in the formulation of this model. The finite element analysis showed that the primary stress concentration was due to the center hole and that the local stress concentrations due to machined fiber ends were secondary. Using the effective strain approach, the conditions under which matrix cracks initiated were predicted.

Damage growth analysis of fiber bridged matrix cracks was developed based on a FB model that accounts for thermal residual stresses. The FB model combines continuum fracture mechanics and micromechanics analyses to derive stress-intensity factor solutions of matrix cracks bridged by fibers. The unknown constant frictional shear stress assumed in this model was used in this study as a curve-fitting parameter to the available experimental data. For a given material system, layup, and applied temperature, a single value of frictional shear stress was sufficient to accurately model the matrix crack growth behavior.

The fatigue life was predicted based on the axial stress in the first intact 0° fiber calculated using the FB model. The critical value of stress at which fiber breakage occurred was first determined for several specimens. For these specimens, the AE technique and the characteristics of the crack growth curve were used to determine the conditions under which fiber breakage occurred. Using these conditions as input parameters, the value of stress in the first intact fiber when fiber breakage occurs was calculated using the FB model and was found to be nearly a constant value. This critical value was then used to predict the onset of fiber breakage during fatigue loading in a subsequent specimen. It was assumed that fatigue life is given by the number of fatigue cycles at which fiber breakage initiates. There was approximately a 22% difference between the observed and predicted fatigue lives.

The post-fatigue residual strength was predicted based on the axial stress in the first intact 0° fiber calculated using either the FB model or a three-dimensional finite element analysis, depending on the classification of fatigue damage. The fatigue damage developed in the SCS-6/TIMETAL-21S laminates can be idealized as a Mode I, self-similar crack. For this case, the fracture mechanics-based approaches are appropriate and the FB model was used. Using the FB model, the predicted applied load causing the onset of fiber breakage was in agreement with AE results. There was approximately a 23% difference in the results.

The virgin static and post-fatigue residual strength was predicted for cross-ply SCS-6/Ti-15-3 laminates containing a center hole using a three-dimensional finite element analysis. The fatigue damage that occurred was quite extensive and widespread consisting of multiple matrix cracks and fiber-matrix debonding. This damage was simulated using a reduced matrix modulus. Using the axial stress in the 0° fiber next to the center hole as the criterion for laminate failure, the predicted strengths were within 8% of the experimental values.

Acknowledgments

The first author gratefully acknowledges the support extended by the National Research Council, Washington, DC, through their Associateship Program, while conducting this research at the NASA Langley Research Center, Hampton VA.

References

[*1*] Hillberry, B. M. and Johnson, W. S., "Prediction of Matrix Fatigue Crack Initiation in Notched SCS-6/Ti-15-3 Metal Matrix Composites," *Journal of Composites Technology & Research,* Vol. 14, No. 4, 1992, pp. 221–224.

[*2*] Bakuckas, J. G., Jr., and Johnson, W. S., "Influence of Thermal Residual Stresses on the Fiber Bridging Analysis of Titanium Matrix Composites," NASA TM 109082, NASA Langley Research Center, Hampton, VA, Feb. 1994.

[*3*] Bakuckas, J. G., Jr., Johnson, W. S., and Bigelow, C. A., "Fatigue Damage in Cross-Ply Titanium Metal Matrix Composites Containing Center Holes," *Journal of Engineering Materials and Technology,* Vol. 115, Oct. 1993, pp. 404–410.

[*4*] Bigelow, C. A., "A Micromechanics-Based Strength Prediction Methodology for Notched Metal-Matrix Composites," *Composites,* Vol. 24, No. 2, 1993, pp. 113–121.

[*5*] Bigelow, C. A. and Johnson, W. S., "Effect of Fiber-Matrix Debonding on Notched Strength of Titanium Matrix Composites," *Fracture Mechanics: Twenty-Third Symposium, ASTM STP 1189,* R. Chona, Ed., American Society for Testing and Materials, Philadelphia, 1993, pp. 696–712.

[*6*] Bakuckas, J. G., Jr., and Johnson, W. S., "Application of Fiber Bridging Models to Fatigue Crack Growth in Unidirectional Titanium Metal Matrix Composites," *Journal of Composites Technology and Research,* Vol. 15, No. 3, Fall 1993, pp. 242–255.

[*7*] Bakuckas, J. G., Jr., and Johnson, W. S., "Modeling Fatigue Damage in Cross Ply Titanium Matrix Composites," NASA TM 108988, NASA Langley Research Center, Hampton, VA, 1993.

[*8*] Bakuckas, J. G., Jr., Prosser, W. H., and Johnson, W. S., "Monitoring Damage Growth in Titanium Matrix Composites Using Acoustic Emission," *Journal of Composite Materials,* Vol. 28, No. 4, 1994, pp. 305–328.

[*9*] Mirdamadi, M., Johnson, W. S., Bahei-El-Din, Y. A., and Castelli, M. G., "Analysis of Thermomechanical Fatigue of Unidirectional Titanium Matrix Composites," NASA TM 104104, NASA Langley Research Center, Hampton, VA, 1991.

[*10*] Mirdamadi, M. and Johnson, W. S., "Prediction of Stress-Strain Response of SCS-6/Timetal-21S Subjected to the Hypersonic Flight Profile," NASA TM 109026, NASA Langley Research Center, Hampton, VA, 1993.

[*11*] Lekhnitskii, S. G., *Theory of Elasticity of an Anisotropic Body,* Holden-Day, San Francisco, 1963.

[*12*] Heywood, R. B., *Designing by Photoelasticity,* Chapman and Hall, Ltd., London, 1952.

[*13*] Marshall, D. B., Cox, B. N., and Evans, A. G., "The Mechanics of Matrix Cracking in Brittle-Matrix Fiber Composites," *Acta Metallurgica,* Vol. 33, No. 11, 1985, pp. 2013–2021.

[*14*] McCartney, L. N., "Mechanics of Matrix Cracking in Brittle-Matrix Fibre-Reinforced Composites," *Proceedings,* Royal Society London, A 409, 1987, pp. 329–350.

[*15*] McMeeking, R. M. and Evans, A. G., "Matrix Fatigue Cracking in Fiber Composites," *Mechanics of Materials*, Vol. 9, 1990, pp. 217–227.

[*16*] Walker, W., "The Effect of Stress Ratio During Crack Propagation and Fatigue for 2024-T3 and 7075-T6 Aluminum," *Effects of Environment and Complex Load History on Fatigue Life, ASTM STP 462*, American Society for Testing and Materials, Philadelphia, 1970, pp. 1–14.

[*17*] Bigelow, C. A. and Bahei-El-Din, Y. A., "Plastic and Failure Analysis of Composites (PAFAC)," LAR-13183, COSMIC, University of Georgia, Athens, 1983.

David M. Harmon,[1] Mark A. Finefield,[1] Charles R. Saff,[1] and James A. Harter[2]

Titanium Matrix Composite Durability Analyses for the Design of Airframe Structures

REFERENCE: Harmon, D. M., Finefield, M. A., Saff, C. R., and Harter, J. A., **"Titanium Matrix Composite Durability Analyses for the Design of Airframe Structures,"** *Life Prediction Methodology for Titanium Matrix Composites, ASTM STP 1253,* W. S. Johnson, J. M. Larsen, and B. N. Cox, Eds., American Society for Testing and Materials, 1996, pp. 520–539.

ABSTRACT: A fatigue life prediction technique is described for use with titanium matrix composite materials. Damage modes specific to this material system are discussed for unnotched and center-notched laminates. Test data for isothermal fatigue and thermomechanical fatigue specimens are compared against predictions made using the analysis method.

KEYWORDS: titanium matrix composites, damage modes, damage initiation, life prediction, titanium, titanium alloys, fatigue (materials), modeling

Metal matrix composites (MMC) are very complex material systems. These materials combine metal alloys, which can have large amounts of plasticity and complex fatigue failure mechanisms, with stiff, brittle ceramic fibers at extremely high temperatures. Depending on the operating temperature for the material, the composite can have a significant residual stress state due to the differences in thermal expansion between the fiber and matrix that can not be ignored during analysis. Additionally, in a multi-directional composite, there may be significant residual stresses between plies. To further appreciate the complexity of this material system, now consider that there is an interface layer between the fiber and matrix that, to some extent, bonds the matrix to the fiber. The composition and properties of the interface are dependent on the constituents and the consolidation process. Both the residual stress state and the interface properties affect fatigue behavior of the composite. Use of these materials in any application requires an understanding of these mechanisms that control damage initiation, growth, and life.

Due to the complex nature of the material and damage modes that may be present, fatigue analyses for metal matrix composites must include damage initiation and growth models for several types of failure modes. The objective of this work was to develop a fatigue life design tool for engineering purposes. Development of this design tool relied on detailed micro-mechanical models to provide an understanding of the material behavior and to subsequently allow for the development of a more simplified analysis technique. However, since not all aspects of the material behavior are thoroughly understood, the model does rely on some lamina data. At a minimum, the model requires matrix fatigue data and manipulates it for lamina life analysis.

[1]Principal technical specialist, project engineer, and MDC fellow, respectively, McDonnell Douglas Aerospace, St. Louis, MO 63166.

[2]Research engineer, Wright Laboratories, WL/FIBEC, Wright Patterson AFB, OH 45433-6553.

The computer program that contains this design tool (MMCLIFE V2.0) was written using a modular format. Therefore, as better understanding is acquired, specific routines within the code may be replaced. The MMCLIFE code serves as a foundation in which future enhancements may be added.

There are three areas to be considered when investigating fatigue in MMCs: (1) the type of damage mode that will occur, (2) when this damage first appears, and (3) how it will grow. This paper concentrates on each of these items. The fatigue analyses and the philosophy behind them are discussed later. The bulk of this work was based on a SCS-6/Ti-15V-3Cr-3Sn-3Al (SCS-6/Ti-15-3) composite. This material was consolidated using an acrylic fiber mat and Ti-15-3 foils. Consolidation was performed at 982°C for 1 h under 103 MPa pressure and then air cooled. No subsequent aging was performed.

Damage Modes in Notched Metal Matrix Composite Coupons

Several distinct damage modes have been observed in center-notched MMCs (Fig. 1). Specifically, three modes have been identified: (1) cracks growing through the net section, (2) matrix cracks growing parallel to fibers in both 0° and off-axis laminates, and (3) matrix cracks growing parallel to fibers with multiple matrix cracks growing across the net section but not breaking fibers. No debonding or delaminations were observed in the MMC material tested under this program, so these modes are not discussed. The type of mode is controlled by many variables, including applied stress level, temperature, lamina stiffness, and interface properties.

Fatigue cracks will initiate at those locations on the notch perimeter at which stresses are highest with respect to the strengths of the matrix or fiber, or the fiber-matrix interface. In unidirectional metal matrix composites that have lower shear strengths, cracks tend to initiate at the notch due to shear stresses and propagate parallel to the fibers. Conversely, unidirectional composites with higher shear strengths initiate cracks at the notch edge 90°

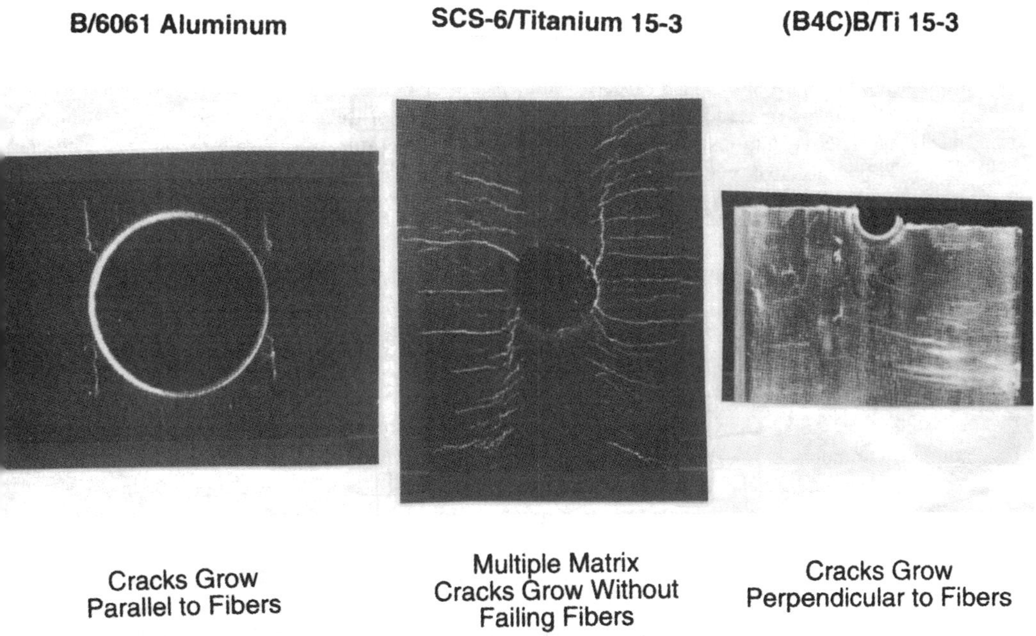

FIG. 1—*Damage modes in three different metal matrix composites.*

from the load direction. The local stress state at each point around the notch can be calculated in terms of longitudinal, transverse, and shear stress components using methods developed by Lekhnitskii [*1*]. When divided by the appropriate strength parameter (that is, longitudinal stress/longitudinal lamina strength or shear stress/shear strength), it is obvious where these critical locations are located (Fig. 2) where θ is measured from the load axis. A failure mode plot has been developed for a longitudinally loaded, 0° laminate with a center open hole. This plot determines the transition from net section cracks to cracks that grow parallel to fibers based on the ratio of lamina longitudinal strength, F^L_{tu}, to shear strength, F_{su}, and the lamina stiffness (Fig. 3). The transition line is determined by equating the gross stress required to initiate a crack at the maximum shear stress location to that required to start a crack at the notch edge. If the gross stress required to initiate failure at the notch edge is less than the gross stress required to fail the lamina in shear, then the crack will propagate through the net section. If the alternative is true, then the crack will propagate parallel to the fibers. The equation that governs the transition line is

$$\frac{F^L_{tu}}{F_{su}} = \frac{K_t}{K_s} \tag{1}$$

where K_s is the ratio of the shear stress parallel to the fibers to the gross applied stress and K_t is the ratio of the longitudinal stress to the gross applied stress. The ratio of K_t/K_s is a function of material stiffness [2].

In aluminum matrix composites, the shear strength is low, driving the strength ratio up and, therefore, results in cracks that initiate on the notch perimeter at an angle of 65° from the load axis, parallel to the fibers. The $(B_4C)B/Ti$ material system was subjected to a secondary heat treatment that degraded the fiber strength and resulted in cracks that grew through the net section failing fibers. At room temperature, the SCS-6/Ti material system resulted in multiple matrix net section cracks that grew throughout the specimen (Fig. 4).

Since the failure mode plot is a function of lamina stiffness (that is, E_1, E_2, G_{12}, ν_{12}) and strength, anything that affects these properties will also affect the mode of damage. Temperature is an obvious example. At high temperatures, the shear strength degrades faster than the longitudinal strength. The effect of this property degradation is to drive the damage mode to one in which cracks initiate away from the edge of the notch and up parallel along the fibers. This is demonstrated using the elevated temperature properties for SCS-6/Ti in Fig. 3. This prediction was verified by testing performed at 427 and 538°C that showed the predicted change in damage mode (Figs. 5 and 6).

Damage Initiation and Life Prediction

Unnotched Life Analysis

Formulation of a fatigue damage initiation analysis for MMCs began with a study of unnotched laminate fatigue data (Fig. 7). Once a model was developed to predict the life of unnotched coupons, then this model could also be applied to predict crack initiation of notched laminates by analyzing the local stresses around the notch perimeter.

A test program was conducted to examine the fatigue life of unnotched $[0°]_6$ unidirectional and $[0°/\pm45°]_s$ SCS-6/Ti-15-3 laminates at room temperature and 260°C. Stress ratios of 0.02 and 0.5 were examined. The specimen dimensions were 0.01 m wide by 0.15 m long. Titanium tabs were welded on to the end of all specimens. Quartz lamps were used for the elevated temperature tests.

A relatively flat stress versus life curve (Fig. 7) was obtained for unidirectional and

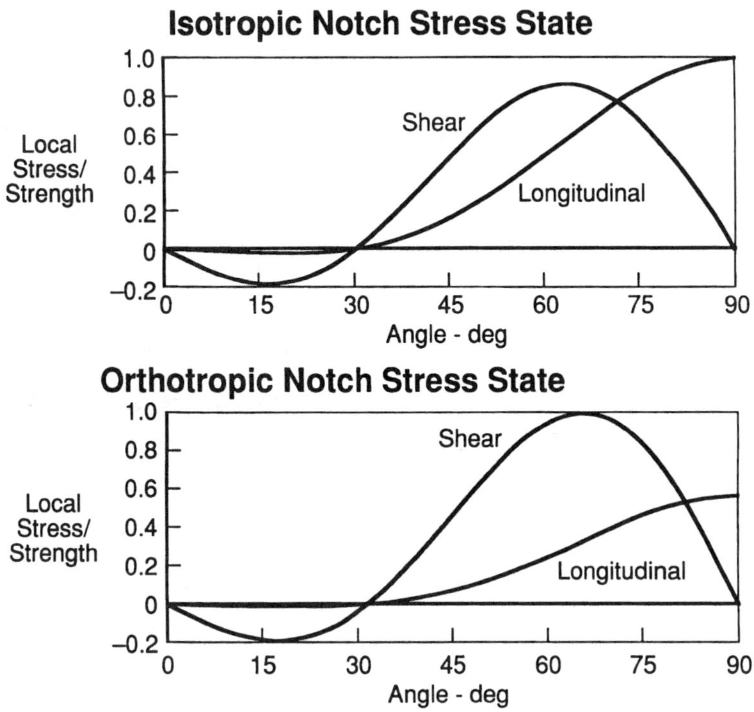

FIG. 2—*Critical locations around notch are easily determined once notch stress state is known.*

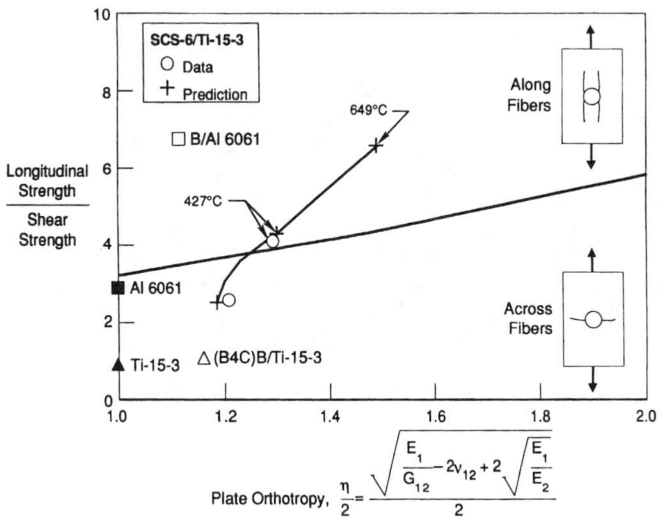

FIG. 3—*Damage modes are a function of stiffness, strength, and temperature.*

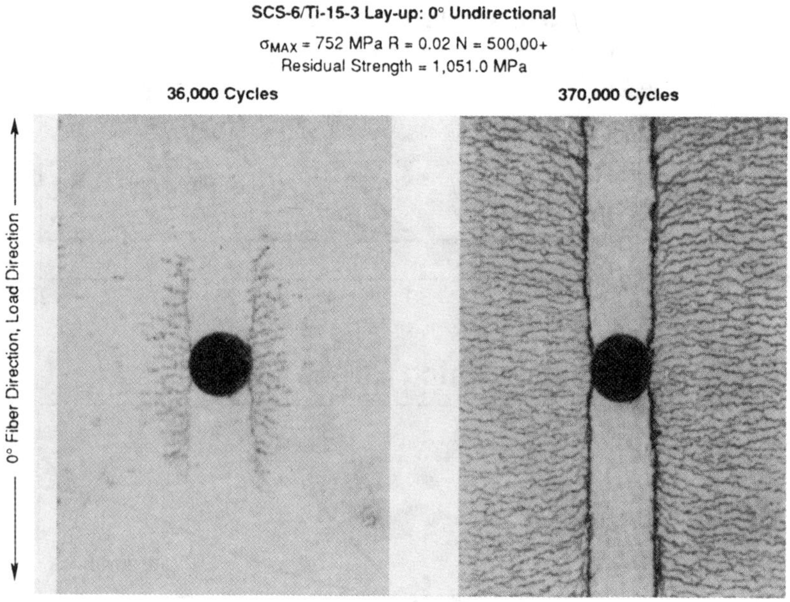

FIG. 4—*A diffuse damage pattern was observed in SCS-6/Ti-15-3 at room temperature.*

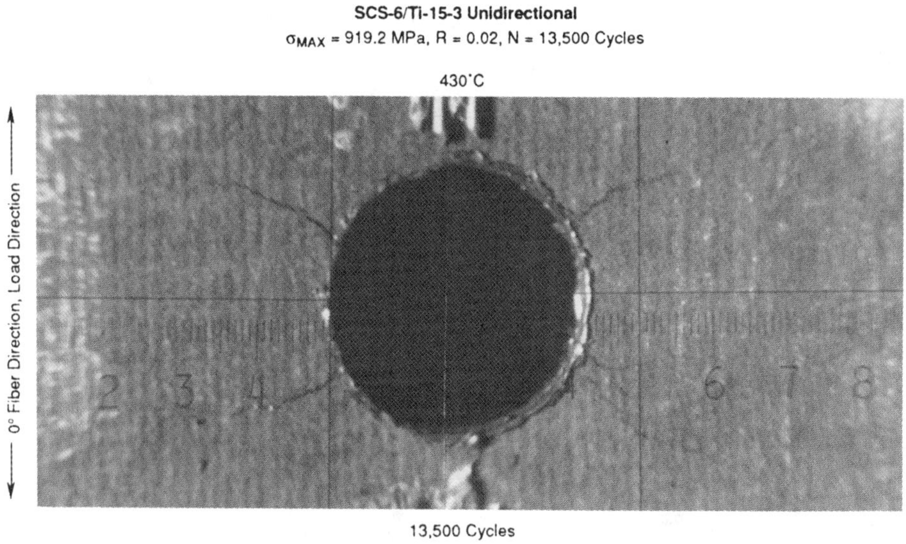

FIG. 5—*Damage mode changes near 427°C in SCS-6/Ti-15-3.*

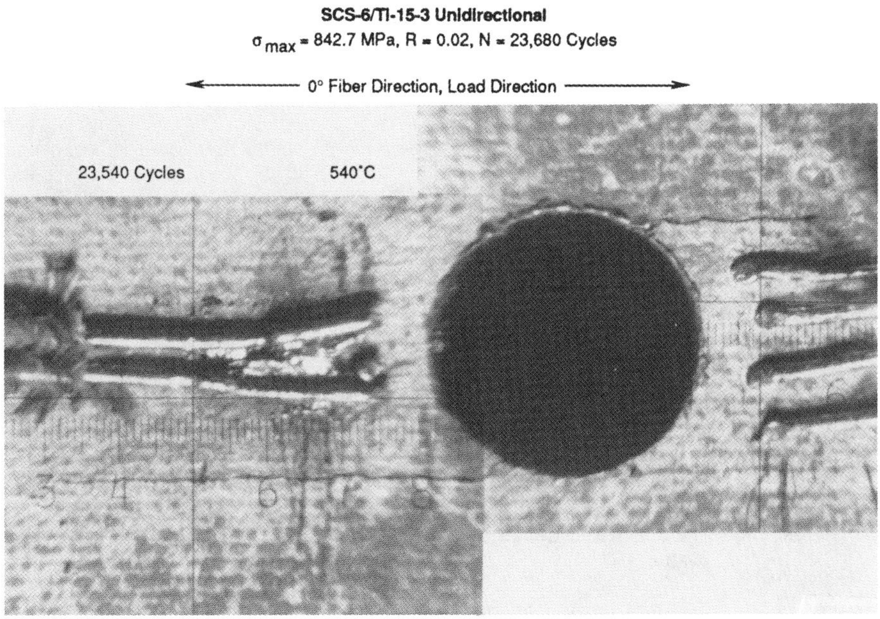

FIG. 6—*Cracks grow parallel to fibers at 538°C in SCS-6/Ti-15-3.*

$[0° \pm 45°]_s$ materials. Little difference was seen between the room temperature and the 260°C tests. The solid symbols represent specimens that were bowed or damaged in some way. The solid lines are straight line curve fits through the data. The bow in the $[0°/\pm 45°]_s$ laminates was a result of improper cool down after consolidation. The deformation occurred during fabrication when one side of the panel was cooled faster than the other, causing it to warp along the fibers.

It was observed that the 0° fiber stress controls the life of these specimens and, therefore, allows the curves from the different laminates to be consolidated to one (Fig. 8). This same

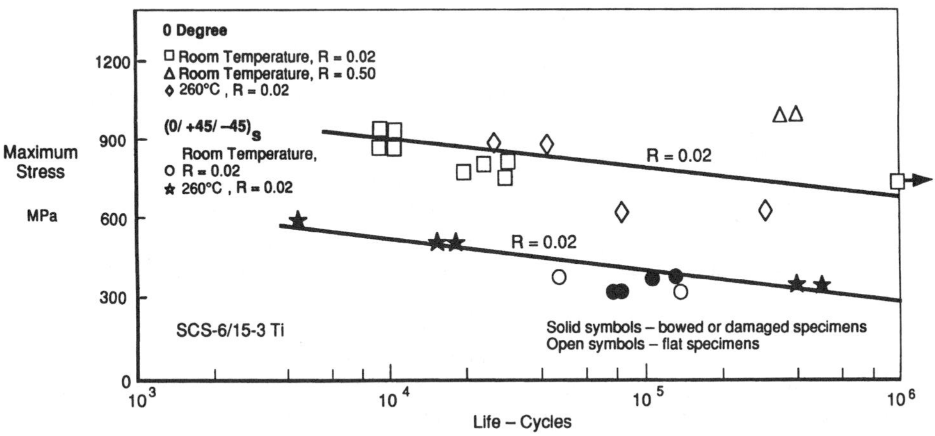

FIG. 7—*Unnotched fatigue life data (SCS-6/Ti-15-3).*

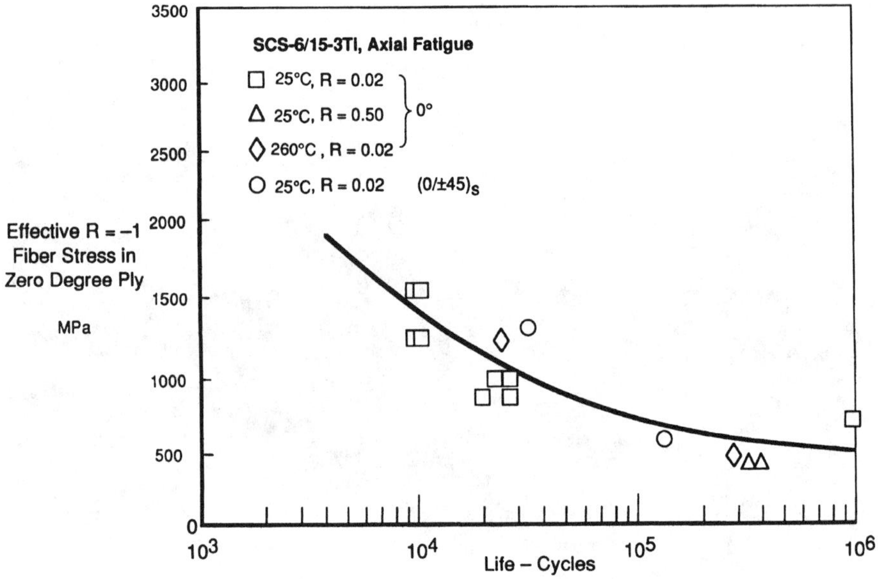

FIG. 8—*0° fiber stress controls life.*

behavior has been observed by Johnson in boron/aluminum coupons [*3*]. The 0° fiber stress was calculated by multiplying the applied stress by the ratio of fiber modulus to laminate modulus. This assumes that the strain is constant across the laminate.

The fatigue life model is based on a damage accumulation scheme that tracks damage in both the fiber and the matrix. Given a generic load and temperature profile, constituent stresses are determined that include the effects of residual stresses and matrix creep relaxation (Fig. 9). The model assumes that the fibers are insensitive to creep.

The material properties required for each constituent include stiffness and strength versus temperature, thermal expansion coefficient, and normalized stress versus life curves. If performing a creep analysis, strain versus time data for several stress levels and temperatures is required. The stress versus life curve for the fibers is obtained from unnotched 0° laminate fatigue tests (Fig. 8).

The applied temperature and load profiles are divided into segments in which the variation of load and temperature within each segment are defined as constant, ramping, or cyclic. Depending on the segment definition, a finite number of points throughout the segment are defined at which the constituent stresses are determined.

The constituent stress analysis performed at each point begins with a temperature-dependent stiffness calculation. Lamina stiffness are determined from rule of mixtures or from available lamina data. Laminated plate theory is used to calculate laminate stiffness. Interpolation schemes are used to find properties for temperatures not included in the input data. Extrapolation of data is not permitted. In the event that a temperature exceeds the range of data provided, then the closest boundary temperature is used.

Based on the applied stress and temperature and using composite mechanics as discussed by Jones [*4*], the longitudinal, transverse, and shear stresses along the lamina local coordinate system are determined for each ply. These lamina stresses are then converted to constituent stresses by inverting the rule of mixtures expressions [2]. The effect of residual stresses is accounted for using superposition. Fiber and matrix residual stresses are calcu-

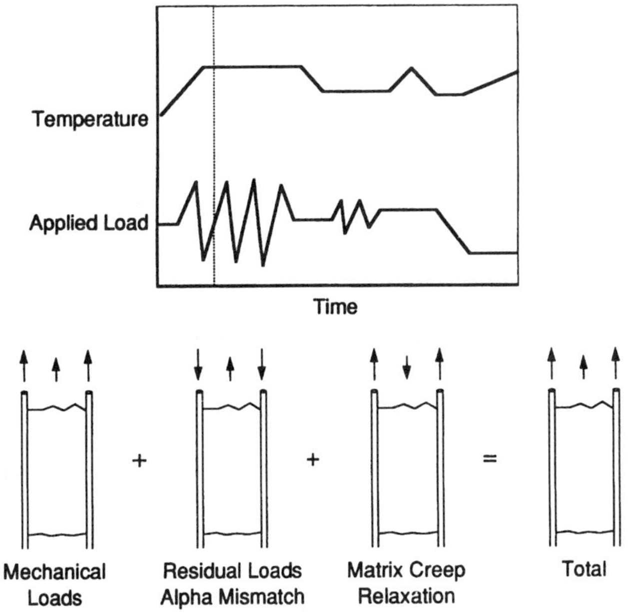

FIG. 9—*Fatigue life analysis includes effects of residual stresses.*

lated assuming a bond has formed between the matrix and fiber at the consolidation temperature, T_c, and that the bond is still intact at the current temperature, T_i. The residual longitudinal fiber and matrix stresses, $\sigma_{\text{res}}^{1,f}$ and $\sigma_{\text{res}}^{1,m}$, are

$$\begin{aligned} \sigma_{\text{res}}^{1,f} &= E_f \epsilon_{\text{res}}^{f} = E_f(\alpha_f - \alpha_{//})(T_c - T_i) \\ \sigma_{\text{res}}^{1,m} &= E_m \epsilon_{\text{res}}^{m} = E_m(\alpha_m - \alpha_{//})(T_c - T_i) \end{aligned} \tag{2}$$

where α_f, α_m, and $\alpha_{//}$ are the thermal expansion coefficients for the fiber, matrix, and lamina. The lamina longitudinal thermal expansion coefficient is calculated simply by summing all forces parallel to the fibers and equating them to zero

$$\Sigma F = [\sigma_{\text{res}}^{1,f} V_f + \sigma_{\text{res}}^{1,m}(1 - V_f)]A_x = 0 \tag{3}$$

where A_x is the cross-sectional area of laminate. Substituting the expressions for the fiber and matrix residual stresses and solving for $\alpha_{//}$, the following expression is obtained

$$\alpha_{//} = \frac{\alpha_f E_f V_f + \alpha_m E_m(1 - V_f)}{E_f V_f + E_m(1 - V_f)} \tag{4}$$

Significant transverse stresses can also exist due to thermal expansion mismatch. Maximum values of matrix residual stress (at the fiber/matrix interface) can be approximated by an analysis based on a concentric cylinder elastic model [5]. The final expression for the transverse matrix residual stress, $\sigma_{\text{res}}^{2,m}$, is

$$\sigma_{\text{res}}^{2,m} = \frac{(\alpha_f - \alpha_m)(T_c - T_i)(1 - V_f)E_m E_f(1 + V_f)/(1 - V_f)}{E_f V_f + E_m(1 - V_f) + E_f/(1 + \upsilon_m(1 - V_f)) - \upsilon_f E_m} \tag{5}$$

The residual stresses in the fiber and matrix act upon the same area. Therefore, the transverse fiber residual stress, $\sigma_{\text{res}}^{2,f}$, is equal in magnitude to the matrix residual stress

$$\sigma_{\text{res}}^{2,f} = \sigma_{\text{res}}^{2,m} \tag{6}$$

If the testing is done at temperature, it is possible that the stress state in the matrix would be partially or completely relieved due to creep relaxation. Methods to accommodate this behavior are significantly more involved and are not included in this discussion. If creep relaxation is ignored, the total fiber and matrix stresses are found by adding the applied and residual stresses

$$\begin{aligned} \sigma^{1,f} &= \sigma_{\text{applied}}^{1,f} + \sigma_{\text{res}}^{1,f} \quad \sigma^{2,f} = \sigma_{\text{applied}}^{2,f} + \sigma_{\text{res}}^{2,f} \\ \sigma^{1,m} &= \sigma_{\text{applied}}^{1,m} + \sigma_{\text{res}}^{1,m} \quad \sigma^{2,m} = \sigma_{\text{applied}}^{2,m} + \sigma_{\text{res}}^{2,m} \end{aligned} \tag{7}$$

Assuming that the fiber and matrix act as separate isotropic materials, an equivalent von Mises stress is calculated for each σ^f and σ^m. This allows the constituent longitudinal, transverse, and shear stress terms to be combined into one equivalent stress term

$$\sigma^m = \sqrt{\frac{1}{2}\left[(\sigma^{1,m} - \sigma^{2,m})^2 + \sigma^{1,m^2} + \sigma^{2,m^2} + 6\tau^{12,m^2}\right]} \tag{8}$$

Since the fiber shear stress is assumed to be zero, the von Mises equation for the fiber simplifies to

$$\sigma^f = \sqrt{\sigma^{1,f^2} + \sigma^{2,f^2} - \sigma^{1,f}\sigma^{2,f}} \tag{9}$$

Matrix and fiber stresses are calculated at every point in the segment and for each segment of the spectrum. Two new stress profiles have been created: one for the fiber and one for the matrix. However, these profiles already include the effects of temperature, residual stresses, and creep relaxation. Due to these effects, new peak and valley stresses exist for both the fiber and matrix. A rainflow counting routine is used to combine the new constituent level stress spectra into peak and valley pairs. Note, however, that these spectra are only valid until the first failure is predicted to occur. When this happens, new fiber and matrix stress profiles must be created using the same procedure described earlier, but with reduced stiffness properties to account for the failed constituent.

Miner's rule is used to track the damage accumulation in both the fiber and matrix (Fig. 10). After combining the peak and valley pairs, a stress ratio, R, is calculated for each pair by dividing the valley stress by the peak stress. A simplified stress ratio correction factor based on the expression developed by Smith, Watson, and Topper [*6*] is applied to obtain effective $R = -1$ stresses

$$\sigma_{\text{eff}}^{f} = \sigma_{\text{peak}}^{f}\sqrt{\frac{1 - R_f}{2}} \qquad \sigma_{\text{eff}}^{m} = \sigma_{\text{peak}}^{m}\sqrt{\frac{1 - R_m}{2}} \tag{10}$$

where

$$R_f = \frac{\sigma_{\text{valley}}^{f}}{\sigma_{\text{peak}}^{f}} \qquad R_m = \frac{\sigma_{\text{valley}}^{m}}{\sigma_{\text{peak}}^{m}} \tag{11}$$

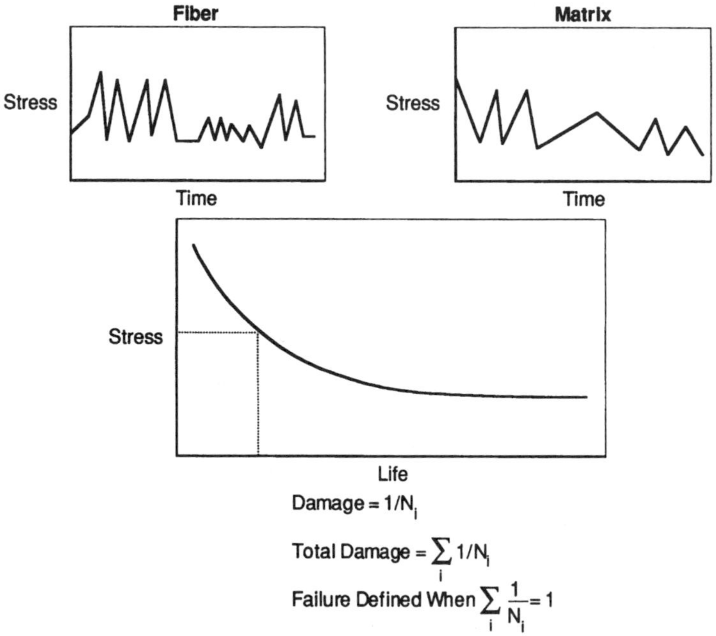

FIG. 10—*A damage accumulation scheme is used to determine constituent life.*

The effective fiber and matrix stresses are normalized by dividing by the corresponding fiber and matrix strengths at temperature, T_i. Then the life to failure is determined from the normalized stress versus life curves. The damage in the fiber and matrix is calculated as

$$D^f_{\text{ply}} = \frac{\#\text{ of cycles @ } \sigma^f_{\text{eff}}}{\text{Life @ } \sigma^f_{\text{eff}}} \qquad D^m_{\text{ply}} = \frac{\#\text{ of cycles @ } \sigma^m_{\text{eff}}}{\text{Life @ } \sigma^m_{\text{eff}}} \tag{12}$$

Damage is tracked in each ply for both the fiber and matrix. The damage is cumulative. When the total damage exceeds 1, failure is predicted to occur.

Lamina and laminate stiffness are recalculated whenever a fiber or matrix failure occurs. If the matrix fails in a ply, the transverse and shear lamina stiffness, E_2 and G_{12}, are reduced. Conversely, if the fibers fail in a ply, the longitudinal stiffness, E_1, is reduced. An arbitrary factor of 1000 was used to reduce the stiffness of the failed constituent. This factor was chosen large enough to reduce the stiffness such that the failed constituent would carry minimal additional load. In theory, the reduced stiffness should be zero, but this led to computational errors. Using a factor of 1000 eliminated these errors in the computer code.

Definition of ultimate laminate failure is dependent on the type of applied loading. If the applied loading is not parallel to any of the fibers in any of the plies in the laminate then the laminate is predicted to fail when the matrix has failed in each ply. However, if the applied loading is parallel to the fibers in at least one ply, then it is assumed that these fibers will control the life of the specimen. In this case, failure will not occur until those fibers have failed. For example, a $[\pm 45°]_s$ loaded longitudinally is predicted to fail when the damage for the matrix in each ply accumulates to one. However, a $[0°/\pm 45°/90°]_s$ laminate loaded transversely will not fail until the fibers in the 90° ply fail.

The complete unnotched fatigue analysis described earlier was used to predict the response of both unidirectional and $[0°/\pm45°]_s$ SCS-6/Ti-15-3 when subject to isothermal constant amplitude loading (Fig. 11). The $R = 0.02$ unidirectional data shown in Fig. 11 was used to create the in situ fiber life curve for this analysis, therefore, it is not surprising that the analysis correlates with this data. However, based on this in-situ fiber life curve, the analysis was also able to predict the $R = 0.50$ unidirectional response and the $[0°/\pm45°]_s$ response.

Crack Initiation in Notched Laminates

The fatigue crack initiation routine for notched laminates combines the unnotched fatigue analysis described earlier with the notched stress analysis developed by Lekhnitskii [*1*] and modified by Harmon et al. [*2*]. Essentially, the unnotched analysis is performed at multiple points around the notch. The same material property data required for the unnotched analysis is used here.

The laminate strain analysis considers the multi-ply, notched laminate as a single orthotropic lamina with a center notch. The properties used for the lamina are those calculated for the multi-ply laminate. The decision to represent the laminate as a single ply was made for two reasons. First, in the vicinity of the notch in an undamaged laminate, the strain field is a global response. Typically, when crack initiation is observed at the surface of the specimen, it tends to be through the thickness. Second, performing the damage analysis at every point around the notch, for every ply in a laminate, is a tedious, time consuming task, unwarranted for the type of design tool being created. The number of calculations are significantly reduced when performing the analysis on a single lamina with smeared laminate properties.

Laminate properties are calculated as described for the unnotched analysis except that a laminate coefficient of thermal expansion is also calculated. This term is calculated using the thermal loads, N_T, induced by the ply-by-ply thermal expansion terms and the inverse of the laminate stiffness matrix, $[\bar{Q}]^{-1}$, for the smeared lamina. $[\bar{Q}]^{-1}$ is determined by taking the

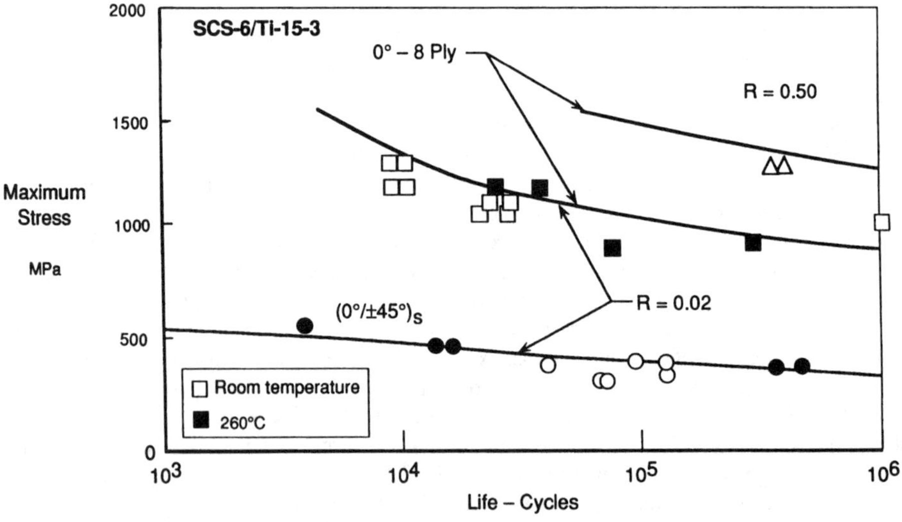

FIG. 11—*Fatigue analysis predicts effects of stress ratio and lay-up on life.*

laminate stiffness terms (E_x, E_y, G_{xy}, and ν_{xy}) and utilizing them as "lamina" properties to calculate a "lamina" stiffness matrix that represents the laminate

$$[\alpha]_{\text{laminate}} = \frac{1}{t\Delta T}[\bar{Q}]^{-1}\{N_T\} \tag{13}$$

In this expression, t is the laminate thickness and ΔT is the temperature change.

Using a closed-form analysis developed by Lekhnitskii, the tangential stress is found at 5° increments around half of the notch. Once the tangential stress is determined, it is broken down into longitudinal, transverse, and shear-ply stresses using Mohr's circle. These terms are then converted to fiber and matrix stresses at each point. The analysis then follows in the same manner as the unnotched routine (residual stresses, creep relaxation, von Mises stresses, Miners rule, and rainflow counting).

The result of this analysis is a prediction of the initiation of a crack at some point around the notch. The analysis can track crack initiation in the matrix, fiber, or both. It assumes the matrix to be elastic-perfectly plastic. As a result, the code will provide similar matrix predictions regardless of whether the matrix is brittle or ductile and, therefore, the analysis predictions for matrix crack initiation tend to be conservative.

Figures 12 and 13 are predictions obtained for a unidirectional SCS-6/Ti-15-3 laminate with a center circular hole. Both of these plots identify crack initiation based on fiber stresses. As a result, the critical point around the notch for the unidirectional laminate is at the hole edge (θ = 90°).

Life Prediction in Laminates with Multiple Matrix Cracks

The analysis developed for unnotched life prediction can also be applied to predict life in notched coupons for the SCS-6/Ti-15-3 material. This material, when fatigue tested at room

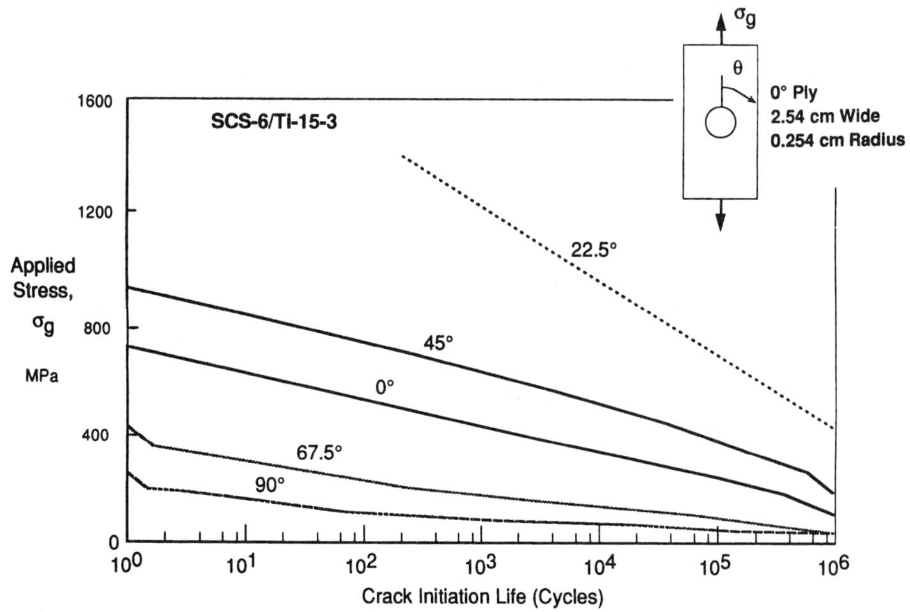

FIG. 12—*Crack initiation as a function of crack location around notch perimeter.*

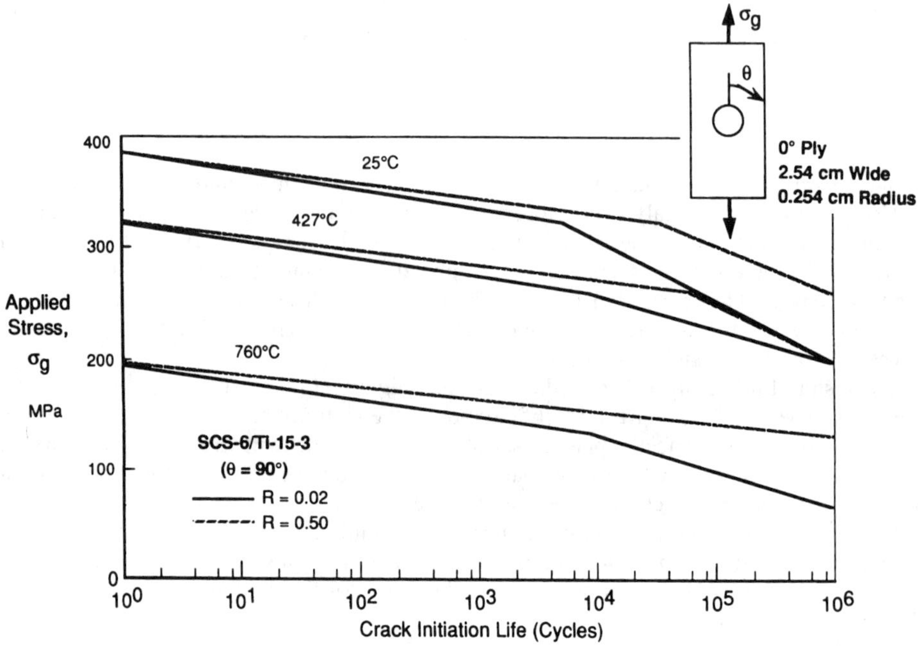

FIG. 13—*Crack initiation life as a function of temperature.*

temperature, was extremely notch insensitive. The specimen shown in Fig. 4 had a residual strength after cycling that was 85% of the virgin notch strength. Cracks initiated from the center hole and grew throughout the specimen without failing fibers. In fact, the residual strength of the coupon is nearly equal to the load carried in the net section fibers when at their ultimate strength.

Proceeding as for the unnotched coupons, an in-situ fiber life curve was created using the notched 0° unidirectional fatigue test coupons. The stress in the net section 0° fibers is calculated as

$$\sigma_{0^\circ\ \text{fiber}} = \sigma_{\text{applied}} \frac{W}{W - D} \frac{1}{V_f} \tag{14}$$

where W is the coupon width and D is the hole diameter.

This expression assumes that the matrix begins to crack very early and, therefore, carries no load. This assumption is based on high residual stresses due to the coefficient of thermal expansion (CTE) mismatch between fiber and matrix. This is demonstrated with the following example for a SCS-6/Ti-15-3 composite. The lamina longitudinal CTE, $\alpha_{//}$, is calculated using a rule of mixtures approach

$$\alpha_{//} = \frac{\alpha_f E_f V_f + \alpha_m E_m (1 - V_f)}{E_f V_f + E_m (1 - V_f)} \tag{15}$$

where α_f = 4.9e-6/°C, α_m = 9.9e-6/°C, E_f = 379 MPa, E_m = 86 MPa [7], and fiber volume fraction, V_f, is 0.4. The laminate CTE is computed as 6.1e-6/°C. The residual stress in the matrix is calculated using the following

$$\sigma_{res}^{1,m} = E_m \epsilon_{res}^{m} = E_m(\alpha_m - \alpha_{//})(T_c - T_i) \tag{16}$$

where T_c = temperature at consolidation and T_i is room temperature. For the SCS-6/Ti-15-3 system, T_c is 871°C and T_i is 25°C. Given these conditions, the residual matrix stress is approximately 276 MPa, which is about 40% of the matrix strength. The stress state imposed on the laminate shown in Fig. 4 causes the stress in the matrix to rapidly approach the failure stress for the matrix. The matrix stress, $\sigma_{applied}^{1,m}$, is found by calculating the load carried by the fiber, $\sigma_{applied}^{1,f}$, and then solving for the matrix stress using rule of mixtures

$$\begin{aligned} \sigma_{applied}^{1,f} &= \sigma_{//}\frac{E_f}{E_fV_f + E_m(1 - V_f)} \\ \sigma_{applied}^{1,m} &= \frac{\sigma_{//} - \sigma_{applied}^{1,f}V_f}{(1 - V_f)} = \sigma_{//}\frac{E_m}{E_fV_f + E_m(1 - V_f)} \end{aligned} \tag{17}$$

where $\sigma_{//}$ is the applied mechanical stress, 752 MPa.

The fatigue test in this example, induces a mechanical stress level of approximately 317 MPa in the matrix. Combining this with the 276 MPa residual stress results in a matrix stress that is about 95% of the ultimate strength of the matrix. This results in the matrix having a short fatigue life and supports the assumption that the matrix contributes almost nothing to the life of the laminate.

To combine data from tests conducted at different stress ratios, R, the Smith, Watson, and Topper stress ratio correction factor [*6*] was applied to the net section fiber stress, $\sigma_{0° fiber}$. The general form of the correction factor is as follows

$$\epsilon_{eff} = \sqrt{\frac{\sigma_{max}\Delta\epsilon}{2E}} \tag{18}$$

where E is the elastic modulus of the material. For a linear elastic material system, the general form can be simplified to

$$\epsilon_{eff} = \epsilon_{max}\sqrt{\frac{1 - R}{2}} \tag{19}$$

and therefore

$$\sigma_{eff} = \epsilon_{eff}E = \sigma_{max}\sqrt{\frac{1 - R}{2}} \tag{20}$$

For the calculations, $\sigma_{max} = \sigma_{0° fiber}$. The effective stress, σ_{eff}, represents an equivalent stress for an $R = -1$ test.

When the effective stress, σ_{eff}, was plotted against the lives obtained in testing for notched specimens, it was observed that a curve fit through this data fell practically on top of the fiber stress curve obtained for unnotched specimens (Fig. 14). This implies that when the material is notch insensitive notched life can be predicted using unnotched data. The effect of the observed fatigue matrix cracks is to direct the load in the specimen to the two net sections as if two unnotched specimens were being tested. Therefore, the notched specimen can be analyzed as an unnotched specimen using the net section stress instead of the applied stress.

A similar comparison was also made for the $[0°/\pm45°]_s$ data (Fig. 15). The agreement is not as good, but the unnotched curve is conservative. This result suggests that the off-axis plies do contribute to the fatigue life of the laminate. The fiber stress formulation ignores

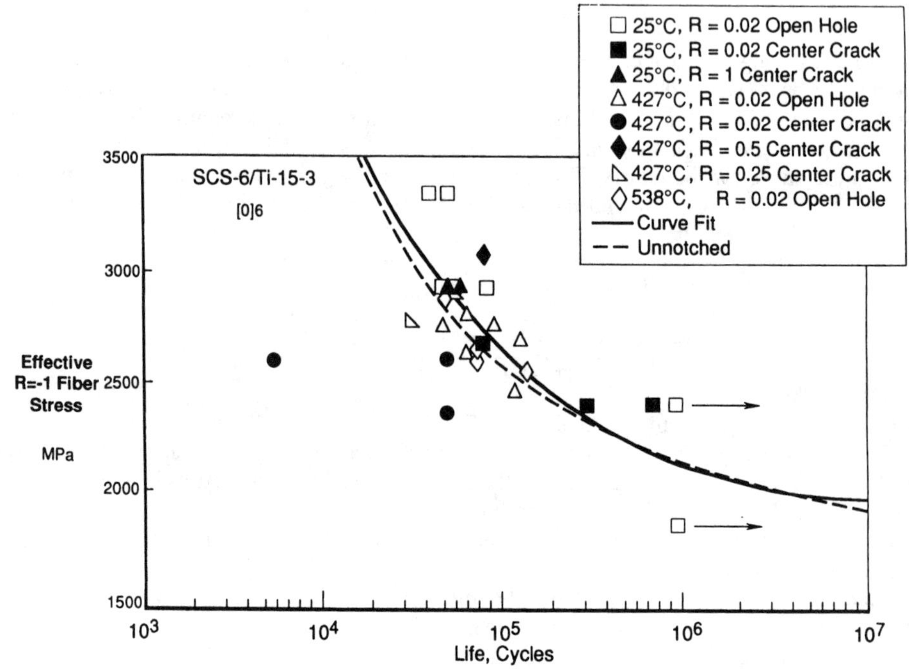

FIG. 14—*0° fiber stress controls fatigue life of notched coupons.*

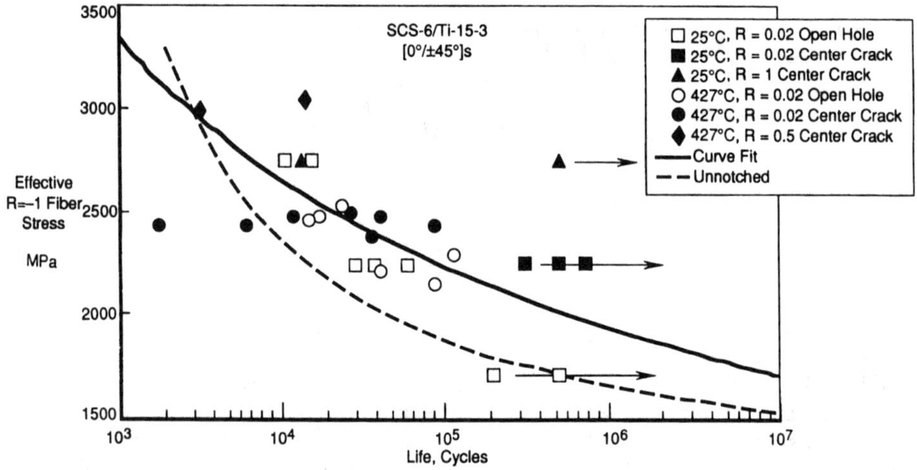

FIG. 15—*Unnotched fatigue life curve provides conservative predictions in* $[0°/\pm 45°]_s$ *laminates.*

this contribution, therefore, the prediction is conservative. In this plot, the 0° fiber stress is computed as

$$\sigma_{0^\circ\ \mathrm{fiber}} = \sigma_{\mathrm{applied}} \frac{W}{W - D} \frac{1}{V_f} \frac{\#\ \text{of plies}}{\#\ \text{of } 0^\circ \text{ plies}} \tag{21}$$

Since the SCS-6/Ti-15-3 material behaves nearly the same in both the notched and unnotched cases, the fatigue life is easily predicted for the notched laminates. The life is obtained by calculating the net section stress and comparing this value to the unnotched stress versus life results predicted by the analysis.

Spectrum Testing

Isothermal and thermomechanical fatigue tests were performed to verify the analysis methodology developed and to provide information on any shortcomings of the program. This information will be used to make future modifications and updates to both the analysis techniques and the computer program.

The SCS-6/Ti-15-3 material system was tested under an advanced fighter spectrum [*2*] shown in Fig. 16. Each of the 16 load blocks consists of 100 load cycles with varying peaks and valleys. This spectrum used in the analysis was simplified and consisted of 16 blocks, each with a single peak and valley that was repeated 100 times. Two analysis spectra were created: one by choosing the highest peak and lowest valley from each load block and the other by choosing the lowest peak and highest valley.

Isothermal spectrum fatigue tests were conducted at 25 and 427°C on center-notched 0° unidirectional and $[0°/\pm 45°]_s$ test coupons. The analysis predictions and the test data are shown in Figs. 17 through 20. The data fell within the range of the predictions in every case except the unidirectional 427°C tests. It is not surprising that tests at 427°C produced shorter lives than predicted, because at this temperature this material appears to be more notch sensitive (Fig. 5) due to the lack of matrix cracks. The stress concentration at the notch edge

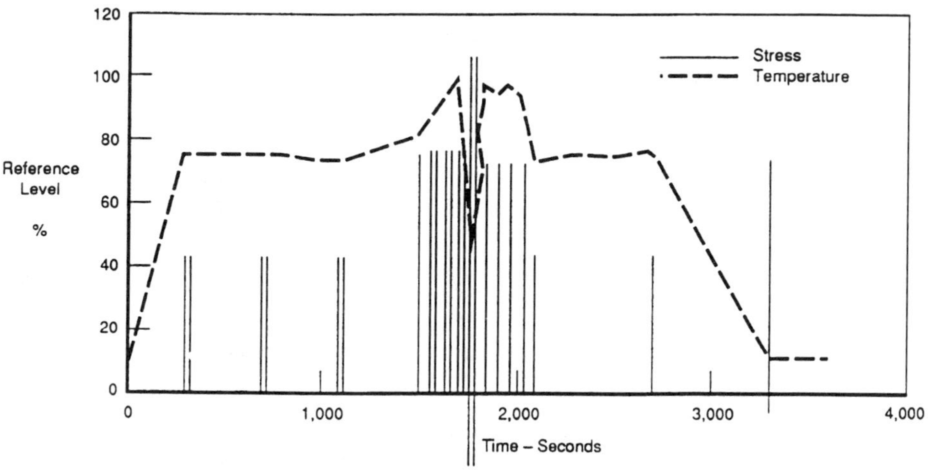

FIG. 16—*Advanced fighter spectrum (100 missions, 1600 load cycles).*

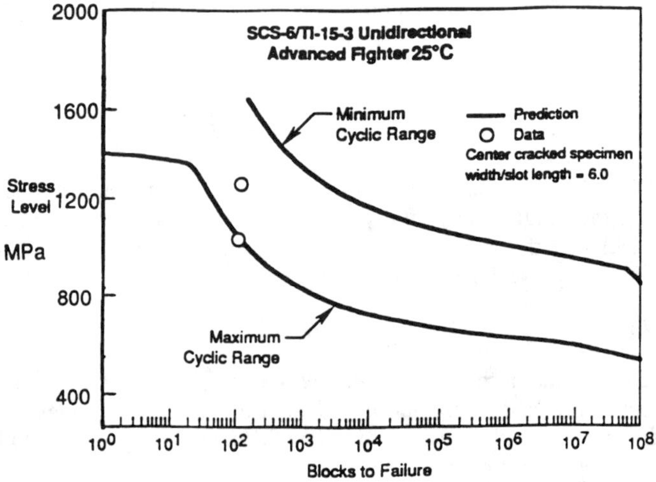

FIG. 17—*Spectrum fatigue life predictions for unidirectional SCS-6/Ti-15-3 at 25°C.*

is not relieved. Therefore, the stress is greater at the notch edge and a fatigue crack initiates and grows through the net section failing fibers.

Thermomechanical fatigue tests were also conducted on center-notched SCS-6/Ti-15-3 using the advanced fighter spectrum. The data points are plotted versus prediction in Figs. 21 and 22. The unidirectional specimens were over-predicted for the lower stress levels while the upper stress levels were closer but were still nonconservative. The predictions for the cross-plied lay-ups were much better for all stress levels.

Discussion of Results

The isothermal and thermomechanical fatigue life predictions were nonconservative for the unidirectional laminates. The fiber life curve used for all of these predictions is based on

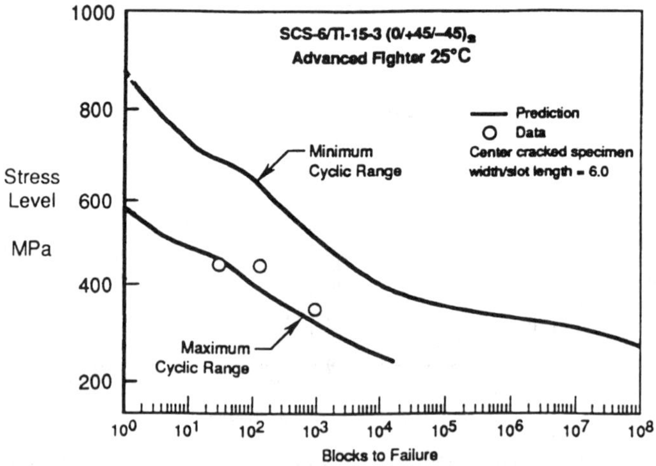

FIG. 18—*Spectrum fatigue life predictions for $[0°/\pm 45°]_s$ SCS-6/Ti-15-3 at 25°C.*

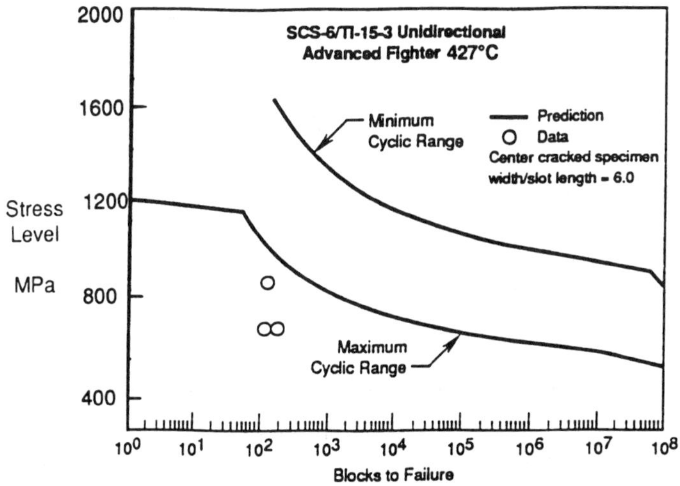

FIG. 19—*Spectrum fatigue life predictions for unidirectional SCS-6/Ti-15-3 at 427°C.*

the fiber life of the isothermal, constant amplitude fatigue, unnotched 0° laminates (Fig. 8). When compared to notched data (Figs. 14 and 15), it is evident that the curve is conservative compared to the cross-plied data. Therefore, the analysis should be under-predicting the lives of the cross-plied laminates. However, the thermomechanical fatigue tests show the analysis to agree very well with the cross-plied data. It is concluded that the 0° fiber life curve is indeed nonconservative in its use here and that some mechanism that was not present in the isothermal, constant amplitude fatigue test is present and unaccounted for in the thermomechanical fatigue tests.

To explain this, lets reexamine the analysis approach. One implication of this approach is that in this SCS-6/Ti-15-3 material system, the fibers appear to be sensitive to fatigue. It is important to realize that the fiber life curve is actually an in situ curve. Any fatigue damage

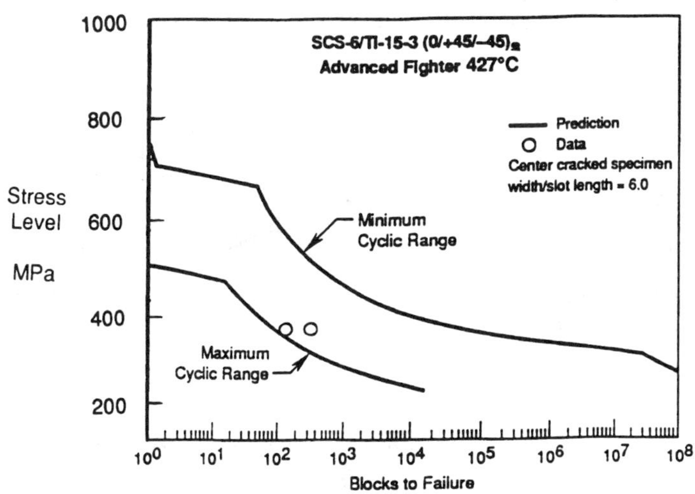

FIG. 20—*Spectrum fatigue life predictions for* $[0°/\pm 45°]_s$ *SCS-6/Ti-15-3 at 427°C.*

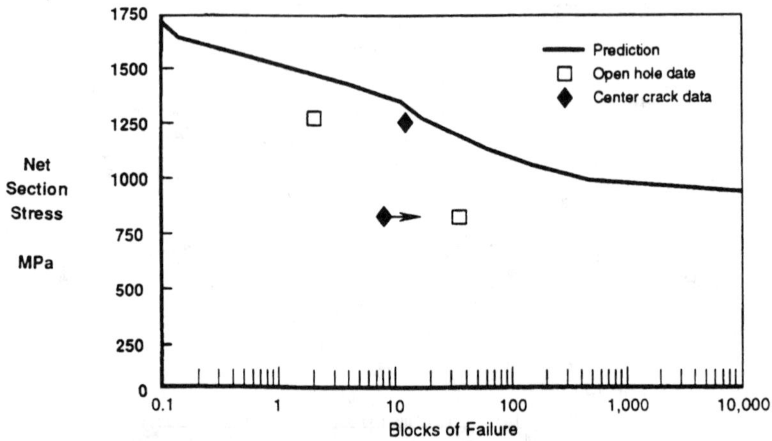

FIG. 21—*Thermomechanical fatigue life predictions for unidirectional SCS-6/Ti-15-3.*

is probably a result of frictional sliding between the fiber and matrix along the interface. This behavior is most evident at room temperature. At higher temperatures, there is less matrix cracking and, therefore, less frictional sliding. The fiber and matrix have the same longitudinal strain as long as the interface bond is intact and there is little matrix cracking. Since matrix cracks are more likely to grow through an intact interface into the fibers, shorter lives than predicted should be expected using the in situ life curve.

Summary

Titanium matrix composites (TMCs) have been considered for several airframe structural components due to their increase in specific stiffness and strength as compared to the metal matrix material alone. Additionally, they show potential for high temperature applications where polymer matrix composites are unusable. However, as with any new material, there are specific durability issues related to the successful application of the system that are still unresolved. This paper summarizes work performed at McDonnell Douglas Aerospace

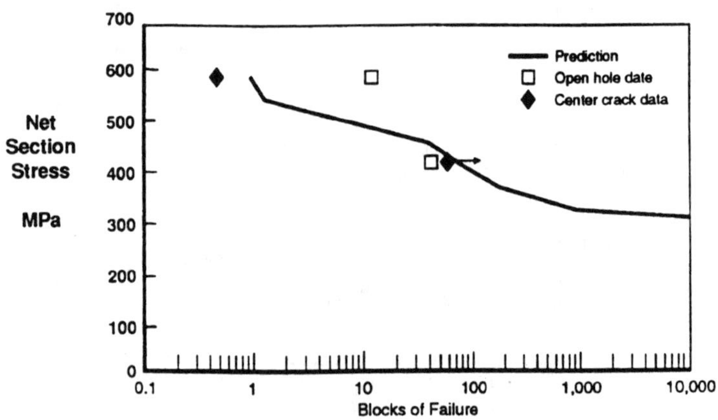

FIG. 22—*Thermomechanical fatigue life predictions for $[0°/\pm 45°]_s$ SCS-6/Ti-15-3.*

(MDA) to characterize and understand damage in TMCs and its effect on life prediction. All work discussed was performed under the Air Force Contract F33615-87-C-3219.

There are many factors that can affect the failure mode in TMCs. These factors include applied stress levels, temperatures, and environments, as well as internal material characteristics, such as residual stress state, and fiber/matrix interface strength. Researchers have investigated several of these parameters by developing detailed micromechanical models that accurately analyze the individual effects. Developing an accurate model that includes all of the pertinent variables to predict damage behavior would be an overwhelming task, and the end result would be a program to large to be practical for design purposes. With this in mind, the objective of the work at MDA has been to develop a design tool that combines a damage model, which accounts for the most important variables, with a limited amount of test data. The result is a computer program that provides fast and reasonably accurate predictions of damage behavior and life.

References

[1] Lekhnitskii, S. G., *Anisotropic Plates,* Gordon and Breach Science Publishers, New York, 1968.
[2] Harmon, D. M. and Finefield, M. A., "Thermomechanical Load History Effects in Metal Matrix Composites, MMCLIFE User's Manual," Report No. WL-TR-94-3016, Wright Laboratory, Cleveland, OH, Dec.
[3] Johnson, W. S., "Modeling Stiffness Loss in Boron/Aluminum Below the Fatigue Limit," TM 83294, NASA Langley Research Center, Hampton, VA, 1982.
[4] Jones, R. M., *Mechanics of Composite Materials,* McGraw Hill, New York, 1975.
[5] Piggot, M. R., *Load Bearing Fibre Composites,* Pergamon Press, Oxford, 1980, p. 205.
[6] Smith, K. N., Watson, P., and Topper, T. H., "A Stress-Strain Function for the Fatigue of Metals," *Journal of Materials,* Vol. 5, No. 4, 1970, pp. 767–778.
[7] *Military Handbook—Metallic Materials and Elements for Aerospace Vehicle Applications,* Mil-Hdbk-5E, Department of Defense, Washington, DC, 1 June 1987.

Mike T. Tong,[1] Suren N. Singhal,[1] Christos C. Chamis,[2] and Pappu L. N. Murthy[2]

Simulation of Fatigue Behavior of High Temperature Metal Matrix Composites

REFERENCE: Tong, M. T., Singhal, S. N., Chamis, C. C., and Murthy, P. L. N., **"Simulation of Fatigue Behavior of High Temperature Metal Matrix Composites,"** *Life Prediction Methodology for Titanium Matrix Composites, ASTM STP 1253,* W. S. Johnson, J. M. Larsen, and B. N. Cox, Eds., American Society for Testing and Materials, 1996, pp. 540–551.

ABSTRACT: A generalized relatively new approach is described for the computational simulation of fatigue behavior of high temperature metal matrix composites (HT-MMCs). This theory is embedded in a specialty-purpose computer code. The effectiveness of the computer code to predict the fatigue behavior of HT-MMCs is demonstrated by applying it to a silicon-fiber/titanium-matrix HT-MMC. Comparative results are shown for mechanical fatigue, thermal fatigue, thermomechanical (in-phase and out-of-phase) fatigue, as well as the effects of oxidizing environments on fatigue life. These results show that the new approach reproduces available experimental data remarkably well.

KEYWORDS: computer codes, computational simulations, fabrication processes, interactions, materials behavior, high temperatures, silicon fibers, titanium matrices, mechanical fatigue, thermal fatigue, thermomechanical fatigue, processing effects, oxidation effects, titanium, titanium matrix composites, life prediction, titanium alloys, fatigue (materials), modeling

In today's globally competitive environment, structural systems need not only perform satisfactorily and reliably, but must also be cost-effective. One of the key questions in highlighting a new structural system is "what material will provide the optimum combination of performance and cost?" The majority of the cost component is usually the cost of extensive test data used to characterize the material, not the material cost. The most often asked question, "how many test data do we need to characterize the material?" should be changed to "how can be quantify the materials behavior with minimum test data?" That is, there is a need to presume and quantify the constituent materials behavior that should be done via computational simulation and by taking advantage of high-speed computer platforms.

As the understanding of different materials and the numerous variables that affect them grows, many different material and service-environment specific models are developed [*1*]. For instance, different models are used for different metallic-matrix composite materials under different in-service static, dynamic, creep, and fatigue loads [*2*]. Despite the sophistication of these models, they usually constitute data fitting to various degrees of success. Further, they are not inclusive of (1) predominant factors contributing to coupled nonlinear material behavior; (2) all material thermal, mechanical, and other relevant properties; and (3) the entire material history from fabrication to sustained load use and to fracture. To

[1]Research engineer, director, Mechanics and Instrumentation revision, and senior aerospace engineer, respectively, NYMA, Inc., Brook Park, OH 44135.

[2] Senior aerospace scientist and senior research engineer, respectively, NASA Lewis Research Center, Cleveland, OH 44135.

understand these specific and limited-in-scope models, investigators have traditionally relied on test data. Clearly, there is a need for one single material characterization model that can quantify all of the requisite effects, starting from primitive variables at the lowest constituent material scale for the whole composite life, with a minimum need for test data.

NASA Lewis Research Center has been, for more than two decades, pursuing alternative general-purpose theories for quantifying coupled nonlinear material behavior [*3–5*]. This approach is based on computational simulation of the entire physical process that governs the coupled material behavior [*6,7*]. The objective of the present investigation is to describe the application of the theory for computationally simulating the thermomechanical cyclic behavior of high temperature metal matrix composites (HT-MMCs) via a single model. The computer code that evolved to perform the computational simulation is described briefly in this paper, with suitable references for more details. The authors hope that the reader will focus on why and what can be done, and will rely on the references for the subtleties on how.

Computational Simulation of Fatigue

In this section, we describe the fundamentals underlying the computational simulation approach along with its respective computer codes. Then, we proceed to describe its effectiveness to simulate mechanical fatigue, thermal fatigue, thermomechanical fatigue, and oxidative effects.

Fundamental Considerations

The computational simulation of HT-MMCs, as described herein, is based on (1) what constituents the composite materials are made from; (2) how their material behavior is manifested at their progressively interactive multiple scales, including the effects of how they are made; (3) how the composite structures respond to service environments; (4) what governs their behavior under a desirable set of design requirements; (5) how the solution of the fundamental governing field equations for all the participating variables is obtained by employing the computer as an integral part of the solution; and (6) how the evolution of the behavior or process is simulated.

By using this simulation, we are able to model various types of effects on the materials behavior such as: (1) service loads, mechanical and thermal; (2) cyclic effects; (3) time effects including creep; (4) temperature effects; (5) chemical and metallurgical effects, such as oxidation; and (6) fabrication-induced stresses.

The simulation starts with constituents and the fabrication-process, and proceeds to unfold the effects induced by the aggressive service loading environments at the structure scale. The structure scale response for a load increment, along with incremental environmental effects, are then used to update the materials behavior at the constituent scale, thus accounting for all the mutual interaction effects. These effects are synthesized up to the structure scale to obtain an up-dated structure deformed position, and the process is repeated until an equilibrium position for the structure is obtained. Next, the loads and environments are updated to their next increments, and the cycle is repeated until the onset of structural fracture is reached. This process is described for a specialty computer code in the next section.

METCAN Computer Code

The METAL matrix Composite Analyzer, METCAN, computationally simulates the materials behavior of HT-MMCs from micromechanics to laminate [*8*]. The structure of METCAN parallels the fabrication process of HT-MMCs. The schematics in Fig. 1 illustrate

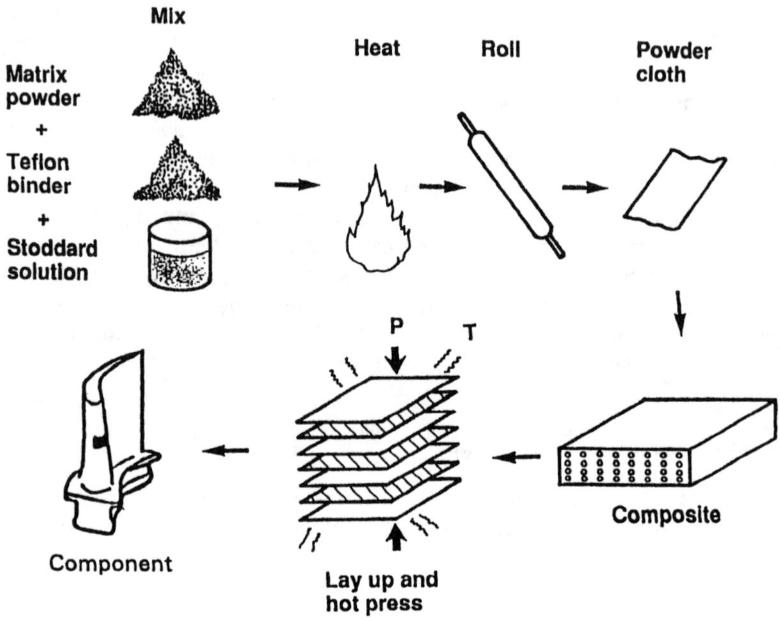

FIG. 1—*Metal-matrix composite fabrication process.*

a typical fabrication process. Figure 2 shows the computational sequence for the simulation of composite behavior including the fabrication process. The sequence consists of: Step I—processing, cool down from processing temperature to room temperature; Step II—heat up to use temperature from room temperature; and Step III—mechanical load to obtain stress-strain data. METCAN is capable of predicting the entire HT-MMC behavior history, including the fabrication process by using only room temperature properties for the fiber, matrix, and interphase.

The METCAN simulation capability is shown in Fig. 3. The simulation starts from describing the material properties at the constituent (fiber/matrix/interphase) scale and synthesizes the ply and laminate scale properties through composite micromechanics, macromechanics, and laminate theories (left-hand side of Fig. 3). The laminate scale

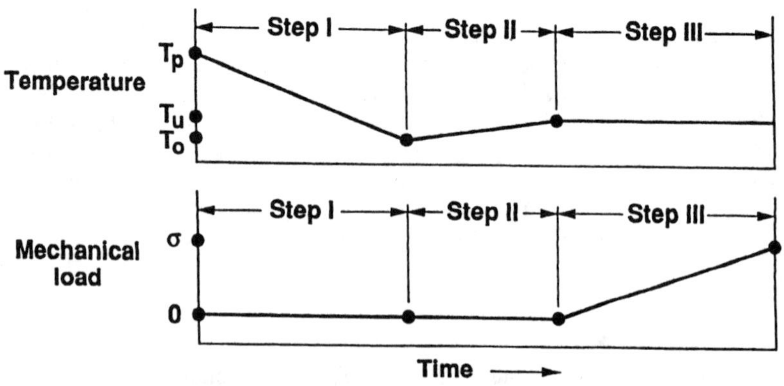

FIG. 2—*METCAN computational simulation sequence.*

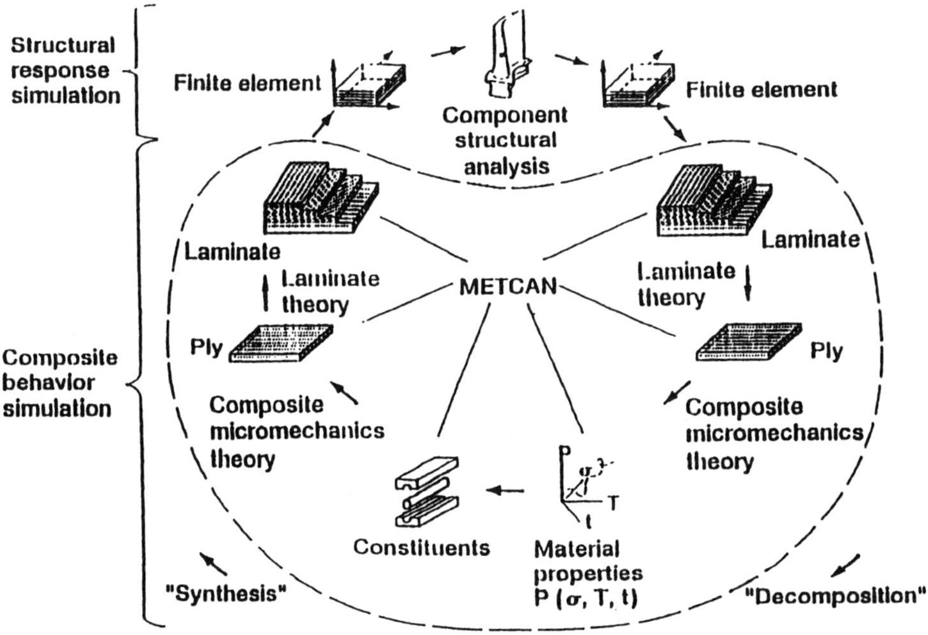

FIG. 3—*METCAN computational simulation of high temperature intermetallic matrix composite behavior.*

properties are used for the global structural analysis (top of Fig. 3). The global structural response is then decomposed to the laminate, ply, and constituent scales (right-hand side of Fig. 3). A detailed description of the micromechanics used for representing the simulation at the constituent materials scale is given in Ref *7*. The room temperature properties of the lowest scale constituent materials are automatically extracted from the METCAN resident data bank that can be augmented for properties of new materials. The data bank is designed to include all room temperature thermal, mechanical, and oxidative properties and their respective exponents for dependence on all types of variables such as temperature, time, and stress. This feature results in a considerable saving of time required for searching and inputting the composite material property data.

Multi-Factor Interaction Model

We start with the premise that if we are to capture all the variables that affect the materials behavior, we need a unified description. Figure 4 shows an n-dimensional material behavior space (MBS) where each dimension represents a specific aspect of material behavior. It is reasonable to assume that the MBS can be described by an assumed interpolation function. One convenient function is a polynomial of product form because mutual interactions can be represented by the overall product to include cross products that are common in higher-degree algebraic polynomials. In this context, the MBS is assumed to be described by a multi-factor interaction model (MFIM) as follows

$$\frac{P}{P_0} = \prod_{i=1}^{n} A_i^{m^i} \tag{1}$$

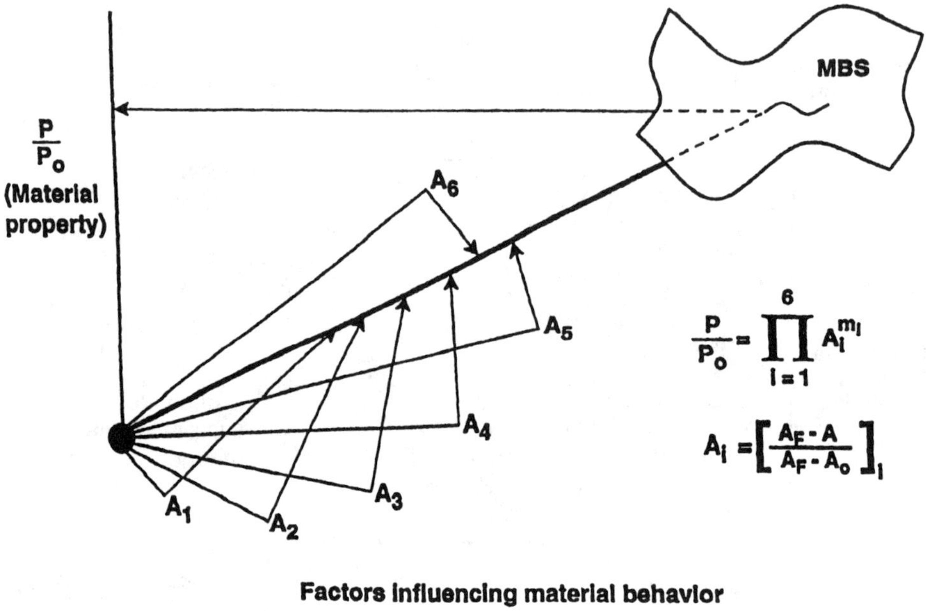

FIG. 4—*Multi-factor interaction model for material behavior space (MBS).*

where P is any material property at current conditions, A_i represents the ith factor that influences the material behavior, m_i denotes the material exponent, and n is the number of factors that affect the material behavior in that specific application. The subscript, 0, refers to the reference condition. The exponent, m_i, is selected from known material behavior. For new materials, m_i can be determined from a few selected tests or from the "best judgment" of a materials expert [9]. The form of the factor, A, is taken to be

$$A_i = \left(\frac{A_F - A}{A_F - A_0}\right)_i \tag{2}$$

The subscripts, 0 and F, refer to the reference and final (or future) conditions, respectively.

Representing the MBS by the MFIM (Eq 2) has another distinct advantage. The factors, A, themselves can also be represented by another level of MFIM or progressive substructuring of Eq 1. This progressive substructuring leads to a multi-tier representation of the MBS that permits intrinsic "lower tier" behaviors to influence more than one factor at the next higher tier. When this is done, n can be limited to a small number but the number of factors influencing the material behavior at the next lower tier will increase exponentially as n^j where j is the number of that tier. For example, when $j = 1$, the number of factors (n) is 4; $j = 2$, $n = 16$; $j = 3$, $N = 64$, etc. This representation is natural for multi-parallel processing machines where the levels of tiers are programmed for different granularities. Obviously, then, the motivation for selecting such a form is for programming convenience and for computational efficiency.

The MFIM general-purpose form is amenable to unlimited extensions for describing the interactive effects of all types of physical (mechanical, thermal, metallurgical, chemical, etc.) variables in one single equation. An expanded form of the MFIM showing typical physical variables and reasons for their selection are summarized in Fig. 5. It is to be noted

$$\frac{P}{P_O} = \left[\frac{T_F - T}{T_F - T_O}\right]^i \left[\frac{S_F - \sigma}{S_F - \sigma_O}\right]^l \left[\frac{\dot{S}_F - \dot{\sigma}}{\dot{S}_F - \dot{\sigma}_O}\right]^m \left[\frac{\dot{T}_F - \dot{\sigma}}{\dot{T}_F - \dot{\sigma}_O}\right]^n \left[\frac{R_F - R}{R_F - R_O}\right]^p \cdots$$

$$\cdots \left[1 - \frac{\sigma_M N_M}{S_F N_{MF}}\right]^r \left[1 - \frac{\sigma_T N_T}{S_F N_{TF}}\right]^s \left[1 - \frac{\sigma t}{S_F t_F}\right]^u \cdots$$

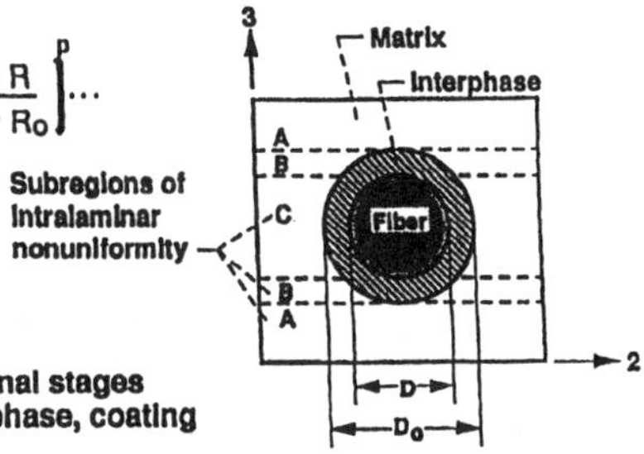

Rationale:

- **Gradual effects during most range, rapidly degrading near final stages**
- **Representative of the in situ behavior for fiber, matrix, interphase, coating**
- **Introduction of primitive variables (PV)**
- **Consistent in situ representation of all constituent properties in terms of PV**
- **Room–temperature values for reference properties**
- **Continuous interphase growth**
- **Simultaneous interaction of all primitive variables**
- **Adaptability to new materials**
- **Amenable to verification inclusive of all properties**
- **Readily adaptable to incremental computational simulation**

Notations:

P – property; T – temperature; S – strength; R – metallurgical reaction; N – number of cycles: t – time; over dot – rate; subscripts: O – reference; F – final; M – mechanical; T – thermal

FIG. 5—*Assumed multi-factor interaction model (MFM) to represent the various factors that influence in situ constituent materials behavior.*

that the exponents in the MFIM represent monotonic shape or the path of the material behavior. For example $-\infty <$ exponent $< \infty$ covers the whole MBS. Obviously, MFIM is amenable to simulating rates-of-change of materials behavior. The inclusiveness of this particular form combined with its simplicity make it very effective for computational simulation. Note that effects not present in the specific application are represented by using "zero" exponent for the respective terms in MFIM. Complete bond was assumed for the simulation described herein. Partial bond is accounted for by voids or low-moduli at the interfaces.

Mechanical Fatigue

In this section, we describe the application of METCAN to predict the mechanical (mechanical load cyclic stress) fatigue for HT-MMCs. Two isothermal cases are described: mechanical fatigue at room temperature and mechanical fatigue at high temperature.

Room Temperature—METCAN simulation results are shown in Fig. 6 where limited data [*9*] are also shown for comparison. The constituent material properties for these comparisons are listed in Table 1. METCAN simulation results reproduce the experimental data (static and cyclic) remarkably well. METCAN simulations for microstresses also help to identify regions of predominant failure mechanisms as is illustrated in Fig. 6. The information in Fig. 6 can be used to assist characterization of fracture surfaces obtained from fractographic studies.

High Temperature—METCAN simulation for this case is shown in Fig. 7 where, for three different frequencies, data [*9–11*] are also shown for comparison purposes. The simulation results follow the test data reasonably well. The authors consider these comparisons remarkable, since the simulation uses only room temperature data for the constituent material properties and an average fiber volume ratio of 0.35, while the data ranged from 0.3 to 0.45. The agreement can be improved by selecting the exponents appropriately for each different material (calibration) [*12*]. The simulation is based on micromechanics and in-

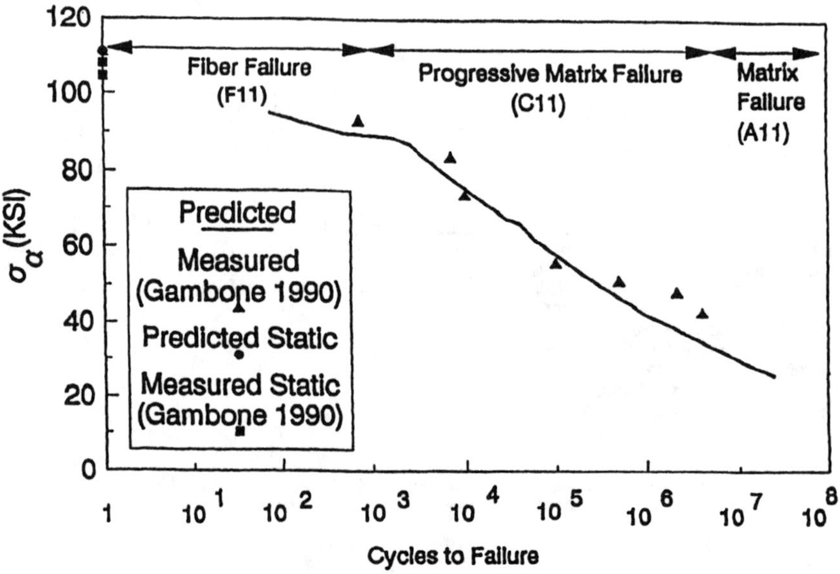

FIG. 6—*Predicted isothermal fatigue life (23°C, R = 0.1).*

TABLE 1—*Constituent (fiber/matrix) material properties used in METCAN.*

	Material Properties			
	Fiber		Matrix	
	SiC	SiC	Ti-15-3	Ti-6-4
ρ, lb/in.^{3}	0.110	0.108	0.172	0.170
T_m, °F	4870	4500	3000	3000
E_{11}, Mpsi	62	60	12	17
E_{22}, Mpsi	62	60	12	17
G_{12}, Mpsi	26	20	5	7
υ_{12}	0.19	0.30	0.32	0.30
α_{11}, ppm/°F	2.7	2.7	4.7	5.2
α_{22}, ppm/°F	2.7	2.7	4.7	5.2
S_{11}, ksi	500	500	60	74
S_{22}, ksi	500	500	60	74
S_{12}, ksi	300	250	36	44
ρ^{gm}/cm^3	3.06	3.0	4.8	4.72
T_m (°C)	2688	2482	1649	1649
E_{11} (Gpa)	427	414	82.7	117
E_{22} (Gpa)	427	414	82.7	117
G_{12} (Gpa)	179.3	138	34.5	48.3
ν_{12}	0.19	0.30	0.32	0.3
α_{11} (ppm/°C)	4.9	4.9	8.5	9.4
α_{22} (ppm/°C)	4.9	4.9	8.5	9.4
S_{11} (Gpa)	3.5	3.5	0.413	0.51
S_{22} (Gpa)	3.5	3.5	0.413	0.51
S_{12} (Gpa)	2.07	1.7	0.25	0.3

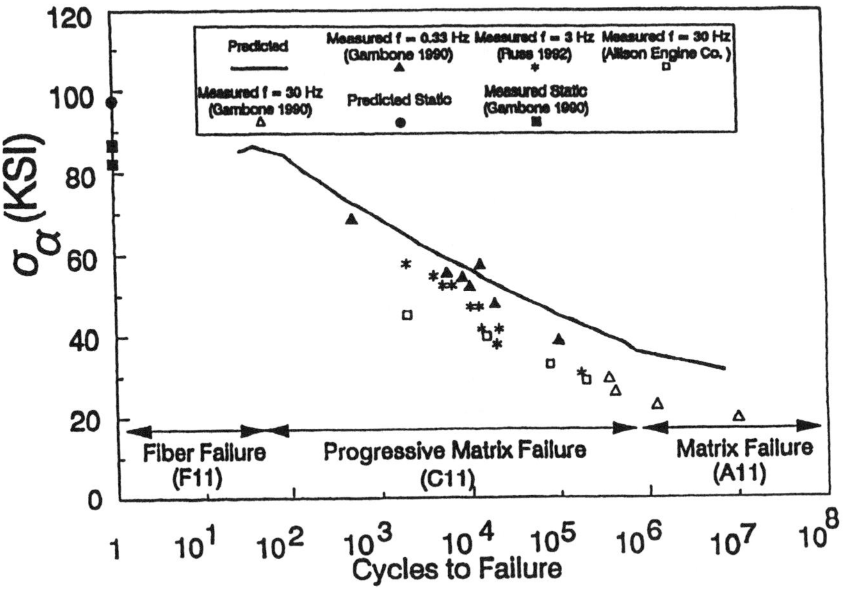

FIG. 7—*Predicted isothermal fatigue life (650°C,* R = *0.1).*

cludes the fabrication process effects and all nonlinear effects through the MFIM. We can confidently conclude that the METCAN simulations are credible.

Note that F11, C11, and A11 in Figs. 6 and 7 refer to the fiber and matrix regions as depicted in Fig. 5.

Thermal Fatigue

METCAN has successfully been used to simulate thermal fatigue. Typical results from such simulations together with limited experimental data [*12*] are shown in Fig. 8. The constituent material properties for these simulations are listed in Table 1. METCAN predicted the composite properties that are compared with measured data in Ref *12*. The simulation was performed by cycling the specimen at the thermal stress in the N_T term in the MFIM (Fig. 5) and then loading it to fracture in tension as depicted in the schematics in Fig. 8. It can be seen in Fig. 8 that the comparisons are in excellent agreement.

This excellent agreement demonstrates that the mechanics embedded in METCAN effectively represent the physics of HT-MMC behavior at all its inherent scales and throughout the load history from the fabrication process to fatigue life at elevated temperatures. Additionally, these comparisons also demonstrate the inclusiveness of the MFIM to describe the various effects on the constituent materials behavior.

Thermomechanical Fatigue

The application of METCAN to computationally simulate thermomechanical fatigue (TMF) is described in this section. Both types of TMF, in-phase and out-of-phase, are simulated. Their respective history sequence is shown in Fig. 9.

METCAN simulation results for 0.35 fiber volume are shown in Fig. 10 for both TMF lives in non-oxidizing environments where limited data [*13*] are also shown. As can be seen,

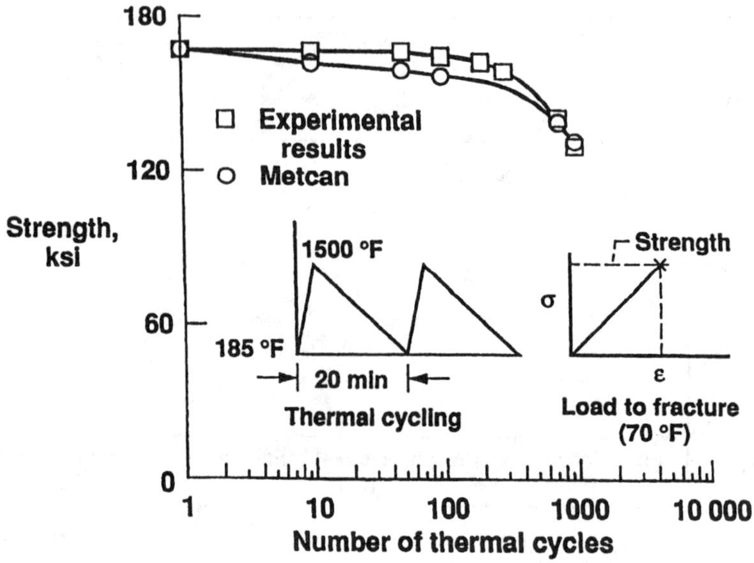

FIG. 8—*METCAN simulated thermal fatigue degradation effects of unidirectional SiC/Ti composite at fiber volume ratio (FVR) 0.36.*

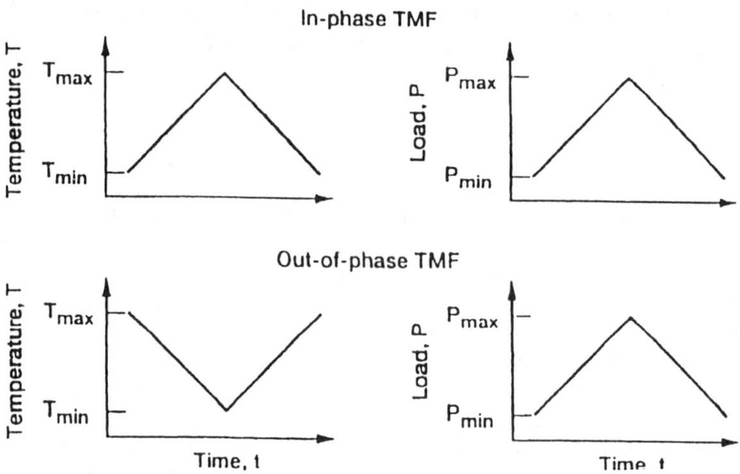

FIG. 9—*In-phase and out-of-phase TMF loading history.*

METCAN simulates the data remarkably well. The oxidizing environment included the N_M and N_T terms in the MFIM in Fig. 5.

Corresponding results for the TMF lives in oxidizing environments are shown in Fig. 11. Here, again, METCAN simulates the data remarkably well. The oxidation effects on TMF were obtained by activating the (R) term in the MFIM in Fig. 5 as was already mentioned.

METCAN simulation results for the same composite but for a different thermal cycle range are shown in Fig. 12 where limited data [*13*] are also shown for comparison purposes. The comparisons between the two are remarkably close, demonstrating the simulation

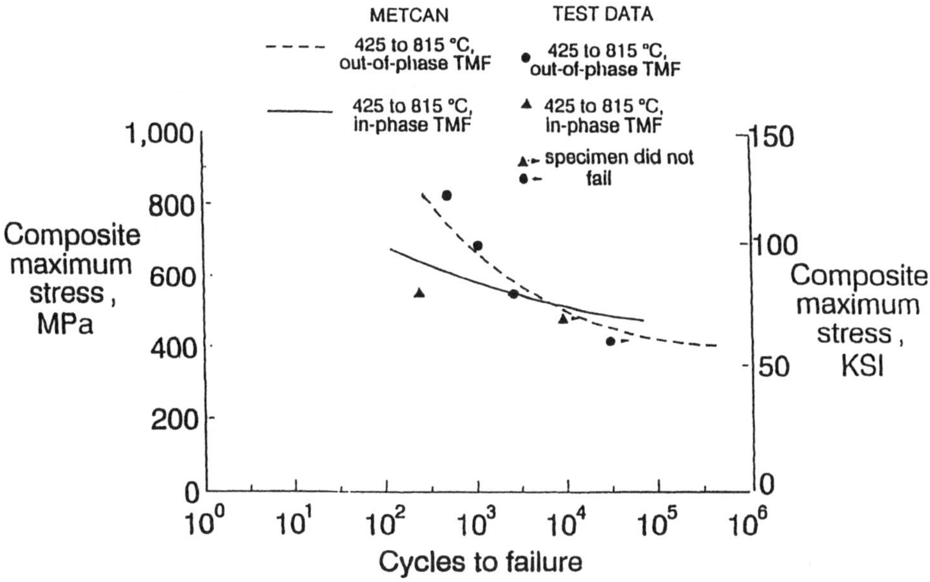

FIG. 10—*TMF lives of [0°] SiC/Ti-24Al-11Nb in flowing argon.*

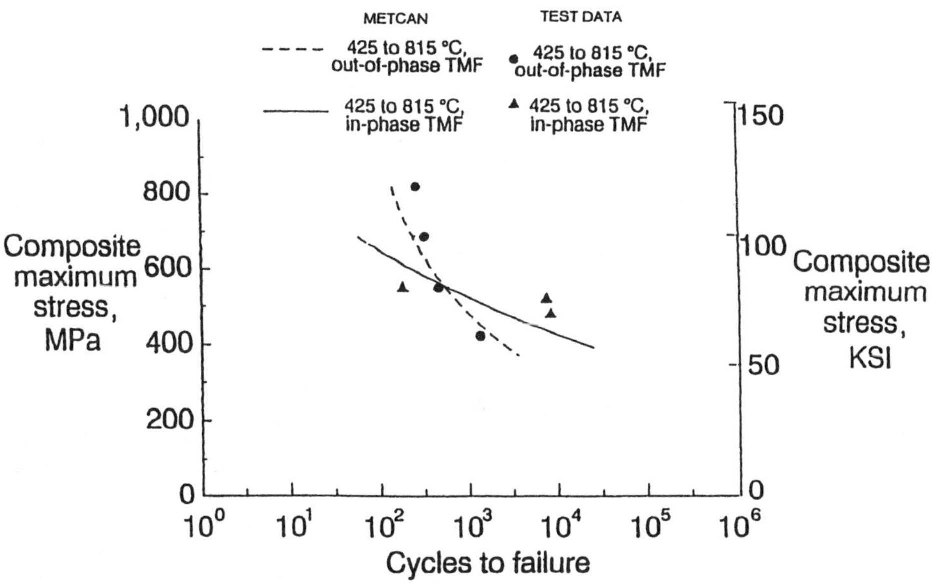

FIG. 11—*TMF lives of [0°] SiC/Ti-24Al-11Nb in air environment.*

versatility of METCAN for oxidizing effects, time dependence, and temperature dependence. It is important to note that the combined cyclic and oxidizing environment degrade the material as predicted by the MFIM. Failure is determined by loading the specimen to fracture in that mode. The local failures are determined by comparing the local micro stress component with its respective uniaxial strength at that region.

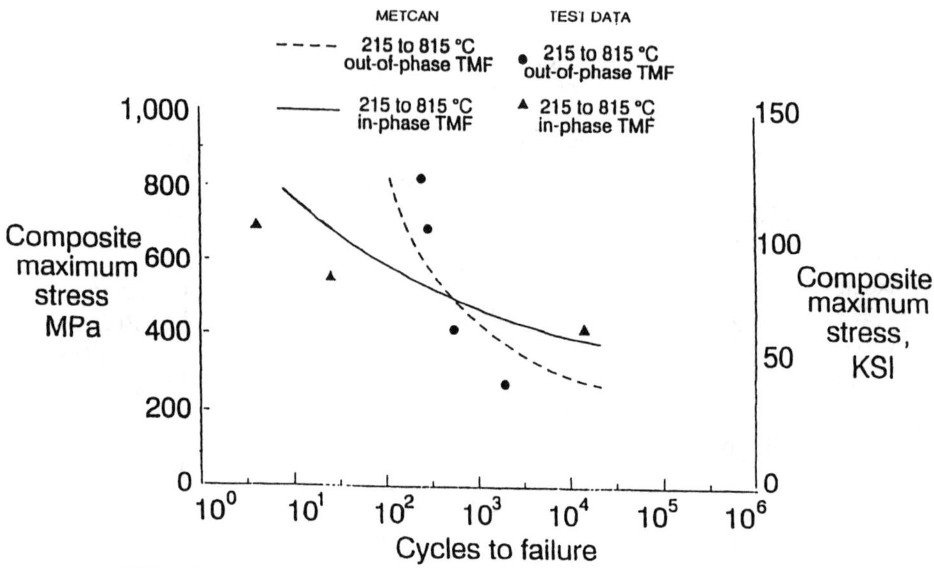

FIG. 12—*TMF lives of [0°] SiC/Ti-24Al-11Nb in air environment.*

General Comments

The previous discussion addressed METCAN simulation of HT-MMC fatigue behavior. Fatigue life is a very important aspect of these composites and, in addition, it serves as a direct means to demonstrate the effectiveness of METCAN simulation, in general. The simulation generality lies in the mathematical expressions (equations) that are used to describe HT-MMC mechanics [*14*]. The effectiveness of METCAN to simulate hot composite structures made from HT-MMCs has also been demonstrated [*15*]. Furthermore, METCAN's effectiveness for static monotonic progressive fracture has been demonstrated [*16*]. The discussion was limited to unidirectional composites because those were the only data we had for comparison. METCAN can handle any laminate configuration including hybrids.

It was hinted throughout the discussion that the effectiveness of the MFIM lies in having credible final values for the various factors in the expanded equation (Fig. 5). Credible final values for various factors may not be available especially for new materials or for long-term effects. The third author recommends an induction-type approach in conjunction with behavior-specific experiments to determine credible final values. The induction-type approach works as follows: (1) to predict the fatigue behavior for known conditions; (2) to simulate the fatigue behavior by extrapolating to a higher number of cycles but lower cyclic stress amplitude; (3) to perform experiments with extrapolated conditions and compare the results; and (4) if the results agree, to repeat Steps 2 and 3 until confidence has been obtained to establish credible final values.

At its present status, the authors consider METCAN a desk-top "virtual test laboratory" to rapidly evaluate the HT-MMC fatigue behavior as well as any other behavior in general. What is more important is that all the simulations are consistent because they are based on the same constituent materials, the same fabrication process, and the same model for all effects.

Conclusions

The important conclusions of this investigation to computationally simulate the fatigue behavior of high temperature metal matrix composites (HT-MMCs) are as follows:

1. Computational simulation, with its attendant computer code, is an effective and general approach to simulate the fatigue behavior of HT-MMCs.
2. The computational simulation is inclusive of the various effects (processing, cycles, stress-amplitude, temperature, oxidative environments, etc.) that affect fatigue behavior or fatigue life, or both.
3. Comparative results for mechanical, thermal, thermomechanical, and oxidative effects demonstrated that computational simulation reproduces experimental data remarkably well.
4. The computational approach, with its attendant computer code, provides a desk-top virtual test laboratory to rapidly evaluate the fatigue behavior of HT-MMCs.

References

[1] *Fatigue of Composite Materials Mechanisms, ASTM STP 569,* American Society for Testing and Materials, Philadelphia, 1975.
[2] Toth, I. J., "Creep and Fatigue Behavior of Unidirectional and Cross-Plied Composites," *Composite Materials: Testing and Design, ASTM STP 460,* American Society for Testing and Materials, Philadelphia, 1969, pp. 236–253.
[3] Chamis, C. C., "Characterization and Design Mechanics for Fiber-Reinforced Metals," NASA TN D-5784, Washington, DC, May 1970.
[4] Chamis, C. C. and Sullivan, T. L., "Nonlinear Response of Boron/Aluminum Angleplied Laminates Under Cyclic Tensile Loading," *Contributing Mechanisms and Their Effects,* NASA TM X-71490, Washington, DC, Nov. 1973.
[5] Chamis, C. C. and Sullivan, T. L., "A Computational Procedure to Analyze Metal Matrix Laminates with Nonlinear Lamination Residual Strains," NASA TM X-71543, Washington, DC, April 1974.
[6] Chamis, C. C. and Hopkins, D. A., "Thermoviscoplastic Nonlinear Constitutive Relationships for Structural Analysis of High Temperature Metal Matrix Composites," NASA TM-87291, Washington, DC, 1985.
[7] Hopkins, D. A. and Chamis, C. C., "A Unique Set of Micromechanics Equations for High Temperature Metal Matrix Composites," NASA TM-87154, Washington, DC, 1985.
[8] Lee, H.-J., Gotsis, P. K., and Murthy, P. L. N., "Metal Matrix Composite Analyzer (METCAN): User's Manual," Version 4.0, NASA TM-105244, Washington, DC, 1991.
[9] Gambone, M. L., "Fatigue and Fracture of Titanium Aluminidcs," WRDC-TR-89-4145, Wright Research and Development Center, Wright-Patterson AFB, OH, 1990.
[10] Deluca, D. P. and Cowles, B. A., "Fatigue and Fracture of Titanium Aluminides," WRDC-TR-89-4136, Wright Research and Development Center, Wright-Patterson AFB, OH, 1989.
[11] Nicholas, T., Russ, S., and Cheney, A., "Frequency and Stress Ratio Effects on Fatigue of Unidirectional SCS/Ti-24Al-11Nb Composite at 650°C," *Fatigue '93,* J. P. Bailson and J. I. Dickson, Eds., Wright Laboratories, Wright-Patterson AFB, OH, 1993.
[12] Chamis, C. C., Caruso, J. J., Lee, H.-J., and Murthy, P. L. N., "METCAN Verification Status," NASA TM-1031119, Washington, DC, 1990.
[13] Bartolotta, P. A. and Verrilli, M. J., "Thermomechanical Fatigue Behavior of SiC/Ti-24Al-11Nb in Air and Argon Environments," NASA TM-105723, Washington, DC, 1992.
[14] Murthy, P. L. N. and Chamis, C. C., *METCAN Theoretical Manual,* NASA TM-106025, Washington, DC, 1993.
[15] Chamis, C. C., Murthy, P. L. N., and Singhal, S. N., "Computational Simulation of Hot Composite Structures," NASA TM-103681, Washington, DC, 1991.
[16] Chamis, C. C., Murthy, P. L. N., and Singhal, S. N., "Simulation of Hot Composite Structures," NASA TM-106200, Washington, DC, 1993.

Brian N. Cox[1]

Life Prediction for Bridged Fatigue Cracks

REFERENCE: Cox, B. N., **"Life Prediction for Bridged Fatigue Cracks,"** *Life Prediction Methodology for Titanium Matrix Composites, ASTM STP 1253,* W. S. Johnson, J. M. Larsen, and B. N. Cox, Eds., American Society for Testing and Materials, 1996, pp. 552–572.

ABSTRACT: The growth rates of matrix fatigue cracks bridged by sliding fibers are now commonly predicted by models based on the micromechanics of frictional interfaces. However, there exist many reasons, both theoretical and experimental, for suspecting that the most popular micromechanical models are probably wrong in detail in the context of fatigue cracks. Furthermore, a review of crack growth data reveals that the validity of the micromechanics-based predictive model has never been tested and may never be tested. In this paper, two alternative approaches are suggested to the engineering problem of predicting the growth rates of bridged cracks without explicit recourse to micromechanics. Instead, it is shown that the material properties required to analyze bridging effects can be deduced directly from crack growth data. Some experiments are proposed to test the validity of the approach.

KEYWORDS: titanium, titanium matrix composites, titanium alloys, life prediction, fatigue (materials), modeling, cracks, crack growth, bridging effects

One of the more promising classes of composites touted for high temperature applications, and certainly the most available,[2] is that of relatively brittle matrices, either ceramic or intermetallic, reinforced by strong, aligned, continuous fibers. Under cyclic loading in the fiber direction, these materials develop matrix cracks that often run perpendicular to the fibers, while the fibers remain intact in the crack wake, supplying bridging tractions across the fracture surfaces [*1–7*]. The bridging tractions shield the crack tip from the applied load, dramatically reducing the crack velocity from that expected in an unreinforced material subjected to the same value, ΔK_a, of the cyclic applied stress intensity factor. An important issue in reliability is the prediction of the growth rates of the bridged cracks.

Many people have worked on this topic in recent years [*5–14*], and it appears that the rudiments of the bridging problem are well understood. In particular, considerable effort has been expended on relating bridging effects to the microstructural characteristics of the composite, focusing especially on the role of the fiber-matrix interface. This research has been invaluable in guiding composite design, both for fatigue crack resistance and enhanced toughness under monotonic loading. Micromechanics is indeed the foundation of rational materials design.

However, predicting the reliability of a given composite in an engineering application is a problem of an entirely different kind. A design engineer requires a predictive methodology

[1]Manager, Design and Reliability Department, Rockwell Science Center, Thousand Oaks, CA 91360.

[2]Cylindrical monofilaments, by the chief virtue of being to hand, have dominated the field of continuously reinforced brittle matrix composites. Yet damage tolerance, transverse and through-thickness strength, and creep resistance are probably better served by reinforcements with different geometry, including three-dimensionally woven or braided filament tows, interpenetrating phases, or even continuously formed laminates. But this is another subject.

that is as simple as possible, that is, containing as few parameters as possible, while remaining true to the material. It is unnecessary and, as argued later, misleading to base predictions on literal applications of micromechanical models.

This paper reviews the uncertainties of applying popular micromechanical models of bridged cracks to the estimation of fatigue lifetime. Two alternative approaches are proposed, based on the simplest parametric models that are likely to capture the significant features of crack growth data, including notch and specimen size effects and rate-dependent bridging phenomena. While both approaches preserve certain functional features that ensue from micromechanical models, they remain noncommittal in regard to the details of such models.

Idealizations from Micromechanics

Early work on fibrous ceramic composites demonstrated the advantages for toughness of weak fiber-matrix interfaces. This condition promotes fiber bridging following matrix cracking and obviates fiber failure and catastrophic cracking [*15*]. Indentation tests on individual fibers in some glass/silicon-carbide (SiC) composites could be fitted very nicely by a simple shear lag model of stress transfer. In this particular case, the data supported the assumption of a uniform friction stress, τ, along the zone of interfacial sliding [*16*]. Both model and experiment showed a quadratic relationship between the indentation load, F, and the displacement of the fiber end (Fig. 1).

An analogous model of fiber sliding was used to develop a relationship between the spatially averaged bridging tractions, p, acting in the wake of a bridged crack and the crack opening displacement, u [*17*]. From Ref *17*

$$p = \beta u^{1/2} \tag{1}$$

where arguments of force balance led to

$$\beta = 2f\left[\frac{\tau E_f E}{(1 - f)\rho E_m}\right]^{1/2} \tag{2}$$

with f the fiber volume fraction, ρ the fiber radius, E_f and E_m the fiber and matrix moduli, and $E = fE_f + (1 - f)E_m$. Conservation of energy or considerations of force balance coupled with a kinematic displacement condition lead alternatively to a coefficient β' in Eq 1, where [*18,19*]

$$\beta' = \frac{E_f}{(1 - f)E_m}\beta \tag{3}$$

From the point of view of theoretical mechanics, Eq 3 should be preferred over Eq 2 for the models posed in Ref *17*; but ignorance of τ and other modeling questions (see following text) can reduce the distinction to a moot matter. In Ref *17*, the dependence of the matrix cracking stress, σ_c, on the matrix crack length, a, was computed for monotonic loading. The criterion for crack advance was assumed to be the attainment of a critical value, K_c, of the net crack tip stress intensity factor, K_{tip}

$$K_{\text{tip}} = K_a - K_b = K_c \tag{4}$$

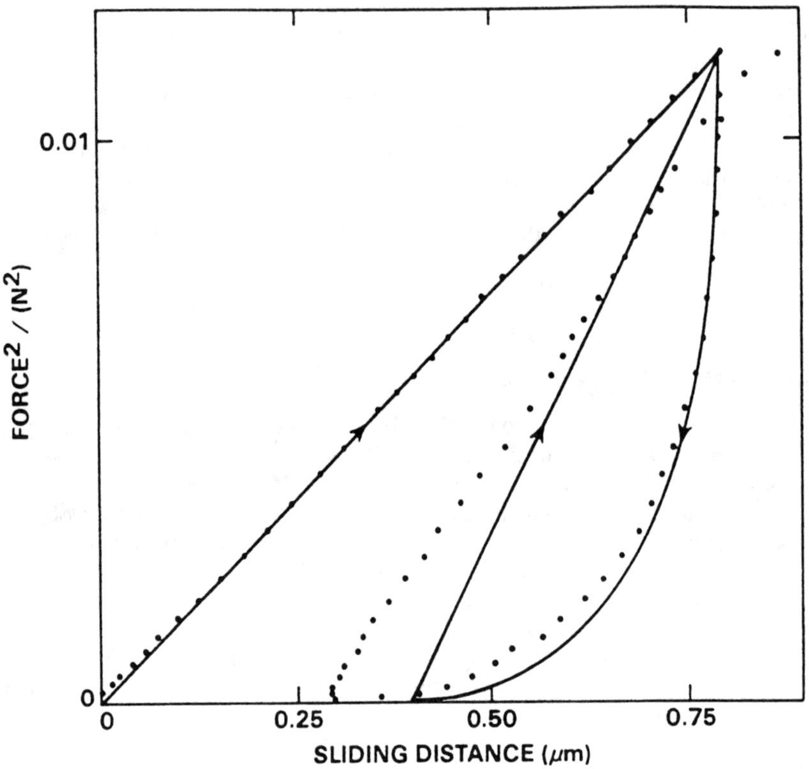

FIG. 1—*Load-displacement data for indentation tests of SiC fibers in a glass matrix and best fit curves from shear lag modeling (from Ref* 16*).*

where K_a and K_b are the applied stress intensity factor and its reduction due to bridging. For fully bridged cracks and increasing functions, $p(u)$, like that of Eq 1, σ_c is a monotonically decreasing function with the asymptote $\sigma_c = \sigma_1$, a constant, at large a [*17,20*]. The condition, $\sigma_c \approx \sigma_1$, signifies the attainment of steady-state cracking, echoing the earlier work of Aveston, Cooper, and Kelly [*21*].

Under cyclic loading, the same shear lag theory that underlies Eqs 1 and 2 leads to

$$|p - p_r| = \bar{\beta}|u - u_r|^{1/2} \tag{5}$$

where $\bar{\beta} = \beta/\sqrt{2}$ and u_r and p_r are the values taken at the last load reversal [*6,20*]. Given a putative relationship between crack growth rate and the cyclic range, ΔK_{tip}, of K_{tip}, for example, a Paris law

$$da/dN = A(\Delta K_{\text{tip}})^m \tag{6}$$

predicting crack growth rates amounts to tracking the evolution of ΔK_{tip} with crack length. As long as fracture surface contact does not occur during reverse loading, computing ΔK_{tip} is simplified by the following theorem: the cyclic ranges of the crack opening displacement, $\Delta u(x)$, the bridging tractions, $\Delta p(x)$, and the stress intensity factor, ΔK_{tip}, for loading under cyclic stress, $\Delta\sigma_a$, are exactly twice the profile, $u(x)$, tractions, $p(x)$, and value, K_{tip}, that would be computed for loading a pristine crack under monotonic loading to stress, $\sigma_a = \Delta K_{tip}/2$ [5].

Following from these basic ideas, numerous papers have demonstrated apparently successful analysis of fatigue crack growth data, particularly in titanium and titanium aluminide alloy matrices reinforced by SiC fibers [3,5,10–14]. Some of the crack growth data analyzed in these works revealed a regime of steady-state crack growth, where, in analogy to the steady state limit of Aveston, Cooper, and Kelly, ΔK_{tip} and da/dN are approximately independent of crack length. Other data and modeling have shown the influence of notches and finite specimen geometries, see Ref *9* and subsequent paragraphs. Extensions of the same modeling precepts to the problem of fiber rupture at maximum load (following Ref *22*; but see also a superior treatment in Ref *23*) have addressed residual composite strength after matrix crack growth [24].

In the analysis of fatigue crack growth rates, several groups have gone so far in applying micromechanical models as to discuss the values of the friction stress, τ, implied by their data [10–12,14]. By and large, good fits of models to data were evidently obtained (for example, Fig. 2), although the best fit values of τ bore only loose correspondence with values measured by independent experiments on individual fibers. Phenomenological analyses of residual strength data have also now appeared [14].

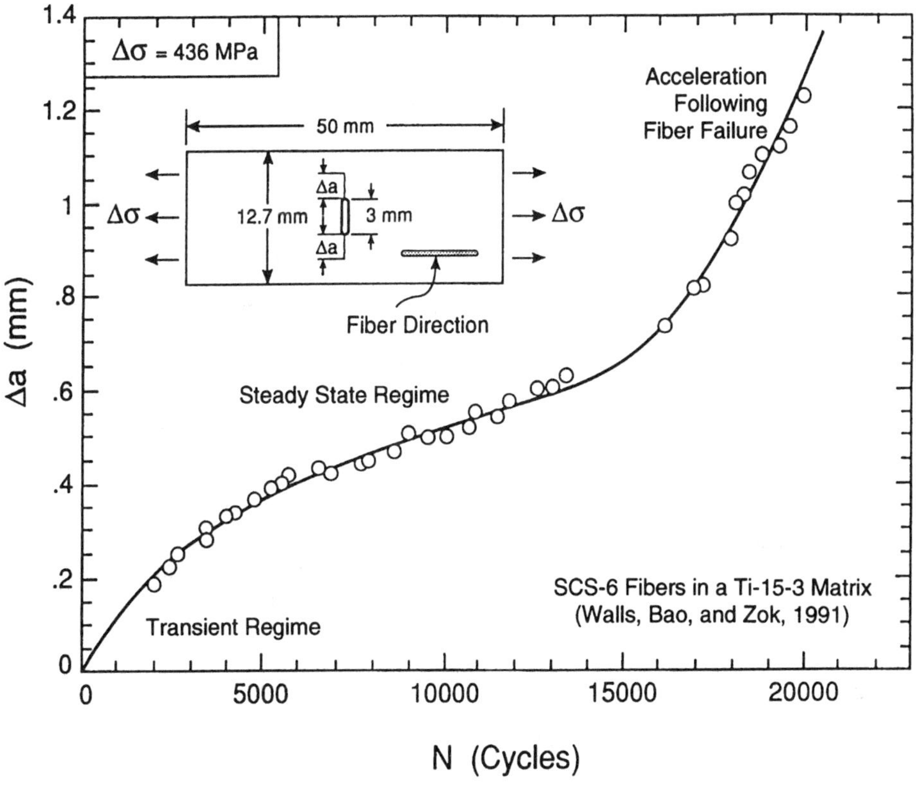

FIG. 2—*Typical fatigue crack growth data for a SiC fiber-reinforced titanium alloy* [12] *showing the characteristic steady-state regime (constant slope) and best fit curves from micromechanics-based modeling (from Ref 24).*

Some Difficulties with the Micromechanical Models

The apparent success of fiber sliding models in calculating crack growth rates is fortuitous. It will be shown in the next section that most fatigue experiments analyzed to date have not been tests of the model. Where a few experiments have provided partial tests, fiber sliding models have fared poorly. Yet even before crack growth data are considered, several fundamental difficulties with fiber sliding models and their relevance to crack bridging raise a strong warning.

Force-Displacement Relationships for Sliding Fibers

Both theoretical and experimental results show that the quadratic relationship between force and displacement (for example, Fig. 1) cannot be generally correct for the fiber sliding phenomenon.

Theoretical studies have dealt chiefly with an ideally cylindrical fiber moving alone in a semi-infinite elastic matrix. Sliding is most often assumed to be opposed by Coulomb friction. The coefficient of friction and radial residual stresses are taken to be uniform along the sliding zone. Even so, Poisson's effect leads to nonuniformity in the net normal interface stresses when the fiber slides. The relationship between the end load on the fiber and its axial displacement is no longer quadratic. In one extreme, Poisson's contraction when the fiber is pulled causes loss of contact between some part of the fiber and the matrix [*25,26*]. The contact-free length of fiber then strains linearly.

Nonuniformity in interfacial stresses will also arise from variations in residual stresses associated with the presence of the traction-free surface of the matrix crack. This complication increases in importance with fiber volume fraction, as does breaking of axial symmetry in the residual stresses due to fiber-fiber interactions. Indeed, in a material with fiber volume fraction high enough to optimize properties, the residual stress field is very difficult to compute or measure in any detail. This is a problem even for composites as received and at room temperature. After thermal cycling or mechanical cycling at constant high temperature, predictions cannot be considered more than qualitative guides.

In reality, the sliding interfaces are much more complicated geometrically than they are ever depicted as being in micromechanical models. The popular SCS6 SiC fiber, for example, contains several distinct interfaces in its carbon-rich coating [*27*]. When composited with titanium-based alloys or intermetallics, debonding and sliding can occur at any one or more of these or at separate interfaces in reaction products formed during consolidation; multiple interfaces can be active around each fiber simultaneously [*8,28*]. The implications for laws of frictional sliding are hard to fathom.

Interfaces are also rough. Figure 3 [*29*] provides a cogent demonstration of how roughness can dominate sliding mechanics. It refers to an experiment in which a short piece of fiber embedded in a slab of matrix was pushed first in one direction and then in the other. Figure 3 shows the load in the reversing step as a function of displacement. The key feature is the large dip in the load when the fiber nears its original position. The load drop corresponds to reseating of matching irregularities on the opposing faces of the rough, sliding interface. Over the width of the dip, that is, over displacements of a few microns, roughness effects dominate the load-displacement curve. Yet this order of magnitude of displacements is precisely that of crack opening displacements in fatigue cracks several mm in length, a critical length in many applications.

The effects of roughness are exceedingly difficult to quantify. Precise profilometry of surfaces of materials similar to those expected at common interfaces reveals the fractal nature of roughness; roughness exists on length scales spanning many orders of magnitude

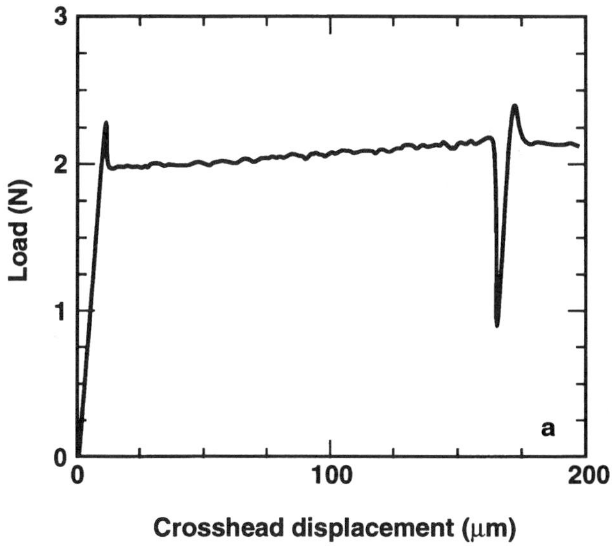

FIG. 3—*Load-displacement data for reversed sliding of a single fiber, exhibiting re-seating phenomenon (from Ref* 29*).*

[*30,31*]. The relationship between contact force and contact area is not linear, but follows a power law that depends on the fractal dimension [*32*]. This implies that the assumption of Coulomb friction may be invalid. Unfortunately, the relationship between contact area and critical shear stress for sliding is not yet understood, preventing completion of the analysis of friction from contact data. Furthermore, characteristic length scales are usually introduced by microstructure, violating the scale invariance assumed in evaluating fractal dimensions. The problem becomes very forbidding.

Sliding Fibers as Bridging Entities

For all their fascinating complexity, none of the models or experiments about fibers sliding in and out of half spaces or slabs solves the problem of sliding fibers as bridging ligaments.

Large-Scale Sliding—Consider the schematic of a bridged crack in Fig. 4. The dashed curves bound the zone of fiber sliding during a loading cycle. Shear lag modeling suggests a relationship similar to Eq 1 between the stress in a fiber and the crack opening displacement for the matrix crack at the fiber's location. In the usual modeling procedure, the problem of Fig. 4*a* is then replaced by that of Fig. 4*b*: the fibrous composite is replaced by a homogeneous (usually orthotropic) continuum; and the bridging fibers are replaced by continuous bridging tractions, p, which act on the fracture surface. The tractions, p, are usually taken to be the stress in the bridging fibers reduced by a factor, f, the fiber volume fraction.

The problems of Figs. 4*a* and *b* are clearly not equivalent for sufficiently small cracks. First, the stress at the fracture plane in the fibers in Fig. 4*a* is transferred to the matrix as a closure stress not just at the fracture surfaces, but over the entire sliding zone and beyond. Now the closure effect (that is, the reduction of K_{tip}) due to a point load of fixed magnitude diminishes as the position at which the load acts moves away from the crack plane [*33*]. Thus the model of Fig. 4*b* overestimates the bridging effect of the fibers in Fig. 4*a* when p follows Eqs 1 to 3 or 5 [*34,35*]. The tractions required in Fig. 4*b* should more appropriately

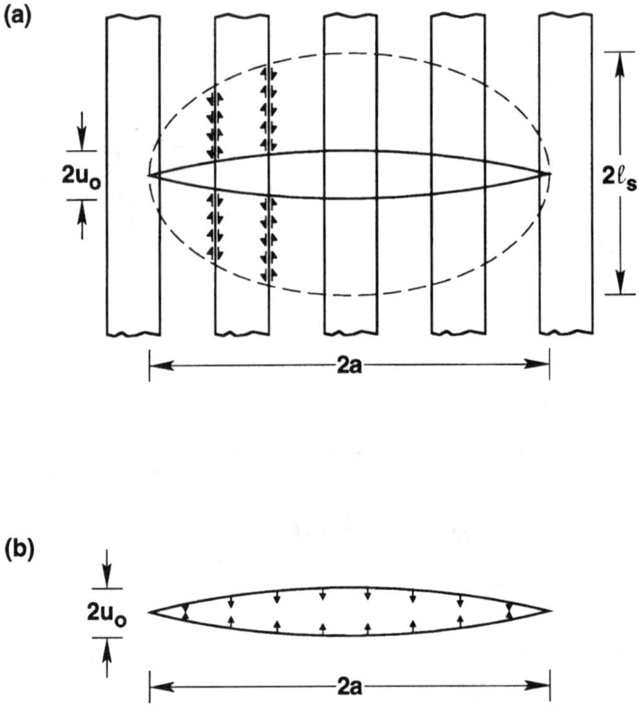

FIG. 4—*Schematic of a bridged crack:* (a) *showing interfacial slip in an inhomogeneous composite, and* (b) *as commonly represented in formulations of the problem as a cracked orthotropic continuum (following Ref* 34*).*

be those required to produce the same crack profile in Fig. 4*b* as the matrix crack has in Fig. 4*a*. The error in modeling Fig. 4*a* by Fig. 4*b* leads to severe overestimates of the shielding effect, ΔK_b, of the bridging fibers unless the sliding distance, ℓ_s, of Fig. 4*a* is much smaller than the crack length, a [*34*]. Fortunately, in this same regime, the bridging zone is immature and $\Delta K_b \ll \Delta K_a$, so that the error in the net ΔK_{tip} remains moderate; nevertheless, it can still be significant. As a becomes very large, the models of Figs. 4*a* and *b* become equivalent, with p in Fig. 4*b* given by Eqs 1 to 3 or 5. For typical modern composites of SiC fibers in titanium alloy matrices, the crack length must exceed ~1 mm for this to be the case.

The limit where $\ell_s \gtrsim a$ has been dubbed "large scale sliding" in recent literature [*36*].

Effects of Discreteness of Bridging Fibers—Other errors arise when tractions supplied by discrete fibers are represented by continuous surface tractions. These errors should be significant only when the crack spans just a few fibers. With 150 μm SiC fibers, for example, this means crack lengths <1 mm. For such small cracks, dips in records of *da/dN* versus a have indeed been attributed to the discreteness of fibers in the crack wake [*11*].

Replacing the fibrous composite by an orthotropic continuum will introduce other errors in describing near-tip fields for cracks of all lengths. The matrix stresses in the composite will always fluctuate as the crack front passes from one fiber to the next, an effect that is lost in the continuum of Fig. 4*b*. The details of the fluctuations in the crack-tip fields are incalculable: they involve elastic heterogeneity, complex residual stresses, deflection of the crack at fiber/matrix interfaces, irregular pinning along the crack front, plasticity constrained by the presence of fibers, etc. In determining the magnitude of resulting fluctuations in *da/dN*, only empirical observations will be credible. Whether fluctuations in tip fields or fluctuations in wake effects due to fiber bridging are responsible for dips in *da/dN* such as those reported in Ref *11* is problematic. However, since large-scale sliding, which diminishes bridging effects, will be in force for fibers very near the crack tip, one might differ with the author of Ref *11* and plump for the former.

Geometrical Irregularity

Fiber sliding models have dealt almost exclusively with axially symmetric stress distributions induced by purely axial applied loads. Given careful specimen alignment, this may well be a fair basis for analyzing experiments consisting of the pushing or pulling of single fibers. However, whether results of such analyses are then valid for a fibrous composite is another question. Fibers are inevitably wavy in a composite. Fluctuations in fiber alignment due to waviness are commonly ~3° in well-manufactured unidirectional composites [*37*]; while much larger variations must occur in woven or braided reinforcement (for example, Refs *38* and *39*). Even in a composite with ideally aligned fibers, misalignment of bridging tractions with the fibers' axes (that is, Mode II tractions) will still occur when a matrix crack fails to remain normal to the fiber direction. In reality, fiber irregularity and non-normal crack surfaces will always generate complex, non-axisymmetric stresses in bridging fibers. While these effects have been recognized and modeled qualitatively [*40*], the unpredictability and complexity of crack and fiber geometry will keep quantitative predictions out of reach.

Another manifestation of fiber misalignment is variance in fiber separation. Residual stresses due to thermal mismatch depend strongly on fiber separation (for example, Ref *41*). In titanium matrix composites, for example, variations of residual hoop stresses with fiber separation are strongly correlated with matrix cracking under transverse loads [*42*]. In the context of sliding fibers, variable residual stresses around individual fibers imply further unknowns in the expected bridging traction law.

Remarks on Prior Analyses of Crack Growth Data

Yet, in spite of all misgivings about the correctness of the underlying micromechanical models, the available crack growth data for matrix fatigue cracks in SiC/titanium alloys often appear to be well fitted by models based on Eqs 5 and 6 (for example, Refs *3–5* and *12–14*). Does this prove that Eqs 5 and 6 are right after all?

The answer lies in understanding the sensitivity of crack growth data to the characteristics of the bridging traction law which can be elicited from an analysis of various contributions to the net stress intensity factor, ΔK_{tip}.

For simplicity, the relevant concepts are developed first for monotonic loading. The crack configuration in the steady state under uniform remote stress, σ_a, is shown schematically in Fig. 5*a*. In the absence of a notch ($c_0 = 0$), p and σ_a must be in equilibrium over the entire far crack wake and therefore the contribution to K_{tip} from the far crack wake must be zero. For $c_0 = 0$, a nonzero value of K_{tip} thus arises only if $|\sigma_a| > |p|$ near the crack tip. If $p(u)$ exists, it must be an increasing function for the steady state to exist. Application of Rice's *J*-integral then yields [*43*]

$$\frac{K_{ss}^2}{E'} = 2\left\{ \sigma_a u_a - \int_0^{u_a} p\,du \right\} \qquad \text{(steady state)} \tag{7}$$

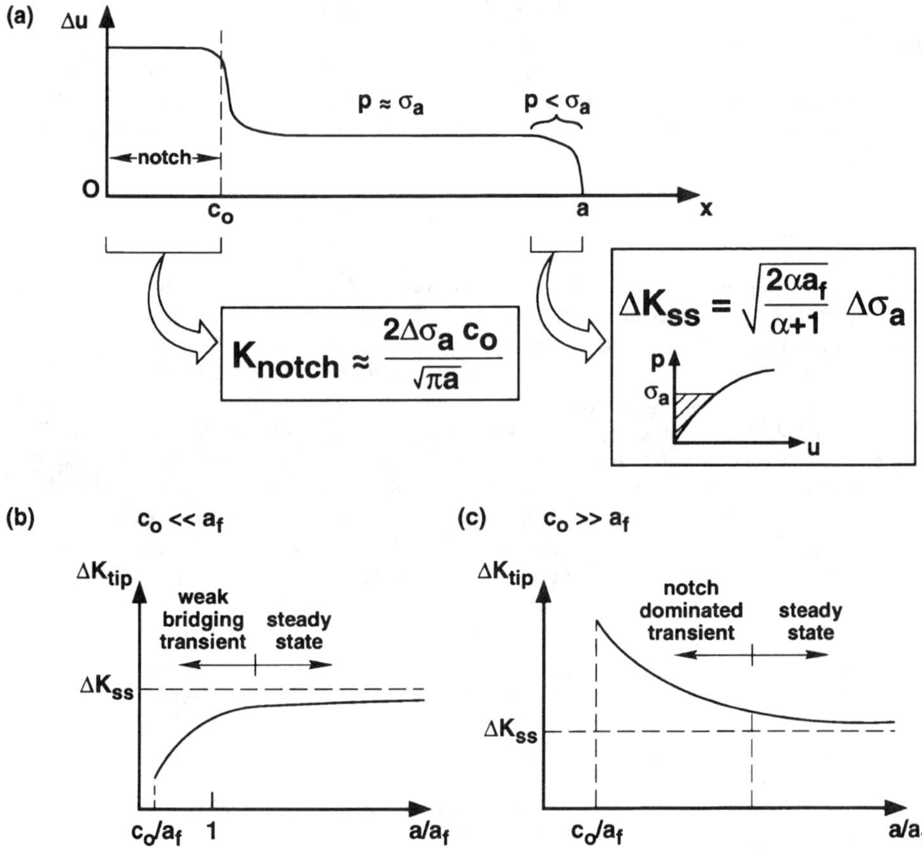

FIG. 5—(a) *Schematic of a bridged crack in the steady-state regime;* (b) *the weak bridging transient regime and steady-state regime for small notches; and* (c) *the notch-dominated transient regime and steady-state regime for large notches.*

for the steady-state value of K_{tip}, where $p(u_a) = \sigma_a$ and E' is a combination of elastic constants defined in Ref *44* (see also Ref *45* and original references therein).

In the absence of a notch, the steady state is preceded by a transient regime in which K_{tip} rises with crack length for fixed applied load. Here crack opening displacements over most of the crack are not yet large enough for p to balance σ_a. This "weak bridging" transient regime persists for a characteristic interval, a_m, of crack growth in monotonic loading, the "bridging length scale," which depends on the shape of $p(u)$.

For cyclic loading following a power law

$$|p - p_r| = \bar{\beta}|u - u_r|^{\alpha} \tag{8}$$

(for which Eq 5 is the special case, $\alpha = \frac{1}{2}$), scale and transformation invariances lead from Eq 7 to [*20,46*]

$$\Delta K_{ss} = \left[\frac{2\alpha^2}{\alpha + 1} \frac{E'}{\bar{\beta}^{1/\alpha}} \Delta\sigma_a^{\frac{\alpha+1}{\alpha}} \right]^{1/2} \tag{9}$$

$$= 2\sqrt{\frac{2\alpha a_f}{\alpha + 1}} \cdot \Delta\sigma_a \tag{10}$$

for the cyclic stress intensity factor in the steady-state regime, where

$$a_f = \frac{\pi\alpha E'}{4\beta^{1/\alpha}} (\Delta\sigma_a)^{\frac{1-\alpha}{\alpha}} \tag{11}$$

is the bridging length scale in fatigue. In the absence of a notch, ΔK_{tip} rises monotonically to ΔK_{ss} at crack length $a \sim a_f$ (Fig. 5*b*).

Now consider the range of influence of a notch of length c_0. Ignoring its perturbation of the steady-state crack profile for $x \gtrsim c_0$, its contribution ΔK_n to ΔK_{tip} is approximately that of the load $\Delta\sigma_a$ acting over the interval $[0, c_0]$ (for example, Ref *45*)

$$\Delta K_n \approx 2\sqrt{\frac{a}{\pi}} \Delta\sigma_a \sin^{-1}\frac{c_0}{a} \approx \frac{2}{\sqrt{\pi a}} \Delta\sigma_a c_0 \tag{12}$$

This function diminishes as $a^{1/2}$ for $a \gg c_0$. As the steady state is approached, the notch loses its influence when $\Delta K_n \ll \Delta K_{ss}$; that is, when

$$\frac{a}{c_0} \gg \frac{\alpha + 1}{2\pi\alpha} \frac{c_0}{a_f} \tag{13}$$

For likely values of α (for energy conserving hysteresis, $\alpha \leq 1$), the factor $(\alpha + 1)/2\pi\alpha$ is ~ 1. If $c_0 \ll a_f$, the notch will have only a small influence on the relationship, $\Delta K_{\text{tip}}(a)$, since bridging will be weak when $a \sim c_0$ (Fig. 5*b*). However, if $c_0 \leq a_f$ and a is too small to satisfy Eq 13, the crack will be in a "notch dominated" transient regime, in which ΔK_{tip} falls with increasing crack length (Fig. 5*c*) (see also Refs *20* and *46*).

In discussing crack growth data, it is imperative to distinguish the regimes of the steady state ($a \gg a_f$ and $a/c_0 \gg c_0/a_f$) from the weak bridging transient regime ($a \lesssim a_f$ and $c_0 \ll a_f$) and the notch-dominated transient regime ($c_0 \gtrsim a_f$ and $a/c_0 \lesssim c_0/a_f$).

Data from the Steady-State Regime

Consider the analysis of crack growth data from the steady-state regime $a \gg a_f$ and $a/c_0 \gg c_0/a_f$ (for example, data in Refs *3* and *12*). The information available about bridging laws from this regime is very limited. More generally than Eq 9, the asymptotically constant value ΔK_{ss} of ΔK_{tip} has the form

$$\Delta K_{ss} = f_k[\Delta\sigma_a, g(p(u))] \tag{14}$$

where g represents the dependence of ΔK_{tip} on the shape of the rising part of the hysteretic bridging relationship, $p(u)$.[3] Given $p(u)$, f_k can be evaluated by application of the J-integral [*43*]; reciprocally, the rising part of $p(u)$ follows from complete knowledge of $\Delta K_{ss}(\Delta\sigma_a)$.

[3]In continuing to write $p(u)$ for the bridging relationship in cyclic loading, it is understood that this is now virtually never a one-to-one correspondence as it often is in monotonic loading, but a multivalued function exhibiting hysteresis, for example, Eq 5.

The problem is to determine ΔK_{ss}. In the absence of load ratio dependence, the velocity is related to ΔK_{ss} by

$$da/dN = f_v(\Delta K_{ss}) \tag{15}$$

where f_v might represent a Paris law and incorporate a fatigue threshold.

Unfortunately, the function, f_v, must in general be regarded as unknown. It cannot be equated to the law deduced from data for the monolithic alloy. Parameters in fatigue crack growth laws, for example, the fatigue exponent and the multiplicative factor in a Paris law or a fatigue threshold, depend on all the nonlinear processes (plasticity, crack deflections, etc.) that act at or near the crack tip. Because of the fiber-related mechanisms listed at the end of the preceding section on sliding fibers as bridging entities, tip processes in a composite are not generally the same as those found in (or hypothesized for) the alloy alone.[4]

Thus, records of da/dN in the steady state for different values of applied load, $\Delta\sigma_a$, cannot test the correctness of a particular form that might be assumed for $p(u)$. To illustrate this point further, consider, for example, a Paris law

$$f_v(\Delta K_{ss}) \equiv A(\Delta K_{ss})^m \tag{16}$$

where the exponent, m, in an unreinforced alloy or intermetallic might range from 2 to 10. Combining this with the power law bridging of Eq 8 gives

$$\left(\frac{da}{dN}\right)_{ss} = A\sqrt{\frac{2\alpha^2}{\alpha+1}\frac{E'}{\bar{\beta}^{1/\alpha}}}(\Delta\sigma_a)^{m\cdot\frac{\alpha+1}{2\alpha}} \tag{17}$$

for the crack velocity in the steady state. If the parameters A and m are unknown, measurements of $(da/dN)_{ss}$ versus $\Delta\sigma_a$ cannot test the correctness of the popular assignment $\alpha = \frac{1}{2}$: the powers α and m appear together.

Data from the Transient Regime

Independent information is potentially entailed in data for the transient regime preceding the steady state.

Weak Bridging Transient Regime—This is the regime where $c_0 \ll a_f$ and $a \lesssim a_f$. The shape of records of da/dN versus a in this regime should in principle be somewhat sensitive to the shape of the relationship $p(u)$. Unfortunately, a_f is usually quite small. It can be estimated via Eq 2 from values of the interface friction stress, τ, deduced from crack growth data by analysis based on Eq 5. For typical titanium alloys reinforced by SiC fibers, a_f ranges from ~0.1 to ~1 mm when the cyclic load amplitude is several hundred MPa [*20*].[5] For such small cracks, the discreteness of fibers leads to relatively large oscillations in da/dN and the limit of large-scale sliding is found. Thus noise and modeling difficulties prevent using data for $a < a_f$ to confirm Eq 5.

[4]Crack growth in a composite in the absence of bridging can be measured if all fibers are broken or cut in the crack wake. In some cases, velocities so measured are within a factor of two of those inferred from crack growth data for the corresponding monolithic alloy [*47*]. However, this simple correspondence cannot be *assumed,* and performing these tests for all temperatures and load levels may not always be practicable.

[5]While the laws of Eqs 2 and 5 might be incorrect, the resulting estimates of a_f are very likely to be representative orders of magnitude.

Notch-Dominated Transient—Uniform Bridging Stress Models—This is the regime where $c_0 \geq a_f$ and $a/c_0 < c_0/a_f$. In the notch-dominated transient, the contribution to ΔK_{tip} of $\Delta\sigma_a$ acting over the notch $[0, c_0]$ is much larger than that of $\Delta\sigma_a - \Delta p$ near the crack tip, while σ_a and p approximately cancel over most of $[0, c_0]$. The crack growth data are therefore insensitive to the near-tip bridging fields and cannot test the traction law, $p(u)$.

This is well illustrated by some of the most recent work on modeling fatigue crack growth data in titanium alloy/SiC (Ti/SiC) composites. In Refs *8*, *10*, and *14*, bridging tractions were simply assumed to be uniform over the entire crack (excluding the notch).[6] In Ref *10*, the bridging tractions were assumed equal and opposite to the applied load, $p = -\sigma_a$. In Ref *14*, higher values ($|p| > |\sigma_a|$) were derived empirically by fitting the resulting predictions of *da/dN* to data. In either case, the measured and predicted growth rate curves followed one another within the (quite considerable) experimental noise.

In all of these tests, *da/dN* was an initially decreasing function, implying that $c_0 > a_f$. Furthermore, *da/dN* was not observed to approach a constant asymptote, implying that the steady state was not reached. Thus the entire recorded crack growth fell in the notch-dominated transient.

The assumption that uniform tractions of magnitude, $|p| \geq |\sigma_a|$, act over the bridged crack will give poor results for crack growth in other regimes. If $c_0 < a_f$, bridging tractions should be weak for small cracks ($a < a_f$); and ΔK_{tip} will be underestimated by the assumption, $|p| \geq |\sigma_a|$. If the crack grows beyond the notch-dominated transient ($a/c_0 \gg c_0/a_f$), ΔK_{tip} should attain its steady-state value, ΔK_{ss}. The dominant contribution to ΔK_{tip} should now come from the near-tip zone where $|p| < |\sigma_a|$, which is not present in a uniform traction model. Thus the assumption of uniform tractions, $|p| \geq |\sigma_a|$, is again nonconservative. Weak shielding for small fully bridged cracks and a positive definite steady-state crack-tip stress intensity factor, ΔK_{ss}, can be predicted only if $|\Delta p|$ is an increasing function of $|\Delta u|$.

Notch-Dominated Transient—Nonuniform Bridging Models—Further confirmation of the insensitivity of the notch-dominated transient to the details of the bridging law is found in modeling work using nonuniform bridging laws. While the quadratic law of Eq 5 has been used most often and generally successfully to fit *da/dN* versus crack length data [*5,12,14*], one recent paper tested out a linear law instead [*50*]. For crack growth extending ~0.6 mm from a 3.5-mm notch in a SiC/Ti-6-4 composite, excellent agreement was obtained between model and experiment. Thus in similar materials quadratic, linear, and uniform traction laws all work well in the notch-dominated transient. Crack growth data taken from this regime alone do not test the correctness of the assumed bridging law.

Analyzing Crack Opening Profiles

Given data with sufficiently low noise levels, the bridging tractions, p (or Δp), can be determined experimentally as a function of position x within the crack from measurements of the crack opening profile. In general, this may be done without assuming any particular form for $p(u)$ or even that the relationship, $p(u)$, is a constant material property [*51*]. The existence and form of a putative load-independent relationship, $p(u)$, can be established by examining $\Delta p(x)$ and $\Delta u(x)$ for various load levels and crack lengths. Unfortunately, this method has been precluded so far by the noise levels in profile data.

Progress has been attempted instead by seeing how well profiles can be fitted if model forms are assumed for $p(u)$. Profile data in Ref *14* appeared to be better fitted when Δp was assumed independent of Δu, rather than following the parabolic law of Eq 5. However, the

[6]These models are very similar to Marissen's earlier studies of fatigue crack growth in "ARALL," a laminate of aluminum and an aramid/epoxy composite [*48,49*].

distinction was unconvincing: in the notch-dominated regime, the crack profile is not especially sensitive to $p(u)$, except near the crack tip. Profile data near the crack tip tend to be noisy. In Ref *10*, profiles predicted for Δp = constant and $|\Delta p| \propto |\Delta u|^{1/2}$ fitted crack opening data equally well.

Collecting crack profile data is certainly worthwhile. The data in Refs *8*, *10*, and *14*, for example, clearly confirm that bridging fibers provide strong crack-tip shielding. However, profile data currently available have not yet provided a test of the bridging law.

Engineering Approaches for Rate-Independent Bridging

Thus neither crack growth data nor crack profile measurements in the literature to date have tested the correctness of the micromechanical model of Eqs 1 to 5. Indeed, the details of the model are unlikely ever to be confirmed.

Therefore, it is highly desirable that a predictive model be formulated that makes no explicit reference to the micromechanics underlying bridging. Instead, the simplest possible representation of bridging effects is sought that will allow crack growth rates to be predicted for any load level or specimen configuration.

It will be assumed at first that the bridging law is rate independent, that is, attrition in the crack wake is negligible. This may not be true, but the assumption is at least helpful in developing concepts and appears in some cases to be supported by experiment. The effects of cycle or other rate dependences are considered in the subsequent section on finite specimens.

Two approaches will be considered. The first is based on evaluating the bridging relationship, $p(u)$, in parametric form from crack growth data; $p(u)$ becomes the defining material property. The second approach is the most empirical possible. Instead of seeking to determine $p(u)$, certain crack growth data are themselves taken as the defining material properties.

Method Based on a Parametric Bridging Law

Some simplifying assumptions appear warranted from prior work.

From the success of existing models in fitting data, it appears that crack growth is governed by ΔK_{tip} and that cyclic bridging tractions can be written as a hysteretic relationship, $p(u)$. In fact, nothing in existing crack growth data suggests that hysteresis in bridging should not be modeled as a one-to-one correspondence between $|p - p_r|$ and $|u - u_r|$; in other words, assuming symmetry between loading and unloading. The power law form of Eq 8 probably offers a sufficiently rich range of hysteresis loops (Fig. 6). The linear law, $p \propto u$ (no hysteresis), and the popular quadratic law, $|p - p_r| \propto |u - u_r|^{1/2}$ are particular cases. These basic assumptions greatly simplify bridging analysis and make available useful analytical and asymptotic results.

The power law for bridging tractions is fully determined by the stiffness parameter, β, and the power, α. In the absence of load ratio effects (see subsequent text), the only other unknown material property is the relationship between da/dN and ΔK_{tip}; that is, the function, f_v, of Eq 15. If β, α, and f_v are known, fatigue crack growth rates can be calculated for any specimen or notch geometry, for example, by integral equation or finite element methods. The necessary computer codes are becoming widely available.

The key to evaluating β, α, and f_v is to acquire data from different crack growth regimes.

1. The existence (or absence) of the steady-state regime must be established. This requires specimens whose width, w, satisfies $w \gg a_f$ and $w/c_0 \gg c_0/a_f$.

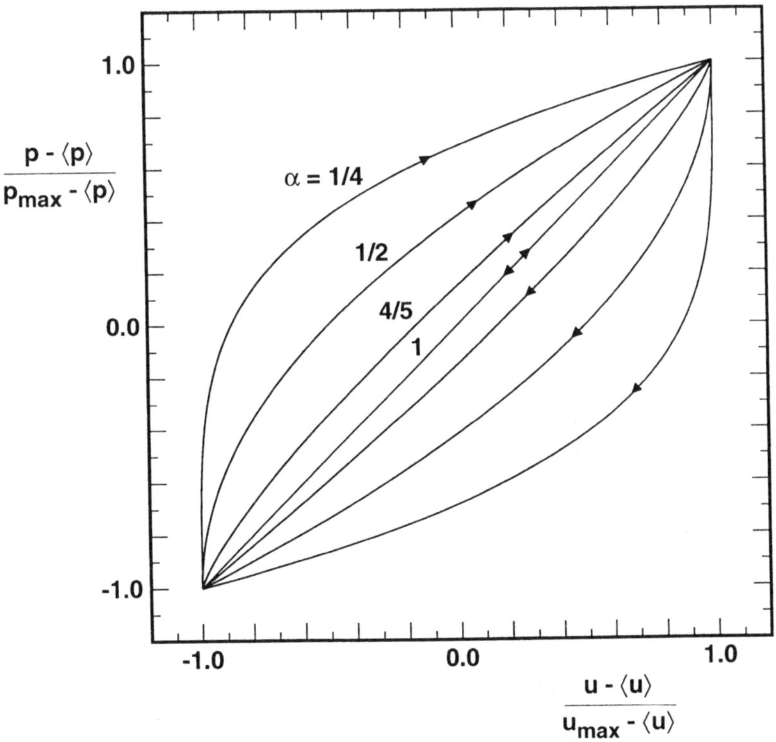

FIG. 6—*Hysteresis loops corresponding to a power law relationship between* $|p - p_r|$ *and* $|u - u_r|$. *The symbols < > indicate average value.*

2. Crack growth rates in the steady-state $(da/dN)_{ss}$ should be measured as a function of applied load amplitude. The resulting relationship

$$(da/dN)_{ss} = f_s(\Delta\sigma_a) \tag{18}$$

is a material property. It is sensitive to the bridging relationship, $p(u)$. Through the analytical result Eq 9, it also links da/dN and ΔK_{tip} via α and β. In other words, if α and β are known from independent data, f_v of Eq 15 is determined by Eq 18 and vice versa.
3. The shape of da/dN versus a in the weak bridging transient regime ($a < a_f$ and $c_0 \ll a_f$) is in principle sensitive to $p(u)$. For example, the length scale, a_f, itself, which characterizes the interval of growth required to reach steady state, depends on $\Delta\sigma_a$ (Eq 11). Comparison of transient crack growth records for different values of $\Delta\sigma_a$ yields an implicit relationship between ΔK_{tip}, f_v, β, and α. This fitting step must involve numerical solutions to the bridged crack problem. Unfortunately, data for small cracks are likely to be quite noisy (see sections covering sliding fibers as bridging entities and data from the transient regime).
4. For the reasons discussed in the preceding section, data from the notch-dominated transient regime ($c_0 > a_f$ and $a/c_0 < c_0/a_f$) show the least dependence on details of $p(u)$. However, the transition from notch-dominated to steady-state crack growth involves long cracks ($a \gg a_f$) and should therefore be relatively noise-free. Determination of this transition for various $\Delta\sigma_u$ will yield the relationship, $a_f(\Delta\sigma_a)$; whence β

and α via Eq 11. Or equivalently, analysis of sets of crack growth data exhibiting both notch-dominated and steady-state behavior at different $\Delta\sigma_a$ will yield β and α. The point of the transition itself need not be explicitly identified. It is enough to know that the data will contain sufficient information if the transition is present.

The precision with which α can be determined will depend on how well the notch-dominated to steady-state transition can be mapped as a function of $\Delta\sigma_a$. If noise in the data makes the uncertainty large enough, a value for α might as well be chosen for convenience. Canonical results allowing simple evaluation of ΔK_{tip} and estimation of the onset of fiber failure have been made available for $\alpha = ½$ [*9,52*]; while analytical approximations serve the same purpose for $\alpha = 1$ (linear springs) [*53–55*]. If α is so set, only β and f_v remain to be determined.

Once α, β, and f_v are evaluated, no interpretation of them in terms of the micromechanics of bridging need be made. Predictions of lifetime follow from these parameters alone by now standard computational methods.

A Highly Empirical Strategy

The second suggested method dispenses altogether with the bridging relationship, $p(u)$. Characterization of steady-state crack growth requires only the function, f_s, given by

$$\left(\frac{da}{dN}\right) = f_s(\Delta\sigma_a) \tag{19}$$

which is measured directly and can be considered a material property. It is not necessary in predicting steady-state crack growth to separate out the relationships, f_k and f_v of Eqs 14 and 15.

It remains to deal with the transient regimes. Transient crack growth for small notches can be treated not too conservatively as being infinitely rapid. In other words, the process of crack initiation could be assumed to create a crack of length a_f, which would grow thenceforth in the steady state. Cycles to initiation are predicted by methods unrelated to bridging phenomena (for example, Ref *56*).

On the other hand, writing $da/dN \rightarrow \infty$ for notch-dominated transient growth will be too conservative, especially if a crack of length $a/c_0 \lesssim c_0/a_f$ is considered fatal. However, in the notch-dominated transient, the assumption of uniform bridging tractions, $p = -\sigma_a$ or $p = -\sigma_{\text{net}}$ (where σ_{net} is the net section stress [*14*]), is probably good enough. Again, knowledge of $p(u)$ is not needed.

Nevertheless, for this most empirical strategy to be complete, some knowledge is required of $a_f(\Delta\sigma_a)$ to bound the weak bridging and notch-dominated transient regimes. Since a_f generally increases with $\Delta\sigma_a$, a conservative estimate of the small notch transient is given by the measured crack length at the attainment of the steady state for the highest load amplitude expected in service. Delimiting the notch-dominated transient is more troublesome. A conservative model would require data showing the attainment of the steady state for every load level. Recall from the section on data from the transient regime that failure to recognize the existence of the steady state leads to nonconservative predictions in uniform traction models when $a/c_0 > c_0/a_f$.

Finite Specimens

When the specimen width, w, fails to satisfy $w \gg a_f$ (uncommon for Ti/SiC composites) or $w/c_0 > c_0/a_f$ (common for many materials), steady-state crack growth cannot be observed. Using such small specimens to qualify a composite for service risks nonconservative estimates for larger structures. On the other hand, once bridging has been analyzed for both the transient and steady-state regimes using large specimens, predictions must also be made for relatively small specimens and various load distributions. For predictions based on the relationship, $p(u)$, no difficulty exists. Calculations can immediately be performed for any case using the same relationship $p(u)$ (for example, Refs *9* and *44*).

In contrast, the highly empirical strategy provides no method of accurately computing finite specimen effects. Nevertheless, this may not be a fatal deficiency. While the applied stress intensity factor, ΔK_a, is a very different function of a for given $\Delta\sigma_a$ in finite and infinite specimens [*57–59*], the distributions of bridging tractions and the crack opening profiles in the two cases are not so dissimilar. Consequently, when the specimen width satisfies $w \gg a_f$, as it must for bridging effects to be significant, da/dN in a specimen of finite width under uniform load $\Delta\sigma_a$ is either close to or below[7] that expected at the same notch size and crack length in an infinite specimen (Fig. 7). Further, for bending that develops a maximum stress, $\Delta\sigma_a$, da/dN is always less than it would be under uniform load, $\Delta\sigma_a$ [*59*]. Thus the empirical strategy will lead to conservative predictions for finite specimens and bending loads—not too conservative for uniform loads, perhaps highly conservative for bending.

Cycle or Other Rate Dependences

Cycling at room temperature degrades the fiber/matrix interface near matrix fatigue cracks. Evidence of degradation includes observation of debris [*8,60*]; load-displacement records of push-pull tests on individual fibers [*61*] and groups of fibers crossing matrix cracks etched right through specimens [*62*]; and some analyses of crack growth data [*14*]. All of this evidence indicates a decrease with cycles in the shear load capacity of the interface in the zone of fiber slip, implying softening of the bridging stiffness parameter, β. Some push-pull tests on groups of fibers at constant load amplitude have suggested that β declines rapidly at first, but then steadies at a constant value over most of the expected fatigue life [*62*]. In contrast, the analysis of crack growth data in Ref *14* led to the conclusion that β declines perceptibly throughout the fatigue record.

Whether bridging softening, $d\beta/dN$, should be regarded as slow or rapid in the context of crack growth depends on how it compares to the rate of crack advance, da/dN. In other words, writing

$$d\beta/dN = R_1 f_\beta(\beta, \Delta p) \tag{20}$$

the shape of da/dN versus a for given notch size and load amplitude depends only on the ratio of the rate constants, R_1 and A, where A is the rate constant of Eq 6 [*63*]. As $R_1/A \rightarrow 0$, bridging degradation becomes undetectable in crack growth; as $R_1/A \rightarrow \infty$, steady-state crack growth cannot be attained even approximately.

To date, no physical model of degradation is available to guide the choice of the function f_β. In view of the difficulty of determining bridging parameters for rate independent bridging, a sensible starting point is a law with a single parameter, for example

$$\frac{d\beta}{dN} = -R_1 \beta \Delta p \tag{21}$$

A case has also been made for replacing Δp in this law by $\Delta p - \Delta p_0$, where Δp_0 is a

[7]On the same grounds, crack growth data for small specimens are unlikely to contribute any information that is not available from larger specimens.

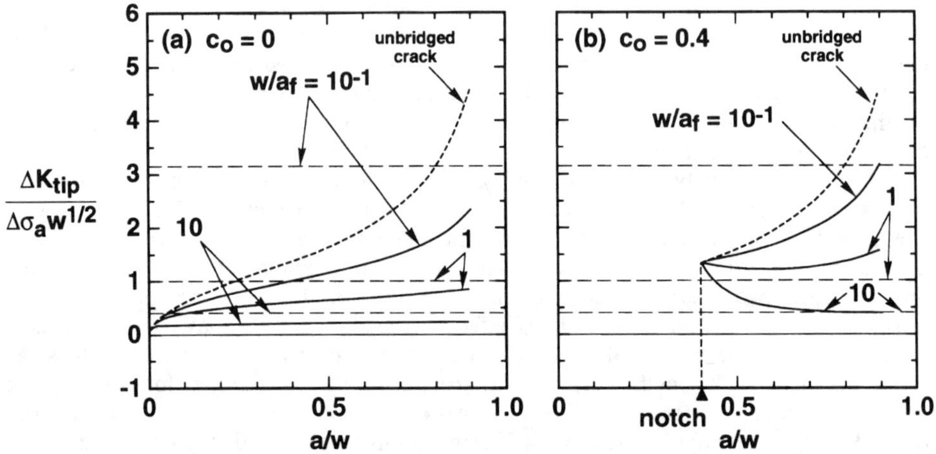

FIG. 7—*The effects of finite specimen geometry shown in predictions of* ΔK_{tip}(a) *(solid curves) for the bridging law of Eq 5 (adapted from Ref* 59*). The long-dashed curves show the steady-state value of* ΔK_{tip} *in an infinite specimen. The short-dashed curves show* ΔK_{tip} (a) *in the finite specimen in the absence of bridging.*

threshold for damage [*63*]. One signature of such a threshold would be long cracks ($a/c_0 \gg c_0/a_f$) for which bridging tractions are uniform while the crack opening profile is not uniform (as for a steady-state crack), but approximately elliptical.

Even when rate dependence exists, the length scale, a_f, remains useful, now defined in terms of the initial value, β_0, of β. (If β declines very rapidly in the first few cycles and slowly thereafter, β_0 would be better defined as the value after the first few cycles.) A second length scale, a_t, is introduced by decay in β. This length scale indicates the interval of crack growth required for the stiffness of newly created bridging ligaments to decay to a relatively small value. It also defines the interval of crack growth over which memory of the initial crack configuration or the effects of a change in load amplitude persist. Further, crack growth approximating the steady state will be found if $a_t \gg a_f$; and not otherwise. Failure to achieve steady state even in a large specimen ($w \gg c_0$) will indicate rapid decay in β.

For given laws for $d\beta/dN$ and da/dN, the length, a_t, depends on ΔK_{tip}; and therefore, like a_f, on $\Delta\sigma_a$. Analytical expressions for a_t have not yet been found in explicit terms of R_1, A, β, and α. However, very good approximations are available for calculating ΔK_{tip} as the only unknown in a transcendental equation when R_1/A is large [*64*]. For general cases, numerical solutions must be employed. An efficient code has been developed for this by rendering an integral equation formulation into vector form [*63*].

Of the two strategies laid out in this article, the approach based on using $p(u)$ as a material property allows the easiest incorporation of degradation in β. Even using this approach, it has not yet been determined what experiments would clearly define a law such as Eq 21. Far more experimental data are required.

Residual Stresses

Residual stresses influence bridging in two ways. First, radial compression at fiber/matrix interfaces influences the stiffness parameter, β. For fatigue at fixed temperature, this effect need not be modeled to predict crack growth rates. In the absence of rate effects, it is incorporated in the stiffness parameter, β. If interfacial attrition and the loss of debris diminish radial compression, the effect will be subsumed in the rate parameter, R_1, of Eq 21. Second, axial tension in the matrix will cause larger mean crack opening displacements. This might lead to greater rates of interface attrition and higher values of R_1; and it promotes fiber failure. However, it has no immediate effect on the cyclic crack opening displacement, $\Delta u(x)$, or on ΔK_{tip}.

If the temperature is cycled, residual stresses have a more prominent role. Relief of compressive interface stresses at higher temperatures will cause the stiffness parameter, β, to cycle with T. Simultaneous cyclic relaxation of axial residual stresses will cause cyclic variation of the crack opening with T, contributing directly to ΔK_{tip}. The resultant effects of these phenomena on crack growth have been modeled [*65*] but not yet measured.

Load Ratio Effects—Fracture Surface Contact

For tension-tension fatigue, fracture surface contact will probably never occur in fibrous composites, except perhaps for relatively long cracks in bending applications [*59*]. However, for load ratios, $R < 0$, it may. Let σ_c be the compressive applied load at which contact occurs. An accurate prediction of ΔK_{tip} will be that estimated for $\Delta\sigma_a = \sigma_{max} - \sigma_c$. The problem is to predict σ_c.

The contact stress, σ_c, depends generally on σ_{max}, α, β, and the magnitude of the axial residual stress, σ_{ax}, in the matrix. It can in principle be predicted once α, β, and σ_{ax} are known. However, uncertainty in values of α and β deduced from crack growth data and the unpredictability of σ_{ax} suggest that σ_c should be measured. For loads other than bending, fracture surface contact will nearly always occur first near the notch or, in the absence of a notch, in the farthest crack wake [*59*]. Closer to the crack tip, the crack is propped open more successfully by hysteresis in $p(u)$. This suggests measuring σ_c for cracks that have been grown into the steady state at various values of $\Delta\sigma_a$ with $\sigma_{min} > 0$. Given values for α and β, a value of σ_{ax} could be deduced and the consistency of the modeling checked by the degree to which it is independent of σ_{max}.

Conclusions

Future experiments on bridged fatigue cracks should focus on verifying the existence of a steady-state regime and measuring the transition to the steady state from the notch-dominated regime. Larger specimens are required than used hitherto.

Methods for predicting lifetime have been proposed that minimize the detail incorporated from micromechanical modeling. While the essential trends in crack growth that are both observed and anticipated from micromechanics are preserved, they are computed with the greatest possible economy in modeling.

Acknowledgments

Work on rate-dependent bridging was supported by the U. S. Department of Energy, Grant No. DE-FG03-89ER45400. Funding by the U. S. Department of Energy does not constitute endorsement of the views expressed herein. Other work was supported by Rockwell Independent Research and Development. The author acknowledges with pleasure discussions with Drs. Fred Leckie and David Marshall.

References

[*1*] Naik, R. A. and Johnson, W. S., "Observations of Fatigue Crack Initiation and Damage Growth in Notched Titanium Matrix Composites," *Composite Materials: Fatigue and Fracture (3rd*

Volume), ASTM STP 1110, T. K. O'Brien, Ed., American Society for Testing and Materials, Philadelphia, 1991, pp. 753–771.

[*2*] Cox, B. N., James, M. R., Marshall, D. B., Morris, W. L., Rhodes, C. G., and Shaw, M., *Proceedings,* 11th International SAMPE Conference, Birmingham, UK, 1989, S. Benson, T. Cook, E. Trewin, and R. M. Turner, Eds., Elsevier, Amsterdam, 1989, pp. 313–320.

[*3*] Sensmeier, M. D. and Wright, P. K., "The Effect of Fiber Bridging on Fatigue Crack Growth in Titanium Matrix Composites," *Fundamental Relationships Between Microstructure and Mechanical Properties of Metal Matrix Composites,* M. N. Gungor and P. K. Liaw, Eds., *Proceedings,* TMS Fall Meeting, Indianapolis, The Metals Society, Warrendale, PA, 1989.

[*4*] Luh, E. Y., Dauskardt, R. H., and Ritchie, R. O., "Cyclic Fatigue-Crack Growth Behavior of Short Cracks in SiC-Reinforced LAS Glass-Ceramic Composite," *Journal of Materials Science Letters,* Vol. 9, 1990, pp. 719–725.

[*5*] McMeeking, R. M. and Evans, A. G., "Matrix Fatigue Cracking in Fiber Composites," *Mechanics of Materials,* Vol. 9, 1990, pp. 217–227.

[*6*] Cox, B. N. and Marshall, D. B., "Crack Bridging in the Fatigue of Fibrous Composites," *Fatigue and Fracture of Engineering Materials and Structures,* Vol. 14, 1991, pp. 847–861.

[*7*] References in *Fatigue of Advanced Materials,* R. O. Ritchie, R. H. Dauskardt, and B. N. Cox, Eds., MCE Publications, Birmingham, UK, 1991.

[*8*] Davidson, D. L., "The Micromechanics of Fatigue Crack Growth at 25°C in Ti-6Al-4V Reinforced with SCS-6 Fibers," *Metallurgical Transactions A,* Vol. 23A, 1992, pp. 865–879.

[*9*] Cox, B. N. and Lo, C. S., "Simple Approximations for Bridged Cracks in Fibrous Composites," *Acta Metallurgica et Materialia,* Vol. 40, 1992, pp. 1487–1496.

[*10*] Ghosn, L. J., Kantzos, P., and Telesman, J., "Modeling of Crack Bridging in a Unidirectional Metal Matrix Composite," *International Journal of Fracture,* Vol. 54, 1992, pp. 345–357.

[*11*] Bowen, P., "Characterization and Micromodeling of Fatigue Crack Growth in Unidirectional Continuous Fiber Reinforced Titanium Alloy Composites," *Proceedings,* IRC '92, Birmingham, UK, Sept. 1992, M. H. Loretto and C. J. Beevers, Eds., MCE Publications, Birmingham, UK, 1992, pp. 697–718.

[*12*] Walls, D. P., Bao, G., and Zok, F. W., "Mode I Fatigue Cracking in a Fiber Reinforced Metal Matrix Composite," *Acta Metallurgica et Materialia,* Vol. 41, 1993, pp. 2061–2071.

[*13*] Bakuckus, J. G., Jr. and Johnson, W. S., "Application of Fiber Bridging Models to Fatigue Crack Growth in Unidirectional Titanium Matrix Composites," *Journal of Composites Technology and Research,* Vol. 15, 1993, pp. 242–255.

[*14*] John, R., Kaldon, S. G., and Ashbaugh, N. E., "Applicability of Fiber Bridging Models to Describe Crack Growth in Unidirectional Titanium Matrix Composites," *Proceedings,* Titanium Metal Matrix Composite Workshop II, La Jolla, CA, June 1993, W. C. Revelos and P. R. Smith, Eds., U.S. Air Force, Dayton, OH, 1993.

[*15*] Evans, A. G. and Marshall, D. B., "Failure Mechanisms in Ceramic-Fiber/Ceramic-Matrix Composites," *Journal,* American Ceramics Society, Vol. 68, 1985, pp. 225–231.

[*16*] Marshall, D. B. and Oliver, W. C., "Measurement of Interfacial Mechanical Properties in Fiber-Reinforced Ceramic Composites," *Journal of Materials Science and Engineering,* Vol. A126, 1987, pp. 95–103.

[*17*] Marshall, D. B., Cox, B. N., and Evans, A. G., "The Mechanics of Matrix Cracking in Brittle-Matrix Fiber Composites," *Acta Metallurgica,* Vol. 33, 1985, pp. 2013–2211.

[*18*] McCartney, L. N., "Mechanics of Matrix Cracking in Brittle-Matrix Fiber-Reinforced Composites," *Proceedings,* The Royal Society, Vol. A409, 1987, pp. 329–350.

[*19*] McCartney, L. N., "Mechanics for the Growth of Bridged Cracks in Composite Materials: Part I. Basic Principles," *Journal of Composites Technology and Research,* Vol. 14, 1992, pp. 133–154.

[*20*] Cox, B. N. and Marshall, D. B., "Concepts for Bridged Cracks in Fracture and Fatigue," *Acta Metallurgica et Materialia,* Vol. 42, 1994, pp. 341–363.

[*21*] Aveston, J., Cooper, G. A., and Kelly, A., "Single and Multiple Fracture," *The Properties of Fiber Composites,* National Physical Laboratory, IPC Science Technology Press, 1971, pp. 15–26.

[*22*] Marshall, D. B. and Cox, B. N., "Tensile Properties of Brittle Matrix Composites. Influence of Fiber Strength," *Acta Metallurgica et Materialia,* Vol. 35, 1987, pp. 2607–2619.

[*23*] Budiansky, B. and Cui, Y. L., "On the Tensile Strength of a Fiber-Reinforced Ceramic Composite Containing a Crack-Like Flaw," *Journal of the Mechanics and Physics of Solids,* Vol. 42, 1994, pp. 1–19.

[*24*] Bao, G. and McMeeking, R. M., "Fatigue Crack Growth in Fiber-Reinforced Metal-Matrix Composites," *Acta Metallurgica et Materialia,* Vol. 42, 1994, pp. 2415–2425.

[25] Gao, Y.-C., Mai, Y.-W., and Cotterell, B., "Fracture of Fiber Reinforced Materials," *Journal of Applied Mathematics and Physics (ZAMP),* Vol. 39, 1988, pp. 550–572.
[26] Hutchinson, J. W. and Jensen, H. M., "Models of Fiber Debonding and Pullout in Brittle Composites with Friction," *Mechanics of Materials,* Vol. 9, 1990, pp. 139–163.
[27] Ning, X. J. and Pirouz, P., "The Microstructure of SCS6 SiC Fiber," *Journal of Materials Research,* Vol. 6, 1991, pp. 2234–2248.
[28] James, M. R., Morris, W. L., and Cox, B. N., "Local Stress Relaxation in Ti-6Al-4V-2Fe/SCS6 Composite During Thermal Cycling," *Proceedings,* ICF7, Houston, March 1989, K. Salama, K. Ravi-Chandar, D. M. R. Taplin, and P. Rama Rao, Eds., Pergamon, New York, 1989, pp. 3029–3037.
[29] Jero, P. D. and Kerans, R. J., "The Contribution of Interfacial Roughness to Sliding Friction of Ceramic Fibers in a Glass Matrix," *Scripta Metallurgica et Materialia,* Vol. 24, 1990, pp. 2315–2318.
[30] Mackin, T. J., Warren, P. D., and Evans, A. G., "Effects of Fiber Roughness on Interface Sliding in Composites," *Acta Metallurgica et Materialia,* Vol. 40, 1992, pp. 1251–1257.
[31] Warren, T. L., Majumdar, A., and Krajcinovic, D., "Atomic Force Microscope Imaging of the Surface Roughness of SCS and TiB_2 Coated SiC Fibers and Uncoated Sapphire Fibers," *Composites,* submitted for publication.
[32] Majumdar, A. and Bhushan, B., "Characterization and Modeling of Surface Roughness for Microtribology," *Microtribology and Its Applications in Magnetic Storage Systems,* B. Bhushan, Ed., CRC Press, Boca Raton, FL, 1994.
[33] Paris, P., McMeeking, R. M., and Tada, H., "The Weight Function Method for Determining Stress Intensity Factors," *Cracks and Fracture, ASTM STP 601,* American Society for Testing and Materials, Philadelphia, 1976, pp. 471–489.
[34] Morris, W. L., unpublished work, Rockwell International Science Center, 1990.
[35] Meda, G. and Steif, P. S., "A Detailed Analysis of Cracks Bridged by Fibers. Part I: Limiting Cases of Short and Long Cracks," *Journal of the Mechanics and Physics of Solids,* submitted for publication.
[36] Xia, Z. C., Hutchinson, J. W., Evans, A. G., and Budiansky, B., "On Large Scale Sliding in Fiber-Reinforced Composites," *Journal of the Mechanics and Physics of Solids,* Vol. 42, 1944, pp. 1139–1158.
[37] Yurgatis, S. W., "Measurement of Small Angle Fiber Misalignment in Continuous Fiber Composites," *Composites Science and Technology,* Vol. 30, 1987, pp. 279–293.
[38] Cox, B. N., Dadkhah, M. S., Kniveton, T., and Morris, W. L., "Simple Models for Triaxially Braided Composites," *Composites,* Vol. 26, 1995, pp. 91–102.
[39] Cox, B. N. and Dadkhah, M. S., "The Macroscopic Elasticity of 3D Woven Composites," *Journal of Composite Materials,* in press.
[40] Li, V. C., Wang, Y., and Backer, S., "Effect of Inclining Angle, Bundling, and Surface Treatment on Synthetic Fiber Pull-Out from a Cement Matrix," *Composites,* Vol. 21, 1990, pp. 132–140.
[41] Kouris, D., "Stress Concentration Due to the Interaction Between Two Imperfectly Bonded Fibers in a Continuous Fiber Composite," *Journal of Applied Mechanics,* Vol. 60, 1993, pp. 203–206.
[42] Marshall, D. B., Morris, W. L., Cox, B. N., Graves, J., Porter, J. R., Kouris, D., and Everett, R. K., "Transverse Strengths and Failure Mechanisms in Ti_3Al Matrix Composites," *Acta Metallurgica et Materialia,* Vol. 42, 1994, pp. 2657–2673.
[43] Marshall, D. B. and Cox, B. N., "A J-Integral Method for Calculating Steady-State Matrix Cracking in Composites," *Mechanics of Materials,* Vol. 7, 1988, pp. 127–133.
[44] Cox, B. N. and Marshall, D. B., "Stable and Unstable Solutions for Bridged Cracks in Various Specimens," *Acta Metallurgica et Materialia,* Vol. 39, 1991, pp. 579–589.
[45] Tada, H., "The Stress Analysis of Cracks Handbook," Paris Productions Inc., St. Louis, MO, 1985, Article 5.13.
[46] Cox, B. N., "Scaling for Bridged Cracks," *Mechanics of Materials,* Vol. 15, 1993, pp. 87–98.
[47] Larsen, J. M., Jira, J. R., John, R., and Ashbaugh, N. E., "Crack Bridging Effects in Notch Fatigue of SCS-6/TIMETAL 21S Composite Laminates," in this volume.
[48] Marissen, R., "Fatigue Crack Growth Predictions in Aramid Reinforced Aluminum Laminates (ARALL)," *Proceedings,* 15th ICAS Conference, London, Sept. 1986, ICAS, 1986, pp. 801–807.
[49] Marissen, R., "Fatigue Crack Growth in Aramid Reinforced Aluminum Laminates (ARALL). Mechanisms and Predictions," Report No. DFVLR-FB-84-37, Wissenschaft Berichtswesen DFVLR, Cologne, Germany, 1984.
[50] Nguyen, T.-H. B. and Yang, J.-M., "Elastic Bridging for Modelling Fatigue Crack Propagation in

a Fiber-Reinforced Titanium Matrix Composite,'' *Fatigue and Fracture of Engineering Materials and Structures,* Vol. 17, 1994, pp. 119–131.

[*51*] Cox, B. N. and Marshall, D. B., ''The Determination of Crack Bridging Forces,'' *International Journal of Fracture,* Vol. 49, 1991, pp. 159–176.

[*52*] Begley, M. R. and McMeeking, R. M., ''Fatigue Crack Growth with Fiber Failure in Metal Matrix Composites,'' *Composites Science and Technology,* in press.

[*53*] Rose, L. R. F., ''Crack Reinforcement by Distributed Springs,'' *Journal of the Mechanics and Physics of Solids,* Vol. 35, 1987, pp. 383–405.

[*54*] Willis, J. R. and Nemat-Nasser, S., ''Singular Perturbation Solution of a Class of Singular Integral Equations,'' *Quarterly Journal of Applied Mathematics,* Vol. 48, 1990, pp. 741–753.

[*55*] Cox, B. N. and Rose, L. R. F., ''A Self-Consistent Approximation for Crack Bridging by Elastic/Perfectly Plastic Ligaments,'' *Mechanics of Materials,* in press.

[*56*] Hillberry, B. M. and Johnson, W. S., ''Prediction of Matrix Fatigue Crack Initiation in Notched SCS6/Ti-15-3 Metal Matrix Composites,'' NASA Technical Memorandum 104141, 1991.

[*57*] Bowling, J. and Groves, G. W., ''The Propagation of Cracks in Composites Consisting of Ductile Wires in a Brittle Matrix,'' *Journal of Materials Science,* Vol. 14, 1979, pp. 443–449.

[*58*] Zok, F. and Hom, C. L., ''Large Scale Bridging in Brittle Matrix Composites,'' *Acta Metallurgica et Materialia,* Vol. 38, 1990, pp. 1895–1904.

[*59*] Cox, B. N. and Lo, C. S., ''Load Ratio, Notch, and Scale Effects for Bridged Cracks in Fibrous Composites,'' *Acta Metallurgica et Materialia,* Vol. 40, 1992, pp. 69–80.

[*60*] Jira, J. M. and Larsen, J., ''Fatigue of Unidirectional SCS-6/Ti-24Al-11Nb Composite Containing a Circular Hole (Part I),'' *Metallurgical Transactions,* in press.

[*61*] Marshall, D. B., Shaw, M. C., and Morris, W. L., ''Measurement of Interfacial Debonding and Sliding Resistance in Fiber Reinforced Intermetallics,'' *Acta Metallurgica et Materialia,* Vol. 40, 1992, pp. 443–454.

[*62*] Walls, D. and Zok, F., ''Interfacial Fatigue in a Fiber-Reinforced Metal Matrix Composite,'' *Acta Metallurgica et Materialia,* Vol. 42, 1994, pp. 2675–2681.

[*63*] Cox, B. N. and Rose, L. R. F., ''Time or Cycle Dependent Crack Bridging,'' *Mechanics of Materials,* Vol. 19, 1994, pp. 39–57.

[*64*] Cox, B. N. and Rose, L. R. F., unpublished work.

[*65*] Bao, G. and McMeeking, R. M., ''Thermomechanical Fatigue Cracking in Fiber Reinforced Metal-Matrix Composites,'' *Journal of the Mechanics and Physics of Solids,* Vol. 43, 1945, pp. 1433–1460.

Massoud Mirdamadi[1] *and W. Stephen Johnson*[2]

Modeling and Life Prediction Methodology for Titanium Matrix Composites Subjected to Mission Profiles

REFERENCE: Mirdamadi, M. and Johnson, W. S., "**Modeling and Life Prediction Methodology for Titanium Matrix Composites Subjected to Mission Profiles,**" *Life Prediction Methodology for Titanium Matrix Composites, ASTM STP 1253,* W. S. Johnson, J. M. Larsen, and B. N. Cox, Eds., American Society for Testing and Materials, 1996, pp. 573–594.

ABSTRACT: Titanium matrix composites (TMCs) are being evaluated as structural materials for elevated temperature applications in future generation hypersonic vehicles. In such applications, TMC components are subjected to complex thermomechanical loading profiles at various elevated temperatures. Therefore, thermomechanical fatigue (TMF) testing, using a simulated mission profile, is essential for evaluation and development of life prediction methodologies. The objective of the research presented in this paper was to evaluate the TMF response of the $[0/90]_{2s}$ SCS-6/TIMETAL-21S subjected to a generic hypersonic flight profile and its portions with a temperature ranging from −130 to 816°C. It was found that the composite modulus, prior to rapid degradation, had consistent values for all the profiles tested. The accumulated minimum strain was also found to be the same for all the profiles tested. A micromechanics-based analysis was used to predict the stress-strain response of the laminate and of the constituents in each ply during thermomechanical loading conditions by using only constituent properties as input. The fiber was modeled as elastic with transverse orthotropic and temperature-dependent properties. The matrix was modeled using a thermoviscoplastic constitutive relationship. In the analysis, the composite modulus degradation was assumed to result from matrix cracking and was modeled by reducing the matrix modulus. Fatigue lives of the composite subjected to the complex generic hypersonic flight profiles were well correlated using the predicted stress in 0° fibers.

KEYWORDS: silicon-carbide fibers, thermomechanical fatigue, residual stresses viscoplasticity theory, thermal strains, titanium, titanium matrix composites, life prediction, titanium alloys, fatigue (materials), modeling

Silicon-carbide fiber-reinforced titanium metal matrix composites (TMCs) are being developed for use as structural materials for elevated temperature applications where low-weight, high-temperature capability, and high stiffness are required. In these applications, TMC components will be subjected to complex flight profiles consisting of a combination of applied thermal and mechanical loads at various elevated temperatures. Due to the large differences in the thermal and mechanical properties of the fiber and the matrix, significant thermal residual stresses develop in the composite. The resulting stress coupled with applied thermal and mechanical loading, can lead to a complex state of stress in these composites. Therefore, thermomechanical fatigue (TMF) testing is critical for the evaluation and devel-

[1]Research scientist, Analytical Services and Materials Inc., Hampton, VA 23666.

[2]Professor, School of Materials Science and Engineering, Georgia Institute of Technology, Atlanta, GA 30332; formerly, senior scientist, Mechanics of Materials Branch, NASA Langley Research Center.

opment of life prediction methodologies. In addition, in order to develop life prediction methodologies for TMC, one must be able to predict the stresses and strains that occur in the constituents of the composite during the complicated TMF loading profiles. Hence, micromechanical-based analyses that can account for the viscoplastic behavior of both the fiber and matrix are helpful in establishing a sound basis for a life prediction methodology.

The objective of this paper is to experimentally evaluate TMF response of the cross-ply SCS-6/TIMETAL-21S[3] subjected to a generic hypersonic flight profile and its portions. Furthermore, a micromechanical-based analysis is used to develop an approach to determine the fatigue life of the laminate under TMF mission profile loading conditions.

Background

A considerable amount of research has been conducted to characterize the fatigue behavior of TMC over the past few years [*1–5*]. At room temperature, the fatigue response of a SCS-6/Ti-15-3 composite, for a variety of lay-ups, was well correlated using the stress range in the 0° fiber [*1*] where the 0° fiber stress was calculated by multiplying the composite stabilized strain range by the fiber modulus (see Fig. 1). The composite's modulus degradation as a result of fatigue cycling was dependent on the composite's lay-up [*1*]. Analysis of the unidirectional SCS-6/Ti-15-3 composite subjected to TMF loading conditions also showed that the stress in the 0° fiber controlled the fatigue lives [*2*]. In Ref *2*, the 0° fiber stress range calculated from a micromechanics analysis was used to compare the in-phase and out-of-phase TMF data reported in Refs 3 and 4 and shown in Fig. 2. The TMF data from Ref *3* had a cycle time of 110 s (0.0091 Hz) and the temperature varied from 300 to 500°C. The TMF data from Ref *4* had a cycle time of 180 s (0.0056 Hz) and the temperature varied from 93 to 593°C. As seen in Fig. 2, the analysis showed that for a given condition, the fatigue strength of the 0° fiber was controlled by a combination of temperature, loading frequency, and time at temperature. Furthermore, for a given temperature, loading frequency, and time at temperature, the stress range in the 0° fiber controlled the fatigue life.

Additionally, the response of $[0/90]_{2s}$ SCS-6/Ti-15-3 laminates subjected to several load temperature profiles was investigated [*5*] and shown in Fig. 3. All profiles shown in the figure had the same sustained stress of 420 MPa applied at the sustained temperature of 427°C. As seen in the figure, the sustained load-sustained temperature test (designated by open circles on Fig. 3) accumulated strain at the fastest rate and resulted in the shortest life. This occurred in spite of the other profiles having cyclic loads and temperatures. The test data also indicated that the total strain to failure decreased with increased time at temperature and is consistent with the previous findings reported in Ref 2. Therefore, this difference in life may be more affected by duration of the exposure at elevated temperature than type of loading. Furthermore, microscopic examination of the $[0/90]_{2s}$ SCS-6/Ti-15-3 composite subjected to the sustained stress of 420 MPa applied at the sustained temperature of 427°C [*6*] revealed that the strain accumulation was due primarily to internal damage in the form of fiber-matrix debonding and matrix cracking and not deformation associated with conventional creep. It was also observed [*6*] that the internal damage initiated and grew from the fiber-matrix interface in the 90° plies and progressed toward the 0° plies. The final failure of the 0° fiber occurred just prior to the failure of the composite. These results indicate that in a cross-ply laminate, the damage and failure of the 90° plies occurs at an early stage, resulting in stress redistribution. However, the final failure of the composite is governed by the stress in the 0° fibers.

Increased exposure at elevated temperature could also produce more reaction between the

[3]TIMETAL-21S is a registered trademark of Timet.

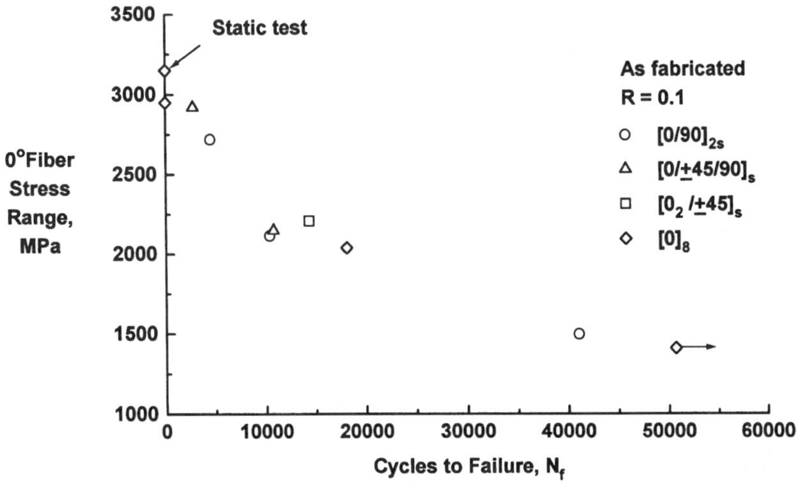

FIG. 1—*Cyclic stress range in the 0° fiber as a function of cycles of failure for the SCS-6/Ti-15-3 composite at room temperature* [*1*].

fiber and the matrix, lowering the static strength [*7*] and fatigue resistance [*8*]. Therefore, the stress (strain) in the 0° fiber stress plays a major role in fatigue strength of TMCs and could be used as a basis for development of life prediction methodologies.

Material and Test Specimens

The material used in this study was SCS-6/TIMETAL-21S that was fabricated into a $[0/90]_{2s}$ laminate by hot isostatic pressing (HIPing) of Ti-15Mo-3Nb-3Al-0.2Si foil between

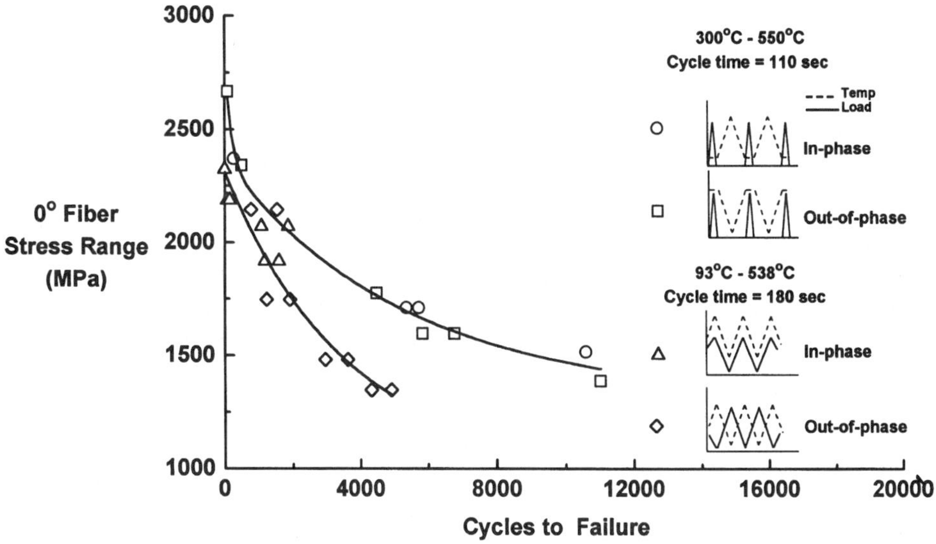

FIG. 2—*The stress range in the 0° fiber as a function of cycles to failure for SCS-6/Ti-15-3 unidirectional composite* [2].

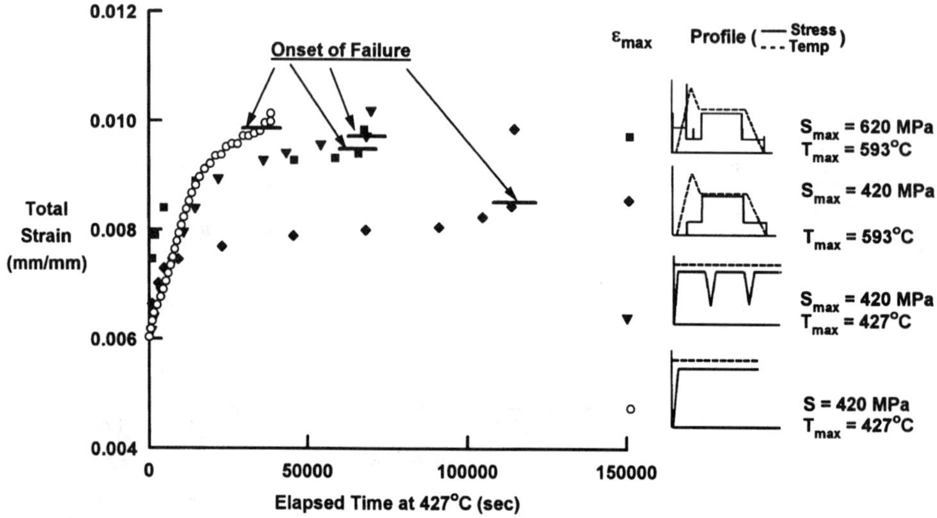

FIG. 3—*Variation of total accumulative strain as a function of elapsed time at 427°C for* $[0/90]_{2s}$ *SCS-6/Ti-15-3 composite with* $v_f = 0.385$ [6].

tapes of unidirectional SCS-6 silicon-carbide fibers having a diameter of 0.14 mm. The fibers were held in place with titanium-niobium (Ti-Nb) crossweave wires. An average volume fraction of 0.365 was measured for this laminate.

Rectangular specimens having a 152-mm length, 12.7-mm width, and 1.69-mm thickness were machined using a diamond wheel saw. Brass tabs (10 by 30 by 1 mm) were placed between the end of the specimen and the grips to avoid specimen failure in the grips. Prior to testing, each specimen was heat treated (aged) at 621°C for 8 h in a vacuum to stabilize the matrix material.

Experimental Procedure

A TMF test capability was developed to conduct a series of generic hypersonic flight profile simulations. The detailed experimental procedure is available in Ref *9*. Briefly, the TMF test setup consists of a 100-kN, closed-loop, servohydraulic test frame equipped with water-cooled hydraulic grips. The test frame is equipped with a load profiler that is capable of generating a variety of waveform segments that can then be combined to construct the required load profiles. The test specimen is heated directly using a 5-kW induction generator with three independent adjustable coils such as developed by Ellis and Bartolotta [*10*] controlled by a temperature profiler. Similar to the load profiler, the temperature profiler can be programmed to generate a variety of waveform segments that can then be combined to construct the required temperature profiles. Type K thermocouple wires were spot welded onto the specimen's surface for the purposes of monitoring and controlling the temperature. In addition, an infrared camera was used to monitor the temperature distribution along the length of the specimen. A temperature variation of ± 10°C (indicated by thermocouple) was maintained throughout the entire applied temperature profiles.

A combination of forced gaseous and liquid nitrogen was used in conjunction with the water-cooled grips to achieve the required cooling rate and low temperature of − 130°C in the temperature profile. A liquid nitrogen tank was modified to supply liquid or gaseous nitrogen or both at constant pressure. Flow of gaseous and liquid nitrogen was controlled by the temperature profiler. With the addition of liquid and gaseous nitrogen, the temperature profiler was used to reach and control cryogenic temperatures as low as −130°C. Rate of cooling is controlled by the temperature profiler through the simultaneous operation of induction heating system and liquid nitrogen cooling system.

Axial strains were measured on the edge of the specimen using a high-temperature, quartz rod water-cooled extensometer with a 25-mm gage length. An eight-channel, analog-digital PC-based data acquisition system using MATE software [*11*] was utilized periodically to record and store the test data. The analog/digital data acquisition system was capable of recording load, strain, time, and temperature data at predetermined levels of these variables. In addition, a strip chart recorder was also used to continuously record the load, temperature, and strain as a function of time. Failure is defined as separation of the test specimen into two pieces and repeat tests were conducted at each test condition.

Generic Hypersonic Profile

The generic hypersonic flight profile used in this study is shown in Fig. 4. All TMF tests were conducted using this flight profile or its portions, as shown in Fig. 5. The generic hypersonic flight profile consists of Mission I and Mission II, as noted in Figs. 4*a* and *b*. The

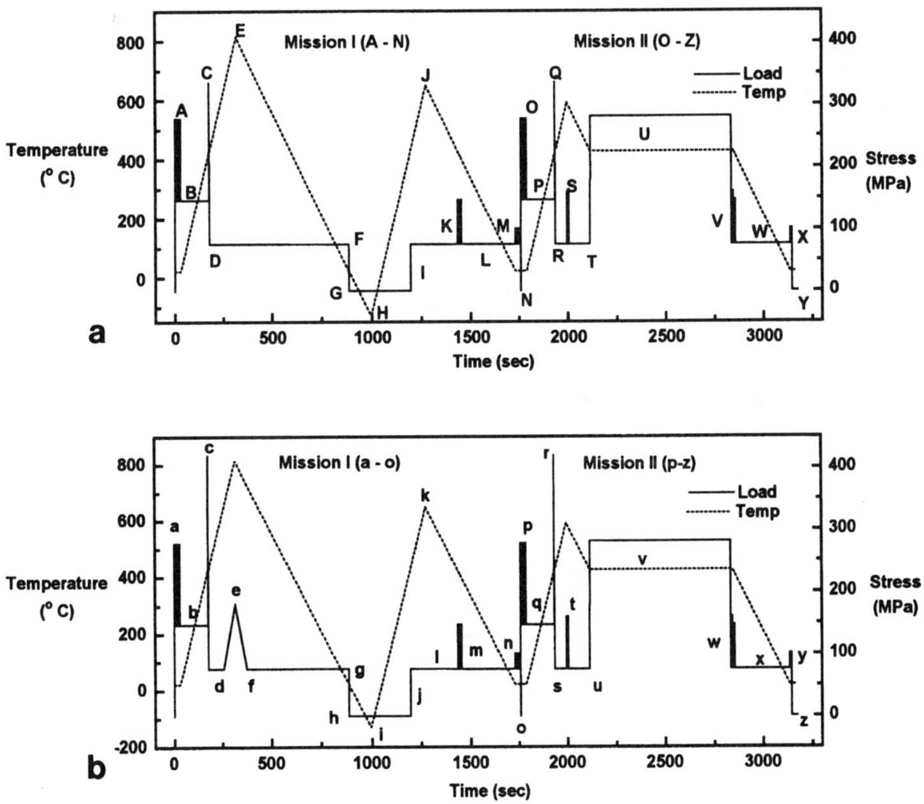

FIG. 4—*Generic hypersonic flight profile;* (a) *used for Flights 1–24, 26–49, 51–75, . . . and* (b) *used for Flights 25, 50, 75, . . .*

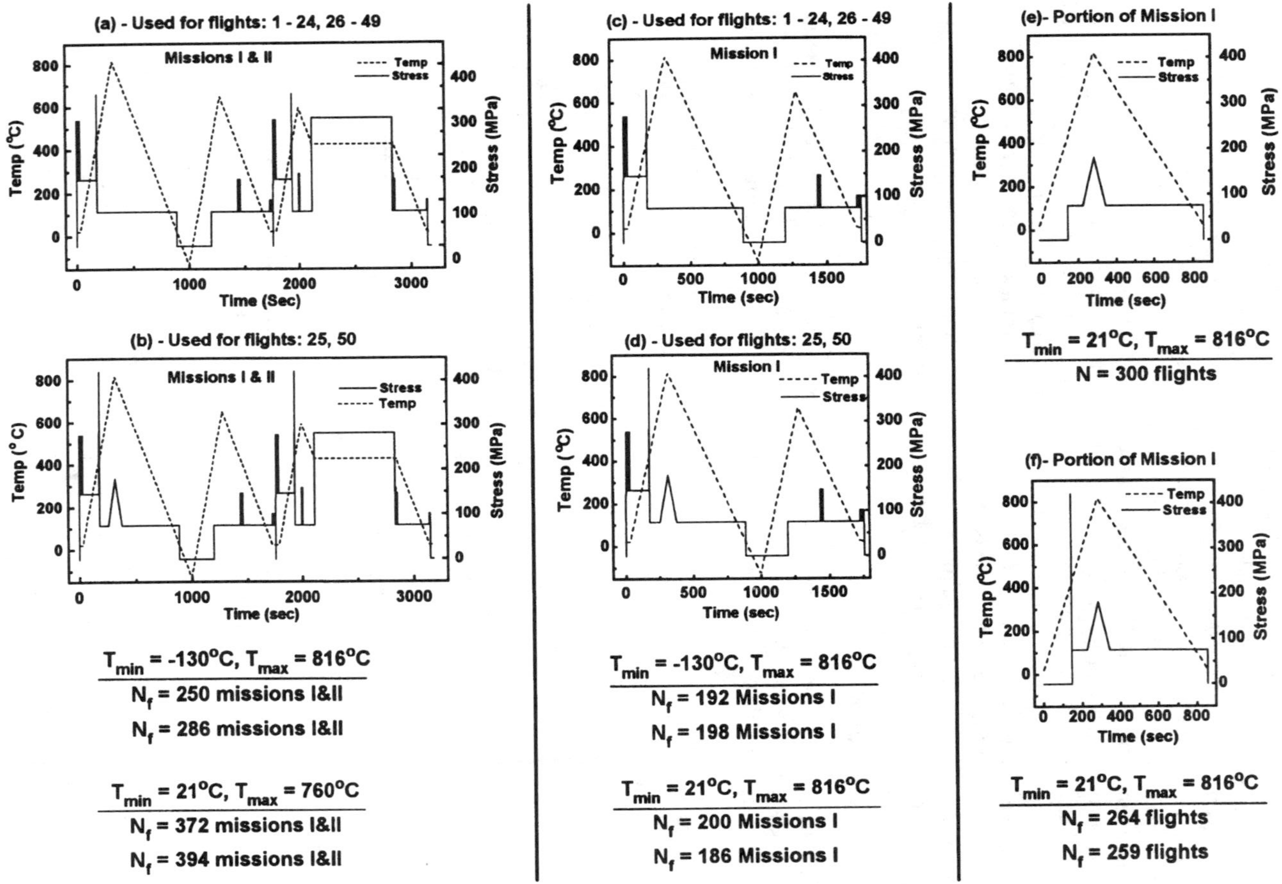

FIG. 5—*Fatigue data of the* $[0/90]_{2s}$ *SCS-6/TIMETAL-21S composite with* $v_f = 0.365$ *subjected to the generic hypersonic flight profile and its portions.*

entire TMF flight profile (Missions I and II), shown in Fig. 4*a*, is applied 24 times followed by the TMF flight profile (Missions I and II), shown in Fig. 4*b*, once. This sequence is repeated until the test specimen is separated into two pieces. The TMF data is reported in terms of the numbers of missions to failure and note that there are two missions (Missions I and II) in the entire flight profile shown in Fig. 4.

The load profile contains segments of continuous fatigue cycling that are shown as heavier lines in the figure. The mechanical fatigue cycles are applied at a frequency of 1 Hz. The temperature profiles shown in Figs. 4*a* and *b* are the same with a maximum and minimum applied temperature of 816°C and − 130°C, respectively. The heating and cooling rates of the temperature profile are 2.8 and 1.4°C/s, respectively. The loading profiles in Figs. 4*a* and *b* have slight differences. In Fig. 4*a*, the maximum applied stress in the loading profile is 336 MPa (Segment D-C) at 427°C, while the maximum applied stress in Fig. 4*b* is 420 MPa (Segment D-C). In addition, Fig. 4*b* contains the segment of the loading profile designated as *d-e-f* that is in-phase with the applied temperature. During this segment, the stress at the maximum applied temperature, 816°C, is 180 MPa.

Analytical Method

A two-dimensional micromechanical model, VISCOPLY, was used to predict the stress-strain response of the composite and of the constituents in each ply during thermomechanical loading conditions by using only constituent properties as input. The VISCOPLY code, developed by Bahei-El-Din, uses the vanishing fiber diameter (VFD) model [*12*] to calculate the orthotropic properties of a ply. The phy properties are then used in a laminated plate theory [*13*] to predict the overall laminate response. Both the fiber and the matrix can be described as thermo-viscoplastic materials.

The viscoplastic constitutive model implemented in the VISCOPLY program was developed by Bahei-El-Din, Shah, and Dvorak [*14*] for high-temperature, isothermal, and nonisothermal applications and is based on the viscoplastic theory of Eisenberg and Yen [*15*]. This constitutive viscoplastic model assumes the existence of an equilibrium stress-strain response that corresponds to the theoretical lower bound of the dynamic response. In addition, this constitutive viscoplastic model assumes that the elastic response is rate-independent and that inelastic rate-dependent deformation takes place if the current stress state is greater than the equilibrium stress. The inelastic strain rate is described using a power law function of overstress [*16*] that is defined as the difference between the current stress and the equilibrium stress.

Complex combinations of thermal and mechanical loads can be modeled using the VISCOPLY program. Sequential jobs can be run to vary the order and rate of load and temperature. Fiber and matrix average stresses and strains and the overall composite response under thermomechanical loading conditions are calculated.

Fiber-matrix interface failure in TMCs occurs early under applied load [*1*] and must be accounted for in the analysis. Previous work [*17*] on $[0/90]_{2s}$ SCS-6/Ti-15-3 composite showed that the fiber-matrix interface failure can be simulated by reducing the fiber transverse modulus in the 90° plies only. Loss of composite stiffness as a result of matrix cracking can also be approximated by reducing the modulus of the matrix in the model. However, it should be pointed out that, due to the assumptions made in the VFD model, local micromechanical effects such as the lateral constraint of the matrix due to the presence of the fibers and stress concentrations in the fibers and matrix are not accounted for in the VISCOPLY program. Nevertheless, the VISCOPLY program has been used to predict the global thermomechanical response within 10% of the experimental results in similar TMCs [*2,17,18*].

Results and Discussion

In this section, the experimental fatigue and analytical results are presented. First, the experimental fatigue results for a series of tests simulating a generic hypersonic flight profile and its portions will be discussed. Second, analytical results from VISCOPLY will be compared to the TMF data.

Fatigue Lives

The experimental fatigue data of cross-ply $[0/90]_{2s}$ SCS-6/TIMETAL-21S subjected to the hypersonic flight profile and portions of the flight profile are shown in Fig. 5. Fatigue life data are presented in terms of completed missions to failure where a mission is the completion of an applied flight profile. Two TMF tests were conducted using the entire flight profile (Missions I and II) as shown in Figs. 5*a* and *b* and a temperature range of −130 to 816°C. The test specimens had fatigue lives of 250 and 286 missions, respectively. In both cases, the test specimen failed during the load spike at 427°C. In order to examine the effect of maximum temperature on fatigue life, two additional tests were conducted using the entire mission profile (Missions I and II) with a temperature range of 21 to 760°C. In these two tests, the flight time and the load profile were kept as before, but the maximum temperature was lowered to 760°C and the minimum temperature was raised to 21°C in Mission I. The fatigue lives under this condition were 372 and 394 missions, respectively. As before, the test specimens failed during the spike load at 427°C. The observed fatigue lives suggest that the fatigue life was considerably increased by reducing the maximum applied temperature from 816 to 760°C.

The maximum temperature of 816°C occurs only during Mission I and was suspected to be the portion of the entire flight profile most detrimental to fatigue life. Therefore, additional TMF tests were conducted using only Mission I as shown in Fig. 5*c* and *d*. Under Mission I, with a temperature range of −130 to 816°C, the test specimens had fatigue lives of 192 and 198 missions during the spike load at 427°C. To examine if the cryogenic temperature had any effect on the fatigue life, two additional tests were conducted using Mission I with the temperature ranging from 21 to 816°C. The fatigue lives were 200 and 186 missions during the spike load at 427°C. These results indicated that the cryogenic temperature of −130 to 816°C did not produce any noticeable effect on fatigue life over specimens tested from 21 to 816°C. As compared to the fatigue results from the entire flight profile (Missions I and II), it was observed that Mission I alone resulted in the failure of the composite.

Since Mission I resulted in fatigue failure, critical portions of Mission I were examined. As indicated in Fig. 4*b*, Mission I contains a load portion (designated as *d-e-f*) that is in-phase with the applied temperature at the maximum temperature of 816°C and occurs once every 25th flight. One TMF test was conducted on this isolated portion of Mission I as shown in Fig. 5*e*. The results of the TMF test under this loading condition did not cause failure after 300 missions, and the test was considered a runout. Recall now that the fatigue results obtained using the entire flight profile (Missions I and II) had an average fatigue life of 268 missions during which the isolated portion (Fig. 5*e*) occurred only five times. It is apparent that this portion by itself is not detrimental to the fatigue life, therefore, this portion including a load spike to a maximum applied stress of 420 MPa at 427°C was included in the testing profile as shown in Fig. 5*f*. Two TMF tests were conducted under this loading condition. The test specimens failed at 264 and 259 flights during the spike load at 427°C. From these results, inclusion of the load spike did not result in a considerable reduction in the fatigue life. Again, comparing these results with those obtained under the Missions I and

II, with the average fatigue life of 268 missions, this portion would only occur about five times. Therefore, even though this portion resulted in fatigue failure, it was not considered as a critical portion of Mission I alone. Although, not shown in Fig. 5, one TMF test was conducted using Mission II of the flight profile, shown in Fig. 4*b* that occurs once every 25 flights. In Mission II, the load profile contains a spike to the maximum applied stress of 420 MPa and 427°C, and a 720-s hold-time segment at the applied stress of 286 MPa and 427°C. The TMF test was conducted under this condition and the test specimen endured 300 flights without failure and was considered a runout.

The fatigue results obtained in this section indicate that the fatigue response is influenced by the applied maximum temperature, applied stress, and time at temperature. Furthermore, these results suggest that the isothermal, simple in-phase and out-of-phase TMF test results may not be extrapolated to describe the TMF flight profile. Therefore, it is essential that the effects of parameters, such as stress, temperature, and time at temperature on fatigue be understood for development of life prediction methodology.

Stress-Strain Behavior

Detailed stress-strain responses and modulus histories for all the TMF mission profiles tested will be presented elsewhere. In this section, only the response under the entire hypersonic flight profile (Figs 5*a* and *b*) and a portion of Mission I (Fig. 5*f*) will be discussed. The stress-strain response was periodically recorded, and the elastic modulus was calculated over the lives of the test specimens. As shown in Fig. 6, the elastic modulus was calculated at room temperature (initial loading), E_i; at 427°C during unloading of the load spike, E_{un}; and during unloading of the in-phase portion of the 25th flight where applicable, E_{in}, for all TMF tests.

The stress-strain responses of the composite subjected to the entire hypersonic flight profile (Missions I and II) for the temperature range of − 130 to 816°C (Flights 1, 2, 75, and

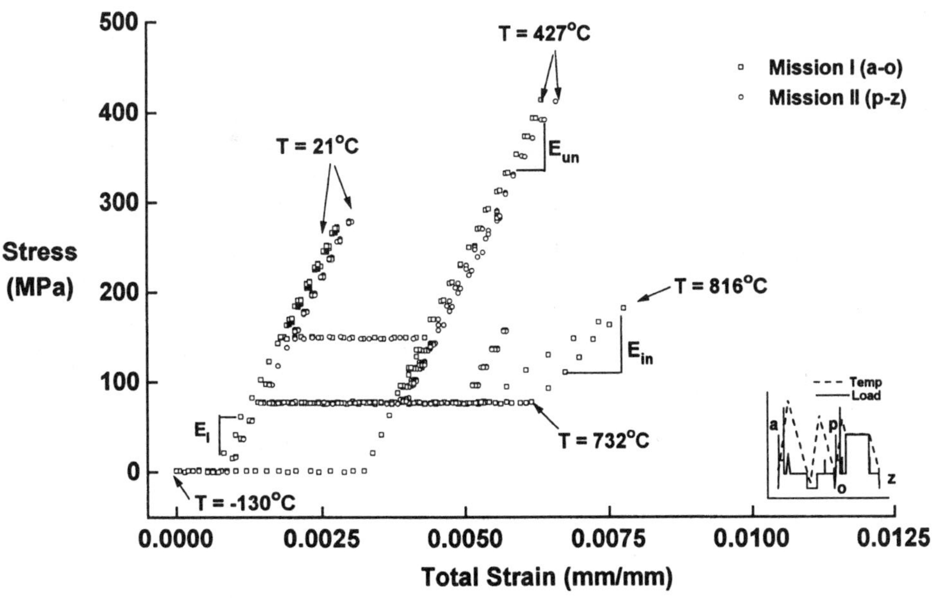

FIG. 6—*Mission profile moduli measurements.*

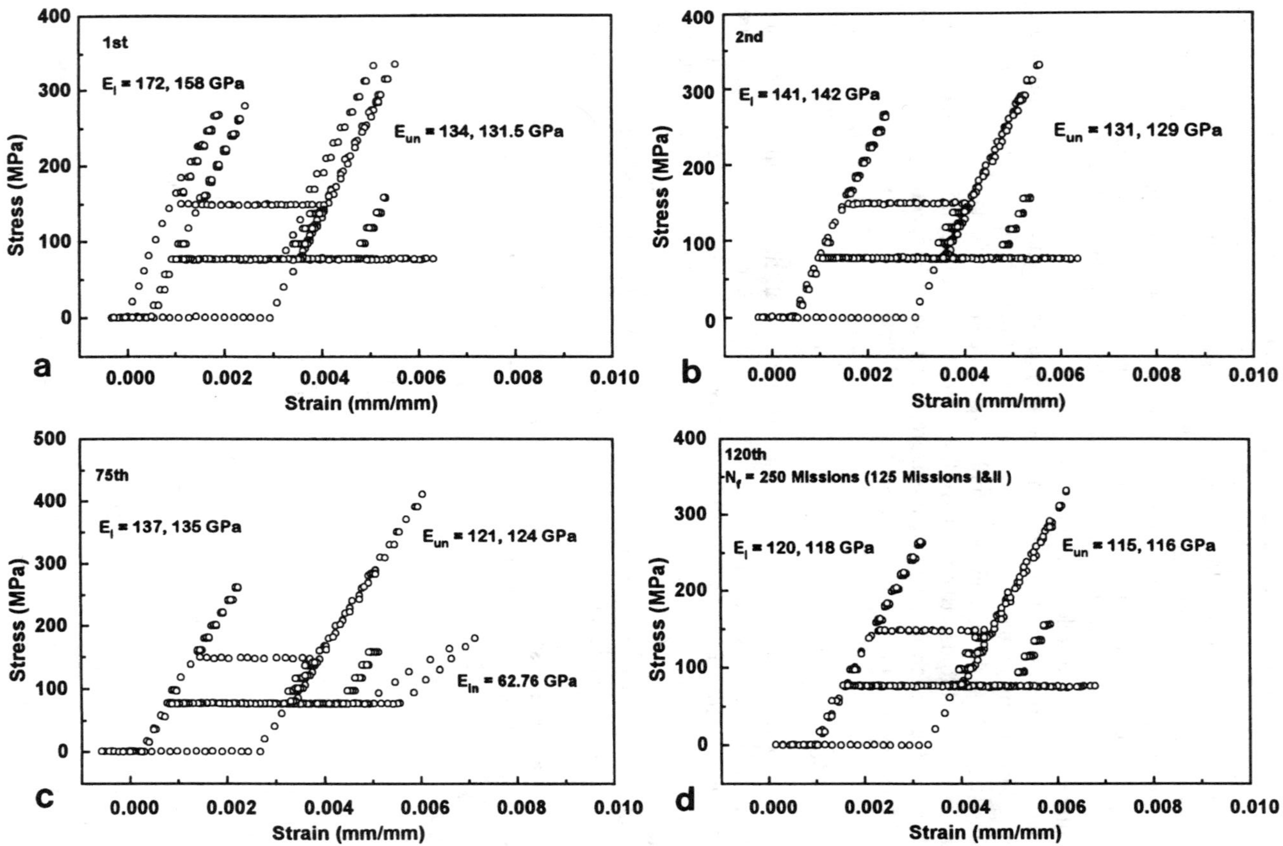

FIG. 7—*Stress-strain response of* $[0/90]_{2s}$ *SCS-6/TIMETAL-21S,* $v_f = 0.365$*, subjected to the entire flight profile and temperature range of* -130 *to* $816°C$.

120) are shown in Fig. 7. Under the entire hypersonic profile (Missions I and II), the elastic modulus is calculated twice, once during Mission I and once during Mission II. As seen in Fig. 7*a*, during the first flight, the initial room temperature modulus, E_i, dropped from 172 GPa in Mission I to 158 GPa in Mission II and is mainly due to fiber/matrix interface failure of the 90° ply [*1*]. Once the interface has failed, only minor variations in the modulus were noted between Mission I and Mission II in each flight. As seen in Fig. 7*b*, the composite stress-strain response stabilized in the second flight. The stress-strain response during the 75th and 120th flights are shown in Figs. 7*c* and *d*. As seen in the figures, the measured modulus at room temperature and at 427°C degraded with applied flights.

The total accumulative strain and normalized modulus for the composite subjected to the entire hypersonic flight profile with a temperature range of −130 to 816°C are given in Figs. 8*a* and *b*, respectively. As seen in Fig. 8*a*, the total minimum and maximum strain gradually increased with applied flights. Comparing with the modulus history, shown in Fig. 8*b*, after initial drop in the modulus, little change to the modulus is observed until the 75th flight. The initial room temperature modulus dropped by 22.5%, and the unloading modulus at 427°C

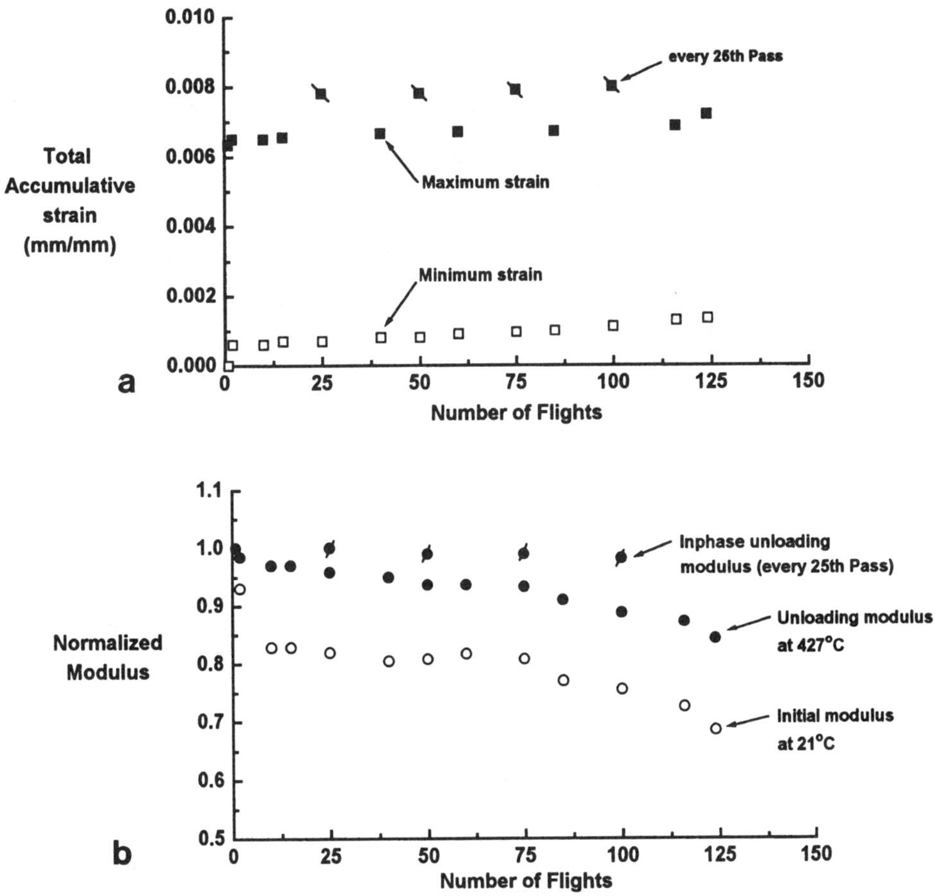

FIG. 8—*Strain history* (a) *and modulus history* (b) *of* $[0/90]_{2s}$ *SCS-6/TIMETAL-21S,* v_f = *0.365, subjected to the entire flight profile and temperature range of −130 to 816°C.*

dropped by 9%. At that point, the moduli decreased at a faster rate until failure. Prior to failure, the unloading modulus at 427°C dropped by 14%. During unloading of the in-phase portion of the 25th flight, the temperature changed from 816 to 732°C. In this temperature range, the matrix modulus is very low. Subsequently, any loss in composite stiffness would be attributable to the fiber breakage. By observing that the unloading modulus of the in-phase portion at the 75th and 100th flight did not change, degradation of the initial and unloading modulus at 427°C is due mainly to the matrix and the fiber/matrix interface cracking. The initial increase in the minimum strain, up the point of stabilization, causes an increase in the stress level in the 0° fibers. This, combined with the matrix cracking, causes additional load to be transferred to the fiber and ultimately results in failure of the laminate.

Figure 9 shows the stress-strain response of the composite subjected to only the portion of Mission I (Flights 1, 2, 200, 259) containing the load spike and in-phase loading, with the applied temperature represented by Fig. 5*f*. As seen in Fig. 9, a larger amount of permanent strain is observed in the first flight than the subsequent flights due to the high applied load and temperature. Note that, in this profile, mechanical loads were applied at elevated temperatures and no initial room-temperature modulus (E_i) was measured. Very little change is noticed in the unloading modulus of the composite at the spike load between the first and second flight. However, as seen in Figs. 9*c* and *d*, the unloading modulus at 427°C degraded in subsequent applied flights. The in-phase unloading modulus (E_{in}) shown in Figs. 9*a* through *d* did not change considerably with the applied flights in comparison. Figures 10*a* and *b* show the variation of the total accumulated strain with flights and the normalized modulus history. The total accumulated strain as shown in Fig. 10*a* gradually increased until 200 flights, and a larger maximum strain was observed to accumulate in this case compared to the previous case. The unloading modulus at 427°C as shown in Fig. 10*b*, after initial drop, gradually degraded up to 200 flights, about 9%, then degraded more rapidly for the remaining life. The unloading modulus at 427°C, prior to failure, dropped by 14%. The in-phase unloading modulus changed little up to 200 flights then decreased at a faster rate for the remaining life. As was mentioned earlier, during unloading of the in-phase portion of the 25th flight, the temperature changed from 816 to 732°C. In this temperature range, the matrix modulus is very low. Subsequently, any loss in composite stiffness would be attributable to the fiber breakage, because the modulus is fiber-dominated.

Comparisons of total accumulative strains and modulus histories, shown in Figs. 8 and 10, indicate a consistent level of total accumulative minimum strain. In addition, comparisons of modulus histories indicated that, prior to rapid modulus degradation, the composite axial modulus had consistent value, independent of applied loading histories. Furthermore, prior to the rapid modulus degradation, the fiber-dominated moduli remain essentially constant, suggesting that the initial modulus degradation is due mainly to the initiation and growth of interfacial and matrix damage (cracks) to a stable level. The modulus degradation as result of fatigue cycling is dependent on the composite's lay-up [*1*]. Therefore, the modulus histories presented in this work are limited to the $[0/90]_{2s}$ SCS-6/TIMETAL-21S.

Theoretical Prediction of Stress-Strain Response

The VISCOPLY program was used to predict the stress-strain response of the cross-ply SCS-6/TIMETAL-21S subjected to the TMF flight profiles. The fiber was modeled as elastic with transverse orthotropic and temperature dependency, and the matrix was modeled using a thermoviscoplastic constitutive relationship that was described previously in the analytical section. A detailed description of the constituent material properties input for use in VISCOPLY are given in Ref *11*. In Ref *12*, a variety of TMF loading profiles of this material system was evaluated and verified. In all of the stress-strain response predictions,

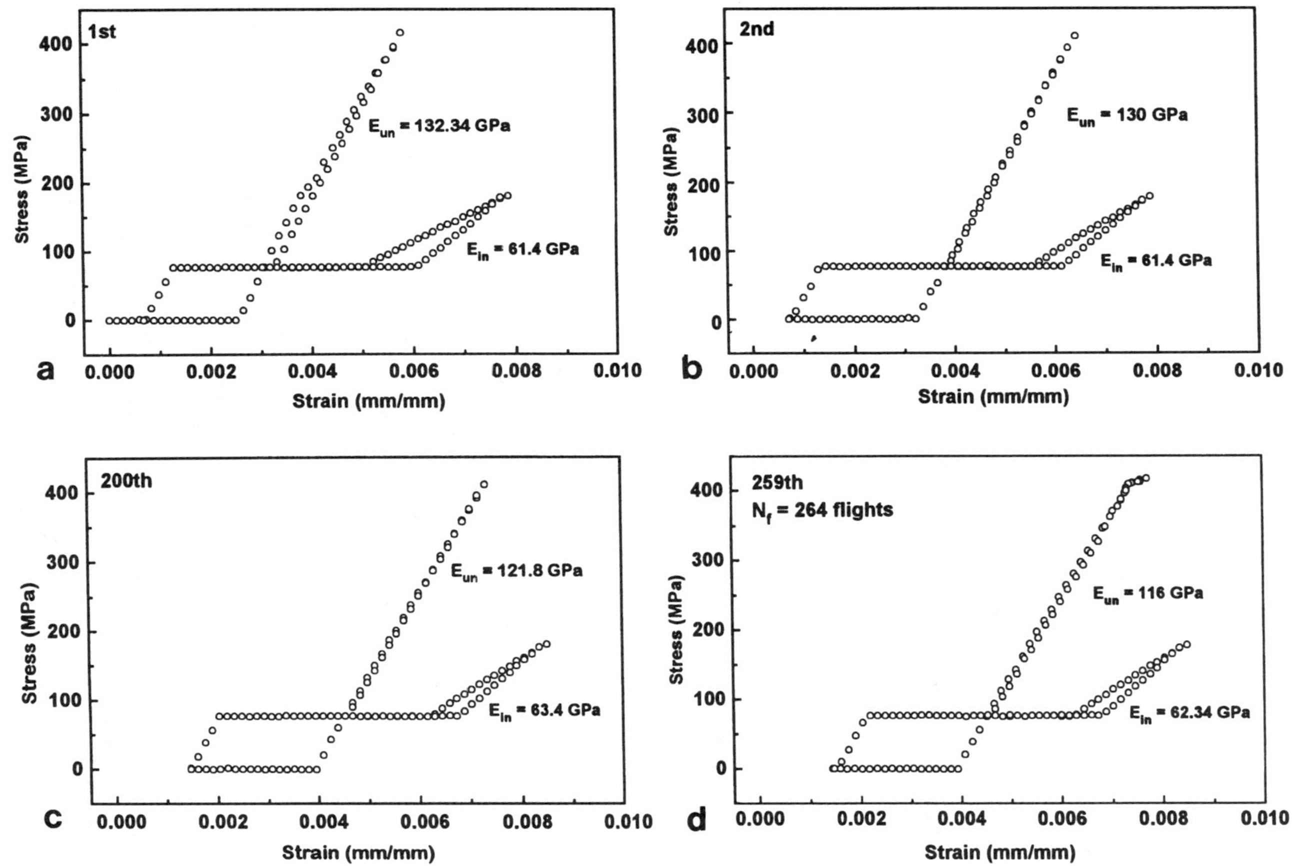

FIG. 9—*Stress-strain response of* $[0/90]_{2s}$ *SCS-6/TIMETAL-21S,* $v_f = 0.365$, *subjected to a portion of Mission I and temperature range of 21 to 816°C.*

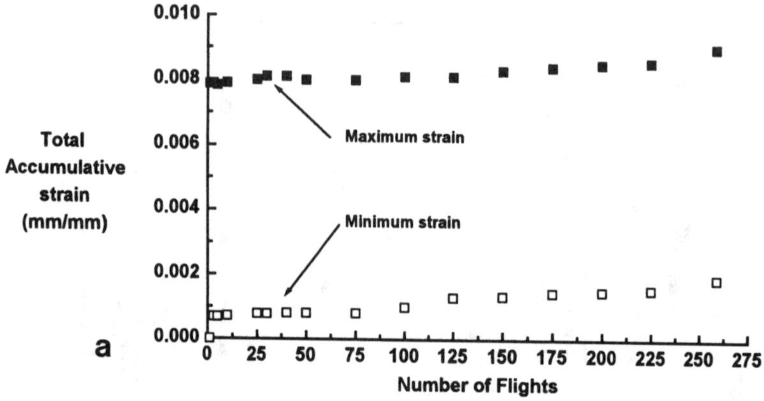

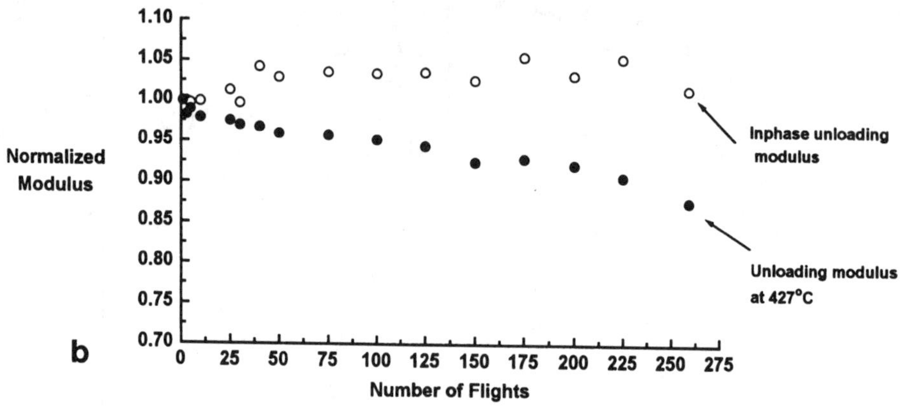

FIG. 10—*Strain history* (a) *and modulus history* (b) *of* $[0/90]_{2s}$ *SCS-6/TIMETAL-21S,* v_f = *0.365, subjected to a portion of Mission I and temperature range of 21 to 816°C.*

the fiber/matrix interface failure was modeled by reducing the transverse modulus of the fiber in the 90° plies. This was shown to be effective in accurately predicting the stress-strain response under a variety of thermomechanical load profiles [*18,19*] as shown in Figs. 11, 12, and 13 for the entire hypersonic flight profile (Missions I and II) and in Fig. 14 for the portion of Mission I. For clarity, different symbols have been used for Mission I and Mission II. The letters placed at various locations on the stress-strain response can be referenced back to Fig. 4 to find the associated segment on the flight profile. As seen in Fig. 11, VISCOPLY reasonably predicted the observed experimental stress-strain response of the first flight. However, a larger permanent strain was observed experimentally (as noted in the figure) than was predicted using VISCOPLY, indicating that there may be local plasticity occurring around the fiber that is not accounted for in the VISCOPLY analysis. The experimental stress-strain response stabilized rather quickly after two flights. However, VISCOPLY predictions did not stabilize until the fifth flight. The experimental and the VISCOPLY predicted stress-strain response of the fifth flight are shown in Fig. 12. In the VISCOPLY analysis, the permanent strain observed experimentally during the fifth flight was matched with the strain at the start of the analysis. As shown in the figure, the VISCOPLY reasonably predicted the stabilized flight stress-strain response. The experimen-

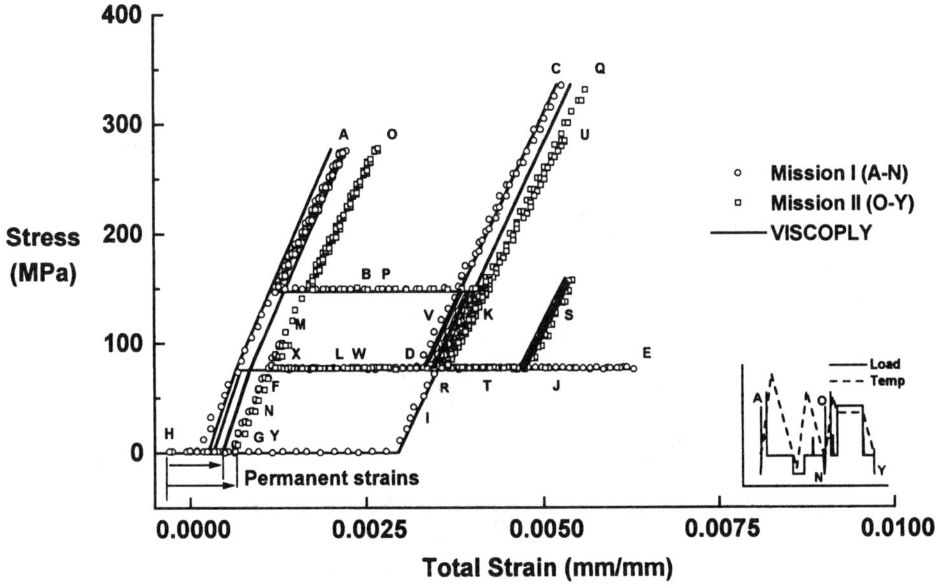

FIG. 11—*Experimental and VISCOPLY prediction of the stress-strain response of $[0/90]_{2s}$ SCS-6/TIMETAL-21S, v_f = 0.365, during the first flight.*

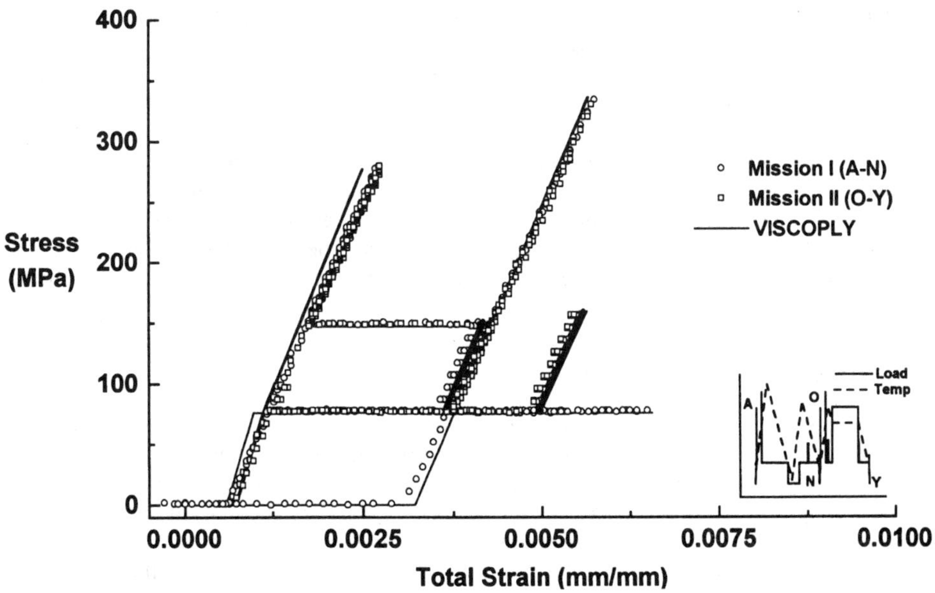

FIG. 12—*Experimental and VISCOPLY prediction of the stress-strain response of $[0/90]_{2s}$ SCS-6/TIMETAL-21S, v_f = 0.365, during the fifth flight.*

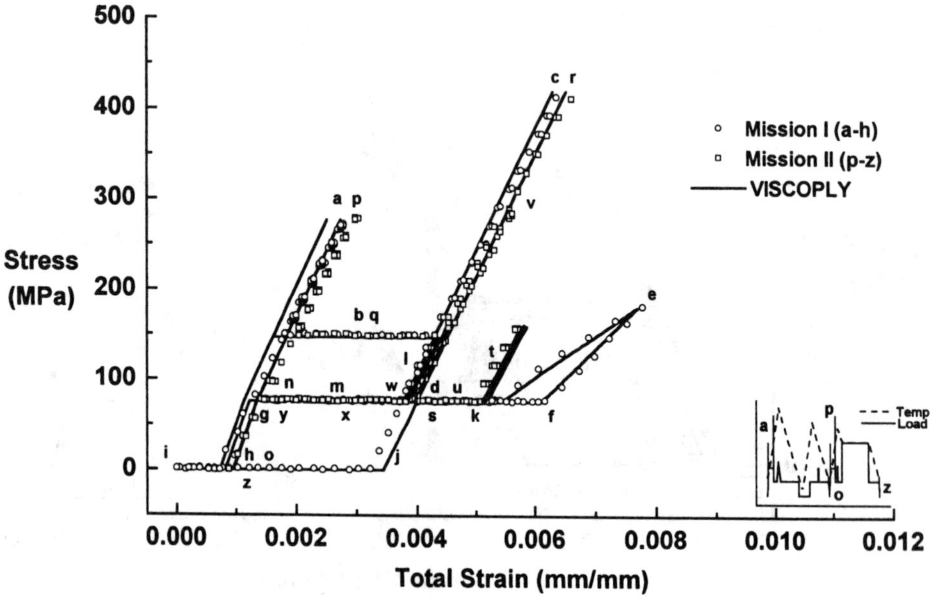

FIG. 13—*Experimental and VISCOPLY prediction of stress-strain response of* $[0/90]_{2s}$ *SCS-6/TIMETAL-21S,* $v_f = 0.365$*, during the 25th flight.*

tal stress-strain response and the VISCOPLY prediction for the 25th flight (using the profile shown in Fig. 4*b*) is shown in Fig. 13. As before, the observed permanent strain as a result of flight cycling was incorporated into the VISCOPLY prediction. As seen in the figure, VISCOPLY accurately predicted the observed experimental results.

The experimental stress-strain response and the VISCOPLY predictions of the first five flights of a portion of Mission I (using the profile shown in Fig. 5*f*) are shown in Fig. 14. As seen in the figure, the experimental stress-strain response stabilized quickly in the second cycle while the VISCOPLY prediction did not stabilize until the fifth flight. As seen in the figure, the VISCOPLY predictions were in excellent agreement with the experimental stress-strain response (the prediction took four flights to stabilize while the actual test took two flights).

Life Prediction Methodology

The VISCOPLY program was used to predict the stress in the 0° fiber for all the TMF profiles tested previously for the stabilized condition and the condition incorporating matrix damage. For the tests subjected to the entire mission profile shown in Figs. 5*a* and *b* and Mission I profile shown in Figs. 5*c* and *d*, VISCOPLY predictions of the stress-strain response initially stabilized during the first five flights. Stabilization after the 25th flight (Fig. 5*b*) was reexamined and only moderate changes were observed in the VISCOPLY prediction of the 0° fiber stress. The predicted 0° fiber stress was then used for analysis of the fatigue data. Using a portion of the Mission I profile shown in Fig. 5*f*, the VISCOPLY prediction of the stress-strain response stabilized after the fifth flight as shown in Fig. 14 and the predicted 0° fiber stress after stabilization was used for the analysis of fatigue data.

The experimental stress-strain behavior indicated that the composite modulus gradually degraded with flights until failure. As was discussed earlier in the stress-strain behavior

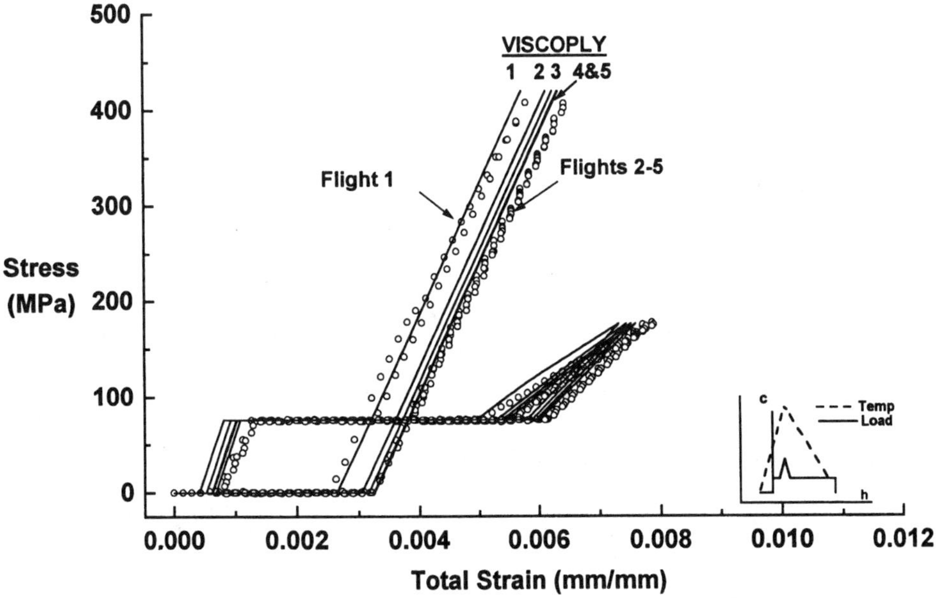

FIG. 14—*Experimental and VISCOPLY prediction of the stress-strain response of* $[0/90]_{2s}$ *SCS-6/TIMETAL-21S,* $v_f = 0.365$*, during the first five flights.*

section, comparisons of modulus histories suggest that the point where the modulus degrades at a faster rate appears to be the onset of 0° fiber failure. The unloading modulus at 427°C at this point (onset of 0° fiber failure) was found to be 120 GPa in all flight profiles tested. This implies that the damage in the composite, in the form of matrix cracks and fiber/matrix interfacial cracks, has become saturated causing more load to be transferred to the 0°

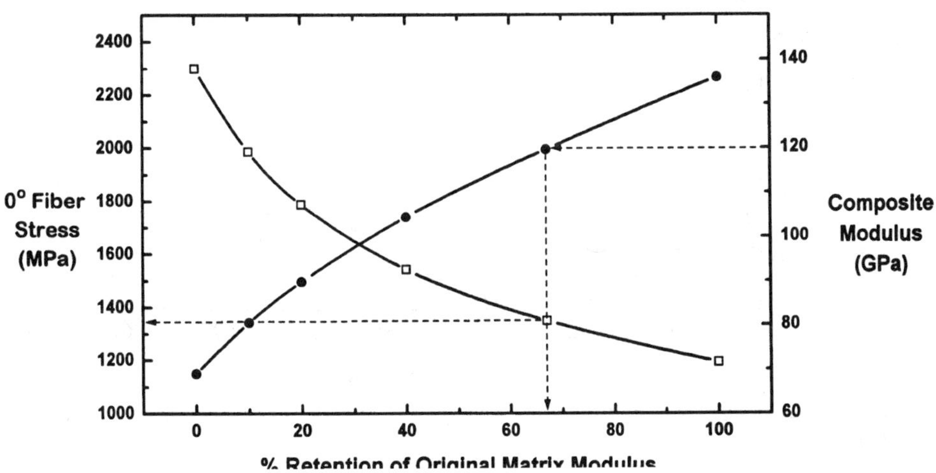

FIG. 15—*VISCOPLY prediction of modulus and 0° fiber stress of* $[0/90]_{2s}$ *SCS-6/TIMETAL-12S,* $v_f = 0.365$*, loaded to 420 MPa at 427°C as a function of percent retention of matrix modulus.*

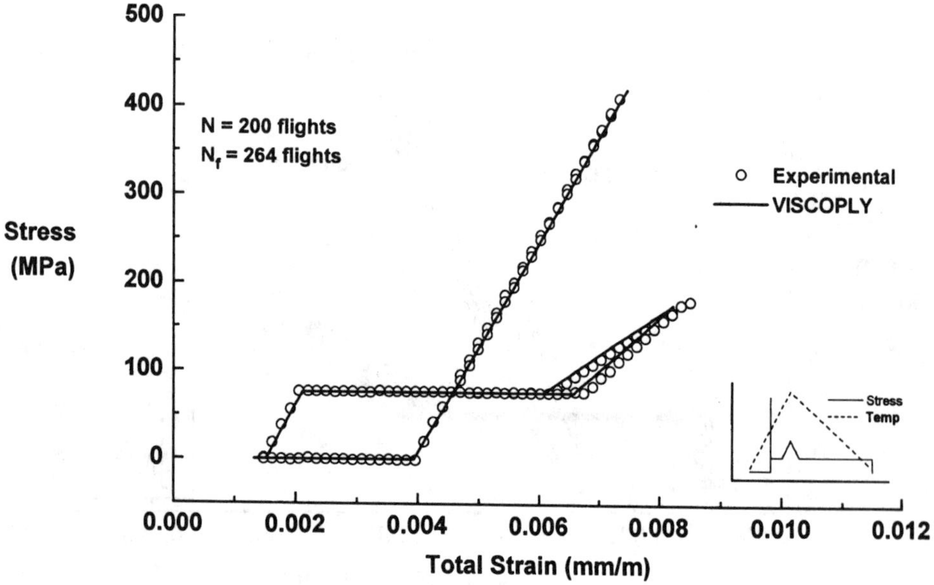

FIG. 16—*VISCOPLY prediction of stress-strain response of* $[0/90]_{2s}$ *SCS-6/TIMETAL-21S,* $v_f = 0.365$*, including matrix damage.*

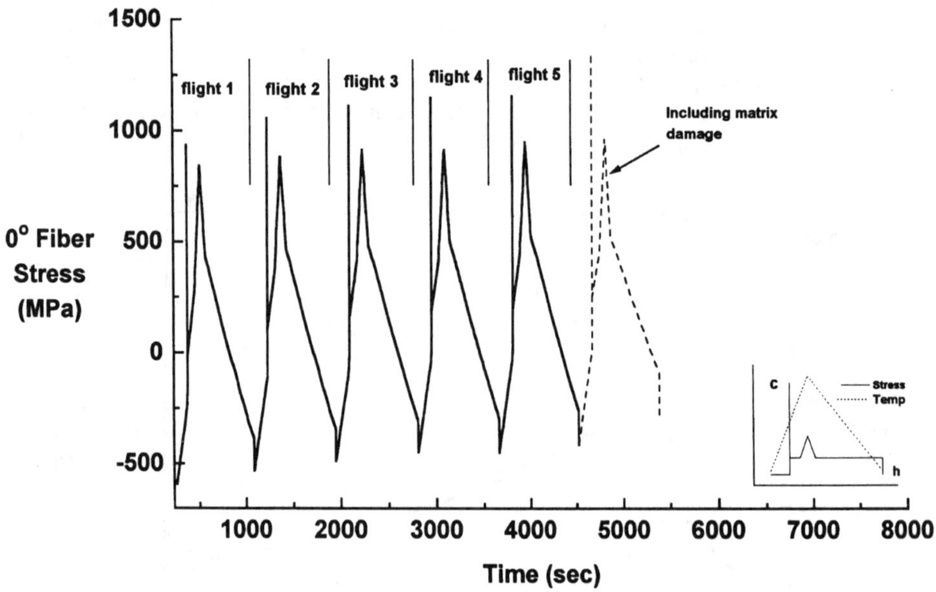

FIG. 17—*VISCOPLY prediction of 0° fiber stress as a function of time for* $[0/90]_{2s}$ *SCS-6/TIMETAL-21S,* $v_f = 0.365$.

fibers leading to eventual failure at points of highly localized stress. As was mentioned earlier, the fiber/matrix interface failure was modeled by reducing the fiber transverse modulus of the 90° plies. To simulate the matrix cracking, the matrix modulus was reduced in the VISCOPLY prediction. Figure 15 shows the VISCOPLY prediction of the composite modulus and stress in the 0° fiber as a function of percent retention of the matrix modulus at 427°C when the composite is loaded to 420 MPa. As seen in the figure, reducing the matrix modulus lowers the composite modulus, resulting in higher stress in the 0° fibers. For all the profiles tested, the unloading modulus at 427°C was found to be about 120 GPa at the point where the modulus begins to decrease rapidly. The composite modulus of 120 GPa was used in Fig. 15 to determine the proper reduction in the matrix modulus and was used in the VISCOPLY program to predict the stress-strain response. Figure 16 shows the stress-strain response of the composite after 200 flights where the unloading modulus was measured to be 120 GPa. The VISCOPLY prediction simulating matrix cracking is also shown in the figure. In the prediction, the matrix modulus was reduced by the same magnitude at all temperatures as was determined for 427°C and the resulting prediction agreed well with the observed experimental results as seen in Fig. 16. Figure 17 shows the predicted 0° fiber stress as a function of time for the first five flights, stabilization, and subsequent prediction simulating matrix cracking. As seen in the figure, the maximum fiber stress increases with each subsequent flight. The increase in fiber stress is due to stress relaxation in the matrix followed by simulated matrix cracking and subsequent load transferal to the fibers as predicted by the VISCOPLY program.

The fatigue data were analyzed using the stress in the 0° fiber as described earlier. Three criterion were considered for analysis of fatigue data, maximum stress (σ_{max}), stress range ($\Delta\sigma$), and Smith-Watson-Topper (SWT) parameter [*19*]. While maximum stress and stress range traditionally are used to describe fatigue response, the SWT effective strain (stress) takes into account the stress ratio effect. The SWT effective strain ($\Delta\epsilon_{eff}$) is given by

$$\Delta\epsilon_{eff} = \left(\frac{\epsilon_{max}\Delta\epsilon}{2}\right)^{1/2} \tag{1}$$

In Eq 1, ϵ_{max} and $\Delta\epsilon$ are the maximum strain and strain range, respectively. Assuming elastic fiber response, the SWT parameter can be written in terms of stress as

$$\Delta\sigma_{eff} = \left(\frac{\sigma_{max}\Delta\sigma}{2}\right)^{1/2} \tag{2}$$

The fatigue data based on the predicted 0° fiber stress as a function of time to failure are shown in Figs. 18 through 20. The time to failure was calculated by multiplying the duration of mission time by the number of missions to failure. The stabilized predictions and predictions including matrix damage are included in Fig. 18. For clarity, Figs. 19 and 20 include predictions simulating matrix damage. As seen in the figures, the fatigue data are well-correlated based on the predicted stress in the 0° fiber. The analysis suggests that the longer exposure at elevated temperature lowered the fatigue strength of the 0° fiber. With the limited available experimental data in this analysis, the SWT parameter appears to collapse the TMF data into a narrower band.

Summary

A cross-ply $[0/90]_{2s}$ SCS-6/TIMETAL-21S composite was subjected to a generic hypersonic flight profile and its portions with a temperature ranging from −130 to 816°C, and

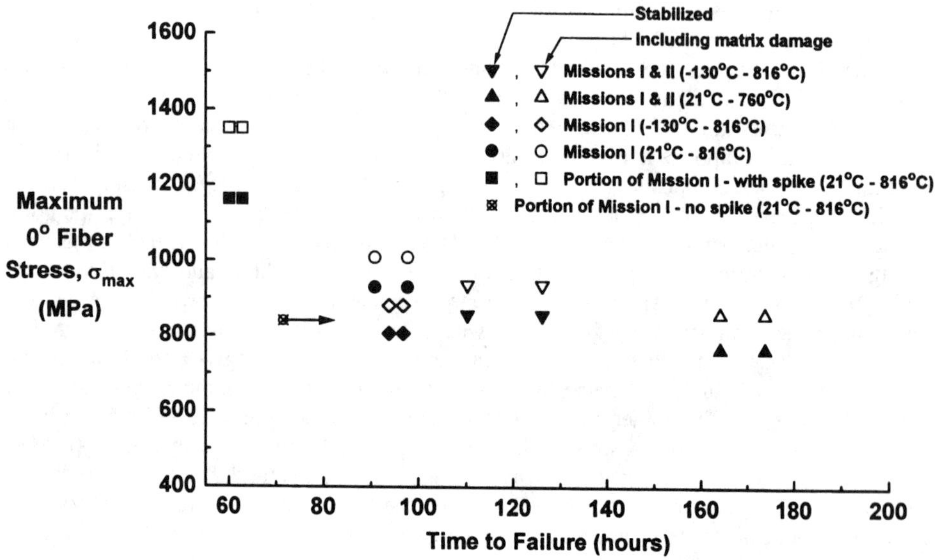

FIG. 18—*Analysis of fatigue data of [0/90]$_{2s}$ SCS-6/TIMETAL-21S,* v_f *= 0.365, based on the maximum stress in the 0° fiber.*

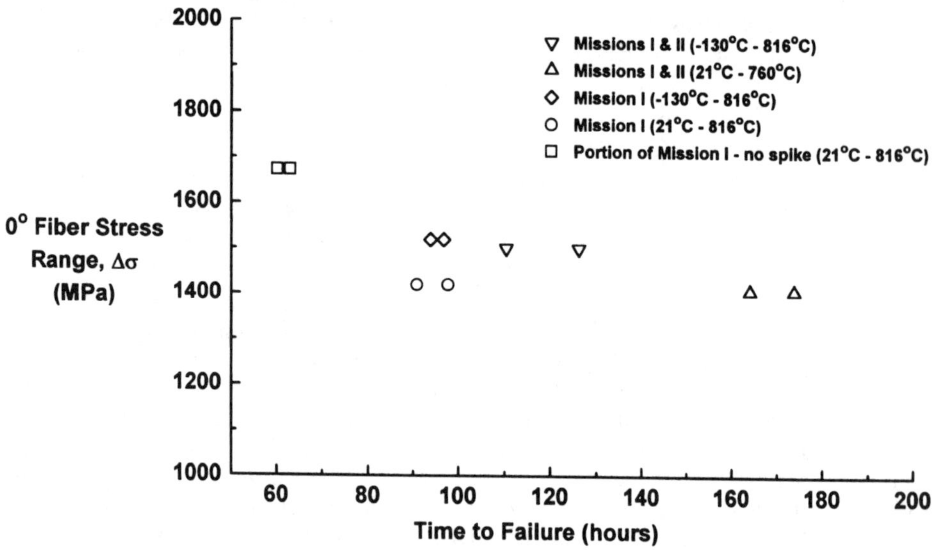

FIG. 19—*Analysis of fatigue data of [0/90]$_{2s}$ SCS-6/TIMETAL-21S,* v_f *= 0.365, based on the stress range of the 0° fiber, including matrix damage.*

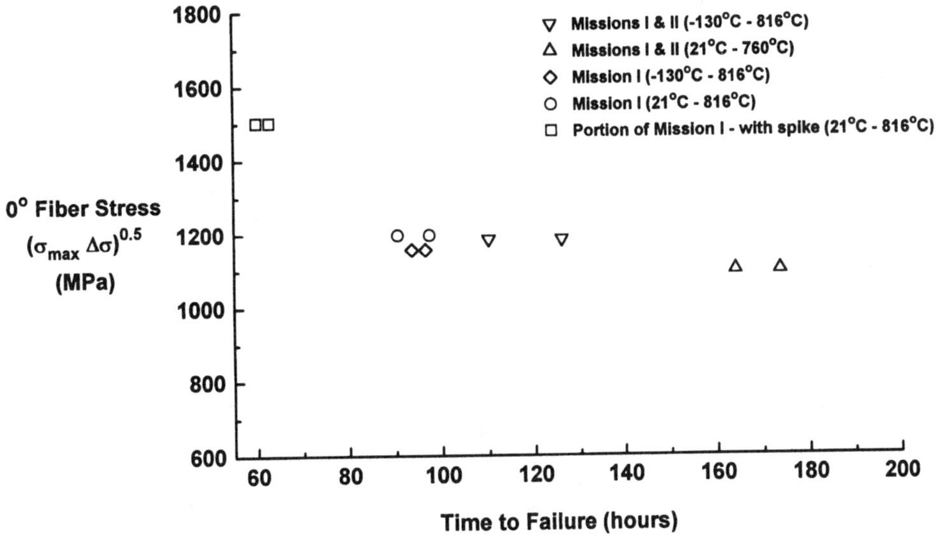

FIG. 20—*Analysis of fatigue data of* $[0/90]_{2s}$ *SCS-6/TIMETAL-21S,* $v_f = 0.365$, *based on the Smith-Watson-Topper effective stress, including matrix damage.*

the TMF fatigue lives were determined. The composite modulus degraded gradually with applied flights and, prior to failure, the degradation accelerated. It was also found that the composite modulus prior to the rapid degradation was consistent in the flight profiles tested. The accumulative minimum strain was shown to reach a consistent level for all of the flight profiles tested. The stress-strain response of the composite of the first flight and the stabilized flight was well predicted using the VISCOPLY program. The loss of modulus was assumed to be a result of matrix cracking and was modeled in the VISCOPLY program by reducing the matrix modulus. The VISCOPLY program was used with the reduced matrix moduli to predict the composite's consistent stresses in the damage state. Fatigue lives of the composite subjected to the generic hypersonic flight profile were well-correlated using the predicted stress in the 0° fibers. This approach can be used to predict fatigue lives for other combinations of TMF loading conditions.

Acknowledgment

Karen Taminger of NASA Langley Research Center, Metallic Materials Branch, provided invaluable assistance in preparation of this report.

References

[1] Johnson, W. S., Lubowinski, S. L., and Highsmith, A. L., "Mechanical Characterization of Unnotched SCS_6/Ti-15-3 Metal Matrix Composites at Room Temperature," *Thermal and Mechanical Behavior of Metal Matrix and Ceramic Matrix Composites, ASTM STP 1080,* J. M. Kennedy, H. H. Moeller, and W. S. Johnson, Eds., American Society for Testing and Materials, Philadelphia, 1990, pp. 193–218.

[2] Mirdamadi, M., Johnson, W. S., Bahei-El-Din, Y. A., and Castelli, M. G., "Analysis of Thermomechanical Fatigue of Unidirectional Titanium Metal Matrix Composite," *Composite Materials: Fatigue and Fracture, Fourth Volume, ASTM STP 1156,* W. W. Stinchcomb and N. E. Ashbaugh, Eds., American Society for Testing and Materials, Philadelphia, 1993, pp. 591–607.

[3] Castelli, M. G., Bartolotta P. A., and Ellis, J. R., "Thermomechanical Fatigue Testing of High

Temperature Composites: Thermomechanical Fatigue Behavior of SiC (SCS-6)/Ti-15-3,'' *Composite Materials: Testing and Design (10th Volume), ASTM STP 1120,* G. C. Grims, Ed., American Society for Testing and Materials, Philadelphia, 1991, pp. 70–86.

[*4*] Gabb, T. P., Gayda, J., and MacKay, R. A. ''Isothermal and Nonisothermal Fatigue Behavior of a Metal Matrix Composite,'' *Journal of Composite Materials,* Vol. 24, 1990, pp. 667–686.

[*5*] Mirdamadi, M. and Johnson, W. S., ''Fatigue of $[0/90]_{2s}$ SCS-6/Ti-15-3 Composite Under Generic Hypersonic Vehicle Flight Simulation,'' *FATIGUE 93,* Vol. II, Bailon and Dickson, Eds., Chameleon Press, London, 1993, pp. 951–956.

[*6*] Mirdamadi, M. Bakuckas, J. G., Jr., and Johnson, W. S., ''Mechanisms of Strain Accumulation in Titanium Matrix Composites at Elevated Temperatures,'' *Mechanics of Composite Materials—Nonlinear Effects, Applied Mechanics Division,* Vol. 159, American Society of Mechanical Engineers, 1993, pp. 245–252.

[*7*] Jeng, S. M., Yang, C. J., Alassoeur, P., and Yang, J. M., ''Deformation and Fracture Mechanisms of Fiber-Reinforced Titanium Alloy Matrix Composites,'' *Proceedings,* Eighth International Conference on Composite Materials (ICCM/8), S. W. Tasi and G. S. Springer, Eds., Society for the Advancement of Material and Process Engineering (SAMPE), Covina, CA, 1991, pp. 25–C–1, 25–C–12.

[*8*] Naik, R. A., Johnson, W. S., and Pollock, W. D., ''Effect of High Temperature Cycle on the Mechanical Properties of Silicon Carbide/Titanium Metal Matrix Composites,'' *Proceedings,* ASC on High Temperature Composites, Technomics, Lancaster, PA, 1989, pp. 94–103.

[*9*] Mirdamadi, M. and Johnson, W. S., ''Experimental Techniques for Hypersonic Flight Simulation Testing,'' *Proceedings,* Society of Experimental Mechanics Conference, 6–8 June 1994, pp. 679–684.

[*10*] Ellis, J. R. and Bartolotta, P. A., ''Adjustable Work Coil Fixture for Facilitating the Use of Induction Heating in Mechanical Testing,'' NASA TM-102461, NASA Lewis Research Center, Cleveland, OH, June 1989.

[*11*] Hartman, G. A. and Ashbaugh, N. E., ''A Fracture Mechanics Test Automation System for a Basic Research Laboratory,'' *Applications of Automation Technology to Fatigue and Fracture Testing, ASTM STP 1092,* A. A. Braun, N. E. Ashbaugh, and F. M. Smith, Eds., American Society for Testing and Materials, Philadelphia, 1990, pp. 95–110.

[*12*] Dvorak, G. J. and Bahei-El-Din, Y. A., ''Plasticity Analysis of Fibrous Composite,'' *Journal of Applied Mechanics,* Vol. 49, 1982, pp. 237–335.

[*13*] Bahei-El-Din, Y. A., ''Plasticity Analysis of Fibrous Composite Laminates Under Thermomechanical Loads,'' *Thermal and Mechanical Behavior of Metal Matrix and Ceramic Matrix Composites, ASTM STP 1080,* J. M. Kennedy, H. H. Moeller, and W. S. Johnson, Eds., American Society for Testing and Materials, Philadelphia, 1990, pp. 20–39.

[*14*] Bahei-El-Din, Y. A., Shah, R. S., and Dvorak, G. J., ''Numerical Analysis of the Rate-Dependent Behavior of High Temperature Fibrous Composites,'' *Mechanics of Composites at Elevated Temperatures,* Vol. 118, S. N. Singhal, W. F. Jones, C. T. Herakovich, and T. Cruse, Eds., Applied Mechanics Division, 1991, pp. 67–78.

[*15*] Eisenberg, M. A. and Yen, C. F., ''A Theory of Multiaxial Anisotropic Viscoplasticity,'' *ASME Journal of Applied Mechanics,* Vol. 48, 1981, pp. 276–284.

[*16*] Malvern, L. E., ''The Propagation of Longitudinal Waves of Plastic Deformation in a Bar of Material Exhibiting a Strain-Rate Effect,'' *ASME Journal of Applied Mechanics,* Vol. 18, 1951, pp. 203–208.

[*17*] Johnson, W. S., Mirdamadi, M., and Bahei-El-Din, Y. A., ''Stress-Strain Analysis of a $[0/90]_{2s}$ Titanium Matrix Laminate Subjected to a Generic Hypersonic Flight Profile,'' *NASA Journal of Composites Technology & Research,* Vol. 15, No. 4, Winter 1993, pp. 297–303.

[*18*] Mirdamadi, M. and Johnson, W. S., ''Prediction of Stress-Strain Response of SCS-6/TIMETAL-21S Subjected to the Hypersonic Flight Profile,'' NASA TM 109026, NASA Langley Research Center, Hampton, VA, Feb. 1994.

[*19*] Smith, K. N., Watson, R., and Topper, T. H., ''A Stress-Strain Function for the Fatigue of Metals,'' *Journal of Materials,* Vol. 5, No. 4, 1970, pp. 767–778.

Theodore Nicholas,[1] Stephan M. Russ,[1] Richard W. Neu,[2] and Norman Schehl[3]

Life Prediction of a [0/90] Metal Matrix Composite Under Isothermal and Thermomechanical Fatigue

REFERENCE: Nicholas, T., Russ, S. M., Neu, R. W., and Schehl, N., **"Life Prediction of a [0/90] Metal Matrix Composite Under Isothermal and Thermomechanical Fatigue,"** *Life Prediction Methodology for Titanium Matrix Composites, ASTM STP 1253,* W. S. Johnson, J. M. Larsen, and B. N. Cox, Eds., American Society for Testing and Materials, 1996, pp. 595–617.

ABSTRACT: Experimental data on cycles to failure are reported for a SCS-6/TIMETAL 21S $[0/90]_s$ composite under isothermal fatigue at 650 and 815°C over a range of frequencies from 0.01 to 200 Hz as well as under in-phase and out-of-phase thermomechanical fatigue from 150 to 650°C. Several models are evaluated for correlating fatigue lives based on computed micromechanical stresses using a life fraction concept that considers both cycle-dependent and time-dependent terms. Fractography is used to illustrate the range of failure modes in this material system as well as to both guide and validate the modeling concepts. Over the range of conditions covered in this investigation, purely time-dependent, purely cycle-dependent, fiber-dominated, and matrix-dominated modes of failure, as well as combinations of these, are observed.

KEYWORDS: titanium, metal matrix composites, fatigue (materials), thermomechanical fatigue, life prediction, elevated temperature, titanium matrix composites, titanium alloys, modeling

Titanium matrix composites continue to attract attention for potential use in advanced aerospace applications such as hypersonic aircraft and advanced turbine engine components where high strength/weight ratios combined with high temperature capability are required. In particular, a great deal of effort has been expended in evaluating the properties of SCS-6/TIMETAL 21S,[4] a beta titanium alloy reinforced with silicon-carbide fibers. Among the challenges to the effective use of such a material is the development of a methodology for predicting the fatigue life under typical mission environments. These include variations in load, stress ratio, temperature, and frequency of loading. It is now well known that cycling of temperature, without any superimposed external loads, can cause significant thermally induced stresses in the fiber and the matrix material because of the large difference in coefficient of thermal expansion between the fiber and the matrix. These stresses, combined

[1]Senior scientist and research engineer, respectively, Wright Laboratory Materials Directorate, Wright-Patterson AFB, OH 45433.

[2]Formerly, NRC associate, Wright Laboratory Materials Directorate, Wright-Patterson AFB, OH 45433; presently, assistant professor, Woodruff School of Mechanical Engineering, Georgia Institute of Technology, Atlanta, GA 30332-0405.

[3]Assistant research engineer, University of Dayton Research Institute, Dayton, OH 45409.

[4]Registered trademark of Timet Corporation, Henderson, NV.

with stresses due to mechanical loading, produce complicated stress histories in the composite constituents and make fatigue life modeling quite difficult.

One concept that has been adopted for the prediction of fatigue life and correlation of fatigue data obtained over a wide variety of conditions has been the linear life-fraction, damage-summation, modeling approach. While thermal fatigue has been shown to lead to degradation of strength in this class of materials [*1–4*], modeling of residual strength or damage accumulation has only recently been attempted [*5*]. Time at temperature, leading to environmentally induced damage, is a major consideration in such analyses. Effects of stress ratio, hold time, and frequency under isothermal conditions have been modeled successfully in SCS-6/Ti-24Al-11Nb, a silicon-carbide fiber-reinforced alpha-2 intermetallic, by separating the cycle-dependent and time-dependent contributions to damage [*6,7*]. In these studies, it has been observed that there is a strong interaction between time-dependent and cycle-dependent mechanisms, and interaction terms have had to be introduced in the modeling to account for these effects. Finally, isothermal as well as thermomechanical fatigue (TMF) lives have been modeled by considering different failure mechanisms that are fiber stress or matrix stress dominated. Considering tests conducted at a single frequency, isothermal fatigue data, both in-phase and out-of-phase TMF data, as well as data under thermal cycling with superimposed sustained loads have been consolidated in a series of studies, first on an alpha-two composite, and then on a beta titanium composite [*8–10*]. In these latter studies, the micromechanical fiber and matrix stresses had to be determined that required the introduction of various stress analysis techniques.

The present study evaluates the behavior of a titanium matrix composite (TMC) laminate consisting of a beta titanium alloy reinforced with silicon-carbide fibers, namely, SCS-6/TIMETAL 21S $[0/90]_s$ for a range of isothermal and thermomechanical loading conditions. Isothermal tests at 650 and 815°C cover a range of frequencies and stress ratios, R, from 0.01 to 200 Hz and 0.1 to 0.8, respectively, while TMF tests involve both in-phase and out-of-phase loading including a few tests at different TMF frequencies. In addition to determining fatigue life as a function of the loading conditions, fractography is conducted to identify the failure mechanisms. This information is then applied in the construction of a life fraction model that accounts for cycle and time-dependent mechanisms while also addressing the roles of micromechanical fiber and matrix stresses in the fatigue process.

Experimental Program

Material

The material used in this study was a titanium matrix composite (TMC) comprised of silicon-carbide fibers (SCS-6) and a beta-titanium matrix (TIMETAL 21S) with a composition of Ti-15Mo-3Nb-3Al-0.2Si% by weight. The composite panels were assembled by alternating layers of rolled titanium foils with fiber mats of SCS-6 woven with a titanium-niobium (Ti-Nb) ribbon, and then consolidated by hot isostatic pressing. The fibers were 142 μm in diameter and were spaced at 126 $\pm$ 3 fibers per inch of fiber mat, the foil thickness was 114 μm, and the thickness of the resulting four-ply composite panels nominally averaged 0.9 mm. Specimens were taken from more than one panel. The fiber volume fraction (V_f) was generally dependent on the panel and varied from 0.32 to 0.41. For all but the high frequency (200 Hz) tests, the panels were cut into straight-sided specimens using abrasive water-jet machining. Specimen widths were nominally 12.7 mm and most specimens were tested as cut. The specimens were heat treated in vacuum at 621°C for 8 h prior to testing to help stabilize the titanium matrix. From previous studies, edge polishing was found not to significantly affect the fatigue life of TMCs [*11*]. For the high frequency tests, a dog bone specimen design was used that had a 50% reduction in cross-sectional area, a length of 72 mm between grips, a 12.7-mm gage section width, and a transition radius of 57 mm.

Equipment

The fatigue machine for all but the very high frequency (200 Hz) tests consisted of a servohydraulic load frame positioned horizontally and quartz lamp heating units [*12,13*]. The heating unit contained two banks of four quartz lamps running perpendicular to the loading axis, one above and one below the specimen. Four thermocouples were attached to the specimen with each actively controlling two lamps, thereby providing four zones of temperature control. A length of about 30 mm in the center of the specimen was heated uniformly, while the ends of the specimen, clamped in the water-cooled grips, remained near room temperature.

The isothermal and TMF tests, excluding 200 Hz, were conducted in load control using MATE software [*14*]. For TMF tests, the software automatically adjusted the temperature setpoints both to keep the temperature uniform throughout the gage section and to account for the lag between the requested and actual temperature in order to ensure correct phasing between stress and temperature. Axial strain was monitored throughout the tests using a high temperature extensometer with quartz extension rods placed on the edge of the specimen. Symmetric triangular wave forms were used for both load and temperature for TMF tests. A triangular wave form was used for load in the 0.1 and 0.01 Hz isothermal tests, while a sinusoidal wave form was used for the 1 Hz testing for more accurate control of peak loads.

For the very high frequency (200 Hz) tests, a unique high frequency test system was utilized. The apparatus, which has been used in an earlier study [*7*] is a fully automated load control machine consisting of an electrodynamic shaker superimposing a sinusoidal high frequency load on a static mean load applied by a pneumatic loading device. A load cell is used to measure mean loads, and a quartz force gage measures the high frequency peak-to-peak load. The dual loading system is capable of producing usable loads in excess of 300 Hz. Specimens are friction gripped in air-cooled titanium grips, resulting in a rigid fixed gripping method rather than a pinned connection, and proper alignment of the specimen and load train is ensured by a translation and rotation-adjustable alignment device clamped with a pneumatic cylinder. The alignment procedure was conducted after each test to verify alignment of the load train. Elevated temperature tests were conducted using three quartz lamps with a two-zone-control temperature monitoring system. Thermocouples placed in close proximity to the specimen provided the necessary feedback to accurately maintain test temperatures.

In all of the tests, failure was defined as complete fracture of the specimen into two pieces. Failure occurred within the heated zone except where noted. A summary of the test conditions and experimental values for cycles to failure is presented in Tables 1 through 3 for the isothermal tests at 650 and 815°C and the TMF tests at 150 to 650°C, respectively. The test data, plotted as cycles to failure against maximum applied stress, are presented in Figs. 1 through 3 for the three data sets. Tests that were terminated prior to failure because of lengthy test times are appropriately indicated in bold in the tables and by arrows in the figures.

Modeling

Model Development

Modeling of fatigue life is conducted to provide a methodology for correlating the experimental data obtained over a range of test conditions. The modeling, which is based on

TABLE 1—*Isothermal fatigue data at 650°C.*

Specimen Number	Max Stress, MPa	Frequency, Hz	Stress Ratio	Cycles to Failure
		1 Hz Data		
92-112	570	1	0.1	3.10E + 03
92-116	470	1	0.1	1.20E + 04
92-114	425	1	0.1	1.84E + 04
92-113	350	1	0.1	5.77E + 04
92-128	300	1	0.1	1.53E + 05
92-120	300	1	0.1	1.53E + 05
92-119	250	1	0.1	**1.00E + 06**
		0.1 Hz Data		
93-444	570	0.1	0.1	1.69E + 02
92-587	470	0.1	0.1	9.18E + 03
		0.01 Hz Data		
92-552	570	0.01	0.1	1.34E + 03
92-553	470	0.01	0.1	3.15E + 03
92-395	400	0.01	0.1	**1.00E + 04**
		NASA 0.01 Hz		
NASA-1	450	0.01	0	2.32E + 03
NASA-2	400	0.01	0	6.98E + 03
NASA-3	350	0.01	0	2.90E + 04
		200 Hz Data		
92-598	200	200	0.1	**7.12E + 06**
92-600	300	200	0.1	1.04E + 06
92-602	400	200	0.1	1.30E + 05
92-606	470	200	0.1	1.87E + 04
92-607	250	200	0.1	**1.50E + 08**
92-608	350	200	0.50	8.79E + 06
92-610	450	200	0.80	2.03E + 05
92-611	425	200	0.80	**7.40E + 07**
93-449	475	200	0.80	4.00E + 07
93-451	337	200	0.10	4.08E + 05

an understanding of the physical processes leading to fatigue, is conducted in three parts. First, a linear-summation life-prediction model, developed previously for correlating isothermal data in a unidirectionally reinforced TMC over a range of frequencies and stress ratios (R), is applied to the 650°C data on the $[0/90]_s$ composite used in this investigation. Second, a modification of the model that accumulates time-dependent damage in a nonlinear fashion is applied to isothermal data at two temperatures, 650 and 815°C. Finally, a model based on micromechanical stresses using the nonlinear damage accumulation concept is applied to all of the 650°C isothermal data as well as the TMF data for the temperature cycle between 150 and 650°C. The details are presented later.

(a) *Linear Damage Summation Model*—The method employed in Refs *6* and *7* was applied to the isothermal data at 650°C and frequencies of 0.01, 0.1, 1.0, and 200 Hz. The majority of the data, presented in Fig. 1, are at $R = 0.1$ or $R = 0$ with only a few at $R = 0.5$ and $R = 0.8$. The model, as before, sums the damage from cycle-dependent effects, time-dependent effects, and mixed-mode or interaction effects in the following manner

$$\frac{N}{N_c} + \frac{N}{N_t} + \frac{N}{N_i} = 1 \tag{1}$$

TABLE 2—*Isothermal fatigue data at 815°C.*

Specimen Number	Max Stress, MPa	Frequency, Hz	Stress Ratio	Cycles to Failure
		1 Hz Data		
92-129	400	1	0.1	7.87E + 03
92-117	275	1	0.1	8.27E + 04
92-115	225	1	0.1	2.46E + 04
92-118	225	1	0.1	2.41E + 04
92-111	347	1	0.1	4.33E + 03
		0.01 Hz Data		
92-588	425	0.01	0.1	6.21E + 02
92-554	350	0.01	0.1	1.01E + 03
92-586	275	0.01	0.1	6.15E + 03
		200 Hz Data		
92-601	200	200	0.1	4.09E + 06
92-604	300	200	0.1	9.00E + 05
92-605	400	200	0.1	6.14E + 04
92-609	290	200	0.50	**7.20E + 07**
92-612	350	200	0.50	3.54E + 06
92-599	440	200	0.50	1.70E + 06
93-160	500	200	0.80	9.64E + 03
93-161	450	200	0.80	6.53E + 05

TABLE 3—*TMF data, 150 to 650°C.*

Specimen Number	Max Stress, MPa	Frequency, Hz	Stress Ratio	Cycles to Failure
		In-Phase		
92-130	550	0.00556	0.1	3.60E + 01
92-108	500	0.00556	0.1	3.43E + 02
92-107	450	0.00556	0.1	4.19E + 03
92-104	400	0.00556	0.1	**5.77E + 03**
92-573	600	0.00556	0.1	6.10E + 01
92-572	550	0.00556	0.1	1.38E + 03
92-574	600	0.00056	0.1	1.10E + 01
92-576	550	0.00056	0.1	2.51E + 02
		Out-of-Phase		
92-106	600	0.00556	0.1	2.64E + 02
92-105	500	0.00556	0.1	4.35E + 02
92-103	400	0.00556	0.1	9.53E + 02
92-109	300	0.00556	0.1	1.29E + 03
92-110	200	0.00556	0.1	2.11E + 03
92-131	150	0.00556	0.1	6.26E + 03
92-551	400	0.00056	0.1	6.54E + 02

where N_c is the number of cycles to failure due to cyclic processes, N_t is the number of cycles to failure due to time-dependent phenomena, and N_i is an interaction term that accounts for the synergistic effect of cyclic- and time-dependent phenomena. The three terms are represented as follows

$$N_c = A(\Delta\sigma_{\text{appl}})^{-m} \tag{2a}$$

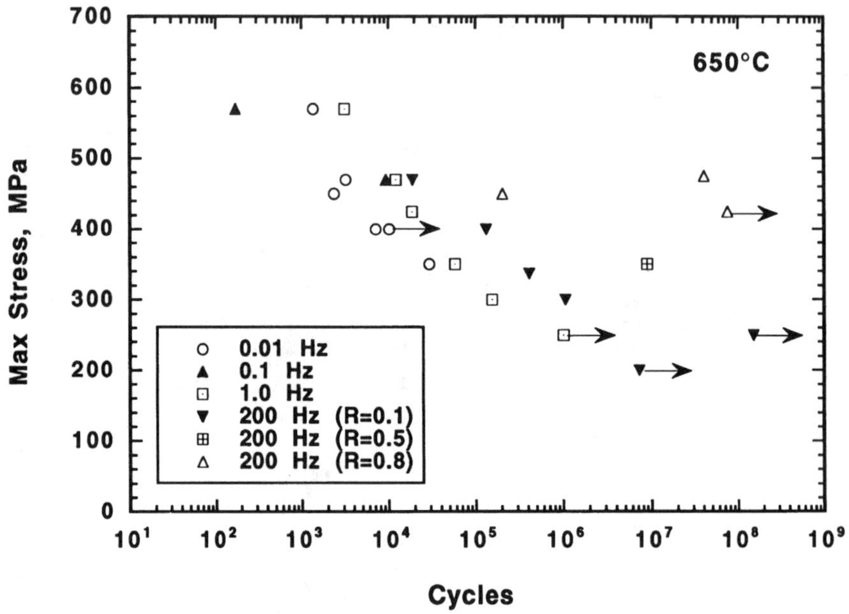

FIG. 1—*Cycles to failure as a function of applied stress for all 650°C data.*

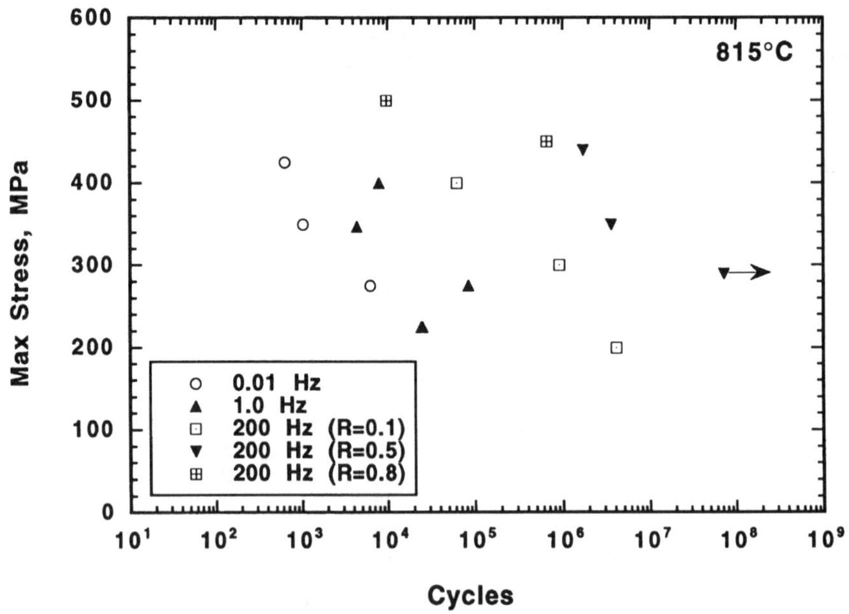

FIG. 2—*Cycles to failure as a function of applied stress for all 815°C data.*

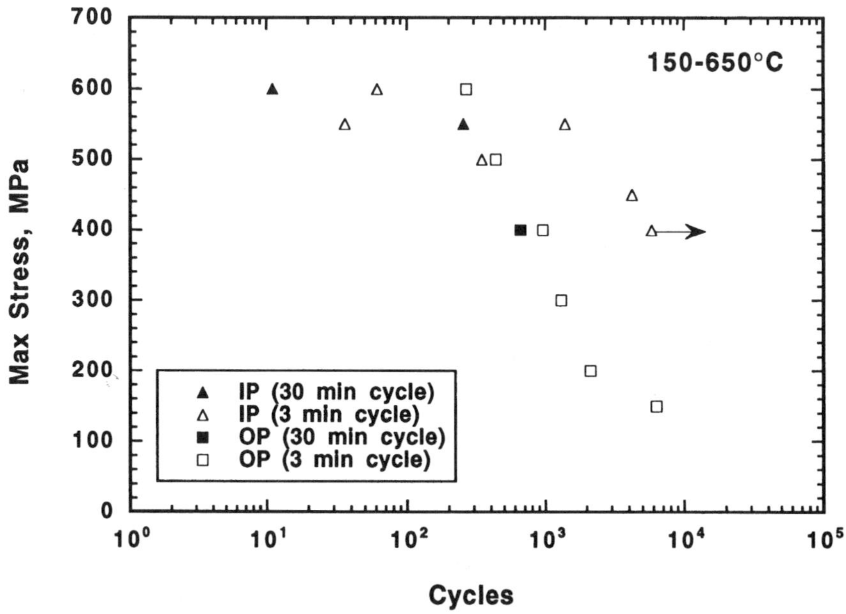

FIG. 3—*Cycles to failure as a function of applied stress for all TMF data.*

$$N_t = \frac{Bf(n+1)(1-R)}{\sigma_{max}^n(1-R^{n+1})} \tag{2b}$$

$$N_i = C\sqrt{N_c N_{tm}} \tag{2c}$$

where f is the cyclic frequency, R is the stress ratio, and A, B, C, m, and n are empirical constants. Equation 2*b* is obtained by integrating the increments of time-dependent damage over an entire cycle, assuming a triangular wave form. Damage accumulation over one cycle is obtained as

$$\frac{1}{N_t} = 2\int_0^{T/2} \frac{dt}{t_c(\sigma)} = \frac{1}{f\sigma_{max}(1-R)}\int_{R\sigma_{max}}^{\sigma_{max}} \frac{d\sigma}{t_c(\sigma)} \tag{3}$$

with damage defined in terms of time to failure under constant stress as

$$t_c = B\sigma^{-n} \tag{4}$$

Equation 2*c*, the interaction term, utilizes the time to failure under the mean stress

$$N_{tm} = ft_c(\sigma_{mean}) \tag{5}$$

In order to account for differences in fiber volume fraction in the various specimens, an effective stress is used in place of the applied stress. The effective stress produces the same stress in the composite as would be obtained in a composite whose fiber volume fraction were 0.35, the average value for this material. For small variations from 0.35, it is easily shown that the effective stress can be accurately represented as

$$\sigma_{eff} = \sigma_{appl}\left(\frac{V_f}{0.35}\right)^{0.718} \tag{6}$$

These equations provide a method for representing a body of isothermal fatigue data obtained at different frequencies and stress ratios, including sustained load tests, and having small variations in fiber volume fraction.

(b) *Nonlinear Damage Summation Model*—The linear damage summation scheme used earlier is modified by elimination of the interaction term and substituting a nonlinear damage accumulation term for the time-dependent portion of the model, thus

$$\frac{N}{N_c} + \left(\frac{N}{N_t}\right)^{0.5} = 1 \tag{7}$$

The form of this equation evolved in an evaluation of thermal fatigue data on the same composite system. An exponent of 0.5 is used to represent damage accumulating with square root of time, typical of diffusion processes. The 0.5 exponent also provides an equation that can be solved directly for N. A further modification to the procedure involves the introduction of an additional constant to account for the possible effect of mean stress as well as stress range. To do this, $\Delta\sigma$ in Eq 2*a* is replaced by $\Delta\sigma_{\text{eff}}$ defined as

$$\Delta\sigma_{\text{eff}} = \sigma_{\text{max}}(1 - R)^p \tag{8}$$

where p can take on values between 0 and 1, $p = 1$ represents $\Delta\sigma$ as the governing term while $p = 0$ produces σ_{max} as the governing term in the computations. As discussed later, for the data at 650°C, the nonlinear model provides substantially the same correlation as the linear model with the interaction term. It was thus decided to continue using the nonlinear model for correlating TMF data.

(c) *General TMF Model*—The TMF modeling utilizes the nonlinear damage-summation approach, Eq 7, but decomposes the cyclic damage into terms governed by the fiber stresses and the matrix stresses separately, thus

$$\frac{1}{N_f} + \frac{1}{N_m} = \frac{1}{N_c} \tag{9}$$

The procedure is an extension of prior work in Refs *8*, *9*, and *10*, based on the computation of the micromechanical stresses in the fiber and the matrix for both isothermal and TMF conditions. Following the prior approach, we take

$$N_f = 10^{N_0\left(1 - \frac{\sigma_{f,\text{max}}(1 - R_f)^0}{\sigma^*}\right)} \tag{10}$$

$$N_m = B_m(\Delta\sigma_m)^{-n_m} \tag{11}$$

where $\sigma_{f,\text{max}}$ and R_f denote the maximum stress and stress ratio in the fiber, respectively, $\Delta\sigma_m$ is the stress range in the matrix, and N_0, σ^*, p, B_m, and n_m are empirical constants for fitting experimental data. The exponent, p, to account for mean stress or stress range effects in the fiber, Eq 10, is the same as that used as the exponent in the applied stress range, Eq 8, where solutions were found to be insensitive to this parameter. For the computations performed herein, the matrix stresses in Eq 11 are the average matrix stresses in the 90° lamina.

The time-dependent term for TMF is obtained in a manner similar to that in the isothermal case, namely, the integration of increments of damage over a triangular wave form cycle as in Eq 3 using the form for time to failure as in Eq 4. For TMF, the time to failure at a given stress under constant load is a function of temperature, thus B and n in Eq 4 are functions of temperature. Since the maximum rate of damage accumulation occurs at the highest temperature, 650°C, in the TMF cycle, and whereas Eq 4 characterizes the behavior at that temperature for the isothermal condition, errors in modeling the damage accumulation at lower temperatures will be less significant than at 650°C. With no data available to characterize the behavior at lower temperatures, we arbitrarily take the temperature-dependent form of Eq 4 as

$$B(T) = \frac{B(T_{\max} - T_{\min})}{T - T_{\min}} \tag{12}$$

that produces the same time to failure at maximum temperature, no damage at minimum temperature, while retaining n as a constant independent of temperature. These assumptions, while somewhat speculative, provide the capability for integrating the damage over a cycle in closed form. Carrying out the procedure, the following equation is obtained

$$\frac{1}{N_t} = \frac{\sigma_{\max}^n}{fB(1 - R)^2} F \tag{13}$$

where the function, F, is given very accurately for $n > 2$ and $R = 0.1$ (all TMF tests were conducted with mechanical cycles of $R = 0.1$) as

$$F = \frac{1}{n+2} - \frac{R}{n+1} \quad \text{for in-phase TMF} \tag{14a}$$

$$F = \frac{1}{n+1} - \frac{1}{n+2} \quad \text{for out-of-phase TMF} \tag{14b}$$

with the terms f, B, n, and R defined as in Eq 2b.

Modeling Observations

Correlation of the experimental data at 650°C, Fig. 1, with computations using the linear life-fraction model with the interaction term is approximately the same as with the nonlinear model. For the linear model with the interaction term, with stress given in MPa and frequency in Hz, a set of constants in Eqs 2 through 5 is determined that gives a best overall fit to the data, not including runout tests, using a least squares error minimization scheme. The constants for this model are summarized in the column, Linear Model, in Table 4. The fit to the same experimental data by the nonlinear model, Eq 7, is obtained by the same error minimization scheme. For this model, a best set of constants is obtained as summarized in Table 4. A slightly better fit is obtained when the constant, p, in Eq 8 is introduced that accounts for the combined effect of maximum stress and stress range. A value of $p = 0.79$ implies that $\Delta\sigma$ is dominant ($p = 1$) while maximum stress has a lesser contribution ($p = 0$). The same procedure is applied to the data obtained at 815°C, resulting in a best set of constants that is summarized in Table 4. In the latter iterations, the results were insensitive to the value of p, thus $p = 1$ was chosen.

The results for the nonlinear model at 650°C, which attempts to fit the data of Fig. 1, are shown in Fig. 4 in terms of predicted life as a function of experimentally determined life for all of the isothermal data at that temperature. A similar plot for the linear model was almost

TABLE 4—*Constants used in fatigue life modeling.*

Constants	Linear Model (Eqs 1–5)	Nonlinear Model, 650°C (Eqs 7 and 8)	Nonlinear Model, 815°C (Eqs 7 and 8)	TMF (Eqs 9 through 11)
A	4.75 E 20	3.78 E 23	3.67 E 21	3.78 E 23
m	6.14	7.34	6.57	7.34
B	1.7 E 12	3.76 E 10	2.21 E 15	9.23 E 09
n	2.92	2.29	4.61	1.48
C	0.33	...	...	...
p	...	0.79	1.0	0.76
N_0	...	...	...	7.34
σ^*	...	...	...	3483
B_m	...	...	...	2.11 E 23
n_m	...	...	...	2.80

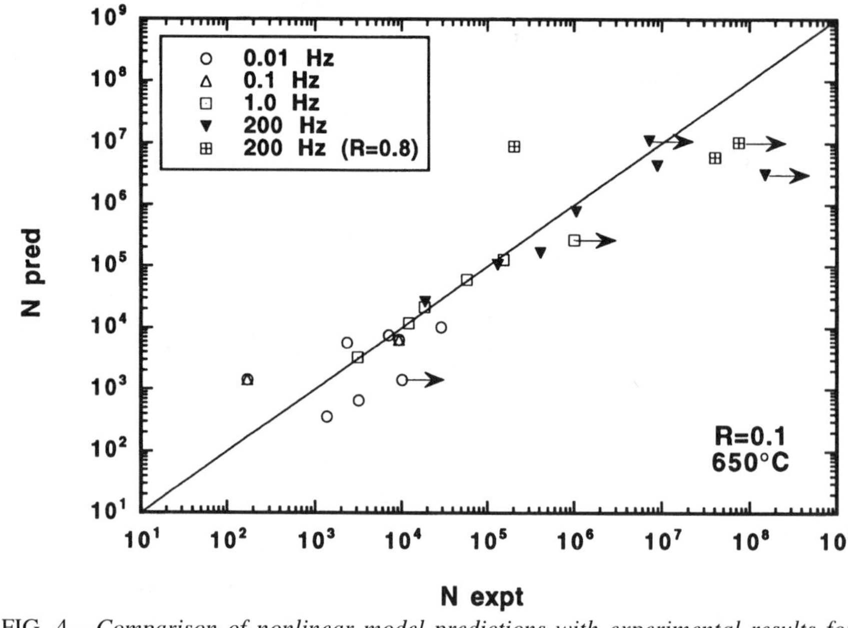

FIG. 4—*Comparison of nonlinear model predictions with experimental results for isothermal fatigue at 650°C.*

identical and, therefore, is not shown. Tests terminated prior to failure are indicated with arrows. The data that are not correlated very well seem to be concentrated at the two extremes of the spectrum. For the shortest life test, the cutoff for static failure is being approached. This cutoff depends on frequency, and varies from approximately 10^3 to 10^4 cycles at 0.01 Hz to over 10^6 at 200 Hz (equivalent to a total test duration of 1.4 h). In this region, life is extremely dependent on stress such that a small error in stress or fiber volume fraction can have an extremely large effect on life. At the high end ($>10^7$ cycles), where the fatigue curve is also expected to be flat, small errors in stress or fiber volume fraction can have a large effect on fatigue life. However, this is the region near runout that is of little practical importance. It can be seen that the model provides a fair correlation of the predictions with the data, especially in the intermediate cycles to failure regime. This is consistent with the concept proposed by Talreja [*15*] that characterizes composite fatigue in three regimes: short life dominated by statistical variability in strength, an intermediate life governed by fatigue processes, and long life or runout that involves statistical scatter. The data in these experiments tend to support such a concept. The data at $R = 0.8$ show the largest amount of scatter and are correlated to a lesser degree than the other data with the model. At these high stress ratios, the cyclic contribution becomes negligible and the behavior is close to that of a pure creep or sustained load test. Such behavior is highly influenced by environmental factors, and slight variations of the material quality can have a large effect on life.

The nonlinear model was also applied to the isothermal data obtained at 815°C, shown in Fig. 2. The results of the model correlation with data are presented in Fig. 5. The number of data points is somewhat limited, but they cover the entire range of frequencies and stress ratios evaluated in this investigation. The nonlinear model is seen to provide a reasonable capability for representing the data, and much of the variability can be attributed to scatter in the experimental data. For the 815°C data, the modeling is insensitive to the value of the exponent, p. This is because at this temperature, the time-dependent behavior appears to dominate, therefore, the cyclic term has little effect on the predicted life.

The model representations of the data for 650 and 815°C are presented in plots of applied stress versus cycles to failure in Figs. 6 and 7, respectively. Symbols are included to identify individual curves that were generated numerically for values of maximum stress at 10 MPa increments. The model representations, which follow the general trends of the actual data, Figs. 1 and 2, are used as the basis for discussion of trends in the data. At 650°C, the cycles to failure differ very little from 1 to 200 Hz. The difference becomes larger at low stress levels (<250 MPa) where the time-dependent mechanisms begin to have more of an effect

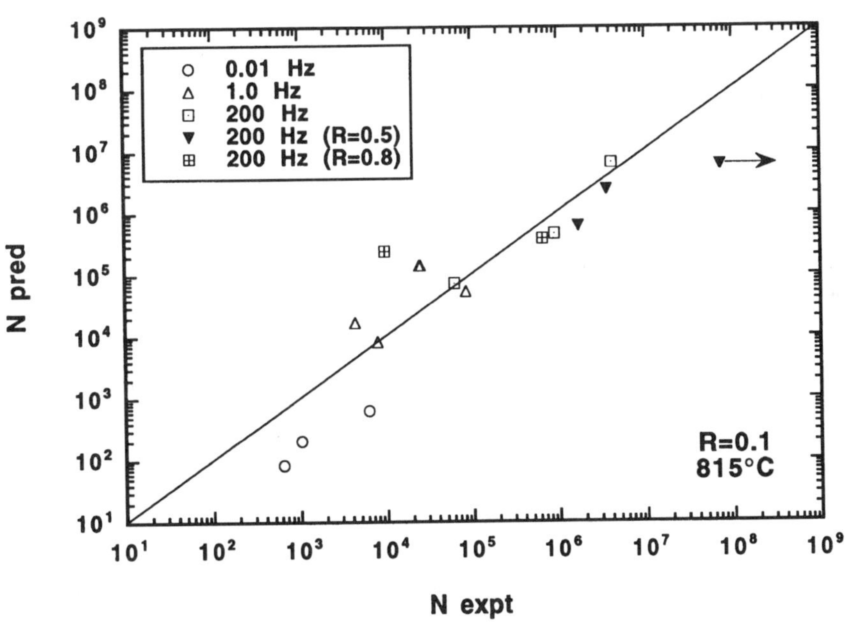

FIG. 5—*Comparison of nonlinear model predictions with experimental results for isothermal fatigue at 815°C.*

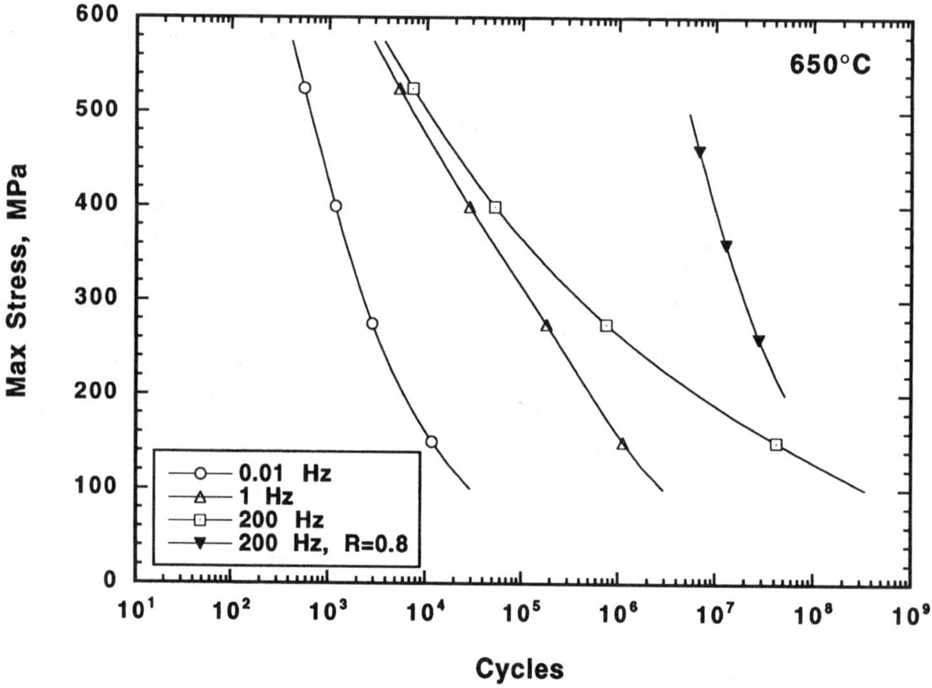

FIG. 6—*Nonlinear model predictions for isothermal fatigue at 650°C.*

at 1 Hz, and times to failure are large (>50 h). This implies that the behavior at this temperature is mainly cycle dependent at frequencies above 1 Hz over the stress levels considered. At $R = 0.8$, the 200 Hz predictions produce lives that are over two decades greater than those at $R = 0.1$, which demonstrates how $\Delta\sigma$ rather than σ_{max} governs life at this high frequency. At the higher temperature of 815°C, on the other hand, the difference in life between $R = 0.1$ and $R = 0.8$ at 200 Hz is only about one decade, demonstrating the much larger effect of σ_{max} and less effect of $\Delta\sigma$. Further, the difference between 1 and 200 Hz is approximately a factor of 10, showing that time dependence is significant even at these high frequencies. Between 0.01 and 1 Hz, a factor of 100 in frequency, the difference in lives is approximately a factor of 100. This indicates a dominance of time-dependent mechanisms in this frequency regime.

For the general TMF model, micromechanical stresses for the fiber and matrix in the axial direction are required. While such computations are relatively simple and straightforward for a unidirectional composite, calculation of stresses for a $[0/90]_s$ composite is quite complicated. For TMCs, it has been verified through combinations of experiments and computations that damage in the form of fiber/matrix interface sliding and separation occurs at very low applied stresses in the [90] lamina [*16–18*]. Further, at elevated temperatures, titanium matrix materials deform in a time-dependent manner that can be characterized using viscoplastic constitutive equations [*19–22*]. To obtain an accurate assessment of the fiber and matrix stresses, a three-dimensional finite element method was employed that incorporates a viscoplastic model for the matrix and accounts for fiber/matrix sliding [*23*]. The matrix stresses used in the fatigue life computations are the average stress in the [90] ply and, further, are those that develop after one complete loading cycle. Residual stresses that develop during cool down from the processing temperature are accounted for in the

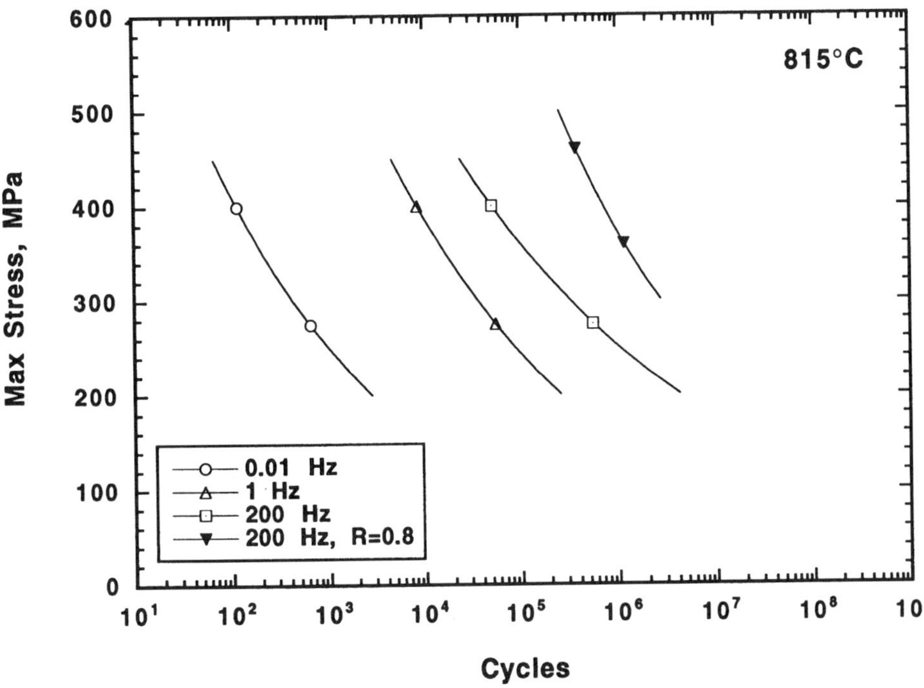

FIG. 7—*Nonlinear model predictions for isothermal fatigue at 815°C.*

analysis. For each of the following conditions: in-phase TMF, out-of-phase TMF, 1-Hz isothermal fatigue, and 0.01-Hz isothermal fatigue, the fiber and matrix stresses at maximum and minimum load after the first cycle are computed for two different applied loads. Linear interpolation is used to determine the stresses at other applied load levels in the modeling computations. Additionally, the 200-Hz isothermal cycles are assumed to develop the same stresses (at 650°C) as the 1-Hz cycles, and the 0.1-Hz cycles are assumed to develop the average of the stresses from the 1- and 0.01-Hz conditions. The composite stress analysis is part of a parallel investigation [*24*]. With this information available, modeling of all of the 650°C isothermal and 150 to 650°C TMF test data is conducted in accordance with the equations presented earlier for the general TMF model.

The ability of the model to consolidate all of the 650°C isothermal as well as the TMF data is shown in Fig. 8, where predicted life is plotted against experimental cycles to failure. As in the isothermal cases, the model shows the overall capability to capture the general trends in the data for each type of test. Because of the relatively large amount of scatter in the data, Figs. 1 and 3, it is difficult to determine whether this is due to inherent material scatter or the inability of the model to capture all of the trends correctly. The model used here is based on computed micromechanical stresses as well as time spent at a given applied stress level. The model is calibrated by numerical techniques to provide the best overall correlation to the experimental data set. Applying a least-squares error minimization technique to the entire set of isothermal and TMF data, a set of constants is obtained to represent the cycles to failure in the general TMF model as summarized in Table 4.

While the model fit to the data is not exceptional, the scatter expected in tests conducted near the static strength of the material as well as those near the fatigue limit tend to make the model look worse than in the intermediate life region. One of the major difficulties in

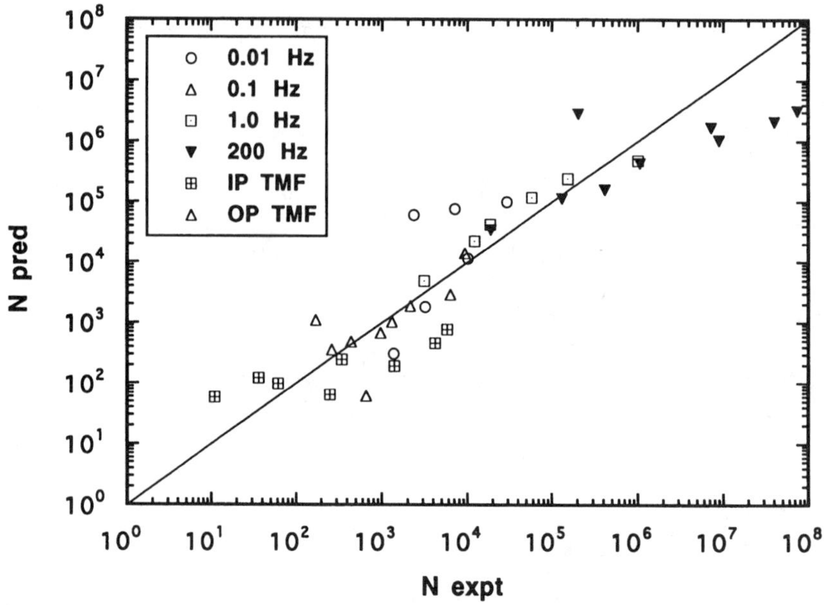

FIG. 8—*Comparison of model predictions with experimental results for all isothermal and TMF data.*

applying a micromechanical stress-based model to a material system such as this one that exhibits creep or stress relaxation is determining which stresses to use. It has been observed, for example, that the difference in lives under in-phase TMF at different frequencies is due, in large part, to the different stresses developed in the fiber because of viscoplastic effects [*25*]. This, in addition to a strong sensitivity of fatigue life to small changes in fiber stress, makes fatigue life predictions very sensitive to the stresses used in the model. While stresses after the first cycle were the only computations available for this investigation, it is speculated that use of saturated steady-state stresses after a number of cycles might provide a better basis for correlating fatigue lives using the modeling concepts presented in this paper. Further, refinements in the model to account for asymptotic limits for short lives as well as a fatigue limit should enhance the life prediction capability. These will be the subjects of a future investigation.

Fractography

An integral part of this investigation involved the evaluation of the fracture surfaces of the specimens tested in order to determine the mechanisms of failure. The fractography, which was conducted in parallel with the modeling in this and prior work leading up to this investigation, is used both to ascertain the failure mechanisms to help guide the model development and to validate some of the trends predicted by the model as well as the data themselves.

Isothermal Fatigue at 650°C

Fracture surfaces of the isothermal fatigue specimens tested at a maximum stress of 470 MPa at different frequencies are compared. Fractography was performed using both an optical and a scanning electron microscope (SEM). An optical microscope at ×30 magnification was used to estimate the percentage of the cross-sectional area that experienced cracking prior to final fracture. The cracked regions are easily discernible from the colors resulting from the oxidation of titanium. The SEM validates the optical observations and is also necessary to identify any cracks that initiated internally, and therefore, were not oxidized because the cracks did not reach the surface of the specimen. Typically, a primary crack initiates at the specimen machined edge. Surface initiation at the 0° ply is not observed. Internal cracks are also observed, initiating at 90° fibers. These are generally small in comparison to the primary crack region.

In general, the extent of fatigue cracking observed in the matrix varies as a function of the frequency. The sample tested at 200 Hz, which experienced the greatest number of mechanical load cycles, exhibits the greatest amount of matrix cracking, while the specimen tested at 0.01 Hz and had the fewest number of cycles, exhibits the least cracking. The percentages of the total fracture surfaces that show evidence of matrix cracking are estimated at 95, 83, 30, and 10% for the 200, 1, 0.1, and 0.01 Hz samples, respectively. These observations are consistent with the fatigue life data and modeling predictions described earlier where the behavior at this temperature is deduced as being mainly cycle dependent over most of the frequency regime, though tending to mixed cycle and time dependent at the low frequency (0.01-Hz) end.

The fracture surfaces typically have more than one fracture plane, suggesting that several fatigue cracks at various locations participated in the final fracture as opposed to one dominant crack. In general, two types of fracture surface morphologies are observed, contrasted in Fig. 9. Figure 9*a* is an SEM photograph of a characteristic region from the specimen tested at 1 Hz and demonstrates matrix cracking through the thickness of the sample. This region was highly oxidized, and only isolated 0° fibers demonstrate significant pull-out. Figure 9*b* is an SEM photograph of one of only two cracked regions from the specimen tested at 0.01 Hz. For this case, the cracking is restricted to the matrix adjacent to the 90° fibers, and oxidation was not significant since the crack never progressed to a free surface. Also noteworthy is the evidence of void formation and coalescence in the matrix surrounding the 0° fibers and near the surface. This dimpling is characteristic of tensile overload. Fiber pull-out is also insignificant in this sample. The fracture surfaces of the higher frequency tests (200 and 1 Hz) are similar in that through-thickness cracking is noted over the vast majority of the sample. It is interesting to note that the predicted lives for these two frequencies are also similar and, therefore, the life is primarily cycle dependent (Fig. 6). The fracture morphology in terms of the amount of matrix cracking significantly differs between 0.01- and 1-Hz tests, and the predicted lives also show an order of magnitude difference indicating that life is both cycle and time dependent at 0.01 Hz (Fig. 6). The fracture surface of the 0.1-Hz specimen has features of both the 0.01- and 1-Hz tests; it exhibits two internal cracks similar to the 0.01-Hz test as well as one through-thickness cracked region similar to the 1-Hz test, Figs. 10*a* and *b*.

Isothermal Fatigue at 815°C

Several specimens tested at 815°C and maximum stress levels of either 400 or 350 MPa were examined. In general, the fracture surfaces show less matrix cracking than the 650°C tests, which might be expected because of the significantly shorter fatigue lives. At this temperature, fiber degradation from environmental exposure is more rapid, and matrix strength is reduced. The 200-Hz test shows the largest amount of matrix fatigue cracking. Regions of both through thickness and 90° ply-only cracking are observed, and nearly 60% of the fracture surface contains matrix fatigue cracks. For the lowest two frequencies, 0.1 and 0.01 Hz, fatigue cracks adjacent to the 90° ply fibers are present; however, the fatigue

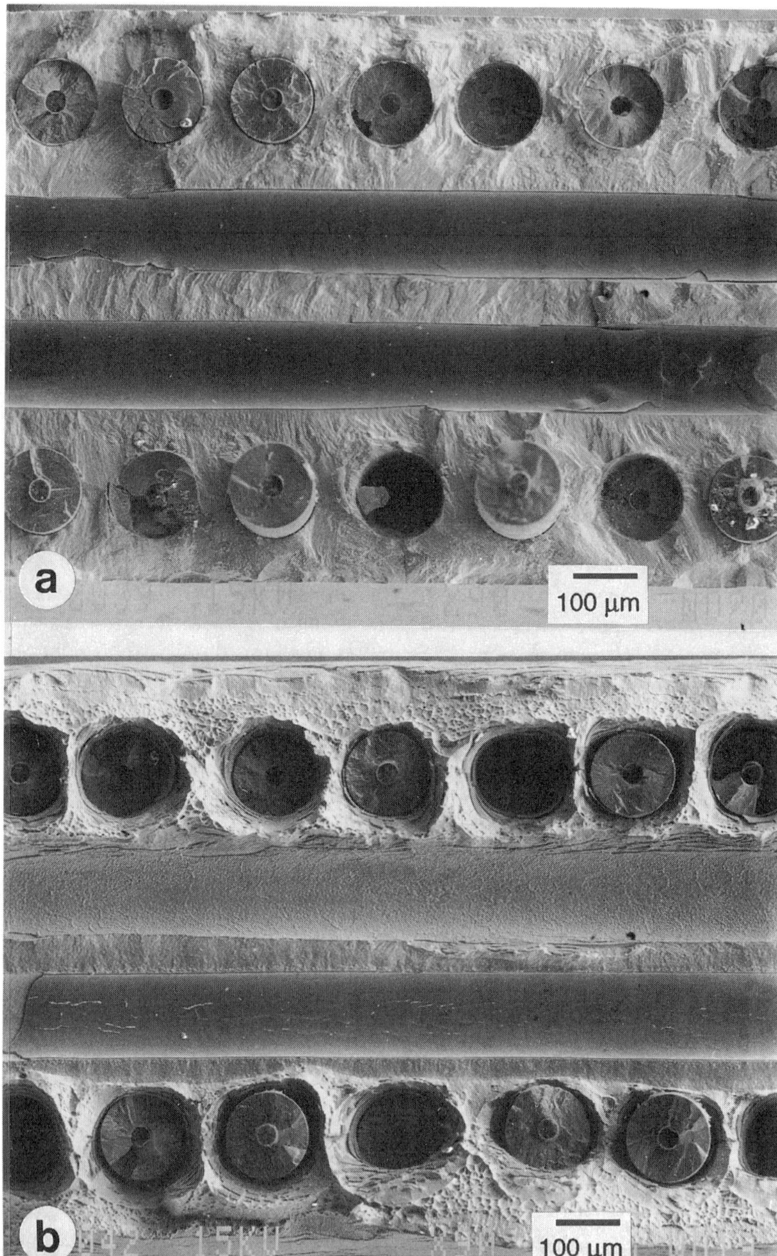

FIG. 9—*Fracture surfaces illustrating matrix cracking for isothermal fatigue specimens tested at 650°C:* (a) *through-thickness cracking at 1 Hz and* (b) *90° ply cracking at 0.01 Hz.*

FIG. 10—*Fracture surfaces of 650°C isothermal fatigue specimens tested at 0.1 Hz showing* (a) *through thickness and* (b) *90° ply matrix cracking.*

crack regions are smaller. Also, relatively small surface cracks can be identified, as seen in Fig. 11 for the 0.1-Hz test sample. The surface-initiated cracking is most likely associated with the increased environmental attack of the surface at 815°C, especially since these specimens were exposed at this temperature for >4 and >27 h for 0.1- and 0.01-Hz tests, respectively. Contrast this with the 200-Hz test at 400 MPa that lasted only 5 min. Of notable exception to the observed pattern of behavior with frequency is the fracture surface of the specimen tested at 1 Hz, where time to failure is less than a few hours at similar stress levels. The specimen tested at 400 MPa that failed after 7866 cycles (2.2 h) experienced matrix cracking over approximately 30% of the cross-sectional area; however, much of the 90° ply cracking is intergranular, Fig. 12. This is in contrast to the other specimens, at both higher and lower frequencies, where the cracking is primarily transgranular in nature. One possible explanation for this phenomena is that creep mechanisms are more active in this particular specimen due to the sinusoidal wave form applied for the 1-Hz test compared to triangular waves at lower frequencies. The 1-Hz frequency also appears to be the dividing line between mainly cycle-dependent behavior at higher frequencies, and time-dependent behavior at lower frequencies, as determined by the data and modeling discussed previously. At this transition frequency, a tradeoff results between sufficient time for creep-fatigue damage and insufficient time for environmental mechanisms to become active. To summarize the fractographic observations for the 815°C tests, the fracture surface morphology is quite dependent on frequency, and as shown earlier in Fig. 7, life predictions are also sensitive to the frequency at this temperature.

FIG. 11—*Fracture surface of isothermal fatigue specimen tested at 815°C, 0.1 Hz, illustrating both 90° ply and surface cracking.*

crack regions are smaller. Also, relatively small surface cracks can be identified, as seen in Fig. 11 for the 0.1-Hz test sample. The surface-initiated cracking is most likely associated with the increased environmental attack of the surface at 815°C, especially since these specimens were exposed at this temperature for >4 and >27 h for 0.1- and 0.01-Hz tests, respectively. Contrast this with the 200-Hz test at 400 MPa that lasted only 5 min. Of notable exception to the observed pattern of behavior with frequency is the fracture surface of the specimen tested at 1 Hz, where time to failure is less than a few hours at similar stress levels. The specimen tested at 400 MPa that failed after 7866 cycles (2.2 h) experienced matrix cracking over approximately 30% of the cross-sectional area; however, much of the 90° ply cracking is intergranular, Fig. 12. This is in contrast to the other specimens, at both higher and lower frequencies, where the cracking is primarily transgranular in nature. One possible explanation for this phenomena is that creep mechanisms are more active in this particular specimen due to the sinusoidal wave form applied for the 1-Hz test compared to triangular waves at lower frequencies. The 1-Hz frequency also appears to be the dividing line between mainly cycle-dependent behavior at higher frequencies, and time-dependent behavior at lower frequencies, as determined by the data and modeling discussed previously. At this transition frequency, a tradeoff results between sufficient time for creep-fatigue damage and insufficient time for environmental mechanisms to become active. To summarize the fractographic observations for the 815°C tests, the fracture surface morphology is quite dependent on frequency, and as shown earlier in Fig. 7, life predictions are also sensitive to the frequency at this temperature.

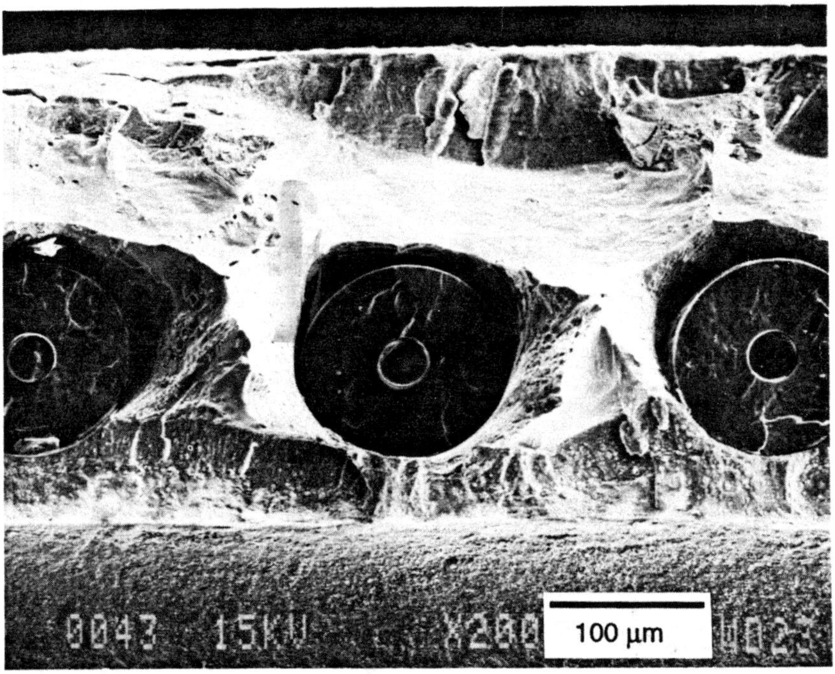

FIG. 11—*Fracture surface of isothermal fatigue specimen tested at 815°C, 0.1 Hz, illustrating both 90° ply and surface cracking.*

cation was used to estimate the percentage of the cross-sectional area that experienced cracking prior to final fracture. The cracked regions are easily discernible from the colors resulting from the oxidation of titanium. The SEM validates the optical observations and is also necessary to identify any cracks that initiated internally, and therefore, were not oxidized because the cracks did not reach the surface of the specimen. Typically, a primary crack initiates at the specimen machined edge. Surface initiation at the 0° ply is not observed. Internal cracks are also observed, initiating at 90° fibers. These are generally small in comparison to the primary crack region.

In general, the extent of fatigue cracking observed in the matrix varies as a function of the frequency. The sample tested at 200 Hz, which experienced the greatest number of mechanical load cycles, exhibits the greatest amount of matrix cracking, while the specimen tested at 0.01 Hz and had the fewest number of cycles, exhibits the least cracking. The percentages of the total fracture surfaces that show evidence of matrix cracking are estimated at 95, 83, 30, and 10% for the 200, 1, 0.1, and 0.01 Hz samples, respectively. These observations are consistent with the fatigue life data and modeling predictions described earlier where the behavior at this temperature is deduced as being mainly cycle dependent over most of the frequency regime, though tending to mixed cycle and time dependent at the low frequency (0.01-Hz) end.

The fracture surfaces typically have more than one fracture plane, suggesting that several fatigue cracks at various locations participated in the final fracture as opposed to one dominant crack. In general, two types of fracture surface morphologies are observed, contrasted in Fig. 9. Figure 9*a* is an SEM photograph of a characteristic region from the specimen tested at 1 Hz and demonstrates matrix cracking through the thickness of the sample. This region was highly oxidized, and only isolated 0° fibers demonstrate significant pull-out. Figure 9*b* is an SEM photograph of one of only two cracked regions from the specimen tested at 0.01 Hz. For this case, the cracking is restricted to the matrix adjacent to the 90° fibers, and oxidation was not significant since the crack never progressed to a free surface. Also noteworthy is the evidence of void formation and coalescence in the matrix surrounding the 0° fibers and near the surface. This dimpling is characteristic of tensile overload. Fiber pull-out is also insignificant in this sample. The fracture surfaces of the higher frequency tests (200 and 1 Hz) are similar in that through-thickness cracking is noted over the vast majority of the sample. It is interesting to note that the predicted lives for these two frequencies are also similar and, therefore, the life is primarily cycle dependent (Fig. 6). The fracture morphology in terms of the amount of matrix cracking significantly differs between 0.01- and 1-Hz tests, and the predicted lives also show an order of magnitude difference indicating that life is both cycle and time dependent at 0.01 Hz (Fig. 6). The fracture surface of the 0.1-Hz specimen has features of both the 0.01- and 1-Hz tests; it exhibits two internal cracks similar to the 0.01-Hz test as well as one through-thickness cracked region similar to the 1-Hz test, Figs. 10*a* and *b*.

Isothermal Fatigue at 815°C

Several specimens tested at 815°C and maximum stress levels of either 400 or 350 MPa were examined. In general, the fracture surfaces show less matrix cracking than the 650°C tests, which might be expected because of the significantly shorter fatigue lives. At this temperature, fiber degradation from environmental exposure is more rapid, and matrix strength is reduced. The 200-Hz test shows the largest amount of matrix fatigue cracking. Regions of both through thickness and 90° ply-only cracking are observed, and nearly 60% of the fracture surface contains matrix fatigue cracks. For the lowest two frequencies, 0.1 and 0.01 Hz, fatigue cracks adjacent to the 90° ply fibers are present; however, the fatigue

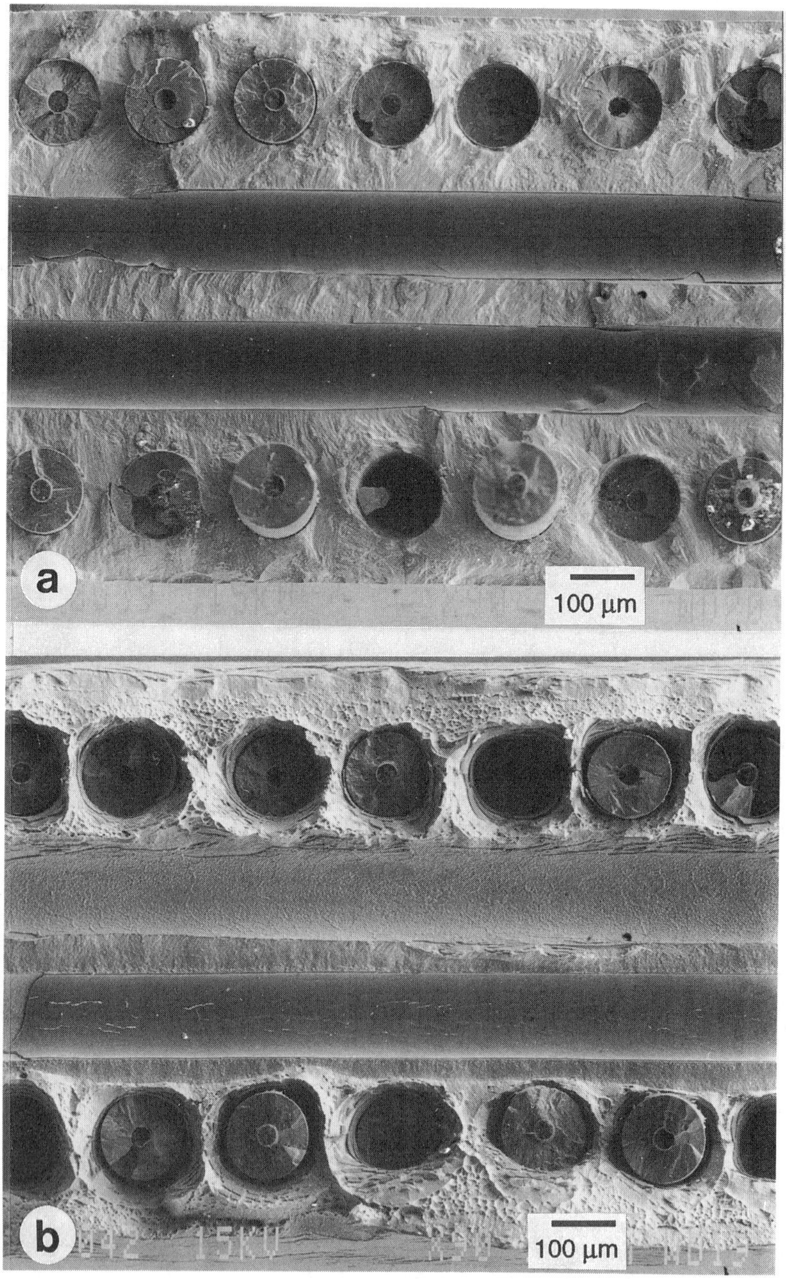

FIG. 9—*Fracture surfaces illustrating matrix cracking for isothermal fatigue specimens tested at 650°C:* (a) *through-thickness cracking at 1 Hz and* (b) *90° ply cracking at 0.01 Hz.*

FIG. 10—*Fracture surfaces of 650°C isothermal fatigue specimens tested at 0.1 Hz showing* (a) *through thickness and* (b) *90° ply matrix cracking.*

crack regions are smaller. Also, relatively small surface cracks can be identified, as seen in Fig. 11 for the 0.1-Hz test sample. The surface-initiated cracking is most likely associated with the increased environmental attack of the surface at 815°C, especially since these specimens were exposed at this temperature for >4 and >27 h for 0.1- and 0.01-Hz tests, respectively. Contrast this with the 200-Hz test at 400 MPa that lasted only 5 min. Of notable exception to the observed pattern of behavior with frequency is the fracture surface of the specimen tested at 1 Hz, where time to failure is less than a few hours at similar stress levels. The specimen tested at 400 MPa that failed after 7866 cycles (2.2 h) experienced matrix cracking over approximately 30% of the cross-sectional area; however, much of the 90° ply cracking is intergranular, Fig. 12. This is in contrast to the other specimens, at both higher and lower frequencies, where the cracking is primarily transgranular in nature. One possible explanation for this phenomena is that creep mechanisms are more active in this particular specimen due to the sinusoidal wave form applied for the 1-Hz test compared to triangular waves at lower frequencies. The 1-Hz frequency also appears to be the dividing line between mainly cycle-dependent behavior at higher frequencies, and time-dependent behavior at lower frequencies, as determined by the data and modeling discussed previously. At this transition frequency, a tradeoff results between sufficient time for creep-fatigue damage and insufficient time for environmental mechanisms to become active. To summarize the fractographic observations for the 815°C tests, the fracture surface morphology is quite dependent on frequency, and as shown earlier in Fig. 7, life predictions are also sensitive to the frequency at this temperature.

FIG. 11—*Fracture surface of isothermal fatigue specimen tested at 815°C, 0.1 Hz, illustrating both 90° ply and surface cracking.*

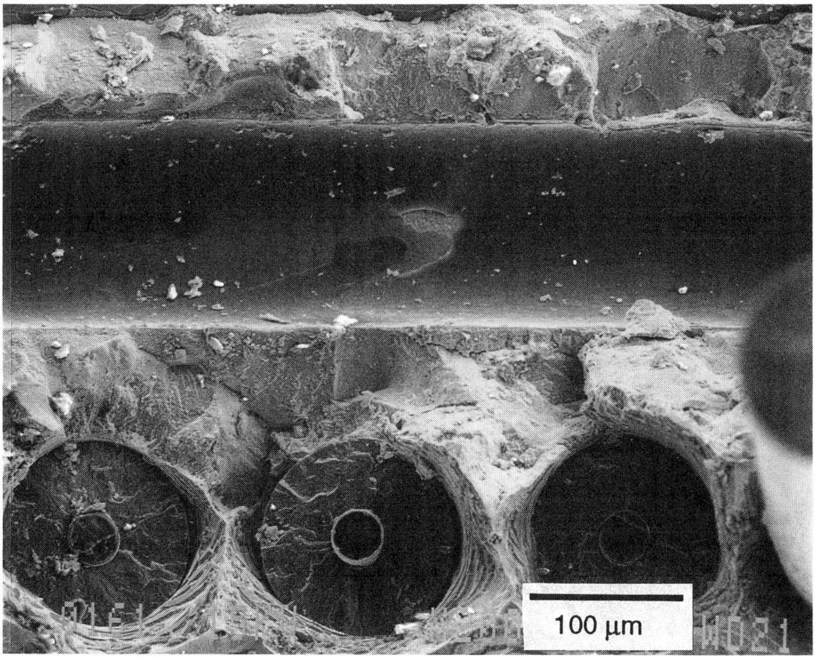

FIG. 12—*Fracture surface of isothermal fatigue specimen tested at 815°C, 1.0 Hz, illustrating 90° ply cracking.*

TMF at 150 to 650°C

Previous work on TMF testing of TMCs showed clear distinctions in the mode of failure between in-phase and out-of-phase TMF loadings [*25–28*]. Notable differences in the fracture surface morphology of cross-ply SCS-6/TIMETAL 21S specimens are observed between those tested in-phase and out-of-phase [*25–27*]. For in-phase TMF, few if any matrix fatigue cracks are observed, whereas for out-of-phase TMF, greater than 50% and sometimes approaching 100% of the matrix on the fracture surface exhibits cracking, with the percentage of matrix fatigue cracks increasing with decreasing maximum applied stress [*25*]. The majority of the matrix cracks are found to initiate at the fiber/matrix interface of the 90° fibers and grow outward toward and into the 0° ply. Fiber pull-out is not generally observed in either the in-phase or out-of-phase loading. The present investigation supports these findings for 0.00556-Hz (3 min/cycle) tests, Figs. 13*a* and *b*. In addition, the fracture surfaces from slower frequency tests, 0.000556 Hz (30 min/cycle), are compared, Figs. 14*a* and *b*. For in-phase TMF, the comparison is made at a maximum stress of 550 MPa, and for out-of-phase, it is made at a maximum stress of 400 MPa. From detailed examination of the fracture surfaces using an SEM, no distinction can be made between the 3 and 30 min/cycle tests. For the out-of-phase TMF test, fatigue cracks are observed in approximately 90% of the matrix for both frequencies (Figs. 13*b* and 14*b*), and the cycles to failure for the 0.000556-Hz test is smaller than the 0.00556-Hz test by less than a factor of two. For in-phase TMF, the life of the 0.000556-Hz test is an order of magnitude lower, yet no differences are observed on the fracture surfaces since both these tests are dominated by a progression of fiber failure during testing and no matrix fatigue cracks are found. In

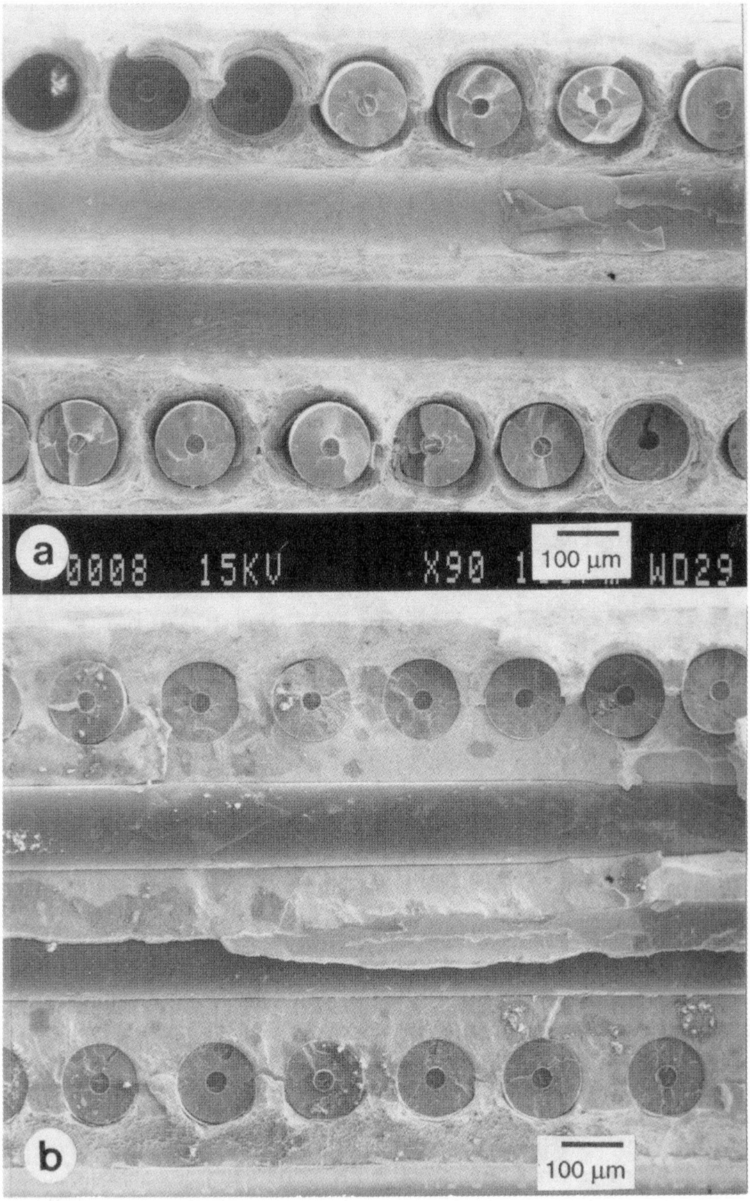

FIG. 13—*Fracture surface of specimens tested under TMF, 150 to 650°C, 3 min cycle:* (a) *in-phase, 550 MPa and* (b) *out-of-phase, 400 MPa.*

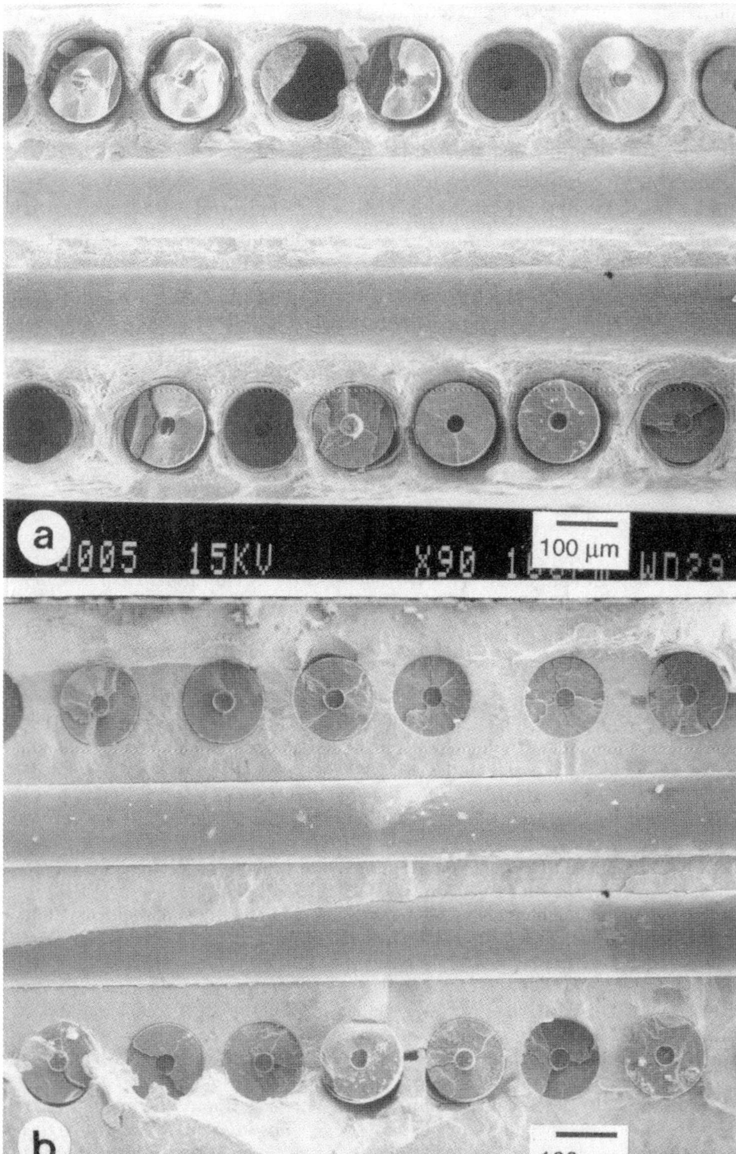

FIG. 14—*Fracture surface of specimens tested under TMF, 150 to 650°C, 30 min cycle:* (a) *in-phase, 550 MPa and* (b) *out-of-phase, 400 MPa.*

summary, an order of magnitude change in the TMF frequency within the range investigated does not significantly change the fracture surface morphology.

Conclusions

Fatigue of a $[0/90]_s$ titanium matrix composite demonstrates a large range of fatigue lives governed by a variety of failure mechanisms under isothermal conditions at 650 and 815°C as well as under in-phase and out-of-phase TMF. Correlation of data from this wide variety of loading conditions can be accomplished to a reasonable extent by consideration of both cycle-dependent and time-dependent terms in a damage summation model. A nonlinear measure of time-dependent damage provides a convenient method for modeling fatigue life, while micromechanical fiber and matrix stresses are demonstrated to be necessary in order to account for the different modes of damage under TMF conditions. Fractography, which illustrates the trend under isothermal conditions from cycle-dependent to time-dependent damage as frequency decreases, validates the concepts in the mathematical modeling. Comparison of in-phase and out-of-phase fracture surfaces illustrates the extremes of fiber-dominated and matrix-dominated modes of failure, respectively.

References

[1] Russ, S. M., "Thermal Fatigue of Ti-24Al-11Nb/SCS-6," *Metallurgical Transactions,* Vol. 21A, June 1990, pp. 1595–1602.

[2] MacKay, R. A., Brindley, P. K., and Froes, F. H., "Continuous Fiber-Reinforced Titanium Aluminide Composites," *Journal of the Minerals, Metals, and Materials Society,* Vol. 43, No. 5, May 1991, pp. 23–29.

[3] Revelos, W. C., "The Thermal Fatigue Response of an SCS-6/Ti-15Mo-3Al-2.6Nb-0.2Si (wt.%) Metal Matrix Composite," *FATIGUE 93, Volume II,* J.-P. Bailon and J. I. Dickson, Eds., Engineering Materials Advisory Services, Ltd., Chameleon Press, London, 1993, pp. 963–968.

[4] Revelos, W. C., Jones, J. W., and Dolley, E. J., "Thermal Fatigue of a SiC/Ti-15Mo-2.7Nb-3Al-0.2Si (wt.%) Composite," *Metallurgical Transactions,* Vol. 26A, May 1995, pp. 1167–1181.

[5] Nicholas, T. and Updegraff, J. J., "Modeling Thermal Fatigue Damage in Metal Matrix Composites," *Composites Engineering,* Vol. 4, 1994, pp. 775–785.

[6] Nicholas, T. and Russ, S. M., "Elevated Temperature Fatigue Behavior of SCS-6/Ti-24Al-11Nb," *Materials Science and Engineering,* Vol. A153, 1992, pp. 514–519.

[7] Nicholas, T., Russ, S., Schehl, N., and Cheney, A., "Frequency and Stress Ratio Effects on Fatigue of Unidirectional SCS-6/Ti-24Al-11Nb Composite at 650°C," *FATIGUE 93, Volume II,* J.-P. Bailon and J. I. Dickson, Eds., Engineering Materials Advisory Services, Ltd., Chameleon Press, London, 1993, pp. 995–1000.

[8] Russ, S. M., Nicholas, T., Bates, M., and Mall, S., "Thermomechanical Fatigue of SCS-6/Ti-24Al-11Nb Metal Matrix Composite," *Failure Mechanisms in High Temperature Composite Materials,* ASME AD-Vol. 22 and AMD-Vol. 122, American Society of Mechanical Engineers, New York, 1991, pp. 37–43.

[9] Mall, S., Hanson, D. G., Nicholas, T., and Russ, S. M., "Thermomechanical Fatigue Behavior of a Cross-Ply SCS-6/β21-S Metal Matrix Composite," *Constitutive Behavior of High Temperature Composites,* MD-Vol. 40, B. S. Majumdar, G. M. Newaz, and S. Mall, Eds., American Society of Mechanical Engineers, New York, 1992, pp. 91–106.

[10] Russ, S. M., Nicholas, T., Hanson, D. G., and Mall, S., "Isothermal and Thermomechanical Fatigue of Cross-Ply SCS-6/β21-S," *Science and Engineering of Composite Materials,* Vol. 3, 1994, pp. 177–189.

[11] Russ, S. M. and Nicholas, T., "Thermal and Mechanical Fatigue of Titanium Aluminide Metal Matrix Composites," *Titanium Aluminide Composites,* WL-TR-91-4020, P. R. Smith, S. J. Balsone, and T. Nicholas, Eds., Wright-Patterson AFB, OH, Feb. 1991, pp. 431–449.

[12] Hartman, G. A. and Russ, S. M., "Techniques for Mechanical and Thermal Testing of Ti_3Al/SCS-6 Metal Matrix Composites," *Metal Matrix Composites: Testing, Analysis, and Failure Modes, ASTM STP 1032,* W. S. Johnson, Ed., American Society for Testing and Materials, Philadelphia, 1989, pp. 43–53.

[13] Hartman, G. A. and Buchanon, D. J., "Methodologies for Thermal and Mechanical Testing of

TMC Materials,'' *Characterization of Fiber Reinforced Titanium Matrix Composites,* AGARD, Report 796, France, 1994, pp. 12-1–12-9.

[*14*] Hartman, G. A. and Ashbaugh, N. E., ''A Fracture Mechanics Test Automation System for a Basic Research Laboratory,'' *Applications of Automation Technology to Fatigue and Fracture Testing, ASTM STP 1092,* A. A. Braun, N. E. Ashbaugh, and F. M. Smith, Eds., American Society for Testing and Materials, Philadelphia, 1990, pp. 95–110.

[*15*] Talreja, R., *Fatigue of Composite Materials,* Technomic Publishing Co., Lancaster, PA, 1987.

[*16*] Johnson, W. S., Lubowinski, S. J., and Highsmith, A. L., ''Mechanical Characterization of Unnotched SCS-6/Ti-15-3 Metal Matrix Composites at Room Temperature,'' *Thermal and Mechanical Behavior of Metal Matrix and Ceramic Matrix Composites, ASTM STP 1080,* J. M. Kennedy, H. H. Moeller, and W. S. Johnson, Eds., American Society for Testing and Materials, Philadelphia, 1990, pp. 193–218.

[*17*] Pollock, W. D. and Johnson, W. S., ''Characterization of Unnotched SCS-6/Ti-15-3 Metal Matrix Composites at 650°C,'' *Composite Materials: Testing and Design (Tenth Volume), ASTM STP 1120,* G. C. Grimes, Ed., American Society for Testing and Materials, Philadelphia, 1992, pp. 175–191.

[*18*] Newaz, G. M. and Majumdar, B. S., ''Deformation and Failure Mechanisms in Mctal Matrix Composites,'' *Failure Mechanisms in High Temperature Composite Materials,* AD-Vol. 22 and AMD-Vol. 122, American Society of Mechanical Engineers, New York, 1991, pp. 55–65.

[*19*] Kroupa, J. L. and Neu, R. W., ''The Nonisothermal Viscoplastic Behavior of a Titanium Matrix Composite,'' *Composites Engineering,* Vol. 4, 1994, pp. 965–977.

[*20*] Sherwood, J. A. and Boyle, M. J., ''Investigation of the Thermomechanical Response of a Titanium-Aluminide/Silicon-Carbide Composite Using a Unified State Variable Model in ADINA,'' *Computers and Structures,* Vol. 40, 1991, pp. 243–255.

[*21*] Johnson, W. S., Mirdamadi, M., and Bahei-El-Din, Y. A., ''Stress-Strain Analysis of a [0/90]2S Titanium Matrix Laminate Subjected to a Generic Hypersonic Flight Profile,'' *Journal of Composites Technology and Research,* Vol. 15, No. 4, Winter 1993, pp. 297–303.

[*22*] Neu, R. W., Coker, D., and Nicholas, T., ''Cyclic Behavior of Unidirectional and Cross-Ply Titanium Matrix Compositics,'' *International Journal of Plasticity,* accepted for publication.

[*23*] Nicholas, T., Kroupa, J. L., and Neu, R. W., ''Analysis of a [0°/90°] Metal Matrix Composite Under Thermomechanical Fatigue Loading,'' *Composites Engineering,* Vol. 3, 1993, pp. 675–689.

[*24*] Kroupa, J. L., Neu, R. W., Nicholas, T., Coker, D., Robertson, D. D., and Mall, S., ''A Comparison of Analysis Tools for Predicting the Inelastic Cyclic Response of Cross-Ply Titanium Matrix Composites,'' in this volume.

[*25*] Neu, R. W. and Nicholas, T., ''Effect of Laminate Orientation on the Thermomechanical Fatigue Behavior of a Titanium Matrix Composite,'' *Journal of Composites Technology and Research,* Vol. 16, 1994, pp. 214–224.

[*26*] Russ, S. M. and Hanson, D. G., ''Fatigue and Thermomechanical Fatigue of a SiC/Titanium $[0/90]_{2s}$ Composite,'' *Fatigue '93,* Vol. II, J. P. Bailon and J. I. Dickson, Eds., Engineering Materials Advisory Services, Ltd., Chameleon Press, London, 1993, pp. 969–974.

[*27*] Castelli, M. G., ''Thermomechanical Fatigue Damage/Failure Mechanisms in SCS-6/TIMETAL 21S $[0/90]_s$ Composite,'' NASA CR-191115, NASA Lewis Research Center, Cleveland, OH, 1994.

[*28*] Mirdamadi, M., Johnson, W. S., Bahei-El-Din, Y. A., and Castelli, M. G., ''Analysis of Thermomechanical Fatigue of Unidirectional Titanium Matrix Composites,'' *Composite Materials: Fatigue and Fracture, Fourth Volume, ASTM STP 1156,* W. W. Stinchcomb and N. E. Ashbaugh, Eds., American Society for Testing and Materials, Philadelphia, 1993, pp. 591–607.

Author Index

R

S

T

W

Y

Z

Subject Index